Lecture Notes in Computer Science 2420

Edited by G. Goos, J. Hartmanis, and J. van Leeuwen

Springer
Berlin
Heidelberg
New York
Barcelona
Hong Kong
London
Milan
Paris
Tokyo

Krzysztof Diks Wojciech Rytter (Eds.)

Mathematical Foundations of Computer Science 2002

27th International Symposium, MFCS 2002
Warsaw, Poland, August 26-30, 2002
Proceedings

Volume Editors

Krzysztof Diks
Wojciech Rytter
Warsaw University, Institute of Informatics
ul. Banacha 2, 02-097 Warszawa, Poland
E-mail: {diks, rytter}@mimuw.edu.pl

Wojciech Rytter
Liverpool University, Department of Computer Science
Peach Street, Liverpool, L69 7ZF, U.K.

Cataloging-in-Publication Data applied for

Die Deutsche Bibliothek - CIP-Einheitsaufnahme

Mathematical foundations of computer science 2002 : 27th international symposium ; proceedings / MFCS 2002, Warsaw, Poland, August 26 - 30, 2002. Krzysztof Diks ; Wojciech Rytter (ed.). - Berlin ; Heidelberg ; New York ; Barcelona ; Hong Kong ; London ; Milan ; Paris ; Tokyo : Springer, 2002
(Lecture notes in computer science ; Vol. 2420)
ISBN 3-540-44040-2

CR Subject Classification (1998): F, G.2, D.3, I.3, E.1

ISSN 0343-2130
ISBN 3-540-44040-2 Springer-Verlag Berlin Heidelberg New York

Springer-Verlag Berlin Heidelberg New York,
a member of BertelsmannSpringer Science+Business Media GmbH

http://www.springer.de

Typesetting: Camera-ready by author, data conversion by Steingräber Satztechnik GmbH, Heidelberg
Printed on acid-free paper SPIN: 10873803 06/3142 5 4 3 2 1 0

Preface

This volume contains papers selected for presentation at the 27th International Symposium on Mathematical Foundations of Computer Science, MFCS 2002, held in Warsaw, Poland, August 26–30, 2002.

MFCS 2002 was organized by the Institute of Informatics of Warsaw University and the Foundation for Information Technology Development. It was supported by the European Association for Theoretical Computer Science. We gratefully acknowledge the support of all these institutions.

The series of MFCS symposia, organized on a rotating basis in the Czech Republic, Poland, and Slovakia, has a well-established tradition. The aim is to encourage high-quality research in all branches of theoretical computer science and bring together specialists who do not usually meet at specialized conferences. The previous meetings took place in: Jablonna, 1972; Štrbské Pleso, 1973; Jadwisin, 1974; Mariánské Lázně, 1975; Gdańsk, 1976; Tatranská Lomnica, 1977; Zakopane, 1978; Olomouc, 1979; Rydzyna, 1980; Štrbské Pleso, 1981; Prague, 1984; Bratislava, 1986; Karlovy Vary, 1988; Porąbka-Kozubnik, 1989; Banská Bystrica, 1990; Kazimierz Dolny, 1991; Prague, 1992; Gdańsk, 1993; Košice, 1994; Prague, 1995; Kraków, 1996; Bratislava, 1997; Brno, 1998; Szklarska Poręba, 1999; Bratislava, 2000; and Mariánské Lázně, 2001.

The MFCS 2002 proceedings consists of five invited papers and 48 contributed papers. We are grateful to all the invited speakers for accepting our invitation and sharing their insights on their research areas. We thank the authors of all submissions for their contribution to the scientific program of the meeting.

The contributed papers were selected by the Program Committee out of a total of 108 submissions. All submissions were evaluated by three members of the committee, with the assistance of referees. After discussions, the final agreement was reached at the selection meeting in Warsaw on April 27, 2002 (the program committee members denoted by * in the list below took part in the meeting). We thank all the program committee members and referees for their work which contributed to the quality of the meeting. We have tried to make the list of referees as complete and accurate as possible and apologize for all omissions and errors.

Special thanks go to Tomasz Waleń and Łukasz Kowalik who provided a reliable software system used for electronic submissions and reviewing. Finally, we would like to thank Mirosław Kowaluk, the chairman of the organizing committee, for his excellent work.

Warszawa, June 2002 Krzysztof Diks and Wojciech Rytter

Preface

This volume contains papers selected for presentation at the 27th International Symposium on Mathematical Foundations of Computer Science, MFCS 2002, held in Warsaw, Poland, August 26–30, 2002.

MFCS 2002 was organized by the Institute of Informatics of Warsaw University and the Foundation for Information Technology Development. It was supported by the European Association for Theoretical Computer Science. We gratefully acknowledge the support of all these institutions.

The series of MFCS symposia, organized on a rotating basis in the Czech Republic, Poland, and Slovakia, has a well-established tradition. The aim is to encourage high-quality research in all branches of theoretical computer science and bring together specialists who do not usually meet at specialized conferences. The previous meetings took place in Jabłonna, 1972; Štrbské Pleso, 1973; Jadwisin, 1974; Mariánské Lázně, 1975; Gdańsk, 1976; Tatranská Lomnica, 1977; Zakopane, 1978; Olomouc, 1979; Rydzyna, 1980; Štrbské Pleso, 1981; Prague, 1984; Bratislava, 1986; Karlovy Vary, 1988; Porąbka-Kozubnik, 1989; Banská Bystrica, 1990; Kazimierz Dolny, 1991; Prague, 1992; Gdańsk, 1993; Košice, 1994; Prague, 1995; Kraków, 1996; Bratislava, 1997; Brno, 1998; Szklarska Poręba, 1999; Bratislava, 2000; and Mariánské Lázně, 2001.

The MFCS 2002 proceedings consists of five invited papers and 48 contributed papers. We are grateful to all the invited speakers for accepting our invitation and sharing their insights on their research areas. We thank the authors of all submissions for their contribution to the scientific program of the meeting.

The contributed papers were selected by the Program Committee out of a total of 108 submissions. All submissions were evaluated by three members of the committee, with the assistance of referees. After discussions, the final agreement was reached at the selection meeting in Warsaw on April 27, 2002 (the program committee members denoted by * in the list below took part in the meeting). We thank all the program committee members and referees for their work which contributed to the quality of the meeting. We have tried to make the list of referees as complete and accurate as possible and apologize for all omissions and errors.

Special thanks go to Tomasz Waleń and Łukasz Kowalik who provided a reliable software system used for electronic submissions and reviewing. Finally we would like to thank Mirosław Kowaluk, the chairman of the organizing committee, for his excellent work.

Warsaw, June 2002 Krzysztof Diks and Wojciech Rytter

Program Committee

M. Crochemore (Marne-la-Valee)
*Krzysztof Diks (Co-chair, Warszawa)
*Costas Iliopoulos (London)
*Jerzy Jaromczyk (Lexington)
Michal Karoński (Poznań)
*Mirosław Kutyłowski (Wrocław)
*Klaus-Jörn Lange (Tübingen)
Thierry Lecroq (Rouen)
Andrzej Lingas (Lund)
*Ernst Mayr (Munich)
*Andrzej Pelc (Hull, Quebec)
*Wojciech Penczek (Warszawa)
*Branislav Rovan (Bratislava)
*Jan Rutten (Amsterdam)
*Wojciech Rytter (Co-chair, Warszawa, Liverpool)
*Jiri Sgall (Prague)
Ugo Vaccaro (Salerno)

Referees

S. Abdeddaim
J. Alber
M. Andersson
J. Andrews
F. Arbab
V. Auletta
M.P. Béal
A. Bałaban
S. Bala
A. Baltag
F. Bartels
T. Bayer
M. Bednarczyk
E. Ben-Sasson
B. Blanchet
M. Bojańczyk
A. Borzyszkowski
T. Borzyszkowski
V. Brattka
V. Bruyére
H. Buhrman
P. Caron
I. Cerna
J.M. Champarnaud
W. Charatonik
B. Chlebus
J. Chomicki
J. Chrząszcz
F. Cicalese
C. Cirstea
B. Codenotti
N. Creignou
J. Czyżowicz
C. Damm
J. Dassow
R. Dąbrowski
P. Degano
A. De Bonis
V. Diekert
G. Di Crescenzo
R. Dowgird
P. Duris
J. Fiala
W. Fokkink
L. Fortnow
P. Fraigniaud
F. Franek
C. Frougny
Z. Fulop
L. Gąsieniec
K. Gołąb
M. Gomułkiewicz
V. Gordon
M. Hammar
U. Hertrampf
V. Heun
J. Honkala
K. Holzapfel
J. Hromkovic
J.M. Jacquet
G. Jakacki
R. Janicki
M. Jantzen
T. Jurdziński
P. Kanarek
J. Karhumäki

J. Kari
R. Klasing
H. Kleine-Buning
P. Kolman
B. Konikowska
W. Kordecki
S. Kosub
Ł. Kowalik
D. Kowalski
M. Kowaluk
D. Kral
E. Kranakis
M. Kurowski
A. Kurucz
A. Kurz
D. Kuske
L. Larmore
S. Lasota
T. Lecroq
P. Lescanne
M. Liśkiewicz
Z. Lonc
E.M. Lundell
J. Lutz
M. Maaß
A. Malinowski
V. Marek
M. Margenstern
B. Masucci
F. Meyer auf der Heide
E. Mayordomo
J. Mazoyer
A. Mazurkiewicz
C. Meinel
M. Mnuk
A. Mora
L. Mouchard
A. Muscholl
M. Napoli
A. Navarra
J. Neraud
R. Niedermeier
D. Niwiński
E. Ochmanski
A. Offtermatt-Souza
D. Parente
K. Park
W. Pawlowski
H. Petersen
Y. Pinzon
M. Piotrów
W. Plandowski
A. Podelski
J.L. Ponty
P. Pudlak
M. Raab
T. Radzik
N. Rahman
I. Rapaport
K. Reinhardt
I. Rohloff
G. Rosu
J. Rutten
U. Rührmair
P. Rychlikowski
M.F. Sagot
P. Samarati
T. Schickinger
B. Smyth
K. Stencel
A. Szałas
M. Szczuka
M. Szreter
W. Szwast
M. Ślusarek
G. Świątek
A. Tarlecki
H. Täubig
D. Therien
W. Thomas
J. Toran
M. Truszczynski
P. Ullrich
P. Valarcher
M. Vardi
Y. Venema
H. Vollmer
K. Wagner
D. Walukiewicz
I. Walukiewicz
G. Wasilkowski
I. Wegener
T. Wierzbicki
J. Winkowski
J. Zhang
U. Zwick
G. Zwoźniak

Table of Contents

Invited Talks

Contributed Talks

Global Development via Local Observational Construction Steps*

Michel Bidoit[1], Donald Sannella[2], and Andrzej Tarlecki[3]

[1] Laboratoire Spécification et Vérification, CNRS & ENS de Cachan, France
[2] Laboratory for Foundations of Computer Science, University of Edinburgh, UK
[3] Institute of Informatics, Warsaw University and
Institute of Computer Science, Polish Academy of Sciences, Warsaw, Poland

Abstract. The way that refinement of individual "local" components of a specification relates to development of a "global" system from a specification of requirements is explored. Observational interpretation of specifications and refinements add expressive power and flexibility while bringing in some subtle problems. The results are instantiated in the context of CASL architectural specifications.

1 Introduction

There has been a great deal of work in the algebraic specification tradition on formalizing the rather intuitive and appealing idea of program development by stepwise refinement, including [EKMP82,Gan83,GM82,Sch87,ST88b]; for a recent survey, see [EK99]). There are many issues that make this a difficult problem, and some of them are rather subtle, one example being the relationship between specification structure and program structure. There are difficult interactions and tradeoffs, an obvious one being between the expressive power of a specification formalism and the ease of reasoning about specifications. Different approaches give more or less prominence to different issues. An overview that covers most of our own contributions is [ST97], with some more recent work addressing the problem of how to prove correctness of refinement steps [BH98], the design of a convenient formalism for writing specifications [ABK+03,BST02], and applications to data refinement in typed λ-calculus [HLST00].

A new angle that we explore here is the "global" effect of refining individual "local" components of a specification. This involves a well-known technique from algebraic specification, namely the use of pushouts of signatures and amalgamation of models to build large systems by composition of separate interrelated components. The situation becomes considerably more subtle when observational interpretation of specifications and refinements is taken into account.

Part of the answer has already been provided, the main references being Schoett's thesis [Sch87,Sch90] and our work on formal development in the EXTENDED ML framework [ST89]; the general ideas go back at least to [Hoa72].

* This work has been partially supported by KBN grant 7T11C 002 21 and European AGILE project IST-2001-32747 (AT), CNRS–PAS Research Cooperation Programme (MB, AT), and British–Polish Research Partnership Programme (DS, AT).

K. Diks et al. (Eds): MFSC 2002, LNCS 2420, pp. 1–24, 2002.

We have another look at these issues here, in the context of the CASL specification formalism [ABK+03] and in particular, its *architectural specifications* [BST02]. Architectural specifications, for describing the modular structure of software systems, are probably the most novel feature of CASL. We view them here as a means of making complex refinement steps, by defining well-structured constructions to be used to build the overall system from implementations of individual units (these also include parametrized units, acting as constructions providing some local construction steps to be used in a more global context).

We begin by introducing in Sect. 2 some details of the underlying logical system we will be working with, and our assumptions concerning specifications built using this system. Our basic view of program development by means of consecutive local refinement steps is presented in Sect. 3. Then, an observational view of specifications is motivated and recalled in Sect. 4. The principal core of the work is in Sect. 5, where we combine the ideas of the previous two sections and discuss program development by local refinement steps with respect to an observational interpretation of the specifications involved. Section 6 introduces a simplified version of CASL architectural specifications, while Sect. 7 sketches their observational semantics and shows how the ideas of Sect. 5 are instantiated in this context. Further work and possible generalizations are discussed in Sect. 8. Due to lack of space we have been unable to include concrete examples that illustrate the definitions and results, but we plan to provide such material in a future extended version.

2 Signatures, Models and Specifications

A basic assumption underpinning algebraic specification and derived approaches to software specification and development is that software systems are modeled as algebras (of some kind) and their static properties are captured by algebraic signatures (again, adapted as appropriate). This leads to quite a flexible framework, which can be tuned as desired to cope with various programming features of interest by selecting the appropriate variation of algebra and signature. This flexibility has been formalized via the notion of *institution* [GB92] and related work on the theory of specifications and formal program development [ST88a,ST97,BH93]. However, rather than exploiting the full generality of institutions, to keep things simple and illustrative we will in this paper base our considerations on a very basic logical framework, leaving to a more extensive presentation elsewhere the required generalization and adaptation to a fully-fledged formalism such as CASL.

So, we will deal here with the usual notions of many-sorted algebraic signatures and signature morphisms; we will assume that all signatures contain a distinguished Boolean part: a sort *bool* with two constants *true* and *false* preserved by all signature morphisms. This yields the category **AlgSig** – it is cocomplete, and we will assume that it comes with some standard construction of pushouts.

For each algebraic signature Σ, $\mathbf{Alg}(\Sigma)$ stands for the usual category of Σ-algebras and their homomorphisms – we restrict attention to algebras with a fixed, standard interpretation of the Boolean part of the signature. As usual, each signature morphism $\sigma: \Sigma \to \Sigma'$ determines a *reduct* functor $\_|_\sigma: \mathbf{Alg}(\Sigma') \to \mathbf{Alg}(\Sigma)$. This yields a functor $\mathbf{Alg}: \mathbf{AlgSig}^{op} \to \mathbf{Cat}$. We refer to [ST99] for a more detailed presentation of the technicalities and for the standard notations we will use in the following.

It can easily be checked that **Alg** is continuous, i.e., maps colimits of algebraic signatures to limits of (algebra) categories (the initial signature, containing the Boolean part only, is mapped to the category having as its only object the algebra providing the fixed interpretation for the Boolean part). In particular, the following *amalgamation property* holds:

Lemma 2.1. *Given a pushout in the category of algebraic signatures* **AlgSig***:*

$$\begin{array}{ccc} \Sigma_1 & \xrightarrow{\iota'} & \Sigma'_1 \\ \uparrow{\scriptstyle\gamma} & & \uparrow{\scriptstyle\gamma'} \\ \Sigma & \xrightarrow[\iota]{} & \Sigma' \end{array}$$

for any algebras $A_1 \in |\mathbf{Alg}(\Sigma_1)|$ and $A' \in |\mathbf{Alg}(\Sigma')|$ such that $A_1|_\gamma = A'|_\iota$ there exists a unique algebra $A'_1 \in |\mathbf{Alg}(\Sigma'_1)|$ such that $A'_1|_{\iota'} = A_1$ and $A'_1|_{\gamma'} = A'$; and similarly for algebra homomorphisms.

Given a signature Σ, terms and first-order formulae with equality are defined as usual. Σ-sentences are closed first-order formulae. Given a Σ-algebra A, a set of variables X and a valuation of variables $v: X \to |A|$, the *value* $t_{A[v]}$ of a term t with variables X in A under v and the *satisfaction* $A[v] \models \phi$ of a formula ϕ with variables X in A under v are defined as usual.

We will also employ a generalized notion of terms, modeling a pretty general idea of how a value may be determined in an algebra. Given a signature Σ, a *conditional term* of sort s with variables X is of the form $p = ((\phi_i, t_i)_{i \geq 0}, t)$, where for $i \geq 0$, ϕ_i are formulae with variables X, and t_i and t are terms of sort s with variables X. Given a Σ-algebra A and a valuation $v: X \to |A|$, the value $p_{A[v]}$ of such a conditional term p is $(t_k)_{A[v]}$ for the least $k \geq 0$ such that $A[v] \models \phi_k$, or $t_{A[v]}$ if no such $k \geq 0$ exists.

This allows for a further generalization of *derived signature morphisms* [SB83], where we allow such a morphism $\delta: \Sigma \to \Sigma'$ to map function symbols $f: s_1 \times \ldots \times s_n \to s$ to conditional terms of sort s with variables $\{x_1: s_1, \ldots, x_n: s_n\}$. Evidently, such a derived signature morphisms $\delta: \Sigma \to \Sigma'$ still determines a reduct function $\_|_\delta: |\mathbf{Alg}(\Sigma')| \to |\mathbf{Alg}(\Sigma)|$ on algebra classes (which in general does *not* extend to a reduct functor between algebra categories).

We will not need to know much about the formalism used for writing specifications. We just assume that some class of specifications is defined, equipped with a semantics that for any specification SP determines its signature $Sig(SP) \in |\mathbf{AlgSig}|$ and its class of *models* $Mod(SP) \subseteq |\mathbf{Alg}(Sig(SP))|$. We also assume that the class specifications is closed under *translation* along signature morphisms, i.e., for any specification SP and signature morphism $\sigma\colon Sig(SP) \to \Sigma'$, we have a specification $\sigma(SP)$ with $Sig(\sigma(SP)) = \Sigma'$ and $Mod(\sigma(SP)) = \{A' \in |\mathbf{Alg}(\Sigma')| \mid A'|_\sigma \in Mod(SP)\}$, and under *unions*, i.e., for any specifications SP_1 and SP_2 with common signature, we have a specification SP_1 **and** SP_2 with $Sig(SP_1$ **and** $SP_2) = Sig(SP_1) = Sig(SP_2)$ and $Mod(SP_1$ **and** $SP_2) = Mod(SP_1) \cap Mod(SP_2)$. So, specifications can for instance be basic specifications, given by a signature and a set of axioms (sentences) over this signature; or structured specifications built over the institution we have implicitly introduced above as defined in [ST88a]; or structured specifications built using more advanced structuring mechanisms such as those of CASL [ABK+03].

3 Program Development and Refinements

In this section we briefly recapitulate our view of the process by means of which software can be formally developed from an algebraic specification of requirements, see [ST88b,ST97]. This is followed by an explanation of the way that development steps can arise from "local" constructions.

Given a requirements specification SP, the programmer's task is to provide a program that correctly implements it. In semantic terms, this amounts to building an algebra $A \in |\mathbf{Alg}(Sig(SP))|$ such that $A \in Mod(SP)$. At this level of generality and abstraction, we will not offer programming techniques for achieving this. We will instead concentrate on the methodological idea that one may proceed in a stepwise fashion by means of successive *refinements*, gradually enriching the original requirements specification with more and more implementation details until a directly implementable specification is obtained:

$$SP_0 \leadsto SP_1 \leadsto \cdots \leadsto SP_n$$

SP_0 is the original requirements specification and $SP_{i-1} \leadsto SP_i$ for $i = 1, \ldots, n$ are individual refinement steps. Joined together, these lead from SP_0 to a specification SP_n which is so detailed that it can be implemented directly – that is, such that an algebra $A_n \in Mod(SP_n)$ can be easily programmed. This "program" A_n correctly implements SP_0 provided we require that refinement steps preserve the specification signature and define:

$$SP \leadsto SP' \iff Mod(SP') \subseteq Mod(SP)$$

Although mathematically simple and quite powerful (in the context of a sufficiently rich specification formalism), this view of the development process may be made more practical by taking into account the fact that successive specifications in the above chain will tend to incorporate more and more details arising from successive design decisions. Some parts thereby become fully determined, and remain fixed until the development process is complete:

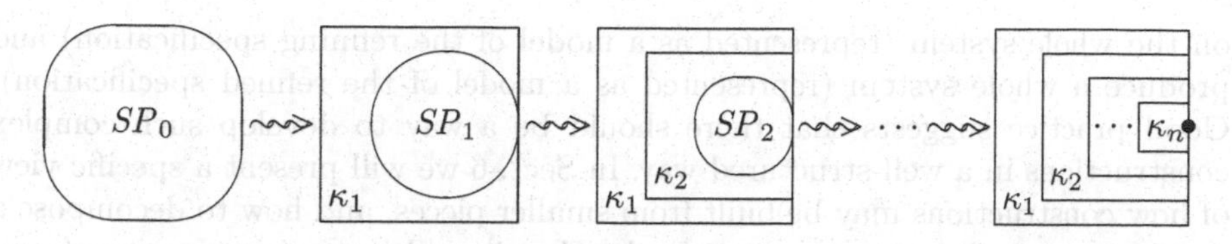

It seems only natural to separate the finished parts from the specification of what remains to be done. This gives the following picture:

$$SP_0 \underset{\kappa_1}{\leadsto} SP_1 \underset{\kappa_2}{\leadsto} SP_2 \underset{\kappa_3}{\leadsto} \cdots \underset{\kappa_n}{\leadsto} \bullet\ SP_n = EMPTY$$

where for $i = 1, \ldots, n$, the specifications SP_i now describe the part of the system that remains to be implemented, while each κ_i is a *parametrized program* [Gog84] which semantically amounts to a (possibly partial) function on algebras $\kappa_i : |\mathbf{Alg}(Sig(SP_i))| \rightharpoonup |\mathbf{Alg}(Sig(SP_{i-1}))|$ which we will call a *construction.* Now, given specifications SP and SP' and a construction $\kappa : |\mathbf{Alg}(Sig(SP'))| \rightharpoonup |\mathbf{Alg}(Sig(SP))|$, we define:

$$SP \underset{\kappa}{\leadsto} SP' \iff Mod(SP') \subseteq dom(\kappa) \text{ and } \kappa(Mod(SP')) \subseteq Mod(SP)$$

This definition captures the correctness requirements we impose on the individual refinement steps, which guarantee that given a successful development sequence:

$$SP_0 \underset{\kappa_1}{\leadsto} SP_1 \underset{\kappa_2}{\leadsto} \cdots \underset{\kappa_n}{\leadsto} SP_n = EMPTY$$

we obtain the algebra:

$$\kappa_1(\kappa_2(\ldots \kappa_n(empty)\ldots)) \in Mod(SP_0)$$

where $EMPTY$ is the empty specification over the "empty" signature (i.e. the initial object in **AlgSig**, containing the Boolean part only) and *empty* is its unique standard realization.

Even though our presentation suggests a "top-down" development process, starting from the requirements specification and proceeding towards a situation where nothing is left to be implemented, this need not be the case in general. We can instead proceed "bottom-up", starting with $EMPTY$ and successively providing constructions which add in bits and pieces in an incremental fashion until an implementation of the original specification is obtained. Or we can combine the two techniques, and proceed in a "middle-out" fashion. What matters is that at the end a chain of correct refinement steps emerges which links the requirements specification with $EMPTY$.

Another point about the above presentation is that it relies on a global view of specifications and their refinement: constructions are required to work on the whole system (represented as a model of the refining specification) and produce a whole system (represented as a model of the refined specification). Good practice suggests that there should be a way to develop such complex constructions in a well-structured way. In Sect. 6 we will present a specific view of how constructions may be built from smaller pieces, and how to decompose a development task into a number of subtasks via multi-argument constructions. For now, let us concentrate on one aspect of this, and discuss how to make refinement steps "local" – that is, how to use only *part* of the system built so far to implement some remaining parts of the requirements specification, and then incorporate the result in the system as a whole.

Technically, this means that we need to look at constructions that map Σ-algebras to Σ'-algebras, but apply them to parts cut out of "larger" Σ_G-algebras, where this "cutting out" is given as the reduct with respect to a signature morphism $\gamma: \Sigma \to \Sigma_G$ that fits the local argument signature into its global context. W.l.o.g. we can assume that constructions are *persistent*: the argument of a construction is always fully included in its result, without modification[1] – note that this assumption holds for all constructions that can be declared and specified in CASL, see Sect. 6. In fact, we generalize this somewhat by considering arbitrary signature morphisms rather than just inclusions.

Throughout the rest of the paper, we will repeatedly refer to the signatures and morphisms in the following pushout diagram:

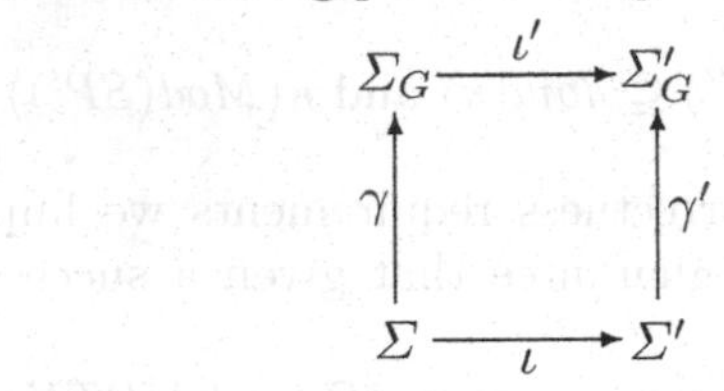

where the local construction is along the bottom of the diagram, "cutting out" its argument from a larger algebra uses the signature morphism on the left, and the resulting global construction is along the top.

Definition 3.1. *Given a signature morphism* $\iota: \Sigma \to \Sigma'$, *a* local construction along ι *is a persistent partial function* $F: |\mathbf{Alg}(\Sigma)| \rightharpoonup |\mathbf{Alg}(\Sigma')|$ *(for each* $A \in dom(F)$, $F(A)|_\iota = A$*). We write* $Mod(\Sigma \xrightarrow{\iota} \Sigma')$ *for the class of all local constructions along* ι.

Given a local construction F *along* $\iota: \Sigma \to \Sigma'$, *a morphism* $\gamma: \Sigma \to \Sigma_G$ *fitting* Σ *into a "global" signature* Σ_G, *and a* Σ_G*-algebra* $\mathcal{G} \in |\mathbf{Alg}(\Sigma_G)|$, *we define the* global result $F_G(\mathcal{G})$ *of applying* F *to* $\mathcal{G}$ *by reference to the pushout diagram above, using the amalgamation property: if* $\mathcal{G}|_\gamma \in dom(F)$ *then* $F_G(\mathcal{G})$ *is the unique* Σ'_G*-algebra such that* $F_G(\mathcal{G})|_{\iota'} = \mathcal{G}$ *and* $F_G(\mathcal{G})|_{\gamma'} = F(\mathcal{G}|_\gamma)$*; otherwise* $F_G(\mathcal{G})$ *is undefined.*

[1] Otherwise we would have to explicitly indicate "sharing" between the argument and result of each construction, and explain how such sharing is preserved by the various ways of putting together constructions, as was painfully spelled out in [ST89]. If necessary, superfluous components of algebras constructed using persistent constructions can be discarded at the end using the reduct along a signature inclusion.

This determines a global construction $F_G: |\mathbf{Alg}(\Sigma_G)| \to |\mathbf{Alg}(\Sigma'_G)|$, *which is persistent along* $\iota': \Sigma_G \to \Sigma'_G$.

This way of "lifting" a persistent function to a larger context via a *"fitting morphism"* using signature pushout and amalgamation is well established in the algebraic specification tradition, going back at least to "parametrized specifications" with free functor semantics, see [EM85].

We will not dwell here on how particular (local) constructions are defined. Free functor semantics for parametrized specifications is one way to proceed, with the persistency requirement giving rise to additional proof obligations [EM85]. Perhaps closer to ordinary programming is to give explicitly a "definitional" derived signature morphism $\delta: \Sigma' \to \Sigma$ that defines Σ'-components in terms of Σ-components. The induced reduct function $\_|_\delta: |\mathbf{Alg}(\Sigma)| \to |\mathbf{Alg}(\Sigma')|$ is a local construction along a signature morphism $\iota: \Sigma \to \Sigma'$ whenever $\iota;\delta = id_\Sigma$.[2]

Suppose now that a local construction F along $\iota: \Sigma \to \Sigma'$ comes with a "semantic" specification of its input/output properties, given as a specification SP with $Sig(SP) = \Sigma$ of the requirements on its arguments together with a specification SP' with $Sig(SP') = \Sigma'$ of the guaranteed properties of its result. Again w.l.o.g. we require that $Mod(SP')|_\iota \subseteq Mod(SP)$, as is indeed ensured for instance in CASL.

Definition 3.2. *A local construction F along $\iota: Sig(SP) \to Sig(SP')$ is* strictly correct w.r.t. SP and SP' *if for all models $A \in Mod(SP)$, $A \in dom(F)$ and $F(A) \in Mod(SP')$. We write $Mod(SP \xrightarrow{\iota} SP')$ for the class of all local constructions along ι that are strictly correct w.r.t. SP and SP'.*

The following theorem shows how such locally correct constructions can be used for global refinement steps.

Theorem 3.3. *Given a local construction $F \in Mod(SP \xrightarrow{\iota} SP')$, specification SP_G with fitting morphism $\gamma: Sig(SP) \to Sig(SP_G)$, and specification SP'_G with $Sig(SP'_G) = \Sigma'_G$, SP_G correctly refines SP'_G via the global construction F_G (i.e., $SP'_G \underset{F_G}{\rightsquigarrow} SP_G$) provided that*

- *$Mod(SP_G) \subseteq Mod(\gamma(SP))$, and*
- *$Mod(\gamma'(SP')$ **and** $\iota'(SP_G)) \subseteq Mod(SP'_G)$.*

Proof. Let $\mathcal{G} \in Mod(SP_G)$. Then $\mathcal{G}|_\gamma \in Mod(SP)$, and so $\mathcal{G}|_\gamma \in dom(F)$ and $F(\mathcal{G}|_\gamma) \in Mod(SP')$. Consequently $F_G(\mathcal{G}) \in Mod(\gamma'(SP')) \cap Mod(\iota'(SP_G))$. □

Informally, this captures directly a "bottom-up" process of building implementations, whereby we start with SP_G, find a local construction $F \in Mod(SP \xrightarrow{\iota} SP')$ with a fitting morphism γ that satisfies the first condition, and define SP'_G such that the second condition is satisfied (e.g. take $SP'_G = \gamma'(SP')$ **and** $\iota'(SP_G)$). When proceeding "top-down", we start with the global requirements specification SP'_G. To use a local construction $F \in Mod(SP \xrightarrow{\iota} SP')$, we have to decide

[2] Composition of derived signature morphisms can be defined in the evident fashion, and equality of two derived signature morphisms is understood here semantically.

which part of the requirements it is going to implement by providing a signature morphism $\gamma'\colon Sig(SP') \to Sig(SP'_G)$, then construct the "pushout complement" $\gamma\colon Sig(SP) \to \Sigma_G$, $\iota'\colon \Sigma_G \to Sig(SP'_G)$ for ι and γ', and finally devise a specification SP_G with $Sig(SP_G) = \Sigma_G$ such that both conditions are satisfied. Then we can proceed with SP_G as the requirements specification for the components that remain to be implemented.

4 Observational Equivalence

So far, we have made few assumptions about the formalism used for writing specifications. Intuitively, it is clear that any such formalism should admit basic specifications given as sets of axioms over some fixed signature. The usual interpretation then is to take as models for such a basic specification all the algebras that satisfy the axioms. However, in many practical examples this turns out to be overly restrictive. The point is that only a subset of the sorts in the signature of a specification are typically intended to be directly observable – the others are treated as internal, with properties of their elements made visible only via *observations* leading to the observable sorts. This calls for a relaxation of the interpretation of specifications, as advocated in numerous "observational" or "behavioural" approaches, going back at least to [GGM76,Rei81]. The starting point is that given an algebraic signature, one has to fix a set of *observable sorts*. Then, roughly, two approaches are possible:

- introduce an internal *observational indistinguishability* relation between algebra elements, and re-interpret equality in the axioms as indistinguishability,
- introduce an external *observational equivalence* on algebras, and re-interpret specifications by closing their class of models under such equivalence.

It turns out that under some acceptable technical conditions, the two approaches are closely related and coincide for most basic specifications [BHW95,BT96]. However, the former approach seems more difficult to extend to structured specifications and parametrization. Hence, we follow here the latter possibility.

Definition 4.1. *Consider a signature Σ with* observable sorts $OBS \subseteq sorts(\Sigma)$. *We always assume that bool* $\in OBS$. *A* correspondence *between two algebras* $A, B \in |\mathbf{Alg}(\Sigma)|$, *written* $\rho\colon A \bowtie B$, *is a relation* $\rho \subseteq |A| \times |B|$ *that is closed under the operations*[3] *and is the identity on* $|A|_{bool} = |B|_{bool}$. *It is* observational *if it is bijective on observable sorts.*

Two algebras $A,B \in |\mathbf{Alg}(\Sigma)|$ *are* observationally equivalent, *written* $A \equiv_{OBS} B$, *if there exists an observational correspondence between them.*

This formulation is due to [Sch87] (cf. "simulations" in [Mil71] and "weak homomorphisms" in [Gin68]) and is equivalent to other standard ways of defining

[3] That is, for $f\colon s_1 \times \ldots \times s_n \to s$, $a_1 \in |A|_{s_1}, \ldots, a_n \in |A|_{s_n}$ and $b_1 \in |B|_{s_1}, \ldots, b_n \in |B|_{s_n}$, if $(a_1, b_1) \in \rho_{s_1}, \ldots, (a_n, b_n) \in \rho_{s_n}$ then $(f_A(a_1, \ldots, a_n), f_B(b_1, \ldots, b_n)) \in \rho_s$.

observational equivalence between algebras, where a special role is played by *observable equalities*, i.e., equalities between terms of observable sorts.

It is easy to check that identities are correspondences and the class of correspondences is closed under composition and reducts w.r.t. signature morphisms.

Correspondences may in fact be identified with certain spans of homomorphisms: a correspondence $\rho: A \bowtie B$ is a span $(h_A: C \to A, h_B: C \to B)$ where, for each sort s distinct from *bool*, $|C|_s$ is a subset of the Cartesian product $|A|_s \times |B|_s$, $|C|_{bool} = |A|_{bool} = |B|_{bool}$, the homomorphisms are the projections for all sorts $s \neq bool$, and the identity on the carrier of the sort *bool*. Such a span is observational if the homomorphisms are bijective on observable sorts. This directly implies that the reduct of a correspondence along a signature morphism is a correspondence. More interestingly, for observational correspondences this extends to derived signature morphisms with observable conditions.

Consider a signature Σ with observable sorts $OBS \subseteq sorts(\Sigma)$. A conditional term $((\phi_i, t_i)_{i \geq 0}, t)$ is *OBS-admissible* if for all $i \geq 0$, ϕ_i are quantifier-free formulae with observable equalities only. A derived signature morphism $\delta: \Sigma' \to \Sigma$ is *OBS*-admissible if it maps Σ'-operations to *OBS*-admissible terms.

Lemma 4.2. *Let $\delta: \Sigma' \to \Sigma$ be an OBS-admissible derived signature morphism, A and B be two Σ-algebras, and $\rho: A \bowtie B$ be an observational correspondence. Then $\rho|_\delta: A|_\delta \bowtie B|_\delta$ is a correspondence as well. Moreover, it is observational for any set $OBS' \subseteq sorts(\Sigma')$ of observable sorts such that $\delta(OBS') \subseteq OBS$.* □

The view of correspondences as spans of homomorphisms also leads to an easy extension to correspondences of the amalgamation property given in Lemma 2.1 for algebras and homomorphisms.

Observational equivalence between algebras can be characterized in terms of the alternative approach based on internal indistinguishability. Consider a signature Σ with observable sorts $OBS \subseteq sorts(\Sigma)$ (with $bool \in OBS$) and an algebra $A \in |\mathbf{Alg}(\Sigma)|$. Let $\langle A \rangle_{OBS}$ be the subalgebra of A generated by the carriers of observable sorts. *Observational indistinguishability* on A, denoted by $\approx_{OBS}$, is the largest congruence on $\langle A \rangle_{OBS}$ that is the identity on observable sorts. The *observational quotient* of A, written $A/{\approx_{OBS}}$, is the quotient of $\langle A \rangle_{OBS}$ by $\approx_{OBS}$.

Theorem 4.3. *Consider a signature Σ with observable sorts $OBS \subseteq sorts(\Sigma)$. Two Σ-algebras are observationally equivalent if and only if their observational quotients are isomorphic.* □

So far we have considered observational equivalence w.r.t. a rather arbitrary set of observable sorts. In practice, however, for any development framework (and programming language), the set of types directly observable to the user is fixed and given in advance – for the framework at hand, the right choice seems to be to take the sort *bool* as the only observable sort. Note that choosing *bool* as the only observable sort is not a restriction, since one can always treat another sort as observable by introducing an "equality predicate" on it. Moreover, this choice will not prevent us from manipulating an explicit set of observable sorts (always keeping *bool* among them though) when considering "local" signatures for verification purposes.

We will consider observational equivalence of "global" models with respect to the single observable sort *bool* – we write $\equiv_{\{bool\}}$ simply as $\equiv$. For any "global" specification SP_G with $Sig(SP_G) = \Sigma_G$, we define its *observational interpretation* by abstracting from the standard interpretation as follows:

$$Abs_{\equiv}(SP_G) = \{\mathcal{G} \in |\mathbf{Alg}(\Sigma_G)| \mid \mathcal{G} \equiv \mathcal{H} \text{ for some } \mathcal{H} \in Mod(SP_G)\}.$$

5 Observational Refinement Steps

The most obvious way to re-interpret correctness of refinement steps $SP' \underset{\kappa}{\rightsquigarrow\!\!>} SP$ to take advantage of the observational interpretation of specifications indicated in the previous section is to relax the earlier definition by requiring $Abs_{\equiv}(SP) \subseteq dom(\kappa)$ and $\kappa(Abs_{\equiv}(SP)) \subseteq Abs_{\equiv}(SP')$. This works, but misses a crucial point: when using a realization of a specification, we should be able to pretend that it satisfies the specification literally, even if when actually implementing it we are permitted to supply an algebra that is correct only up to observational equivalence. This leads to a new notion of *observational refinement*: given specifications SP and SP' and a construction κ: $|\mathbf{Alg}(Sig(SP'))| \rightharpoonup |\mathbf{Alg}(Sig(SP))|$, we define:

$$SP \overset{\equiv}{\underset{\kappa}{\rightsquigarrow\!\!>}} SP' \iff Mod(SP') \subseteq dom(\kappa) \text{ and } \kappa(Mod(SP')) \subseteq Abs_{\equiv}(SP)$$

This relaxation has a price: observational refinements do not automatically compose! The crucial insight to resolve this problem comes from [Sch87], who noticed that well-behaved constructions satisfy the following *stability* property.

Definition 5.1. *A construction* κ: $|\mathbf{Alg}(\Sigma)| \rightharpoonup |\mathbf{Alg}(\Sigma')|$ *is* stable *if it preserves observational equivalence of algebras, i.e., for any algebras* $A,B \in |\mathbf{Alg}(\Sigma)|$ *such that* $A \equiv B$, *if* $A \in dom(\kappa)$ *then* $B \in dom(\kappa)$ *and* $\kappa(A) \equiv \kappa(B)$.

Now, if all the constructions involved are stable then from a successful chain of observational refinements

$$SP_0 \overset{\equiv}{\underset{\kappa_1}{\rightsquigarrow\!\!>}} SP_1 \overset{\equiv}{\underset{\kappa_2}{\rightsquigarrow\!\!>}} \cdots \overset{\equiv}{\underset{\kappa_n}{\rightsquigarrow\!\!>}} SP_n = EMPTY$$

we obtain:

$$\kappa_1(\kappa_2(\ldots\kappa_n(empty)\ldots)) \in Abs_{\equiv}(SP_0).$$

The rest of this section is devoted to an analysis of conditions that ensure stability of constructions and observational correctness of refinement steps when the constructions arise via the use of local constructions, as in Sect. 3. The problem is that we want to restrict attention to conditions that are essentially local to the local constructions involved, rather than conditions that refer to all the possible global contexts in which such a construction can be used.

Let us start with the stability property.

Definition 5.2. *A local construction F along $\iota\colon \Sigma \to \Sigma'$ is* locally stable *if for any Σ-algebras $A, B \in |\mathbf{Alg}(\Sigma)|$ and correspondence $\rho\colon A \bowtie B$, $A \in dom(F)$ if and only if $B \in dom(F)$ and moreover, if this is the case then there exists a correspondence $\rho'\colon F(A) \bowtie F(B)$ that extends ρ (i.e., $\rho'|_\iota = \rho$).*

Proposition 5.3. *The composition of locally stable constructions is a locally stable construction.* □

Lemma 5.4. *If F is a locally stable construction along $\iota\colon \Sigma \to \Sigma'$ then for any signature Σ_G and fitting morphism $\gamma\colon \Sigma \to \Sigma_G$, the induced global construction $F_G\colon |\mathbf{Alg}(\Sigma_G)| \rightharpoonup |\mathbf{Alg}(\Sigma'_G)|$ along $\iota'\colon \Sigma_G \to \Sigma'_G$ is locally stable as well.*

Proof. Consider a correspondence $\rho_G\colon \mathcal{G} \bowtie \mathcal{H}$ between algebras $\mathcal{G}, \mathcal{H} \in |\mathbf{Alg}(\Sigma_G)|$. Its reduct is a correspondence $\rho_G|_\gamma\colon \mathcal{G}|_\gamma \bowtie \mathcal{H}|_\gamma$, so $\mathcal{G}|_\gamma \in dom(F)$ iff $\mathcal{H}|_\gamma \in dom(F)$, and consequently $\mathcal{G} \in dom(F_G)$ iff $\mathcal{H} \in dom(F_G)$. Suppose $\mathcal{G}|_\gamma \in dom(F)$. Then there exists a correspondence $\rho'\colon F(\mathcal{G}|_\gamma) \bowtie F(\mathcal{H}|_\gamma)$ with $\rho'|_\iota = \rho_G|_\gamma$. Amalgamation of ρ_G and ρ' yields a correspondence $\rho'_G\colon F_G(\mathcal{G}) \bowtie F_G(\mathcal{H})$ such that $\rho'_G|_{\iota'} = \rho_G$. □

Corollary 5.5. *If F is a locally stable construction along $\iota\colon \Sigma \to \Sigma'$ then for any signature Σ_G and fitting morphism $\gamma\colon \Sigma \to \Sigma_G$, the induced global construction $F_G\colon |\mathbf{Alg}(\Sigma_G)| \rightharpoonup |\mathbf{Alg}(\Sigma'_G)|$ along $\iota'\colon \Sigma_G \to \Sigma'_G$ is stable.*

Proof. Let $\mathcal{G}, \mathcal{H} \in |\mathbf{Alg}(\Sigma_G)|$ be such that $\mathcal{G} \equiv \mathcal{H}$. Then there is a correspondence $\rho_G\colon \mathcal{G} \bowtie \mathcal{H}$. By Lemma 5.4, if $\mathcal{G} \in dom(F_G)$ then $\mathcal{H} \in dom(F_G)$ and there is a correspondence $\rho'_G\colon F_G(\mathcal{G}) \bowtie F_G(\mathcal{H})$, which proves $F_G(\mathcal{G}) \equiv F_G(\mathcal{H})$. □

This establishes a sufficient local condition which ensures that a local construction induces a stable global construction in every possible context of use.

The following is a corollary of Lemma 4.2.

Corollary 5.6. *Let $\delta\colon \Sigma' \to \Sigma$ be a $\{bool\}$-admissible derived signature morphism and $\iota\colon \Sigma \to \Sigma'$ be a signature morphism such that $\iota;\delta = id_\Sigma$. Then the reduct $F = \_|_\delta\colon Mod(\Sigma) \to Mod(\Sigma')$) is a local construction that is locally stable.* □

The above corollary supports the point put forward in [Sch87] that stable constructions are those that respect modularity in the software construction process. That is, such constructions can use the components provided by their imported parameters, but they cannot take advantage of their particular internal properties. This is the point of the requirement that δ be $\{bool\}$-admissible: any branching in the code must be governed by directly observable properties. This turns (local) stability into a directive for language design, rather than a condition to be checked on a case-by-case basis: in a language with good modularization facilities, all constructions that one can code should be locally stable.

Let us turn now to the issue of correctness w.r.t. given specifications.

Definition 5.7. *A local construction F along $\iota\colon Sig(SP) \to Sig(SP')$ is* observationally correct w.r.t. *SP* and *SP'* *if for every model $A \in Mod(SP)$, $A \in dom(F)$*

and there exists a model $A' \in Mod(SP')$ and correspondence $\rho': A' \bowtie F(A)$ such that $\rho'|_\iota$ is the identity.

We write $Mod_{lc}(SP \xrightarrow{\iota} SP')$ for the class of all locally stable constructions along ι that are observationally correct w.r.t. SP and SP'.

The requirement above implies that $A'|_\iota = A$ and $A' \equiv_{\iota(sorts(\Sigma))} F(A)$, which in turn is in general stronger than $F(A) \in Abs_{\equiv_{\iota(sorts(\Sigma))}}(SP')$. It follows that if $F \in Mod_{lc}(SP \xrightarrow{\iota} SP')$ then there is some $F' \in Mod(SP \xrightarrow{\iota} SP')$ such that $dom(F') = dom(F)$ and for each $A \in Mod(SP)$, $F'(A) \equiv_{\iota(sorts(\Sigma))} F(A)$. However, in general $Mod(SP \xrightarrow{\iota} SP') \not\subseteq Mod_{lc}(SP \xrightarrow{\iota} SP')$, as strictly correct local constructions need not be stable. Moreover, it may happen that there are no stable observationally correct constructions, even if there are strictly correct ones: that is, we may have $Mod_{lc}(SP \xrightarrow{\iota} SP') = \emptyset$ even if $Mod(SP \xrightarrow{\iota} SP') \neq \emptyset$. This was perhaps first pointed out in [Ber87], in a different framework.

Counterexample 5.8. Let SP_1 include a non-observable sort s with two constants $a, b: s$, and let SP_2 enrich SP_1 by an observable sort o, two constants $c, d: o$ and axiom $c \neq d \iff a = b$. Then $Mod(SP_1 \to SP_2)$ is non-empty, with any construction in it mapping models satisfying $a = b$ to those that satisfy $c \neq d$, and models satisfying $a \neq b$ to those that satisfy $c = d$. But none of these constructions is stable!

Lemma 5.9. *Consider a local construction F along $\iota: Sig(SP) \to Sig(SP')$ that is observationally correct w.r.t. SP and SP'. Then, for every global signature Σ_G and fitting morphism $\gamma: Sig(SP) \to \Sigma_G$, for every $\mathcal{G} \in Mod(\gamma(SP))$ we have $\mathcal{G} \in dom(F_G)$ and there is some $\mathcal{G}' \in Mod(\gamma'(SP'))$ such that $\mathcal{G}'|_{\iota'} = \mathcal{G}$ and $\mathcal{G}' \equiv F_G(\mathcal{G})$.*

Proof. We have $\mathcal{G}|_\gamma \in Mod(SP)$, and so $\mathcal{G}|_\gamma \in dom(F)$ and there exist $A' \in Mod(SP')$ and a correspondence $\rho': A' \bowtie F(\mathcal{G}|_\gamma)$ with identity reduct $\rho'|_\iota$. Consider the unique Σ'_G-algebra $\mathcal{G}'$ such that $\mathcal{G}'|_{\iota'} = \mathcal{G}$ and $\mathcal{G}'|_{\gamma'} = A'$. Then the identity $id_{\mathcal{G}}: \mathcal{G} \bowtie \mathcal{G}$ and $\rho': A' \bowtie F(\mathcal{G}|_\gamma)$ amalgamate to a correspondence $\rho'_G: \mathcal{G}' \bowtie F_G(\mathcal{G})$, which proves that $F_G(\mathcal{G}) \equiv \mathcal{G}' \in Mod(\gamma'(SP'))$. □

If $F \in Mod_{lc}(SP \xrightarrow{\iota} SP')$ and $\gamma: Sig(SP) \to \Sigma_G$, then by Lemma 5.9 we obtain $\gamma'(SP') \underset{F_G}{\overset{\equiv}{\leadsto}} \gamma(SP)$, and since F_G is stable by Cor. 5.5, we can use this in the observational development process. Given two "global" specifications SP_G with $Sig(SP_G) = \Sigma_G$ and SP'_G with $Sig(SP'_G) = \Sigma'_G$, we have $SP'_G \underset{F_G}{\overset{\equiv}{\leadsto}} SP_G$ whenever $Mod(SP_G) \subseteq Abs_{\equiv}(\gamma(SP))$ and $Mod(\gamma'(SP')) \subseteq Abs_{\equiv}(SP'_G)$. But while the former requirement is quite acceptable, the latter is in fact impossible to achieve in practice since it implicitly requires that all the global requirements must follow (up to observational equivalence) from the result specification for the local construction. More practical requirements are obtained by generalizing Thm. 3.3 to the observational setting:

Theorem 5.10. *Given a local construction* $F \in Mod_{lc}(SP \xrightarrow{\iota} SP')$, *specification* SP_G *with fitting morphism* $\gamma: Sig(SP) \to Sig(SP_G)$, *and specification* SP'_G *with* $Sig(SP'_G) = \Sigma'_G$, *if*

(i) $Mod(SP_G) \subseteq Abs_{\equiv}(SP_G \text{ **and** } \gamma(SP))$, *and*
(ii) $Mod(\gamma'(SP') \text{ **and** } \iota'(SP_G)) \subseteq Abs_{\equiv}(SP'_G)$

then for every $\mathcal{G} \in Mod(SP_G)$, *we have* $\mathcal{G} \in dom(F_G)$ *and* $F_G(\mathcal{G}) \in Abs_{\equiv}(SP'_G)$. *Consequently:*

$$SP'_G \underset{F_G}{\overset{\equiv}{\rightsquigarrow}} SP_G.$$

Proof. Let $\mathcal{G} \in Mod(SP_G)$. Then $\mathcal{G} \equiv \mathcal{H}$ for some $\mathcal{H} \in Mod(SP_G) \cap Mod(\gamma(SP))$ by (i). By Lemma 5.9, $F_G(\mathcal{H}) \equiv \mathcal{H}'$ for some $\mathcal{H}' \in Mod(\gamma'(SP'))$ with $\mathcal{H}'|_{\iota'} = \mathcal{H} \in Mod(SP_G)$. Hence $\mathcal{H}' \in Abs_{\equiv}(SP'_G)$ by (ii). By stability of F_G (Cor. 5.5), $\mathcal{G} \in dom(F_G)$ and $F_G(\mathcal{G}) \equiv F_G(\mathcal{H}) \equiv \mathcal{H}'$, and so $F_G(\mathcal{G}) \in Abs_{\equiv}(SP'_G)$. □

Requirement (i) is perhaps the only surprising assumption in this theorem. Note though that it straightforwardly follows from the inclusion of strict model classes $Mod(SP_G) \subseteq Mod(\gamma(SP))$ (or equivalently, $Mod(SP_G)|_{\gamma} \subseteq Mod(SP)$), which is often easiest to verify. However, (i) is strictly stronger in general than the perhaps more expected $Mod(SP_G) \subseteq Abs_{\equiv}(\gamma(SP))$. This weaker condition turns out to be sufficient (and in fact, equivalent to (i)) if we additionally assume that the two specifications involved are *behaviourally consistent* [BHW95], that is, closed under observational quotients. When this is not the case, then the use of this weaker condition must be paid for by a stronger version of (ii):

$$Abs_{\equiv}(\gamma'(SP')) \cap Mod(\iota'(SP_G)) \subseteq Abs_{\equiv}(SP'_G),$$

which seems even less convenient to use than (i). Overall, we need a way to pass information on the global context from SP_G to SP'_G independently from the observational interpretation of the local construction and its correctness, and this must result in some inconvenience of verification on either the parameter or the result side.

6 Architectural Specifications

Using local constructions for global implementations of specifications, we have moved only one step away from the monolithic global view of specifications and constructions used to implement them. The notion of *architectural specification* [BST02] as introduced for CASL takes us much further. An architectural specification *prescribes* a decomposition of the task of implementing a requirements specification into a number of subtasks to implement specifications of "modular components" (called *units*) of the system under development. The units may be parametrized, and then we can identify them with local constructions; non-parametrized units are modeled as algebras. Another essential part of an architectural specification is a prescription of how the units, once developed, are to be put together using a few simple operators. One of these is an application of a parametrized unit which corresponds exactly to the lifting of a local construction to a larger context studied above. Thus, an architectural specification may be thought of as a definition of a complex construction to be used in a development process to implement a requirements specification by a number of specifications (of non-parametrized units), where the construction uses a number of specified local constructions to be developed as well.

For the sake of readability, we will discuss here only a very simple version of CASL architectural specifications, with a limited (but representative) number of constructs, shaped after a somewhat less simplified fragment used in [SMT+01]; a generalization of the work presented here to full architectural specifications of CASL would be tedious but rather straightforward, except perhaps for the "unguarded import" mechanism, see [Hof01]. Our version of architectural specifications is defined as follows.

Architectural Specifications: ASP ::= **arch spec** Dcl^* **result** T
An architectural specification consists of a list of unit declarations followed by a unit result term.

Unit Declarations: $Dcl ::= U{:}\, SP \mid U{:}\, SP_1 \xrightarrow{\iota} SP_2$
A unit declaration introduces a unit name with its type, which is either a specification or a specification of a parametrized unit, determined by a specification of its parameter and its result, which extends the parameter via a signature morphism ι.

Unit Terms: $T ::= U \mid U[T$ **fit** $\sigma] \mid T_1$ **and** T_2
A unit term is either a (non-parametrized) unit name, or a unit application with an argument that fits via a signature morphism σ, or an amalgamation of units.

Following the semantics of full CASL [CoFI02], see also [SMT+01], we give the semantics of this CASL fragment in two stages: first we give its *extended static semantics* and then the *strict model semantics.*

An *extended static context* $\mathcal{C}_{st} = (P_{st}, \mathcal{B}_{st}, \Sigma_G)$ in which CASL phrases are elaborated, consists of a static context for parametrized units P_{st} mapping parametrized unit names to signature morphisms (from the parameter to the result signatures), a global context signature Σ_G, and an extended static context for non-parametrized units $\mathcal{B}_{st}$ mapping non-parametrized unit names to morphisms from the unit signature to Σ_G. From any such extended static context we can extract a *static context* $ctx(\mathcal{C}_{st}) = (P_{st}, B_{st})$ by forgetting the global context signature and restricting the information about non-parametrized units to their signatures only (sources of the morphisms given by $\mathcal{B}_{st}$).

Given a morphism $\theta{:}\, \Sigma_G \to \Sigma'_G$, we write $\mathcal{B}_{st};\theta$ for the extended static context $\mathcal{B}'_{st}$ with the same domain as $\mathcal{B}_{st}$ and such that for any name $U \in dom(\mathcal{B}_{st})$, $\mathcal{B}'_{st}(U) = \mathcal{B}_{st}(U);\theta$. Then the extended static context $\mathcal{C}_{st};\theta$ is $(P_{st}, (\mathcal{B}_{st};\theta), \Sigma'_G)$. $\mathcal{C}^{\emptyset}_{st}$ stands for the "empty" extended static context that consists of the empty parametrized and non-parametrized unit contexts and the initial signature.

Figure 1 gives rules to derive semantic judgments of the following forms:

$$\frac{\vdash Dcl^* \rhd \mathcal{C}_{st} \qquad \mathcal{C}_{st} \vdash T \rhd (\theta\colon \Sigma_G \to \Sigma'_G, i\colon \Sigma \to \Sigma'_G)}{\vdash \textbf{arch spec } Dcl^* \textbf{ result } T \rhd (ctx(\mathcal{C}_{st}), \Sigma)}$$

$$\frac{\begin{array}{c}\mathcal{C}^{\emptyset}_{st} \vdash Dcl_1 \rhd (\mathcal{C}_{st})_1 \\ \dots \\ (\mathcal{C}_{st})_{n-1} \vdash Dcl_n \rhd (\mathcal{C}_{st})_n\end{array}}{\vdash Dcl_1 \dots Dcl_n \rhd (\mathcal{C}_{st})_n}$$

$$\frac{\begin{array}{c}U \notin (dom(P_{st}) \cup dom(\mathcal{B}_{st})) \\ \Sigma'_G \text{ is the coproduct of } \Sigma_G \text{ and } Sig(SP) \\ \text{with injections } \theta\colon \Sigma_G \to \Sigma'_G, i\colon Sig(SP) \to \Sigma'_G\end{array}}{(P_{st}, \mathcal{B}_{st}, \Sigma_G) \vdash U\colon SP \rhd (P_{st}, (\mathcal{B}_{st};\theta) + \{U \mapsto i\}, \Sigma'_G)}$$

$$\frac{\begin{array}{c}\iota : Sig(SP_1) \to Sig(SP_2) \\ U \notin (dom(P_{st}) \cup dom(\mathcal{B}_{st}))\end{array}}{(P_{st}, \mathcal{B}_{st}, \Sigma_G) \vdash U\colon SP_1 \xrightarrow{\iota} SP_2 \rhd (P_{st} + \{U \mapsto \iota\}, \mathcal{B}_{st}, \Sigma_G)}$$

$$\frac{U \in dom(\mathcal{B}_{st})}{(P_{st}, \mathcal{B}_{st}, \Sigma_G) \vdash U \rhd (id_{\Sigma_G}, \mathcal{B}_{st}(U))}$$

$$\frac{\begin{array}{c}(P_{st}, \mathcal{B}_{st}, \Sigma_G) \vdash T \rhd (\theta\colon \Sigma_G \to \Sigma'_G, i\colon \Sigma_T \to \Sigma'_G) \\ P_{st}(U) = \iota\colon \Sigma \to \Sigma' \qquad \sigma\colon \Sigma \to \Sigma_T \\ (\iota'\colon \Sigma_T \to \Sigma'_T, \sigma'\colon \Sigma' \to \Sigma'_T) \text{ is the pushout of } (\sigma, \iota) \\ (\iota''\colon \Sigma'_G \to \Sigma''_G, i'\colon \Sigma'_T \to \Sigma''_G) \text{ is the pushout of } (i, \iota')\end{array}}{(P_{st}, \mathcal{B}_{st}, \Sigma_G) \vdash U[T \textbf{ fit } \sigma] \rhd (\theta;\iota'', i'\colon \Sigma'_T \to \Sigma''_G)}$$

$$\frac{\begin{array}{c}(P_{st}, \mathcal{B}_{st}, \Sigma_G) \vdash T_1 \rhd (\theta_1\colon \Sigma_G \to \Sigma^1_G, i_1\colon \Sigma_1 \to \Sigma^1_G) \\ (P_{st}, \mathcal{B}_{st}, \Sigma_G) \vdash T_2 \rhd (\theta_2\colon \Sigma_G \to \Sigma^2_G, i_2\colon \Sigma_2 \to \Sigma^2_G) \\ \Sigma = \Sigma_1 \cup \Sigma_2 \text{ with inclusions } \iota_1\colon \Sigma_1 \to \Sigma, \iota_2\colon \Sigma_2 \to \Sigma \\ (\theta'_2\colon \Sigma^1_G \to \Sigma'_G, \theta'_1\colon \Sigma^2_G \to \Sigma'_G) \text{ is the pushout of } (\theta_1, \theta_2) \\ \text{there is a (unique) morphism } j\colon \Sigma \to \Sigma'_G \text{ such that } \iota_1;j = i_1;\theta'_2 \text{ and } \iota_2;j = i_2;\theta'_1\end{array}}{(P_{st}, \mathcal{B}_{st}, \Sigma_G) \vdash T_1 \textbf{ and } T_2 \rhd (\theta_1;\theta'_2, j)}$$

Fig. 1. Extended static semantics

- $\vdash ASP \rhd (\mathcal{C}_{st}, \Sigma)$: the architectural specification ASP yields a static context describing the units declared and the signature of the result unit;
- $\mathcal{C}_{st} \vdash Dcl \rhd \mathcal{C}'_{st}$: the unit declaration Dcl in the extended static context $\mathcal{C}_{st}$ yields a new extended static context $\mathcal{C}'_{st}$; similarly for a sequence of unit declarations;
- $(P_{st}, \mathcal{B}_{st}, \Sigma_G) \vdash T \rhd (\theta\colon \Sigma_G \to \Sigma'_G, i\colon \Sigma \to \Sigma'_G)$: the unit term T in the extended static context $(P_{st}, \mathcal{B}_{st}, \Sigma_G)$ extends the global context signature Σ_G to a new one Σ'_G along a signature morphism $\theta\colon \Sigma_G \to \Sigma'_G$ and yields the signature Σ of the unit built, indicating how the unit resides in the global context using the morphism $i\colon \Sigma \to \Sigma'_G$.

In the strict model semantics we work with *contexts* $\mathcal{C}$ that are sets of *unit environments* E. Environments map unit names to either local constructions (for parametrized units) or to individual algebras (for non-parametrized units). *Unit evaluators* UEv map unit environments to algebras.

Given an extended static unit context $\mathcal{C}_{st} = (P_{st}, \mathcal{B}_{st}, \Sigma_G)$, an environment E *fits* $\mathcal{C}_{st}$ if

- for each $U \in dom(P_{st})$, $E(U)$ is a local construction along $P_{st}(U)$, and
- there exists an algebra $\mathcal{G} \in |\mathbf{Alg}(\Sigma_G)|$ such that for each $U \in dom(\mathcal{B}_{st})$, $E(U) = \mathcal{G}|_{\mathcal{B}_{st}(U)}$; we say then that $\mathcal{G}$ *witnesses* E.

We write $ucx(\mathcal{C}_{st})$ for the class of all environments that fit $\mathcal{C}_{st}$. $\mathcal{C}^{\emptyset} = ucx(\mathcal{C}^{\emptyset}_{st})$ is the context which constrains no unit name. Given a unit context $\mathcal{C}$, a unit name U and a class of units $\mathcal{V}$, we write $\mathcal{C} \times \{U \mapsto \mathcal{V}\}$ for $\{E + \{U \mapsto V\} \mid E \in \mathcal{C}, V \in \mathcal{V}\}$, where $E + \{U \mapsto V\}$ maps U to V and otherwise behaves like E.

Figure 2 gives rules to derive semantic judgments of the following forms:

- $\vdash ASP \Rightarrow (\mathcal{C}, UEv)$: the architectural specification ASP yields a context $\mathcal{C}$ with environments providing interpretations for the units declared and the unit evaluator that for each such environment determines the result unit;
- $\mathcal{C} \vdash Dcl \Rightarrow \mathcal{C}'$: the unit declaration Dcl in the context $\mathcal{C}_{st}$ yields a new context $\mathcal{C}'_{st}$; similarly for a sequence of unit declarations;
- $\mathcal{C} \vdash T \Rightarrow UEv$: the unit term T in the context $\mathcal{C}$ yields a unit evaluator UEv that when given an environment (in $\mathcal{C}$) yields the unit resulting from the evaluation of T in this environment.

The rules rely on a successful run of the extended static semantics; this allows us to use the static concepts and notations introduced there. Moreover, the following invariants link the extended static semantics and model semantics and are maintained by the rules:

- $\vdash ASP \rhd\!\!\!> (\mathcal{C}_{st}, \Sigma)$ and $\vdash ASP \Rightarrow (\mathcal{C}, UEv)$: there is an extended static context $\mathcal{C}_{st}$ such that $ctx(\mathcal{C}_{st}) = C_{st}$ and $\mathcal{C} \subseteq ucx(\mathcal{C}_{st})$, $\mathcal{C} \subseteq dom(UEv)$, and for each $E \in \mathcal{C}$, $UEv(E) \in |\mathbf{Alg}(\Sigma)|$;
- $\mathcal{C}_{st} \vdash Dcl \rhd\!\!\!> \mathcal{C}'_{st}$ and $\mathcal{C} \vdash Dcl \Rightarrow \mathcal{C}'$: if $\mathcal{C} \subseteq ucx(\mathcal{C}_{st})$ then $\mathcal{C}' \subseteq ucx(\mathcal{C}'_{st})$; similarly for a sequence of unit declarations;
- $\mathcal{C}_{st} \vdash T \rhd\!\!\!> (\theta\colon \Sigma_G \to \Sigma'_G, i\colon \Sigma \to \Sigma'_G)$ and $\mathcal{C} \vdash T \Rightarrow UEv$: if $\mathcal{C} \subseteq ucx(\mathcal{C}_{st})$ then for each unit environment $E \in \mathcal{C}$ and each algebra $\mathcal{G}$ that witnesses E, there exists a model $\mathcal{G}' \in |\mathbf{Alg}(\Sigma'_G)|$ such that $\mathcal{G}'|_{\theta} = \mathcal{G}$ and $UEv(E) = \mathcal{G}'|_i$.

The invariants ensure that the crossed out premise of the unit amalgamation rule of the model semantics follows from the premises of the corresponding rule of the extended static semantics.

7 Observational Interpretation of Architectural Specifications

In this section we discuss an observational interpretation of the architectural specifications introduced in Sect. 6. The extended static semantics remains unchanged – observational interpretation of specifications does not affect their

$$\frac{\vdash Dcl^* \Rightarrow \mathcal{C} \qquad \mathcal{C} \vdash T \Rightarrow UEv}{\vdash \textbf{arch spec } Dcl^* \textbf{ result } T \Rightarrow (\mathcal{C}, UEv)}$$

$$\frac{\begin{array}{c}\mathcal{C}^{\emptyset} \vdash Dcl_1 \Rightarrow \mathcal{C}_1 \\ \cdots \\ \mathcal{C}_{n-1} \vdash Dcl_n \Rightarrow \mathcal{C}_n\end{array}}{\vdash Dcl_1 \ldots Dcl_n \Rightarrow \mathcal{C}_n}$$

$$\frac{}{\mathcal{C} \vdash U{:}\, SP \Rightarrow \mathcal{C} \times \{U \mapsto Mod(SP)\}}$$

$$\frac{}{\mathcal{C} \vdash U{:}\, SP_1 \xrightarrow{\iota} SP_2 \Rightarrow \mathcal{C} \times \{U \mapsto Mod(SP_1 \xrightarrow{\iota} SP_2)\}}$$

$$\frac{}{\mathcal{C} \vdash U \Rightarrow \lambda E \in \mathcal{C} \cdot E(U)}$$

$$\frac{\begin{array}{c}\mathcal{C} \vdash T \Rightarrow UEv; \quad \textbf{for each } E \in \mathcal{C},\, UEv(E)|_{\sigma} \in dom(E(U)) \\ UEv' = \{E \mapsto A \mid E \in \mathcal{C}, A|_{\iota'} = UEv(E), A|_{\sigma'} = E(U)(UEv(E)|_{\sigma})\}\end{array}}{\mathcal{C} \vdash U[T \textbf{ fit } \sigma] \Rightarrow UEv'}$$

$$\frac{\begin{array}{c}\mathcal{C} \vdash T_1 \Rightarrow UEv_1 \qquad \mathcal{C} \vdash T_2 \Rightarrow UEv_2 \\ \sout{\textbf{for each } E \in \mathcal{C}, \textbf{ there is a unique } A \in |\textbf{Alg}(\Sigma)| \textbf{ such that}} \\ \sout{A|_{\iota_1} = UEv_1(E), A|_{\iota_2} = UEv_2(E)} \\ UEv = \{E \mapsto A \mid E \in \mathcal{C}, A|_{\iota_1} = UEv_1(E), A|_{\iota_2} = UEv_2(E)\}\end{array}}{\mathcal{C} \vdash T_1 \textbf{ and } T_2 \Rightarrow UEv}$$

Fig. 2. Strict model semantics

static properties. We provide, however, a new *observational model semantics*, with judgments written as $\_ \vdash \_ \overset{\equiv}{\Longrightarrow} \_$.

To begin with, the effect of unit declarations has to be modified, taking into account observational interpretation of the specifications involved, as discussed in Sects. 4 and 5. The new rules follow in Fig. 3.

$$\frac{}{\mathcal{C} \vdash U{:}\, SP \overset{\equiv}{\Longrightarrow} \mathcal{C} \times \{U \mapsto Abs_{\equiv}(SP)\}}$$

$$\frac{}{\mathcal{C} \vdash U{:}\, SP_1 \xrightarrow{\iota} SP_2 \overset{\equiv}{\Longrightarrow} \mathcal{C} \times \{U \mapsto Mod_{lc}(SP_1 \xrightarrow{\iota} SP_2)\}}$$

Fig. 3. Observational model semantics – the modified rules

No other modifications are necessary: all the remaining rules are the same for observational and strict model semantics. This should not be surprising: the interpretation of the constructs on unit terms remains the same, all we change is the interpretation of unit specifications.

Moreover, the observational model semantics can be linked to the extended static semantics in exactly the same way as in the case of the strict model semantics: the invariants stated in Sect. 6 carry over without change.

This does not mean that the two semantics quite coincide: there is one point in the model semantics where verification is performed, and the resulting verification conditions for strict and observational model semantics differ. Namely, in the rule for parametrized unit application, the premise

$$\text{for each } E \in \mathcal{C},\ UEv(E)|_{\sigma} \in dom(E(U))$$

checks whether what we can conclude about the argument ensures that it is indeed in the domain of the parametrized unit. Suppose the corresponding unit declaration was $U{:}\, SP \xrightarrow{\iota} SP'$. Then in the strict model semantics this requirement reduces to

$$\text{for each } E \in \mathcal{C},\ UEv(E)|_{\sigma} \in Mod(SP).$$

Now, in the observational model semantics, this is in fact replaced by a more permissive condition:

$$\text{for each } E \in \mathcal{C},\ UEv(E)|_{\sigma} \in Abs_{\equiv}(SP).$$

Of course, the situation is complicated by the fact that the contexts $\mathcal{C}$ from which environments are taken are different in the two semantics. In the simplest case, where the argument T is simply given as a unit name previously declared with a specification SP_T, for the strict model semantics the above verification condition is

$$Mod(SP_T) \subseteq Mod(SP)$$

while for the observational model semantics we get, as expected,

$$Mod(SP_T) \subseteq Abs_{\equiv}(SP).$$

In particular, it follows that there are statically correct architectural specifications ASP (i.e., $\vdash ASP \rhd\!\!\!\rhd (\mathcal{C}_{st}, \Sigma)$ for some extended static context $\mathcal{C}_{st}$ and signature Σ) that are observationally correct (i.e., $\vdash ASP \stackrel{\equiv}{\Longrightarrow} (\mathcal{C}_{obs}, UEv_{obs})$ for some unit context $\mathcal{C}_{obs}$ and evaluator UEv_{obs}) but *are not* strictly correct (i.e., for *no* unit context $\mathcal{C}$ and evaluator UEv can we derive $\vdash ASP \Rightarrow (\mathcal{C}, UEv)$).

A complete study of verification conditions for architectural specifications is beyond the scope of this paper; we refer to [Hof01] for work in this direction, which still has to be combined with the observational interpretation as given by the semantics here and presented in a simpler setting in Sect. 5. In the rest of this paper we will concentrate on some aspects of the relationship between the strict and observational model semantics and on stability of unit constructions as introduced in Sect. 6.

Our first aim is to show that the constructions that can be defined by architectural specifications are (locally) stable. To state this precisely, we need some more notation and terminology, as the constructions are captured here by unit evaluators operating on environments rather than on individual units.

Local constructions F_1, F_2 along $\iota: \Sigma \to \Sigma'$ are *observationally equivalent*, written $F_1 \equiv F_2$, if $dom(F_1) = dom(F_2)$ and for each $A \in dom(F_1)$ there exists a correspondence $\rho: F_1(A) \bowtie F_2(A)$ such that its reduct $\rho|_\iota$ is the identity on A.

Proposition 7.1. *Let F_1 and F_2 be observationally equivalent local constructions along $\iota: \Sigma \to \Sigma'$. Then if F_1 is locally stable then so is F_2.* □

Environments E_1, E_2 are *observationally equivalent*, written $E_1 \equiv E_2$, if $dom(E_1) = dom(E_2)$ and for each $U \in dom(E_1)$, $E_1(U) \equiv E_2(U)$.

A unit environment is *stable* if all the parametrized units it contains are locally stable. By Prop. 7.1, the class of stable environments is closed under observational equivalence. Given an extended static context $\mathcal{C}_{st}$, we write $ucx_{obs}(\mathcal{C}_{st})$ for the class of those unit environments in $ucx(\mathcal{C}_{st})$ that are stable. Then, given a unit context $\mathcal{C}$, we write $Abs_\equiv(\mathcal{C})$ for the class of all stable unit environments equivalent to a unit environment in $\mathcal{C}$; clearly, if $\mathcal{C} \subseteq ucx(\mathcal{C}_{st})$ for some static context $\mathcal{C}_{st}$ then $Abs_\equiv(\mathcal{C}) \subseteq ucx_{obs}(\mathcal{C}_{st})$.

Back to the stability of the constructions defined by architectural specifications: we want to show that if $\vdash ASP \rhd\!\!\!\rhd (\mathcal{C}_{st}, \Sigma)$ and $\vdash ASP \stackrel{\equiv}{\Longrightarrow} (\mathcal{C}_{obs}, UEv_{obs})$ then the unit evaluator UEv_{obs} is stable, i.e., maps observationally equivalent environments to observationally equivalent algebras. Unfortunately, this cannot be proved by a simple induction on the structure of the unit terms involved. The trouble is with amalgamation, since in general amalgamation is not stable – informally, joining the signatures of two algebras may introduce new observations for either or both of them.

Counterexample 7.2. Let Σ_1 and Σ_2 be signatures containing the Boolean part and a sort s (the same in both signatures). Moreover, let Σ_1 contain constants $a, b: s$; and let Σ_2 contain a function $f: s \to bool$. Since in either of the signatures there are no observations for the non-observable sort s, all algebras in $\mathbf{Alg}(\Sigma_1)$ are observationally equivalent, and similarly for algebras in $\mathbf{Alg}(\Sigma_2)$. However, observational equivalence between $(\Sigma_1 \cup \Sigma_2)$-algebras is non-trivial; for instance, algebras with $f(a) = f(b)$ are not equivalent to those where $f(a) \neq f(b)$. Consequently, given an algebra $A \in |\mathbf{Alg}(\Sigma_1)|$ with $a_A \neq b_A$, it is easy to indicate algebras $B, B' \in |\mathbf{Alg}(\Sigma_2)|$, with the same carrier of sort s as A and such that $B \equiv B'$, while the amalgamation of A with B and B', respectively, yields algebras in $\mathbf{Alg}(\Sigma_1 \cup \Sigma_2)$ that are not observationally equivalent.

However, the key point here is that amalgamation in unit terms in architectural specifications is not used as a construction on its own, but it just identifies a new part of the global context that has been constructed earlier. Since the "essential" constructions used to build new components of the global context are locally stable, such use of amalgamation can cause no harm.

To demonstrate this, we introduce a more detailed form of the semantics for unit terms, which carries more information about the construction of the global context performed on the way. Given $\mathcal{C}_{st} \vdash T \rhd\!\!\!\rhd (\theta\colon \Sigma_G \to \Sigma'_G, i\colon \Sigma \to \Sigma'_G)$, we derive judgments of the form $\mathcal{C}_{obs} \vdash T \overset{\equiv}{\Longrightarrow\!\!\!\!\Rightarrow} (\langle F_E \rangle_{E \in \mathcal{C}}, UEv_{obs})$, where for each $E \in \mathcal{C}_{obs}$, $F_E\colon |\mathbf{Alg}(\Sigma_G)| \rightharpoonup |\mathbf{Alg}(\Sigma'_G)|$ is a construction along $\theta\colon \Sigma_G \to \Sigma'_G$. The rules are given in Fig. 4 (*ID* in the first rule is the family of identities, appropriately indexed); as before, the rules rely on the notation introduced by the corresponding rules of the extended static semantics, see Fig. 1.

Lemma 7.3. *If* $\mathcal{C}_{st} \vdash T \rhd\!\!\!\rhd (\theta\colon \Sigma_G \to \Sigma'_G, i\colon \Sigma \to \Sigma'_G)$ *and* $\mathcal{C}_{obs} \vdash T \overset{\equiv}{\Longrightarrow} UEv_{obs}$ *with* $\mathcal{C}_{obs} \subseteq ucx_{obs}(\mathcal{C}_{st})$, *then* $\mathcal{C}_{obs} \vdash T \overset{\equiv}{\Longrightarrow\!\!\!\!\Rightarrow} (\langle F_E \rangle_{E \in \mathcal{C}}, UEv_{obs})$ *for some family* $\langle F_E \rangle_{E \in \mathcal{C}_{obs}}$ *such that*

- *for* $E \in \mathcal{C}_{obs}$, $F_E\colon |\mathbf{Alg}(\Sigma_G)| \rightharpoonup |\mathbf{Alg}(\Sigma'_G)|$ *is persistent along* $\theta\colon \Sigma_G \to \Sigma'_G$;
- *for* $E \in \mathcal{C}_{obs}$, *if* $\mathcal{G} \in |\mathbf{Alg}(\Sigma_G)|$ *witnesses* E *then* $\mathcal{G} \in dom(F_E)$ *and* $UEv_{obs}(E) = F_E(\mathcal{G})|_i$;
- *the family* $\langle F_E \rangle_{E \in \mathcal{C}_{obs}}$ *is locally stable in the following sense: for* $E_1, E_2 \in \mathcal{C}_{obs}$ *such that* $E_1 \equiv E_2$, $\mathcal{G}_1, \mathcal{G}_2 \in |\mathbf{Alg}(\Sigma_G)|$ *that witness* E_1 *and* E_2, *respectively, and correspondence* $\rho\colon \mathcal{G}_1 \bowtie \mathcal{G}_2$, *if* $\mathcal{G}_1 \in dom(F_E)$ *then* $\mathcal{G}_2 \in dom(F_E)$ *as well and there exists a correspondence* $\rho'\colon F_{E_1}(\mathcal{G}_1) \bowtie F_{E_2}(\mathcal{G}_2)$ *with* $\rho'|_\theta = \rho$.

Proof. By induction on the structure of the unit term. In each case, the first two properties follow easily from the construction and Lemma 2.1. Prop. 5.3 and Lemma 5.4 imply the last property for parametrized unit application and unit amalgamation. □

Since reducts preserve observational equivalence, Lemma 7.3 directly implies stability of unit constructions definable by architectural specifications:

$$\frac{}{\mathcal{C} \vdash U \overset{\equiv}{\Longrightarrow\!\!\!\!\Rightarrow} (ID, \lambda E \in \mathcal{C} \cdot E(U))}$$

$$\frac{\begin{array}{c}\mathcal{C} \vdash T \overset{\equiv}{\Longrightarrow\!\!\!\!\Rightarrow} (\langle F_E \rangle_{E \in \mathcal{C}}, UEv) \\ \text{for each } E \in \mathcal{C}, UEv(E)|_\sigma \in dom(E(U)) \\ UEv' = \{E \mapsto A \mid E \in \mathcal{C}, A|_{\iota'} = UEv(E), A|_{\sigma'} = E(U)(UEv(E)|_\sigma)\} \\ \text{for } E \in \mathcal{C}, F'_E = \{\mathcal{G} \mapsto \mathcal{G}' \mid \mathcal{G} \in |\mathbf{Alg}(\Sigma_G)| \text{ witnesses } E, \\ \qquad \mathcal{G}'|_{\iota''} = \mathcal{G}, \mathcal{G}'|_{\sigma';i'} = E(U)(UEv(E)|_\sigma)\}\end{array}}{\mathcal{C} \vdash U[T \textbf{ fit } \sigma] \overset{\equiv}{\Longrightarrow\!\!\!\!\Rightarrow} (\langle F_E ; F'_E \rangle_{E \in \mathcal{C}}, UEv')}$$

$$\frac{\begin{array}{c}\mathcal{C} \vdash T_1 \overset{\equiv}{\Longrightarrow\!\!\!\!\Rightarrow} (\langle F^1_E \rangle_{E \in \mathcal{C}}, UEv_1) \quad \mathcal{C} \vdash T_2 \overset{\equiv}{\Longrightarrow\!\!\!\!\Rightarrow} (\langle F^2_E \rangle_{E \in \mathcal{C}}, UEv_2) \\ UEv = \{E \mapsto A \mid E \in \mathcal{C}, A|_{\iota_1} = UEv_1(E), A|_{\iota_2} = UEv_2(E)\} \\ \text{for } E \in \mathcal{C}, F_E = \{\mathcal{G} \mapsto \mathcal{G}' \mid \mathcal{G} \in |\mathbf{Alg}(\Sigma_G)| \text{ witnesses } E, \\ \qquad \mathcal{G}'|_{\theta_1} = F^1_E(\mathcal{G}), \mathcal{G}'|_{\theta_2} = F^2_E(\mathcal{G})\}\end{array}}{\mathcal{C} \vdash T_1 \textbf{ and } T_2 \overset{\equiv}{\Longrightarrow\!\!\!\!\Rightarrow} (\langle F_E \rangle_{E \in \mathcal{C}}, UEv)}$$

Fig. 4. Modified observational model semantics

Corollary 7.4. *If* $\vdash ASP \mathrel{\rhd\!\!\!\rhd} (\mathcal{C}_{st}, \Sigma)$ *and* $\vdash ASP \stackrel{\equiv}{\Longrightarrow} (\mathcal{C}_{obs}, UEv_{obs})$ *then for any unit environments* $E_1, E_2 \in \mathcal{C}_{obs}$ *such that* $E_1 \equiv E_2$*, we have* $UEv_{obs}(E_1) \equiv UEv_{obs}(E_2)$. □

As already mentioned, the observational semantics is more permissive than the strict model semantics: there existence of a successful derivation of an observational meaning for an architectural specification does not in general imply that its strict model semantics is defined as well. Moreover, the observational semantics may "lose" some results permitted by the strict model semantics, which follows from Counterexample 5.8. However, if an architectural specification has a strict model semantics then its observational semantics is defined as well and up to observational equivalence, nothing new is added:

Theorem 7.5. *If* $\vdash ASP \mathrel{\rhd\!\!\!\rhd} (\mathcal{C}_{st}, \Sigma)$ *and* $\vdash ASP \Rightarrow (\mathcal{C}, UEv)$ *then* $\vdash ASP \stackrel{\equiv}{\Longrightarrow} (\mathcal{C}_{obs}, UEv_{obs})$*, where for every* $E_{obs} \in \mathcal{C}_{obs}$ *there exists* $E \in \mathcal{C}$ *such that* $E_{obs} \equiv E$ *and* $UEv_{obs}(E_{obs}) \equiv UEv(E)$.

Proof. The following can be proved inductively:

1. $\vdash ASP \mathrel{\rhd\!\!\!\rhd} (\mathcal{C}_{st}, \Sigma)$ and $\vdash ASP \Rightarrow (\mathcal{C}, UEv)$: then $\vdash ASP \stackrel{\equiv}{\Longrightarrow} (\mathcal{C}_{obs}, UEv_{obs})$ with $\mathcal{C}_{obs} = Abs_{\equiv}(\mathcal{C})$ and for each stable $E \in \mathcal{C}$ (then necessarily $E \in \mathcal{C}_{obs}$) we have $UEv_{obs}(E) = UEv(E)$;
2. $\mathcal{C}_{st} \vdash Dcl \mathrel{\rhd\!\!\!\rhd} \mathcal{C}'_{st}$ and $\mathcal{C} \vdash Dcl \Rightarrow \mathcal{C}'$, where $\mathcal{C} \subseteq ucx(\mathcal{C}_{st})$: then $Abs_{\equiv}(\mathcal{C}) \vdash Dcl \stackrel{\equiv}{\Longrightarrow} Abs_{\equiv}(\mathcal{C}')$, and similarly for sequences of unit declarations;
3. $\mathcal{C}_{st} \vdash T \mathrel{\rhd\!\!\!\rhd} (\theta\colon \Sigma_G \to \Sigma'_G, i\colon \Sigma \to \Sigma'_G)$ and $\mathcal{C} \vdash T \Rightarrow UEv$ with $\mathcal{C} \subseteq ucx(\mathcal{C}_{st})$: then $Abs_{\equiv}(\mathcal{C}) \vdash T \stackrel{\equiv}{\Longrightarrow} UEv_{obs}$ where for each stable $E \in \mathcal{C}$ (then $E \in Abs_{\equiv}(\mathcal{C})$) we have $UEv_{obs}(E) = UEv(E)$.

The only potential difficulty is in the proof of item 3 for parametrized unit application, where to deduce the premise of the observational semantics rule that captures verification that the argument is in the domain of the parametrized unit, we need to rely on the corresponding premise of the strict model semantics, on the stability of the parametrized unit and on Lemma 7.3.

The theorem follows easily now: given the assumptions, item 1 implies that $\vdash ASP \stackrel{\equiv}{\Longrightarrow} (\mathcal{C}_{obs}, UEv_{obs})$ with $\mathcal{C}_{obs} = Abs_{\equiv}(\mathcal{C})$, and so for each $E_{obs} \in \mathcal{C}_{obs}$ there is a stable environment $E \in \mathcal{C}$ such that $E_{obs} \equiv E$. Thus, by item 1 and Cor. 7.4, $UEv(E) = UEv_{obs}(E) \equiv UEv_{obs}(E_{obs})$. □

8 Conclusions and Further Work

Apart from the preliminaries, this paper consists of two parts. Sects. 3, 4, and 5 recall a now rather standard and quite general view of the software development process, paying special attention to observational interpretation of the specifications involved and discussing in more detail than usual how "global" developments proceed using "local" constructions. We point out how observational interpretation of specifications leads to the crucial – and quite natural, Cor. 5.6 – stability requirement on the constructions, and how this in turn helps to establish correctness of development steps, Thm. 5.10. Then, Sects. 6 and 7 study how these general ideas may be instantiated in the context of architectural specifications as borrowed from CASL in a simplified version. We view here architectural specifications as means to build complex constructions to be used in the software development process. Observational interpretation of specifications brings out rather non-trivial issues. We study stability of the constructions involved, with the expected positive result in Cor. 7.4, and link the results under observational interpretation with those for the standard interpretation of architectural specifications, Thm. 7.5. Clearly, as mentioned in Sect. 7, this must be augmented with an analysis of the internal correctness of architectural specifications under observational interpretation.

Although formally we have worked in a specific – and simple – logical framework, it should be clear that much of the above applies to a wide range of institutions of interest. Rather than trying to embark on an exercise of formally spelling out the appropriate notion of "institution with extra structure", let us just remark that surprisingly little is required. A special notion of *observational model morphisms* that must be closed under composition and reducts, plus some extra categorical structure to identify "correspondences" as certain spans of such morphisms, seems necessary and sufficient to formulate most of the material presented. Notice that we have in effect not referred to the set of observable sorts in the technical development. The trick is, however, to study a sufficient number of special cases to demonstrate that observational intuitions in various institutions may well be captured by such a simple structure. Further justification may be provided via links with indistinguishability relations (via factorization properties, like Thm. 4.3, which in turn may require a richer context of *concrete institutions*, with model categories equipped with concretization structure subject to a number of technical requirements as in [BT96]).

On the other hand, to transfer the present work to the specific framework of CASL we need a precise and convincing definition of observational equivalence between CASL models (many-sorted algebras with predicates, partial operations and subsorting). In terms of the institutional structure hinted at above, our first attempts dictate to simply use *closed homomorphisms* as observational morphisms – but the resulting notion of equivalence needs a more detailed analysis, from both the methodological and technical point of view.

The semantics for our simplified architectural specifications made reference to the cocompleteness of the category of signatures and to the amalgamation property of the underlying institution. Many institutions enjoy these properties, including the many-sorted versions of various standard logics. However, the amalgamation property fails for full CASL with subsorts, as discussed in detail in [SMT+01]. There are at least two ways to circumvent this problem. One is to present the global context as a diagram of signatures and a compatible family of models over this diagram, as in [SMT+01]. The other possibility is to use an extension of the CASL institution to "enriched" signatures (where multiple embeddings between subsorts are allowed) and their corresponding models, where the amalgamation property holds, again presented in [SMT+01].

References

[ABK+03] E. Astesiano, M. Bidoit, H. Kirchner, B. Krieg-Brückner, P.D. Mosses, D. Sannella and A. Tarlecki. CASL: The Common Algebraic Specification Language. *Theoretical Computer Science*, to appear (2003). See also the CASL Summary at `http://www.brics.dk/Projects/CoFI/Documents/CASL/Summary/`.

[AKBK99] E. Astesiano, B. Krieg-Brückner and H.-J. Kreowski, eds. *Algebraic Foundations of Systems Specification.* Springer (1999).

[Ber87] G. Bernot. Good functors ... are those preserving philosophy! *Proc. 2nd Summer Conf. on Category Theory and Computer Science CTCS'87*, Springer LNCS 283, 182–195 (1987).

[BH93] M. Bidoit and R. Hennicker. A general framework for modular implementations of modular systems. *Proc. 4th Int. Conf. Theory and Practice of Software Development TAPSOFT'93*, Springer LNCS 668, 199–214 (1993).

[BH98] M. Bidoit and R. Hennicker. Modular correctness proofs of behavioural implementations. *Acta Informatica* 35(11):951–1005 (1998).

[BT96] M. Bidoit and A. Tarlecki. Behavioural satisfaction and equivalence in concrete model categories. *Proc. 20th Coll. on Trees in Algebra and Computing CAAP'96*, Linköping, Springer LNCS 1059, 241–256 (1996).

[BHW95] M. Bidoit, R. Hennicker and M. Wirsing. Behavioural and abstractor specifications. *Science of Computer Programming* 25:149–186 (1995).

[BST02] M. Bidoit, D. Sannella and A. Tarlecki. Architectural specifications in CASL. *Formal Aspects of Computing*, to appear (2002). Available at `http://www.lsv.ens-cachan.fr/Publis/PAPERS/BST-FAC2002.ps`. Extended abstract: *Proc. 7th Intl. Conf. on Algebraic Methodology and Software Technology, AMAST'98.* Springer LNCS 1548, 341–357 (1999).

[CoFI02] The CoFI Task Group on Semantics. Semantics of the Common Algebraic Specification Language CASL. Available at `http://www.brics.dk/Projects/CoFI/Documents/CASL/Semantics/` (2002).

[EK99] H. Ehrig and H.-J. Kreowski. Refinement and implementation. In: [AKBK99], 201–242.

[EKMP82] H. Ehrig, H.-J. Kreowski, B. Mahr and P. Padawitz. Algebraic implementation of abstract data types. *Theoretical Comp. Sci.* 20:209–263 (1982).

[EM85] H. Ehrig and B. Mahr. *Fundamentals of Algebraic Specification I: Equations and Initial Semantics.* Springer (1985).

[Gan83] H. Ganzinger. Parameterized specifications: parameter passing and implementation with respect to observability. *ACM Transactions on Programming Languages and Systems* 5:318–354 (1983).

[Gin68] A. Ginzburg. *Algebraic Theory of Automata.* Academic Press (1968).

[Gog84] J. Goguen. Parameterized programming. *IEEE Trans. on Software Engineering* SE-10(5):528–543 (1984).

[GB92] J. Goguen and R. Burstall. Institutions: abstract model theory for specification and programming. *J. of the ACM* 39:95–146 (1992).

[GM82] J. Goguen and J. Meseguer. Universal realization, persistent interconnection and implementation of abstract modules. *Proc. 9th Intl. Coll. on Automata, Languages and Programming.* Springer LNCS 140, 265–281 (1982).

[GGM76] V. Giarratana, F. Gimona and U. Montanari. Observability concepts in abstract data type specifications. *Proc. 5th Intl. Symp. on Mathematical Foundations of Computer Science*, Springer LNCS 45, 576–587 (1976).

[Hoa72] C.A.R. Hoare. Proofs of correctness of data representations. *Acta Informatica* 1:271–281 (1972).

[Hof01] P. Hoffman. Verifying architectural specifications. *Recent Trends in Algebraic development Techniques, Selected Papers, WADT'01*, Springer LNCS 2267, 152-175 (2001).

[HLST00] F. Honsell, J. Longley, D. Sannella and A. Tarlecki. Constructive data refinement in typed lambda calculus. *Proc. 2nd Intl. Conf. on Foundations of Software Science and Computation Structures.* Springer LNCS 1784, 149–164 (2000).

[KHT+01] B. Klin, P. Hoffman, A. Tarlecki, L. Schröder and T. Mossakowski. Checking amalgamability conditions for CASL architectural specifications. *Proc. 26th Intl. Symp. Mathematical Foundations of Computer Science MFCS'01*, Springer LNCS 2136, 451–463 (2001).

[Mil71] R. Milner. An algebraic definition of simulation between programs. *Proc. 2nd Intl. Joint Conf. on Artificial Intelligence*, London, 481–489 (1971).

[Rei81] H. Reichel. Behavioural equivalence – a unifying concept for initial and final specification methods. *Proc. 3rd Hungarian Comp. Sci. Conference*, 27–39 (1981).

[Sch87] O. Schoett. *Data Abstraction and the Correctness of Modular Programming.* Ph.D. thesis, report CST-42-87, Dept. of Computer Science, Univ. of Edinburgh (1987).

[Sch90] O. Schoett. Behavioural correctness of data representations. *Science of Computer Programming* 14:43–57 (1990).

[SB83] D. Sannella and R. Burstall. Structured theories in LCF. *Proc. Colloq. on Trees in Algebra and Programming.* Springer LNCS 159, 377–391 (1983).

[ST87] D. Sannella and A. Tarlecki. On observational equivalence and algebraic specification. *J. of Computer and System Sciences* 34:150–178 (1987).

[ST88a] D. Sannella and A. Tarlecki. Specifications in an arbitrary institution. *Information and Computation* 76:165–210 (1988).

[ST88b] D. Sannella and A. Tarlecki. Toward formal development of programs from algebraic specifications: implementations revisited. *Acta Informatica* 25:233–281 (1988).

[ST89] D. Sannella and A. Tarlecki. Toward formal development of ML programs: foundations and methodology. *Proc. Colloq. on Current Issues in Programming Languages, Intl. Joint Conf. on Theory and Practice of Software Development TAPSOFT'89*, Barcelona. Springer LNCS 352, 375–389 (1989).

[ST97] D. Sannella and A. Tarlecki. Essential concepts of algebraic specification and program development. *Formal Aspects of Computing* 9:229–269 (1997).

[ST99] D. Sannella and A. Tarlecki. Algebraic preliminaries. In: [AKBK99], 13–30.

[SMT+01] L. Schröder, T. Mossakowski, A. Tarlecki, P. Hoffman and B. Klin. Semantics of architectural specifications in CASL. *Proc. 4th Intl. Conf. Fundamental Approaches to Software Engineering FASE'01*, Genova. Springer LNCS 2029, 253–268 (2001).

Edge-Colouring Pairs of Binary Trees: Towards a Concise Proof of the Four-Colour Theorem of Planar Maps

Alan Gibbons and Paul Sant

Department of Computer Science, King's College London, UK,
amg/sant@dcs.kcl.ac.uk

Abstract. The famous *Four-colour Problem* (*FCP*) of planar maps is equivalent, by an optimally fast reduction, to the problem of *Colouring Pairs of Binary Trees* (*CPBT*). Extant proofs of *FCP* lack conciseness, lucidity and require hours of electronic computation. The search for a satisfactory proof continues and, in this spirit, we explore two approaches to *CPBT*. In the first, we prove that a satisfactory proof exists if the *rotational path* between the two trees of the problem instance always satisfies a specific condition embodied in our *Shortest Path Conjecture*. In our second approach, we look for *patterns of colourability* within *regular forms* of tree pairs and seek to understand all instances of *CPBT* as a perturbation of these. In this *Colouring Topologies* approach, we prove, for instance, that concise proofs to *CPBT* exist for instances contained within many infinite-sized sets of trees.

1 Introduction

One of the most famous and fascinating problems in graph theory is the *Four-Colour Problem of Planar Maps* (see [13], for example) which is to show that every planar graph is vertex-colourable using at most four colours. The conjecture that four colours are sufficient was made by Francis Guthrie in about 1852 but it was not until 1977 when Appel, Haken and Koch [5,6,7] provided a solution that the four colour theorem was (controversially) established. Their proof was extremely long and could not be checked without the use (at the time) of thousands of hours of mainframe computing. Moreover, the proof did not provide concise and lucid comprehension of why the theorem is true as a more traditional style of proof might have done. Much more recently, in 1996, Robertson, Sanders, Seymour and Thomas [11] published a new proof, based on similar mathematical principles but with greatly improved computational aspects. The computational requirements of this new proof are less than a day on a PC. Moreover, rather than requiring $O(n^4)$ time to colour a graph with n nodes, as Appel, Haken and Koch would need, only $O(n^2)$ time is required. However, the discomfort initially associated with the first proof still remains.

The Four-Colour Problem is equivalent, by an optimally fast reduction, to the problem of *Colouring Pairs of Binary Trees* (*CPBT*) [2,4]. This problem

K. Diks et al. (Eds): MFSC 2002, LNCS 2420, pp. 25–39, 2002.

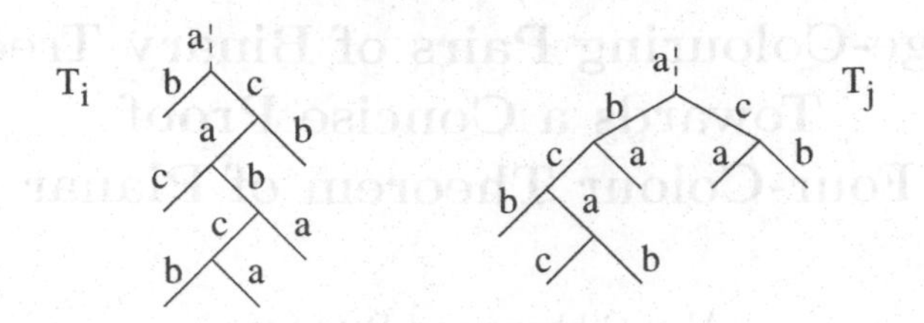

Fig. 1. A solution to an instance of *CPBT*.

is defined as follows. Let (T_i, T_j) be any two binary trees of the same size, the problem is to show that there exists a 3-edge-colouring of T_i and a 3-edge-colouring of T_j such that for every k, the edge adjacent to the kth leaf (from the left) is the same in both trees. Figure 1 illustrates a solution to *CPBT* for one pair of trees.

This paper describes some recent work on *CPBT*. In [2] the problem was related to others involving the intersection of regular languages, finding an integer solution of a set of linear equations and finding a common evaluation of two algebraic expressions. Here, we are concerned with the colouring problem *per se.* Our ultimate goal is a succinct and concise proof of the Four-Colour Problem by this means. In this spirit, we describe two approaches to *CPBT*. In Section 3 we show how the colouring problem can be coupled to *rotational paths* between the two trees in question and prove that there is always a solution if a particular condition prevails. We believe this condition is always valid and is expressed in the *Shortest Sequence Conjecture* of that section. In Section 4 we briefly describe an approach through *Colouring Topologies* or patterns of colourability for regular forms of trees, and, for example, we provide short proofs that solutions to *CPBT* exist for certain infinitely large classes of tree pairs. Both approaches to the problem of *CPBT* described have rich potential for further exploration and, in this sense, the work here lays the foundations for that.

2 Preliminaries

Here we establish some basic terminology and lemmas specific to this paper. Otherwise we use basic graph-theoretic definitions (to be found, for example, in [3]). A *free binary tree* is a non-rooted undirected tree in which every internal node has degree 3. Edges with leaves as endpoints are called *leaf-edges.* The following, easy-to-prove, lemma will be of much use later.

Lemma 1 (Colour-Parity Lemma). *In a 3-edge-colouring of a free binary tree, using colours from the set $\{a, b, c\}$, if n_a, n_b and n_c are, respectively, the number of occurences of the colours a, b and c at leaf-edges, then:*

$$n_a \bmod 2 = n_b \bmod 2 = n_c \bmod 2$$

By a *rooted* tree, we mean a free binary tree rooted at a *non-leaf* vertex. The next lemma, easy-to-prove, will also be useful. Here, the parity of a leaf-edge is *odd* or *even* according to its distance from the root; edges with the root as an endpoint are at a distance one.

Lemma 2 (Leaves-Parity Lemma). *Let L_o and L_e respectively denote the number of leaf-edges of a non-leaf-rooted, 3-regular, tree that have odd and that have even parity, then:*

$$L_o \bmod 3 = L_e \bmod 3$$

From Lemma 2, a rooted tree, with $L = L_o + L_e$ leaves, can only have certain values for (L_o, L_e). For example, if $L \bmod 3 = 1$, then $L_o \bmod 3 = L_e \bmod 3 = 2$ and (L_o, L_e) may only take values $(2, L-2), (5, L-5), \ldots, (L-2, 2)$.

From now on, we shall be concerned with trees that are rooted at an internal node adjacent to at least one leaf. The particular leaf-edge joining the root to this specific leaf will be called the *root-edge.* In all our figures, root-edges appear above the root and other leaf-edges below. By *even-parity* (*odd-parity*) trees we will mean trees for which every *non-root-leaf* leaf-edge is of even-parity (odd-parity). For any 3-edge-colouring of a particular tree T, a concise description of the colouring is given by the sentence, S, formed by the colours (read from left-to-right in our figures) assigned to leaf-edges excluding the root-edge. Notice that all other edge colours of T follow from S, because each colour must be present at every vertex. From Lemma 1, we see that the colour of the root-edge is *independent* of the topology of the tree and can be deduced directly from S. Thus, in Figure 1, the (dashed) root-edges are similarly coloured. We say that *S colours T* and refer to T as being *colourable* by S. We also say that T is *a-coloured* if a is the colour assigned to the root-edge. The notation $C(e)$ denotes the colour of edge e. An *instance* of *CPBT* is a pair of trees (T_i, T_j) and a *solution* is a string S that colours both T_i and T_j.

3 Rotations and Colour-Constrained Trees

As is well-known [14,1,10], any binary tree may be transformed into any other by a series of so-called *rotations.* Figure 2 shows how the locality of a tree may be transformed by rotating an edge e through the vertex v. Figure 3 shows the complete *rotational space*, R_5, for binary trees with 5 leaves. R_n is a graph in which two nodes (trees) are connected by an edge if and only if either tree can be transformed into the other by a single rotation. R_n is exponentially large in the number of leaves (more precisely, the number of nodes of the rotational

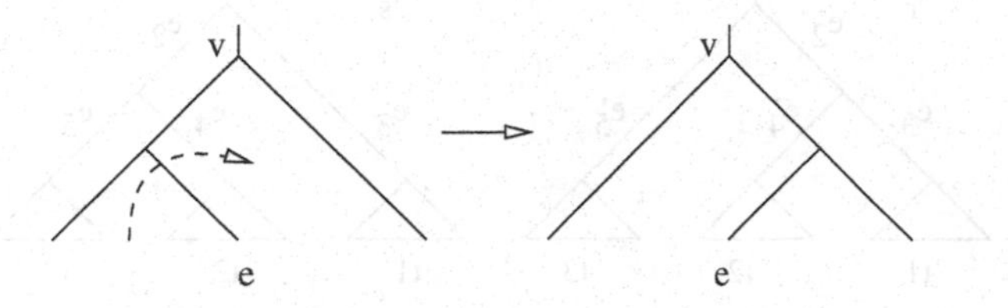

Fig. 2. A (clockwise) rotation of an edge e through a vertex v.

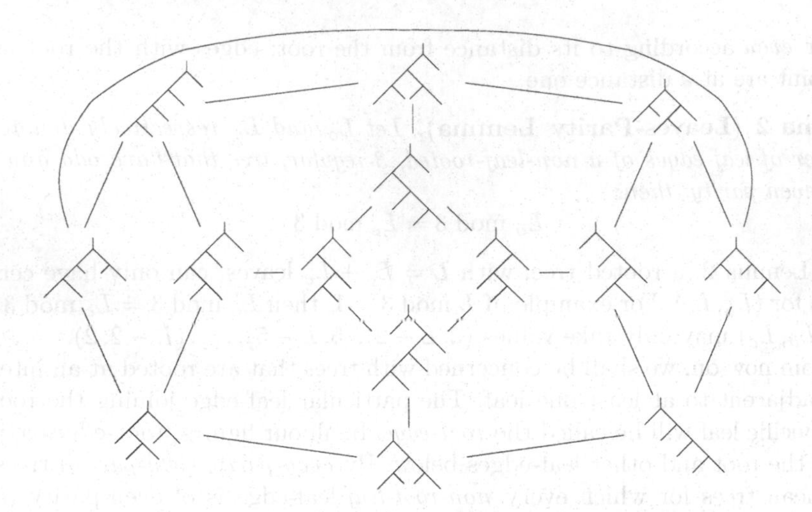

Fig. 3. The *rotational space* for binary trees with 5 leaves.

space is given by the nth *Catalan number*). Given (T_1, T_i), we consider a path, $(T_1, T_2, T_3, \ldots, T_i)$, in the rotational space R_n. In general, there are many such paths. At this point we need not be specific. The idea is to see how the series of rotations associated with the path might be used to find a solution to $CPBT$ for the tree pair (T_1, T_i). We first consider how we might generate the solution to $CPBT$ for a pair of binary trees which are a distance one apart in rotational space.

Lemma 3. *Let (T_r, T_s) be a pair of binary trees which differ by a single rotation (Figure 4) and let S colour T_s. Then S is a solution to $CPBT$ for (T_r, T_s) if and only if $C(e_1) = C(e_4)$. Also $C(e_1) = C(e_4)$ implies that $C(e'_1) = C(e'_4)$.*

Proof. There are only three colours available and so adjacency gives us:

$$C(e_1) = C(e_4) \text{ or } C(e_1) = C(e_5) \quad (C(e_4) \neq C(e_5))$$

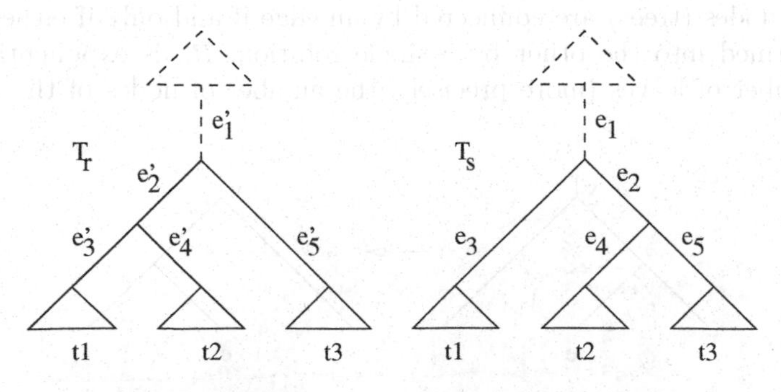

Fig. 4. Solutions to $CPBT$ for (T_r, T_s) require that $C(e_1) = C(e_4) = C(e'_1) = C(e'_4)$.

- If S is a solution to $CPBT$ for (T_r, T_s), then S colours T_r as well as T_s and the *Colour-Parity Lemma* (section 2) gives:

$$C(e_1) = C(e_1'), C(e_3) = C(e_3'), C(e_4) = C(e_4'), C(e_5) = C(e_5')$$

  But $C(e_1) = C(e_1') \neq C(e_5')(adjacency) = C(e_5)$ and so we must have that $C(e_1) = C(e_4)(= C(e_1') = C(e_4'))$
- If S is not a solution to $CPBT$ for (T_r, T_s), then S does not colour T_r. However, since S colours T_s it must colour $t1$, $t2$ and $t3$ (identical in T_r and T_s). The *Colour-Parity Lemma* then gives:

$$C(e_3) = C(e_3'), C(e_4) = C(e_4'), C(e_5) = C(e_5')$$

  Assume that $C(e_1) = C(e_4)$, then $C(e_3) = C(e_5)$ and there would be no colour clashes in the region of e_2' in T_r. It would follow that $C(e_1) = C(e_1')$, and that, contrary to our assumption, S would colour T_r. It must therefore be the case that $C(e_1) = C(e_5)$ and that S does not colour T_r because it would cause $C(e_3') = C(e_4')$.

Notice that Lemma 3 also implies (Figure 4) that if S is a solution to $CPBT$ for (T_r, T_s) then:

$$C(e_3) = C(e_5) = C(e_3') = C(e_5') \neq C(e_2) = C(e_2') \quad (1)$$

It also follows from the Lemma that the number of solutions to $CPBT$ for (T_r, T_s) is exactly half the number of strings that colour either T_r or colour T_s; this is because in T_s, for example, exactly half the colouring strings have $C(e_1) = C(e_4)$ and the other half have $C(e_1) \neq C(e_4)$.

We define a *colour-constrained tree* to be a binary tree in which *some* of the edges are partitioned into *colour-constrained* subsets. All the edges in any one of these subsets are required to be of the same (unspecified) colour and any two edges in such a subset cannot be adjacent. Lemma 3 provides our first illustration of a colour-constrained tree. Given (T_r, T_s) of that lemma, we see that all proper colourings of T_s, for example, in which $C(e_1) = C(e_4)$ give solutions to $CPBT$. Here, the edge subset $\{e_1, e_4\}$ is a colour-constrained subset.

Provided that no colour-clash is induced by the rotation (more on this later), *we notice that the proof of Lemma 3 works for colour-constrained trees also*: we require the same colour-constraints to apply in both trees and that in T_r, e_4' and e_5' must not be in the same colour-constrained subset, otherwise there would be a colour-clash in T_s after the rotation.

We now extend the idea of colour-constraining across a series of rotations which produce the sequence $(T_1, T_2, T_3, \ldots, T_i)$. The idea is that starting from T_1, we build up a series of colour-contrained trees, adding more constraints as we approach T_i. From the colour-constrained version of T_j, $1 < j \leq i$, we can obtain (as we shall see) a solution to $CPBT$ for all (T_1, T_j). Lemma 3 describes how to proceed and this is illustrated in Figure 5 in which edges belonging to the same colour-constrained subsets are similarly represented (as bold or dashed lines). This small example illustrates a number of points:

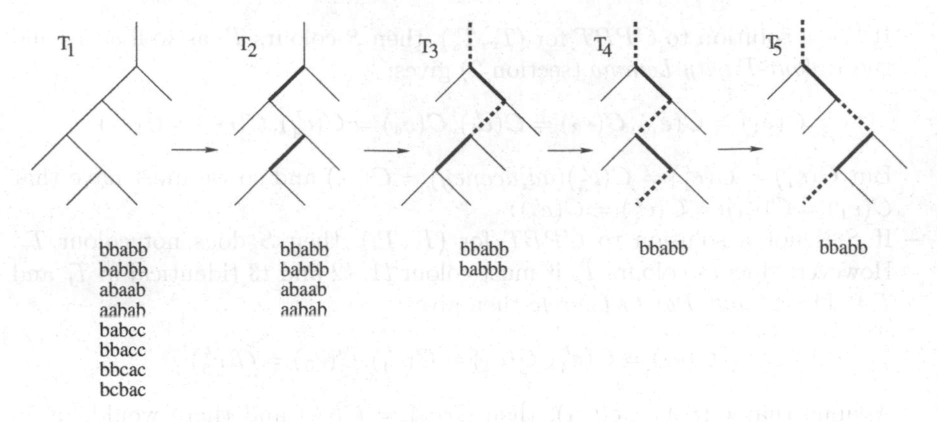

Fig. 5. Colour-constraining along a rotational sequence.

- In such a sequence, a colour-constraint demanded by a particular rotation might already be in place before the rotation occurs. This is illustrated, for example, in going from T_4 of Figure 5 to T_5. If, however, the rotation demands a *new* colour constraint in going from T_j to T_{j+1}, say, then the number of solutions to $CPBT$ for (T_1, T_j), given by the process, is twice the number for (T_1, T_{j+1}). Again, this is illustrated in Figure 5 by the solutions being listed below each T_j; for the sake of brevity only distinct a-colourings are listed (that is, solutions obtained by interchanging b's and c's are ignored).
- At an arbitrary rotation, from T_j to T_{j+1}, a new colour-constrained subset can be obtained in one of three ways: by merging two constrained edge-sets (one containing e_1 and the other containing e_4), by merging a constrained subset with an unconstrained edge, or by combining two unconstrained edges (e_1 and e_4). For example, consider the dashed-edges in going from T_3 to T_4, here a constrained subset of two edges is merged with an unconstrained edge.
- If the parent-edge of a rotated edge is colour-constrained just before a rotation (in T_j, say), then the parent-edge of the rotated edge after rotation (in T_{j+1}) is similarly colour-constrained. This follows from equation 1 and is first illustrated in Figure 5 in going from T_2 to T_3.
- It is possible for an individual edge to undergo several rotations in the same (clockwise or anticlockwise) direction without experiencing *any* rotations in the opposite sense. Even the small example of Figure 5 shows this: the edge that is first rotated (clockwise) in going from T_1 to T_2 is further rotated (clockwise) in going from T_4 to T_5 without, at other times, experiencing any rotations in the opposite (anticlockwise) sense.

Before we can claim that the colour-constrained T_i will provide a solution to $CPBT$ for (T_1, T_i), we need to prove that any colour-constrained tree produced by the process is indeed 3-edge-colourable. This is not immediately obvious because not every colour-constrained tree is 3-edge-colourable, even if no two members of the same subset of colour-constrained edges are adjacent. For example, the colour-constrained tree on the left of Figure 6 is not 3-edge-colourable.

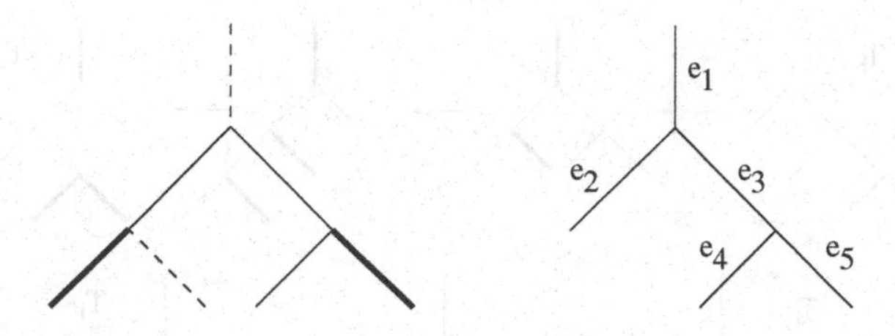

Fig. 6. The colour-constrained tree on the left is not 3-edge-colourable

Lemma 4. *Any colour-constrained tree, generated by a series of rotations, is 3-edge-colourable in linear time.*

Proof. We colour the edges in linear-time by conducting a *depth-first* traversal of the tree, T. At the start of the traversal the root-edge and all edges in the same colour-constrained subset are coloured with the same colour (arbitrarily) chosen from the set $\{a, b, c\}$. As each edge, e, is encountered for the first time, provided e is not yet coloured, we set $C(e)$ to be a colour in $\{a, b, c\}$ not yet used by its neighbours and colour every edge in the same colour-constrained subset with the same colour. A colour-clash is only possible if every colour in $\{a, b, c\}$ has already been used by edges adjacent to e. This, however, is not possible for a colour-constrained tree generated by a rotational sequence as we shall see.

A subgraph T_v of the colour-constrained tree T *induced* by the edges in the same colour-constrained subset V is defined as follows: the vertices of T_v are the endpoints of the edges in V and its edges are those edges of T which have both endpoints in T_v. It is very easy to see that if T is generated by a series of rotations, then T_v is a *connected* subgraph of T.

Suppose that the right-hand-side of Figure 6 shows a locality of T just as the edge e_3 is about to be coloured and after e_1 has been coloured so that $C(e_1) = a$. Either or both of the edges e_4 and e_5 can only have become coloured if individually they had belonged to the same colour-constrained subset as some edge that had already been coloured. Notice that any such previously coloured edge must belong to one component of T defined by the cut-edge e_3 and (e_4 and e_5) must belong to the other. Because the subgraph of T induced by any colour-constrained subset is connected, it follows that any colour used at this step at (e_4 and e_5) must also be present at (e_1 and e_2). There are a small number of ways in which we can colour all or some of the edges (e_2, e_4, e_5) so that all of a, b and c are adjacent to e_3, however, an exhaustive check easily shows that none of these is compatible with the requirement that colours present at (e_4 and e_5) must also be present at (e_1 and e_2). For example, this is the case for $C(e_2) = C(e_4) = b$ and $C(e_5) = c$ because c is not present at e_1 or e_2.

We have, until now, put aside any considerations of what might be a best path to follow in rotational space from T_1 to T_i. It is easy to see that some paths can yield a sequence of proper colour-constrained trees, whereas other paths (with the same endpoints) may encounter colour-clashes. This is illustrated by the small example of Figure 7. The upper and the lower portions of the Figure

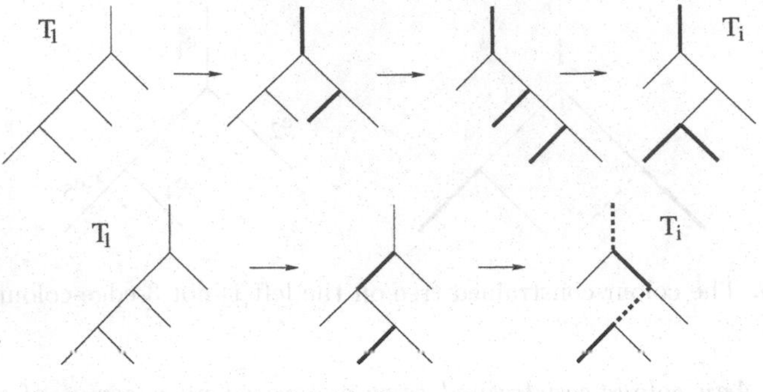

Fig. 7. A colour-clashing rotational sequence (top) and a non-clashing sequence.

illustrate different rotational routes between the same initial tree, T_1, and the same final tree, T_i. In the upper case a colour clash is encountered whereas no such clash occurs in the lower route.

The following theorem follows from the preceding considerations:

Theorem 1. *If there is a series of rotations defining the sequence of colour-constrained trees* $(T_1, T_2, \ldots, T_i)$ *such that no colour clashes occur, then there is a solution to* $CPBT$ *for each pair* (T_1, T_j), *for* $1 < j \leq i$, *given by a 3-edge-colouring of the colour-constrained tree* T_j.

We observe that the successful rotational route of Figure 7 is a *shortest* rotational route from T_1 to T_i. We also observe, by inspection of Figure 3, that the rotational path taken in Figure 5 is one of four possible shortest paths. Each of these shortest paths yields no colour clashes, whereas all the longer rotational paths do. We have been unable to construct a counterexample to the following (Shortest Sequence) Conjecture which we believe it to be true.

Conjecture (Shortest Sequence Conjecture) *For any pair of binary trees,* (T_1, T_i), *there exists at least one rotational sequence inducing no colour-clashes. Such a sequence is defined by a shortest path in the rotational space.*

Of course, a proof of the Shortest Sequence Conjecture would be a proof of the Four-Colour Problem for Planar Maps. If such a proof also yielded a linear-time algorithm to find such a shortest path, then we would have a linear-time algorithm for four-colouring planar maps. This is because the reductions between the Four-Colouring Problem and $CPBT$ are linear-time [2] and because of Lemma 4. It does not seem likely that an efficient algorithm to find a shortest rotational path will exist because of the exponential size of the rotational space. It may be that the problem is NP-complete although we know of no proof. For some specific tree pairs it is known how to construct a shortest path in linear-time. We illustrate this in what follows.

A *spine* is a binary tree in which every node has at least one child which is a leaf. For instance, in a *right* spine, the left-child of every node is always a leaf. Let T_{rsp} denote the right spine with n leaves and T_{arb} an arbitrary binary tree of the same size. The following is typical of a number of lemmas in which we have short proofs of the existence of solutions to $CPBT$ for pairs (T_{arb}, T), where T is spine-like (the definition of T might, for example, relax the requirement that the root has to have a leaf for a child).

Lemma 5. *There exists a solution to* $CPBT$ *for any pair of the form* (T_{arb}, T_{rsp}).

Proof. Starting from T_{arb}, the simple strategy of rotating an edge onto the spine (in other words, increasing the length of the path from the root to the right-most leaf by one) at each rotational step will, after at most $n-2$ steps, produce a proper colour restricted T_{rsp}. To see this, we only have to observe that at any step in the execution of the algorithm, all the colour restricted edges are either on or are adjacent to the path from the root to the right-most leaf in the partially-constructed right spine. This excludes all edges yet to be rotated. Thus each edge, just before being rotated is not colour-restricted and so no colour clash can occur at rotation.

We observe that, in the execution of the algorithm of Lemma 5, a shortest rotational path from T_{arb} to T_{rsp} is taken. This follows from the fact that the path length from root to right-most leaf is increased by one at every step and that no algorithm employing rotations can do better.

4 Colouring Topologies

Our motivation within this section is to look for *patterns of colourability* within *regular forms* of tree pairs, and to seek to understand colourings of all tree pairs as a modification or perturbation of these. This is best understood by a number of examples. Before coming to these, we note that it will be convenient occasionally to use a regular expression to specify a string colouring a tree. In this we adopt a particular convention: if the tree in question has n (non-root) leaf-edges, then we take the *first* n characters of a sufficiently-long string that can be generated by the expression. Thus, in this abuse of normal convention, for 4 leaf-edges, $(abc)*$ would mean *abca* and $(bac)*$ would mean *bacb*.

As a first example, consider the three different colourings of the four-leaved tree shown in Figure 8. Notice that the leaf-edge colours in each case spell out a subsequence of $(abc)*$. We can stack copies of this small graph, starting with

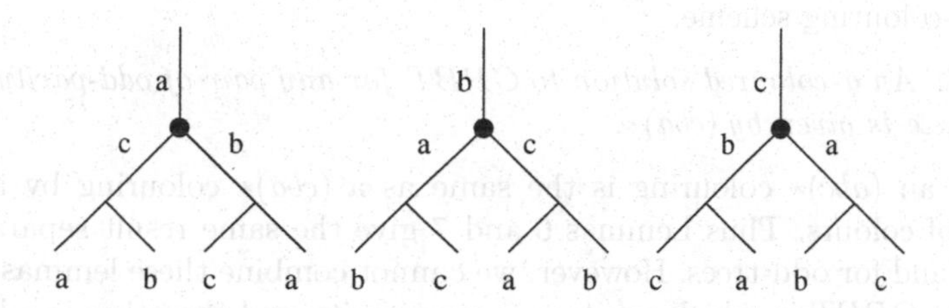

Fig. 8. Stackable colourings for even-parity trees.

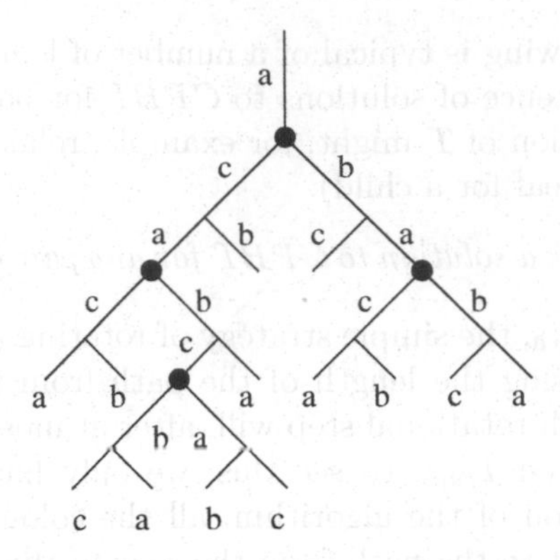

Fig. 9. Stacked 4-leaf trees producing an $(abc)*$ colouring of an even-parity tree.

the leftmost, for a-coloured trees, in such a way that the root-edge of each additional copy is made to coincide with a similarly coloured leaf-edge of another copy. Figure 9 shows an example of the construction. Clearly we can construct arbitrarily large coloured graphs by the same means and in every case the tree will be a-coloured by a string $(abc)*$ of length $(3k+1)$, where k is the number of coloured four-leaved tree used. It is easy to see that any *even-parity* tree can be constructed in this way giving the first lemma of this section.

Lemma 6. *An a-coloured solution to CPBT for any pair of even-parity trees of the same size is given by* $(abc)*$.

What is interesting about this simple result is that it provides a *pattern of colourability* for any pair of trees drawn from an infinite-sized set of trees of particular *regular form.* It is convenient here to define the *locked-colouring scheme.* Any binary tree T is coloured according to this scheme by completely embedding it in a larger a-coloured tree, T_{lc},constructed from the coloured components of Figure 8. The embedding is the natural one of making the root-edges of T and T_{lc} coincide, then their respective left and right child-edges and so on. The colour of each edge of T is then just the colour of the corresponding edge of T_{lc}. Notice that no edge of a tree coloured by this scheme can be rotated without a colour-clash occuring (see Lemma 3, here every rotatable edge does not have the colour of its grandparent) and hence the name *locked-colouring* scheme. The next lemma follows immediately from colouring odd-parity trees according to the locked-colouring scheme.

Lemma 7. *An a-coloured solution to CPBT for any pair of odd-parity trees of the same size is given by* $(cba)*$.

Of course, an $(abc)*$ colouring is the same as a $(cba)*$ colouring by a simple renaming of colours. Thus Lemmas 6 and 7 give the same result separately for even-trees and for odd-trees. However, we cannot combine these lemmas to solve instances of $CPBT$ in which one tree is even-parity and the other is odd-parity because no such pair of trees can be of the same size. This follows from the

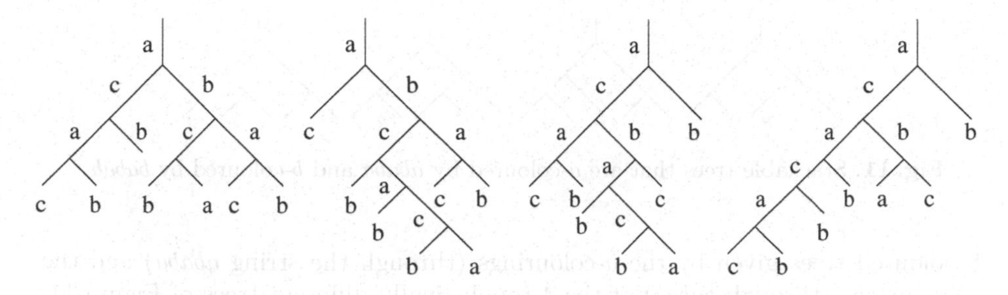

Fig. 10. A set of same-parity sequence trees all coloured by *cbbbacb*.

Leaves-Parity Lemma (Section 2). However, the Lemmas can be combined in the interesting sense of Theorem 2 which is illustrated in Figure 10. The four trees shown there all have the same *parity sequence*. That is, in all these trees the ith leaf-edge from the left, $1 \leq i \leq 7$, has the same parity. If we denote odd parity by o and even parity by e, then the parity sequence for all these trees is: $ooeooo = o^2eo^4$.

Theorem 2. *Let* (T_i, T_j) *denote any pair of binary trees of the same size and same parity sequence, then a solution to CPBT for* (T_i, T_j) *is given by the locked-colouring scheme.*

Proof. (Sketch). Without loss of generality, let the parity sequence of both T_i and T_j be $o^{n_1}e^{n_2}o^{n_3}\dots e^{n_k}$. Then for each sequence of odd-parity leaf-edges: o^{n_1}, o^{n_3}, ..., the leaf-edges will be coloured by a subsequence of $(cba)*$ (as for Lemma 7). Similarly, for each sequence of even-parity nodes e^{n_2}, e^{n_4}, ..., the leaf-edges will be coloured by a subsequence of $(abc)*$ (as for Lemma 6). The question then remains as to whether these colour patterns are *in phase* in both trees. To see that this must be so, we observe that the locked-colouring scheme always ensures that the first leaf-edge of an odd (or even) sequence repeats the colour of the last leaf-edge of the preceding even (or odd) sequence. In this way, reading from left to right, we can see that colours of leaf-edges in both trees must always agree.

Theorem 2 followed naturally from initial considerations about infinite sets of graphs (the even-parity and odd-parity graphs) of which every member could be coloured by $(abc)*$. We specifically observe here that the *Colour-Parity Lemma* (Section 2) requires that any $(abc)*$ colourable tree cannot have $3k$ consecutive leaves, for any $k \geq 1$, which form the leaves of a subtree of the tree. This is because that subtree would have equal numbers (and therefore parities) of the colours a, b and c present in its leaves so that any colouring of the root edge would violate the Colour-Parity Lemma. We briefly consider trees which are $(ab)*$ colourable in the same light. In this case, the Colour-Parity Lemma requires that the trees do not have $4k$ leaves, for any $k \geq 1$, that form the leaves of a subtree of the graph. Here, instead of the three different colourings of the even-tree with four leaves of Figure 8, we can consider, for example, a set of

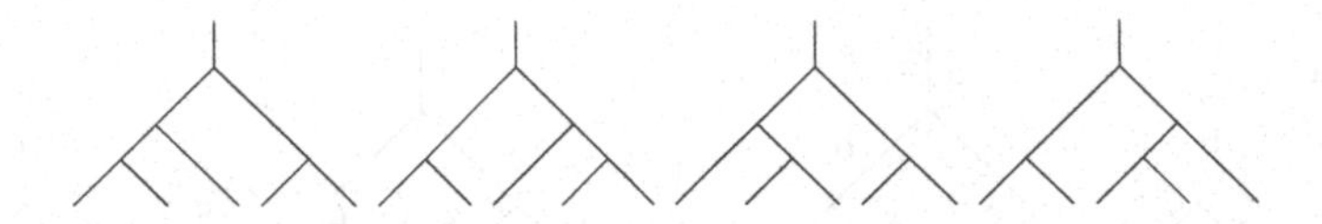

Fig. 11. Stackable trees that are a-coloured by *ababa* and b-coloured by *babab*.

8 coloured trees given by the a-colourings (through the string *ababa*) and the b-colourings (through *babab*) of the 4 topologically different trees of Figure 11. These trees, all with 5 leaves, can be stacked, in a manner entirely analogous to the stacking of the 4-leaved trees of Figure 9, to give an infinite-sized set of $(ab)*$ colourable trees. It follows that there is a solution to $CPBT$ for any pair of these trees of the same size. Such a solution follows in linear-time. Here, it is not so easy to provide a succinct description of the trees in this set (like *even-parity*, for example), but what is interesting is that any such pair of trees may have very different parity sequences unlike the $(abc)*$ case described previously.

Before concluding this section, we note that Lemma 6 is a special case of the following Lemma.

Lemma 8. *Any even-parity tree is coloured by any string from* $((abc) \vee (bac))*$.

Proof. (Sketch). Any string, S, generated by $((abc) \vee (bac))*$ has c's occuring regularly at positions 3 apart and these are separated by either *ab* or *ba*. It is easy to see that, in any such S, we can replace *any* 4-character sequence, by the letter that uniquely occurs twice in that sequence, and obtain a shorter string that is also generated by $((abc) \vee (bac))*$. For example, *bcba* can occur in L and would be replaced by b. Since *bcba* must be preceded by a and followed by c in L, we see that the effect of this 4-letter replacement is to replace *abcbac* in L by *abc*. Thus the sequence generated from L is still a sequence of 3-character subsequences each of which is either *abc* or *bac*. Recursive application of replacements, such as we have described, models the colouring of *any* even-parity tree by any string S generated by $((abc) \vee (bac))*$.

Lemma 8 shows, for example, that any even-parity tree is coloured by the sequence $(abcbac)*$. Each of the family of trees represented by Figure 12 (one tree with $(3k+1)$ leaves, for each $k \geq 1$) is also coloured by $(abcbac)*$. Thus we have solutions to $CPBT$ for a tree from Figure 12 and any even-parity tree of the same size. Notice each tree of Figure 12 is a mix of even-parity and odd-parity leaves. Earlier no such concise solution was provided for $CPBT$ in which only one of the trees was an even-parity tree.

It is clear that many other results can be generated by the methodology of this section. The real interest of this material though lies in whether insights provided by this approach may eventually yield a solution to $CPBT$ for *any* pair of trees.

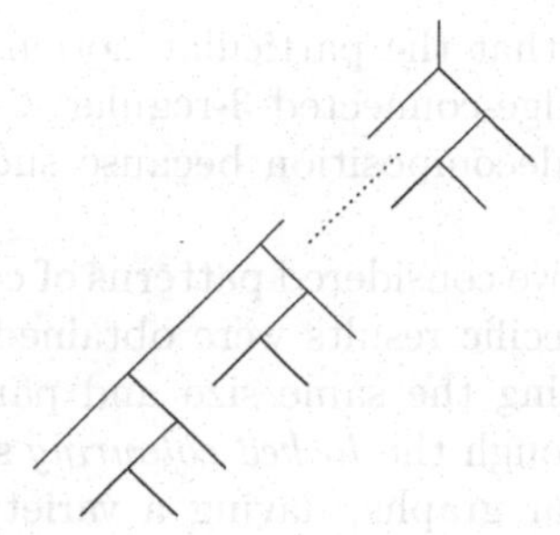

Fig. 12. Trees that are $(abcbac)*$ colourable.

5 Summary and Further Work

We have described two novel approaches in the search for a concise and lucid proof of the problem of *Colouring Pairs of Binary Trees* (*CPBT*). Such a discovery would, in turn, provide such a proof for the *Four-Colour Problem* of planar maps. Both approaches have rich potential for further exploration and, in that sense, the work described here lays the foundations.

In our first approach, we proved that a concise solution to *CPBT* is possible, for any instance, provided that a rotational path exists between the two trees of the instance which invokes no colour clashes. We believe that a shortest rotational path always guarantees that no such clash will occur and this is embodied in the *Shortest Path Conjecture* of Section 3. It seems that a colour clash can occur when an individual edge experiences rotations in both a clockwise and an anticlockwise sense in the course of transforming one tree into the other along the rotational path. Intuitively, we believe that both types of rotations should not be necessary for any individual edge along a shortest rotational path. This view could be a starting point for further work. At this time, it is not known how to find a shortest path in polynomial-time, indeed the problem might be NP-complete. In [9], the authors prove that the problem of finding shortest paths using rotations and so-called *twists* is NP-complete but the proof does not carry over, as the authors state, for rotations alone. In [12], the author describes an $O(n \log n)$-time algorithm to find a short path between any two vertices in the rotational space R_n. However, the length of the path is only guaranteed to be less than twice that of an optimally short path. This is not tight enough for our purposes as an example in the text illustrates. Even if this shortest path problem is NP-complete, it may still be possible to find a satisfactory proof of the problem of *CPBT* through the rotational approach that we have described. Its NP-completeness would be likely, however, to deny an efficient algorithm by these means. An aspect of exploring a solution to *CPBT* would be opened up by knowledge that every 3-regular, 2-connected planar graph always contained some standard form for one of the trees in a (T_i, T_j) decomposition. If, for example, one of the trees could always be made to be a *spine*, then we would have solved the *Four-Colour Problem* through one of the lemmas described in the text (it is easy to observe that the particular notion of the spine here doesn't itelf work because any 4-edge-connected 3-regular, 2-connected planar graphs, G, cannot have any such decomposition because such decompositions require 3-edge-connectivity of G).

In our second approach we considered patterns of colourability in trees of regular form. A number of specific results were obtained. For example, we proved that any pair of trees having the same size and parity sequence have a concise solution to $CPBT$ through the *locked-colouring* scheme. We also described classes of generically similar graphs, having a variety of parity sequences, for which there is a ready solution. One approach to $CPBT$ is to recognise that the set of strings colouring any particular binary tree is a regular language (described by a regular expression without *Kleene closure*). See, for example, [2]. Then $CPBT$ reduces to showing that the intersection of any two such languages is non-empty. There is a well-known construction for the finite automaton which accepts the intersection of two regular languages (see [8]). Proof of the problem of $CPBT$ by this route is fraught with tedious detail and it doesn't appear to offer potential for clarity and conciseness. Using the colouring topologies approach, we attempt to show that there might be an easy way to identify colouring patterns, for generically similar trees, within this intersection. The arguments used, although related to finite automata theory, are not really of that domain. It is not clear how far the approach may be usefully extended while retaining its lucidity and conciseness. The patterns of 3-edge-colourability of binary trees appear to be intimately tied up with the notion of parity patterns in sequences of leaves. In this sense, we do not believe, for example, that the potentials of the *Colour-Parity* and *Leaves-Parity* Lemmas (Section 2) have been fully explored. One aspect of the colouring topologies approach is that it can solve $CPBT$ for generically similar, but rotationally-distant, trees (for example, same parity sequence trees), (T_i, T_j); a solution to another pair (T_i', T_j'), such that T_i is close to T_i' in rotational space and T_j is close to T_j', might then follow by some small, well-understood (but as yet unexplored), perturbation. A different topological-colouring scheme from *locked-colouring* would be needed if the perturbation was to be handled by *rotations*. However, other topological-colourings are available. For example, the two rotationally-distant trees given by the bracket structures ((((.(..)).)((..)((..).))).) and (.((((..)(..))(..))(.(..)))) are coloured by *babbcabcbc* in what we might call the *unlocked colouring* scheme. Here every rotatable edge is assigned the colour of its grandparent so that many near neighbours of each tree can be similarly coloured.

There are clearly many lines of further enquiry possible in this work.

References

1. Karel Culik II and Derick Wood. A note on some tree similarity measures. *Information Processing Letters*. Volume 15:1, (1982), pages 39-42.
2. Artur Czumaj and Alan Gibbons, Guthrie's problem: new equivalences and rapid reductions, *Theoretical Computer Science*, Volume 154, Issue 1, 3-22, January 1996.
3. Alan Gibbons. Algorithmic Graph Theory. Cambridge University Press (1985).

4. Alan Gibbons. Problems on pairs of trees equivalent to the four colour problem of planar maps. Second British Colloquium for Theoretical Computer Science, Warwick, March 1986.
5. K. Appel and W. Haken. Every planar map is four colourable. Part I. Discharging. Illinois Journal of Mathematics 21 (1977), 429-490.
6. K. Appel, W. Haken and J. Koch. Every planar map is four colourable. Part II. Reducibility. Illinois Journal of Mathematics 21 (1977), 491-567.
7. K. Appel and W. Haken. Every planar map is four colourable. Comtemporary Mathematics 98 (1989) entire issue.
8. John E. Hopcroft and Jeffrey D. Ullman. Introduction to Automata Theory, Languages and Computation. Addison-Wesley Publishing Company (1979).
9. Ming Li and Louxin Zhang. Twist-rotation transformations of binary trees and arithmetic expressions. *Journal of Algorithms*, Volume 32:2,(1999), pages 155-166.
10. Jean Pallo. On the rotation distance in the lattice of Binary Trees. *Information Processing Letters.* Volume 25:6, (1987), pages 369-373.
11. N. Robertson, D. P. Sanders, P.D.Seymour and R. Thomas. A new proof of the four-colour theorem. Electron. Res. Announc.Amer. Math. Soc. 2 (1996), no.1, 17-25 also: The four colour theorem, J. Combin. Theory Ser. B. 70 (1997), 2-44.
12. Rodney O. Rogers. On finding shortest paths in the rotation graph of binary trees. Congressus Numerantium 137 (1999), 77-95.
13. T. L. Saaty and P. C. Kainen. The Four-Colour Problem: Assaults and Conquest. McGraw-Hill (1977).
14. Daniel D. Sleator, Robert E. Tarjan and William P. Thurston. Rotation Distance, Triangulations and Hyperbolic Geometry. *Journal of the American Mathematical Society.* Volume 1 (1988), pages 647-682.

Applications of Finite Automata*

Juhani Karhumäki

Department of Mathematics & TUCS, University of Turku,
FIN-20014 Turku, Finland,
karhumak@cs.utu.fi

Abstract. We consider three different recent applications of finite automata. They are chosen to emphasize the diversified applicapity of the theory.

1 Introduction

The theory of finite automata is almost half a century old. It started from a seminal paper of Kleene [Kl], and within a few years developed into a rich mathematical research topic. This is much due to several influencial papers like that of [RS]. From the very beginning research on finite automata was very intensive, so that fundamental results and the beauty of the theory was revealed in a short period, in a decade or so, cf. e.g. [HU]. However, some challenging problems were solved only much later, or are still open at the moment.

Problems of the first type were, for example, the equivalence problems for deterministic multitape finite automata and for deterministic pushdown automata, cf. [HK] and [Se], respectively. Problems of the latter type are the generalized star height problem and the comparision of the number of states of equivalent deterministic and nondeterministic 2-way finite automata, cf. [Br] and [Mi], respectively. Also many problems on rational formal power series, i.e. finite automata with multiplicities are still unanswered. For example, famous *Skolem's Problem* can be stated as a question to decide whether two unary finite automata accept some word equally many times, cf. [SS].

From the very beginning finite automata constituted a core of computer science. Part of the reason is that they capture something very fundamental as is witnessed by a numerous different characterizations of the family of rational languages, that is languages defined by finite automata. More important, however, is that they model a fundamental intuitive notion: what can be computed by a finite memory? A third, and I believe, very important reason is the usefulness of finite automata in many applications of computer science.

In fact, the interrelation of finite automata and their applications in computer science is a splendid example of a really fruitful connection of theory and practice. Finite automata played a crucial role in the theory of programming languages and compiler constructions just to mention two applications, cf. e.g.

* Supported under the grant 44087 of the Academy of Finland

K. Diks et al. (Eds): MFSC 2002, LNCS 2420, pp. 40–58, 2002.

[ASU]. On the other hand, applications had a definite impact on automata theory by introducing important questions and problems. The above mentioned equivalence problems are concrete examples in this respect.

Later on, and in particular over the past ten years, the applications of finite automata have been more diversified, and I would not be surprised, if this development would continue in the future. This is the point we want to make here.

Our goal is not to discuss, or even not to list, all recent applications of finite automata. For example, we do not consider the use of finite automata in problems on model checking or program verifications, cf. e.g. [Bo], [BCMS], [T] and [V], although these seem to be among the most promising and widely used applications at the moment. Instead, we concentrate on three concrete examples of applications of finite automata. They are chosen to support a view that the theory of finite automata can be used essentially in many different areas, not only in computer science, but also in combinatorics and algebra.

The first application is to algebra. Theory of automatic groups, that is groups the presentations of which can be computed by a finite transducer, is an excellent example of such applications, cf. [ECHLPT]. We point out here a much more specific, but, we believe, a very illustrative example. Namely, how finite automata can be used to decide the isomorphism of two finitely generated subsemigroups of a free semigroup, cf. [ChHK].

The second application is to combinatorics, or more precisely to tilings of the plane. A remarkable result of Berger [Berg] says that there exist finite sets of tiles which can cover the plane only in a nonperiodic way, that is there exist nonperiodic tile sets. A completely new, and so far the best with respect to the cardinality of the tile set, proof of this result was based on a beautiful idea of using finite automata (and number theory) to solve the problem, cf. [Kari].

Finally, as the third application, we use finite automata to generate, compress and transform (two) dimensional images. Here finite automata with multiplicities, that is rational formal power series, come into the play. An interesting, and potentially important feature is that many transformations can be done directly on the level of the compressed automata representation of the image. For example, integration of the image is such a property. This research was initiated in [BM] and [DC], and later continued for example in [CKarh], [CKariI] and [DKLTI]. A related application is to use automata, again with multiplicities, for speech recognition, for more see [Mo]. Another methods based on automata to compress texts were introduced in [CMRS].

2 Notation

In this section we recall briefly basic definitions of different variants of finite automata, mainly in order to fix the terminology, for more details see e.g. [HU], [Bers] and [E].

For a finite alphabet A let A^* (resp. A^ω) denote the set of all finite (resp. one way infinite) words over A. Denoting the empty word by 1 we set $A^+ = A^* \setminus \{1\}$.

A finite automaton FA over A is a quituple $\mathcal{A} = (Q, A, E, I, T)$ where Q is a finite set of states, $I \subseteq Q$ is a set of initial states, $T \subseteq Q$ is a set of final states, and $E \subseteq Q \times A \times Q$ is a set of transitions. A transition $(p, a, q) \in E$ is usually denoted by

$$p \xrightarrow{a} q. \tag{1}$$

A finite automaton is *deterministic* if for each pair (p, a) there exists at most one state q such that $(p, a, q) \in E$ and I is a singleton, say $I = \{i_0\}$. The language accepted by $\mathcal{A}$, in symbols $L(\mathcal{A})$, for a deterministic FA, can be computed as follows: Let $Q = \{1, \ldots, n\}$ and define, for each a, an $n \times n$ matrix by the condition

$$(M_a)_{i,j} = 1 \text{ if } i \xrightarrow{a} j \in E, \tag{2}$$

and otherwise $(M_a)_{i,j} = 0$. Then, as is well known, for $w = a_1 \ldots a_n$, with $a_i \in A$, we have

$$w \in L(\mathcal{A}) \Longleftrightarrow \pi_i \, M_{a_1} \cdots M_{a_n} \pi_t = 1,$$

or equivalently, $\pi_i M_{a_i} \cdots M_{a_n} \pi_t \neq 0$, where π_i is the n-dimensional vector having 1 in the position corresponding to the initial state and 0 elsewhere, and π_t is the n-dimensional (column) vector having 1 in the positions corresponding to final states and 0 elsewhere.

We preferred the above slightly unusual formalism, since it immediately extends to formal power series, that is to finite automata with multiplicities. Indeed, if $\mathcal{A}$ is a *nondeterministic* FA we have

$$w \in L(\mathcal{A}) \Longleftrightarrow \pi_i M_{a_1} \cdots M_{a_n} \pi_t \neq 0,$$

and moreover, the value of the right hand side tells how many times w is accepted, i.e. tells the *multiplicity* of w.

In formal power series we go a step further. Now, transitions (1) are with multiplicities, i.e. of the form

$$p \xrightarrow{a,\alpha} q,$$

where α is the multiplicity of the transition. Here it will be a real number, but in general an element of a semiring, cf. [SS]. Then (2) gets a form

$$(M_a)_{ij} = \alpha \text{ if } i \xrightarrow{a,\alpha} j \in E,$$

and the multiplicity of w in $\mathcal{A}$ is computed by a formula

$$\pi_i M_{a_1} \cdots M_{a_n} \pi_j, \tag{3}$$

where now M_{a_j}'s are matrices over real numbers and the vectors π_i and π_j have real entries defined by the multiplicities of initial and final states. Consequently, a finite automaton $\mathcal{A}$ with multiplicities defines a function

$$A^* \longrightarrow \mathbb{R}$$

referred to as the formal power series defined by $\mathcal{A}$. We call finite automata with multiplicities *weighted finite automata*.

Example 1. A nondeterministic automaton

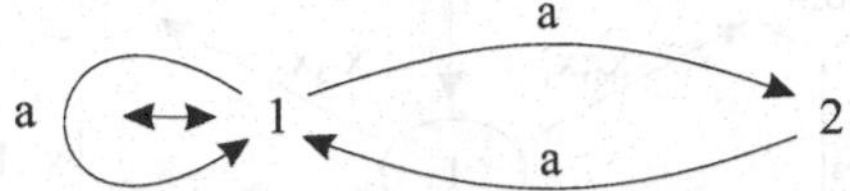

accepts the word a^n exactly F_n times where F_n is the nth Fibonacci number. This is seen, for example, from the fact that a^n is accepted, by (3), $(1,\ 0)\binom{1\,1}{1\,0}^n\binom{1}{0}$ times.

Besides finite automata with multiplicities finite transducers, i.e. finite automata with outputs, are useful for our applications. For definitions and basic results we refer to [Bers]. Here we specify only the minimum needed for this paper.

A *finite transducer* is a finite automata capable of producing an output at any step of the computation. Consequently, transitions of (1) are now of the form

$$p \xrightarrow{\alpha,\ \beta} q,$$

where α is the input read and β the output produced. A transducer might be with or without final states. A transducer is *sequential*, sometimes called *generalized sequential machine*, if the input α is always a single letter. In general it can be a word including the empty one. A finite transducer computes or defines a relation from Σ^* into Δ^*.

The following two examples are crucial in our applications and illustrate the power of finite transducers.

Example 2. Let $X = \{a, ab, ba\}$. Then X is not a code, that is there exist words having two different X-factorizations, in other words, words that can be expressed as products of X in two different ways. For example, we have $a.ba = aba = ab.a$. All double X-factorizations can be computed by a finite transducer as follows. First, define a bijection $\Theta \to X$, say $\Theta = \{x_1, x_2, x_3\}$ with $x_1 \leftrightarrow a$, $x_2 \leftrightarrow ab$ and $x_3 \leftrightarrow ba$.Then a finite transducer of Fig. 1 computes all *minimal* double X-factorizations, where minimal means that the initial parts of these are not factorizations of a same word. Here we use state ① as both the initial and final state. It is easy to modify the above to a transducer computing all double X-factorizations.

Note that the above construction is very illustrative. The machine remembers in its states which of the factorizations is ahead and by how much. After this observation it is obvious how the above can be extended to arbitrary finite sets X. The conclusion is that *the set of all double X-factorizations of a finite set X can be computed by a finite transducer*, i.e. it is *rational.*

In the next example we define two rather peculiar looking sequential transducers, defined by J. Kari to obtain a beautibul solution to an important problem in combinatorics.

Example 3. Consider a sequential transducer depicted in Fig. 2. The inputs for the machine are numbers 1 and 2, while the outputs are numbers 0,1 and 2. In its

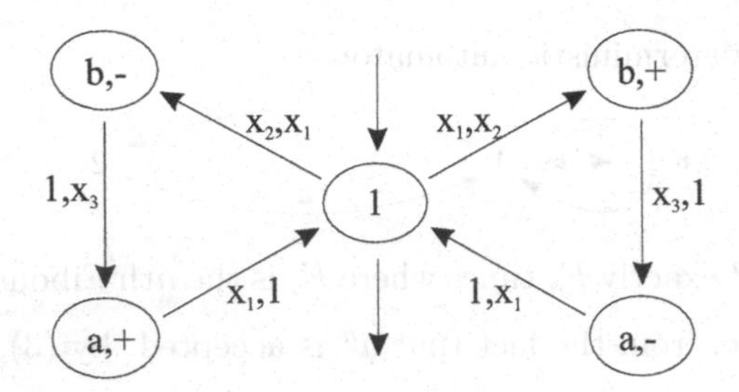

Fig. 1. A transducer for minimal double X-factorizations

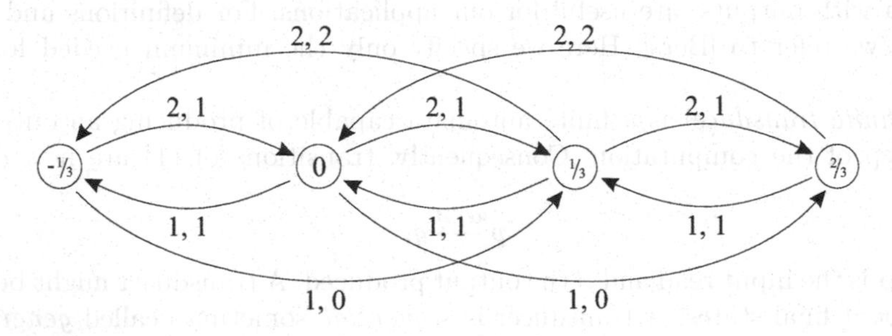

Fig. 2. Sequential transducer $\mathcal{M}_{2/3}$

states it remembers a remainder which is either $-1/3$, 0, $1/3$, or $2/3$. Now, in a step the automaton multiplies the input a by $2/3$, adds the current remainder q and represents it in the form $2/3 \cdot a + q = b + s$. Then the output produced is b and the remainder s tells the new state. Note, however, that due to the choices of the remainders and the alphabets, it is no reason that the above representation is unique. Indeed, the machine $\mathcal{M}_{2/3}$ is nondeterministic, cf. e.g. state 0.

Another similarly defined machine, where the multiplication is by 2, is shown in Fig. 3. Note here that the machine is not complete, it cannot read 1 at state (-1) - according to the rules that would require a new input letter 0. We do not want to make it complete, since a crucial idea in the application is to have as few transitions as possible!

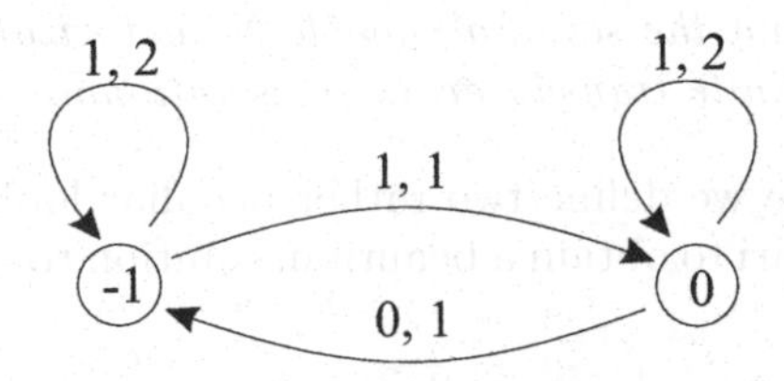

Fig. 3. Sequential transducer $\mathcal{M}_2$

3 Isomorphisms of F-Semigroups

In this chapter we consider one central problem of algebra, namely the question of deciding whether two finitely generated algebraic structures are isomorphic. As is well known, cf. e.g. [KS], many decision questions in algebra are undecidable. Indeed, first such examples were discovered at the very beginning of the theory of undecidability, see [P] and [Ma]. Here we show, as an application of automata theory (and combinatorics on words) how we can prove an important decidability result.

Let us call a subsemigroup of a free semigroup A^* *F-semigroup*. From the point of view of our considerations we can assume that the generating set A is fixed. The *isomorphism problem* for an algebraic structure asks to decide whether two elements of this structure are isomorphic. We shall show, see [ChHK], the following result.

Theorem 1. *The isomorphism problem for finitely generated F-semigroups is decidable.*

Proof. We repeat the proof in order to emphasize the role of automata theory in the solution. So let $S_1 = X^*$ and $S_2 = Y^*$ for finite sets $X, Y \subseteq A^*$. Clearly, the necessary condition for the isomorphism is that X and Y are of the same cardinality, say $X = \{x_1, \ldots, x_n\}$ and $Y = \{y_1, \ldots, y_n\}$. Further X and Y are isomorphic if and only if there is a bijection $\varphi : X \to Y$ such that whenever elements of X satisfy a relation the corresponding elements (via φ) of Y satisfy the same relation, and vice versa. Since there exist only finitely many different bijections φ it is enough to consider just one, which can be assumed to be the mapping $x_i \mapsto y_i$.

Next we introduce a copy of X and Y, say Θ, and interpret any nontrivial relation satisfied by X as an equation over variables Θ, as was done in Example 2. Hence the set of all nontrivial relations satisfied by X can be interpreted as an infinite system of equations, say $R(X) \subseteq \Theta^* \times \Theta^*$. Similarly, we obtain $R(Y)$.

The first observation is the one made already in Example 2.

Fact 1. *$R(X)$ and $R(Y)$ are rational relations, and moreover, finite transducers realizing these can be effectively found.*

Hence our problem is reduced to decide the equivalence of two rational relations. But this is a well known *undecidable* problem! A way to go around this is revealed when we notice that we ask in our problem much less than the equivalence of $R(X)$ and $R(Y)$. Namely, we ask only whether X is a solution of $R(Y)$ and Y is a solution of $R(X)$.

It is still not clear how to test these questions since $R(X)$ and $R(Y)$ are infinite systems of equations. Part of a solution comes from the fundamental compactness result of word equations, cf. e.g. [CKarh]:

Fact 2. *$R(X)$ and $R(Y)$ as systems of equations over A^* are equivalent to some of their finite subsystems $R_0(X)$ and $R_0(Y)$, respectively.*

Now, clearly our problem becomes decidable if these subsystems can be effectively found. In general, the compactness result of Fact 2 is only existential. But since $R(X)$ and $R(Y)$ are rational relations the subsystems can be effectively found. This follows straightforwardly from the pumping lemma for finite automata and the following easily provable implication on combinatorics on words: for any words α, β, γ, δ, $\bar{\alpha}$, $\bar{\beta}$, $\bar{\gamma}$ and $\bar{\delta}$ we have

$$\left.\begin{array}{l} \alpha\beta = \bar{\alpha}\bar{\beta} \\ \alpha\gamma\beta = \bar{\alpha}\bar{\gamma}\bar{\beta} \\ \alpha\delta\beta = \bar{\alpha}\bar{\delta}\bar{\beta} \end{array}\right\} \Rightarrow \alpha\gamma\delta\beta = \bar{\alpha}\bar{\gamma}\bar{\delta}\bar{\beta}.$$

So we can formulate

Fact 3. *Finite equivalent subsystems $R_0(X)$ and $R_0(Y)$ of $R(X)$ and $R(Y)$, respectively, can be effectively found.*

This completes our presentation of the proof of Theorem 1.□

Theorem 1 proposes a number of remarks. First, the proof method actually implies even a stronger result:

Theorem 2. *It is decidable whether a given finitely generated F-semigroup is embeddable into another one.*

Second, and more interestingly, our proof method breaks completely if instead of finitely generated, rationally generated F-semigroups are considered. So we formulate

Open Problen 1. *Is the isomorphism problem for F-semigroups having rational generating sets decidable?*

Third, for more complicated semigroups the isomorphism problem becomes easily undecidable. This is already the case for multiplicative matrix semigroups over nonnegative integers, cf. [KBS] or [CaHK]. This is essentially due to embeddings $A^* \hookrightarrow \mathcal{M}_{2\times 2}(\mathbb{N})$, which make Post Correspondence Problem applicable.

To conclude this section we recall that there are much deeper and more involved connections of automata theory and different parts of algebra, for example in automatic groups, cf. [ECHLPT]. Here we wanted to pick up just one, and we believe, very illustrative example.

4 Nonperiodic Tilings

In this section we recall a marvelous construction of J. Kari, see [Kari], to reprove, and also to improve, a result of Berger [Berg] saying that there exist aperiodic *tile sets.* A set of squares with coloured edges is a *tile set.* A *proper tiling* of the plane by a tile set T is the arrangement of elements of T such that the whole plane is covered and the neighboring edges are always of the same colour. Here the tiles are not allowed to be rotated and, of course, there are infinitely many tiles of each type of T available. A tiling might be *periodic*, i.e. invariant under some translation, or *aperiodic*.

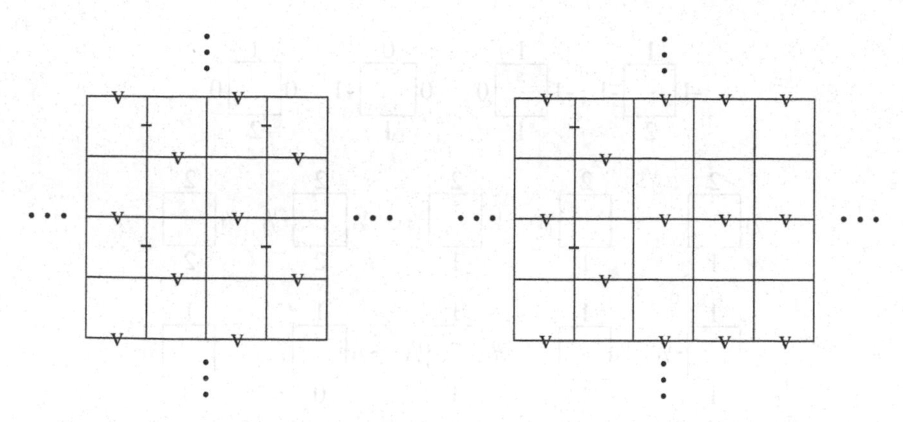

Fig. 4. Two tilings by $\mathcal{S}$

Example 4. Consider the set of the following four tiles

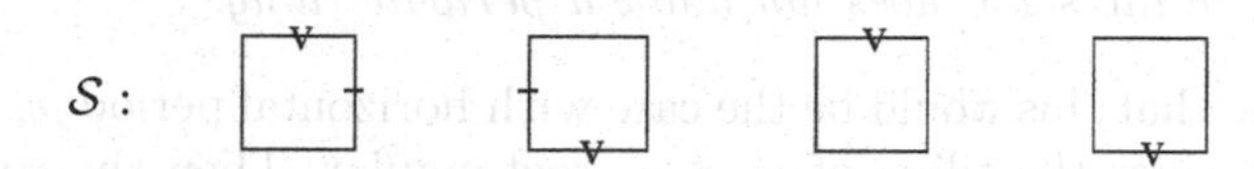

Clearly, it allows several tilings of the plane, for example those shown in Fig. 4. The first one is periodic in both directions (with period 2) while the second one is only vertically periodic. In this case it would also be easy to construct an aperiodic tiling of the plane.

A challenging combinatorial question was whether there exist finite tile sets allowing only aperiodic tilings. This was answered positively by Berger in his seminal work [Berg] using more than 20000 tiles! Later the number of needed tiles has been reduced, the best results being achieved by the method of J. Kari using rational relations. He constructed a aperiodic tile set of size 14, later K. Culik II modified the construction to only 13 tiles, cf. [Kari] and [Cu], respectively.

A tile set of 14 tiles is obtained from the sequential transducer of Example 3. Each transition of these both machines defines a tile. For example from transition $1/3 \xrightarrow{2,1} 2/3$ we create a tile, where horizontal edges are labeled by the input and output symbols and the vertical edges by the states, i.e. a tile

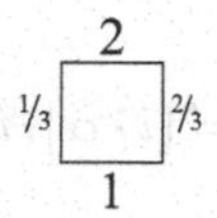

This way we obtain the tile set T:

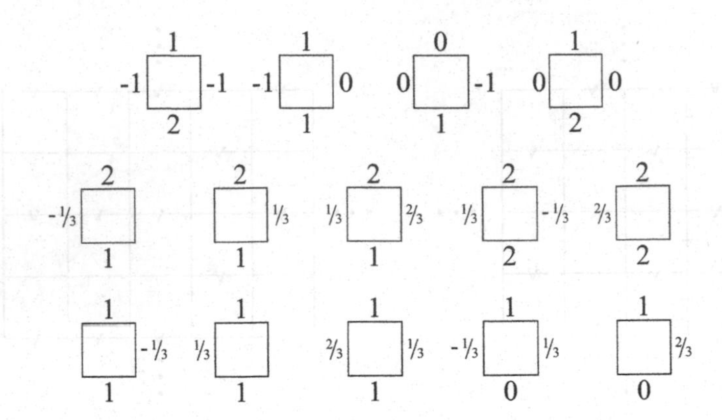

Note here that the states 0 of the machines $\mathcal{M}_2$ and $\mathcal{M}_{2/3}$ are replaced in the tiles by different symbols 0 and θ. This guarantees that the tile sets obtained from $\mathcal{M}_2$ and $\mathcal{M}_{2/3}$, respectively, cannot be mixed on any horizontal line of a proper tiling.

What is now easy, is the following:

Lemma 1. *The tile set T does not allow a periodic tiling.*

Proof. Assume that this would be the case with horizontal period p_h and vertical period p_v. Consider the tiling of $p_h \times p_v$ rectangular. Then the corresponding labels on the top and on the bottom are the same, as well as those on the left and on the right, respectively. Let us denote by n_i, for $i = 0, \ldots, p_v$, the sum of labels on the ith horizontal line of the rectangular. Due to the horizontal periodicity, and the definition of the machines $\mathcal{M}_2$ and $\mathcal{M}_{2/3}$, we have

$$n_{i+1} = q_i n_i \quad \text{for } i = 0, \ldots, p_v - 1,$$

where $q_i = 2$ or $q_i = 2/3$ depending on whether the ith line comes from the tiles obtained from $\mathcal{M}_2$ or $\mathcal{M}_{2/3}$, respectively. Now, the vertical periodicity implies that $n_0 = n_{p_v}$, and hence

$$n_0 = q_0 \ldots q_{p_v - 1} n_0,$$

which, since $n_0 \neq 0$, means that $1 = q_0 \ldots q_{p_v-1}$. This, however, is not possible since q_i's are either 2's or (2/3)'s. □

To prove that T allows an aperiodic tiling is more difficult but still easily explainable. It is based on so-called Beatty sequences and balanced representations of numbers, introduced in 1920's, cf. [Bea]. Much studied Sturmian words, cf. [Lo], are special cases of these. For a given irrational $\alpha \in \mathbb{R}$ and integer $i \in \mathbb{Z}$ we define the *Beatty sequence*

$$\mathcal{A}_i(\alpha) = \lfloor i \cdot \alpha \rfloor, i \in \mathbb{Z},$$

where the brackets denote the integer part of the number. Then the *balanced representation* of α is defined as the sequence $\mathcal{B}(\alpha) = (\mathcal{B}(\alpha))_{i \geq 0}$:

$$\mathcal{B}(\alpha)_i = \mathcal{A}(\alpha)_i - \mathcal{A}(\alpha)_{i-1}.$$

This sequence have many remarkable properties. What is important here is how the machines $\mathcal{M}_2$ and $\mathcal{M}_{2/3}$ are related to balanced representations of numbers. This is the main part of the construction: For suitably, but rather freely, chosen α, if the automaton $\mathcal{M}_2$ or $\mathcal{M}_{2/3}$ is fed by an infinite balanced representation of α, i.e. $\mathcal{B}(\alpha)$, then they can output $\mathcal{B}(2\alpha)$ or $\mathcal{B}(^2/_3\alpha)$, respectively. And the procedure can be repeated, as well as reversed. This implies, see [Kari] for details,

Lemma 2. *The tile set T allows a proper tiling.*

From Lemmas 1 and 2 we obtain:

Theorem 3. *(J. Kari) There exists an aperiodic tile set of size 14.*

As we already set in [Cu] the above ideas were modified to obtain an aperiodic tile set of size 13.

5 Image Manipulation

A finite automaton is a compact description of a typically infinite language. Consequently, it is no surprise that finite automata can be used to compress images, properly encoded into languages. Especially useful they are in compressing fractal type of images. Another particularly interesting phenomenon is that many types of transformations of images can be done directly on their compressed automata-theoretic representations.

The idea of using finite automata and rational languages to compress real images seems to be first introduced in [BM]. Soon after that [DC] introduced finite automata as alternative to mutually recursive function systems (MRFS) defined in [Ba] for the compression of the fractal type images. The mathematical formalism of weighted automata for image compression was developed in [CKarh], while their practical usefulness was considered e.g. in [CKariI]. Articles [CF], [CKariII] are important sources dealing with image manipulation using weighted finite automata.

In what follows we present the basic definitions and try to illustrate some features of the applicability of finite automata in the image compression and manipulation.

An $n \times n$ real image I can be defined as a language $L_I \subseteq A^n$, or as a formal power series $s_I : A^n \to S$ as follows. Choose $A = \{0, 1, 2, 3\}$ and $n = 2^p$ for $p \geq 0$. Then words in L of length n define a pixel with colour "black" in I. All other pixel get colour "white". This corresponds black and white pictures. On the other hand, formal power series define a colour or graytone picture by associating an element of S (its darkness) to each of the pixels via words of length n.

More formally, the correspondence with words in L_I and pixels in I is defined as follows. We divide the unit square iteratively smaller and smaller units by horizontal and vertical lines:

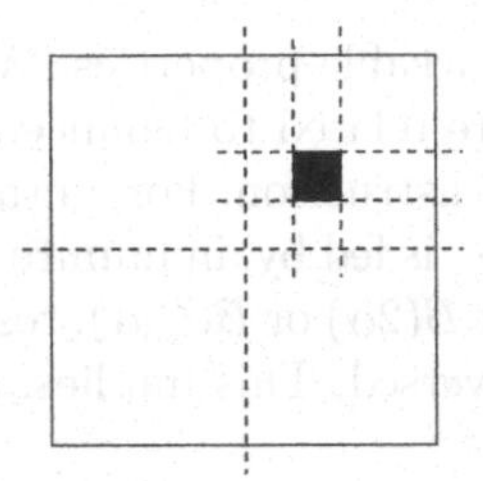

We denote the four quadrants by 0, 1, 2 and 3 when going from the left bottom quadrant clockwise. Hence the address of the black square in the above figure of the third level of the iteration is 202. Consequently, the language L_I or formal power series s_I gives an approximation of the image I on the resolution level p.

Example 5. Consider the image:

To find an automaton representing (or approximating) it we divide the image to four quadrants:

Denoting these by states of the automaton and analyzing how they are related in the above addressing of the pixels we obtain the automaton

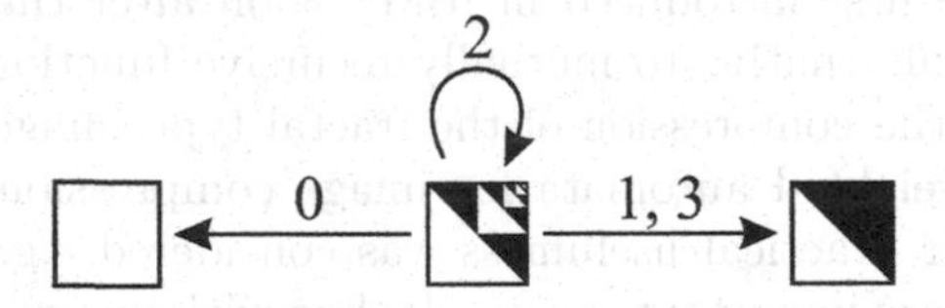

Repeating the procedure we obtain

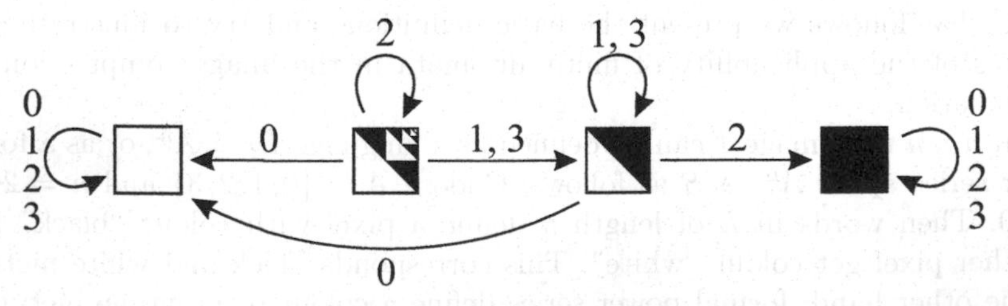

The state ◩ is, of course, the initial state, while the set of final states might be all except the state □, or only the black state. In these choices the words of length $n = 2^p$ accepted by the automata would give slightly different approximations of the image - however, when p goes to infinite they approach to the same original image.

In above approximations of the image we consider a pixel black if its address is an accepted word of the automaton. Intuitively, however, pixel ■ is darker than pixel ◪, for example. This can be taken into account by generalizing the above automaton to a finite automaton with weights.

Example 6. Simple computations show how much of darkness goes from any state to its daughter states. For example 2 takes half of the darkness of state ◪ into state ■, while inputs 1 and 3 take just one quarter into itself. Hence, we obtain a weighted automaton

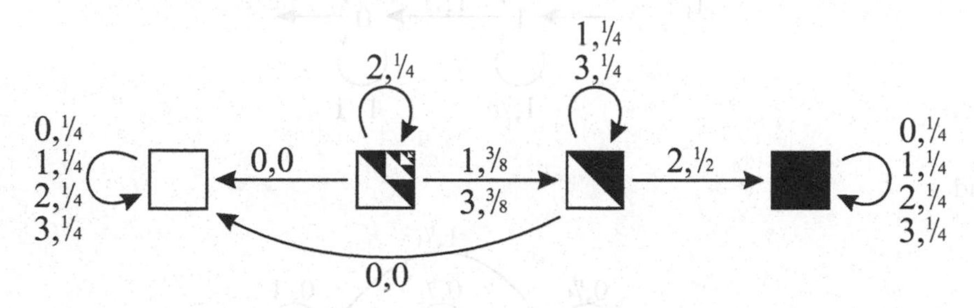

Of course, the above method of iterative subdivisions does not lead in general to a finite number of states, i.e. to a finite automaton.

Many problems on rational formal power series, i.e. on weighted automata, are very difficult or even undecidable. This was part of the reason why the article [CKarh] concentrated only to special cases of these, so-called level automata. The other reason was that even these are quite powerful to generate complex images.

We recall that a *level automaton* is weighted finite automaton where (i) the automaton is acyclic except that loops from a state into itself are allowed, (ii) all weights are nonnegative real numbers, and (iii) all weights of the loops are smaller than or equal to one, and moreover, only loops in the state of level zero (i.e. in the state from which there is no outgoing transitions) are allowed, and required, to be equal to 1.

Note that the automaton of Example 6 can be viewed as a level automaton since the illegal loop is labeled by the weight zero. Some reasons for the above restrictions are as follows. Condition (ii) guarantees that the series defined are $\mathbb{N}$-rational, which together with (i) makes many questions much easier to consider. Condition (iii) is essential since the automata are used to specify also infinite resolution images, for that we need the convergence on infinite inputs. Concerning the formal definitions and further details we refer to [CKarh].

Actually, instead of using weighted finite automata to generate real 2-dimensional pictures, they were mainly used in [CKarh] to generate 1-dimensional pictures, i.e. to compute functions $[0,1]^* \to \mathbb{R}_+$. However, already here many challenging features can be revealed, and connections to 2-dimensional pictures are close.

In this approach the alphabet is, of course, binary $A = \{0, 1\}$, 0 representing the left half and 1 the right one. Then numbers in the interval $[0, 1]$ are identified with the binary decimal representations. The value computed by a weighted automaton $\mathcal{A}$ on an infinite word w (or real number it represents) is the multiplicity of w in $\mathcal{A}$. The ambiguity that the words $w10^\omega$ and $w01^\omega$ represent the same number is a source of many problems, but also a reason for interesting results. For more see again in [CKarh].

Example 7. Consider the level automata

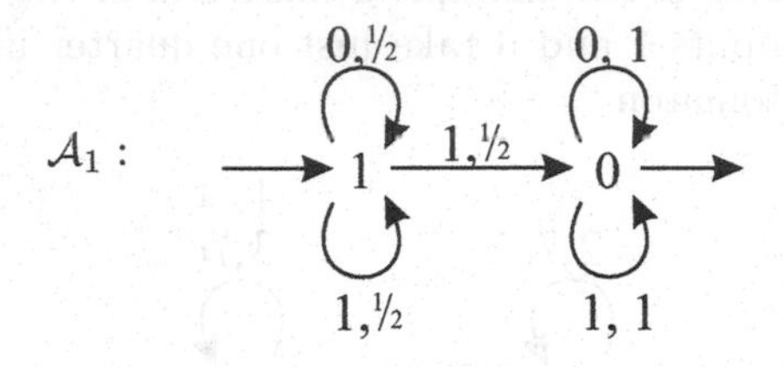

and

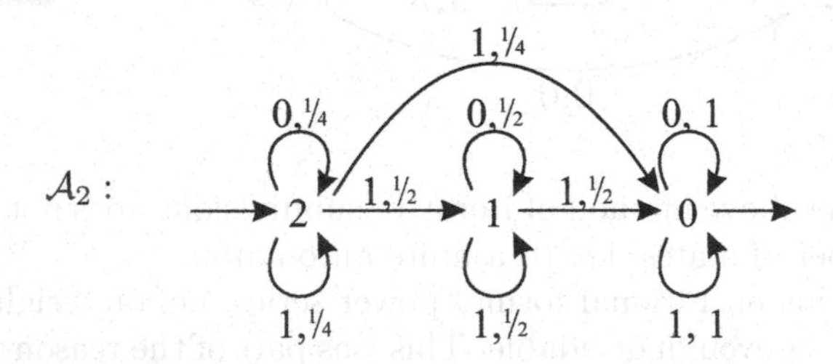

The first automaton computes, as is easy to see, the linear function $f(x) = x$. For example, the weight of $w = 0101^\omega$, as well as that of $w = 0110^\omega$ is

$$\frac{1}{4} + \frac{1}{16}\sum_{i=0}^{\infty}\frac{1}{2^i} = \frac{1}{4} + \frac{1}{8}.$$

Accordingly, the automaton $\mathcal{A}_2$ computes the parabola $f(x) = x^2$.

Automata $\mathcal{A}_1$ and $\mathcal{A}_2$ deserve a few comments. First, $\mathcal{A}_1$ is the unique two state level automaton computing the function x. Similarly, the automaton $\mathcal{A}_2$ is (up to certain simple transformations) the unique three state level automaton computing the parabola. It is also interesting to note that if we decompose the automaton $\mathcal{A}_2$ into two automata according to two paths from state 2 to state 0, then both of these components computes a noncontinuous function. However, their sum have very nice behaviour!

Example 7 can be extended to , cf. [CKarh] and [DKLTII], where the *smoothness* means that the function has all derivatives:

Theorem 4. *Each polynomial of degree k on interval $[0,1]$ can be computed by a level automaton with $k+1$ states (if negative values are allowed in the initial vector). Moreover, polynomials are the only smooth functions computable by level automata.*

Typically, weighted finite automata compute noncontinuous functions. This is demonstrated also in the next example. The other points of the example are to show that weighted automata, even very small weighted automata, can compute very complicated continuous functions, as well as to point out that small changes in the automata change the function drastically.

It can be shown that the upper part of the automaton, i.e. the automaton where the state $1'$ is omitted, computes, for each t, a continuous function only on one value of $x(t)$. The same, of course, holds for the lower part of the automaton. Consequently, on these values the whole automaton computes a continuous function, for details cf. [DKLTI].

The functions computed on values $t = 1/4$, $t = 2/3$ and $t = 3/4$ are shown in Figures 5, 6 and 7. Interestingly, the function in Figure 7 is an example of a continuous function having no derivative at any point!

Example 8. Consider the automaton

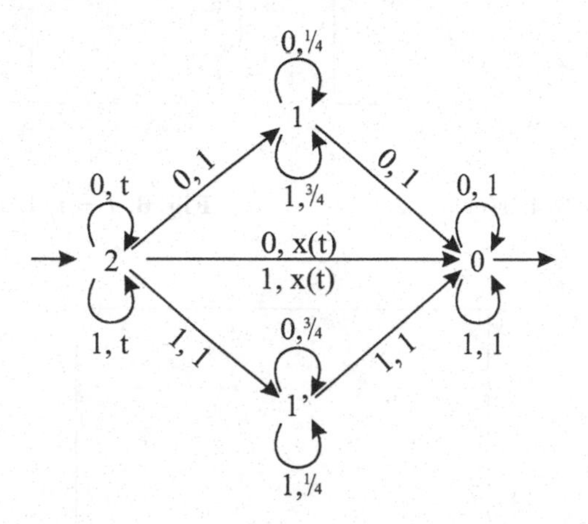

We conclude by discussing about particular advantages of the above automata-theoretic image generation and compression. What is characteristic for finite automata is their strong closure properties. Here this means that many transformations of images can be done directly based on their compressed automata representations. Examples of simple operations are *scaling*, *rotating* and *zooming* of the images. A less intuitive transformation is that we can *integrate* the image canonically from its automata representation, cf. [CKarh]. For more details we refer to e.g. [CF] and [CKariII].

We consider two operations more concretely. A *cartesian product* of two automata is one of the standard operations of the automata theory. In classical automata theory this corresponds to intersection. Here, in the weighted automata case, it corresponds the pointwise product. That is if automata $\mathcal{A}$ and $\mathcal{B}$ compute the functions $f_{\mathcal{A}}(w)$ and $f_{\mathcal{B}}(w)$ then their product $\mathcal{A} \times \mathcal{B}$ computes the function $f_{\mathcal{A}}(w) \cdot f_{\mathcal{B}}(w)$, that is $f_{\mathcal{A}\times\mathcal{B}}(w) = f_{\mathcal{A}}(w) \cdot f_{\mathcal{B}}(w)$. We illustrate that in Example 10.

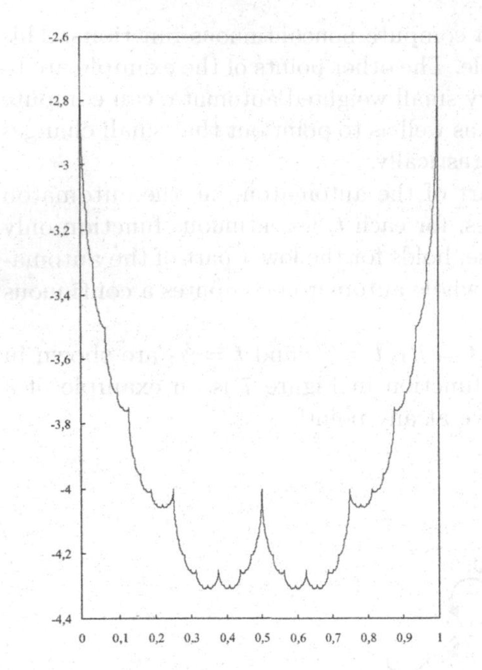

Fig. 5. $t = 3/4$; $x(t) = -2$

Fig. 6. $t = 1/4$; $x(t) = 2/3$

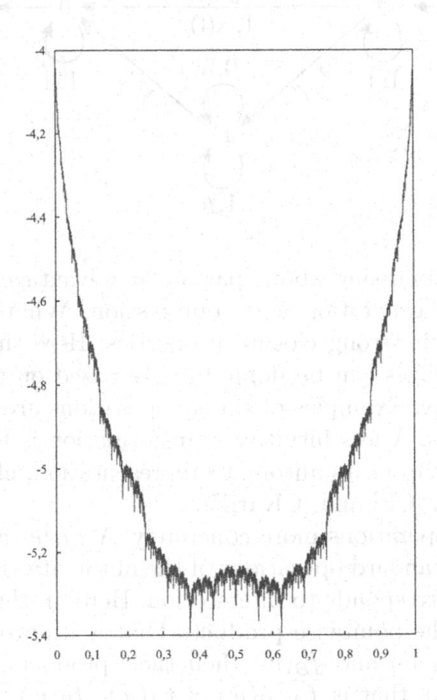

Fig. 7. $t = 2/3$; $x(t) = -8/3$

Another natural, but much less used product of two automata, is their *complete cartesian product.* This turns out to be very useful here. In this product not only the product of states is considered, but also the product of inputs is used. More specifically, if automata $\mathcal{A}$ and $\mathcal{B}$ have transitions

$$p \xrightarrow{a,\alpha} q \quad \text{and} \quad r \xrightarrow{b,\beta} s,$$

then in their complete cartesian product $\mathcal{A} \times_c \mathcal{B}$ we have transition

$$(p,q) \xrightarrow{(a,b),\alpha\beta} (r,s).$$

The size of the alphabet of $\mathcal{A} \times_c \mathcal{B}$ is the product of the sizes of the alphabets of $\mathcal{A}$ and $\mathcal{B}$, respectively. This means that from automata computing functions of the interval $[0,1]$ we obtain automata for real pictures on the unit square. This is illustrated in the following example.

Example 9. Consider the automata $\mathcal{A}_1$ and $\mathcal{A}_2$ of Example 8 computing functions. The complete cartesian product of these automata is:

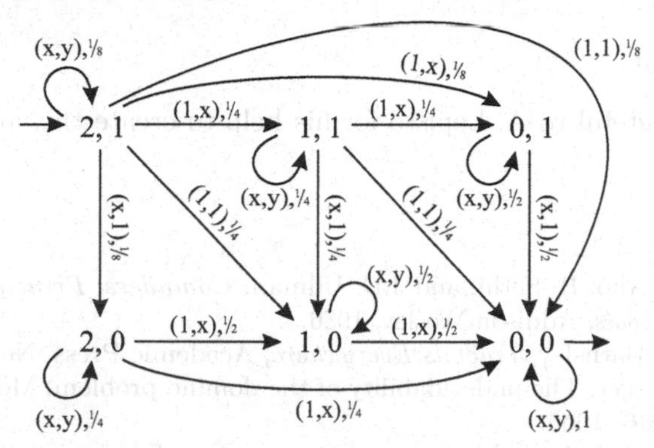

Here x and y ranges over $\{0,1\}$. Now, identifying inputs as

$$(0,0) \leftrightarrow 0, (0,1) \leftrightarrow 1, (1,0) \leftrightarrow 3 \text{ and } (1,1) \leftrightarrow 2$$

we can view the automaton $\mathcal{A}_1 \times_c \mathcal{A}_2$ as a level automaton computing a real picture. In fact, it computes the picture representing the function

$$f(x,y) = x^2 y.$$

Now, the above two products allow a lot of possibilities to modify pictures.

Example 10. Let $\mathcal{A}$ be the automaton of Example 6, and $\mathcal{A}_1$ and $\mathcal{A}_2$ as in Example 8. Then the automaton

$$\mathcal{A} \times (\mathcal{A}_1 \times_c \mathcal{A}_2)$$

transforms the picture of Example 5 such that in the horizontal direction it will be quadratically darkening and vertically linearly darkening. The automaton contains 24 states. The image is shown in Fig. 8.

Fig. 8. The picture for $\mathcal{A} \times (\mathcal{A}_1 \times_c \mathcal{A}_2)$

Acknowlegment

The author is grateful to A. Lepistö for his help to create this article.

References

[ASU] A.V. Aho, R. Sethi, and J.D. Ullman, *Compilers: Principles, Techniques, and Tools*, Addison-Wesley, 1986.

[Ba] M.F. Barnsley, *Fractals Everywhere*, Academic Press, New York, 1988.

[Berg] R. Berger, The undecidability of the domino problem, Mem. Amer. Math. Soc. **66**, 1966.

[Bers] J. Berstel, Transductions and Context-Free Languages, Teubner 1979.

[BM] J. Berstel and M. Morcrette, Compact representation of patterns by finite automata, Proceedings Pixim'89 Paris, 387–402, 1998.

[Bea] S. Beatty, Problem 3173, Amer. Math. Monthly **33**, 159, 1926, solution in **34**, 159, 1927.

[Bo] A. Bouajjani, Languages, rewriting systems, and verification of infinite-state systems, in: F. Orejas et al. (eds.), *Proc. ICALP 2001*,Springer LNCS **2076**, 24–39, Springer-Verlag, 2001.

[BCMS] O. Burkart, D. Caucal, F. Moller, and B. Steffen, Verification over Infinite States, in: J. Bergstra, A. Ponse, and S.A. Smolka (eds.), *Handbook of Process Algebra*, 545–623, Elsevier Publishers, 2001.

[Br] J. Brzozowski, Open problems about regular languages, in: R.V. Book (ed.), *Formal language theory - Perspectives and open problems*, Academic Press, New York, 1980

[CaHK] J. Cassaigne, T. Harju, and J. Karhumäki, On the undecidability of freeness of matrix semigroups, Inter. J. Alg. and Comput. **9**, 295–305, 1999.

[ChHK] C. Choffrut, T. Harju, and J. Karhumäki, A note on decidability questions on presentations of word semigroups, Theoret. Comput. Sci. **183**, 83–92, 1997.

[CK] C. Choffrut and J. Karhumäki, Combinatorics of Words, in: G. Rozenberg and A. Salomaa (eds.), *Handbook of Formal Languages*, Vol. 1, Springer, 329–438, 1997.

[CMRS] M. Crochemore, F. Mignosi, A. Restivo, and S. Salemi, Text compression using anti-dictionaries, Springer LNCS **1644**, 261–270, 1999.

[Cu] K. Culik II, An aperiodic set of 13 tiles, Discr. Math. **160**, 245–251, 1996.

[CKarh] K. Culik II and J. Karhumäki, Finite automata computing real functions, SIAM J. Comput. **23**, 789–814, 1994.

[CKariI] K. Culik II and J. Kari, Image compression using weighted finite automata, Computer and Graphics **13**, 305–313, 1993.

[CKariII] K. Culik II and J. Kari, Digital Images and Formal Languages, in: A.Salomaa and G.Rozenberg (eds.), *Handbook of Formal Languages*, Vol. 3, 599–616, Springer-Verlag, 1997.

[CF] K. Culik II and I. Fris, Weighted finite transducers in image processing, Discr. Appl. Math. **58**, 223–237, 1995.

[DKLTI] D. Derencourt, J. Karhumäki, M. Latteux, and A. Terlutte, On continuous functions computed by finite automata, Theor. Inform. and Appl. **28**, 387–403, 1994.

[DKLTII] D. Derencourt, J. Karhumäki, M. Latteux, and A. Terlutte, On the computational power of weighted finite automata, Fundamenta Informaticae **25**, 285–293, 1996.

[DC] S. Dube and K. Culik II, Affine automata and related techniques for generating complex images, Proceedings of MFCS'90, Springer LNCS **452**, 224–231, 1990.

[E] S. Eilenberg, *Automata, Languages and Machines*, Vol. A, Academic Press, New York, 1974.

[ECHLPT] D. Epstein, J. Cannon, D. Holt, S. Levy, M. Paterson, and W. Thurston, *Word processing in groups*, Jones and Bartlett, Boston, 1992.

[HK] T. Harju and J. Karhumäki, The equivalence problem for multitape automata, Theoret. Comput. Sci. **78**, 347–355, 1991.

[HU] J.E. Hopcroft and J.D. Ullman, *Introduction to Automata Theory, Languages and Computation*, Addison-Wesley, 1979.

[Kari] J. Kari, A small aperiodic set of Wang tiles, Discr. Math. **160**, 259–264, 1996.

[KS] O. Kharlampovich and M. Sapir, Algorithmic problems in varieties, Int. J. Alg. and Comput. **5**, 379–602, 1995.

[KBS] D. Klarner, J.C. Birget, W. Satterfield, On the undecidability of the freeness of integer matrix semigroups, Int. J. Alg. and Comput. **1**, 223-226, 1991.

[Kl] S. Kleene, Representation of events in nerve nets and finite automata, in: C.E. Shannon and J. McCarthy (eds.), *Automata Studies*, Princeton University Press, 3–42, 1956.

[Lo] M. Lothaire, *Algebraic Combinatorics on Words*, Encyclopedia of Mathematics and its Applications **90**, Cambridge University Press, 2002.

[Ma] A.A. Markov, On the Impossibility of Certain Algorithm in the Theory of Associative Systems I+II(Russian), Doklady Akademii Nauk S.S.S.R.,n.s., vol. **55**, 587–590, vol. **58**, 353–356, 1947. English translation for I: Comptes rendus de l'acadêmie des sciences de l'U.R.S.S., n.s., vol. **55**, 583–586, 1947.

[Mi] S. Micali, Two-way deterministic finite automata are exponentially more succinct than sweeping automata. Inform. Proc. Letters **12**(2), 103–105, 1981.

[Mo] M. Mohri, Finite-State Transducers in Language and Speech Processing, Computational Linguistics **23**(2), 269–311, 1997.

[P] E. Post, Recursive unsolvability of a problem of Thue, J. Symbolic Logic **12**, 1–11, 1947.

[RS] M. Rabin and D. Scott, Finite Automata and their decision problems, IBM J. Res. **3**, 115–125, 1959.

[SS] A. Salomaa and M. Soittola, *Automata-Theoretic Aspects of Formal Power Series*, Springer, 1978.

[Se] G. Senizergues, L(A)=L(B)? decidability results from formal systems, Theoret. Comput. Sci. **251**, 1–166, 2001.

[T] W. Thomas, Automata on infinite objects, in: J. van Leeuwen (eds.), *Handbook of Theoretical Computer Science*, Vol. B, 135–191, Elsevier, The MIT Press, 1990.

[V] M. Vardi, Nontraditional Applications of Automata Theory, TACS'94, Springer LNCS **789**, 575–597, Springer-Verlag, 1994.

[Y] S. Yu, Regular languages, in: G. Rozenberg and A. Salomaa (eds.), *Handbook of Formal Languages*, 41–110, Springer, 1997.

Approximability of the Minimum Bisection Problem: An Algorithmic Challenge

Marek Karpinski*

University of Bonn,
marek@cs.uni-bonn.de

Abstract. We survey some recent results on the complexity of computing approximate solutions for instances of the *Minimum Bisection* problem and formulate some very intriguing and still open questions about the approximability status of that problem.

1 Introduction

The problem of approximating the minimum bisection of a graph, i.e., the problem of partitioning a given graph into two equal halfs so as to minimize the number of edges with exactly one end in each half, belongs to the most intriguing problems currently in the area of combinatorial optimization and the approximation algorithms. The reason being that we are not able to cope at the moment with the global conditions imposed on the vertices of a graph like the condition that the two parts of a partition are of equal size. The MIN-BISECTION problem arises profoundly in several contexts, either explicitly or implicitly, which range from problems of statistical physics and combinatorial optimization to computational geometry and various clustering problems, (cf., e.g., [MPV87], [JS93], [H97]). We refer also to [PY91] and [AL97] for the background on approximation algorithms and approximation hardness of optimization problems.

2 Instances of MIN-BISECTION Problem

We are going to define the instances of the MIN-BISECTION problem studied in our paper.

- *MIN-BISECTION:* Given an undirected graph, partition the vertices into two equal halves so as to minimize the number of edges with exactly one endpoint in each half.
- *Paired MIN-BISECTION:* Given an undirected graph, and a set of pairs of its vertices, partition the vertices into two equal halves so as to split each given pair of vertices and to minimize the number of edges with exactly one endpoint in each half.

* Research supported in part by DFG grants, DIMACS, PROCOPE project, IST grant 14036 (RAND-APX), and Max-Planck Research Prize. Research partially done while visiting Department of Computer Science, Yale University.

K. Diks et al. (Eds): MFSC 2002, LNCS 2420, pp. 59–67, 2002.

- *Weighted MIN-BISECTION:* Given a *weighted* undirected graph, partition the vertices into two equal halves so as to minimize the sum of weights of the edges with exactly one endpoint in each half.

We refer to a graph $G = (V, E)$ as a *dense* graph if its *minimal* degree is $\Theta(n)$. We call a graph G *planar*, if G can be embedded into a plane graph. A *weighted* complete graph G is called *metric*, if G can be embedded into a finite metric space.

- *Dense MIN-BISECTION* is the MIN-BISECTION problem restricted to the dense graphs.
- *Dense Paired MIN-BISECTION* is the Paired MIN-BISECTION problem restricted to the dense graphs.
- *Planar MIN-BISECTION* is the MIN-BISECTION problem restricted to the planar graphs.
- *Metric MIN-BISECTION* is the Weighted MIN BISECTION problem restricted to the metric graphs.

We define, in a similar way, the dual MAX-BISECTION problems for the general, dense, planar and metric graphs, respectively.

It is not difficult to see that the *dense* and *metric* instances of MIN-BISECTION and Paired MIN-BISECTION are both NP-hard in exact setting (cf. [AKK95], [BF99], [FK98b]). It was proven recently that the Planar MAX-BISECTION ([J00], see also [JKLS01]) is *NP-hard* in *exact* setting, however the status of the Planar MIN-BISECTION remains still an intriguing open problem.

We refer to [K01a] and [K01b] for a survey on approximability of dense and sparse instances of some other NP-hard combinatorial optimization problems.

3 Dense Instances of MIN-BISECTION, Paired MIN-BISECTION, and MIN-2SAT

We consider here also the following minimization problems.

- *MIN-2SAT:* Given a 2CNF formula, construct an assignment as to minimize the number of clauses satisfied.

We refer to the 2CNF formula to be *dense* if the number of occurrences of each variable is $\Theta(n)$.

- *Dense MIN-2SAT* is the MIN-2SAT problem restricted to the dense formulas.

It is known that the large fragments of Minimum Constraint Satisfaction (MIN-CSP) problems do not have polynomial time approximation schemes even if restricted to the dense instances (see [CT96], [KZ97], [BFK01]). It has turned out however, a bit surprisingly, that the dense instances of MIN-BISECTION do have a PTAS [AKK95].

Theorem 1. *([AKK95])* There exists a PTAS for Dense MIN-BISECTION.

The method used in [AKK95] depended on a new technique of approximating Smooth Polynomial Integer Programs for large values of objective functions, and a biased *radical* placement method for the small values. The variant of that technique was used in Bazgan and Fernandez de la Vega [BF99] to prove that dense instances of Paired MIN-BISECTION possess a PTAS.

Theorem 2. *([BF99])* There exists a PTAS for Dense Paired MIN-BISECTION.

The above result was used to derive the existence of a PTAS for dense instances of MIN-2SAT. It has turned however out that the proof in [BF99] contained an error. The corrected proof was established in Bazgan, Fernandez de la Vega and Karpinski [BFK01].

Theorem 3. *([BFK01])* There exists a PTAS for Dense MIN-2SAT.

We notice that both Paired MIN-BISECTION and MIN-2SAT are both provably MAX-SNP-hard (cf. [BF99], [KKM94]), and thus not having PTASs under usual complexity theoretic assumptions. Intriguingly, all attempts to establish a connection between the approximation hardness of Paired MIN-BISECTION and MIN-2SAT and the approximation hardness of MIN-BISECTION have failed utterly up to now. The approximation hardness status of MIN-BISECTION remains an outstanding open problem. At the moment we are not even able to exclude a possibility of existence of a PTAS for that problem.

Open Problem 1. *Is MIN-BISECTION NP-hard to approximate to within a constant factor?*

On the positive side, there was recent substantial improvement on approximation ratio for MIN-BISECTION, cf. Feige, Krautghamer and Nissim [FKN00], and Feige and Krautghamer [FK00].

Theorem 4. *([FK00])* MIN-BISECTION can be approximated in polynomial time to within $O(log^2 n)$ factor.

[FK00] gives also an improved approximation factor for planar instances of MIN-BISECTION.

Theorem 5. *([FK00])* Planar MIN-BISECTION can be approximated in polynomial time to within $O(logn)$ factor.

4 Planar Instances of MIN-BISECTION

There has been a very recent progress on the approximability status of planar MAX-BISECTION resulting in design of the first PTAS for that problem, and also in the first proof of its NP-hardness in exact setting [J00], [JKLS01].

The status of planar MAX-BISECTION was an open problem for a long time. An intriguing context for that problem is the fact that planar MAX-CUT can be computed exactly in polynomial time [H75]. An additional paradigm connected to it was based on analysis of cut polytops, and the fact that the value of the planar MAX-CUT semidefinite relaxation with *triangle constraints* is just equal to the value of the optimal cut (cf. [BM86] for the background). The corresponding problem for *bisectional* polytops however remains still open.

The *exact* computation status for planar MAX-BISECTION was resolved recently by Jerrum [J00] (cf. also [JKLS01]) in proving its NP-hardness. The technique of his proof is similar to the method used by Barahona [B82] for the planar *spin glass* problem within a *magnetic field*, and is based on the NP-hardness of the maximum independent set on 3-regular planar graphs.

Theorem 6. *([J00])* Planar MAX-BISECTION is NP-hard in exact setting.

Soon after Jansen, Karpinski, Lingas and Seidel [JKLS01] were able to design the first PTAS for planar MAX-BISECTION, and for some special cases of planar MIN-BISECTION. The method of solution depended on a new method of finding maximum partitions of bounded treewidth graphs, combined with the tree-type dynamic programming method of dividing planar graph into k-outerplanar graphs [B83].

Theorem 7. *([JKLS01])* There exists a PTAS for Planar MAX-BISECTION.

We notice that Theorem 6 and 7 do not entail readily any corresponding result for planar MIN-BISECTION. The reason being that the operation of *complementing* an instance of planar MAX-BISECTION does not result in a planar instance of MIN-BISECTION (alike some other situations).

However, the results of [JKLS01] entail also the following.

Theorem 8. *([JKLS01])* There exists a PTAS for instances of the Planar MIN-BISECTION with a size of minimum bisection Ω(n log logn/logn).

The proof of Theorem 8 depends on the fact that the PTAS of Theorem 7 works for the partitionings of the treewidth up to $O(logn)$. We observe, by the planar separator theorem [LT79] for bounded degree planar graphs, that the size of minimum bisection is $O(\sqrt{n})$. This fact yields also

Theorem 9. *([JKLS01])* Given an instance G of Planar MIN-BISECTION of size n and maximum degree d, a minimum bisection of G of size $O(d\sqrt{n})$ can be computed in time $O(nlogn)$.

The problem on whether Planar MIN-BISECTION admits PTAS, or perhaps even polynomial time exact algorithms, remains open.

Open Problem 2. *Is Planar MIN-BISECTION NP-hard in exact setting?*

Open Problem 3. *Does Planar MIN-BISECTION have a PTAS?*

We will study in the next section the case of metric MIN-BISECTION, and connected problems of metric MIN-CLUSTERING.

5 Metric Instances of MIN-BISECTION

Metric (and more restrictively, geometric) instances of combinatorial optimization problems occur in a number of realistic scenarios and are strongly motivated by various applications (cf.[H97]). The instances of such problems are given by embeddings in finite metric spaces.

We consider first two dual metric instances of MAX-CUT, and MIN-2CLUSTERING (also known as MIN-UNCUT, cf. [KST97]).

- Metric MIN-2CLUSTERING: Given a finite metric space (X, d), partition X into two sets C_1 and C_2 so as to minimize the sum $\sum_{i=1}^{2} \sum_{x,y \in C_i} d(x, y)$ (called the sum of *intra-cluster distances*).

Fernandez de la Vega and Kenyon [FK98b] were the first to design a PTAS for the Metric MAX-CUT. Their method followed the earlier work of Fernandez de la Vega and Karpinski [FK98a] on existence of a PTAS on dense weighted instances of MAX-CUT.

Theorem 10. *([FK98b])* There exists a PTAS for Metric MAX-CUT.

Metric MIN-2CLUSTERING was left open in [FK98b], and solved later by Indyk in [I99]. The main difficulty in Indyk's solution was coping with the situations were the value of max-cut was much higher than the value of the 2-clustering.

Theorem 11. *([I99])* There exists a PTAS for Metric MIN-2CLUSTERING.

In a very recent work, Fernandez de la Vega, Karpinski and Kenyon [FKK02] resolved finally the status of the Metric MIN-BISECTION problem by proving an existence of a PTAS for that problem. The method of solution depends on a new kind of biased sampling and a new type of rounding. This method could be also of independent interest.

Theorem 12. *([FKK02])* There exists a PTAS for Metric MIN-BISECTION.

The method of [FKK02] gave rise also to the consecutive solution for Metric MIN-kCLUSTERING problem (number of clusters k being now an arbitrary constant).

Fernandez de la Vega, Karpinski, Kenyon and Rabani [FKKR02] proved the following general result.

Theorem 13. *([FKKR02]) There exists a PTAS for Metric MIN-kCLUSTERING for each fixed k.*

6 Sparse Instances of MIN-BISECTION

We turn our attention to the dual class, i.e. to the class of *sparse* instances. For the representative of this class we choose the class of 3-regular graphs.

We introduce first some notation. We will call an approximation algorithm A for an optimization problem P, an $(r(n), t(n))$*-approximation algorithm*, if A

approximates P within a factor $r(n)$, and running time of A is $O(t(n))$ for n the size of an instance.

The following result has being proven recently by Berman and Karpinski [BK01].

Theorem 14. *([BK01])* Suppose there exists an $(r(n), t(n))$-approximation algorithm for 3-regular instances of MIN-BISECTION. Then there exists an $(r(n^3), t(n^3))$-approximation algorithm for MIN-BISECTION on general instances.

The construction of [BK01] can be modified as to yield a similar result on 3-regular planar graphs. In such a modification we use a slightly larger piece of hexagonal mesh, and replace each edge between a pair of nodes with a pair of edges between meshes that replaced those nodes (cf. [BK01]).

Theorem 15. Suppose there exists an $(r(n), t(n))$-approximation algorithm for 3-regular instances of Planar MIN-BISECTION. Then there exists an $(r(n^3), t(n^3))$-approximation algorithm for Planar MIN-BISECTION.

Theorem 14, and 15 give a relative hardness of 3-regular instances of MIN-BISECTION. The approximation lower bound status of MIN-BISECTION, and, in fact, the exact computation lower bound for Planar MIN-BISECTION, remain important and intriguing open problems.

It is also interesting to notice that the recent improvements in approximation ratios for 3-regular instances of MAX-BISECTION (cf. e.g. [FKL00], [KKL00]) were not paralleled by the analogous improvements on the 3-regular instances of MIN-BISECTION. Theorem 14, and 15 give good reasons for this development.

7 Summary of Known Approximation Results for MIN-BISECTION

We present here (Table 1) the best up to now approximation upper and lower bounds for the instances of MIN-BISECTION problem.

8 Further Research

The most challenging and intriguing open problem remains the status of the MIN-BISECTION on general graphs. The known so far PCP-techniques do not seem to yield any approximation lower bounds for that problem. The same holds for the known techniques to approximate MIN-BISECTION. They do not seem to allow us to break the approximation factor at any level below $O(logn)$, and this even for 3-regular planar graphs. It seems that the improved upper and lower approximation bounds for MIN-BISECTION will require essentially new techniques. This holds not only for the general MIN-BISECTION, but also for the very restricted Planar MIN-BISECTION instances, for which we are not able to prove at the moment even NP-hardness for the exact computation, nor are we able to give any better than $O(logn)$ approximation factors.

Table 1. Approximation Upper and Lower Bounds for MIN-BISECTION

Instances	Approx. Upper	Approx. Lower
General	$O(\log^2 n)$	Not known
Dense	PTAS	–
Sparse	$O(\log^2 n)$	Equal to MIN-BISECTION
Planar	$O(\log n)$	Not known to be NP-Hard even in exact setting
Sparse Planar	$O(\log n)$	Equal to Planar MIN-BISECTION
Metric	PTAS	–

Acknowledgments

My thanks go to Mark Jerrum, Piotr Berman, Uri Feige, W. Fernandez de la Vega, Ravi Kannan, and Claire Kenyon for many stimulating discussions.

References

[AKK95] S. Arora, D. Karger, and M. Karpinski, *Polynomial Time Approximation Schemes for Dense Instances of NP-Hard Problems*, Proc. 27th ACM STOC (1995), pp.284-293; the full version appeared in J. Comput. System Sciences 58 (1999), pp. 193-210.

[AL97] S. Arora and C. Lund, *Hardness of Approximations, in Approximation Algorithms for NP-Hard Problems* (D. Hochbaum, ed.), PWS Publ. Co. (1997), pp. 399-446.

[B83] B. S. Baker, *Approximation algorithms for NP-complete problems on planar graphs*, Proceedings of the 24th IEEE Foundation of Computer Science, 1983, pp. 265-273.

[B82] F. Barahona, *On the Computational Complexity of Ising Spin Glass Models*, J. Phys. A. Math. Gen. 15 (1982) pp. 3241-3253.

[BM86] F. Barahona and A. R. Mahjoub, *On the Cut Polytope*, Mathematical Programming 36 (1986), pp. 157-173.

[BF99] C. Bazgan and W. Fernandez de la Vega, *A Polynomial Time Approximation Scheme for Dense MIN 2SAT*, Proc. Fundamentals of Computation Theory, LNCS 1684, Springer, 1999, pp. 91-99

[BFK01] C. Bazgan, W. Fernandez de la Vega and M. Karpinski, *Polynomial Time Approximation Schemes for Dense Instances of Minimum Constraint Satisfaction*, ECCC Technical Report, TR01-034, submitted to Random Structures and Algorithms.

[BK99] P. Berman and M. Karpinski, *On Some Tighter Inapproximability Results*, Proc. 26th ICALP (1999), LNCS 1644, Springer, 1999, pp. 200-209.

[BK01] P. Berman and M. Karpinski, *Approximation Hardness of Bounded Degree MIN-CSP and MIN-BISECTION*, ECCC Technical Report, TR01-026, 2001, to appear in Proc. 29th ICALP (2002).

[CT96] A. E. F. Clementi and L.Trevisan, *Improved Non-approximability Results for Vertex Cover with Density Constraints*, Proc. of 2nd Conference on Computing and Combinatorics, COCOON'96, Springer, 1996, pp. 333-342.

[FKL00] U. Feige, M. Karpinski and M. Langberg, *A Note on Approximation MAX-BISECTION on Regular Graphs*, ECCC Technical Report TR00-043, 2000, also in Information Processing Letters 79 (2001), pp, 181-188.

[FK00] U. Feige and R. Krauthgamer, *A Polylogarithmic Approximation of the Minimum Bisection*, Proc. 41st IEEE FOCS (2000), pp. 105–115.

[FKN00] U. Feige, R. Krauthgamer and K. Nissim, *Aproximating the Minimum Bisection Size*, Proc. 32nd ACM STOC (2000), pp. 530-536.

[FK98a] W. Fernandez de la Vega and M. Karpinski, *Polynomial Time Approximation of Dense Weighted Instances of MAX-CUT*, ECCC Technical Report TR98-064 (1998), final version Random Structures and Algorithms 8 (2000), pp. 314-332.

[FKK02] W. Fernandez de la Vega, M. Karpinski and C. Kenyon, *Polynomial Time Approximation Scheme for Metric MIN-BISECTION*, Manuscript, 2002.

[FKKR02] W. Fernandez de la Vega, M. Karpinski, C. Kenyon and Y. Rabani, *Ploynomial Time Approximation Schemes for Metric Min-Sum Clustering*, ECCC Tech. Report TR02-025, 2002.

[FK98b] W. Fernandez de la Vega and C. Kenyon, *A Randomized Approximation Scheme for Metric MAX-CUT*, Proc. 39th IEEE FOCS (1998), pp. 468-471, final version Journal of Computer and System Sciences 63 (2001), pp. 531-534.

[H75] F. Hadlock, *Finding a Maximum Cut of a Planar Graph in Polynomial Time*, SIAM Journal on Computing 4 (1975), pp. 221-225.

[H97] D. S. Hochbaum (ed.), *Approximation Algorithms for NP-Hard Problems*, PWS, 1997.

[I99] P. Indyk, *A Sublinear Time Approximtion Scheme for Clustering in Metric Spaces*, Proc. 40th IEEE FOCS (1999), 154-159.

[JKLS01] K. Jansen, M. Karpinski, A. Lingas, and E. Seidel, *Polynomial Time Approximation Schemes for MAX-BISECTION on Planar and Geometric Graphs*, Proc. 18th STACS (2001), LNCS 2010, Springer, 2001, pp. 365-375.

[J00] M. Jerrum, Personal communication, 2000.

[JS93] M. Jerrum and G. B. Sorkin, *Simulated Annealing for Graph Bisection*, Proc. 34th IEEE FOCS (1993), pp. 94-103

[K01a] M. Karpinski, *Polynomial Time Approximation Schemes for Some Dense Instances of NP-Hard Optimization Problems*, Algorithmica 30 (2001), pp. 386-397.

[K01b] M. Karpinski, *Approximating bounded degree instances of NP-hard problems*, Proc. 13th Symp. on Fundamentals of Computation Theory, LNCS 2138, Springer, 2001, pp. 24-34

[KKL00] M. Karpinski, M. Kowaluk, and A. Lingas, *Approximation Algorithms for MAX-BISECTION on Low Degree Regular Graphs and Planar Graphs*, ECCC Technical Report TR00-051 (2000).

[KZ97] M. Karpinski and A. Zelikovsky, *Approximationg Dense Cases of Covering Problems*, ECCC Technical Report TR 97-004,1997, appeared also in DIMACS Series in Discrete Mathematics and Theoretical Computer Science, Vol.40, 1998, pp. 169-178.

[KST97] S. Khanna, M. Sudan and L. Trevisan, *Constraint Satisfaction: the approximability of minimization problems*, Proc. of 12th IEEE Computational Complexity, 1997, pp. 282-296.

[KKM94] R. Kohli, R. Krishnamurti and P. Mirchandani, *The Minimum Satisfiability Problem*, SIAM Journal on Discrete Mathematics 7 (1994), pp. 275-283.

[LT79] R. J. Lipton and R. E. Tarjan. *A separator theorem for planar graphs*, SIAM Journal of Applied Mathematics, 36 (1979), pp.177-189.

[MPV87] M. Mezard, G. Parisi and M. A. Virasoro, *Spin Glass Theory and Beyond*, World Scientific, 1987.

[PY91] C. Papadimitriou and M. Yannakakis, *Approximation and Complexity Classes*, J. Comput. System Sciences 43 (1991), pp. 425-440.

Low Stretch Spanning Trees

David Peleg*

Department of Computer Science and Applied Mathematics,
The Weizmann Institute of Science,
Rehovot 76100, Israel,
peleg@wisdom.weizmann.ac.il

Abstract. The paper provides a brief review of problems and results concerning low stretch and low communication spanning trees for graphs.

1 Introduction

This paper concerns spanning trees that (approximately) preserve the distance properties of the graph they span. A common example is the shortest paths spanning tree, which provides optimal paths from its root to any other vertex. Its obvious drawback is that the paths it provides between other pairs of vertices may be rather long, compared to their distance in the original graph. It is thus natural to look for other types of spanning trees that may provide a better approximation for the original distances. The paper reviews some known results concerning a number of spanning tree types related to distance preservation.

Consider a weighted graph $G = (V, E, \omega)$, where the (nonnegative and symmetric) weight of an edge represents its length. The length of a path in G is the sum of the weights of its edges, and the distance between two vertices is the length of the shortest path connecting them. For a spanning tree T of G, the distance between two vertices in T is the length of the unique path connecting them in T.

A natural way to evaluate the quality of a spanning tree T is based on the *stretch* parameter. The stretch of a pair of vertices is the ratio between their distance in T and their distance in the spanned graph G. Either the maximum or the average stretch over all vertex pairs may serve as a measure of interest. A tree minimizing the maximum stretch is referred to as a *minimum max stretch tree (MMST)* [14], and a tree minimizing the average stretch is referred to as a *minimum average stretch tree (MAST)* [1].

Another commonly used distance-related measure for the goodness of a spanning tree T is the average distance in T between a pair of vertices. A tree minimizing this average is called a *minimum routing cost tree (MRCT)* [42] or *shortest total path length tree* [40]. In fact, the MRCT can be viewed as a "uniform" version of the *minimum communication tree (MCT)* [20], presented more formally later on, which is defined in a more general setting taking into account

* Supported in part by a grant from the Israel Science Foundation.

K. Diks et al. (Eds): MFSC 2002, LNCS 2420, pp. 68–80, 2002.

the possibility of attributing different significance levels (representing, say, different traffic requirements) to different vertex pairs. It turns out that MCTs and MASTs are in fact equivalent notions.

Distance preserving trees have found natural applications in diverse contexts. For instance, in a communication network or a distributed system based on a point-to-point message passing model, spanning trees often serve as a basis for certain control structures and information gathering and dissemination mechanisms. In these contexts, the weight of an edge may reflect the transmission delay or the communication cost over the corresponding link. Consequently, a number of network design problems involve constructing a spanning tree of a given network so as to minimize the delays or costs related to communication over the tree between the network vertices [20,22,4,19,15,29,38,39,40,41,42]. In the realm of distributed systems, low-stretch trees were used for improving the complexity of distributed directories [30]. Low average stretch trees and low stretch probability distributions of trees were also used in the design of competitive online algorithms for a variety of problems. The MAST problem was studied in [1] in the context of devising a randomized competitive online algorithm for the k-server problem. Randomized online algorithms were presented in [7] for metrical task systems, distributed paging, the (centralized and distributed) k-server problem, distributed resource management, file allocation and more. In computational biology, MRCTs have been used for the construction of good multiple sequence alignments [42].

In what follows we define the tree types and problems to be discussed, and review some of the results known for these problems.

2 Low Stretch and Low Communication Trees

2.1 Low Stretch Trees

Consider a weighted connected graph $G = (V, E, \omega)$, where $\omega : E \mapsto \mathbb{R}$ assigns a nonnegative weight $\omega(e)$ to each edge e, representing its length. For a subgraph H of G, let $dist_H(u, v)$ denote the distance between u and v in H. The *stretch* of a pair $u, v \in V$ in a spanning tree T, denoted $\textsc{Str}_T(u, v)$, is the ratio between their distance in the tree T and their distance in the original graph G, i.e.,

$$\textsc{Str}_T(u, v) = \frac{dist_T(u, v)}{dist_G(u, v)} .$$

We are interested in two measures. The first is the *maximum stretch* over the tree T,

$$\textsc{MaxStr}(T) = \max_{u,v}\{\textsc{Str}_T(u, v)\}.$$

Given a graph G, the spanning tree T minimizing $\textsc{MaxStr}(T)$ is referred to as the *minimum max stretch tree (MMST)* of G. The *MMST problem*, introduced in [14], is to find the MMST of a given graph G.

The second measure is the *average stretch* over the tree T,

$$\text{AvStr}(T) = \frac{1}{\binom{n}{2}} \sum_{u,v} \text{Str}_T(u,v).$$

The spanning tree T minimizing $\text{AvStr}(T)$ is referred to as the *minimum average stretch tree (MAST)* of G. The *MAST problem,* introduced in [1], is to find the MAST of a given graph G.

To illustrate the differences between maximum and average stretch, let us look at two examples. The n-vertex unweighted ring R_n demonstrates the widest gap between the two measures. A spanning tree T for R_n is obtained by discarding any single edge. The maximum stretch of T is $\text{MaxStr}(T) = n-1$, and by symmetry the maximum stretch of R_n is $n-1$. In contrast, the average stretch of T is $\text{AvStr}(T) = 2(n-1)/n < 2$, and hence the average stretch of R_n is less than 2. As a second example for a wide gap, the 2-dimensional n-vertex grid has average stretch $\Theta(\log n)$ but maximum stretch $\Theta(\sqrt{n})$ [1].

MAST on Multigraphs: A more general version of the MAST problem, allowing us to average the stretch over partial collections of pairs (e.g., the edges of E), concerns a multigraph G, with a *multiplicity* $m(u,v)$ associated with each pair of vertices. The selected tree should minimize

$$\text{AvStr}_m(T) = \frac{1}{\sum_{u,v} m(u,v)} \sum_{u,v} m(u,v) \cdot \text{Str}_T(u,v).$$

Spanners and Tree Spanners: An attractive and often viable alternative to spanning trees as a structure for approximating the distances in a general graph is using (possibly non-tree) *sparse spanners,* introduced in [33,31] and studied extensively since then. A *t-spanner* is a spanning subgraph that approximately maintains the distances between pairs of vertices up to factor t. Our previous definitions apply to t-spanners, namely, given a (possibly non-tree) subgraph $H = (V, E')$ of a weighted graph $G = (V, E, \omega)$ (where $E' \subseteq E$), we define the notions of stretch and maximum stretch in the same manner as for a spanning tree. We then say that the subgraph H is a *t-spanner* of G if $\text{MaxStr}(H) \leq t$.

Cast in this terminology, the MMST problem asks for finding a *tree* t-spanner minimizing t. Tree t-spanners were introduced in [14], and their theoretical properties, as well as their construction in various special graph classes, were studied in [12,13,14,24].

In general, the number of edges needed for a (non-tree) spanner depends on the desired stretch bound t and may be rather low (cf. [28]), making it a useful alternative when spanning trees turn out to have very high stretch. Nevertheless, some applications may exclude solutions based on using a non-tree spanner. Moreover, the use of a single tree as a spanning structure may in some cases be preferred due to practical advantages such as structural simplicity, the existence of a unique path between each pair of vertices, and having the fewest edges among all possible spanners.

2.2 Low Communication Trees

Let us now consider a wider setting in which, in addition to the weighted graph $G = (V, E, \omega)$, we are given also a (symmetric) requirement matrix specifying a nonnegative *communication requirement* $r(u, v)$ for every pair of vertices u and v in V. For a spanning tree T of G, the total *communication cost* of T is defined as

$$\mathrm{CC}(T) = \sum_{u,v} (r(u, v) \cdot dist_T(u, v)).$$

A trivial lower bound for this value is the *communication volume* in the original network, $\mathrm{CC}(G) = \sum_{u,v}(r(u, v) \cdot dist_G(u, v))$. The *minimum communication tree (MCT)* of G is a spanning tree T minimizing $\mathrm{CC}(G)$. The *MCT problem,* introduced in [20], is to find the MCT of a given graph G.

A well-studied special case of MCT is the *uniform MCT* problem, in which the communication requirements between any two sites are equal (i.e., $r(u, v) = 1$ for every u and v). Hence this variant is equivalent to the problem of finding the MRCT of the given graph.

MCT vs. MAST: In [27] is was shown that the MAST problem on multigraphs is equivalent to the MCT problem, in the sense that an instance I of MAST can be transformed into an instance I' of MCT with the same underlying graph G (and vice versa), so that for every spanning tree T, $\mathrm{AvStr}(T)$ in I equals $\mathrm{CC}(T)/\mathrm{CC}(G)$ in I'. This connection implies that every result concerning MAST on multigraphs applies to MCTs as well, and vice versa. In what follows, we will usually state each result in the context in which it was originally presented.

3 Basic Properties

3.1 Hardness

The MMST problem was proven to be NP-hard in [14]. In [18] it was shown that determining the optimal maximum stretch is NP-hard even for unweighted planar graphs. In [30] it was shown that the MMST problem cannot be approximated by a factor better than $(1 + \sqrt{5})/2$ unless $P = NP$. Thus the problem does not admit a polynomial-time approximation scheme (PTAS), unless P=NP.

The NP-hardness of the MCT problem was established in [22], even for the restricted case of uniform requirements over arbitrary unweighted graphs ($\omega(u, v) \in \{1, \infty\}$). In [42] it was shown that the problem remains NP-hard even in the case of uniform requirements over complete weighted graphs with weights satisfying the triangle inequality. In [29] it was shown that even the simplified case of 2-*uniform* MCT, where there are only two designated vertices with uniform requirements to all the vertices in the network, is still NP-hard. Moreover, the MCT problem was shown in [29] to be MaxSNP-hard [26,2], and thus it too does not admit a PTAS unless P=NP [3].

3.2 Constructions and Bounds for General Graph Classes

The MMST Problem: We are unaware of nontrivial approximation algorithms for the MMST problem on arbitrary weighted graphs. The problem was studied on planar graphs. A polynomial algorithm with fixed parameter t, that for any given planar unweighted graph G with bounded face length decides whether G has a spanning tree with stretch t, is presented in [18]. Furthermore, it is proved therein that it can be decided in polynomial time whether a given unweighted planar graph has a spanning tree with maximum stretch 3. Polynomial time algorithms are given in [32] for the MMST problem on outerplanar graphs and for the class of $EF2$ graphs, namely, 1-face depth graphs with no vertices of degree 2.

The MAST/MCT Problem: The MAST problem was studied in [1], where it is shown how to deterministically construct, for any n-vertex weighted graph, a spanning tree with average stretch $\exp(O(\sqrt{\log n \log\log n}))$. This result applies also to $\textsc{AvStr}_m$ on multigraphs. (In fact, the average stretch measure used in [1], $\textsc{AvStr}'$, is slightly different from the one given here, in that it averages the stretch over edges $e \in E$ only, rather than over all vertex pairs. This makes no essential difference since, as shown in [35], any algorithm for constructing a spanning tree approximating the minimum $\textsc{AvStr}'$ can be translated into one for constructing a tree approximating the minimum $\textsc{AvStr}_m$ or $\textsc{AvStr}$, with the same ratio.)

In the opposite direction, a lower bound of $\Omega(\log n)$ is established in [1] on the average stretch for certain n-vertex graphs (and in particular the 2-dimensional grid). This leaves a wide gap betwen the upper and lower bounds. It is conjectured in [1] that the lower bound is tight, namely, that every n-vertex graph has average stretch $O(\log n)$.

For the MCT problem, it follows from the construction of [1] that the problem enjoys a polynomial-time approximation algorithm on arbitrary weighted graphs with approximation ratio $\exp(O(\sqrt{\log n \log\log n}))$. Conversely, it follows from the lower bound of [1] that there are graphs in which the *gap* between the communication cost of the optimal tree and the communication volume of the graph is $\Omega(\log n)$, i.e., the communication cost of the MCT is $\Omega(\log n)$ larger than $\mathrm{CC}(G)$. This implies that techniques comparing the communication cost to the communication volume cannot yield an approximation ratio better than $\Omega(\log n)$ for the MCT problem.

The planar case is studied further in [32]. Polynomial time algorithms are given for the MAST problem on outerplanar and $EF2$ graphs. For general planar graphs, an approximation algorithm with ratio $2k$ is given for the MAST problem, where k is the *face depth* of the graph.

The problem of finding a light-weight tree approximating the MAST is studied in [5]. The paper presents a polynomial time algorithm for constructing a spanning tree of average stretch $O(\log n \cdot \log diam \cdot \exp(O(\sqrt{\log n \log\log n})))$, where *diam* denotes the diameter of the graph G, that approximates the total weight of the minimum-weight spanning tree for the given graph up to a factor of $O(\log n)$.

4 Low Stretch Trees in Metric Spaces

4.1 Low-Distortion Embeddings in Metric Spaces

Our problems are related to the intensively studied area of low-distortion embeddings of metric spaces in structurally simpler metric spaces, such as tree (or additive) metrics, normed spaces or Euclidean spaces (see, e.g., [11,25,21]), and can be cast as embedding problems in this framework.

Metric Spaces: A *metric space* is a pair $\mathcal{M} = (V, \delta)$ where V is a set of points and $\delta : V \times V \mapsto \mathbb{R}$ is a nonnegative and symmetric distance function satisfying the triangle inequality, i.e., $\delta(u, v) \leq \delta(u, x) + \delta(x, v)$ for every $u, v, x \in V$. The triangle inequality can be defined also for arbitrary graphs: the edge weights of a weighted graph $G = (V, E, \omega)$ are said to satisfy the triangle inequality if every edge $e = (u, v)$ is a shortest path between its endpoints u and v, i.e., $dist_G(u, v) = \omega(e)$. Thus a *metric space* can be thought of as a complete weighted graph satisfying the triangle inequality[1]. An arbitrary weighted graph $G = (V, E, \omega)$ induces a corresponding metric space, denoted $\mathcal{M}_G = (V, \delta)$, defined by setting $\delta(u, v) = dist_G(u, v)$ for every $u, v \in V$. Note that every metric space is induced by some weighted graph. A metric space $\mathcal{M} = (V, \delta)$ is said to be a *tree metric* if there exists a weighted *tree* T inducing it, i.e., such that $\mathcal{M}_T = \mathcal{M}$.

Embeddings: Consider two (finite) metric spaces $\mathcal{M}_i = (V, \delta_i)$, $i = 1, 2$, over the same set of points, where intuitively $\mathcal{M}_1$ plays the role of the simpler metric and $\mathcal{M}_2$ is the metric to be approximated. We require that $\mathcal{M}_1$ dominates $\mathcal{M}_2$, i.e., $\delta_2(u, v) \leq \delta_1(u, v)$ for every $u, v \in V$. Let us imagine that $\mathcal{M}_2$ is embedded into $\mathcal{M}_1$. The *distortion* of this embedding is defined as $\max_{u,v \in V}\{\delta_1(u, v)/\delta_2(u, v)\}$. We say that $\mathcal{M}_1$ *ρ-approximates* $\mathcal{M}_2$ if this distortion is at most ρ.

The problem of finding a spanning tree T satisfying $\text{MaxStr}(T) \leq \rho$ for a given weighted graph G can now be reformulated as follows: Given a weighted graph G, find a spanning tree T whose induced metric, $\mathcal{M}_T$, ρ-approximates the metric induced by the graph, $\mathcal{M}_G$. Similarly, the MMST problem can be formulated as looking for a spanning tree T such that the embedding of $\mathcal{M}_G$ into $\mathcal{M}_T$ has minimum distortion. The MAST problem can be cast in this framework as well, in terms of average distortion.

Tree Metrics vs. Spanning Trees: The typical questions asked in the area of low-distortion embeddings concern metric spaces and not graphs. Namely, given an arbitrary weighted graph $G = (V, E, \omega)$, one may be concerned with looking for a low-distortion embedding of the metric space $\mathcal{M}_G$ into a tree metric $\mathcal{M}_T$, but disregard the issue of whether the corresponding tree T is a spanning tree of the original G. In particular, the resulting tree T may contain both edges and vertices that do not exist in the original graph G.

This additional freedom in selecting the tree may generally yield better results. Nevertheless, in some cases it does not help. For instance, returning to the

[1] Indeed, in what follows we may occasionally use the terms "metric space" and "complete metric graph" interchangeably.

n-vertex ring R_n, it is shown in [34] that for any tree T (possibly including new vertices and edges), the distortion of embedding $\mathcal{M}_{R_n}$ in $\mathcal{M}_T$ is $\Omega(n)$.

As pointed out in [23], in network design applications the graph G represents a *physical* entity (the communication network), and therefore an approximating tree metric that is not a spanning tree may prove infeasible, as its edges and vertices might not correspond to ones that physically exist in G. For application types where G is a *logical* graph, such as developing competitive online algorithms, this may make no difference.

4.2 Probabilistic Approximation by Tree Metrics

The inadequacy of individual trees as approximations for general graphs, in terms of their maximum stretch, has led to the idea of replacing the use of a single spanning tree by a *probability distribution* of trees. This idea, originated in [1] in a restricted form focusing on the use of spanning trees, was taken one step further in [7], which explicitly introduced the notion of *probabilistic approximation* of metric spaces by general tree metrics.

Probabilistically Approximating Distributions (PADs): A probability distribution $\mathcal{D}$ of tree metrics over $V' \supseteq V$ is said to be a *ρ-probabilistically approximating distribution (ρ-PAD)* for a metric space $\mathcal{M} = (V, \delta)$ if for every $u, v \in V$, $dist_T(u,v) \geq \delta(u,v)$, and the expected stretch of trees in the distribution $\mathcal{D}_G$ satisfies $\mathbb{E}_{\mathcal{D}}[\text{STR}_T(u,v)] \leq \rho$. Intuitively, the maximum stretch $\text{MAXSTR}(T)$ of each tree T in the distribution may be high, and therefore the expected maximum stretch of a tree in the distribution, $\mathbb{E}_{\mathcal{D}}[\text{MAXSTR}(T)] = \mathbb{E}_{\mathcal{D}}[\max_{u,v}\{\text{STR}_T(u,v)\}]$, may still be high. However, the maximum expected stretch, $\max_{u,v}\{\mathbb{E}_{\mathcal{D}}[\text{STR}_T(u,v)]\}$, is guaranteed to be low.

To appreciate the way in which using a distribution of tree metrics instead of a single tree can help overcome the obstacle of high maximum stretch, let us once again return to the n-vertex ring R_n, and note that it has a 2-PAD, namely, the uniform distribution on the collection $\mathcal{T}$ of all n possible spanning trees. In other words, for every pair of vertices u and v, if one picks a tree T from the collection $\mathcal{T}$ uniformly at random, then the expected stretch will satisfy $\text{STR}_T(u,v) < 2$.

From Low Average Stretch Trees to PADs: It turns out that there exists a useful connection between spanning trees of average stretch ρ and ρ-PADs. Namely, given an algorithm that for every graph G constructs a spanning tree of average stretch ρ, one can (deterministically) construct a ρ-PAD for G. This transformation is implicitly used in [1], cast in game-theoretic terminology, for the application of low average stretch trees to online algorithms (see [35]). As the construction of [1] yields a spanning tree of average stretch $\exp(O(\sqrt{\log n \log\log n}))$ for any n-vertex graph G, it follows that any such graph has an $\exp(O(\sqrt{\log n \log\log n}))$-PAD of spanning trees.

Hierarchically Well-Separated Trees (HSTs): It was subsequently shown in [7] that any graph has an $O(\log^2 n)$-PAD of tree metrics. The trees of the distribution are of a special type called *hierarchically well-separated trees (HSTs)*.

A k-HST is a rooted tree with the following two properties: (1) the edges from an internal node to its children are all of the same length, and (2) the weights of consecutive edges along any root-to-leaf path decrease by a factor of k or more.

Each HST corresponds to a hierarchical partition of the graph into low-diameter clusters. A *probabilistic partition* of the graph with parameter r is a distribution over partitions of the graph into clusters of diameter $O(r \log n)$ such that the probability that two vertices u and v are separated (i.e., fall in different clusters) is proportional to $\omega(u, v)/r$. Picking an HST at random from the distribution is performed by constructing it recursively together with its corresponding randomly chosen partition. The process starts with the entire graph, partitions it into clusters, and then recursively constructs an HST for each cluster and glues them together to yield an HST for the entire graph. Hence the construction can be thought of as a probabilistic version of the recursive tree construction of [1], based on the sparse partitions of [6].

The results of [7] were later improved in [8] using the techniques of [36,17], showing that any metric space has an $O(\log n \log\log n)$-PAD of tree metrics. This is already close to the lower bound on the best possible distortion, shown in [7] to be $\Omega(\log n)$.

It is shown in [23] that a *planar* metric $\mathcal{M}_G$ (i.e., the metric induced by a planar graph G) has an $O(\log diam)$-PAD of tree metrics, and that this is the best possible.

Applications: The probabilistic approximation technique was initially used in [1,7] for the design of randomized competitive online algorithms for a variety of problems. A randomized online algorithm with ratio $\exp(O(\sqrt{\log n \log\log n}))$ for the k-server problem was given in [1]. A randomized online algorithm with sublinear competitive ratio for metrical task systems was presented in [7]. Randomized online algorithms with logarithmic competitive ratio were also given in [7] for distributed paging, the (centralized and distributed) k-server problem, distributed resource management, file allocation and more.

In the opposite direction, a near-logarithmic lower bound on the competitive ratio of randomized online algorithms for metrical task systems (including the k-server problem) was shown in [9], by establishing a Ramsey-type theorem stating that every metric space contains a large subspace which is approximately a hierarchically well-separated tree (HST).

The probabilistic approximation technique was then used extensively to derive randomized approximation algorithms with polylogarithmic ratio for a variety of optimization problems. The technique applies to optimization problems that share two key characteristics. First, the input is a weighted graph in which one must find a minimum weight subset of edges satisfying certain constraints. Second, the problem should be easier to solve (exactly or approximately) on a tree. A problem satisfying these properties can be solved by embedding the given graph in a tree metric that approximates its weights with low distortion, and solving the problem on the tree. The resulting approximation ratio for the problem on the original graph will depend on the distortion of the embedding.

In particular, this approach was used for approximating the *group Steiner tree* problem, that given a weighted graph and a collection of k sets of vertices requires to find a minimum weight Steiner tree connecting at least one element from each set. Using PADs of tree metrics, the group Steiner tree problem was given an $O(\log^4 n)$-approximation algorithm in [19] and then an $O(\log^2 n \log k \log\log n)$-approximation algorithm in [15]. The *k-median* problem requires to find, in a given weighted graph, k locations so that if every vertex is assigned to the nearest location, the total sum of distances from vertices to their assigned location is minimized. This problem was given an $O(\log k \log\log k)$-approximation algorithm in [15] using PADs of tree metrics. Other problems approximated in this technique include the buy-at-bulk network design problem [4] and the min-sum k-clustering problem in metric spaces [10].

Deterministic Constructions: The probabilistic approximation technique of [7,8] may yield an exponentially large probability distribution. Therefore, the resulting algorithm for constructing a tree with low average stretch for a given complete metric graph is inherently *randomized*.

In [16], it is shown that for every n-point metric space it is possible to construct (in polynomial time) a rather small $O(\log n \log\log n)$-PAD; in fact, a collection of $O(n \log n)$ trees will do. This, in turn, implies that the randomized approximation algorithms for the problems discussed above can be efficiently derandomized, since it is possible to efficiently generate and test each of the trees in the distribution in polynomial time. Hence we get *deterministic* approximation algorithms for these problems.

PADs of Spanning Trees: As discussed earlier, the fact that the approximating tree metrics are not spanning trees of the given graph may be problematic for certain applications, such as network design. As partial remedy, it is shown in [23] that the HST constructions of [7] can be modified, dispensing with the external vertices and thus remaining with a spanning tree of the original metric space. Still, this transformation does not handle the case of an arbitrary weighted graph G, as the tree generated by it is a spanning tree of the induced metric space $\mathcal{M}_G$ but might contain edges that do not exist in G itself. Hence for arbitrary graphs, the currently best construction for PADs of spanning trees is still that of [1].

From PADs to Low Average Stretch Trees: Note that the connection between spanning trees of average stretch ρ and ρ-PADs for complete metric graphs, discussed earlier, is in fact bidirectional. As shown in [42], the existence of a (constructible) ρ-PAD of a complete metric graph G implies a randomized approximation algorithm for MCT, returning a spanning tree whose expected communication cost is $O(\rho) \cdot \mathrm{CC}(G)$. Equivalently, it yields a randomized algorithm for selecting a spanning tree of average stretch $O(\rho)$ for G. This is achieved by sampling the tree distribution for a suitable number of times, taking the best resulting HST, and using the transformation of [23] to modify this HST into a spanning tree for G.

Combined with the result of [8], this yields an $O(\log n \operatorname{loglog} n)$-approximation for the MCT problem, or a randomized algorithm for selecting a spanning tree of average stretch $O(\log n \log\log n)$, for every complete metric graph G. Again, for arbitrary graphs the currently best construction for low average stretch spanning trees is that of [1].

Finally, these algorithms can be derandomized by applying them on the small distributions of [16]. This yields a deterministic algorithm for the construction of a spanning tree average stretch $O(\log n \log\log n)$, based on generating and testing every tree in the distribution. Similarly, we get a deterministic algorithm for MCT on complete metric graphs which returns a spanning tree with communication cost $O(\log n \log\log n) \cdot \mathrm{CC}(G)$, improving on the deterministic algorithm of [29] whose resulting communication cost was $O(\log^2 n) \cdot \mathrm{CC}(G)$.

5 MCT on Restricted Instances

We conclude this review with a discussion of results obtained for various special cases of the MCT problem.

Uniform Requirements on Arbitrary Weighted Graphs: The special case of uniform MCT on arbitrary weighted graphs was studied in [37], where it was shown that there exists a vertex v such that the shortest paths tree from v provides a 2-approximation for problem. Some two decades later, a PTAS was established for the uniform MCT problem in [42]. This was established in two stages. First, a PTAS was proposed for the problem on a complete metric graph. The PTAS is based on the notion of a k-star, which is a tree with at most k internal (nonleaf) vertices. The central observation is that for a uniform instance and for every fixed k, there exists some k-tree whose communication cost is at most $1 + 2/(k + 1)$ times that of the MCT. The paper then shows how on a complete metric graph, the minimum communication cost k-tree can be found in time $O(n^{2k})$, which is polynomial for fixed k. This yields a PTAS for the problem on complete metric graphs. In the second stage, the uniform MCT problem on arbitrary weighted graphs is reduced to the uniform MCT problem on complete metric graphs, by showing that for an arbitrary weighted graph G, given any spanning tree $\bar{T}$ of the induced metric space $\mathcal{M}_G$, it is possible to gradually transform $\bar{T}$ into a spanning tree T for G using no "bad" edges (i.e., edges that violate the triangle inequality), such that $\mathrm{CC}(T) \leq \mathrm{CC}(\bar{T})$. This subsequently provides a PTAS for the problem on arbitrary weighted graphs as well.

A more restricted special case of the uniform case, studied in [20], imposes the additional constraint that the edge weights satisfy that for every $1 \leq x, y, z \leq n$ such that $\omega(x, y) \leq \omega(x, z) \leq \omega(y, z)$, we have $(\omega(y, z) - \omega(x, y))/\omega(x, z) \leq (n - 2)/(2n - 2)$. (This can be thought of as a stronger version of the triangle inequality.) It was shown in [20] that for this case, the optimal tree is a star (composed of a root and $n - 1$ leaves), hence the algorithm simply checks all n possible star trees.

The problem of finding a light-weight tree approximating the uniform MCT is studied in [38]. The paper presents a polynomial time algorithm that given a uniform instance constructs a tree approximating up to a constant factor both the communication cost of the MCT and the total weight of the minimum-weight spanning tree for the given graph.

Other Requirements on Arbitrary Weighted Graphs: A number of papers studied intermediate settings between the uniform MCT problem and the MCT problem on arbitrary inputs. In [29] it is shown that the shortest paths heuristic of [37] may be extended to provide a 3-approximation for k-uniform MCT, the variant where there are k designated vertices that have uniform requirements to all the vertices of the network. In another setting, studied in [39,41], each vertex v is assigned a requirement coefficient $p(v)$, determining its requirements. In the sum-requirement (SR) case, the communication requirements are set to $r(u,v) = p(u) + p(v)$, and in the product-requirement (PR) case, the requirements are set to $r(u,v) = p(u) \cdot p(v)$. Both settings were studied for complete metric graphs. In [39], both problems were given constant approximation algorithms (with the constant being 2 and 1.577 respectively). In [41], the PR problem was given a PTAS, based on extending the technique of [42]. The observation that some k-star forms a $(1+1/k)$-approximate solution is still valid, but for the PR problem it seems more difficult to compute the optimum k-star. Instead, the algorithm is based on finding a k-star approximating the optimum k-star via a rounding and scaling technique.

General Requirements on Geometric Graphs: The MCT problem becomes somewhat easier for *geometric* graphs, namely, graphs whose vertices are given as coordinates in $\mathbb{R}^d$ and the distance function is the Euclidean distance between the coordinates. In particular, a *deterministic* algorithm was presented in [29] that given a geometric graph, finds a spanning tree T with communication cost $O(d \cdot \log n) \cdot \mathrm{CC}(G)$. This yields an $O(d \log n)$-approximation for the MCT problem (or $O(\log n)$-approximation when d is fixed). This was then improved in [16] to an approximation ratio of $O(\sqrt{d} \cdot \log n)$.

References

1. N. Alon, R.M. Karp, D. Peleg, and D. West. A graph-theoretic game and its application to the k-server problem. *SIAM J. on Computing*, pages 78–100, 1995.
2. S. Arora and C. Lund. Hardness of approximation. In Dorit Hochbaum, editor, *Approximation Algorithms for NP-Hard Problems*, pages 399–446. PWS Publishing Company, Boston, MA, 1997.
3. S. Arora, C. Lund, R. Motwani, M. Sudan, and M. Szegedy. Proof verification and hardness of approximation problems. In *Proc. 33rd IEEE Symp. on Foundations of Computer Science*, 1992.
4. B. Awerbuch and Y. Azar. Buy-at-bulk network design. In *Proc. 38th IEEE Symp. on Foundations of Computer Science*, pages 542–547, 1997.
5. B. Awerbuch, A. Baratz, and D. Peleg. Efficient broadcast and light-weight spanners. Unpublished manuscript, November 1991.
6. B. Awerbuch and D. Peleg. Sparse partitions. In 31st *IEEE Symp. on Foundations of Computer Science*, pages 503–513, October 1990.
7. Y. Bartal. Probabilistic approximation of metric spaces and its algorithmic applications. In *Proc. 37th IEEE Symp. on Foundations of Computer Science*, pages 184–193, 1996.

8. Y. Bartal. On approximating arbitrary metrics by tree metrics. In *Proc. 30th ACM Symp. on Theory of Computing*, pages 161–168, 1998.
9. Y. Bartal, B. Bollobas, and M. Mendel. Ramsey-type theorems for metric spaces with applications to online problems. In *Proc. 42nd IEEE Symp. on Foundations of Computer Science*, 2001.
10. Y. Bartal, M. Charikar, and D. Raz. Approximating min-sum k-clustering in metric spaces. In *Proc. 33rd ACM Symp. on Theory of Computing*, 2001.
11. J. Bourgain. On Lipschitz embeddings of finite metric spaces in Hilbert spaces. *Israel J. Math.*, pages 46–52, 1985.
12. L. Cai. Tree 2-spanners. Technical Report TR 91-4, Simon Fraser University, Burnaby, B.C., Canada, 1991.
13. L. Cai and D. Corneil. Isomorphic tree spanner problems. *Algorithmica*, 14:138–153, 1995.
14. L. Cai and D. Corneil. Tree spanners. *SIAM J. on Discr. Math.*, 8:359–387, 1995.
15. M. Charikar, C. Chekuri, A. Goel, and S. Guha. Rounding via trees: deterministic approximation algorithms for group steiner trees and k-median. In *Proc. 30th ACM Symp. on Theory of Computing*, pages 114–123, 1998.
16. M. Charikar, C. Chekuri, A. Goel, S. Guha, and S. Plotkin. Approximating a finite metric by a small number of tree metrics. In *Proc. 39th IEEE Symp. on Foundations of Computer Science*, pages 379–388, 1998.
17. G. Even, J. Naor, S. Rao, and B. Schieber. Divide-and-conquer approximation algorithms via spreading metrics. In *Proc. 36th IEEE Symp. on Foundations of Computer Science*, pages 62–71, 1995.
18. S.P. Fekete and J. Kremer. Tree spanners in planar graphs. In *Proc. 24th Int. Workshop on Graph-Theoretic Concepts in Computer Science*. Springer-Verlag, 1998.
19. N. Garg, G. Konjevod, and R. Ravi. A polylogarithmic approximation algorithm for the group Steiner tree problem. In *Proc. 9th ACM-SIAM Symp. on Discrete Algorithms*, 1998.
20. T.C. Hu. Optimum communication spanning trees. *SIAM J. on Computing*, pages 188–195, 1974.
21. P. Indyk. Algorithmic applications of low-distortion geometric embeddings. In *Proc. 42nd IEEE Symp. on Foundations of Computer Science*, 2001.
22. D.S. Johnson, J.K. Lenstra, and A.H.G. Rinnooy-Kan. The complexity of the network design problem. *Networks*, 8:275–285, 1978.
23. G. Konjevod, R. Ravi, and F.S. Salman. On approximating planar metrics by tree metrics. *Information Processing Letters*, 80:213–219, 2001.
24. H.-O. Le and V.B. Le. Optimal tree 3-spanners in directed path graphs. *Networks*, 34:81–87, 1999.
25. N. Linial, E. London, and Y. Rabinovich. The geometry of graphs and some of its algorithmic applications. *Combinatorica*, 15:215–245, 1995.
26. C.H. Papadimitriou and M. Yannakakis. Optimization, approximation, and complexity classes. *J. Comput. and Syst. Sci.*, 43:425–440, 1991.
27. D. Peleg. Approximating minimum communication spanning trees. In *Proc. 4th Colloq. on Structural Information & Communication Complexity*, pages 1–11, July 1997.
28. D. Peleg. *Distributed Computing: A Locality-Sensitive Approach.* SIAM, 2000.
29. D. Peleg and E. Reshef. Deterministic polylog approximation for minimum communication spanning trees. In *Proc. 25th Int. Colloq. on Automata, Languages & Prog.*, pages 670–681, 1998.

30. D. Peleg and E. Reshef. A variant of the arrow distributed directory protocol with low average-case complexity. In *Proc. 26th Int. Colloq. on Automata, Languages & Prog.*, pages 615–624, 1999.
31. D. Peleg and A.A. Schäffer. Graph spanners. *J. of Graph Theory*, 13:99–116, 1989.
32. D. Peleg and D. Tendler. Low stretch spanning trees for planar graphs. Technical Report MCS01-14, The Weizmann Institute of Science, 2001.
33. D. Peleg and J.D. Ullman. An optimal synchronizer for the hypercube. *SIAM J. on Computing*, 18(2):740–747, 1989.
34. Y. Rabinovich and R. Raz. Lower bounds on the distortion of embedding finite metric spaces in graphs. *Discrete and Computational Geometry*, 19:79–94, 1998.
35. E. Reshef. Approximating minimum communication cost spanning trees and related problems. Master's thesis, The Weizmann Institute of Science, Rehovot, Israel, 1999.
36. P.D. Seymour. Packing directed circuites fractionally. *Combinatorica*, 15:281–288, 1995.
37. R. Wong. Worst-case analysis of network design problem heuristics. *SIAM J. on Alg. and Discr. Meth.*, 1:51–63, 1980.
38. B.Y. Wu, K.-M. Chao, and C.Y. Tang. Constructing light spanning trees with small routing cost. In *Proc. 16th Symp. on Theoretical Aspects of Computer Science*, pages 334–344, 1999.
39. B.Y. Wu, K.-M. Chao, and C.Y. Tang. Approximation algorithms for some optimum communication spanning tree problems. *Discrete Applied Mathematics*, 102:245–266, 2000.
40. B.Y. Wu, K.-M. Chao, and C.Y. Tang. Approximation algorithms for the shortest total path length spanning tree problem. *Discrete Applied Mathematics*, 105:273–289, 2000.
41. B.Y. Wu, K.-M. Chao, and C.Y. Tang. A polynomial time approximation scheme for optimal product-requirement communication spanning trees. *J. of Algorithms*, 36:182–204, 2000.
42. B.Y. Wu, G. Lancia, V. Bafna, K.M. Chao, R. Ravi, and C. Y. Tang. A polynomial-time approximation scheme for minimum routing spanning trees. In *Proc. 9th ACM-SIAM Symp. on Discrete Algorithms*, pages 21–32, San Francisco, California, January 1998.

On Radiocoloring Hierarchically Specified Planar Graphs: $\mathcal{PSPACE}$-Completeness and Approximations

Maria I. Andreou[1], Dimitris A. Fotakis[2], Sotiris E. Nikoletseas[1], Vicky G. Papadopoulou[1], and Paul G. Spirakis[1,*]

[1] Computer Technology Institute (CTI) and Patras University, Greece, Riga Fereou 61, 26221 Patras, Greece. Fax: +30-61-222086, {mandreou,nikole,viki,spirakis}@cti.gr

[2] Max-Planck-Institute für Informatik, Stuhlsatzenhausweg 85, 66123 Saarbrücken, Germany, fotakis@mpi-sb.mpg.de

Abstract. Hierarchical specifications of graphs have been widely used in many important applications, such as VLSI design, parallel programming and software engineering. A well known hierarchical specification model, considered in this work, is that of Lengauer [9, 10] referred to as *L-specifications*. In this paper we discuss a restriction on the L-specifications resulting to graphs which we call Well-Separated (WS). This class is characterized by a polynomial time (to the size of the specification of the graph) testable combinatorial property.

In this work we study the Radiocoloring Problem (RCP) on WS L-specified hierarchical planar graphs. The optimization version of RCP studied here, consists in assigning colors to the vertices of a graph, such that any two vertices of distance at most two get different colors. The objective here is to minimize the number of colors used. This problem is equivalent to the problem of vertex coloring the square of a graph G, G^2, where G^2 has the same vertex set as G and there is an edge between any two vertices of G^2 if their distance in G is at most 2.

We first show that RCP is $\mathcal{PSPACE}$-complete for WS L-specified hierarchical planar graphs. Second, we present a polynomial time 3-approximation algorithm as well as a more efficient 4-approximation algorithm for RCP on graphs of this class.

We note that, the best currently known approximation ratio for the RCP on ordinary (non-hierarchical) planar graphs of general degree is 2 ([6, 1]). Note also that the only known results on any kind of coloring problems have been shown for another special kind of hierarchical graphs (unit disk graphs) achieving a 6-approximation solution [13].

* This work has been partially supported by the EU IST/FET projects ALCOM-FT, FLAGS, CRESCCO and EU/RTN Project ARACNE. Part of the last author's work was done during his visit at Max-Planck-Institute für Informatik (MPI).

K. Diks et al. (Eds): MFSC 2002, LNCS 2420, pp. 81–92, 2002.

1 Introduction, Our Results and Related Work

1.1 Motivation

Many practical applications of graph theory and combinatorial optimization in CAD systems, VLSI design, parallel programming and software engineering involve the processing of large (but regular) objects constructed in a systematic manner from smaller and more manageable components. As a result, the graphs that abstract such circuits (designs) also have a regular structure and are defined in a systematic manner using smaller graphs. The methods for specifying such large but regular objects by small specifications are referred to as *succinct specifications*. One way to succinctly represent objects is to specify the graph hierarchically. Hierarchical specifications are more concise in describing objects than ordinary graph representations. A well known hierarchical specification model, considered in this work, is that of Lengauer, introduced in [9,10], referred to as *L-specifications*.

In modern networks, Frequency Assignment Problems (FAP) have important applications in efficient bandwidth utilization, by trying to minimize the number (or the range) of frequencies used, in a way that however keeps the interference of nearby transmitters at an acceptable level. Problems of assigning frequencies in networks are usually abstracted by variations of coloring graphs. An important version of Frequency Assignment Problems is the Radiocoloring Problem (RCP). The optimization version of RCP studied here, consists in assigning colors (frequencies) to the vertices (transmitters) of a graph (network), so that any two vertices of distance at most two get different colors. The objective here is to minimize the number of distinct colors used.

In this work we study RCP on L-specified hierarchical graphs. Note that RCP is equivalent to the problem of vertex coloring the square of a graph G, G^2, where G^2 has the same vertex set as G and there is an edge between any two vertices of G^2 if their distance in G is at most 2. We study here planar hierarchical graphs.

Also, our interest in coloring the square of a hierarchical planar graph is inspired by real communication networks, especially wireless and large ones, that may be structured in a hierarchical way and are usually planar.

1.2 Summary of Our Results

We investigate the computational complexity and provide efficient approximation algorithms for the RCP on a class of L-specified hierarchical planar graphs which we call Well-Separated (WS) graphs. In such graphs, levels in the hierarchy are allowed to directly connect *only* to their immediate descendants. In particular:

1. We prove that the decision version of the RCP for Well-Separated L-specified hierarchical planar graphs is $\mathcal{PSPACE}$-complete.
2. We present two approximation algorithms for RCP for this class of graphs. These algorithms offer alternative trade-offs between the quality and the

efficiency of the solution achieved. The first one is a simple and very efficient 4-approximation algorithm, while the second one achieves a better solution; it is a 3-approximation algorithm, but is less efficient, although polynomial.

We note that the class of WS L-specified hierarchical graphs considered here can lead to graphs that are exponentially large in the size of their specification. The WS class is a subclass of the class of L-specified hierarchical graphs considered in [11], called k-level-restricted graphs.

1.3 Related Work and Comparison

In a fundamental work, Lengauer and Wagner [10] proved that the following problems are $\mathcal{PSPACE}$-complete for L-specified hierarchical graphs: 3-coloring, hamiltonian circuit and path, monotone circuit value, network flow and independent set. For L-specified graphs, Lengauer ([9]) have given efficient algorithms to solve several important graph theoretic problems including 2-coloring, min spanning forest and planarity testing.

Marathe et al in [12,11] studied the complexity and provided approximation schemes for several graph theoretic problems for L-specified hierarchical planar graphs including maximum independent set, minimum vertex cover, minimum edge dominating set, max 3SAT and max cut.

We remark that the best currently known approximation ratio for the RCP on ordinary (non-hierarchical) planar graphs of general degree is 2 ([6,1]). Approximations for various classes of graphs presented in [5]. Also, the only known results on any kind of coloring problems have been shown for the vertex coloring for a special kind of hierarchical graphs (k-level-restricted unit disk graphs) achieving a 6-approximation solution ([13]).

2 Preliminaries

In this work we study an optimization version of the Radiocoloring problem ([6]), where the objective is to minimize *the number* of colors used:

Definition 1. Min Order RCP*: Given a graph $G(V,E)$, find an assignment of G, i.e. a coloring function $\Lambda : V \rightarrow N^*$ assigning integers (colors) to the vertices of G such that $\Lambda(u) \neq \Lambda(v)$ if $d(u,v) \leq 2$, where $d(u,v)$ is the distance between u and v in G, that uses a minimum number of colors. The number of different integers in such an assignment, is called the* order *of RCP on G and is denoted here by $\lambda(G)$, i.e. $\lambda(G) = |\Lambda(V)|$.*

For simplicity reasons, in the sequel we refer to it as the RCP. Remark that:

Proposition 1. *The min order RCP of a given graph G is equivalent to the problem of coloring the square of the graph G, G^2. G^2 has the same vertex set as G and there is an edge between any two vertices of G^2 if their distance in G is at most 2.*

We study the RCP on hierarchical graphs as specified by Lengauer [9].

Definition 2. (**L-specifications,** [9]) *An L-specification* $\Gamma = (G_1, \cdots, G_i, \cdots, G_n)$, *where* n *is the number of levels in the specification, of a graph* G *is a sequence of labeled undirected simple graphs* G_i *called* cells. *The graph* G_i *has* m_i *edges and* n_i *vertices. The* p_i *of the vertices are called pins. The other* $(n_i - p_i)$ *vertices are called inner vertices. The* r_i *of the inner vertices are called nonterminals. The* $(n_i - r_i)$ *vertices are called terminals. The remaining* $n_i - p_i - r_i$ *vertices of* G_i *that are neither pins nor nonterminals are called* explicit *vertices.*

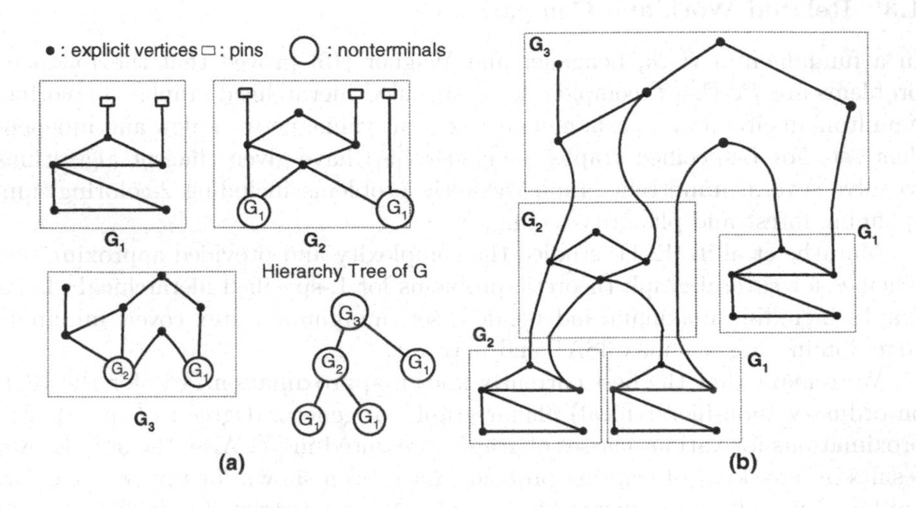

Fig. 1. **(a)** An L-specification $\Gamma = (G_1, G_2, G_3)$ of a graph G and its hierarchy tree $HT(G)$. **(b)** The expansion of the graph G.

Each pin of G_i *has a unique label, its name. The pins are assumed to be numbered from 1 to* p_i. *Each nonterminal in* G_i *has two labels* (v, t), *a name and a type. The type* t *of a nonterminal in* G_i *is a symbol from* $G_1, \cdots, G_{i-1}$. *The neighbours of a nonterminal vertex must be terminals. If a nonterminal vertex* v *is of type* G_j *in* G_i, $1 \leq j \leq i-1$, *then* v *has degree* p_j *and each terminal vertex that is a neighbor of* v *has a distinct label* (v, l) *such that* $1 \leq l \leq p_j$. *We say that the neighbor of* v *labeled* (v, l) *matches the* l*-th pin of* G_j.

See Figure 1(a) for an example of an L-specification. Note that a terminal vertex may be a neighbor of several nonterminal vertices. Given an L-specification Γ, $N = \sum_{1 \leq i \leq n} n_i$ denotes the *vertex number* and $M = \sum_{1 \leq i \leq n} m_i$ denotes the *edge number* of Γ. The size of Γ, denoted by $size(\Gamma)$, is $N + M$. For simplicity reasons, we assume that $n = \max_i\{n_i\}$.

Definition 3. (**Expansion of an L-specificied Hierarchical Graph,** [9]) *Let any L-specified hierarchical graph, given by* $\Gamma = (G_1, \cdots, G_n)$. *The* expanded graph $E(\Gamma)$ *(i.e. the graph associated with* Γ*) is iteratively obtained as follows:*

$k = 1$: $E(\Gamma) = G_1$.

$k > 1$: *Repeat the following step for each nonterminal* v *of* G_k *say of the type* G_j: *delete* v *and the edges incident on* v. *Insert a copy of* $E(\Gamma_j)$ *by identifying the* l*-th pin of* $E(\Gamma_j)$ *with the node in* G_k *that is labeled* (v, l). *The inserted copy of* $E(\Gamma_j)$ *is called a subcell of* G_k.

For example, the expansion of the hierarchical graph G of Fig. 1(a) is shown in Fig. 1(b). To each L-specification $\Gamma = (G_1, \cdots, G_n)$, we associate a labeled rooted unoriented tree $HT(\Gamma)$ depicting the insertions of the copies of the graphs $E(\Gamma_j)$ $(1 \leq j \leq n-1)$, made during the construction of $E(\Gamma)$ as follows:

Definition 4. *(***Hierarchy Tree of an L-specification,** **[9]***) Let* $\Gamma = (G_1, \cdots, G_n)$, *be an L-specification of the graph* $E(\Gamma)$. *The hierarchy tree of* Γ, *denoted by* $HT(\Gamma)$, *is a labeled rooted unordered tree defined as follows:*

1. *Let* r *the root of* $HT(\Gamma)$. *The label of* r *is* G_n. *The children of* r *in* $HT(\Gamma)$ *are in one-to-one correspondence with the nonterminal vertices of* G_n *as follows: The label of the child* s *of* r *in* $HT(\Gamma)$ *corresponding to the nonterminal vertex* (v, G_j) *of* G_n *is* (v, G_j).
2. *For all other vertices* s *of* $HT(\Gamma)$ *and letting the label of* $s = (v, G_j)$, *the children of* s *in* $HT(\Gamma)$ *are in one-to-one correspondence with the nonterminal vertices of* G_j *as follows: The label of the child* t *of* s *in* $HT(\Gamma)$ *corresponding to the nonterminal vertex* (w, G_l) *of* G_j *is* (w, G_l).

We consider hierarchical planar graphs as studied in [9]:

Definition 5. *(***Strongly Planar Hierarchical Graph, [9]***) An L-specified hierarchical graph* G *given by* $\Gamma = (G_1, \ldots, G_n)$ *is strongly planar if* $E(\Gamma)$ *has a planar embedding such that for each* $E(\Gamma_i)$ *all pins of it occur around a common face and the rest of* $E(\Gamma_i)$ *is completely inside this face.*

In fact, we here study a subclass of strongly planar hierarchical graphs, where additionally to the above condition, *all graphs* G_i, $1 \leq i \leq n$, *are planar.* We call this class as **fully planar hierarchical graphs**. In the sequel, and when there is no ambiguity, we refer to such graphs simply as *hierarchical planar graphs.*

Moreover, we concentrate on a class of L-specified hierarchical graphs which we call Well-Separated (WS) graphs, defined in the sequel using the followings:

Consider an L-specified hierarchical graph G, given by $\Gamma = (G_1, \ldots, G_n)$. For each graph G_i $(1 \leq i \leq n)$, we define the following subgraphs:

Definition 6. Inner Subgraph of Graph G_i, $\mathbf{G_{in\,i}}$: *is induced by the explicit vertices of* G_i *not connected to any pin or nonterminal of* G_i.

Definition 7. Outer-Up Subgraph of Graph G_i, $\mathbf{G_{outUp\,i}}$: *is induced by the explicit vertices of* G_i *connected to at least one pin of* G_i.

Definition 8. Outer-Down Subgraph of Graph G_i, $\mathbf{G_{outDown\,i}}$: *is induced by the explicit vertices of* G_i *connected to at least one nonterminal of* G_i.

Remark 1. Generally, an explicit vertex of G_i might belong to both outer-up, outer-down subgraphs of G_i. In this work we study the following class of graphs:

Definition 9. *(***Well-Separated, WS***) We call Well-Separated graphs the class of L-specified hierarchical graphs of which any explicit vertex of G_i, $1 \leq i \leq n$, belongs either to $G_{outDown\ i}$ or $G_{outUp\ i}$ or none of them, but not to both of them. Moreover, any vertex of $G_{outDown\ i}$ is located at distance at least 3 from any vertex of $G_{outUp\ i}$.*

Observe that the WS class of hierarchical graphs is testable in time polynomial in the size of the L-specification of a hierarchical graph. Note also that the WS class is a subclass of k-level-restricted graphs, which is another class of L-specified hierarchical graphs, studied by Marathe et al. [11] (see full version [2] for a proof).

Definition 10. *The* **Maximum Degree of a Hierarchical Graph** *G, $\Delta(G)$, is the maximum degree of a vertex in the expansion of the graph, $E(\Gamma)$.*

The following definitions and results are needed by our approximation algorithms.

Definition 11. *(***k-Outerplanar Graph** *G* **[3]***) A k-outerplanar graph G is defined recursively by taking an embedding of the planar graph G, finding the vertices in the exterior face of the graph and removing those vertices and the edges incident to them. Then, the remaining graph should be a $(k-1)$-outerplanar graph. A 1-outerplanar graph is an outerplanar graph.*

Theorem 1. *[4] Any k-outerplanar graph is a $3k-1$ bounded treewidth graph.*

Theorem 2. *[16] Let $G(V, E)$ be a k-tree of n vertices given by its tree-decomposition, let C be a set of colors, and let $\alpha = |C|$. Then, it can be determined in polynomial time $T(n, k)$, whether G has a radiocoloring that uses the colors of set C, and if such a radiocoloring exists, it can found in the same time, where $T(n, k) = O(n(2\alpha + 1)^{2^{2(k+1)(l+2)+1}} + n^3)$, $l = 2$ and $n = |V|$.*

3 The Complexity of the Radiocoloring Problem

In this section, we study the complexity of RCP on L-specified hierarchical planar graphs. A critical observation about the constructions utilized in the $\mathcal{PSPACE}$-completeness proofs, is that the transformations are local ([7]). I.e, given any hierarchical graph G, the graph G'_i obtained from each G_i, is the same for all appearances of G_i in the hierarchy tree of G. Thus, the resulting hierarchical graph G' can be computed in time polynomial in the size of the L-specification of the graph G.

Another important issue for the $\mathcal{PSPACE}$-completeness reductions is whether an already known $\mathcal{NP}$-completeness proof for the same problem, that fulfills such locality characteristics, can be modified so that to apply for a hierarchical graph G. This technique has been used in previous papers to get $\mathcal{PSPACE}$-completeness results for a number of problems considered (e.g. [13]).

In our case, there was no such 'local' $\mathcal{NP}$-completeness reduction available. The corresponding $\mathcal{NP}$-completeness reductions that *both* could be adapted for the hierarchical case are the reductions of [15,14]. However, although the reduction of [15] is local that of [14] is not. Henceforth, they can not be used to get the $\mathcal{PSPACE}$-completeness of L-specified hierarchical planar graphs.

For these reasons, we have developed a new $\mathcal{NP}$-completeness proof for the RCP of ordinary planar graphs which reduces it from the problem of 3-coloring planar graphs. The construction satisfies the desired locality characteristics and thus, we can utilize it to get the $\mathcal{PSPACE}$-completeness proof of the RCP for L-specified hierarchical planar graphs.

3.1 The $\mathcal{NP}$-Completeness of RCP for Planar Graphs

In this section we provide a new $\mathcal{NP}$-completeness proof for the problem of radiocoloring for ordinary (non-hierarchical) planar graphs, which is 'local'. We remark that this reduction is the only one that works for the cases where $\Delta(G) < 7$ ($\Delta(G) \geq 3$), in contrast to the only known $\mathcal{NP}$-completeness proof of [15].

Theorem 3. *The following decision problem is $\mathcal{NP}$-complete:*
Input: A planar graph $G(V, E)$.
Question: Is $\lambda(G) \leq 4$?

Proof. It can be easily seen that the problem belongs in $\mathcal{NP}$. Let any planar graph G. We reduce RCP from the 3-COLORING problem of planar graphs, which is known to be $\mathcal{NP}$-complete ([7]). I.e. we will construct in polynomial time a new graph G', which is 4-radiocolorable if and only if G is 3-colorable. The reduction employs the component design technique.

The construction replaces every vertex u of degree d_u of the initial graph G with a component, called 'cycle node'. The cycle node obtained by a vertex of degree d_u is said to be 'a cycle node of degree d_u'. An instance of it is shown in Figure 2(a). A cycle node of degree d_u is constructed as follows:

Add a cycle of $3d_u$ vertices, called *outer cycle*, as shown in Figure 2(a). Call the vertices of each triad as *first, second* and *third.* For each triad, add two more vertices (called *fourth* and *fifth*) and connect them to the triad as shown in the Figure. Now, group together every five such vertices into pentads and number them as shown in the Figure. Next, add another cycle of $3d_u$ vertices, called *inner cycle.* For each triad, add a fourth vertex and connect it to the triad as shown in the Figure. Now, group together every four such vertices into quadruples and call them as in Figure 2(a). Finally, connect the i-th pentad of the outer cycle to the i-th quadruplet of the inner cycle as in the Figure. In the sequel, we explain how the cycle node is used to construct in polynomial time from any planar graph G a new planar graph G', with the desired properties. We consider a planar embedding of graph G. The new graph G' is constructed as follows (See Figure 2(b) for an example):

1. Replace each vertex of degree d_u of the initial graph G, with a cycle node of degree d_u.
2. For each vertex u of graph G, number the edges incident to u, in increasing, clockwise, order. For every edge of the initial graph $e = (u, v)$ connecting u

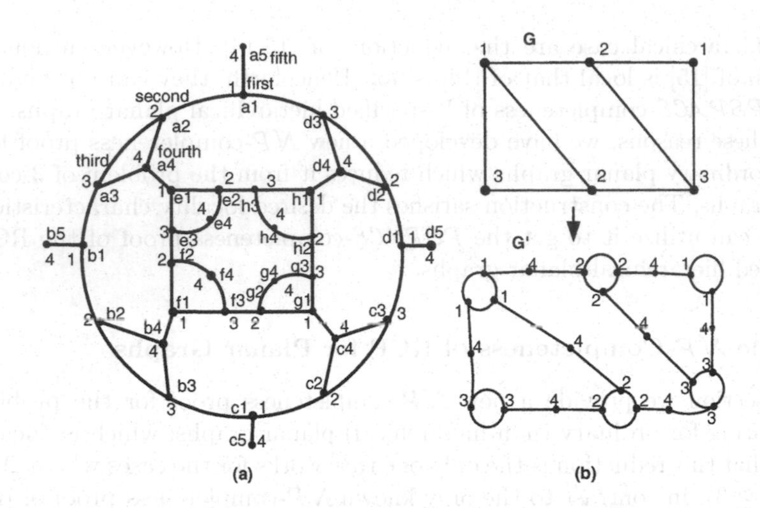

Fig. 2. (a) The 'cycle node' for a vertex of degree 4 and a 4-radiocoloring of it. (b) A graph G with a 3-coloring and the graph G' obtained with the resp. 4-radiocoloring.

and v, let u_i be the number of edge e given by vertex u and let v_j be the number of the edge e given by vertex v. Then, take the fifth vertex of the u_i-th group of the cycle node of vertex u and the fifth vertex of the v_j-th group of the cycle node of vertex v and collapse them to a single vertex uv.

Obviously, the new graph G' is a planar graph.

Lemma 1. $\lambda(G') \leq 4$ *if and only if* $\chi(G) \leq 3$.

(See the full version [2] of the paper for the proof of the Lemma) □

3.2 The $\mathcal{PSPACE}$-Completeness of RCP for Hierarchical Planar Graphs

The $\mathcal{PSPACE}$-completeness reduction for the RCP of L-specified hierarchical planar graphs, utilizes the 'local' construction of the $\mathcal{NP}$-completeness proof of the radiocoloring problem for ordinary planar graphs, given in section 3.1, Theorem 3, in order to be of polynomial time to the size of the L-specification.

The $\mathcal{PSPACE}$-Completeness of 3-Coloring for Hierarchical Planar Graphs. In order to be able to utilize the $\mathcal{NP}$-completeness reduction of Theorem 3 to prove the $\mathcal{PSPACE}$-completeness of RCP for hierarchical planar graphs, we need to prove that:

Theorem 4. *The following decision problem is $\mathcal{PSPACE}$-complete:*

Input: A fully planar hierarchical graph G, given by the L-specification $\Gamma = (G_1, \ldots, G_n)$.

Question: Is $\chi(G) \leq 3$?

Proof. We adapt the $\mathcal{NP}$-completeness construction of [7]. See full version of the paper [2] for the proof of the Theorem. □

The $PSPACE$-Completeness of RCP for Hierarchical Planar Graphs. Before introducing the $\mathcal{PSPACE}$-completeness Theorem, the following observation is needed. We denote by $d(u)_H$ the degree of vertex u in graph H. Observe that, for an L-specified hierarchical graph G, for any G_i, $i = 1, \ldots n$, $d(u)_{G_i} = d(u)_{E(\Gamma)}$, $u \in G_i$.

Theorem 5. *The following decision problem is $\mathcal{PSPACE}$-complete:*
Input: A WS fully planar hierarchical graph G, given by the L-specification $\Gamma = (G_1, \ldots, G_n)$.
Question: Is $\lambda(G) \leq 4$?

Proof. **Membership in $\mathcal{PSPACE}$:** Similar to the $\mathcal{PSPACE}$-membership of 3-coloring an L-specified hierarchical graph ([10]).

$\mathcal{PSPACE}$-Completeness: We reduce the RCP for WS fully planar hierarchical graphs from the 3-COLORING of fully planar hierarchical graphs, proved to be $\mathcal{PSPACE}$-complete in Theorem 4 using the construction of Theorem 3.

Let any L-specified hierarchical planar graph G, given by $\Gamma = (G_1, \ldots, G_n)$. For each graph $G_i = (V_i, E_i)$ of its L-specification we construct a new graph $G'_i = (V'_i, E'_i)$ using the rules of Theorem 3:

1. Apply Rule 1 of Theorem 3 on the explicit vertices of G_i. Apply Rules 2,3 of the Theorem on the edges of G_i connecting any two explicit vertices of it.
2. All nonterminal vertices of G_i are present in G'_i. However, the type of a nonterminal, assume one of type G_j is changed to G'_j.
3. Take each $e = (u, t)$ of G_i connecting an explicit vertex u to a nonterminal vertex t of type G_j, matching the l-th pin of the graph G_j. Then, connect the corresponding fifth vertex of the cycle node of u to the nonterminal t matching the l-th pin of G_j.
4. All pins vertices of G_i are present in G'_i. Take each edge $e = (u, p)$ of G_i connecting an explicit vertex u to a pin p of G_i, numbered as the l-th. Then, remove the corresponding fifth vertex of the cycle node of vertex u and connect the corresponding first vertex of the cycle node to the pin p.

Observe that graphs $G'_1, G'_2, \ldots, G'_n$ obtained G, define an L-specified hierarchical graph G', given by $\Gamma' = (G'_1, G'_2, \ldots, G'_n)$. To see why note that the graph G'_i obtained by G_i is the same for all appearances of G_i in the hierarchical tree $HT(\Gamma)$ of G. Moreover, the graph G'_i has the same pins and calls the same terminals as the initial G_i.

Lemma 2. $\lambda(G') \leq 4$ *if and only if* $\chi(G) \leq 3$. □

The lemma is proved using the same arguments as those in the proof of Lemma 1 and the observation that the expansion of the graph G' obtained, $E(\Gamma')$, is the same graph as the graph obtained by the construction of Theorem 3 when applied on the expansion of the initial hierarchical graph G.

Finally, it can be easily proved (see full version [2]) that the resulting graph is a WS fully planar hierarchical graph. □

4 Approximations to RCP for WS Fully Planar Graphs

In this section we present two approximation algorithms for the min order RCP on WS fully planar hierarchical graphs: a simple and fast algorithm, that achieves an approximation ratio of 4, and a more sophisticated one which, being still polynomial, achieves a 3-approximation ratio. These algorithms offer alternative options that trade-off the efficiency of the algorithm and the quality of the solution achieved. Both algorithms utilize a bottom up methodology of radiocoloring an L-specified hierarchical planar graph G, given by $\Gamma = (G_1, \ldots, G_n)$. Actually, they compute at most $n-i$ radiocolorings for a subgraph of each graph G_i, $1 \leq i \leq n$, and use these radiocolorings for all copies of G_i in the expansion of G, $E(\Gamma)$. This enables them to run in time only polynomial to the size of the L-specification of G.

More analytically, we wish to compute only one radiocoloring assignment for each G_i, and use this in all appearances of G_i in the expansion of $G, E(\Gamma)$. However, due to the structure of L-specified hierarchical graphs, the distance two neighborhood of the outer vertices of each G_i, may differentiate for every call of G_i by other graphs G_j. Henceforth, a radiocoloring for such a vertex (the outer ones) may becomes invalid due to a change of the distance two neighborhood of the vertex. Since each graph G_i may be called by at most $n-i$ other graphs, we need to compute at most $n-i$ radiocolorings of those (outer) vertices.

Moreover, we need to guarantee that the different radiocolorings of the outer part of G_i do not introduce any implication in the radiocoloring of its inner part. By having only one radiocoloring for the inner part of G_i, we manage to have also no implications to the radiocoloring of the subtree of G_i, $HT(G_i)$.

Based on this design approach, both algorithms partition appropriately, each G_i into three parts: (1) *the inner part*, (2) *the outer up* and (2) *the outer down part*. Remark, that these subgraphs might be different from the inner, outer up, outer down subgraphs of G_i defined in the Definitions 6, 7 and 8.

Then, the algorithms radiocolor the inner part of G_i only once, using, each of them, a different method. Both of them, they group and radiocolor the outer down part of it together with the outer up parts of the graphs called by it, using a known 2-approximation algorithm.

4.1 A 4-Approximation Algorithm HRC_1

We first provide a simple and efficient algorithm (HRC_1) that achieves a 4-approximation for RCP on WS fully planar hierarchical graphs. Let A_1, B_1 two disjoint sets of colors of size $2\Delta(G) + 25$ each, where $\Delta(G)$ is the maximum degree of G.

Overview of the HRC_1 Algorithm: First, the algorithm defines for each G_i its inner, outer up and outer down parts to be the inner, outer up and outer down subgraphs $G_{in\ i}$, $G_{outUp\ i}$ and $G_{outDown\ i}$, respectively.

Then, it radiocolors the inner part of the graph G_i using a known 2-approximation algorithm e.g. [6,1] using the color set A_1. Also, it radiocolors the outer down part of the graph G_i together with the outer up parts of its children using the 2-approximation algorithm with the color set B_1.

Theorem 6. *Algorithm $HRC_1(G)$ produces a radiocoloring of a WS fully planar hierarchical graph G in time $O(n^5)$ and achieves a 4-approximation ratio.*

For, a detailed description of the HRC_1 algorithm and the proof of Theorem 6, see the full version of the paper [2]. □

4.2 A 3-Approximation Algorithm HRC_2

Overview of the Algorithm: We provide a more sophisticated radiocoloring algorithm, that achieves a 3-approximation ratio rather than 4, for fully planar hierarchical graphs of class WS. The basic idea of the algorithm, called HRC_2, is to partition the vertices of each graph G_i into outerplanar *levels* using a BFS (similar to [3,8]) and define the three parts of each G_i based on this search.

The *outer up part of* G_i consists of the first level of the BFS tree obtained. Thus, it is the outer up subgraph of G_i, $G_{outUp\,i}$. The *outer down part of* G_i consists of the graph induced by the vertices of the BFS tree of levels D up to the end of the tree, where D is the first level of the tree having an outer down vertex of G_i. The *inner part of* G_i *is the rest of the graph* G_i.

More analytically, the *inner part of* G_i *is radiocolored* as follows: Radiocolor every two successive levels of the inner part of G_i optimally interchanging color sets A, B, where $|A| = |B| = OPT(G)$ and $OPT(G)$ is the optimal number of colors needed to radiocolor G. This can be achieved in polynomial time and without any conflicts as we prove: We first show that any two successive levels, call them a double level, is a 4-outerplanar graph. Thus, by Theorem 1, it is a bounded treewidth graph. Consequently, applying Theorem 2, each double level can be radiocolored optimally in polynomial time. Moreover, by the BFS partitioning procedure, there is no conflict between double levels colored using the same color set.

The *outer down part of* G_i *is radiocolored* together with the outer up parts of its children using a known (2-approximation) radiocoloring algorithm for ordinary planar graphs using color sets A and C, where $|C| = |A| = OPT(G)$. Since the algorithm uses only color sets A, B, C, it has a 3-approximation algorithm.

Theorem 7. *HRC_2 Algorithm produces a valid radiocoloring of a WS fully planar hierarchical graph G using at most $3\ OPT(G) + 50$ colors, achieving a 3 approximation ratio. It runs in $O(n^2 \cdot T(n,k) + n^5)$ time, where $T(n,k)$ is a polynomial time function for the optimal radiocoloring of a k-tree of size n, specified in [16] ($k = 11$), (see Theorem 2).*

For, a detailed description of the HRC_2 algorithm and the proof of Theorem 7, see the full version of the paper [2].

References

1. Geir Agnarsson, Magnus M.Halldórsson: Coloring Powers of Planar Graphs. ACM Symposium on Discrete Algorithms (SODA) (2000).

2. M.I. Andreou, D.A. Fotakis, S.E. Nikoletseas, V.G. Papadopoulou and P.G. Spirakis: On Radiocoloring Hierarchically Specified Planar Graphs: $\mathcal{PSPACE}$-completeness and Approximations (full version). CTI Technichal Report 2002, URL *http : //students.ceid.upatras.gr/~viki*.
3. B. S. Baker: Approximation algorithms for NP-complete problems on planar graphs. Journal of the ACM 41:153–180, 1994.
4. Bodlaender, H.L.: Planar graphs with bounded treewidth. TR RUU-CS-88-14, Department of Computer Science, University of Utrecht, The Netherlands, March, 1988.
5. Bodlaender, H.L., T. Kloks, R.B. Tan and J. van Leeuwen: Approximations for λ-coloring of graphs. In Proc. 17th Annual Symp. on Theoretical Aspects of Computer Science (STACS). LNCS Vol. 1770, pp. 395-406, 2000.
6. D.A. Fotakis, S.E. Nikoletseas, V.G. Papadopoulou and P.G. Spirakis: $\mathcal{NP}$-completeness Results and Efficient Approximations for Radiocoloring in Planar Graphs. In Proceedings of the 25th International Symposium on Mathematical Foundations of Computer Science (MFCS), Editors Mogens Nielsen, Branislav Rovan, LNCS 1893, pp 363-372, 2000.
7. M. R. Garey, D. S. Johnson: "Computers and Intractability: A guide to the Theory of NP-completeness", W. H./ Freeman and Company, 1979.
8. Krumke, M.V. Marathe and S. S. Ravi: Approximation algorithms for channel assignment in radio networks. In DIALM for Mobility, 2nd International Workshop on Discrete Algorithms and methods for Mobile Computing and Communications, Dallas, Texas, 1998.
9. T. Lengauer: Hierarchical Planarity Testing. Journal of ACM, Vol 36, No 3, pp. 474-509, (1989).
10. T. Lengauer and K.W. Wagner: Correlation between the Complexities of the Of hierarchical and Hierarchical Versions of graph Problems. Journal of Computer and System Sciences, 44, pp. 63-93 (1992).
11. M.V. Marathe, H.B. Hunt III, R.E. Stearns and V. Radhakrishnan: Approximation Algorithms for PSPACE-Hard Hierarchically and Periodically Specified Problems. Proc. 26th Annual ACM Symposium on the Theory of Computing (STOC), pp. 468-478, May 1994. A complete version appears in SIAM Journal on Computing , Vol. 27, No 5, Oct. 1998, pp. 1237–1261.
12. H. Marathe, H. Hunt III, R. Stearns and V. Radhakrishnan: "Complexity of hierachically and 1-dimensioned periodically specified problems". DIMACS Workshop on Satisfiability Problem: Theory and Applications (1996).
13. M.V. Marathe, V. Radhakrishnan, H.B. Hunt III and S.S. Ravi: Hierarchically Specified Unit Disk Graphs. Proc. 19th International Workshop on Graph-Theoretic Concepts in Computer Science (WG) , Utrecht, Netherlands, LNCS 790, pp. 21-32, Springer Verlag, (June 1993). Journal version appears in Theoretical Computer Science , 174(1-2), pp. 23-65, (March 1997).
14. S. T. McCormick: Optimal approximation of sparse hessians and its equivalence to a graph coloring problem. Technical Report SOL 81-22, Dept. of Operations Research, Standford University, 1981.
15. S. Ramanathan, E. R. Loyd: The complexity of distance2-coloring. 4th International Conference of Computing and information, (1992) 71-74.
16. X. Zhou, Y. Kanari. T. Nishizeki: Generalized vertex-coloring of partial k-trees. IEICE Trans. Foundamentals, EXX-A(1), 2000.

Finite Domain Constraint Satisfaction Using Quantum Computation

Ola Angelsmark*, Vilhelm Dahllöf**, and Peter Jonsson***

Department of Computer and Information Science,
Linköping University, SE-581 83 Linköping, Sweden,
Fax: +46/(0)13/284499,
{olaan,vilda,petej}@ida.liu.se

Abstract. We present a quantum algorithm for finite domain constraint solving, where the constraints have arity 2. It is complete and runs in $O((\lceil d/2 \rceil)^{n/2})$ time, where d is size of the domain of the variables and n the number of variables. For the case of $d = 3$ we provide a method to obtain an upper time bound of $O(8^{n/8}) \approx O(1.2968^n)$. Also for $d = 5$ the upper bound has been improved. Using this method in a slightly different way we can decide 3-colourability in $O(1.2185^n)$ time.

1 Introduction

Several algorithms for quantum computers have been presented recently, among them Shor's celebrated algorithm for finding prime factors of composite integers [12]. Another interesting result is the search algorithm presented by Grover [9] and improved by Boyer *et al.* [4]. Its input consists of a polynomial-time computable oracle function $f : \{0, \ldots, N-1\} \rightarrow \{0, 1\}$ and the algorithm decides whether there exists an x such that $f(x) = 1$ by only performing $O(\sqrt{N})$ evaluations of f. Obviously, any deterministic algorithm would need $O(N)$ evaluations of f to solve this problem. The algorithm has a constant probability of error, but it can easily be reduced to an arbitrarily low level by repeating the algorithm. The idea to use it for solving NP-complete problems has been proposed before – already Grover [9] discusses its possible use for solving propositional satisfiability (SAT) problems and comes up with an $O(2^{n/2})$ time algorithm. Another more elaborate example on Hamiltonian paths in graphs is given by Nielsen and Chuang [11]. We will focus on the well-known NP-complete *constraint satisfaction problem* (CSP) where we are given a set of variables that take their values from a finite domain and a set of constraints (*e.g.* relations) that constrain the values different variables can take – the question is whether the variables can

* The research is supported in part by CUGS – National Graduate School in Computer Science, Sweden.

** The research is supported by CUGS – National Graduate School in Computer Science, Sweden.

*** Partially supported by the *Swedish Research Council* (VR) under grant 221-2000-361.

K. Diks et al. (Eds): MFSC 2002, LNCS 2420, pp. 93–103, 2002.

be assigned values such that all constraints are satisfied. By (d,l)-CSP, we mean the CSP problem restricted to domains of size d and constraint arities less than or equal to l. We will exclusively consider *binary* problems where $l = 2$.

Now, assume that A is an $O(c^n)$-time ($c > 1$) deterministic algorithm for some NP-complete problem L. It is obvious that A cannot immediately be combined with Grover's algorithm in order to obtain a $O(c^{n/2})$ quantum algorithm for L; after all, Grover's algorithm can only decide whether a given polynomial-time computable function $f : \{0,\ldots,N-1\} \to \{0,1\}$ returns the value 1 for some input. In fact, it is typically as hard to construct a $O(c^{n/2})$ time quantum algorithm as constructing a $O(c^n)$ time deterministic algorithm for a given problem. Our basic idea is that given a problem instance, we construct a function f where N is 'small' and there exists an x such that $f(x) = 1$ if and only if the given instance is a member of the problem L. Consider, for example, a SAT problem over n variables $x_1,\ldots,x_n$. A straightforward proof of satisfiability is a string containing n bits where the ith bit corresponds to the value of variable x_i in some satisfying assignment. As was noted above, we can find such a string (if it exists) in $O(2^{n/2}) \approx O(1.4142^n)$ time on a quantum computer. If there was a shorter proof, say containing only $0.9n$ bits, we would immediately get a considerably faster $O(2^{0.45n}) \approx O(1.3660^n)$ time algorithm. Our algorithm for $(d,2)$-CSP (Section 3) is a basic application of this 'short proof' approach where we use $\log_2(\lceil d/2 \rceil)$ bits for proving that the given instance is satisfiable. Our algorithm for 3-colourability (Section 5) elaborates upon this idea by adapting the proof system to the given instance; we prove that approximately $0.5702n$ bits are enough in this case.

In this paper, we show that $(d,2)$-CSP can be solved in $O(\lceil d/2 \rceil^{n/2})$ time using quantum computation, where n is the number of variables in the instance. Furthermore, we give improved upper bounds for the cases $d = 3$ and $d = 5$ ($O(8^{n/8})$ and $O(8^{n/4})$, respectively) together with an algorithm for 3-colourability running in $O(1.2185^n)$ time. The best known deterministic algorithm for the 3-colourability problem [3] runs in $O(1.3289^n)$ so our algorithm has a considerably lower time complexity. Our results on $(d,2)$-CSP are shown in Table 1 and compared with those found by others. Feder and Motwani [7] have a probabilistic algorithm, which runs in $O((d!)^{n/d})$ time. The algorithms given by Eppstein [6] runs in $O(1.3645^n)$ time for $d = 3$, and $O((0.4518d)^n)$ time for $d > 3$ (this algorithm is deterministic for $d \leq 4$ and probabilistic for $d > 4$). Comparing our results with those of Eppstein and Feder & Motwani, it seems that our results are close to the square-root of the old results as d grows. These algorithms are, of course, based on an entirely different computational model.

Cerf *et al.* [5] present a quantum algorithm for CSP and it runs in $O(d^{0.2080n})$ time for $(d,2)$-CSP. Note, however, that this time complexity is not an upper bound for the worst case. Their analysis is made in the region where the number of constraints per variable is close to $d^l \cdot \log d$, *i.e.*, where the problem exhibits a *phase transition*. In Gent & Walsh [8], it is shown that many NP-complete problems have a different region where the problems are under-constrained and satisfiable but they can be orders of magnitude harder than those in the middle

Table 1. Upper time bound for various algorithms for binary CSPs

	Feder & Motwani	Eppstein	Cerf *et al.*	New results
(3, 2)-CSP	$O(1.8171^n)$	$O(1.3645^n)$	$O(1.2567^n)$	$O(1.2968^n)$
(4, 2)-CSP	$O(2.2134^n)$	$O(1.8072^n)$	$O(1.3342^n)$	$O(1.4142^n)$
(5, 2)-CSP	$O(2.6052^n)$	$O(2.2590^n)$	$O(1.3976^n)$	$O(1.6818^n)$
(6, 2)-CSP	$O(2.9938^n)$	$O(2.7108^n)$	$O(1.4516^n)$	$O(1.7321^n)$
(7, 2)-CSP	$O(3.3800^n)$	$O(3.1626^n)$	$O(1.4989^n)$	$O(2^n)$
(8, 2)-CSP	$O(3.7644^n)$	$O(3.6144^n)$	$O(1.5411^n)$	$O(2^n)$
(9, 2)-CSP	$O(4.1472^n)$	$O(4.0662^n)$	$O(1.5794^n)$	$O(2.2361^n)$
(10, 2)-CSP	$O(4.5287^n)$	$O(4.5180^n)$	$O(1.6144^n)$	$O(2.2361^n)$
(11, 2)-CSP	$O(4.9092^n)$	$O(4.9698^n)$	$O(1.6467^n)$	$O(2.4495^n)$
(12, 2)-CSP	$O(5.2889^n)$	$O(5.4216^n)$	$O(1.6768^n)$	$O(2.4495^n)$
(15, 2)-CSP	$O(6.4234^n)$	$O(6.7770^n)$	$O(1.7564^n)$	$O(2.8284^n)$
(20, 2)-CSP	$O(8.3044^n)$	$O(9.0360^n)$	$O(1.8647^n)$	$O(3.1623^n)$

of the phase transition. Also note that the very existence of phase transitions has been questioned, cf. [1]. Thus the complexity given in Cerf *et al.* cannot be directly compared with ours.

The structure of the paper is as follows: Section 2 contains some basic information on constraint satisfaction and quantum computation needed for the algorithm, which is presented and described in Section 3. Here we also find the main theorem of the paper, as well as a conjecture on the time complexity of the problem. In Section 4 we study two special cases of $(d, 2)$-CSP, namely domain sizes 3 and 5, in order to corroborate our conjecture and to lay the foundations for the algorithm for deciding 3-colourability presented in Section 5.

2 Preliminaries

A *binary constraint satisfaction problem ((d,2)-*CSP*)* is a triple $\langle V, D, C\rangle$ with V a finite set of variables, D a finite domain of values, with $|D| = d$, and C a set of constraints $\{c_1, c_2, \ldots, c_q\}$. Each constraint $c_i \in C$ is a triple xRy where $x, y \in V$ and $R \subseteq D \times D$. A *solution* to a CSP instance, is a function $f : V \to D$, such that for each constraint xRy, $(f(x), f(y)) \in R$. Given a CSP instance, the computational problem is to decide whether the instance has a solution or not.

For a good introduction to quantum computation, we recommend [11]. In this paper, however, we only need to point out a few things. Given a polynomial-time computable function $f : \{0, \ldots, N-1\} \to \{0, 1\}$, we can (in polynomial time) construct a polynomial size classical circuit representing the function f. This circuit can (once again, in polynomial time) be converted into a quantum circuit suitable for the quantum search algorithm as described by [9] and [4]. We will denote this search algorithm $QuantFind(A, N)$ where A is a description of the polynomial-time algorithm computing f. $QuantFind(A, N)$ will, with constant probability of error and in $O(\sqrt{N})$ time, return a number m such that $f(m) = 1$ (if such a number exists) and otherwise answer 'no'.

The algorithm presented in the next section relies on propositional logic. A *propositional variable* is a variable which represents the truth or falsehood of a proposition. A *2-SAT formula* is a sentence in conjunctive normal form consisting of the conjunction of a number of clauses, where each clause contains at most two literals. The *2-SAT problem* is to decide whether a given 2-SAT formula is satisfiable or not. The problem of finding a satisfying assignment for 2-SAT formulae is solvable in linear time [2].

3 Algorithm for $(d, 2)$-CSP

The main idea of our algorithm is to construct a polynomial-time computable function $f : \{0, \ldots, (d/2)^n - 1\} \to \{0, 1\}$ which answers 1 for some input iff there is a solution to the given CSP instance. Once f is found we employ it as an oracle for *QuantFind*. Our method builds on grouping the domain values into subsets of size 2 in order to obtain a shorter proof of satisfiability. This approach resembles Eppstein's randomized algorithm for $(d, 2)$-CSP; by dividing the domain into 4-element sets a number of $(4, 2)$-CSP instances are spawned for which he has an efficient algorithm. Since it is unlikely that k-SAT (assuming $P \neq NP$) is polynomial time solvable for $k > 2$, we are limited to subsets of size 2.

Theorem 1. *The $(d, 2)$-CSP can be solved in $O((\lceil d/2 \rceil)^{n/2})$ time, where d is the size of the domain of the variables, and n is the number of variables.*

Proof. Let $\Theta = (V, D, C)$ be a CSP instance such that $d = |D|$ is even; the case when d is odd will be considered below. The problem can be transformed to a number of instances of 2-SAT, $I_0, \ldots, I_{(d/2)^n - 1}$, such that there exists an instance I_k that has a solution iff Θ is satisfiable.

Each I_k contains $|V| \cdot d$ propositional variables x_d where $x \in V$ and $d \in D$. Our intended intepretation of these variables is that x_d is true if and only if x is assigned the value d.

The clauses are of two types: first, we have clauses for representing the constraints and these are the same for all instances I_k. Given a constraint $xRy \in C$, we represent it with a formula C_{xRy} as follows:

$$C_{xRy} = \bigwedge_{a,b \in D,\ (a,b) \notin R} (\neg x_a \vee \neg y_b).$$

For example, given the constraint $x \neq y$ where '$\neq$' $= \{(a, b) \mid a \neq b \wedge a, b \in D\}$, then the corresponding set of clauses prevents x and y from being assigned the same value.

The rest of the clauses will ensure that each variable takes exactly one value, *i.e.* that exactly one of x_a, $a \in D$, is assigned true for every $x \in V$. Assume that we want variable x to take either value a or value b; this can be guaranteed by the clauses

$$(x_a \vee x_b) \wedge (\neg x_a \vee \neg x_b) \wedge \bigwedge_{c \in D,\ c \neq a,\ c \neq b} (\neg x_c).$$

By using this construction, there are $d/2$ alternatives to consider per variable and we can construct $(d/2)^{|V|}$ instances that cover all possible assignments of the variables. If we combine these 'variable restricting' clauses with the clauses representing the constraints, it is obvious that if one of these instances is satisfiable, then Θ is satisfiable and vice-versa. It should be clear that we can enumerate these instances $I_0, \ldots, I_{(d/2)^n-1}$ in such a way that given a number m, we can in polynomial time construct I_m.

Finally, we construct a function $f : \{0, \ldots, (d/2)^n - 1\} \rightarrow \{0, 1\}$, computable in polynomial-time, such that $f(m) = 1$ if and only if I_m is satisfiable. By computing $QuantFind(f, (d/2)^n - 1)$, we can find out whether Θ is satisfiable or not in $O((d/2)^{n/2})$ time. Note that we tacitly ignore all polynomial factors in the analysis of exponential time algorithms.

If d is odd, then we add a dummy element to the domain and the algorithm will run in $O(((d+1)/2)^{n/2}))$ time. Consequently, the worst-case running time of the algorithm is in $O((\lceil d/2 \rceil)^{n/2})$. □

Section 4 shows that there are cases when the upper limit on the time complexity of Q can be improved. It seems counter-intuitive that a domain with odd cardinality should be so much harder to solve than one with even cardinality. This leads us to the following plausible conjecture.

Conjecture 1. There exists a quantum algorithm which solves $(d, 2)$-CSP in time $O((d/2)^{n/2})$.

4 The Special Cases $|D| = 3$ and $|D| = 5$

We will now consider $(3, 2)$- and $(5, 2)$-CSP and construct algorithms faster than those presented in the previous section. Assume that $D = \{1, 2, 3\}$ and let us consider two variables at a time (instead of one at a time as in Section 3). Table 2 depicts this situation, with the variables x and y. It is easily seen that the second, fourth, and ninth columns, with boxed entries, *i.e.*, the clauses $(x_1 \vee x_2) \wedge (y_1 \vee y_3)$, $(x_1 \vee x_3) \wedge (y_1 \vee y_2)$ and $(x_2 \vee x_3) \wedge (y_2 \vee y_3)$, cover all assignments of values to the variables. Thus, by arbitrarily putting the variables in groups of two, we need only consider three different 2-SAT instances instead of four. By using ideas similar to those used in Section 3, this would yield an upper time bound of $O(1.3161^n)$.

However, we can also consider four variables at a time, the result of which is found in Proposition 1. This result will be used in Section 5 for constructing an algorithm for 3-colourability.

Proposition 1. $(3, 2)$-CSP *can be solved in* $O(8^{n/8}) \approx O(1.2968^n)$ *time.*

Table 2. The special case of $|D| = 3$, with $D = \{1, 2, 3\}, V = \{x, y\}$. ($x_i, x_j$ is short for $x_i \vee x_j$)

		x_1, x_2	x_1, x_2	x_1, x_2	x_1, x_3	x_1, x_3	x_1, x_3	x_2, x_3	x_2, x_3	x_2, x_3
x	y	y_1, y_2	y_1, y_3	y_2, y_3	y_1, y_2	y_1, y_3	y_2, y_3	y_1, y_2	y_1, y_3	y_2, y_3
1	1	X	X	-	X	X	-	-	-	-
1	2	X	-	X	X	-	X	-	-	-
1	3	-	X	X	-	X	X	-	-	-
2	1	X	X	-	-	-	-	X	X	-
2	2	X	-	X	-	-	-	X	-	X
2	3	-	X	X	-	-	-	-	X	X
3	1	-	-	-	X	X	-	X	X	-
3	2	-	-	-	X	-	X	X	-	X
3	3	-	-	-	-	X	X	-	X	X

Proof. Assume $D = \{1, 2, 3\}$ and $V = \{p, q, r, s\}$. Analogously to the previous situation with two variables, we observe that we need the following eight cases to cover all possible assignments of values:

1. $(p = 1 \vee p = 2)$, $(q = 1 \vee q = 2)$, $(r = 1 \vee r = 2)$, $(s = 1 \vee s = 2)$
2. $(p = 1 \vee p = 2)$, $(q = 1 \vee q = 2)$, $(r = 1 \vee r = 3)$, $(s = 1 \vee s = 3)$
3. $(p = 1 \vee p = 2)$, $(q = 1 \vee q = 3)$, $(r = 2 \vee r = 3)$, $(s = 2 \vee s = 3)$
4. $(p = 1 \vee p = 3)$, $(q = 1 \vee q = 2)$, $(r = 2 \vee r = 3)$, $(s = 2 \vee s = 3)$
5. $(p = 1 \vee p = 3)$, $(q = 1 \vee q = 3)$, $(r = 1 \vee r = 2)$, $(s = 1 \vee s = 2)$
6. $(p = 1 \vee p = 3)$, $(q = 1 \vee q = 3)$, $(r = 1 \vee r = 3)$, $(s = 1 \vee s = 3)$
7. $(p = 2 \vee p = 3)$, $(q = 2 \vee q = 3)$, $(r = 1 \vee r = 2)$, $(s = 1 \vee s = 3)$
8. $(p = 2 \vee p = 3)$, $(q = 2 \vee q = 3)$, $(r = 1 \vee r = 3)$, $(s = 1 \vee s = 2)$

By performing a computer-assisted enumeration of all possible assignments, we found that at least 8 clauses are needed. Obviously, we can modify the function f (described in the previous section) so that it considers the variables in groups of four using the method described above. Thus we get 8 possibilities per four variables, and the call $QuantFind(f, 8^{n/4})$ will have a time complexity of $O(\sqrt{8^{n/4}})$. □

To further bolster Conjecture 1, we look at the case $|D| = 5$.

Proposition 2. $(5, 2)$-CSP *can be solved in* $O(8^{n/4}) \approx O(1.6818^n)$ *time*

Proof. Assume $D = \{1, 2, 3, 4, 5\}$, and $V = \{p, q\}$. The following eight cases are needed to cover all possible assignments of values, and it cannot be done with fewer:

1. $(p = 1 \vee p = 2)$, $(q = 1 \vee q = 2)$
2. $(p = 1 \vee p = 2)$, $(q = 1 \vee q = 3)$
3. $(p = 1 \vee p = 2)$, $(q = 4 \vee q = 5)$
4. $(p = 1 \vee p = 3)$, $(q = 1 \vee q = 2)$
5. $(p = 3 \vee p = 4)$, $(q = 3 \vee q = 4)$
6. $(p = 3 \vee p = 5)$, $(q = 3 \vee q = 5)$
7. $(p = 4 \vee p = 5)$, $(q = 1 \vee q = 4)$
8. $(p = 4 \vee p = 5)$, $(q = 2 \vee q = 5)$

Analogous to the proof of Proposition 1, we can change function f to reflect this new situation, and with the call $QuantFind(f, \sqrt{8^{n/2}})$, we get a time complexity of $O(8^{n/4}) \approx O(1.6818^n)$. □

In theory, we could examine cases with more variables and/or larger domains with the aid of a computer. However, the search space of the problem grows rapidly and it quickly becomes infeasible to enumerate the instances. For example, the case with 4 variables and domain size 5, gives 625 possible assignments. Each clause would satisfy 16 assignments, so we need at least 40 (possibly more) clauses to cover all possibilities. This implies that we need to consider at least $\binom{625}{40} \approx 2.34 \cdot 10^{63}$ possible choices of clauses, making the enumerative approach very time consuming.

5 Algorithm for 3-Colourability

We will now present an $O(1.2185^n)$ time quantum algorithm for the 3-colourability problem. In order to obtain this time complexity (which is significantly lower than that for (3,2)-CSP), the algorithm begins by analysing the given graph and choose how to construct the oracle function depending on this analysis. Before we present our algorithm we need to give some more preliminaries:

An undirected *graph* G consists of a set $V(G)$ of *vertices*, and a set $E(G)$ of *edges*, where each element of $E(G)$ is an unordered pair of vertices. The *size* of a graph G, denoted $|G|$, is the number of vertices. An *independent set* S, is a set $S \subseteq V(G)$, such that for every pair $v, w \in S$, $\{v, w\} \notin E(G)$. A *3-colouring* of a graph G is a function $C : V(G) \to \{R, G, B\}$, such that, for all pairs $v, w \in V(G)$, if $C(v) = C(w)$, then $\{v, w\} \notin E(G)$; *viz.*, no adjacent vertices have the same colour. If G is a graph, and $S \subseteq V(G)$, the *subgraph of G induced by S* has the vertex set S and edge set $E(G|S) = \{\{u, v\} \in E(G) \mid u, v \in S\}$, and is denoted $G|S$. We write $G - S$ to denote the graph $G|(V - S)$.

A *matching* in a graph, is a set M of vertices such that each vertex $v \in M$ has an edge to one, and only one, other vertex in M. The maximum matching (wrt the number of matched vertices) of a graph is computable in polynomial time [10]. Let *Match*(G) be a function which computes a maximum matching of the graph G and returns a pair (U, M), where $U \subseteq V(G)$ contains the unmatched vertices and $M \subseteq E(G)$ the matched pairs.

We consider an arbitrary graph G with n vertices (where each vertex have an associated colour) and assume, without loss of generality, that it is connected. Our algorithm $3C$ works as follows: First we start by setting $(G_u, G_m) = Match(G)$. The algorithm then considers two cases: (1) $|G_u| < C \cdot |G|$; and (2) $|G_u| \geq C \cdot |G|$ where $C \approx 0.2806$ is the unique real solution to the equation $8^{C/4} \cdot 2^{(1-C)/2} = 3^{(1-C)/2}$ (the choice of C will be clear at the complexity analysis).

In the first case, we construct a polynomial-time computable function $a : \{0, \ldots, 8^{|G_u|/4} \cdot 2^{|G_m|/2} - 1\} \to \{0, 1\}$ such that there exists a value x such that $a(x) = 1$ if and only-if G is 3-colourable. The construction of a is described in Subsection 5.1. The second case is similar; we construct a function

$b : \{0, \ldots, 3^{|G_m|/2}\} \to \{0, 1\}$ (see Subsection 5.2) with properties similar to function a. In both cases, *QuantFind* is employed to check whether the function returns 1 for some input. In Subsection 5.3, we show that the time complexity of the resulting algorithm is $O(1.2185^n)$.

5.1 Case 1: $|G_u| < C \cdot |G|$

Assume one transforms the graph into a (3,2)-CSP instance as described in Section 4 (*i.e.*, the vertices are treated as variables over the domain $\{R, G, B\}$ and an edge (x, y) corresponds to the constraint $x \neq y$). Solving this instance directly using Proposition 1 would lead to an algorithm running in $O(\sqrt{8^{|G|/4}}) = O(1.2968^{|G|})$ time. We improve this bound by treating the matched and unmatched vertices differently.

Consider a pair $x - y$ in G_m. Since x and y are connected by an edge we know that $x \neq y$ over the domain $\{R, G, B\}$. There are six allowed tuples (x, y) : $(R, G), (R, B), (G, R), (G, B), (B, R)$ and (B, G); and they are in fact covered by two cases: (1) $x = R \vee x = G$, $y = G \vee y = B$; and (2) $x = G \vee x = B$, $y = R \vee y = G$.

For the vertices in G_u, we arbitrarily group the vertices in sets of size 4. By Proposition 1, we need eight cases to cover all possibilities of such a set. It is now straightforward to construct a polynomial-time computable function $a : \{0, \ldots, 8^{|G_u|/4} \cdot 2^{|G_m|/2} - 1\} \to \{0, 1\}$ such that $a(m) = 1$ for some m iff G is 3-colourable.

5.2 Case 2: $|G_u| \geq C \cdot |G|$

We will now consider the function b (see Figure 1) that is used in the second case of $3C$. We show that $b(m) = 1$ iff G is 3-colourable.

Similarly to the vertices, we assume that each pair $p \in G_m$ have an associated colour. Let $H = G|G_m$. We define an $R\{G/B\}$ *assignment* of G as an assignment of colours to matched pairs such that any pair $x - y$ from G_m shall be assigned either $R - G/B$, $G/B - R$ or $G/B - G/B$. We say that we *extend* such an assignment to the whole of G iff for each of the vertices $v \in G_u$, we assign $v := R$ if it is not adjacent to any vertex with colour R; else we assign

```
1  algorithm b(m)
2  assign colours to H according to m
3  extend the R{G/B} assignment of H to G
4  V_R := all elements of G having colour R
5  if V_R is not an independent set then return 0
6  V_{G/B} := all elements of G having colour G/B
7  if G|V_{G/B} is not 2-colourable then return 0
8  else return 1
```

Fig. 1. The function b

$v := G/B$. Observe that every vertex will have a colour after an extension since G is connected. We say that an extended R{G/B} assignment for G is *refineable* to a 3-colouring of G iff for each of the vertices v having colour G/B by the extended R{G/B} assignment, we can assign $v := G$ or $v := B$ in such a manner that we obtain a 3-colouring of G. We note that saying that we have an extended R{G/B} assignment for G which is refineable to a 3-colouring of G, is equivalent of saying that the assignment has the following properties:

P1. the vertices with colour R form an independent set;

P2. the induced subgraph of vertices with colour G/B is 2-colourable.

By the following lemma we establish that it suffices to investigate each R{G/B} assignment for H to find a 3-colouring of G.

Lemma 1. *G is 3-colourable iff there is an R{G/B} assignment for H which, after being extended to the whole of G, is refineable to a 3-colouring of G.*

Proof. One direction of the proof is trivial: if we have an R{G/B} assignment for H, extend it to the whole of G, and find that it is refineable to a 3-colouring then clearly G is 3-colourable.

The other direction: Assume $T : V(G) \to \{R, G, B\}$ is a 3-colouring of G. For each $h \in V(H)$, let h have the colour R if $T(h) = R$ and G/B otherwise; obviously, this a R{G/B} assignment of H. Extend this assignment to the whole of G. We continue to show that this R{G/B} assignment satisfies properties **P1** and **P2** and, in other words, that this assignment can be refined to a 3-colouring of G.

P1. Assume $(p, q) \in E(G)$ and both p and q have the colour R. If $p, q \in V(H)$, then $T(p) = T(q)$ which is a contradiction. If $p \in V(H)$ and $q \in G_u$ (or vice versa), then q would have been given the colour G/B in the extension of the initial assignment – a contradiction. Finally, both p and q cannot be members of G_u since G_u is an independent set.

P2. We show that if a vertex $p \in V(G)$ is assigned the colour G/B, then $T(p) \in \{G, B\}$. This proves that the induced subgraph of vertices with colour G/B is 2-colourable since T is a 3-colouring of G. Assume that p is coloured G/B and note that the claim trivially holds if $p \in V(H)$. If $p \in G_u$, then p is adjacent to a node q such that $T(q) = R$ which implies that $T(p) \in \{G, B\}$. □

Now, the idea behind function b is to carry out the tests described in Lemma 1 given a number that represents the R{G/B} assignment of the vertices of H. An enumeration of all assignments is trivial to construct and we have thus shown that there exists an m such that $b(m) = 1$ if and only if G is 3-colourable.

5.3 Complexity Analysis

We are now ready to analyse the complexity of algorithm $3C$ when applied to a graph containing n vertices. We will show that both Case 1 and 2 takes $O(1.2185^n)$ time which proves that $3C$ runs in $O(1.2185^n)$ time.

Case 1. When $|G_u| < C \cdot |G|$, the maximum number of assignment appears when $|G_u|$ is as large as possible, *i.e.*, $|G_u|$ is close to $C \cdot n$ (since $8^{C/4} \cdot 2^{(1-C)/2}$ is a strictly increasing function for $C > 0$). Thus, $8^{(C \cdot n)/4} \cdot 2^{(1-C)n/2} = 1.4847^n$ is an upper bound to the number of assignments, and the algorithm will run in $O(\sqrt{1.4847^n}) = O(1.2185^n)$ time.

Case 2. Given a graph of size $n = |G_u| + 2 \cdot |G_m|$, the worst case will arise when G_u is as small as possible, *i.e.*$|G_u| = C \cdot n$, since the number of possible R{G/B} assignments equals $3^{|G_m|} = 3^{(n-|G_u|)/2}$. So, the number of possible R{G/B} assignments is at most $3^{(1-C)n/2} \approx 1.4847^n$, and our algorithm runs in $O(1.2185^n)$ time in this case as well.

Acknowledgments

The authors wish to thank Johan Thapper for his useful comments.

References

1. D. Achlioptas, L. M. Kirousis, E. Kranakis, D. Krizanc, M. Molloy, and Y. Stamatiou. Random constraint satisfaction: A more accurate picture. *Constraints*, 6(4):329–344, 2001.
2. B. Aspvall, M. F. Plass, and R. E. Tarjan. A linear time algorithm for testing the truth of certain quantified Boolean formulas. *Inf. Process. Lett.*, 8:121–123, 1979.
3. R. Beigel and D. Eppstein. 3-coloring in time $O(1.3289^n)$. ACM Computing Research Repository, June 2000.
4. M. Boyer, G. Brassard, P. Høyer, and A. Tapp. Tight bounds on quantum searching. In *Proc. 4th Workshop on Physics and Computation*, pages 36–43, 1996.
5. N. J. Cerf, L. K. Grover, and C. P. Williams. Nested quantum search and NP-hard problems. *Applicable Algebra in Engineering, Communication and Computing*, 10(4/5):311–338, 2000.
6. D. Eppstein. Improved algorithms for 3-coloring, 3-edge-coloring, and constraint satisfaction. In *Proc. 12th Annual Symposium on Discrete Algorithms (SODA)*, pages 329–337, 2001.
7. T. Feder and R. Motwani. Worst-case time bounds for coloring and satisfiability problems. Unpublished manuscript.
8. I. P. Gent and T. Walsh. The satisfiability constraint gap. *Artificial Intelligence*, 81(1–2):59–80, 1996.
9. L. K. Grover. A fast quantum mechanical algorithm for database search. In *Proc. 28th Annual ACM Symposium on the Theory of Computing (STOC)*, pages 212–219, 1996.

10. S. Micali and V. V. Vazirani. An $O(\sqrt{|v|} \cdot |e|)$ algorithm for finding maximum matching in general graphs. In *21st Annual Symposium on Foundations of Computer Science (FOCS)*, pages 10–16, IEEE Computer Society, 1980.
11. M. A. Nielsen and I. L. Chuang. *Quantum Computation and Quantum Information*. Cambridge University Press, 2000.
12. P. W. Shor. Polynomial-time algorithms for prime factorization and discrete logarithms on a quantum computer. *SIAM Journal on Computing*, 26(5):1484–1509, 1997.

Fast Algorithms with Algebraic Monge Properties*

Wolfgang W. Bein[1,**], Peter Brucker[2],
Lawrence L. Larmore[1,*], and James K. Park[3***]

[1] Department of Computer Science, University of Nevada,
Las Vegas, Nevada 89154, USA,
bein@cs.unlv.edu larmore@cs.unlv.edu

[2] Universität Osnabrück, Fachbereich Mathematik/Informatik
D-49069 Osnabrück, Germany,
peter@mathematik.uni-osnabrueck.de

[3] Bremer Associates, Inc.,
215 First Street, Cambridge, Massachusetts 02142, USA,
james.park@bremer-inc.com

Abstract. When restricted to cost arrays possessing the sum Monge property, many combinatorial optimization problems with sum objective functions become significantly easier to solve. The more general algebraic assignment and transportation problems are similarly easier to solve given cost arrays possessing the corresponding algebraic Monge property. We show that Monge-array results for two sum-of-edge-costs shortest-path problems can likewise be extended to a general algebraic setting, provided the problems' ordered commutative semigroup satisfies one additional restriction. In addition to this general result, we also show how our algorithms can be modified to solve certain bottleneck shortest-path problems, even though the ordered commutative semigroup naturally associated with bottleneck problems does not satisfy our additional restriction. We show how our bottleneck shortest-path techniques can be used to obtain fast algorithms for a variant of Hirschberg and Larmore's optimal paragraph formation problem, and a special case of the bottleneck traveling-salesman problem.

1 Introduction

In an algebraic combinatorial optimization problem, we are given a collection $\mathcal{S}$ of subsets of a finite nonempty set E as well as a cost function $\phi : E \to H$, where $(H, *, \preceq)$ is an ordered commutative semigroup. It is further assumed that the internal composition $*$ is *compatible* with the order relation $\preceq$, i.e. for

* Conference Version.
** Research of these authors supported by NSF grant CCR-9821009.
*** This author's work was done while the author was at Sandia National Laboratories, and supported by the U.S. Department of Energy under Contract DE-AC04-76DP00789.

K. Diks et al. (Eds): MFSC 2002, LNCS 2420, pp. 104–117, 2002.

all $a, b, c \in H$, $a \prec b$ implies $c * a \preceq c * b$. Then an algebraic combinatorial problem is given by

$$\min_{S \in \mathcal{S}} \bigotimes_{e \in S} \phi(e)$$

with $\bigotimes_{e \in S} := \phi(e_1) * \phi(e_2) * \cdots * \phi(e_k)$ for $S = \{e_1, e_2, \ldots, e_k\} \subset E$. When the operation "$*$" is the "$+$" operation we have a regular sum objective. Often objectives more general than sum objectives accomodate practical optimization problems more easily. Of particular interest are bottleneck objectives, where the operation "$*$" is replaced by the "max" operation. Many combinatorial optimization problems with sum objectives have efficient algorithms for algebraic objective functions (see Burkard and Zimmermann [7] for a survey of classical results, as well as Burkard [6,8] and Seiffart [16]).

These generalizations are especially elegant for shortest path problems and assignment problems. In this case the set E is a subset of $U \times V$, where $U = \{1, \ldots, m\}$ and $V = \{1, \ldots, n\}$ for some $m, n \in N$. The cost function ϕ is then expressed in terms of a cost array $A = \{a[i, j]\}$ with $\phi(e) = a[i, j]$ if $e = (i, j)$. We say that an $m \times n$ array $A = \{a[i, j]\}$ possesses the *algebraic Monge property* if for all $1 \leq i < k \leq m$ and $1 \leq j < \ell \leq n$, $a[i, j] * a[k, \ell] \preceq a[i, \ell] * a[k, j]$. If operation "$*$" is the "$+$" operation we just say that array A has the *Monge property*, if the operation "$*$" is replaced by the "max" operation we say that array A has the *bottleneck Monge property.*

One of the earliest results concerning optimization problems with (sum) Monge arrays goes back to Hoffman [13]. He showed the now classical result that the transportation problem is solved by a greedy $O(n)$ algorithm if the underlying cost array is a Monge array and $m = n$.

For path problems, consider the complete directed acyclic graph $G = (V, E)$, i.e. G has vertices $V = \{1, \ldots, n\}$ and edges $(i, j) \in E$ iff $i < j$. The *unrestricted shortest-path problem* is the problem of finding the shortest path from vertex 1 to vertex n whereas the *k-edge shortest-path problem* is the problem of finding such a path that has exactly k edges. Larmore and Schieber [15] have shown that, for (sum) Monge distances, the unrestricted shortest path problem on a complete directed acyclic graph can be solved in $O(n)$, whereas the k-shortest path problem can be solved in $O(kn)$. Aggarwal, Schieber, and Tokuyama [3] present an alternate $O(n\sqrt{k \lg n})$ algorithm using parametric search.

In this paper we will derive such results for two path problems with algebraic cost arrays.

Consider the complete directed acyclic graph G as above. Associated with the edges are costs $a[i, j]$, which are drawn from the ordered commutative semigroup $(H, *, \preceq)$. An essential requirement of our algorithms will be that the internal composition $*$ be strictly compatible with the order relation $\preceq$ (the operation $*$ is *strictly compatible* with the order relation $\preceq$ if for all $a, b, c \in H$, $a \prec b$ implies $c * a \prec c * b$).

As mentioned earlier, an important case for practical applications of algebraic objective functions is bottleneck objective functions. Gabow and Tarjan [10] give results concerning bottleneck shortest path problems. In [5], Burkard and Sandholzer identify several families of cost arrays in which the bottleneck traveling-salesman problem can be solved in polynomial time; their results include bottleneck Monge arrays as an important special case. Klinz, Rudolf, and Woeginger [14] have developed an algorithm to recognize bottleneck Monge matrices in linear time. We derive here an $O(n)$ algorithms for the unrestricted shortest path bottleneck problem, as well as an $O(kn)$ algorithm and an alternative $O(n^{3/2} \lg^{5/2} n)$ for the k-shortest path bottleneck problems.

Consider the ordered commutative subgroup $(\Re, \max, \leq)$ naturally associated with bottleneck combinatorial optimization problems. Note that the composition max is compatible with the order relation $\leq$ but not strictly compatible with it. (For example, $5 < 7$ but $\max\{8,5\} \not< \max\{8,7\}$.) For an example of an ordered commutative semigroup $(H, *, \preceq)$ whose internal composition $*$ is strictly compatible with its order relation $\preceq$, consider the set T of ordered tuples $(r_1, r_2, \ldots, r_n)$ such that $n \geq 0$, $r_i \in \Re$ for $1 \leq i \leq n$, and $r_1 \geq r_2 \geq \cdots \geq r_n$. Furthermore, suppose we define "$\oplus$" so that

$$(q_1, q_2, \ldots, q_m) \oplus (r_1, r_2, \ldots, r_n) = (s_1, s_2, \ldots, s_{m+n}) ,$$

where $s_1, s_2, \ldots, s_{m+n}$ is the sorted sequence obtained by merging the sequences $(q_1, q_2, \ldots, q_m)$ and $(r_1, r_2, \ldots, r_n)$, and we define $\prec$ so that $(q_1, q_2, \ldots, q_m) \prec (r_1, r_2, \ldots, r_n)$ if and only if there exists an i in the range $1 \leq i \leq m$ such that $q_i < r_i$ and $q_j = r_j$ for $1 \leq j < i$ *or* $m < n$ and $q_j = r_j$ for $1 \leq j \leq m$. It is not hard to see that $(T, \oplus, \preceq)$ is an ordered commutative semigroup and $\oplus$ is strictly compatible with $\preceq$. As we will see later this semigroup can be used to model strict bottleneck Monge conditions.

In Section 2 we derive the general algorithm for algebraic shortest path problems with the Monge property. The results follows fairly directly from the fact that if matrix A possesses the algebraic Monge property, then A is totally monotone. Section 3 develops the algorithm for the bottleneck case. As discussed in the previous paragraph, the results for the bottleneck case is more intricate. Section 4 contains an alternate algorithm that in some sense generalizes Aggarwal, Schieber, and Tokuyama's [3] algorithm. In Section 5 we apply our results to a variant of Hirschberg and Larmore's optimal-paragraph-formation problem [11] and in Section 6 we obtain a fast algorithms for a special case of the bottleneck traveling-salesman problem.

2 Algorithms for Algebraic Shortest-Path Problems

We will now show that both the unrestricted and the k-edge variants of the algebraic shortest-path problem for an ordered commutative semigroup $(H, *, \preceq)$ are significantly easier to solve given edge costs with the algebraic Monge property, provided the internal composition $*$ is strictly compatible. An $m \times n$ array $A = \{a[i,j]\}$ is *totally monotone* if for all $1 \leq i < k \leq m$ and $1 \leq j < \ell \leq n$, either $a[i,j] \preceq a[i,\ell]$ or $a[k,j] \succ a[k,\ell]$. (We say that A is *transpose totally monotone* if the transpose of A is totally monotone; *i.e.*, if the same condition holds with the role of rows and columns reversed.) Strict compatibility is necessary to insure that every array possessing the algebraic Monge property is also totally monotone, the crucial property exploited by our algorithms. The following lemma makes this last claim precise.

Lemma 2.1. *Let* $(H, *, \preceq)$ *denote an ordered commutative semigroup whose internal composition* $*$ *is* strictly *compatible with its order relation* $\preceq$*, and let* $A = \{a[i,j]\}$ *denote an array whose entries are drawn from* $(H, *, \preceq)$*. If* A *possesses the algebraic Monge property, then* A *is totally monotone and also transpose totally monotone.*

Proof. By contradiction. Suppose that $(H, *, \preceq)$ is an ordered commutative semigroup whose internal composition $*$ is strictly compatible with its order relation $\preceq$, $A = \{a[i,j]\}$ is an algebraic Monge array whose entries are drawn from $(H, *, \preceq)$, and A is not totally monotone. Then for some i, j, k, and ℓ satisfying $1 \leq i < k \leq m$ and $1 \leq j < \ell \leq n$, $a[i,j] \succ a[i,\ell]$ *and* $a[k,j] \preceq a[k,\ell]$. Because the composition $*$ is strictly compatible with $\preceq$, we have that $a[i,j] * a[k,\ell] \succ a[i,\ell] * a[k,\ell]$ and $a[i,\ell] * a[k,\ell] \succeq a[i,\ell] * a[k,j]$. By transitivity of the order relation, we have $a[i,j] * a[k,\ell] \succ a[i,\ell] * a[k,j]$, which contradicts the Monge property of A. The proof that A is transpose totally monotone is similar, as the algebraic Monge property is invariant under tansposition.

Note that if the semigroup's composition $*$ is compatible with its order relation $\preceq$ but not strictly compatible with it, then an array whose entries are drawn from the semigroup may possess the algebraic Monge property without being totally monotone. For example, consider again the ordered commutative subgroup associated with bottleneck combinatorial optimization problems. The array

$$\begin{bmatrix} 1 & 0 \\ 1 & 1 \end{bmatrix}$$

satisfies the inequality $\max\{a[i,j], a[k,\ell]\} \leq \max\{a[i,\ell], a[k,j]\}$ for all $i < k$ and $j < \ell$, but it is not totally monotone.

The total monotonicity of arrays possessing the algebraic Monge property allows us to locate these arrays' smallest entries using the array-searching algorithms of Aggarwal et al. [1] (called the SMAWK algorithm) and Larmore and Schieber [15]. However, before we can obtain the desired shortest-path algorithms, we need one more lemma. (Note that this lemma does not require the strict-compatibility assumption.)

Lemma 2.2. *Let* $(H, *, \preceq)$ *denote an ordered commutative semigroup whose internal composition* $*$ *is compatible with its order relation* $\preceq$*, and let* $C = \{c[i,j]\}$ *denote an array whose entries are drawn from* $(H, *, \leq)$*. Furthermore, let* $B = \{b[i]\}$ *denote any vector, and let* $A = \{a[i,j]\}$ *denote the array given by* $a[i,j] = b[i] * c[i,j]$*. If* C *possesses the algebraic Monge property, then so does* A*.*

Proof. If C is algebraic Monge, then for all $i < k$ and $j < \ell$, $c[i,j] * c[k,\ell] \preceq c[i,\ell] * c[k,j]$. Since the order relation is compatible, and the composition $*$ is commutative, we have

$$c[i,j] * c[k,\ell] * b[i] * b[k] \preceq c[i,\ell] * c[k,j] * b[i] * b[k]$$
$$b[i] * c[i,j] * b[k] * c[k,\ell] \preceq b[i] * c[i,\ell] * b[k] * c[k,j]$$
$$a[i,j] * a[k,\ell] \preceq a[i,\ell] * a[k,j] .$$

Theorem 2.3. *Let $(H, *, \preceq)$ denote an ordered commutative semigroup whose internal composition $*$ is strictly compatible with its order relation $\preceq$, and let G denote a complete directed acyclic graph on vertices $1, \ldots, n$ whose edge costs $C = \{c[i,j]\}$ are drawn from H. If C possesses the algebraic Monge property, then the algebraic k-edge shortest-path problem for G can be solved in $O((t_a + t_c)kn)$ time, where t_a is the worst-case time required for performing a composition $*$ and t_c is the worst-case time required for comparing two elements of H.*

Proof. Let $1 \hookrightarrow j$ denote a path from vertex 1 to vertex j. Define $a^\ell[i,j]$ to be the length of the shortest ℓ-edge $1 \hookrightarrow j$ path that contains the edge (i,j), and $d^\ell[i]$ to be the length of the shortest ℓ-edge $1 \hookrightarrow i$ path. Then

$$a^\ell[i,j] = \begin{cases} d^{\ell-1}[i] * c[i,j] & \text{if } i < j \\ \infty & \text{otherwise} \end{cases} \quad \text{and} \quad d^\ell[j] = \begin{cases} \min_{1 \le i < j} a^\ell[i,j] & \text{if } \ell > 1 \\ c[i,j] & \text{if } \ell = 1 \end{cases}$$

where the min operation is performed over the order relation $\preceq$. By Lemma 2.2, for $2 \le \ell \le k$, the array $A^\ell = \{a^\ell[i,j]\}$ is algebraic Monge.

Our algorithm contains k phases. In phase ℓ, we use $d^{\ell-1}[1]$ through $d^{\ell-1}[n]$ to compute $d^\ell[1]$ through $d^\ell[n]$. Note that $d^\ell[1]$ through $d^\ell[n]$ are simply the minima of column 1 through column n of array A^ℓ. In phase ℓ, any entry $a^\ell[i,j]$ of array A^ℓ can be computed in t_a time, since $d^{\ell-1}[i]$ is already known. Thus, using the algorithm of Aggarwal et al. [1], the column minima of A^ℓ can be found in $O((t_a + t_c)n)$ time. Since our algorithm has k phases, the total running time is $O((t_a + t_c)kn)$.

Theorem 2.4. *Let $(H, *, \preceq)$ denote an ordered commutative semigroup whose internal composition $*$ is strictly compatible with its order relation $\le$, and let G denote a complete directed acyclic graph on vertices $1, \ldots, n$ whose edge costs are drawn from H. If G's edge costs possess the algebraic Monge property, then the algebraic unrestricted shortest-path problem for G can be solved in $O((t_a + t_c)n)$ time, where t_a is the worst-case time required for computing $d[i] * c[i,j]$ and t_c is the worst-case time required for comparing two entries of A.*

Proof. This proof is very similar to the proof of Theorem 2.3 and is given in the journal version.

3 Algorithms for the Bottleneck Shortest-Path Problems

In this section, we show how our two algebraic shortest-path algorithms can be modified to handle the bottleneck shortest-path problems. As mentioned in

Section 1, the ordered commutative subgroup $(\Re, \max, \leq)$ naturally associated with bottleneck combinatorial optimization problems has a composition that is not strictly compatible with its order relation. Thus, we need to model the bottleneck problems with a different semigroup.

To obtain our results we assume that the cost array possesses what we call the *strict bottleneck Monge property*, which requires that for all $i < k$ and $j < \ell$, either $\max\{c[i,j], c[k,\ell]\} < \max\{c[i,\ell], c[k,j]\}$ or both $\max\{c[i,j], c[k,\ell]\} = \max\{c[i,\ell], c[k,j]\}$ and $\min\{c[i,j], c[k,\ell]\} \leq \min\{c[i,\ell], c[k,j]\}$.

Define the *cost* of the bottleneck shortest-path to be an ordered tuple containing the costs of all the edges on this path sorted into non-increasing order. Define the *bottleneck-cost* of the bottleneck shortest-path to be the first (i.e. the largest) entry in the cost of the bottleneck shortest-path. For example, if the bottleneck shortest $1 \hookrightarrow j$ path consists of edges $(1, i_1)$, (i_1, i_2), (i_2, i_3), (i_3, j) with the costs $c[1, i_1] = 6$, $c[i_1, i_2] = 3$, $c[i_2, i_3] = 9$, $c[i_3, j] = 5$, then the cost of this path is $(9, 6, 5, 3)$ and its bottleneck-cost is 9. We model the bottleneck shortest-path problems using the ordered commutative semigroup $(T, \oplus, \preceq)$ that was defined in Seciton 1, where the set T contains the costs of the bottleneck shortest-paths. To be able to use our results for the algebraic Monge property, we need the following lemma.

Lemma 3.1. *Let $C = \{c[i,j]\}$ be the array of edge costs. Let $C_T = \{c_T[i,j]\}$ denote an array where each entry $c_T[i,j]$ is a tuple consisting of a single element $c[i,j]$. If C possesses the strict bottleneck Monge property, then C_T possesses the algebraic Monge property under $(T, \oplus, \preceq)$.*

Proof. We need to prove that $(c_T[i,j]) \oplus (c_T[k,\ell]) \preceq (c_T[i,\ell]) \oplus (c_T[k,j])$. Let $L_1 = \max\{c[i,j], c[k,\ell]\}$, $L_2 = \min\{c[i,j], c[k,\ell]\}$, $R_1 = \max\{c[i,\ell], c[k,j]\}$ and $R_2 = \min\{c[i,\ell], c[k,j]\}$. Using this notation, $(c_T[i,j]) \oplus (c_T[k,\ell]) = (L_1, L_2)$ and $(c_T[i,\ell]) \oplus (c_T[k,j]) = (R_1, R_2)$. Thus, we need to prove that $(L_1, L_2) \preceq (R_1, R_2)$.

If C possesses the strict bottleneck Monge property, then for all $i < k$ and $j < \ell$, we have one of two cases

Case 1: $\max\{c[i,j], c[k,\ell]\} < \max\{c[i,\ell], c[k,j]\}$

Case 2: both $\max\{c[i,j], c[k,\ell]\} = \max\{c[i,\ell], c[k,j]\}$ and $\min\{c[i,j], c[k,\ell]\} \leq \min\{c[i,\ell], c[k,j]\}$

In Case 1, we have $L_1 < R_1$, which implies that $(L_1, L_2) \prec (R_1, R_2)$. In Case 2, we have $L_1 = R_1$ and $L_2 \leq R_2$, which implies that $(L_1, L_2) \preceq (R_1, R_2)$.

As was discussed in Section 1, the composition $\oplus$ is strictly compatible with the order relation $\preceq$. Thus, given Lemma 3.1, if we use array C_T as our cost array, all the lemmas and theorems of Section 2 apply to the bottleneck shortest-path problems modelled by $(T, \oplus, \preceq)$. Note that once the algorithms in Section 2 return their answers for the cost of the shortest-path, the bottleneck-cost of the shortest-path can be determined by taking the largest element in the cost tuple. We now state our results:

Theorem 3.2. *The bottleneck k-edge shortest-path problem for an n-vertex directed acyclic graph whose edge costs possess the strict bottleneck Monge property can be solved in $O(kn)$ time.*

We only give the motivation for the proof of Theorem 3.2 here: Adapting the proof of Theorem 2.3 to our semigroup, recall that we called the SMAWK algorithm to determine the column minima of the array $A^\ell = \{a^\ell[i,j]\}$, where

$$a^\ell[i,j] = \begin{cases} d^{\ell-1}[i] \oplus c_T[i,j] & \text{if } i < j \\ \infty & \text{otherwise} \end{cases} \quad \text{and} \quad d^\ell[j] = \begin{cases} \min_{1 \le i < j} a^\ell[i,j] & \text{if } \ell > 1 \\ c_T[i,j] & \text{if } \ell = 1 \end{cases}$$

In order to improve the running time of Theorem 2.3 on the semigroup $(T, \oplus, \preceq)$, we need to take a close look at the kinds of comparisons performed in the proof of that theorem. All the work in that proof is done by the SMAWK algorithm (Aggarwal et al. [1]). The key idea is that instead of storing and manipulating an entire tuple, we can simply keep first two element of $a^\ell[i,j]$. The resulting matrix (in the ordered semi-group consisting of unordered pairs, instead of k-tuples) is still algebraically Monge. Thus, the times t_a and t_c are reduced to $O(1)$, and using the proof of Theorem 2.3, we can solve the bottleneck k-edge shortest-path problem in $O(kn)$ time.

By analyzing the algorithm of Larmore and Schieber [15], instead of SMAWK, we similarly obtain:

Theorem 3.3. *The bottleneck unrestricted shortest-path problem for an n-vertex directed acyclic graph whose edge costs possess the strict bottleneck Monge property can be solved in $O(n)$ time.*

4 An Alternate Algorithm for the Bottleneck k-Edge Shortest-Path

In this section we present a second algorithm for the bottleneck k-edge shortest-path problem that in some sense generalizes Aggarwal, Schieber, and Tokuyama's [4] algorithm. Our algorithm is based on a $O(n)$-time query subroutine for determining whether the graph contains a k-edge $1 \hookrightarrow n$ path using only edges whose costs are less than or equal to some threshold T. To create the query subroutine, we need two technical lemmas. We say that a path *satisfies the threshold* if every edge on the path is less than or equal to T.

Define the following two graphs with the same vertex and edge sets at G, but with different cost functions: a graph $G'_T = (V, E)$ with edge costs

$$c'[i,j] = \begin{cases} 1 & \text{if } c[i,j] \le T \\ +\infty & \text{otherwise} \end{cases}$$

and a graph $G''_T = (V, E)$ with edge costs

$$c''[i,j] = \begin{cases} -1 & \text{if } c[i,j] \le T \\ +\infty & \text{otherwise} \end{cases}$$

Lemma 4.1. *If the cost array $C = \{c[i,j]\}$ is bottleneck Monge, then the arrays $C' = \{c'[i,j]\}$ and $C'' = \{c''[i,j]\}$ are Monge.*

Proof. The proof is routine and is given in the journal version.

Lemma 4.2. *Suppose the unrestricted shortest $1 \hookrightarrow n$ path P in G'_T contains k' edges and has length that is less than $+\infty$. Also, suppose the unrestricted shortest $1 \hookrightarrow n$ path Q in G''_T contains k'' edges and has length that is less than $+\infty$. Then there exists a k-edge $1 \hookrightarrow n$ path in G that satisfies the threshold if $k' \leq k \leq k''$.*

Proof. The proof is ommited here and is given in the journal version.

We are now ready to design the query subroutine for determining whether the given graph contains a k-edge $1 \hookrightarrow n$ path that uses only edges whose costs are less than or equal to some threshold T. Given Lemmas 4.2, such a path exists if $k' \leq k \leq k''$, where k' is the length of the unrestricted shortest $1 \hookrightarrow n$ path in G'_T and k'' is the length of the unrestricted shortest $1 \hookrightarrow n$ path in G''_T. Using the algorithm of Larmore and Schieber [15], we can determine the length of the unrestricted shortest $1 \hookrightarrow n$ path in a graph with Monge property in $O(n)$ time. Since the cost arrays of both G'_T and G''_T satisfy the Monge property (Lemma 4.1), our query subroutine runs in $O(n)$ time.

Theorem 4.3. *The bottleneck k-edge shortest-path problem for an n-vertex graph whose edge costs possess the strict bottleneck Monge property can be solved in $O(n^{3/2} \lg^2 n)$ time (or in $O(n \lg^2 n)$ time if the problem's cost array is also bitonic[1]).*

Proof. We use the result of Agarwal and Sen [2], who show how to find the d-th smallest entry in an $m \times n$ totally monotone array in $O((m+n)\sqrt{n} \lg n)$ time. For our $n \times n$ totally monotone array, this translates into $O(n^{3/2} \lg n)$ time. There are n^2 entries in the $n \times n$ cost array C_T of the bottleneck shortest-path problem. We perform a binary search on these n^2 entries by calling the procedure Binary-Search(C_T, 1, n^2). Binary-Search(C_T, i, j) does the following. We use Agarwal and Sen [2] to find the $\lfloor \frac{j+i}{2} \rfloor$-th smallest entry in C_T, which we call τ. We use our query algorithm to test if the graph contains a k-edge $1 \hookrightarrow n$ path that uses only edges whose costs are less than or equal to τ. If the query answers "yes," then we call Binary-Search(C_T, i, $\lfloor \frac{j+i}{2} \rfloor$); otherwise we call Binary-Search(C_T, $\lfloor \frac{j+i}{2} \rfloor$, j). The binary search returns the smallest entry τ^* in C_T for which there exists a k-edge $1 \hookrightarrow n$ path that uses only edges whose costs are less than or equal to τ^*. Thus, τ^* is the bottleneck-cost of the bottleneck k-edge shortest-path.

To find the actual path, we consider Lemma 4.2. We compute the unrestricted shortest $1 \hookrightarrow n$ paths P and Q, $|P| = k'$, $|Q| = k''$, in the graphs G'_{τ^*} and G''_{τ^*}.

[1] An n-entry vector $B = \{b[i]\}$ is called bitonic if there exists an i satisfying $1 \leq i \leq n$ such that $b[1] \geq \cdots \geq b[i-1] \geq b[i] \leq b[i+1] \leq \cdots \leq b[n]$. We call a 2-dimensional array bitonic if its rows or its columns are bitonic.

Scanning vertices in increasing order starting at 1, we then look for an edge (i, t) in P and an edge (s, j) in Q that satisfy $i < s < j \leq t$ and $|1 \overset{P}{\hookrightarrow} t| = |1 \overset{P}{\hookrightarrow} s|$. Let Q' consist of $1 \overset{P}{\hookrightarrow} i$, followed by the edge (i, j), followed by $j \overset{Q}{\hookrightarrow} n$; and P' consist of $1 \overset{Q}{\hookrightarrow} s$, followed by the edge (s, t), followed by $t \overset{P}{\hookrightarrow} n$. From Lemma 4.2, we know that P' and Q' are both $1 \hookrightarrow n$ paths that satisfy τ^* and that $|P'| = k' + 1$ and $|Q'| = k'' - 1$. If $k = k' + 1$ or $k = k'' - 1$, we are done. Otherwise, we repeat this procedure, except that we start our search for the new edges where we want to switch the paths not from vertex 1, but from vertices i and s. Thus, the total running time to find the path given τ^* is $O(n+k) = O(n)$. The total running time to find the bottleneck k-edge shortest-path is $O(\lg n^2(n^{3/2} \lg n + n)) = O(n^{3/2} \lg^2 n)$.

In the case of a bitonic cost array, the selection problem is simpler. The selection algorithm of Frederickson and Johnson [9] computes the n largest elements overall in $O(n)$ sorted lists in $O(n)$ time. When the array is bitonic, we can easily decompose it into $2n$ sorted lists in $O(n \lg n)$ time. Applying Frederickson and Johnson [9], we can compute the d-th smallest entry in an $n \times n$ bitonic array in $O(n \lg n)$ time. Thus, for cost arrays that satisfy the string bottleneck Monge property and are bitonic, the k-edge shortest-path can be found in $O(n \lg^2 n)$ time.

Using a similar query technique, we can also obtain the following result for unbalanced assignment problem.

Theorem 4.4. *The bottleneck assignment problem for an $m \times n$ bipartite graph, where $m \leq n$ and the edge costs possess the strict bottleneck Monge property can be solved in $O((m\sqrt{n} \lg m + n) \lg^2 n)$ time (or in $O(m \lg^2 n + n \lg n)$ time if the problem's cost array is also bitonic).*

Proof. We say that a matching *satisfies the threshold* T if all the edges of the matching have weight T or less. First, we design a query algorithm analogous to the one used in Theorem 4.3. The query algorithm, given a bipartite graph G and a threshold T, determines if there is a perfect matching of G that satisfies the threshold. The query algorithm is greedy. It works by finding the minimum value j_1 such that $w[1, j_1] \leq T$, then the minimum value j_2 such that $w[2, j_2] \leq T$ and $j_2 > j_1$, then minimum value j_3 such that $w[3, j_3] \leq T$ and $j_3 > j_2$, and so forth. This algorithm takes $O(n)$ time. If this algorithm produces a perfect matching, then clearly this matching satisfies the threshold. Furthermore, it is not difficult to see that if there exists a perfect matching that satisfies the threshold, then our algorithm finds one such matching. To see this, consider any perfect matching M that satisfies the threshold. Consider the first vertex i such that $(i, j_i) \notin M$. We show how to get another perfect matching M' in which the first vertex i' such that $(i', j_{i'}) \notin M'$ is greater than i. Suppose that i is matched to ℓ in M. Our computation of j_i guarantees that $\ell > j_i$. If j_i is unmatched in M, then we can simply remove (i, ℓ) and add (i, j_i) to get another perfect matching M' that satisfies the threshold and contains (i, j_i). Hence, assume that j_i is matched in M to s, $s > i$. By the strict bottleneck Monge property, we have that $\max\{c[i,j_i], c[s,\ell]\} < \max\{c[i,\ell], c[s,j_i]\}$ or both $\max\{c[i,j_i], c[s,\ell]\} = \max\{c[i,\ell], c[s,j_i]\}$ and $\min\{c[i,j_i], c[s,\ell]\} \leq \min\{c[i,\ell], c[s,j_i]\}$. Thus, if M satisfies the threshold, then $M' = M - \{(i,\ell),(s,j_i)\} + \{(i,j_i),(s,\ell)\}$ is another perfect matching and it contains (i, j_i).

To complete the proof, we use the query algorithm just as in the proof of Theorem 4.3. We use the algorithm of Agarwal and Sen [2] in our binary search, and for each d-th smallest value T in the $m \times n$ cost array, we query to see if there exists a perfect matching that satisfies the threshold T. The total running time is $O(\lg(mn)((m+n)\sqrt{n}\lg n + n)) = O(n^{3/2}\lg^2 n)$. Similarly, if the cost array is bitonic, then the running time is $O(\lg(mn)(m\lg n + n)) = O(m\lg^2 n + n\lg n)$.

5 A Paragraph-Formation Problem

We discuss a variant of Hirschberg and Larmore's optimal-paragraph-formation problem [11]. A slightly simplified version of their problem is to break a sequence of words $w_1, \ldots, w_n$ into lines in order to form a paragraph. Define $c[i,j]$ to be the cost of a line consisting of words w_i through w_{j-1}. Let L be the optimal line width and $|w_i|$ be the length of word w_i plus one for the cost of the space after word w_i. Then, following the ideas of Hirschberg and Larmore, the cost function is $c[i,j] = (|w_{i+1}| + |w_{i+2}| + \cdots + |w_j| - 1 - L)^2$. Their objective is to construct a paragraph minimizing the sum of the paragraph's line costs. This problem is easily transformed into an instance of the sum unrestricted shortest-path problem. Consider a directed acyclic graph $G = (V, E)$, where the vertices are numbered 0 through n and $(i,j) \in E$ if and only if $i < j$. The cost of edge (i,j) is $c[i,j]$ defined above. A $0 \hookrightarrow n$ path $p = \langle (0,i_1),(i_1,i_2),\ldots,(i_s,n)\rangle$ corresponds to putting words w_1 through w_{i_1} into the first line, words w_{i_1+1} through w_{i_2} into the second line, and so forth, with words w_{i_s} through w_n forming the last line. The shortest $0 \hookrightarrow n$ path in this graph corresponds to the minimum sum of the paragraph's line costs. Hirschberg and Larmore prove that the above cost function satisfies the sum Monge property, and thus it can be solved in $O(n)$ time. (Credit for the linear-time algorithm belongs to Wilber [17], as well as to Larmore and Schieber [15].)

If we instead seek to minimize the maximum line cost, we obtain an instance of the bottleneck unrestricted shortest-path problem. The following two lemmas prove that the edge costs in this problem possess the strict bottleneck Monge property. We call a line cost function $f[i,j]$ *strictly bitonic* if any sequence of costs $f[i_1,j_1], f[i_2,j_2], \ldots, f[i_s,j_s]$, which satisfies the following conditions for every $1 \leq \ell < s$:

1. either $i_\ell = i_{\ell+1}$ and $j_\ell < j_{\ell+1}$,
2. or $j_\ell = j_{\ell+1}$ and $i_\ell > i_{\ell+1}$

is strictly decreasing then strictly increasing. Note that either the strictly decreasing subsequence or the strictly increasing subsequence may have length zero.

Lemma 5.1. *Let $F = \{f[i,j]\}$, where $f[i,j]$ is a line cost function that is strictly bitonic. Then F satisfies the strict bottleneck Monge property.*

Proof. The proof is routine.

Lemma 5.2. *Cost function $c[i,j] = (|w_{i+1}| + |w_{i+2}| + \ldots + |w_j| - 1 - L)^2$ is strictly bitonic.*

Proof. The proof is routine and is given in the full paper version.

From these two lemmas, we conclude that any strictly bitonic line cost function $f(i,j)$ satisfies the strict bottleneck Monge property, and thus, by Theorem 3.3, a variant of Hirschberg and Larmore's problem which uses $f(i,j)$ can also be solved in $O(n)$ time. In particular, the variant with cost function $c[i,j]$ can be solved in $O(n)$ time.

We have thus far considered a simplified version of Hirshberg and Larmore cost function. There are three other factors about paragraph formation which they include in their cost. One is the ability to hyphenate a word at a fixed cost per hyphenation. Another is the fact that in addition to penalizing quadratically lines that differ too much from the ideal line length, there may be some lower and upper limits beyond which the line length is simply not allowed. And finally, the last line in the paragraph should not be penalized for being too short. We now attempt to incorporate these factors into our cost function.

The second factor is fairly easy to incorporate. Suppose that that the maximum and the minimum allowed length for a line are *maxlen* and *minlen*. Then define the cost function to be

$$c[i,j] = \begin{cases} (p[i,j] - 1 - L)^2 & \text{if } minlen \le p[i,j] \le maxlen \\ M^{i+j} & \text{otherwise} \end{cases}$$

where M is a very large number (say the sum of all the word lengths). It is not hard to see that the proof that $c[i,j]$ is bitonic holds with this new definition of the cost function.

To take into account the last factor, we would need to define a new cost function

$$c'[i,j] = \begin{cases} 0 & \text{if } j = n \text{ and } p[i,j] \le L \\ (p[i,j] - 1 - L)^2 & \text{if } minlen \le p[i,j] \le maxlen \\ M^{i+j} & \text{otherwise} \end{cases}$$

Unfortunately, this new cost function $c'[i,j]$ is not strictly bitonic and furthermore not even bottleneck Monge. Thus, we stick to $c[i,j]$ as our cost function, but we modify the algorithm. Instead of computing the $1 \hookrightarrow n$ bottleneck shortest path in Theorem 3.3, we compute the bottleneck cost of the $1 \hookrightarrow n-1$ bottleneck shortest path. The algorithm in the proof of this theorem actually computes all $d[i]$'s for $1 \le i \le n-1$, where $d[i]$ is the bottleneck cost of the bottleneck shortest $1 \hookrightarrow i$ path. To find the bottleneck shortest $1 \hookrightarrow n$ path, we evaluate the following, which takes $O(n)$ time:

$$d[n] = \min_{1 \leq i \leq n-1} \left\{ \begin{array}{ll} d[i] & \text{if } p[i,n] \leq L \\ \max\{d[i], (p[i,n] - 1 - L)^2\} & \text{if } L \leq p[i,n] \leq maxlen \\ \infty & \text{otherwise} \end{array} \right\}$$

Unfortunately, it is not possible to incorporate the first factor into the bottleneck framework. If we are to assign a fixed penalty function B for breaking a word in the middle and hyphenating it, the cost function is no longer bottleneck Monge. More specifically, we assume that each w_i is now a syllable. Suppose we have $i < k < j < \ell$ and j-th syllable is not the last syllable of its word, but the ℓ-th syllable is. Then for the cost function to be bottleneck Monge, we would need to have

$$\begin{aligned} &\max\left\{(p[i,j] - 1 - L)^2 + B, (p[k.\ell] - 1 - L)^2\right\} \\ &\leq \max\left\{(p[k,j] - 1 - L)^2 + B, (p[i.\ell] - 1 - L)^2\right\} \end{aligned}$$

It is easy to come up with a numerical example when this does not hold.

We also consider one variation of the cost function. In some circumstances, a more accurate portrayal of a typical text formatting application would be that instead of having an ideal line length and penalizing for both running under and running over this ideal length, the application has L as the available line width. In this case, there is a very large penalty for running over this available line width and a quadratic penalty for running under this width. The cost function in this case is

$$c[i,j] = \left\{ \begin{array}{ll} (p[i,j] - 1 - L)^2 & \text{if } p[i,j] \leq L \\ M^{i+j} & \text{otherwise} \end{array} \right.$$

where M is a very large number (say the sum of all the word lengths). It is not hard to see that this $c[i,j]$ is strictly bitonic.

6 A Special Case of the Bottleneck Traveling-Salesman Problem

For our final application, we consider a special case of the bottleneck traveling-salesman problem. Given a complete directed graph G on vertices $\{1, \ldots, n\}$ and a cost array $C = \{c[i,j]\}$ assigning cost $c[i,j]$ to the edge (i,j), we seek a tour of G that visits every vertex of G exactly once and minimizes the maximum of the tour's edges' costs. We call such a tour the *bottleneck shortest-tour.* In [5], Burkard and Sandholzer identified several families of cost arrays corresponding to graphs containing at least one bottleneck-optimal tour that is pyramidal. A tour T is called *pyramidal* if (1) the vertices on the path T starting from vertex n and ending at vertex 1 have monotonically decreasing labels, and (2) the vertices on the path T starting from vertex 1 and ending at vertex n have monotonically increasing labels. For example, a tour $T = \langle 4, 2, 1, 3, 6, 8, 7, 5, 4 \rangle$

is pyramidal, but a tour $T = \langle 4, 2, 1, 6, 3, 8, 7, 5, 4 \rangle$ is not. Thus, since there is a simple $O(n^2)$-time dynamic-programming algorithm for computing a pyramidal tour whose maximum edge cost is minimum among all pyramidal tours, the bottleneck traveling-salesman problem for any graph whose cost array is a member of one of Burkard and Sandholzer's families can be solved in $O(n^2)$ time. We show that if a graph's edge cost array possesses the strict bottleneck Monge property, then it is possible to find the graph's bottleneck-shortest pyramidal tour in $O(n \lg^2 n)$ time.

Theorem 6.1. *Given a graph G whose cost array C satisfies the strict bottleneck Monge property, the bottleneck pyramidal shortest-tour of G can be found in $O(n \lg^2 n)$ time.*

Proof. The proof is given in the full paper version. The basic idea is to reduce the problem to two instances of the on-line monotone matrix searching problem, and solve both of them using the algorithm of [15]. The two on-line processors, each using the on-line matrix searching algorithm, run simultaneously, passing information to each other at each step. The total number of queries needed is $O(n)$.

The reduction is not constant cost, however. A query in the reduced problem may require the computation of the cost of a path from i to j which passes through every intermediate point. We can use $O(n \lg n)$ preprocessing time to create a data structure and then satisfy each query in $O(\lg^2 n)$ time, as given in [12].

References

1. A. Aggarwal, M. M. Klawe, S. Moran, P. Shor, and R. Wilber. Geometric applications of a matrix-searching algorithm. *Algorithmica*, 2(2):195–208, 1987.
2. P. K. Agarwal and S. Sen. Selection in monotone matrices and computing k^{th} nearest neighbors. In *Proceedings of the 4th Scandinavian Workshop on Algorithm Theory*, 1994.
3. A. Aggarwal, B. Schieber, and T. Tokuyama. Finding a minimum weight k-link path in graphs with Monge property and applications. In *Proc. 9th Annu. ACM Sympos. Comput. Geom.*, pages 189–197, 1993.
4. A. Aggarwal, B. Schieber, and T. Tokuyama. Finding a minimum-weight k-link path in graphs with the concave Monge property and applications. *Discrete Comput. Geom.*, 12:263–280, 1994.
5. R. E. Burkard and W. Sandholzer. Efficiently solvable special cases of bottleneck travelling salesman problems. *Discrete Applied Mathematics*, 32(1):61–76, 1991.
6. R. E. Burkard. Remarks on some scheduling problems with algebraic objective functions. *Operations Research Verfahren*, 32:63–77, 1979.
7. R. E. Burkard and U. Zimmermann. Combinatorial optimization in linearly ordered semimodules: A survey. In B. Korte, editor, *Modern Applied Mathematics: Optimization and Operations Research*, pages 391–436. North-Holland Publishing Company, Amsterdam, Holland, 1982.

8. R. E. Burkard and B. Klinz and R. Rudolf. Perspectives of monge properties in optimization. *Discrete Applied Mathematics*, 70:95–161, 1996.
9. G. N. Frederickson and D. B. Johnson. The complexity of selection and ranking in X + Y and matrices with sorted columns. *Journal of Computer and System Sciences*, 24(4):197–208, 1982.
10. H. N. Gabow and R. E. Tarjan. Algorithms for two bottleneck optimization problems. *Journal of Algorithms*, 9(3):411–417, 1988.
11. D. S. Hirschberg and L. L. Larmore. The least weight subsequence problem. *SIAM Journal on Computing*, 16(4):628–638, 1987.
12. D. S. Hirschberg and D. J. Volper. Improved update/query algorithms for the interval valuation problem. *Information Processing Letters*, 24:307–310, 1987.
13. A.J. Hoffman. On simple linear programming problems. In *Convexity, Proc. Symposia in Pure Mathematics*, volume 7, pages 317 – 327, Providence, RI, 1961. American Mathematical Society.
14. B. Klinz, R. Rudolf, and G.J. Woeginger. On the recognition of bottleneck monge matrices. *Discrete Applied Mathematics*, 63:43–74, 1995.
15. L. L. Larmore and B. Schieber. On-line dynamic programming with applications to the prediction of RNA secondary structure. *Journal of Algorithms*, 12(3):490–515, 1991.
16. E. Seiffart. Algebraic transportation and assignment problems with "Monge-property" and "quasi-convexity". *Discrete Applied Mathematics*, 1993.
17. R. Wilber. The concave least-weight subsequence problem revisited. *Journal of Algorithms*, 9(3):418–425, 1988.

Packing Edges in Random Regular Graphs

Mihalis Beis[2], William Duckworth[1], and Michele Zito[2]

[1] Department of Computing, Macquarie University, Sydney, NSW 2109, Australia
[2] Department of Computer Science, University of Liverpool, Liverpool, L69 7ZF, UK

Abstract. A k-separated matching in a graph is a set of edges at distance at least k from one another (hence, for instance, a 1-separated matching is just a matching in the classical sense). We consider the problem of approximating the solution to the maximum k-separated matching problem in random r-regular graphs for each fixed integer k and each fixed $r \geq 3$. We prove both constructive lower bounds and combinatorial upper bounds on the size of the optimal solutions.

1 Introduction

In this paper we consider graphs generated according to the $\mathcal{G}(n,r\text{-reg})$ model (see, for example, [11, Chapter 9]). Given n urns, each containing r balls (with rn even), a set of $rn/2$ distinct pairs of balls is chosen *uniformly at random* (u.a.r.). Then a (random) n-vertex graph, $G = (V, E)$, is obtained by identifying the n urns with the vertices of the graph and letting $\{i, j\} \in E$ if and only if there is at least one pair with one ball belonging to urn i and the other ball belonging to urn j. The maximum degree of a vertex in G (i.e. the maximum number of edges incident to a vertex) is at most r. Moreover, for every integer $r > 0$, there is a positive fixed probability that the random pairing contains neither pairs with two balls from the same urn nor couples of pairs with balls coming from just two urns (see, for example, [18, Section 2.2]). In this case the graph is r-regular (all vertices have degree r). Notation $G \in \mathcal{G}(n,r\text{-reg})$ will signify that G is selected according to the $\mathcal{G}(n,r\text{-reg})$ model. An event, $\mathcal{E}_n$, describing a property of a random graph depending on a parameter n, holds *asymptotically almost surely* (a.a.s.), if the probability that $\mathcal{E}_n$ holds tends to one as n tends to infinity.

The distance between two vertices in a graph is the number of edges in a shortest path between the two vertices. The distance between two edges $\{u_1, u_2\}$ and $\{v_1, v_2\}$ is the minimum distance between any two vertices u_i and v_j. For any positive integer k, a *k-separated matching* of a graph, is a set of edges, $\mathcal{M}$, such that the minimum distance between any two edges in $\mathcal{M}$ is at least k (the qualifier "separated" will normally be omitted in the rest of this paper). Let $\nu_k(G)$ be the size of a largest k-matching in G. The *maximum k-matching* (MkM) problem asks for a k-matching of size $\nu_k(G)$. For $k = 1$ this is the classical maximum matching problem [13]. Stockmeyer and Vazirani [16] introduced the generalisation for $k \geq 2$, motivating it (for $k = 2$) as the "risk-free marriage problem" (find the maximum number of married couples such that each person is compatible

K. Diks et al. (Eds): MFSC 2002, LNCS 2420, pp. 118–130, 2002.

only with the person (s)he is married to). The M2M problem (also known as the *maximum induced matching* problem) stimulated much interest in other areas of theoretical computer science and discrete mathematics as finding a maximum 2-matching of a graph is a sub-task of finding a strong edge-colouring of a graph (a proper colouring of the edges such that no edge is incident with more than one edge of the same colour as each other, see (for example) [8,9,12,15]). The separation constraint imposed on the matching edges when $k > 2$ is a distinctive feature of the MkM problem and the main motivation for our algorithmic investigation of such problems.

MkM is NP-hard [16] (polynomial time solvable [7]), for each $k \geq 2$ (for $k = 1$). Improved complexity results are known for M1M [14] on random instances. In particular it has been proven that simple greedy heuristics a.a.s. produce sets of $\frac{n}{2} - o(n)$ independent edges [1] in dense random graphs and random regular graphs. A number of results are known on the approximability of an optimal 2-matching [2,4,19]. Zito [20] presented some simple results on the approximability of an optimal 2-matching in dense random graphs.

In this paper, we consider several natural (and simple) strategies for approximating the solution to the MkM problem, for each positive integer k, and analyse their performance on graphs generated according to the $\mathcal{G}(n,r\text{-reg})$ model. We also prove combinatorial upper bounds on $\nu_k(G)$ which hold a.a.s. if $G \in \mathcal{G}(n,r\text{-reg})$. The algorithm we present for M2M was analysed deterministically in [4] where it was shown to return a 2-matching of size at least $r(n-2)/2(2r-1)(r-1)$ in a connected r-regular graph on n vertices, for each $r \geq 3$. Furthermore, it was shown that there exist infinitely many r-regular graphs on n-vertices for which the algorithm only achieves this bound. For the case $r = 3$, the cardinality of a largest 2-matching $\mathcal{M}$ of a random 3-regular graph a.a.s. satisfies $0.26645n \leq |\mathcal{M}| \leq 0.282069n$ [6] (unfortunately the optimistic $0.270413n$ lower bound claimed in the paper is not correct). We analyse the performances of such an algorithm when the input is distributed according to $\mathcal{G}(n,r\text{-reg})$. Generalisations of this algorithm to the case $k > 2$ fails. Any greedy process based on a sequence of local choices/updates has no permanent record of the original neighbourhood structure of each vertex. Hence, an edge chosen by the algorithm to add to $\mathcal{M}$ may cause the matching not to be k-separated. For $k > 2$, different strategies, based on a more selective updating, must be used. The following Theorem encompasses the results of this paper. Its proof appears in subsequent sections.

Theorem 1. *For each fixed positive integer k and fixed integer $r \geq 3$ there exist two positive real numbers $\lambda_k = \lambda_k(r)$ and $\mu_k = \mu_k(r)$ such that if $G \in \mathcal{G}(n,r\text{-reg})$, then $\lambda_k n \leq \nu_k(G) \leq \mu_k n$ a.a.s..*

Table 1 reports the values of λ_k and μ_k for the first few values of r and k.

2 Lower Bounds

In this section we describe the greedy heuristics used to construct a large k-matching. The algorithms are quite general and may be applied to any graph.

Table 1.

r	M1M		M2M		M3M		M4M	
	λ_1	μ_1	λ_2	μ_2	λ_3	μ_3	λ_4	μ_4
3	0.5	0.5	0.26645	0.28207	0.11239	0.15605	0.03943	0.09455
4	0.5	0.5	0.22953	0.25	0.07314	0.10757	0.01672	0.05007
5	0.5	0.5	0.20465	0.22695	0.05071	0.07922	0.01856	0.02933
6	0.5	0.5	0.18615	0.2091	0.03801	0.0611	0.00478	0.01859

The analyses presented at the end of each sub-section give lower bounds on the size of the resulting k-matching if the input graph is selected according to the $\mathcal{G}(n, r\text{-reg})$ model.

2.1 Dense Matchings

Next we describe the algorithm that will be used to find a large k-matching in a random r-regular graph, when[1] $k \leq 2$. Let $\Gamma(u) = \{v \in G : \{u, v\} \in E\}$ be the *neighbourhood of vertex u*.

Algorithm DegreeGreedy(G, k)
Input: a graph $G = (V, E)$ on n vertices.
  $\mathcal{M} \leftarrow \emptyset$;
  while $E \neq \emptyset$
    let u be a vertex of minimum positive degree in $V(G)$;
    if $|\Gamma(u)| > 0$
      let v be a vertex of minimum positive degree in $\Gamma(u)$;
    $\mathcal{M} \leftarrow \mathcal{M} \cup \{\{u, v\}\}$;
    shrink$(G, \{u, v\}, k)$;

Ties in the selection of u and v may be broken by any reasonable rule, e.g. making selections at random or according to some predefined ordering on the graph vertices. For each iteration of the algorithm, procedure *shrink* updates G by removing u, v, all vertices within distance $k-1$ from $\{u, v\}$ and all edges incident with these vertices.

Let $V_i = \{v : |\Gamma(v)| = i\}$. In the following discussion G always denotes the subgraph of the input graph still to be dealt with at some point during the execution of DegreeGreedy. In each iteration of the main while loop a further portion of G is dealt with.

The analysis below is based on the fact that if $G \in \mathcal{G}(n, r\text{-reg})$ the state of the dynamic process associated with the execution of DegreeGreedy may be described by the vector $(|V_1|, |V_2|, \ldots, |V_r|)$. Furthermore, the evolution of such a vector is (approximately) Markovian and quite peculiar. The algorithm essentially runs for $r-1$ *phases*: Phase 1, Phase 2, ..., Phase $r-1$. In Phase j (for $j < r-1$) there are essentially no vertices of degree less than $r-j-1$, and

[1] We believe that the values reported in Section 1 justify the attribute "dense" in the title of this section.

few vertices of degree $r-j-1$ which are used up as soon as they are created. Therefore, the vertex u is selected predominantly from V_{r-j}, and occasionally (but rarely) from V_{r-j-1}. However, towards the end of the phase, the rate at which vertices of degree $r-j-1$ are generated becomes larger than the rate at which they are consumed. This determines the transition to Phase $j+1$. We refer to one iteration of the main while loop that selects u from V_j as a *selection of a vertex of degree* j. A *clutch* (or *step*) in a given Phase j is defined as a sequence of iterations of the main while loop. The first of these iterations selects u from V_{r-j}, and the remainder selects u from V_{r-j-1}. The last iteration in the clutch is the last one before the next u of degree $r-j$ is selected.

For $1 \le j \le r-1$, let $Y_i = Y_i^j(t)$ denote $|V_{r-i}|$ after step t of Phase j. Let $Y_0^1(0) = n$ and $Y_i^1(0) = 0$ for all $i > 0$. Furthermore, let $Y_i^{j+1}(0)$ be equal to the final value of Y_i in Phase j for all $i \ge 0$ and $1 \le j \le r-2$. Finally, let $X^j(t) = \sum_{i=0}^{r-1}(r-i)Y_i^j(t)$. From now on the dependency on t and j will be omitted unless ambiguity arises. The key ingredient in the analysis of the algorithm above is the use (in each Phase) of Theorem 6.1 from [18] which provides tight asymptotics for the most likely values of Y_i (for each $i \in \{0, \dots, r-1\}$), as the algorithm progresses through successive Phases.

Note that it would be fairly simple to modify the graph generation process described in Section 1 to incorporate the decisions made by the algorithm DegreeGreedy: the input random graph and the output structure $\mathcal{M}$ would be generated at the same time. In this setting one would keep track of the degree sequence of the so called *evolving graph* H_t (see for instance [17]). H_0 would be empty, and a number of edges would be added to H_t to get H_{t+1} according to the behaviour of algorithm DegreeGreedy during step $t+1$. The random variables Y_i also denote the number of vertices of degree i in the evolving graph H_t. We prefer to give our description in terms of the original regular graph. In this way the algorithm is easier to understand, and all details about the analysis of its performances are kept away from the programmer.

Let $\mathrm{E}_j(\Delta Y_i)$ denote the expected change of Y_i during step $t+1$ in some given Phase j, conditioned to the history of the algorithm execution from the start of the phase until step t. This is asymptotically the sum of two terms: the expected change due to the updating following the selection of a vertex of degree $r-j$ at the beginning of a clutch and the one due to the updating in the remaining part of the clutch. We denote by $\boldsymbol{d} \equiv (d_1, d_2, \dots, d_{|\Gamma(u)|})$ the degrees of u's neighbours in $G \setminus \{u\}$. In what follows we assume that a.a.s. $r-j-1 \le d_1 \le d_2 \le \dots \le d_{|\Gamma(u)|} \le r-1$. If Births_{j+1} denotes the expected number of vertices of degree $r-(j+1)$ generated during a clutch when $j < r-1$ (and $\text{Births}_r = 0$) then the asymptotic expression for $\mathrm{E}_j(\Delta Y_i)$ is

$$\sum_{\boldsymbol{d}} \mathrm{E}_j(\Delta Y_i \mid \boldsymbol{d}) \Pr[\boldsymbol{d}] + \text{Births}_{j+1} \sum_{\boldsymbol{d}'} \mathrm{E}_{j+1}(\Delta Y_i \mid \boldsymbol{d}) \Pr[\boldsymbol{d}'] \qquad (1)$$

where $\mathrm{E}_j(\Delta Y_i \mid \boldsymbol{d})$ is the expected change of Y_i conditional to the degree(s) of u's neighbour(s) in $G \setminus \{u\}$ being described by $\boldsymbol{d}$. The expression $\Pr[\boldsymbol{d}]$ denotes (asymptotically) the probability that configuration $\boldsymbol{d}$ occurs conditioned to the history of the algorithm so far. The first (second) sum is over all possible configurations $\boldsymbol{d}$ (resp. $\boldsymbol{d}'$) with $|\boldsymbol{d}| = r-j$ (resp. $|\boldsymbol{d}'| = r-j-1$). Also the expected change in the size of the structure output by the algorithm, $\mathrm{E}_j(\Delta|\mathcal{M}|)$, is asymptotically

$$1 + \mathrm{Births}_{j+1} \tag{2}$$

since an edge is added to $\mathcal{M}$ following each selection. Setting $x = t/n$, $y_i^j(x) = Y_i^j/n$ (again dependency on j will be usually omitted), and $\lambda = \lambda_k(x) = |\mathcal{M}|/n$, the following system of differential equations is associated with each Phase j,

$$\frac{dy_i}{dx} = \tilde{\mathrm{E}}_j(\Delta Y_i)\Big|_{t=xn, Y_i=y_i n} \quad \frac{d\lambda}{dx} = \tilde{\mathrm{E}}_j(\Delta|\mathcal{M}|)\Big|_{t=xn, Y_i=y_i n} \tag{3}$$

where $\tilde{\mathrm{E}}_j(\Delta Y_i)$ and $\tilde{\mathrm{E}}_j(\Delta|\mathcal{M}|)$ denote the asymptotic expressions for the corresponding expectations obtained from (1) and (2) using the estimates on $\Pr[\boldsymbol{d}]$ and $\mathrm{E}_j(\Delta Y_i \mid \boldsymbol{d})$ given later on. Since x does not occur in $\tilde{\mathrm{E}}_j(\Delta Y_i)$ and $\tilde{\mathrm{E}}_j(\Delta|\mathcal{M}|)$, we actually solve, one after the other, the systems

$$\frac{dy_i}{d\lambda} = \frac{\tilde{\mathrm{E}}_j(\Delta Y_i)}{\tilde{\mathrm{E}}_j(\Delta|\mathcal{M}|)}\Big|_{Y_i=y_i n} \tag{4}$$

where differentiation is w.r.t. λ, setting $y_0^0(0) = 1$ and $y_i^0(0) = 0$ for all $i \geq 1$ and using the final conditions of Phase j, as initial conditions of Phase $j+1$ for each $j > 0$. The value of λ_k reported in the table in Section 1 is the sum of the final values of λ in each Phase.

We use Theorem 6.1 (see Appendix) from [18] to show that, during each phase, the values of $|\mathcal{M}|/n$ and Y_i/n a.a.s. remain within $o(1)$ of the corresponding deterministic solutions to the differential equations (4). To apply such result to the multi-dimensional random process $(t/n, |\mathcal{M}|/n, Y_0/n, \ldots, Y_{r-1}/n)$ we need to define a bounded connected open set in $\mathbb{R}^{r+2}$ containing the closure of the set

$$\{(0, z^{(1)}, \ldots, z^{(r+1)}) : \Pr[Y_i(t) = z^{(i+2)}n, 0 \leq i \leq r-1] \neq 0 \text{ for some } n\},$$

and then verify that the boundedness, trend, and Lipschitz hipotheses hold uniformly for each t smaller than the minimum t such that $(t/n, |\mathcal{M}|/n, Y_0/n, \ldots, Y_{r-1}/n) \notin \mathcal{D}$. For Phase j, where $j < r-1$, and for an arbitrarily small $\epsilon > 0$, define $\mathcal{D}_{\epsilon,j}$ to be the set of all $(x, z^{(1)}, \ldots, z^{(r+1)})$ for which $x > -\epsilon$, $\sum_{i=2}^{r+1} z^{(i)} > \epsilon$, $\mathrm{Births}_{j+1} < (1-\epsilon)n$, $z^{(1)} > -\epsilon$ and $z^{(i)} < 1 + \epsilon$ where $1 \leq i \leq r+1$. It is fairly simple, if tedious, to verify that the functions under consideration satisfy the three main hypotheses of Wormald's theorem in the domain $\mathcal{D}_{\epsilon,j}$. A similar argument applies to Phase $r-1$ as well except that, here, a clutch is equivalent to processing just one vertex.

Some additional work is required to ensure that once the end of Phase j is reached, the process proceeds as described informally into the next phase. This involves ensuring that the expected number of births in a clutch (divided by n) rises strictly above 1. This may be achieved by computing partial derivatives of the equations representing the expected changes in the variables Y_i for processing a vertex (not a clutch) in Phase j. For reasons of brevity, we do not include these details here.

From the point in Phase $r-1$ after which Wormald's analysis tool does not apply until the completion of the algorithm, the change in each variable per step is bounded by a constant. Hence, the change in the random variables Y_i and $|\mathcal{M}|$ is $o(n)$. This completes the proof of the lower bounds in Theorem 1.

In the remainder of this section details are given on how to compute asymptotic expressions for $\Pr[\boldsymbol{d}]$, $\mathrm{E}_j(\Delta Y_i \mid \boldsymbol{d})$, for each $\boldsymbol{d}$, and Births_{j+1}, and some comments are made on how to solve the systems in (4).

Probability of a Configuration. The formula for $\Pr[\boldsymbol{d}]$ is better understood if we think of the algorithm DegreeGreedy as embedded in the graph generation process. Each configuration $\boldsymbol{d} \equiv (d_1, d_2, \ldots, d_{|\Gamma(u)|})$ occurs at the neighbourhood of a given u if the $|\Gamma(u)|$ balls still available in the urn U associated with u are paired up with balls from random urns containing respectively $d_1+1, d_2+1, \ldots, d_{|\Gamma(u)|}+1$ free balls. The probability of pairing up one ball from U with a ball from an urn with $r-i$ free balls is $P_i = \frac{(r-i)Y_i}{X}$ for $i \in \{0, \ldots, r\}$ at the beginning of a step, and this only changes by a $o(1)$ factor during each step due to the multiple edge selections which are part of a step. Let $\Pr[\boldsymbol{d}]$ be $\binom{|\Gamma(u)|}{m_1, \ldots, m_{|\Gamma(u)|}} P_{r-(d_1+1)} \cdot P_{r-(d_2+1)} \cdot \ldots \cdot P_{r-(d_{|\Gamma(u)|}+1)}$ where m_z are the multiplicities of the possibly $|\Gamma(u)|$ distinct values occurring in $\boldsymbol{d}$.

Conditional Expectations. Note that the sequence $\boldsymbol{d}$ contains all the information needed to compute asymptotic expressions for all the conditional expected changes of the variables Y_i. We define

$$\mathrm{E}_j(\Delta Y_i \mid \boldsymbol{d}) = -\mathrm{rm}(r, j, i, \boldsymbol{d}) - \partial(r, j, i, \boldsymbol{d})$$

where $\mathrm{rm}(r, j, i, \boldsymbol{d})$, gives, for each $\boldsymbol{d}$ in a given Phase j, the number of vertices of degree $r-i$ removed from the subgraph of G induced by $\{u\} \cup \Gamma(u)$ and ∂ is a function accounting for the degree changes occurring outside such a subgraph. Let $Q_i(0) = P_i - P_{i-1}$ for $i \in \{0, \ldots, r\}$ (with $P_{-1} = 0$) and $Q_i(1) = \sum_{z=0}^{r-1} P_z(\delta_{i,z} + (r-z-1)Q_i(0))$, for $i \in \{0, 1, \ldots, r\}$ (where $\delta_{x,y} = 1$ if $x = y$ and zero otherwise). The expression $-Q_i(0)$ describes asymptotically the contribution to $\mathrm{E}_j(\Delta Y_i \mid \boldsymbol{d})$ given by the removal from G of one edge incident to a vertex whose degree is decreased by one during one execution of *shrink*. Expression $-Q_i(1)$ is (asymptotically) the contribution to $\mathrm{E}_j(\Delta Y_i \mid \boldsymbol{d})$ given by the removal from G of the vertices in $\Gamma(v)$ (and all their incident edges). We have

$$\begin{aligned}
\mathrm{rm}(r, j, i, \boldsymbol{d}) &= \delta_{i,j} + \delta_{i,r-(d_1+1)} + \delta_{k,2} \textstyle\sum_{h=2}^{|\Gamma(u)|} \delta_{i,r-(d_h+1)}, \\
\partial(r, j, i, \boldsymbol{d}) &= d_1 Q_i(k-1) + \textstyle\sum_{h=2}^{|\Gamma(u)|} (\delta_{k,1}(\delta_{i,r-(d_h+1)} - \delta_{i,r-d_h}) + \delta_{k,2} d_h Q_i(0)).
\end{aligned}$$

Vertices of Degree $r-j-1$ in Phase j. We remind the reader that Births_{j+1} denotes the expected number of vertices of degree $r-j-1$ generated during a clutch. This quantity can be computed by modelling the generation of these vertices as a discrete branching process. The first selection in a clutch will, asymptotically, generates $\mathrm{E}_j(\Delta Y_{j+1})$ vertices of degree $r-j-1$: this can be considered as the first generation of the branching process. In general, the bth generation will contain approximately

$$\mathrm{E}_j(\Delta Y_{j+1})(\mathrm{E}_{j+1}(\Delta Y_{j+1}))^{b-1}$$

vertices, and therefore, provided $\mathrm{E}_{j+1}(\Delta Y_{j+1})$ is strictly smaller than one, the asymptotic expression for Births_{j+1} is

$$\frac{\mathrm{E}_j(\Delta Y_{j+1})}{1-\mathrm{E}_{j+1}(\Delta Y_{j+1})}$$

(and after all these vertices have been removed from G, there is no vertex of degree $r-j-1$ left).

Computational Aspects. The systems of differential equations in this paper have been solved using a method developed by Cash and Karp [3]. A basic 5th order Runge-Kutta solver is repeatedly run twice, with different step sizes, to ensure better accuracy. Error control is performed using local extrapolation [10]. In all cases the equations are quite stable and the numbers provided in the table in Section 1 are accurate to the fifth decimal digit.

2.2 Sparse Matchings

Any obvious generalisation of the algorithm DegreeGreedy to the case $k > 2$ fails. The DegreeGreedy process, which repeatedly picks sparsely connected edges $\{u, v\}$ and removes their neighbourhood within distance $k-1$, has no permanent record of the original neighbourhood structure of each vertex. Hence, an edge chosen to add to $\mathcal{M}$ may cause the matching not to be k-separated. For $k > 2$ we therefore resort to a different heuristic to approximate ν_k. Such an algorithm is based on a more selective process that repeatedly picks vertices of degree r from the given graph, explores their neighbourhood at distance at most $\lfloor k/2 \rfloor$ and performs different updatings depending on the degree structure around the chosen vertex. We treat the k even and k odd cases separately, as some extra care is needed to define a large k-matching when k is odd.

Let $t_0(r)$ be the trivial tree formed by a single vertex. Let $t_d(r)$ be the (rooted) tree obtained by taking r copies of $t_{d-1}(r)$ and joining their roots to a new vertex u. For any integer $k \geq 2$, the tree $T_k(r)$ is a rooted tree whose root u has a child v which is the root of a copy of $t_{\lfloor k/2 \rfloor -1}(r-1)$ and, when $k \geq 4$, $r-1$ other children $v_2, \ldots, v_r$ which are roots of copies of $t_{\lfloor k/2 \rfloor -2}(r-1)$. It can be easily verified that $T_k(r)$ contains $\mathrm{rem}(r,k) = \frac{2(r-1)^{\lfloor k/2 \rfloor}-2}{r-2}$ vertices. The k-matching algorithm may be described as follows:

Algorithm Sparse(G, k)
Input: an r-regular graph $G = (V, E)$ on n vertices.
  $\mathcal{M} \leftarrow \emptyset$;
  while $V_r \neq \emptyset$
    let u be a vertex of degree r in $V(G)$;
    if (u is the root of a copy of $T_k(r)$ such that the
      vertex v is chosen at u.a.r. in $\Gamma(u)$ and

$\mathcal{M} \cup \{\{u,v\}\}$ is a k-matching)
$\mathcal{M} \leftarrow \mathcal{M} \cup \{\{u,v\}\}$;
remove $T_k(r)$ from G;
else
remove u from G;

It would be quite simple to modify the description above so that the algorithm could handle any graph on n vertices. For clarity of exposition we preferred to state the process for r-regular graphs only.

Analysis for Even k. The dynamics of this algorithm can again be described by looking at the evolution of the (random) vector $(|V_1|, \ldots, |V_r|)$. Furthermore the analysis in such case is much simpler than in the case $k \leq 2$. During each iteration of the main while loop one of two things can happen. If a copy of $T_k(r)$ is found "around" u, then the matching is updated and $\frac{2(r-1)^{\frac{k}{2}}-2}{r-2} = r^{O(k)}$ vertices are removed from G. Otherwise we simply remove u (along with its incident edges) from G. In other words the algorithm is often mimicking a simple process which builds an independent dominating set of vertices in a regular graph by repeatedly stripping off vertices of degree r. Such process has been analysed in [17]. Therefore it is fairly easy to verify that $\mathrm{E}(\Delta Y_i)$ is asymptotically

$$-\delta_{i,0} - rQ_i(0) + ((r+1-\mathrm{rem}(r,k))\delta_{i,0} + r\delta_{i,1} - 2(r-1)^{\frac{k}{2}}Q_i(0)) \times \mathrm{E}(\Delta\mathcal{M})$$

where $\mathrm{E}(\Delta\mathcal{M})$ can be approximated (asymptotically) by $\left(\frac{rY_0}{X}\right)^{\mathrm{rem}(r,k)-1}$. Solving numerically the associated differential equations leads, for $k = 4$ to the values reported in the table in Section 1.

Analysis for Odd k. When k is odd the analysis needs some additional book-keeping to model the process that ensures that the matching is actually k-separated. Let V_i^+ (V_i^-), for each $i \in \{1, \ldots, r-1\}$, be the collection of vertices of degree i in G which are at distance at least two (resp. one) from a matching edge, and define $Y_{r-i}^{\mathrm{sgn}} = |V_i^{\mathrm{sgn}}|$ for each sgn $\in$ {"+", "−"} (with $Y_0 = |V_r|$ as usual). Let $\boldsymbol{D}$ denote an arbitrary configuration around u. For arbitrary k, configurations describe, essentially, the sequence of degrees of the vertices at distance at most $\lceil k/2 \rceil$ from u. *Good* configurations are those leading to an increas in $|\mathcal{M}|$. The expected change in Y_i^{sgn} can be expressed in the "usual" way in terms of a linear combination of conditional expectations. Function $\mathrm{E}(\Delta Y_i^{\mathrm{sgn}} \mid \boldsymbol{D})$ takes different expressions depending on whether $\boldsymbol{D}$ is a good configuration or not. If $\boldsymbol{D}$ is not good then the conditional expectation is a.a.s.

$$-\delta_{i,0} - \sum(\delta_{i,r-(|d_h|+1)} - \delta_{i,r-|d_h|})$$

(the sum being over all d_h referring to vertices in $\Gamma(u)$ with sign sgn). If $\boldsymbol{D}$ is good then $\mathrm{E}(\Delta Y_i^{\mathrm{sgn}} \mid \boldsymbol{D})$ has a contribution $-\delta_{i,0}$ for each vertex of $T_k(r)$, a second one of the form $-\sum \delta_{i,r-(|d_h|+1)}$ for each vertex not in $T_k(r)$ adjacent to a leaf of such tree and a third one, $\sum \delta_{i,r-|d_h|}$, present only if sgn $\equiv$ "−". In other words in the case of a good configuration each of the Y_i^+ is decreased, whereas the Y_i^- can be increased due to constributions from either Y_{i-1}^+ or Y_{i-1}^-.

3 The Upper Bounds

Consider a random n-vertex r-regular graph G generated using the pairing model given in Section 1. Let Q_k be the event "G contains at least one k-matching of size y". We will show that there exists a positive real number μ_k such that when $\mu > \mu_k$ and $y = \mu n$, then $\Pr[Q_k] = o(1)$, thus proving the upper bound in Theorem 1. Let Q'_k be the corresponding event defined on pairings that correspond to n-vertex r-regular graphs. It is well known [18, Section 2.2] that there is a constant c_r such that $\Pr[Q_k]$ is, asymptotically $c_r \Pr[Q'_k]$, hence we can estimate $\Pr[Q_k]$ by performing all our calculations using the pairing model.

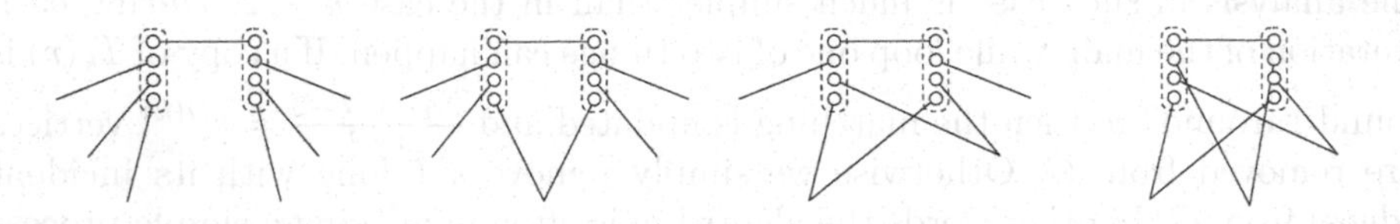

Fig. 1. Different graphs in $\mathcal{H}_{4,3}$. The vertices u_1 and u_2 are represented as a collection of points (in the pairing model).

Let $N(i) = \frac{(2i)!}{i!2^i}$. Let $X_k = X_k(G, r, y)$ be the number of k-matchings of size y in a random pairing. We calculate an asymptotic expression for $\mathrm{E}(X_k)$ and show that when $y = \mu_k n$, then $\mathrm{E}(X_k) = o(1)$, thus proving the upper bound in Theorem 1 by Markov's inequality. Such number can be computed using linearity of expectation. If $X_{\mathcal{M}}$ is a random indicator equal to one if $\mathcal{M}$ is a k-matching of size y in a random pairing, then $\mathrm{E}(X_k) = \sum_{\mathcal{M}} \mathrm{E}(X_{\mathcal{M}})$. Notice that $\mathrm{E}(X_{\mathcal{M}})$ is simply the probability that $\mathcal{M}$ actually occurs in the random pairing, viz. $N(rn/2)^{-1}$ times the number of ways in which a pairing can be built with $\mathcal{M}$ embedded in it. Such pairings have a very special structure. If $\mathcal{M} = \{e_1, \ldots, e_y\}$ is a k-matching in the given pairing G, then it is possible to define y unlabelled vertex-disjoint induced pairings, $H_1, \ldots, H_y$ such that:

1. e_j belongs to (a labelled version of) H_j for each $j \in \{1, \ldots, y\}$;
2. each such copy of H_j is connected[2] to the rest of G through a set of *socket* urns;
3. all non-socket urns of H_j have degree r in H_j.

Let ψ be a mapping that associates a tuple $(H_{i_1}, \ldots, H_{i_y})$ to each k-matching of size y, where for each $j \in \{1, \ldots, y\}$, (a labelled copy of) H_{i_j} is the largest sub-pairing of G satisfying the three properties described above. It is possible to count the number of k-matchings of a given size by finding the size of the range

[2] Of course the term *connected* is quite ambiguous in this context. A pairing G is a collection of pairs on the given set of points, therefore no part of it can be connected in the classical graph-theoretic sense. Clearly we are referring here to a property of the regular graph corresponding to G.

of ψ and multiplying such number by the number of ways in which H_{i_j} can be linked to the rest of G.

Let $\mathcal{H}_{r,k}$ be the collection of all distinct unlabelled pairings H whose labelled versions may occur in $\psi(\mathcal{M})$ for some k-matching $\mathcal{M}$. It should be remarked that $\mathcal{H}_{r,k} = \mathcal{H}_{r,k+1}$, for every odd k and consequently each sequence $(H_{i_1}, \ldots, H_{i_y})$ will characterise a number of k-matchings and $k+1$-matchings. In the case of a k-matching edges are allowed to connect pairs of sockets in distinct subgraphs, whereas in the case of $k+1$-matchings this is not allowed. For instance, $\mathcal{H}_{r,3}$ is formed by the r distinct graphs H_i, for $i \in \{0, \ldots, r-1\}$, having two vertices of degree r, u_1 and u_2, joined by an edge (this is the edge added to the 3-matching), i vertices of degree two each joined to both u_1 and u_2, and $2(r-1)-i$ vertices of degree one, each joined to either u_1 or u_2 (see Fig. 1). The subgraph H_i has $2r-i$ vertices and it consists of a pairing of $2r+2(r-1)-i+2i = 4r+i-2$ points.

A final additional approximation is needed to make the counting "viable". It is well known [11] that w.h.p. random regular graphs contain very few short cycles. This implies that, w.h.p. most of the k-matchings are described by tuples $(H_0, \ldots, H_0)$.

These remarks in particular imply that

$$\mathrm{E}(X_3) \sim \binom{n}{2y, 2y(r-1)} N(y) r^{2yr} (2y(r-1))! \frac{N((rn/2 - y(2r-1))}{N(rn/2)}$$
$$\mathrm{E}(X_4) \sim \binom{n}{2y, 2y(r-1)} N(y) r^{2yr} (2y(r-1))! \frac{(rn-2yr^2) N((rn/2 - y(2r-1) - 2y(r-1)^2)}{(rn-2yr^2-2y(r-1)^2) N(rn/2)}$$

and a formula valid for arbitrary k can be easily deduced by generalising the reasoning above. Setting $y = \mu n$, and using standard Stirling's approximations to the various factorials, the expressions above have both the form $f_r(\mu)^n$, with f_r being a continuous unimodal function of μ. Therefore, for each r, one can find a value $\mu_k(r)$ such that $f_r(\mu) < 1$ for $\mu > \mu_k(r)$.

Maximal vs. Non-maximal k-Matchings. It should be remarked that slightly smaller values of $\mu_k(r)$ than those reported in the table in Section 1 can be numerically computed counting *maximal* k-matchings (a stronger 0.28206915 value for $\mu_2(3)$ is reported in [6]). However we preferred to keep the simpler exposition presented above as the magnitude of the improvements (less than 10^{-5} for each $r \geq 3$ and $k = 2$ and even smaller for larger k) makes the more convoluted analysis required to perform the counting correctly rather uninteresting.

References

1. J. Aronson, A. Frieze, and B. G. Pittel. Maximum matchings in sparse random graphs: Karp-sipser revisited. *Random Structures and Algorithms*, 12:111–178, 1998.
2. K. Cameron. Induced matchings. *Discrete and Applied Mathematics*, 24(1-3):97–102, 1989.
3. J. R. Cash and A. H. Karp. A variable order runge-kutta method for initial value problems with rapidly varying right-hand sides. *ACM Transactions on Mathematical Software*, 16(3):201–222, 1990.

4. W. Duckworth, D. Manlove, and M. Zito. On the approximability of the maximum induced matching problem. Technical Report TR-2000-56, Department of Computing Science of Glasgow University, April 2000.
5. W. Duckworth, N. C. Wormald, and M. Zito. Maximum induced matchings of random cubic graphs. In D. Z. Du, P. Eades, V. Estivill-Castro, X. Lin, and A. Sharma, editors, *Computing and Combinatorics; 6th Annual International Conference, COCOON'00*, volume 1858 of *Lecture Notes in Computer Science*, pages 34–43. Springer-Verlag, 2000.
6. W. Duckworth, N. C. Wormald, and M. Zito. Maximum induced matchings of random cubic graphs. *Journal of Computational and Applied Mathematics*, 142(1):39–50, 2002. Preliminary version appeared in [5].
7. J. Edmonds. Paths, trees and flowers. *Canadian Journal of Mathematics*, 15:449–467, 1965.
8. P. Erdős. Problems and results in combinatorial analysis and graph theory. *Discrete Mathematics*, 72:81–92, 1988.
9. R. J. Faudree, A. Gyárfas, R. H. Schelp, and Z. Tuza. Induced matchings in bipartite graphs. *Discrete Mathematics*, 78(1-2):83–87, 1989.
10. C. W. Gear. *Numerical Initial Value Problems in Ordinary Differential Equations.* Prentice-Hall, 1971.
11. S. Janson, T. Łuczak, and A. Rucínski. *Random Graphs.* John Wiley and Sons, 2000.
12. J. Liu and H. Zhou. Maximum induced matchings in graphs. *Discrete Mathematics*, 170:277–281, 1997.
13. L. Lovász and M. D. Plummer. *Matching Theory*, volume 29 of *Annals of Discrete Mathematics.* North Holland, 1986.
14. R. Motwani. Average-case analysis of algorithms for matchings and related problems. *Journal of the Association for Computing Machinery*, 41(6):1329–1356, November 1994.
15. A. Steger and M. Yu. On induced matchings. *Discrete Mathematics*, 120:291–295, 1993.
16. L. J. Stockmeyer and V. V. Vazirani. NP-Completeness of some generalizations of the maximum matching problem. *Information Processing Letters*, 15(1):14–19, August 1982.
17. N. C. Wormald. Differential equations for random processes and random graphs. *Annals of Applied Probability*, 5:1217–1235, 1995.
18. N. C. Wormald. The differential equation method for random graph processes and greedy algorithms. In M. Karoński and H. J. Prömel, editors, *Lectures on Approximation and Randomized Algorithms*, pages 73–155. PWN, Warsaw, 1999.
19. M. Zito. Induced matchings in regular graphs and trees. In P. Widmayer, G. Neyer, and S. Eidenbenz, editors, *Graph Theoretic Concepts in Computer Science; 25th International Workshop, WG'99*, volume 1665 of *Lecture Notes in Computer Science*, pages 89–100. Springer-Verlag, 1999.
20. M. Zito. *Randomised Techniques in Combinatorial Algorithmics.* PhD thesis, Department of Computer Science, University of Warwick, 1999.

Appendix

A *(discrete time) random process* is a probability space Ω defined over sequences of values $(q_0, q_1, \ldots)$ of some set S. We call $H_t = (q_0, q_1, \ldots, q_t)$ the *history* of the random process up to time t. If y is a function defined on histories, the random variable $y(H_t)$ will be denoted simply by $Y(t)$. The following theorem deals with sequences of random processes Ω_n for $n = 1, 2, 3, \ldots$. Hence, for instance, $q_i = q_i(n)$, and $S = S_n$, although the dependency on n will be usually dropped for simplicity. A function $f(x_1, \ldots, x_j)$ satisfies a *Lipschitz condition* on $\mathcal{D} \subseteq \mathbb{R}^j$ if there is a positive constant L such that

$$|f(x_1, \ldots, x_j) - f(y_1, \ldots, y_j)| \leq L \max_{1 \leq i \leq j} |x_i - y_i|$$

for all $(x_1, \ldots, x_j), (y_1, \ldots, y_j) \in \mathcal{D}$. In such a case f is said to be Lipschitz on $\mathcal{D}$. A useful sufficient condition for f to be Lipschitz on a given domain $\mathcal{D}$ is that all partial derivatives of f are continuous and bounded in $\mathcal{D}$.

Theorem 2. [18] *Let a be fixed and let $\mathcal{D}^* \subseteq \mathbb{R}^{a+1}$. For $1 \leq i \leq a$, let $y^{(i)} : \bigcup_n S_n^+ \to \mathbb{R}$ and $f_i : \mathbb{R}^{a+1} \to \mathbb{R}$, such that for some constant C_0 and for all i, $|Y_i(t)| < C_0 n$ for all $H_t \in S_n^+$ for all n. Define the* stopping time $T_\mathcal{D} = T_\mathcal{D}(Y_1, \ldots, Y_a)$ *to be the minimum t such that $(t/n, Y_1(t)/n, \ldots, Y_a(t)/n) \notin \mathcal{D}$ and let $T_{\mathcal{D}^*}$ be defined analogously w.r.t. $\mathcal{D}^*$. Assume the following three conditions hold, where in (ii) and (iii) $\mathcal{D}$ is some bounded connected open set containing the closure of $\{(0, z^{(1)}, \ldots, z^{(a)}) : \Pr[Y_i(t) = z^{(i)}n, 1 \leq i \leq a] \neq 0$ for some $n\}$.*

(i) (Boundedness hypothesis.) For some functions $\beta = \beta(n) \geq 1$ and $\gamma = \gamma(n)$, the probability that $\max_{1 \leq i \leq a} |Y_i(t+1) - Y_i(t)| < \beta$ conditional upon H_t, is at least $1 - \gamma$ for $t < \min\{T_{\mathcal{D}^*}, T_\mathcal{D}\}$.

(ii) (Trend hypothesis.) For some function $\lambda_1 = \lambda_1(n) = o(1)$, for all $i \leq a$

$$|\mathrm{E}(Y_i(t+1) - Y_i(t) \mid H_t)| - f_i(t/n, Y_1(t)/n, \ldots, Y_a(t)/n)| \leq \lambda_1$$

uniformly for $t < \min\{T_{\mathcal{D}^*}, T_\mathcal{D}\}$.

(iii) (Lipschitz hypothesis.) Each function f_i is continuous and Lipschitz on the set $\mathcal{D} \cap \{(\xi, z^{(1)}, \ldots, z^{(a)}) : \xi \geq 0\}$ with the same Lipschitz constant for each i.

Then the following are true

1. *For $(0, \hat{z}^{(1)}, \ldots, \hat{z}^{(a)}) \in \mathcal{D}$ the system of differential equations*

$$\frac{dy_i}{ds} = f_i(s, y_1, \ldots, y_a), \qquad i = 1, \ldots, a, \tag{5}$$

has a unique solution in $\mathcal{D}$ for $y_i : \mathbb{R} \to \mathbb{R}$ passing through

$$y_i(0) = \hat{z}^{(i)}, \qquad 1 \leq i \leq a,$$

and which extends to points arbitrarily close to the boundary of $\mathcal{D}$.

2. *Let $\lambda > \lambda_1 + C_0 n\gamma$ with $\lambda = o(1)$. For a sufficiently large constant C, with probability $1 - O(n\gamma + \frac{\beta}{\lambda}\exp(-\frac{n\lambda^3}{\beta^3}))$,*

$$Y_i(t) = ny_i(t/n) + O(\lambda n) \tag{6}$$

uniformly for $0 \leq t \leq \min\{\sigma n, T_{\mathcal{D}^*}$ *and for each* i, *where* $yz_i(x)$ *is the solution in (5) with* $\hat{z}^{(i)} = Y_i(0)/n$, *and* $\sigma = \sigma(n)$ *is the supremum of those* x *to which the solution can be extended before reaching within* l^∞*-distance* $C\lambda$ *the boundary of* $\mathcal{D}$.

To prove the results for $k \leq 2$, Theorem 2 is applied to the variables $|\mathcal{M}|$ and $Y_0, \ldots, Y_{r-1}$, defined in Section 2.1, independently in each phase. The functions $f_1, f_2, \ldots$ are $\tilde{\mathrm{E}}(\Delta|\mathcal{M}|)$, $\tilde{\mathrm{E}}(\Delta Y_0), \ldots$ respectively. They clearly satisfy the trend hypothesis and simple calculations of their partial derivatives imply that all of them satisfy the Lipschitz hypothesis in $\mathcal{D}_{\epsilon,j}$.

Simpler versions of the same results [18] are sufficient to prove the results in Section 2.2.

A Lower Bound Technique for Nondeterministic Graph-Driven Read-Once-Branching Programs and Its Applications
(Extended Abstract)

Beate Bollig* and Philipp Woelfel*

FB Informatik, LS2, Univ. Dortmund, 44221 Dortmund, Germany,
Fax: +49 (0)231 755-2047,
{bollig,woelfel}@Ls2.cs.uni-dortmund.de

Abstract. We present a new lower bound technique for a restricted Branching Program model, namely for nondeterministic graph-driven read-once Branching Programs (g.d.-BP1s). The technique is derived by drawing a connection between ω-nondeterministic g.d.-BP1s and ω-nondeterministic communication complexity (for the nondeterministic acceptance modes $\omega \in \{\vee, \wedge, \oplus\}$). We apply the technique in order to prove an exponential lower bound for integer multiplication for ω-nondeterministic well-structured g.d.-BP1s. (For $\omega = \oplus$ an exponential lower bound was already obtained in [5] by using a different technique.) Further, we use the lower bound technique to prove for an explicitly defined fnction which can be represented by polynomial size ω-nondeterministic BP1s that it has exponential complexity in the ω-nondeterministic well-structured g.d.-BP1 model for $\omega \in \{\vee, \oplus\}$. This answers an open question from Brosenne, Homeister, and Waack [7], whether the nondeterministic BP1 model is in fact more powerful than the well-structured graph-driven variant.

1 Introduction and Results

Branching Programs (BPs) or equivalently Binary Decision Diagrams (BDDs) belong to the most important nonuniform models of computation. (For a history of results on Branching Programs see, e.g., the monograph of Wegener [19].)

Definition 1.1. *A* Branching Program *on the variable set* $\mathcal{X}_n = \{x_1, \ldots, x_n\}$ *is a directed acyclic graph with one source and two sinks. The internal nodes are marked with variables in* $\mathcal{X}_n$ *and the sinks are labeled with the Boolean constants 0 and 1. Further, each internal node has two outgoing edges, marked with 0 and 1, respectively.*

Let B_n *denote the set of Boolean functions* $f_n : \{0,1\}^n \rightarrow \{0,1\}$. *A Branching Program on* $\mathcal{X}_n$ *represents at each node* v *a function* $f_v \in B_n$ *in the following way. If* v *is a c-sink,* $c \in \{0,1\}$*, then* $f_v = c$ *and if* v *is an internal node*

* Supported in part by DFG grant We 1066

K. Diks et al. (Eds): MFSC 2002, LNCS 2420, pp. 131–142, 2002.

with 0-successor v_0 *and 1-successor* v_1*, then* $f_v = \overline{x_i} f_{v_0} \vee x_i f_{v_1}$. *The function represented by the Branching Program itself is the function represented at the source. The* size *of a Branching Program G is the number of its nodes, denoted by* $|G|$*, and the* Branching Program complexity *of a Boolean function f is the size of the smallest Branching Program representing f.*

Nondeterminism is one of the most powerful concepts in complexity theory. In analogy to the definition of Turing machines, different modes of acceptance have been studied for Branching Programs. The following definition is due to Meinel [16].

Definition 1.2. *Let* Ω *be a set of binary operations. An* Ω-nondeterministic *Branching Program is a Branching Program of which some internal nodes are labeled with an operation* $\omega \in \Omega$ *instead of a variable. Such nodes are called* nondeterministic *nodes, and the function represented at the nondeterministic node* v*, labeled with* ω *and with 0-successor* v_0 *and 1-successor* v_1*, is* $f_v = f_{v_0}\, \omega\, f_{v_1}$. *As in the deterministic case, a nondeterministic Branching Program represents the function which is represented at the source. The* size *of an* Ω*-nondeterministic Branching Program is the number of its deterministic nodes.*

For the ease of notation, we write ω instead of $\{\omega\}$ if the considered set Ω of binary operations is a singleton. In this paper, we investigate the most common acceptance modes OR, AND, and PARITY, denoted by $\vee$, $\wedge$, and $\oplus$, respectively (although our lower bound technique is not limited to these acceptance modes). For certain acceptance modes ω, an alternative way to determine the function value of a function represented by an ω-nondeterministic Branching Program is to count the number of computation paths of an input a which lead to the 1-sink. (A source-to-sink path is a computation path of the input $a = (a_1 \dots a_n)$ if it leaves any deterministic node labeled by x_i over the edge labeled by a_i and any nondeterministic node over an arbitrary edge.) E.g., a $\oplus$-nondeterministic BP accepts an input a if and only if an odd number of computation paths of a lead to the 1-sink.

Deterministic and nondeterministic BPs can be simulated by the corresponding Turing machines, and the BP complexity of a Boolean function is a measure for the space complexity of the corresponding model of sequential computation. Therefore, one is interested in large lower bounds for BPs. Until today, no superpolynomial lower bounds for general BPs representing an explicitly defined function are known. Therefore, various types of restricted BPs have been investigated, and one is interested in refining the proof techniques in order to obtain lower bounds for less restricted BPs. (For the latest breakthrough see e.g. [1], [2], and [3].) There are several reasonable possibilities to restrict BPs, among them restrictions concerning the multiplicity of variable tests or the ordering in which variables may be tested.

Definition 1.3. (i) *A (nondeterministic)* read-once Branching Program *(short:* BP1*) is a (nondeterministic) Branching Program where each variable appears on each computation path at most once.*

(ii) *A (nondeterministic) Branching Program is called* s-oblivious, *for a sequence of variables* $s = (s_1, \ldots, s_l)$, $s_i \in X_n$, *if the set of decision nodes can be partitioned into disjoint sets* V_i, $1 \le i \le l$, *such that all nodes from* V_i *are labeled with* s_i *and the edges which leave* V_i*-nodes reach a sink or a* V_j*-node where* $j > i$.

Besides the theoretical viewpoint people have used BPs in applications. Oblivious BP1s, introduced by Bryant [8] under the term OBDDs, have found a large variety of applications, e.g. in circuit verification. Obliviousness, though, is a very strong restriction. Gergov and Meinel [12] and Sieling and Wegener [18] have independently generalized the concept of obliviousness in the deterministic read-once case in order to show how to use BP1s for verification.

Definition 1.4. *A* graph ordering *is a Branching Program with a single sink, where on each path from the source to the sink all variables appear exactly once. A (nondeterministic)* graph-driven BP1 *(short:* g.d.-BP1*) is a (nondeterministic)* BP1 G *for which there exists a graph ordering* G_0 *with the following property: If for an input* a*, a variable* x_i *appears on the computation path of* a *in* G *before the variable* x_j*, then* x_i *also appears on the unique computation path of* a *in* G_0 *before* x_j.

A (nondeterministic) g.d.-BP1 G *with graph ordering* G_0 *is called* well-structured, *if there exists a mapping* α *from the node set of* G *to the node set of* G_0 *such that for every node* v *in* G *the node* $\alpha(v)$ *is labeled with the same variable as* v*, and such that if a computation path of an input* a *passes through* v*, then the computation path of* a *in* G_0 *passes through* $\alpha(v)$.

The main idea is that in g.d.-BP1s with the graph ordering G_0 for each input the variables are tested in the same ordering, whereas (different from OBDDs) for different inputs different orderings may be used. The stronger structural property of the well-structured model leads to the design of simpler and faster algorithms in the deterministic case [18]. The difference between the two models is the following one. In a general graph-driven Branching Program it is possible that the computation paths of two different inputs pass through the same node labeled by x_i, whereas in the graph ordering they pass through different nodes labeled by x_i. This is not allowed in the well-structured case. For the parity case Brosenne, Homeister, and Waack [7] were the first ones realizing that the property of beeing well-structured can be used to determine the minimal number of nodes which are necessary to represent a Boolean function f. Until now well-structured $\oplus$-BP1s are the most general parity Branching Programs (without any restriction on the number of nondeterministic nodes) for which exponential lower bounds for explicitly defined functions are known.

It is easy to see that any BP1 is in fact a well-structured g.d.-BP1 for a suitably chosen graph ordering but for nondeterministic BP1s graph orderings do not exist in general. Hence, it is an intriguing question, whether nondeterministic (well-structured) g.d.-BP1s are in fact significantly more restricted than general nondeterministic BP1s. One of the main contributions of this paper is that we answer this question for the well-structured case in an affirmative way for the most important nondeterministic acceptance modes. This is done by presenting a function called $n/2$-MRC$_n$, which can be represented in polynomial size by ω-nondeterministic BP1s but has exponential complexity in the ω-nondeterministic well-structured g.d.-BP1 model (for $\omega \in \{\vee, \oplus\}$). Note that an analogous separation result for $\omega = \wedge$ follows right away from de Morgan's rules for the complement of $n/2$-MRC$_n$.

In order to prove the separation result, we derive a new lower bound technique. Until now, there was only one general lower bound technique known for nondeterministic well-structured g.d.-BP1s, which in addition worked only for the parity-acceptance mode [7]. We follow a more general approach by drawing connections to communication complexity. Hence, our lower bound technique can be applied to all acceptance modes, where corresponding lower bounds for communication complexity can be proven.

As another application of our lower bound technique, we prove an exponential lower bound for integer multiplication. Lower Bounds for integer multiplication are motivated by the general interest in the complexity of important arithmetic functions and the insight into the structure of such functions which is often gained by lower bound proofs. Furthermore, since exponential lower bounds are often proven for functions which are "designed" in such a way that they fit to a given lower bound technique, the lower bound proofs for important functions can lead to refinements of the proof techniques.

Definition 1.5. *The Boolean function* $\mathrm{MUL}_{i,n} \in B_{2n}$ *maps two* n*-bit integers* $x = x_{n-1} \dots x_0$ *and* $y = y_{n-1} \dots y_0$ *to the* i*th bit of their product, i.e.* $\mathrm{MUL}_{i,n}(x, y) = z_i$, *where* $x \cdot y = z_{2n-1} \dots z_0$.

Since the middle bit (the bit $n-1$) of integer multiplication is the hardest bit to compute, one is interested mainly in the complexity of $\mathrm{MUL}_n := \mathrm{MUL}_{n-1,n}$. Bryant [9] has proven an exponential lower bound of $2^{n/8}$ for the function MUL_n in the OBDD model, and Gergov has presented an exponential lower bound for nondeterministic linear-length oblivious Branching Programs [11]. Later Ponzio has shown that the complexity of this function is $2^{\Omega(\sqrt{n})}$ for BP1s [17], and Bollig [4] has proven an exponential lower bound for nondeterministic tree-driven BP1s (i.e. g.d.-BP1s where the graph ordering is a tree of polynomial size).

Recently, progress in the analysis of MUL_n has been achieved by a new approach using universal hashing. Woelfel [20] has improved Bryant's lower bound to $\Omega(2^{n/2})$ and Bollig and Woelfel [6] have presented a lower bound of $\Omega(2^{n/4})$ for BP1s. Finally, Bollig, Waack, and Woelfel [5] have proven a lower bound of $2^{(n-46)/12}/n$ for $\oplus$-nondeterministic well-structured g.d.-BP1s. Their proof, though, is limited to this type of acceptance mode.

Until now exponential lower bounds for MUL_n for unrestricted nondeterministic BP1s are unknown. One step towards proving such bounds might be to investigate BP models "inbetween" the deterministic and the nondeterministic BP1s. This was also the motivation behind a result in [21] where an exponential lower bound has been proven for nondeterministic BP1s which have only a restricted number of nondeterministic nodes at the top of the BP1.

The lower bound for integer multiplication presented here is $2^{n/12-4} \cdot n^{-1/3}$ and is valid for all ω-nondeterministic well-structured g.d.-BP1s where $\omega \in \{\vee, \wedge, \oplus\}$. Comparing with the algebraic approach of [5], one advantage is that using methods from communication complexity, all important types of nondeterminism can be handled simultaneously.

Due to the lack of space we have to omit some of the proofs.

2 A Lower Bound Technique for Nondeterministic Graph-Driven BP1s

Methods from communication complexity have been used to prove lower bounds in several Branching Program models, e.g. for OBDDs. (See e.g. [13,15] for the theory of communication complexity.) Consider a Boolean function $f \in B_n$ which is defined on the variables in $\mathcal{X}_n = \{x_1, \ldots, x_n\}$, and let $\Pi = (\mathcal{X}_A, \mathcal{X}_B)$ be a partition of $\mathcal{X}_n$. Assume that Alice has access only to the input variables in $\mathcal{X}_A$ and Bob has access only to the input variables in $\mathcal{X}_B$. In a one-way communication protocol, upon a given input x, Alice is allowed to send a single message (depending on the input variables in $\mathcal{X}_A$) to Bob who must then be able to compute the answer $f(x)$. In an ω-nondeterministic communication protocol, $\omega \in \{\vee, \wedge, \oplus\}$, Alice is allowed to "guess" a message. The function value is one if the number of guesses upon which Bob accepts the input matches the corresponding acceptance mode ω (e.g. is at least one in the case of $\omega = \vee$ or odd in case of $\omega = \oplus$). The *ω-nondeterministic one-way communication complexity* of the function f is the number of bits of communication which need to be transmitted by such a protocol that computes f. It is denoted by $\mathrm{ND}_\omega^{A\to B}(f, \Pi)$.

In order to state the lower bound technique for nondeterministic g.d.-BP1s, we have to introduce some further notation, first. A *filter* of a set X is a closed upward subset of 2^X (i.e. if $S \in \mathcal{F}$, then all supersets of S are in $\mathcal{F}$). Let $\mathcal{F}$ be a filter of $\mathcal{X}_n = \{x_1, \ldots, x_n\}$. A subset $B \subseteq \mathcal{X}_n$ is said to be in the *boundary* of $\mathcal{F}$ if $B \notin \mathcal{F}$ but $B \cup \{x_i\} \in \mathcal{F}$ for some $x_i \in \mathcal{X}_n$.

Let f be a function in B_n defined on the variables in $\mathcal{X}_n$ and $\mathcal{F}$ be a filter of $\mathcal{X}_n$. For a subset $Z \subseteq \mathcal{X}_n$, we denote by $\mathcal{A}(Z)$ the set of all possible assignments to the variables in Z. Let $\Pi = (\mathcal{X}_A, \mathcal{X}_B)$ be a partition of $\mathcal{X}_n$. If $\mathcal{X}_B$ is in the boundary of $\mathcal{F}$, then Π is called *$\mathcal{F}$-partition* of $\mathcal{X}_n$. Finally, a function $f' \in B_n$ is called *(ϵ, Π)-close* to f, if there exists a set $R \subseteq \mathcal{A}(\mathcal{X}_A)$ with $|R| \geq \epsilon \cdot 2^{|\mathcal{X}_A|}$, such that f and f' coincide on all inputs in $R \times \mathcal{A}(\mathcal{X}_B)$.

Theorem 2.1. *Let $\mathcal{F}$ be a filter on $\mathcal{X}_n$, $f \in B_n$, $0 < \epsilon \leq 1$ and $\ell \in \mathbb{N}$. If for every $\mathcal{F}$-partition Π of $\mathcal{X}_n$ and for every function f' which is (ϵ, Π)-close to f it is $\mathrm{ND}_\omega^{A\to B}(f', \Pi) > \ell$, then any graph-driven ω-nondeterministic* BP1 *representing f either has a size of at least $2^\ell + 1$ or its graph ordering has a size of more than $1/\epsilon$ (for $\omega \in \{\vee, \wedge, \oplus\}$).*

The above technique does not yield lower bounds for nondeterministic g.d.-BP1s directly, because the size of the graph ordering of such a Branching Program is not part of the nondeterministic g.d.-BP1 size. Until now it is unknown whether there exists a class of functions f_n which has polynomial complexity in the nondeterministic g.d.-BP1 model whereas the size of every graph ordering of a polynomial size nondeterministic g.d.-BP1 for f_n is exponential. The situation is different in the well-structured case as Bollig, Waack, and Woelfel [5] have shown by the following proposition.

Proposition 2.2 ([5]). *For any well-structured nondeterministic graph driven* BP1 *G on n variables, there exists a graph ordering G_0 such that G is G_0-driven and $|G_0| \le 2n|G|$.*

Corollary 2.3. *Let $f \in B_n$ be a function satisfying the conditions of Theorem 2.1 for some filter $\mathcal{F}$ on $\mathcal{X}_n$ and the parameters ϵ and ℓ. Then any well-structured ω-nondeterministic graph driven* BP1 *for f has a size of more than* $\min\{2^\ell, (\epsilon \cdot 2n)^{-1}\}$.

3 An Exponential Lower Bound for Integer Multiplication

As a first application of the lower bound technique, we prove a lower bound for integer multiplication. We consider here the Boolean function $\mathrm{MUL}_n^* \in B_{2n-2}$; this is the subfunction of MUL_n, which takes as inputs only odd integers (i.e. the least significant bits of the two n-bit factors are fixed to 1). Obviously, a lower bound on the (nondeterministic) communication complexity of MUL_n^* implies the same lower bound for MUL_n.

The following lemma describes the connection between integer multiplication and nondeterministic communication complexity, which we need to apply Corollary 2.3. It is well known that a large nondeterministic communication complexity can be shown by proving that the communication matrix according to a given partition Π contains a large triangular submatrix (this follows e.g. from the methods in [10]). Note that we use the term *submatrix* here in the common combinatorial sense, which means that each submatrix is obtained from a matrix M by selecting an arbitrary set of rows and columns of M and ordering them arbitrarily.

Lemma 3.1. *Let $A, B \subseteq \mathbb{Z}_{2^n}$ and $Y \subseteq \mathbb{Z}_{2^n}^* := \{1, 3, \ldots, 2^n - 1\}$ and assume that $|B| = 2^\beta$ and $|Y| = 2^\mu$. Consider the $|A| \times |B \times Y|$-matrix M, where each row is identified with an integer $a \in A$ and each column is identified with a pair $(b, y) \in B \times Y$, and the entry of the matrix in row a and column (b, y) equals $\mathrm{MUL}_n^*(a + b,\ y)$. Then M contains a triangular $s \times s$-submatrix where* $s = \min\{|A|/2 - 1,\ 2^{(3\mu+\beta-3n-10)/4} - 1\}$.

In order to prove Lemma 3.1, we need to recall some properties about integer multiplication which have been derived by Bollig and Woelfel [6] and Bollig, Woelfel, and Waack [5] using universal hashing. Let $\mathbb{Z}_{2^n}^*$ be the set of odd n-bit integers.

Lemma 3.2 ([6,5]). *Let $X \subseteq \mathbb{Z}_{2^n}$ and $Y \subseteq \mathbb{Z}_{2^n}^*$. If $|X| \cdot |Y| \ge 2^{n+2r+1}$, $r \ge 0$, then there exists an element $y \in Y$ such that*

$$\forall q \in \{0, \ldots, 2^r - 1\}\ \exists x \in X:\ q \cdot 2^{n-r} \le (xy) \bmod 2^n < (q+1) \cdot 2^{n-r}.$$

Lemma 3.3 **([5]).** *Let* $Y \subseteq \mathbb{Z}_{2^n}^*$, $1 \le r \le n-1$ *and* $(z_i, z_i') \in \mathbb{Z}_{2^n} \times \mathbb{Z}_{2^n}$, *where* $z_i \ne z_i'$ *for* $1 \le i \le t$. *Then there exists a subset* $Y' \subseteq Y$, $|Y'| \ge |Y| - t \cdot 2^{n-r+1}$, *such that for all pairs* (z_i, z_i'), $1 \le i \le t$,

$$\forall y \in Y' : \; 2 \cdot 2^{n-r} \le ((z_i - z_i')y) \bmod 2^n \le 2^n - 2 \cdot 2^{n-r}.$$

Proof (of Lemma 3.1). We show below that there exist an element $y \in Y$, a subset $\{a_1, \dots, a_{s+1}\} \subseteq A$ and a subset $\{b_1, \dots, b_s\} \subseteq B$ such that for all $1 \le j \le s+1$ and $1 \le i \le s$

$$\mathrm{MUL}_n^*(a_j + b_i, y) = \begin{cases} 0 & \text{if } i \ge j \\ 1 & \text{if } i < j. \end{cases} \tag{1}$$

This means that the $s \times s$-submatrix of M consisting of the rows $a_2, \dots, a_{s+1}$ and of the columns $(b_1, y), \dots, (b_s, y)$ is triangular.

Let $r = (\mu + \beta - n)/2 - 1$. If $|A| \le 2^{(3\mu+\beta-3n-6)/4}$, then we let $A' = A$. Otherwise, we let A' be an arbitrary subset of A containing exactly $2^{(3\mu+\beta-3n-6)/4}$ elements.

Consider now the $t = |A'|(|A'| - 1)$ pairs (z_i, z_i'), $1 \le i \le t$, with $z_i, z_i' \in A'$ and $z_i \ne z_i'$. Applying Lemma 3.3, we obtain a subset $Y' \subseteq Y$, $|Y'| \ge |Y| - |A'|^2 \cdot 2^{n-r+1}$, such that for all different $a, a' \in A'$ it holds

$$\forall y \in Y' : \; 2 \cdot 2^{n-r} \le ((a - a')y) \bmod 2^n \le 2^n - 2 \cdot 2^{n-r}. \tag{2}$$

Then

$$\begin{aligned} |B| \cdot |Y'| &\ge |B| \cdot |Y| - |B| \cdot |A'|^2 \cdot 2^{n-r+1} \ge 2^{\beta+\mu} - 2^{\beta+(3\mu+\beta-3n-6)/2+n-r+1} \\ &= 2^{\beta+\mu} - 2^{\beta+\mu+(\mu+\beta-n)/2-1-r-1} = 2^{\beta+\mu} - 2^{\beta+\mu-1} = 2^{\beta+\mu-1} \\ &= 2^{n+2r+1}. \end{aligned}$$

Therefore, we may apply Lemma 3.2 (with $X = B$) in order to see that there exists an element $y \in Y'$ such that

$$\forall q \in \{0, \dots, 2^r - 1\} \, \exists b \in B : \; q \cdot 2^{n-r} \le (by) \bmod 2^n < (q+1) \cdot 2^{n-r}. \tag{3}$$

We let this element $y \in Y'$ be fixed from now on. Further, let

$$A'_< = \left\{ a \in A' \mid (ay) \bmod 2^n < 2^{n-1} \right\}$$

and

$$A'_\ge = \left\{ a \in A \mid (ay) \bmod 2^n \ge 2^{n-1} \right\}.$$

We choose A^* to be the set which has at least as many elements as the other one. Hence,

$$|A^*| \ge |A'|/2 \ge \min\left\{ |A|, 2^{(3\mu+\beta-3n-6)/4} \right\}/2 = s+1.$$

We consider only the case where A^* equals $|A'_<|$; the other case is symmetric and can be proven analogously. We label the elements in A^* by $a_1, \ldots, a_{s+1}$ in such a way that

$$0 \le (a_1 y) \bmod 2^n \le \ldots \le (a_{s+1} y) \bmod 2^n < 2^{n-1}. \quad (4)$$

Then we obtain by (2) that

$$\forall 1 \le i \le s: \ (a_i y) \bmod 2^n + 2 \cdot 2^{n-r} \le (a_{i+1} y) \bmod 2^n. \quad (5)$$

For $1 \le i \le s$ we let now

$$q_i := \left\lfloor \frac{2^{n-1} - (a_i y) \bmod 2^n}{2^{n-r}} \right\rfloor - 1 \quad (6)$$

and choose $b_i \in B$ such that

$$q_i \cdot 2^{n-r} \le (b_i y) \bmod 2^n < (q_i + 1) \cdot 2^{n-r}. \quad (7)$$

(Such a b_i exists because of (3)). Hence, we get for $1 \le j \le i$

$$(a_j y) \bmod 2^n + (b_i y) \bmod 2^n \overset{(4),(7)}{<} (a_i y) \bmod 2^n + (q_i + 1) \cdot 2^{n-r} \overset{(6)}{\le} 2^{n-1}. \quad (8)$$

Thus, $\big((a_j + b_i) y\big) \bmod 2^n < 2^{n-1}$, which implies $\mathrm{MUL}_n^*(a_j + b_i, y) = 0$. This already proves the claim (1) for the case $i \ge j$.

We consider now the case $i < j$. First of all, we have

$$\begin{aligned}(a_{i+1} y) \bmod 2^n + (b_i y) \bmod 2^n &\overset{(5),(7)}{\ge} (a_i y) \bmod 2^n + 2 \cdot 2^{n-r} + q_i \cdot 2^{n-r} \overset{(6)}{\ge} \\ (a_i y) \bmod 2^n + 2 \cdot 2^{n-r} + 2^{n-1} - (a_i y) \bmod 2^n - 2 \cdot 2^{n-r} &= 2^{n-1}. \quad (9)\end{aligned}$$

Hence, by (4) we also obtain $(a_j y) \bmod 2^n + (b_i y) \bmod 2^n \ge 2^{n-1}$. Thus,

$$\begin{aligned}2^{n-1} &\le (a_j y) \bmod 2^n + (b_i y) \bmod 2^n \\ &= (a_j y) \bmod 2^n - (a_i y) \bmod 2^n + (a_i y) \bmod 2^n + (b_i y) \bmod 2^n \\ &\overset{(4),(8)}{<} 2^{n-1} + 2^{n-1} = 2^n.\end{aligned}$$

These inequalities tell us that $\big((a_j + b_i) y\big) \bmod 2^n \ge 2^{n-1}$, and hence $\mathrm{MUL}_n^*(a_j + b_i) = 1$. Altogether, we have shown (1). □

In order to derive a lower bound for integer multiplication by the use of Theorem 2.1, we need to define an appropriate filter on the input variables. We use the filters $\mathcal{F}_k(Z)$ which are defined on an m-element variable set Z for $1 \le k < m$ as $\mathcal{F}_k(Z) = \{M \subseteq Z \mid |M| \ge m - k + 1\}$. This definition ensures that (Z_A, Z_B) is an $\mathcal{F}_k$-partition if and only if $|Z_A| = k$.

In the following let $\mathcal{X}_{n-1} = \{x_1, \ldots, x_{n-1}\}$ and $\mathcal{Y}_{n-1} = \{y_1, \ldots, y_{n-1}\}$ be the input variables for the odd x- and the y-integer, which are multiplied by MUL_n^*.

Lemma 3.4. *Let* $k = \lceil n/3 + 2/3 \log(n-1) - 9/2 \rceil$ *and* $\epsilon = 2^{n/4-k-5/2}$. *Further, let* $\mathcal{X}_A, \mathcal{X}_B \subseteq \mathcal{X}_{n-1}$ *and* $\mathcal{Y}_A, \mathcal{Y}_B \subseteq \mathcal{Y}_{n-1}$. *If* $\Pi = (\mathcal{X}_A \cup \mathcal{Y}_A, \mathcal{X}_B \cup \mathcal{Y}_B)$ *is an* $\mathcal{F}_k(\mathcal{X}_{n-1} \cup \mathcal{Y}_{n-1})$*-partition of* $\mathcal{X}_{n-1} \cup \mathcal{Y}_{n-1}$ *and* f' *is* (ϵ, Π)*-close to* MUL_n^*, *then* $\mathrm{ND}_\omega^{A\to B}(f', \Pi) \geq n/12 - \log(n-1)/3 - 3$ *for any* $\omega \in \{\vee, \wedge, \oplus\}$.

A simple calculation using the parameters from the lemma above shows that $(\epsilon \cdot 4(n-1))^{-1} \geq 2^{n/12-\log(n-1)/3-4}$. Using Corollary 2.3, this yields the following exponential lower bound for well-structured g.d.-BP1s representing MUL_n.

Corollary 3.5. *Let* $\omega \in \{\vee, \wedge, \oplus\}$. *The size of any well-structured* ω*-nondeterministic graph-driven* BP1 *for* MUL_n *is larger than* $2^{n/12-4} \cdot (n-1)^{-1/3}$.

4 Separating Well-Structured Nondeterministic Graph-Driven BP1s from Nondeterministic BP1s

Here we answer an open question from Brosenne, Homeister, and Waack [7] whether the class of all Boolean functions representable in polynomial size by ω-nondeterministic well-structured graph-driven BP1s is a proper subclass of all Boolean functions representable in polynomial size by ω-nondeterministic BP1s in an affirmative way.

The function $n/2$-MRC_n is defined on an $n \times n$ Boolean matrix X on the variables $\mathcal{X}_{n\times n} = \{x_{1,1}, \ldots, x_{n,n}\}$. Its function value is 1 if and only if the following two conditions are fulfilled (for the sake of readability we assume that n is an even number.)

1. The number of ones in the matrix is at least $n^2/4+n$ and at most $(3/4)n^2-n$.
2. The matrix either contains exactly $n/2$ monochromatic rows and each non-monochromatic row contains exactly $n/2$ ones, or it contains exactly $n/2$ monochromatic columns and each non-monochromatic column contains exactly $n/2$ ones.

Note that because of condition 1, there cannot be $n/2$ monochromatic rows and $n/2$ monochromatic columns for a satisfying input. Furthermore, if condition 2 is satisfied, then condition 1 is fulfilled if and only if at least one of the monochromatic rows (columns) satisfying condition 2 consists only of ones, and at least one of the monochromatic rows (columns) consists only of zeros.

The Branching Program model for which we show the upper bound is even more restricted than the general ω-nondeterministic BP1 model.

Definition 4.1. *An* (ω, k)*-PBDD* G *consists of* k *OBDDs* $G_1, \ldots, G_k$ *whoses variable orderings may be different. If* $f_1, \ldots, f_k$ *are the functions represented by* $G_1, \ldots, G_k$, *then* G *represents the function* $f_1 \,\omega\, f_2 \,\omega \cdots \omega\, f_k$. *The size of* G *is* $|G| = |G_1| + \ldots + |G_k|$.

Note that we can regard an (ω, k)-PBDD as an ω-nondeterministic BP1 which has $k-1$ nondeterministic nodes at the top, which generate k paths leading to the disjoint OBDDs $G_1, \ldots, G_k$. Motivated by applications, the model of $(\vee, k)$-PBDDs has been introduced in [14].

Theorem 4.2. *For $\omega \in \{\vee, \oplus\}$, the function $n/2$-MRC$_n$ can be represented by $(\omega, 2)$-PBDDs with size $O(n^4)$, but its complexity is $\Omega(2^{n/4}/n)$ for well-structured ω-nondeterministic graph-driven* BP1*s.*

A rowwise (columnwise) variable ordering is an ordering, where all variables of one row (column) are tested one after another. The existence of an $(\omega, 2)$-PBDD for $n/2$-MRC$_n$ can be proven in a straight forward way by realizing that there exists an OBDD of size $O(n^4)$, testing the variables in a rowwise (columnwise) ordering and computing $n/2$-MRC$_n$ correctly if the input contains exactly $n/2$ monochromatic rows (columns), and returning 0 otherwise. Because any satisfying input contains either $n/2$ monochromatic rows or $n/2$ monochromatic columns, but not both, this is also sufficient for the parity case. Due to space restrictions, we omit the full proof, but instead focus in the rest of this section on proving the lower bound of the theorem.

We apply again the technique from Corollary 2.3. In order to do so, we have to define an appropriate filter $\mathcal{F}_{\mathrm{M}}$ on the variable set $\mathcal{X}_{n\times n}$. A set $T \subseteq \mathcal{X}_{n\times n}$ is in the filter $\mathcal{F}_{\mathrm{M}}$, if T contains all variables from $n/2+1$ arbitrary rows and $n/2+1$ arbitrary columns. If $\Pi = (\mathcal{X}_A, \mathcal{X}_B)$ is an $\mathcal{F}_{\mathrm{M}}$-partition, then by definition $\mathcal{X}_B \notin \mathcal{F}_{\mathrm{M}}$ and there exists a variable $x_{i,j}$ such that $\mathcal{X}_B \cup \{x_{i,j}\} \in \mathcal{F}_{\mathrm{M}}$. Hence, $\mathcal{X}_A$ contains variables from exactly $n/2$ different rows and from at most $n/2$ different columns or vice versa.

The lower bound of Theorem 4.2 follows right away from the following lemma and Corollary 2.3 by the choice $\epsilon = 1/\left(n \cdot 2^{n/4}\right)$.

Lemma 4.3. *Let $0 < \epsilon \le 1$ and Π be an arbitrary $\mathcal{F}_{\mathrm{M}}$-partition of $\mathcal{X}_{n\times n}$. Then for every function f' which is (ϵ, Π)-close to $n/2$-MRC$_n$, it is $\mathrm{ND}_\omega^{A\to B}(f', \Pi) \ge n/2 + \log \epsilon$.*

Proof. Let $\Pi = (\mathcal{X}_A, \mathcal{X}_B)$ be an $\mathcal{F}_{\mathrm{M}}$-partition and f' be (ϵ, Π)-close to $n/2$-MRC$_n$. We may assume w.l.o.g. that $\mathcal{X}_A$ contains variables from exactly the rows $1, \ldots, n/2$, whereas there are at most $n/2$ columns from which variables are contained in $\mathcal{X}_A$. Since f' is (ϵ, Π)-close to $n/2$-MRC$_n$, there exists a subset $R \subseteq \mathcal{A}(\mathcal{X}_A)$, $|R| \ge \epsilon \cdot 2^{|\mathcal{X}_A|}$, such that f' coincides with $n/2$-MRC$_n$ on all inputs in $R \times \mathcal{A}(\mathcal{X}_B)$. For $1 \le i \le n/2$ let k_i be the number of variables in row i which are contained in $\mathcal{X}_A$. We consider the mapping

$$\mu : \mathcal{A}(\mathcal{X}_A) \to \{0, \ldots, k_1\} \times \ldots \times \{0, \ldots, k_{n/2}\},$$

which maps a partial assignment α to the tuple $\mu(\alpha) = (z_1, \ldots, z_{n/2})$, where z_i is the number of bits in row i being fixed to 1 by α.

Let $\mu(R) = \{\mu(\alpha) \mid \alpha \in R\}$. Below, we show the following two inequalities from which the lemma follows right away.

(I1) $\mathrm{ND}_\omega^{A\to B}(f', \Pi) \ \ge \ \log |\mu(R)|$.
(I2) $|\mu(R)| \ \ge \ \epsilon \cdot 2^{n/2}$.

Proof of (I1): We show that the communication matrix contains a diagonal $s \times s$-submatrix, where $s = |\mu(R)|$. For an arbitrary partial assignment $\alpha \in$

R let $\mu(\alpha) = \big(\mu_1(\alpha), \ldots, \mu_{n/2}(\alpha)\big)$. We fix for each such α a corresponding partial assignment $\beta \in \mathcal{A}(\mathcal{X}_B)$ as follows. In row i, $1 \le i \le n/2$, β sets exactly $n/2 - \mu_i(\alpha)$ variables to 1 and the other variables to zero. (Recall that $\mathcal{X}_A$ contains variables from at most $n/2$ columns, and hence at least $n/2$ variables from each row are in $\mathcal{X}_B$.) All the variables in the rows $n/2+1, \ldots, n-1$ are fixed to 0 and the variables in row n are all set to 1. Then $(\alpha\beta)$ contains exactly $n/2$ rows with exactly $n/2$ ones each (the rows $1, \ldots, n/2$), and it contains $n/2-1$ 0-monochromatic rows and one 1-monochromatic row. Hence, $n/2\text{-MRC}_n(\alpha\beta) = 1$.

We consider now s arbitrary partial assignments $\alpha_1, \ldots, \alpha_s \in R$ such that $\mu(\alpha_i) \ne \mu(\alpha_j)$ for $i \ne j$. Let $\beta_1, \ldots, \beta_s$ be the corresponding partial assignments in $\mathcal{A}(\mathcal{X}_B)$. (It is obvious that also $\beta_i \ne \beta_j$ for $i \ne j$.) Clearly, the $s \times s$-matrix which has in row i and column j the entry $n/2\text{-MRC}_n(\alpha_i\beta_j)$ is a submatrix of the communication matrix of $n/2\text{-MRC}_n$. Hence, for the claim (I1), it suffices to show that this matrix is a diagonal matrix. For the diagonal elements, we have already proven above that $n/2\text{-MRC}_n(\alpha_i\beta_i) = 1$. Consider now an element in row i and column j, $i \ne j$. Since $\alpha_i \ne \alpha_j$, there exists an index $1 \le t \le n/2$ for which $\mu_t(\alpha_i) \ne \mu_t(\alpha_j)$. Hence, by construction the matrix X defined by the input $\alpha_j\beta_i$ contains in row t not exactly $n/2$ ones. But the construction also ensures that none of the rows $n/2+1, \ldots, n$ of X contains exactly $n/2$ ones, thus there exist less than $n/2$ rows with exactly $n/2$ ones. Finally, the property that row n is 1-monochromatic and the row $n-1$ is 0-monochromatic ensures that there exists no monochromatic column. Altogether, this yields that $n/2\text{-MRC}_n(\alpha_i, \beta_j) = 0$.

Proof of (I2): Recall that $\mathcal{X}_A$ contains k_i variables in row i of the matrix X $(1 \le i \le n/2)$. Hence, there are exactly 2^{k_i} possible settings of those variables in row i and among these, there are $\binom{k_i}{z_i}$ settings for which row i contains exactly z_i ones. Hence, for every tuple $z = (z_1, \ldots, z_{n/2}) \in \{0, \ldots, k_1\} \times \ldots \times \{0, \ldots, k_{n/2}\}$ we obtain that

$$\frac{|\mu^{-1}(z)|}{|\mathcal{A}(\mathcal{X}_A)|} = \frac{\binom{k_1}{z_1} \cdot \ldots \cdot \binom{k_{n/2}}{z_{n/2}}}{2^{k_1} \cdot \ldots \cdot 2^{k_{n/2}}} \le \frac{2^{k_1-1} \cdot \ldots \cdot 2^{k_{n/2}-1}}{2^{k_1} \cdot \ldots \cdot 2^{k_{n/2}}} = 2^{-n/2}. \tag{10}$$

Since R is the union of all $\mu^{-1}(z)$ for $z \in \mu(R)$, there exists by the pigeonhole principle an element $z \in \mu(R)$ for which $|\mu^{-1}(z)| \ge |R|/|\mu(R)|$. Using the precondition that $|R| \ge \epsilon \cdot 2^{|\mathcal{X}_A|}$ together with inequality (10) yields

$$|\mu(R)| \ge \frac{|R|}{|\mu^{-1}(z)|} \ge \frac{\epsilon \cdot 2^{|\mathcal{X}_A|}}{2^{-n/2} \cdot |\mathcal{A}(\mathcal{X}_A)|} = \epsilon \cdot 2^{n/2}.$$

This finally proves (I2). □

Acknowledgment

We would like to thank Martin Sauerhoff and Ingo Wegener for fruitful discussions about the subject of this paper and helpful comments.

References

1. M. Ajtai. A non-linear time lower bound for boolean branching programs. In *Proc. of 40th FOCS*, pp. 60–70. 1999.
2. P. Beame, M. Saks, X. Sun, and E. Vee. Super-linear time-space tradeoff lower bounds for randomized computation. In *Proc. of 41st FOCS*, pp. 169–179. 2000.
3. P. Beame and E. Vee. Time-space tradeoffs, multiparty communication complexity, and nearest neighbor problems. In *Proc. of 34th ACM STOC*. 2002. To appear.
4. B. Bollig. Restricted nondeterministic read-once branching programs and an exponential lower bound for integer multiplication. *RAIRO*, 35:149–162, 2001.
5. B. Bollig, S. Waack, and P. Woelfel. Parity graph-driven read-once branching programs and an exponential lower bound for integer multiplication. In *Proc. of 2nd TCS*. 2002. To appear.
6. B. Bollig and P. Woelfel. A read-once branching program lower bound of $\Omega(2^{n/4})$ for integer multiplication using universal hashing. In *Proc. of 33rd ACM STOC*, pp. 419–424. 2001.
7. H. Brosenne, M. Homeister, and S. Waack. Graph-driven free parity BDDs: Algorithms and lower bounds. In *Proc. of 26th MFCS*, pp. 212–223. 2001.
8. R. E. Bryant. Graph-based algorithms for boolean function manipulation. *IEEE Trans. on Comp.*, C-35:677–691, 1986.
9. R. E. Bryant. On the complexity of VLSI implementations and graph representations of boolean functions with applications to integer multiplication. *IEEE Trans. on Comp.*, 40:205–213, 1991.
10. C. Damm, M. Krause, C. Meinel, and S. Waack. Separating counting communication complexity classes. In *Proc. of 9th STACS*, pp. 281–292. 1992.
11. J. Gergov. Time-space tradeoffs for integer multiplication on various types of input oblivious sequential machines. *Information Processing Letters*, 51:265–269, 1994.
12. J. Gergov and C. Meinel. Efficient Analysis and Manipulation of OBDDs can be extended to FBDDs. *IEEE Trans. on Comp.*, 43:1197–1209, 1994.
13. J. Hromkovič. *Communication Complexity and Parallel Computing.* Springer, 1997.
14. J. Jain, J. Bitner, D. S. Fussell, and J. A. Abraham. Functional partitioning for verification and related problems. In *Brown MIT VLSI Conf.*, pp. 210–226. 1992.
15. E. Kushilevitz and N. Nisan. *Communication Complexity.* Cambridge University Press, 1997.
16. C. Meinel. The power of polynomial size Ω-branching programs. In *Proc. of 5th STACS*, pp. 81–90. 1988.
17. S. Ponzio. A lower bound for integer multiplication with read-once branching programs. *SIAM Journal on Computing*, 28:798–815, 1998.
18. D. Sieling and I. Wegener. Graph driven BDDs – a new data structure for Boolean functions. *Theor. Comp. Sci.*, 141:283–310, 1995.
19. I. Wegener. *Branching Programs and Binary Decision Diagrams - Theory and Applications.* SIAM, 2000.
20. P. Woelfel. New bounds on the OBDD-size of integer multiplication via universal hashing. In *Proc. of 18th STACS*, pp. 563–574. 2001.
21. P. Woelfel. On the complexity of integer multiplication in branching programs with multiple tests and in read-once branching programs with limited nondeterminism. In *Proc. of 17th Conf. on Comp. Compl.*. 2002. To appear.

Matroid Intersections, Polymatroid Inequalities, and Related Problems*

Endre Boros[1], Khaled Elbassioni[2], Vladimir Gurvich[1], and Leonid Khachiyan[2]

[1] RUTCOR, Rutgers University,
640 Bartholomew Road, Piscataway NJ 08854-8003,
{boros,gurvich}@rutcor.rutgers.edu

[2] Department of Computer Science, Rutgers University,
110 Frelinghuysen Road, Piscataway NJ 08854-8003,
elbassio@paul.rutgers.edu,leonid@cs.rutgers.edu

Abstract. Given m matroids $M_1, \ldots, M_m$ on the common ground set V, it is shown that all maximal subsets of V, independent in the m matroids, can be generated in quasi-polynomial time. More generally, given a system of polymatroid inequalities $f_1(X) \geq t_1, \ldots, f_m(X) \geq t_m$ with quasi-polynomially bounded right hand sides $t_1, \ldots, t_m$, all minimal feasible solutions $X \subseteq V$ to the system can be generated in incremental quasi-polynomial time. Our proof of these results is based on a combinatorial inequality for polymatroid functions which may be of independent interest. Precisely, for a polymatroid function f and an integer threshold $t \geq 1$, let $\alpha = \alpha(f,t)$ denote the number of maximal sets $X \subseteq V$ satisfying $f(X) < t$, let $\beta = \beta(f,t)$ be the number of minimal sets $X \subseteq V$ for which $f(X) \geq t$, and let $n = |V|$. We show that $\alpha \leq \max\{n, \beta^{(\log t)/c}\}$, where $c = c(n,\beta)$ is the unique positive root of the equation $2^c(n^{c/\log\beta} - 1) = 1$. In particular, our bound implies that $\alpha \leq (n\beta)^{\log t}$. We also give examples of polymatroid functions with arbitrarily large t, n, α and β for which $\alpha = \beta^{(1-o(1))\log t/c}$.

1 Introduction

Given m matroids $M_1, \ldots, M_m$ on the common ground set V of cardinality n, Lawler, Lenstra and Rinnooy Kan [14] in 1980 asked the question of the complexity of generating all maximal sets independent in all the matroids, and gave an exponential-time algorithm whose running time is $O(n^{m+2})$ per each generated maximal independent set. This matroid intersection problem has interesting applications in a variety of fields including combinatorial optimization [13,19] and symbolic analysis of electrical circuits [10]. In this paper, we show that all

* This research was supported by the National Science Foundation (Grant IIS-0118635), and by the Office of Naval Research (Grant N00014-92-J-1375). The second and third authors are also grateful for the partial support by DIMACS, the National Science Foundation's Center for Discrete Mathematics and Theoretical Computer Science.

K. Diks et al. (Eds): MFSC 2002, LNCS 2420, pp. 143–154, 2002.

maximal sets independent in m matroids can be generated in incremental quasi-polynomial time. More precisely, assume that each matroid M_i is described by an independence oracle, i.e., an algorithm that, given a set $X \subseteq V$, determines whether or not X is independent in M_i.

Theorem 1. *Let $M_1, \ldots, M_m$ be m matroids on the common ground set V, $|V| = n$, and let $\mathcal{F} \subseteq 2^V$ be the family of all maximal sets independent in all the matroids. Given a partial list $\mathcal{H} \subseteq \mathcal{F}$, either a new element in $\mathcal{F} \setminus \mathcal{H}$ can be computed, or $\mathcal{F} = \mathcal{H}$ can be recognized, in $k^{o(\log k)}$ time and poly(k) calls to the independence oracles, where $k \stackrel{\text{def}}{=} \max\{m, n, |\mathcal{H}|\}$.*

In fact, we shall consider a wider class of problems of which matroid intersection is a special case. Let V be a finite set of cardinality $|V| = n$, let $f : 2^V \mapsto \mathbb{Z}_+$ be a set-function taking non-negative integral values, and let $r = r(f)$ denote the *range* of f, i.e., $r(f) = \max\{f(X) \mid X \subseteq V\}$. The set-function f is called *monotone* if $f(X) \leq f(Y)$ whenever $X \subseteq Y$, and *submodular* if

$$f(X \cup Y) + f(X \cap Y) \leq f(X) + f(Y)$$

holds for all subsets $X, Y \subseteq V$. Finally, f is called a *polymatroid function* if it is monotone, submodular and $f(\emptyset) = 0$. Given a system of *polymatroid* inequalities:

$$f_i(X) \geq t_i, \quad i = 1, \ldots, m, \tag{1}$$

where each of the polymatroid functions $f_i : 2^V \mapsto \mathbb{Z}_+$ is defined via an *evaluation oracle*, and $t_1, \ldots, t_m$ are given positive integral thresholds, let $\mathcal{A}$ and $\mathcal{B}$, respectively, denote the family of all maximal infeasible and minimal feasible sets for (1). It is easy to see that $\mathcal{A} = \mathcal{I}(\mathcal{B})$, where $\mathcal{I}(\cdot)$ denotes the family of all maximal independent sets for the hypergraph $(\cdot)$. Consider the following problem:

$GEN(\mathcal{B}, \mathcal{H})$: Given a system of polymatroid inequalities (1) and a collection $\mathcal{H} \subseteq \mathcal{B}$ of minimal feasible sets for (1), either find a new minimal feasible set $H \in \mathcal{B} \setminus \mathcal{H}$ for (1), or show that $\mathcal{H} = \mathcal{B}$.

Clearly, the matroid intersection problem can be described as a system of polymatroid inequalities (1). [Indeed, let $\rho_i : 2^V \mapsto \{0, 1, \ldots, n\}$ be the rank function of M_i. Then the rank function of the dual matroid

$$f_i(X) = \rho_i(V \setminus X) + |X| - \rho_i(V) : 2^V \mapsto \{0, 1, \ldots, n\}$$

is a polymatroid function. Furthermore, a set $X \subseteq V$ is independent in M_i if and only if $f_i(V \setminus X) \geq n - \rho_i(V)$. Letting $\mathcal{B} \stackrel{\text{def}}{=} \{V \setminus X \mid X \in \mathcal{F}\}$, we conclude, therefore, that $\mathcal{B}$ is the family of minimal solutions for the system of polymatroid inequalities $f_i(X) \geq n - \rho_i(V)$, $i = 1, \ldots, m$.]

The main result of this paper, Theorem 2 below, generalizes Theorem 1 to systems of polymatroid inequalities (1). Let, as before, $\mathcal{B}$ denote the family of all minimal solutions to (1). A generation algorithm for $\mathcal{B}$ is said to run in *incremental quasi-polynomial* time if it can solve problem $GEN(\mathcal{B}, \mathcal{H})$ in $2^{\text{polylog}k}$ operations and calls to the evaluation oracles for $f_1, \ldots, f_m$, where $k = \max\{m, n, |\mathcal{H}|\}$.

Theorem 2. *Consider a system of polymatroid inequalities (1) in which the right-hand sides are bounded by a quasi-polynomial in the dimension of the system:*

$$\max\{t_1, \ldots, t_m\} \leq 2^{\text{polylog}(nm)}.$$

Then all minimal solutions to (1) can be generated in incremental quasi-polynomial time.

Theorem 2 can be complemented with the following negative result.

Proposition 1. *There exist polymatroid inequalities $f(X) \geq t$, with polynomial-time computable left-hand side, for which problem $GEN(\mathcal{B}, \mathcal{H})$ is NP-hard for exponentially large t.*

The paper is organized as follows. In section 2, we present a combinatorial inequality bounding the number of maximal infeasible sets $\mathcal{A}$ by a quasi-polynomial in n and the number of minimal feasible sets $\mathcal{B}$. We then use this inequality in Section 4 to reduce problem $GEN(\mathcal{B}, \mathcal{H})$, in quasi-polynomial time, to the well-known hypergraph dualization problem, that is, the generation of all maximal independent sets of an explicitly given hypergraph. Since the hypergraph dualization problem can be solved in incremental quasi-polynomial time [9], this will prove Theorem 2 and allow for the efficiently incremental solution of a number of applications, in addition to matroid intersections. Some of these applications are briefly discussed in Section 3, including the generation of minimal feasible solutions to a system of non-negative linear inequalities in Boolean variables (integer programming), minimal infrequent sets of a database (data mining), minimal connectivity ensuring collections of subgraphs from a given list (reliability theory), and minimal spanning collections of subspaces from a given list (linear algebra). The proof of the polymatroid inequality will be given in Sections 5 and 6.

2 An Inequality for Polymatroid Functions

Given a polymatroid function $f : 2^V \mapsto \{0, 1, \ldots, r\}$ and an integral threshold $t \in \{1, ..., r\}$, let us denote by $\mathcal{B}_t = \mathcal{B}_t(f)$ the family of all minimal subsets $X \subseteq V$ for which $f(X) \geq t$, and analogously, let us denote by $\mathcal{A}_t = \mathcal{A}_t(f)$ the family of all maximal subsets $X \subseteq V$ for which $f(X) < t$. Throughout the paper we shall use the notation $\alpha = |\mathcal{A}_t(f)|$ and $\beta = |\mathcal{B}_t(f)|$.

Theorem 3. *For every polymatroid function f and threshold $t \in \{1, \ldots, r(f)\}$ such that $\beta \geq 2$ we have the inequality*

$$\alpha \leq \beta^{(\log t)/c(n,\beta)}, \tag{2}$$

where $c(n,\beta)$ is the unique positive root of the equation[1]

$$2^c(n^{c/\log\beta}-1)=1. \tag{3}$$

In addition, $\alpha \le n$ holds if $\beta = 1$.

Let us first remark that by (3), $1 = n^{-c/\log\beta} + (n\beta)^{-c/\log\beta} \ge 2(n\beta)^{-c/\log\beta}$, and hence $\beta^{1/c(n,\beta)} \le n\beta$. Consequently, for $\beta \ge 2$ (in which case $n \ge 2$ is implied, too) we can replace (2) by the simpler but weaker inequality

$$\alpha \le (n\beta)^{\log t}. \tag{4}$$

In fact, (4) holds even in case of $\beta = 1$, because if the hypergraph $\mathcal{B}_t$ consists only of a single hyperedge $X \subseteq V$, then $|\mathcal{A}_t| \le |X| \le n$ follows immediately by the relation $\mathcal{A}_t = \mathcal{I}(\mathcal{B}_t)$. On the other hand, for large β the bound of Theorem 3 becomes increasingly stronger than (4). For instance, $c(n,n) = \log(1+\sqrt{5})-1 >$ .694, $c(n,n^2) > 1.102$, and $c(n,n^\sigma) \sim \log\sigma$ for large σ.

Let us remark next that the bound of Theorem 3 is reasonably sharp. For instance, given positive integers k, l, and d, let $V = V_1 \cup \cdots \cup V_k$ be the disjoint union of k sets of l vertices each, and for $X \subseteq V$, define $f(X) = d^k$ if $|X \cap V_i| \ge d$ for some $i \in \{1,\ldots,k\}$, and $f(X) = d^k - \prod_{i=1}^k (d - |X \cap V_i|)$ otherwise. Then f is a polymatroid function of range $r = d^k$ for which $n = kl$, $|\mathcal{A}_r| = \binom{l}{d-1}^k$, and $|\mathcal{B}_r| = k\binom{l}{d}$. Thus, letting $t = r$, $d = k$, and $l = 2^k$, we obtain an infinite family of polymatroid functions for which $c(n,\beta) = (1-o(1))\log k$ and

$$\alpha = \beta^{(1-o(1))\log t/c(n,\beta)},$$

as $k \to \infty$.

Let us finally note that for many classes of polymatroid functions, β cannot be bounded by a quasi-polynomial estimate of the form $(n\alpha)^{\operatorname{poly}\log r}$. Let us consider for instance, a graph $G = t \times K_2$ consisting of t disjoint edges, and let $f(X)$ be the number of edges X intersects, for $X \subseteq V(G)$. Then f is a polymatroid function of range $r = t$, and we have $n = 2t$, $\alpha = |\mathcal{A}_t| = t$ and $\beta = |\mathcal{B}_t| = 2^t$.

Given a non-empty hypergraph $\mathcal{H}$ on the vertex set V, a polymatroid function $f : 2^V \mapsto \mathbb{Z}_+$, and a integral positive threshold t, the pair (f,t) is called a *polymatroid separator* for $\mathcal{H}$ if $f(H) \ge t$ for all $H \in \mathcal{H}$. We can further strengthen Theorem 3 as follows.

Theorem 4. *Let (f,t) be a polymatroid separator for a hypergraph $\mathcal{H}$ of cardinality $|\mathcal{H}| \ge 2$. Then*

$$|\mathcal{A}_t(f) \cap \mathcal{I}(\mathcal{H})| \le |\mathcal{H}|^{(\log t)/c(n,|\mathcal{H}|)}, \tag{5}$$

where $\mathcal{I}(\mathcal{H})$ is the family of all maximal independent sets for $\mathcal{H}$.

Clearly, Theorem 3 is a special case of Theorem 4 for $\mathcal{H} = \mathcal{B}_t(f)$. Since the right-hand side of (5) monotonically increases with $|\mathcal{H}|$, we can assume without loss of generality that $\mathcal{H}$ is Sperner, i.e., none of the hyperedges of $\mathcal{H}$ contains another hyperedge of $\mathcal{H}$.

[1] All logarithms in this paper are assumed to have base 2

3 Applications

Before proving Theorems 2 and 4, let us consider first some applications.

Monotone Systems of Linear Inequalities in Binary and Integer Variables: Consider a system $Ax \geq b$ of m linear inequalities in n Boolean variables, where A is a given non-negative integer $m \times n$-matrix and b is given integer m-vector. Since a linear inequality with non-negative integer coefficients is clearly polymatroid, all minimal Boolean vectors x feasible for the system can be generated in quasi-polynomial time by Theorem 2, provided that the right-hand side b is bounded by a quasi-polynomial in n and m. In fact, for linear systems the latter condition can be dropped and the bound of Theorem 3 can be strengthened to a linear bound valid even for real A and b and integer x in an arbitrary box $0 \leq x \leq c$. This gives an incremental quasi-polynomial algorithm for enumerating minimal solutions to an arbitrary nonnegative system of linear inequalities in Boolean or integer variables (see [4,6] for more details). Thus knapsack, generalized knapsack, and set covering problems are all included as special cases. The quasi-polynomial generation of all maximal feasible solutions to a generalized knapsack problem improves on known results, since for instance Lawler, Lenstra and Rinnooy Kan [14] conjectured that the generation of the maximal binary feasible solutions of a generalized knapsack problem cannot be done in incremental polynomial time, unless $P = NP$.

Minimal Infrequent Sets for a Database: Given a hypergraph $\mathcal{H} \subseteq 2^V$ (or equivalently, a database with binary attributes), and an integer threshold t, a set $X \subseteq V$ is called *t-frequent* if it is contained in at least t hyperedges of $\mathcal{H}$, and is called *t-infrequent* otherwise. The generation of maximal frequent and minimal infrequent sets for are important tasks in knowledge discovery and data mining applications (see, for instance, [1,2,18]). Since the function $f(X) \stackrel{\text{def}}{=} |\{H \in \mathcal{H} \mid H \not\supseteq X\}|$ is polymatroid of range $|\mathcal{H}|$, Theorems 3 and 2 imply respectively that the number of maximal frequent sets can be bounded by a quasi-polynomial in the number of minimal infrequent sets and the sizes of $V, \mathcal{H}$, and that the minimal infrequent sets can be generated in quasi-polynomial time. In fact, the bound of Theorem 4 can be strengthened to a sharp linear bound in this case, see [7].

Connectivity Ensuring Collections of Subgraphs: Let R be a finite set of r vertices and let $E_1, \dots, E_n \subseteq R \times R$ be a collection of n graphs on R. Given a set $X \subseteq \{1, \dots, n\}$ define $k(X)$ to be the number of connected components in the graph $(R, \bigcup_{i \in X} E_i)$. Then $k(X)$ is an anti-monotone supermodular function and hence for any integral threshold t, the inequality $f(X) = r - k(X) \geq t$ is polymatroid. In particular, $\mathcal{B}_{r-1}(f)$ is the family of all minimal collections of the input graphs $E_1, \dots, E_n$ which interconnect all vertices in R. (If the n input graphs are just n disjoint edges, then $\mathcal{B}_{r-1}$ is the set of all spanning trees in the graph $E_1 \cup \dots \cup E_n$, see [17].) Since $k(X)$ can be evaluated at any set X in polynomial time, Theorem 2 implies that for each $t \in \{1, \dots, r\}$, all elements of

$\mathcal{B}_t$ can be enumerated in incremental quasi-polynomial time. This problem has applications in reliability theory [8,16].

Spanning a Linear Space by Linear Subspaces: Given a collection $\mathcal{V} = \{\mathcal{V}_1, \ldots, \mathcal{V}_n\}$ of n linear subspaces of $\mathbf{F}^r$, for some field $\mathbf{F}$, consider the problem of enumerating all minimal sub-collections X of $V = \{1, \ldots, n\}$ such that $\text{Span}\langle \bigcup_{i \in X} \mathcal{V}_i \rangle = \mathbf{F}^r$. More generally, consider the polymatroid inequality

$$f(X) = \dim(\bigcup_{i \in X} \mathcal{V}_i) \geq t, \tag{6}$$

where $t \in \{1, \ldots, r\}$ is a given threshold. Then the set $\mathcal{B}_t(f)$ of minimal solutions to (6) is the collection of all minimal subsets of $\mathcal{V}$ the dimension of whose union is at least t. Theorem 4 then states that for all $t \in \{1, \ldots, r\}$, the size of $\mathcal{A}_t(f)$ can be bounded by a $\log t$-degree polynomial in n and $|\mathcal{B}_t(f)|$, and thus all sets in $\mathcal{B}_t(f)$ can be enumerated in incremental quasi-polynomial time.

It is worth mentioning that in all of the above examples, generating all maximal infeasible sets for (1) turns out to be NP-hard, see [7,11,15].

4 Proof of Theorem 2

In this Section we show that Theorem 2 follows from Theorem 4.

Let $\mathcal{B}$ be the set of minimal feasible sets for (1). Clearly, we can incrementally generate all sets in $\mathcal{B}$ by initializing $\mathcal{H} = \emptyset$ and then iteratively solving problem $GEN(\mathcal{B}, \mathcal{H})$ a number of $|\mathcal{B}| + 1$ times. It is easy to see that the first minimal feasible set $H \in \mathcal{B}$ can be found (or $\mathcal{B} = \emptyset$ can be recognized) by evaluating (1) $n + 1$-times. Furthermore, since $\mathcal{I}(\{H\}) = \{V \setminus \{x\} \mid x \in H\}$, the second minimal feasible set can also be identified (or $\mathcal{B} = \{H\}$ can be recognized) in another $n + |H|$ evaluations of (1). Thus, in what follows we can assume without loss of generality that the current set $\mathcal{H} \subseteq \mathcal{B}$ of minimal solutions to (1) has cardinality of at least 2.

By definition, each pair (f_i, t_i) is a polymatroid separator for $\mathcal{H}$, and therefore Theorem 4 implies the inequalities

$$|\mathcal{A}_{t_i}(f_i) \cap \mathcal{I}(\mathcal{H})| \leq |\mathcal{H}|^{(\log t_i)/c(n, |\mathcal{H}|)}, \quad i = 1, \ldots, m.$$

Let $\mathcal{A} = \mathcal{I}(\mathcal{B})$ be the hypergraph of all maximal infeasible sets for (1), then $\mathcal{A} \subseteq \bigcup_{i=1}^m \mathcal{A}_{t_i}(f_i)$. Hence we arrive at the following bound:

$$|\mathcal{I}(\mathcal{B}) \cap \mathcal{I}(\mathcal{H})| \leq m |\mathcal{H}|^{(\log t)/c(n, |\mathcal{H}|)},$$

where $t = \max\{t_1, \ldots, t_m\}$. Now, since $t_1, \ldots, t_m$ are bounded by a quasi-polynomial in n and m, we conclude that

$$|\mathcal{I}(\mathcal{B}) \cap \mathcal{I}(\mathcal{H})| \leq 2^{\text{polylog} k} \text{ where } k = \max\{n, m, |\mathcal{H}|\}. \tag{7}$$

By definition, the family $\mathcal{B} \subseteq 2^V$ of all minimal feasible sets for (1) is a Sperner hypergraph. Furthermore, the hypergraph $\mathcal{B}$ has a simple superset oracle: given a set $X \subseteq V$, we can determine whether or not X contains some set $H \in \mathcal{B}$ by checking the feasibility of X for (1), i.e., by evaluating $f_1(X), \ldots, f_m(X)$. As observed in [3,11], for any Sperner hypergraph $\mathcal{B}$ defined via a superset oracle, problem $GEN(\mathcal{B}, \mathcal{H})$ reduces in quasi-polynomial time to $|\mathcal{I}(\mathcal{B}) \cap \mathcal{I}(\mathcal{H})|$ instances of the *hypergraph dualization problem*: *Given two explicitly listed Sperner families* $\mathcal{H} \subseteq 2^V$ *and* $\mathcal{G} \subseteq \mathcal{I}(\mathcal{H})$, *either find a new maximal independent set* $X \in \mathcal{I}(\mathcal{H}) \setminus \mathcal{G}$ *or show that* $\mathcal{G} = \mathcal{I}(\mathcal{H})$. (To see this reduction, consider an arbitrary hypergraph $\mathcal{H} \subseteq \mathcal{B}$. Start generating maximal independent sets for $\mathcal{H}$ checking, for each generated set $X \in \mathcal{I}(\mathcal{H})$, whether or not X is feasible for (1). If X is feasible for (1) then X contains a new minimal solution to (1) which can be found by querying the superset oracle at most $|X| + 1$ times. If $X \in \mathcal{I}(\mathcal{H})$ is infeasible for (1), then it is easy to see that $X \in \mathcal{I}(\mathcal{B})$, and hence the number of such infeasible sets X is bounded by $|\mathcal{I}(\mathcal{B}) \cap \mathcal{I}(\mathcal{H})|$.)

Combining the above reduction with (7) and the fact that the hypergraph dualization problem can be solved in quasi-polynomial time $poly(n) + (|\mathcal{H}| + |\mathcal{G}|)^{o(\log(|\mathcal{H}|+|\mathcal{G}|))}$ (see [9]), we readily obtain Theorem 2.

5 Proper Mappings of Independent Sets into Binary Trees

Our proof of Theorem 4 makes use of a combinatorial construction which may be of independent interest. Theorem 4 states that for any polymatroid separator (f, t) of a hypergraph $\mathcal{H}$ we have

$$r(f) \geq t \geq |\mathcal{S}|^{c(n,|\mathcal{H}|)/\log(|\mathcal{H}|)},$$

where $\mathcal{S} = \mathcal{I}(\mathcal{H}) \cap \{X \mid f(X) < t\}$, i.e., the range of f must increase with the size of $\mathcal{S} \subseteq \mathcal{I}(\mathcal{H})$. Thus, to prove the theorem we must first find ways to provide lower bounds on the range of a polymatroid function. To this end we shall show that the number of independent sets which can be organized in a special way into a binary tree structure provides such a lower bound.

Let $\mathbf{T}$ denote a rooted binary tree, $V(\mathbf{T})$ denote its node set, and let $L(\mathbf{T})$ denote the set of its leaves. For every node $v \in V(\mathbf{T})$, let $\mathbf{T}(v)$ be the binary sub-tree rooted at v. Obviously, for every two nodes u, v of $\mathbf{T}$ either the sub-trees $\mathbf{T}(u)$ and $\mathbf{T}(v)$ are disjoint, or one of them is a sub-tree of the other. The nodes u and v are called *incomparable* in the first case, and *comparable* in the second case.

Given a Sperner hypergraph $\mathcal{H}$ and a binary tree $\mathbf{T}$, let us consider mappings $\phi : L(\mathbf{T}) \mapsto \mathcal{I}(\mathcal{H})$ assigning maximal independent sets $I_l \in \mathcal{I}(\mathcal{H})$ to the leaves $l \in L(\mathbf{T})$. Let us associate furthermore to every node $v \in V(\mathbf{T})$ the intersection $S_v = \bigcap_{l \in L(\mathbf{T}(v))} I_l$. Let us call finally the mapping ϕ *proper* if it is injective, i.e., assigns different independent sets to different leaves, and if the sets $S_u \cup S_v$ are not independent whenever u and v are incomparable nodes of $\mathbf{T}$. Let us point out that the latter condition means that the set $S_u \cup S_v$, for incomparable nodes

u and v, must contain a hyperedge $H \in \mathcal{H}$, as a subset. Since the intersection of independent sets is always independent, it follows, in particular that both S_v and S_u are non-empty independent sets (otherwise their union could not be non-independent.) Finally, since all non-root nodes $u \in V(\mathbf{T})$ have at least one incomparable node $v \in V(\mathbf{T})$, we conclude that the sets S_u are non-empty and independent, for all non-root nodes u.

Lemma 1. *Let us consider a Sperner hypergraph $\mathcal{H}$ and a polymatroid separator (f,t) of it, and let us denote by $\mathcal{S}$ the subfamily of maximal independent sets, separated by (f,t) from $\mathcal{H}$, as before. Let us assume further that $\mathbf{T}$ is a binary tree for which there exists a proper mapping $\phi : L(\mathbf{T}) \mapsto \mathcal{S}$. Then, we have*

$$r(f) \geq t \geq |L(\mathbf{T})|. \tag{8}$$

Let us note that if a proper mapping exists for a binary tree $\mathbf{T}$, then we can associate a hyperedge $H_u \in \mathcal{H}$ to every node $u \in V(\mathbf{T}) \setminus L(\mathbf{T})$ in the following way: Let v and w be the two successors of u in $\mathbf{T}$. Since v and w are incomparable, the union $S_v \cup S_w$ must contain a hyperedge from $\mathcal{H}$. Let us choose such a hyperedge, and denote it by H_u. Let us observe next that if $l \in L(\mathbf{T}(v))$ and $l' \in L(\mathbf{T}(w))$, then $S_v \subseteq I_l$ and $S_w \subseteq I_{l'}$, and thus $H_u \subseteq I_l \cup I_{l'}$. In other words, to construct a large binary tree for which there exists a proper mapping, we have to find a way of splitting the family of independent sets, repeatedly, such that the union of any two independent sets, belonging to different parts of the split contains a hyperedge of $\mathcal{H}$. We shall show next that indeed, such a construction is possible.

Lemma 2. *For every Sperner hypergraph $\mathcal{H} \subseteq 2^V$, $|\mathcal{H}| \geq 2$, and for every subfamily $\mathcal{S} \subseteq \mathcal{I}(\mathcal{H})$ of its maximal independent sets there exists a binary tree $\mathbf{T}$ and a proper mapping $\phi : L(\mathbf{T}) \mapsto \mathcal{S}$, such that*

$$|L(\mathbf{T})| \geq |\mathcal{S}|^{c(|V|,|\mathcal{H}|)/\log|\mathcal{H}|}. \tag{9}$$

Clearly, Lemmas 1 and 2 imply Theorem 4, which in turn implies Theorem 3. The proof of Lemmas 1 and 2 is given in the next Section.

6 Proof of Main Lemmas

In this section we prove Lemmas 1 and 2, which are the key statements needed to prove our main results.

Proof of Lemma 1. Let us recall that (f,t) is a polymatroid separator of the hypergraph $\mathcal{H}$, separating the maximal independent sets $\mathcal{S} = \mathcal{S}(\mathcal{H},f,t)$ from $\mathcal{H}$, and that to every node v of $\mathbf{T}$ we have associated an independent set $S_v = \bigcap_{l \in L(\mathbf{T}(v))} I_l$, where $I_l \in \mathcal{S}$ denotes the maximal independent set assigned to the leaf $l \in L(\mathbf{T})$ by the proper assignment ϕ.

To prove the statement of the lemma, we shall show by induction that

$$f(S_w) \leq t - |L(\mathbf{T}(w))| \tag{10}$$

holds for every node w of the tree $\mathbf{T}$. Since f is non-negative, it follows that

$$|L(\mathbf{T}(w))| \leq t \leq r(f)$$

which, if applied to the root of $\mathbf{T}$, proves the lemma. To see (10), let us apply induction by the size of $L(\mathbf{T}(w))$. Clearly, if $w = l$ is a leaf of $\mathbf{T}$, then $|L(\mathbf{T}(l))| = 1$, $S_w = I_l \in \mathcal{S}$, and (10) follows by the assumption that (f,t) is separating $\mathcal{H}$ from $\mathcal{S}$. Let us assume now that w is a node of $\mathbf{T}$ with u and v as its immediate successors. Then $|L(\mathbf{T}(w))| = |L(\mathbf{T}(u))| + |L(\mathbf{T}(v))|$, and $S_w = S_u \cap S_v$. By our inductive hypothesis, and since f is submodular, we have the inequalities

$$\begin{aligned} f(S_u \cup S_v) + f(S_w) \leq f(S_u) + f(S_v) &\leq t - |L(\mathbf{T}(u))| + t - |L(\mathbf{T}(v))| \\ &= 2t - |L(\mathbf{T}(w))|. \end{aligned}$$

Since ϕ is a proper mapping, the set $S_u \cup S_v$ contains a hyperedge $H \in \mathcal{H}$, and thus $f(S_u \cup S_v) \geq f(H) \geq t$ by the monotonicity of f, and by our assumption that (f,t) is a separator for $\mathcal{H}$. Thus, from the above inequality we get $t + f(S_w) \leq f(S_u \cup S_v) + f(S_w) \leq 2t - |L(\mathbf{T}(w))|$, from which (10) follows. □

For a hypergraph $\mathcal{H}$ and a vertex $v \in V = V(\mathcal{H})$ let us denote by $d_{\mathcal{H}}(v)$ the *degree* of vertex v in $\mathcal{H}$, i.e., $d_{\mathcal{H}}(v)$ is the number of hyperedges of $\mathcal{H}$ containing v.

Lemma 3. *For every Sperner hypergraph $\mathcal{H} \subseteq 2^V$ on $n = |V| > 1$ vertices, with $m = |\mathcal{H}| \geq n$ hyperedges, there exists a vertex $v \in V$ for which*

$$m\frac{1}{n} \leq d_{\mathcal{H}}(v) \leq m\left(1 - \frac{1}{n}\right).$$

Proof. Let us define

$$X = \{v \in V \mid d_{\mathcal{H}}(v) < m\frac{1}{n}\} \quad \text{and} \quad Y = \{v \in V \mid d_{\mathcal{H}}(v) > m(1 - \frac{1}{n})\},$$

and let us assume indirectly that $X \cup Y = V$ forms a partition of the vertex set.

Let us observe first that $|X| < n$ must hold, since otherwise a contradiction

$$m \leq \sum_{H \in \mathcal{H}} |H| = \sum_{v \in X} d_{\mathcal{H}}(v) < n\frac{m}{n} = m,$$

would follow. Let us observe next that $|X| > 0$ must hold, since otherwise

$$\sum_{H \in \mathcal{H}} |H| = \sum_{v \in V} d_{\mathcal{H}}(v) = \sum_{v \in Y} d_{\mathcal{H}}(v) > n \times m(1 - \frac{1}{n}) = m(n-1)$$

follows, implying the existence of a hyperedge $H \in \mathcal{H}$ of size $|H| = n$, i.e., $V \in \mathcal{H}$. Since $\mathcal{H}$ is Sperner, $1 = m < n$ would follow, contradicting our assumptions.

Let us observe finally that the number of those hyperedges which avoid some points of Y cannot be more than $|Y|m/n$, and since $|Y| < n$ by our previous observation, there must exist a hyperedge $H \in \mathcal{H}$ containing Y. Thus, all other hyperedges must intersect X, and hence we have

$$m - 1 \leq \sum_{H \in \mathcal{H}} |H \cap X| = \sum_{v \in X} d_{\mathcal{H}}(v) < |X|\frac{m}{n} \leq m\frac{n-1}{n}$$

by our first observation. From this $m < n$ would follow, contradicting again our assumption that $m \geq n$. This last contradiction hence proves X and Y cannot cover V, and thus follows the lemma. □

For a subset $X \subseteq V$ let $\mathcal{H}^X \stackrel{\text{def}}{=} \{H \in \mathcal{H} \mid H \supseteq X\}$, and let us simply write $\mathcal{H}^v$ if $X = \{v\}$.

Lemma 4. *Given a hypergraph $\mathcal{H}$ and a subfamily $\mathcal{S} \subseteq \mathcal{I}(\mathcal{H})$ of its maximal independent sets, $|\mathcal{S}| \geq 2$, there exists a hyperedge $H \in \mathcal{H}$ and a vertex $v \in H$ such that*

$$|\mathcal{S}^v| \geq \frac{|\mathcal{S}|}{n} \text{ and } |\mathcal{S}^{H \setminus v}| \geq \frac{|\mathcal{S}|}{n|\mathcal{H}|}.$$

Proof. Let us note first that if $2 \leq |\mathcal{S}| < n$, then the statement is almost trivially true. To see this, let us choose two distinct maximal independent sets S_1 and S_2 from $\mathcal{S}$, and a vertex $v \in S_2 \setminus S_1$. Since $S_1 \cup \{v\}$ is not independent, there exists a hyperedge $H \in \mathcal{H}$ for which $v \in H \cap S_2$ and $H \setminus \{v\} \subseteq S_1$, implying thus that both $|\mathcal{S}^v|$ and $|\mathcal{S}^{H \setminus v}|$ are at least 1, and the right-hand sides in the claimed inequalities are not more than 1.

Thus, we can assume in the sequel that $|\mathcal{S}| \geq n$. Let us then apply Lemma 3 for the Sperner hypergraph $\mathcal{S}^c \stackrel{\text{def}}{=} \{V \setminus I \mid I \in \mathcal{S}\}$, and obtain that

$$\frac{|\mathcal{S}|}{n} \leq d_{\mathcal{S}^c}(v) \leq |\mathcal{S}|(1 - \frac{1}{n})$$

holds for some $v \in V$, since $|\mathcal{S}| = |\mathcal{S}^c|$ obviously. Thus, from the second inequality we obtain

$$|\mathcal{S}^v| \geq \frac{|\mathcal{S}|}{n}.$$

To see the second inequality of Lemma 4, let us note that members of $\mathcal{S}^c$ are minimal transversals of $\mathcal{H}$, and thus for every $T \in \mathcal{S}^c$, $T \ni v$ there exists a hyperedge $H \in \mathcal{H}$ for which $H \cap T = \{v\}$, by the definition of minimal transversals. Thus,

$$\bigcup_{H \in \mathcal{H}: H \ni v} \{T \in \mathcal{S}^c \mid T \cap H = \{v\}\} \supseteq \{T \in \mathcal{S}^c \mid T \ni v\}$$

holds, from which

$$\sum_{H \in \mathcal{H}: H \ni v} |\mathcal{S}^{H \setminus v}| \geq d_{\mathcal{S}^c}(v) \geq \frac{|\mathcal{S}|}{n}$$

follows. Therefore, since $|\{H \in \mathcal{H} \mid H \ni v\}| = d_{\mathcal{H}}(v) \leq |\mathcal{H}|$ holds obviously, there must exist a hyperedge $H \in \mathcal{H}$, $H \ni v$, for which

$$|\mathcal{S}^{H \setminus v}| \geq \frac{|\mathcal{S}|}{n|\mathcal{H}|}$$

holds, implying thus the lemma. □

Proof of Lemma 2. Let us denote by $L(\alpha)$ the maximum number of leaves of a binary tree $\mathbf{T}$ with a proper mapping $\phi : V(\mathbf{T}) \to \mathcal{S}$, where $\mathcal{S} \subseteq \mathcal{I}(\mathcal{H})$ is an arbitrary subfamily of maximal independent sets of $\mathcal{H}$. To simplify notation, let us write $\alpha = |\mathcal{S}|$ and $\beta = |\mathcal{H}|$. To prove the statement, we need to show that

$$L(\alpha) \geq \alpha^{c/\log\beta} \tag{11}$$

where $c = c(n, \beta)$ is as defined in (3).

Let us prove this inequality by induction on α. Clearly, if $\alpha = 1$, then $L(1) = 1$ holds, and we have equality in (11).

Let us assume next that we already have verified the claim for all subfamilies of size smaller than α, and let us consider a subfamily $\mathcal{S} \subseteq \mathcal{I}(\mathcal{H})$ of size $\alpha = |\mathcal{S}|$. According to Lemma 4, we can choose two disjoint subfamilies $\mathcal{S}', \mathcal{S}'' \subseteq \mathcal{S}$ such that $|\mathcal{S}'| \geq \frac{\alpha}{n}$ and $|\mathcal{S}''| \geq \frac{\alpha}{n\beta}$, and such that for any pair of sets $S' \in \mathcal{S}'$ and $S'' \in \mathcal{S}''$ the union $S' \cup S''$ contains a member of $\mathcal{H}$. Thus, building binary trees with proper mappings separately for $\mathcal{S}'$ and $\mathcal{S}''$, and joining them as two siblings of a common root, we obtain a binary tree with a proper mapping for $\mathcal{S}$. Since the right-hand side of our claim is a monotone function of α, we can conclude for the number of leaves in the obtained binary tree that

$$L(\alpha) \geq L(\frac{\alpha}{n}) + L(\frac{\alpha}{n\beta}). \tag{12}$$

Applying now our inductive hypothesis, we get

$$L(\alpha) \geq \left(\frac{\alpha}{n}\right)^{\frac{c}{\log\beta}} + \left(\frac{\alpha}{n\beta}\right)^{\frac{c}{\log\beta}} = \alpha^{\frac{c}{\log\beta}} \left[n^{\frac{-c}{\log\beta}} + (n\beta)^{\frac{-c}{\log\beta}}\right] = \alpha^{c/\log\beta},$$

where the last equality holds by (3). This proves (11), and hence the lemma follows. □

Note that the right-hand side of (11) is the least possible solution of the recursion (12).

References

1. R. Agrawal, T. Imielinski and A. Swami, Mining associations between sets of items in massive databases, *Proc. 1993 ACM-SIGMOD Int. Conf. on Management of Data*, pp. 207-216.
2. R. Agrawal, H. Mannila, R. Srikant, H. Toivonen and A. I. Verkamo, Fast discovery of association rules, in U. M. Fayyad, G. Piatetsky-Shapiro, P. Smyth and R. Uthurusamy eds., *Advances in Knowledge Discovery and Data Mining*, pp. 307-328, AAAI Press, Menlo Park, California, 1996.
3. J. C. Bioch and T. Ibaraki, Complexity of identification and dualization of positive Boolean functions, *Information and Computation* 123 (1995) pp. 50-63.
4. E. Boros, K. Elbassioni, V. Gurvich, L. Khachiyan and K. Makino, On generating all minimal integer solutions for a monotone system of linear inequalities, in *ICALP 2001*, LNCS 2076, pp. 92-103. An extended version is to appear in *SIAM Journal on Computing*.

5. E. Boros, K. Elbassioni, V. Gurvich and L. Khachiyan, An inequality for polymatroid functions and its applications, DIMACS Technical Report 2001-14, Rutgers University, http://dimacs.rutgers.edu/TechnicalReports/2001.html.
6. E. Boros, V. Gurvich, L. Khachiyan and K. Makino, Dual bounded generating problems: partial and multiple transversals of a hypergraph. *SIAM Journal on Computing* 30 (6) (2001) pp. 2036-2050.
7. E. Boros, V. Gurvich, L. Khachiyan and K. Makino, On the complexity of generating maximal frequent and minimal infrequent sets, in *STACS 2002*, LNCS 2285, pp. 133-141.
8. C. J. Colbourn, *The combinatorics of network reliability*, Oxford Univ. Press, 1987.
9. M. L. Fredman and L. Khachiyan, On the complexity of dualization of monotone disjunctive normal forms, *Journal of Algorithms*, 21 (1996) pp. 618-628.
10. M. Galań, I. Garciá-Vargas, F.V. Fernańdez and A. Rodriǵuez-Vaźquez, A new matroid intersection algorithm for symbolic large circuit analysis, in *Proc. 4th Int. Workshop on Symbolic Methods and Applications to Circuit Design*, Oct. 1996.
11. V. Gurvich and L. Khachiyan, On generating the irredundant conjunctive and disjunctive normal forms of monotone Boolean functions, *Discrete Applied Mathematics*, 96-97 (1999) pp. 363-373.
12. T. Helgason, Aspects of the theory of hypermatroids, in Hypergraph Seminar, Lecture Notes in Math. 411 (1975) Springer, pp. 191-214.
13. E. L. Lawler, *Combinatorial Optimization: Networks and Matroids*, Holt, Rinehart and Winston, New York, 1976.
14. E. Lawler, J. K. Lenstra and A. H. G. Rinnooy Kan, Generating all maximal independent sets: NP-hardness and polynomial-time algorithms, *SIAM Journal on Computing*, 9 (1980) pp. 558-565.
15. K. Makino and T. Ibaraki, Interior and exterior functions of Boolean functions, *Discrete Applied Mathematics*, 69 (1996) pp. 209-231.
16. K. G. Ramamurthy, *Coherent Structures and Simple Games*, Kluwer Academic Publishers, 1990.
17. R. C. Read and R. E. Tarjan, Bounds on backtrack algorithms for listing cycles, paths, and spanning trees, *Networks*, 5 (1975) pp. 237-252.
18. R. H. Sloan, K. Takata and G. Turan, On frequent sets of Boolean matrices, *Annals of Mathematics and Artificial Intelligence* 24 (1998) pp. 1-4.
19. D.J.A. Welsh, Matroid Theory (Academic Press, London, New York, San Francisco 1976).

Accessibility in Automata on Scattered Linear Orderings

Olivier Carton

Institut Gaspard Monge, Université de Marne-la-Vallée, 5 boulevard Descartes, F-77454 Marne-la-Vallée Cedex 2, France,
Olivier.Carton@univ-mlv.fr, http://www-igm.univ-mlv.fr/~carton/

Abstract. In a preceding paper, automata have been introduced for words indexed by linear orderings. These automata are a generalization of automata on transfinite words introduced by Büchi. In this paper, we show that if only words indexed by scattered linear orderings are considered, the accessibility and the emptiness in these automata can be checked in time nm^2 where n and m are the number of states and the number of transitions. This solves the problem for automata on transfinite words.

1 Introduction

The first result in automata theory and formal languages is the Kleene theorem which establishes the equivalence between sets of words accepted by automata and sets of words described by rational expressions. Since the seminal paper of Kleene [8], automata accepting different kinds of objects have been introduced. This includes the automata on infinite words introduced by Büchi [3], automata on bi-infinite words, automata on finite and infinite trees [12], automata on traces, automata on pictures and automata on transfinite words introduced again by Büchi [4].

In [1], we have considered linear structures in a general framework, *i.e.*, words indexed by a linear ordering. This approach allows us to treat in the same way finite words, left- and right-infinite words, bi-infinite words, ordinal words which are studied separately in the literature. We have introduced a new notion of automaton accepting words on linear orderings, which is simple, natural and includes previously defined automata. We have also defined rational expressions for such words. We have proved the related Kleene-like theorem when the orderings are restricted to countable scattered linear orderings. This result extends Kleene's theorem for finite words [8], infinite words [2,10], bi-infinite words [7, 11] and ordinal words [4,5,18].

The automata we have defined in [1] are a natural generalization of those introduced by Büchi for transfinite words. These latter automata are usual automata on finite words with additional limit transitions of the form $P \to q$ where P is a subset of states and q is a state. These limit transitions are needed for the limit ordinals. The automata we have introduced have limit transitions of

K. Diks et al. (Eds): MFSC 2002, LNCS 2420, pp. 155–164, 2002.

the form $P \to q$ but they also have limit transitions of the form $q \to P$. To some extend, they are more symmetrical.

Compared to other formalisms like rational expressions or logic formulas, automata give more tractability. First, automata have been used to show the decidability of some logics [2, 12]. Second, automata are in many problems the key to efficient algorithms. This has been heavily used in model-checking [15] or in language processing [13]. Although most operations on automata can be performed in polynomial time, there are some pitfalls when dealing when automata on infinite words. The complexity may highly depend on the kind of automata that are considered. For instance, the computation of the Rabin index which roughly measures the nesting of accepting and non-accepting loops is NP-complete for deterministic Rabin automata [9] whereas it is polynomial for deterministic Muller automata [16].

Automata can be considered as finite state transitions systems. One important issue in the study of transitions systems is the accessibility. The accessibility consists in deciding whether a state p of the system is reachable from another state q. This issue is especially important because it corresponds to safety properties of the system.

The purpose of this paper is to show that under some natural restriction, the accessibility in automata on linear orderings can be efficiently checked. These automata can accept words indexed by any linear orderings. In this paper, we only consider countable linear orderings that do not contain a dense subordering like $\mathbb{Q}$. These orderings are called scattered in the literature [14]. This restriction is motivated by the fact that these orderings look discrete. Furthermore, the path of a non scattered word is non countable even if the word is countable.

We prove that if only paths labeled by words indexed by countable scattered orderings are considered, the accessibility can be checked in time nm^2 where n and m are the number of states and the number of transitions of the automaton. Since ordinals are scattered orderings, this result also holds for automata on transfinite words introduced by Büchi. Even those simpler automata have limit transitions of the form $P \to q$ and these transitions may be nested. Indeed, if $q' \in P$ and if $P' \subset P$, a transition $P' \to q'$ may be used in a loop which in turn allows the transition $P \to q$ to be used. For that reason, the result is not straightforward since a naive approach would consider all subsets of Q. Since the emptiness can easily be reduced to the accessibility, we also solve the emptiness in the same time.

The result is achieved in two steps. We first prove a kind of pumping lemma which essentially states that is suffices to consider simple paths which are rational. Then, we reduce the problem of accessibility in an automaton on linear orderings to the accessibility in an usual automaton of finite words which can be computed in time nm^2.

The paper is organized as follows. In Sections 2 and 3, we recall the definitions of words and automata on linear orderings. In Section 4, we state and we prove the pumping lemma for these automata. Section 5 is finally devoted to the accessibility in these automata. The main result of the paper is stated in that section. The proofs are given in the Appendix.

2 Words on Linear Orderings

A *linear ordering* J is an ordering $<$ which is total, that is, for any $j \neq k$ in J, either $j < k$ or $k < j$ holds. We refer the reader to [14] for a complete introduction to linear orderings. Given a finite alphabet A, a *word* $(a_j)_{j \in J}$ is a function from J to A which maps any element j of J to a letter a_j of A. We say that J is the *length* $|x|$ of the word x. For instance, the *empty word* ε is indexed by the empty linear ordering $J = \varnothing$. Usual finite words are the words indexed by finite orderings $J = \{1, 2, \ldots, n\}$, $n \geq 0$. A word of length $J = \omega$ is a word usually called an ω-word or an infinite word. A word of length $J = \zeta$ is a sequence $\ldots a_{-2}a_{-1}a_0a_1a_2\ldots$ of letters which is usually called a bi-infinite word.

Given a linear ordering J, we denote by $-J$ the *backwards* linear ordering obtained by reversing the ordering relation. For instance, $-\omega$ is the backwards linear ordering of ω which is used to indexed the so-called left-infinite words.

Given two linear orderings J and K, the linear ordering $J + K$ is obtained by juxtaposition of J and K, i.e., it is the linear ordering on the disjoint union $J \cup K$ extended with $j < k$ for any $j \in J$ and any $k \in K$. For instance, the linear ordering ζ can be obtained as the sum $-\omega + \omega$. More generally, let J and K_j for $j \in J$, be linear orderings. The linear ordering $\sum_{j \in J} K_j$ is obtained by juxtaposition of the orderings K_j in respect of J. More formally, the *sum* $\sum_{j \in J} K_j$ is the set L of all pairs (k, j) such that $k \in K_j$. The relation $(k_1, j_1) < (k_2, j_2)$ holds iff $j_1 < j_2$ or $j_1 = j_2$ and $k_1 < k_2$ in K_{j_1}.

The sum operation on linear orderings leads to a notion of product of words as follows. Let J and K_j for $j \in J$, be linear orderings. Let $x_j = (a_{k,j})_{k \in K_j}$ be a word of length K_j, for any $j \in J$. The *product* $\prod_{j \in J} x_j$ is the word z of length $L = \sum_{j \in J} K_j$ equal to $(a_{k,j})_{(k,j) \in L}$. For instance, the word $a^\zeta = a^{-\omega} \cdot a^\omega$ of length $\zeta = -\omega + \omega$ is the product of the two words $a^{-\omega}$ and a^ω of length $-\omega$ and ω respectively.

In this paper as in [1], we only consider linear orderings which are *countable* and *scattered*, i.e., without any dense subordering. This class is denoted by $\mathcal{S}$ and its elements are shortly called *orderings*. We use notation $\mathcal{N}$ for the subclass of $\mathcal{S}$ of finite linear orderings and $\mathcal{O}$ for the subclass of countable ordinals. Recall that an *ordinal* is a linear ordering which is well-ordered, that is, without the subordering $-\omega$.

The following characterization of the class $\mathcal{S}$ is due to Hausdorff [14]. The finite ordering with one element is denoted by $\mathbf{1}$.

Theorem 1 (Hausdorff). $\mathcal{S} = \bigcup_{\alpha \in \mathcal{O}} V_\alpha$ *where the classes* V_α *are inductively defined by*

1. $V_0 = \{\varnothing, \mathbf{1}\}$;
2. $V_\alpha = \{\sum_{j \in J} K_j \mid J \in \mathcal{N} \cup \{\omega, -\omega, \zeta\}$ *and* $K_j \in \bigcup_{\beta < \alpha} V_\beta\}$.

For any $\alpha \in \mathcal{O}$, we also define the class $V_{\alpha+\frac{1}{2}}$ of orderings as follows.

$$V_{\alpha+\frac{1}{2}} = \{\sum_{j \in J} K_j \mid J \in \mathcal{N} \text{ and } K_j \in V_\alpha\}$$

The notation $V_{\alpha+\frac{1}{2}}$ is justified by the following inclusions which hold for any ordinal $\alpha \in \mathcal{O}$.

$$V_\alpha \subset V_{\alpha+\frac{1}{2}} \subset V_{\alpha+1}.$$

All these inclusions are strict. The ordinal ω^α belongs to V_α but does not belong to any V_β for $\beta < \alpha$. The ordinal $\omega^\alpha \cdot 2 = \omega^\alpha + \omega^\alpha$ belongs to $V_{\alpha+\frac{1}{2}}$ but not to V_α.

The *rank* of a scattered ordering J is the smallest ordinal α such that $J \in V_{\alpha+\frac{1}{2}}$. In some sense, the rank measured the size of an ordering. Since the ranks of countable scattered orderings range over all countable ordinals, the orderings of finite rank can be considered as very small orderings. By extension, the *rank* of a word is the rank of its length.

Note that the orderings of rank 0 are the finite ones. More generally, the rank of an ordinal α is also its degree (see [14, p. 61]). This is the first exponent α_1 in its Cantor normal $\alpha = \omega_1^\alpha \cdot n_1 + \cdots + \omega^{\alpha_k} \cdot n_k$. Therefore, the ordinals of finite rank are exactly the ordinals smaller that ω^ω. This justifies why the rank is defined from the classes $V_{\alpha+\frac{1}{2}}$ but not from the classes V_α. Otherwise, the rank of ω^2 would have been 2 while the rank of $\omega^2 \cdot 2$ would have been 3.

2.1 Rational Words

We define here the rational words. A word x is said to be *rational* if the singleton $\{x\}$ is rational. The class R of rational words is the smallest class which contains the letters and which is closed under concatenation and the operations ω and $-\omega$. This means that R is the smallest class such that:

- if $a \in A$ then $a \in R$,
- if $x \in R$ and $y \in R$ then $xy \in R$,
- if $x \in R$, then $x^\omega \in R$ and $x^{-\omega} \in R$.

The rational words are the generalization in our framework of ultimately periodic words. Indeed, a set $\{x\}$ containing a unique infinite word x is rational if and only if this word is ultimately periodic, that is equal to uv^ω for finite words u and v.

Note that the rank of a rational word is always an integer. For that reason, rational words can be seen as very small words. The rank of a rational word x is equal to the maximum number of nested ω and $-\omega$ in any rational expressions denoting the word x. There are many expression denoting the same word x but this number is independent of the expression.

3 Automata

Automata accepting words on linear orderings are a natural extension of finite automata. As above they are defined as $\mathcal{A} = (Q, A, E, I, F)$. The set E is composed with three types of transitions: the usual *successor* transitions in $Q \times A \times E$, the *left limit* transitions which belong to $\mathcal{P}(Q) \times Q$ and the *right limit* transitions which belong to $Q \times \mathcal{P}(Q)$.

Example 1. The automaton depicted in Figure 1 has one left limit transition $\{2\} \to 0$ and one right limit transition $0 \to \{1\}$.

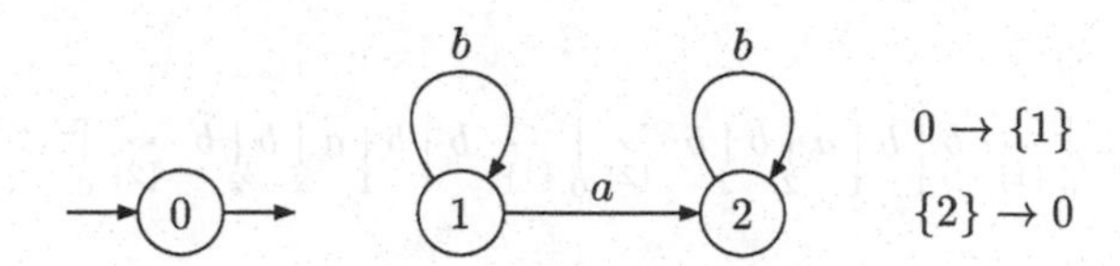

Fig. 1. An automaton on linear orderings

The notion of cut is needed to define a path in such an automaton. A *cut* of an ordering J is a pair (K, L) of intervals such that $J = K \cup L$ and for any $k \in K$ and $l \in L$, $k < l$. The two subsets must be disjoint and they form a partition of the set J. The set of all cuts of the ordering J is denoted by $\hat{J}$. The set $\hat{J}$ can be linearly ordered as follows. For any cuts $c_1 = (K_1, L_1)$ and $c_2 = (K_2, L_2)$, define the relation $c_1 < c_2$ iff $K_1 \subsetneq K_2$. Note that $\hat{J}$ has always a least cut $(\varnothing, J)$ denoted $c_{\min}$ and a greatest cut $(J, \varnothing)$ denoted $c_{\max}$.

See Figure 2 where each element of J is represented by a bullet, and each cut by a vertical bar.

$$|\cdots|\bullet|\bullet|\bullet|\cdots|\cdots|\bullet|\bullet|\bullet|\cdots|$$

Fig. 2. Ordering $J \cup \hat{J}$ for $J = \zeta + \zeta$.

A word $x = (a_j)_{j \in J}$ of length J is accepted by $\mathcal{A}$ if it is the label of a successful path. A *path* γ is a sequence of states $\gamma = (q_c)_{c \in \hat{J}}$ of length $\hat{J}$ verifying the following conditions. For two consecutive states in γ, there must be a successor transition labeled by the letter in between. For a state $q \in \gamma$ which has no predecessor on γ, there must be a left limit transition $P \to q$ where P is the limit set of γ on the left of q. Right limit transitions are used similarly when q has no successor on γ. A path is *successful* if its first state $q_{c_{\min}}$ is initial and its last state $q_{c_{\max}}$ is final.

More precisely, for any cut $c \in \hat{J}$, define the sets $\lim_{c^-} \gamma$ and $\lim_{c^+} \gamma$ as follows:

$$\lim_{c^-} \gamma = \{q \in Q \mid \forall c' < c \, \exists k \quad c' < k < c \text{ and } q = q_k\},$$

$$\lim_{c^+} \gamma = \{q \in Q \mid \forall c < c' \, \exists k \quad c < k < c' \text{ and } q = q_k\}.$$

For any consecutive cuts c_j^- and c_j^+ of $\hat{J}$, $q_{c_j^-} \xrightarrow{a_j} q_{c_j^+}$ must be a successor transition. For any cut $c \neq c_{\min}$ in $\hat{J}$ which has no predecessor, $\lim_{c^-} \gamma \rightarrow q_c$ must be a left limit transition. For any cut $c \neq c_{\min}$ in $\hat{J}$ which has no successor, $q_c \rightarrow \lim_{c^+} \gamma$ must be a right limit transition.

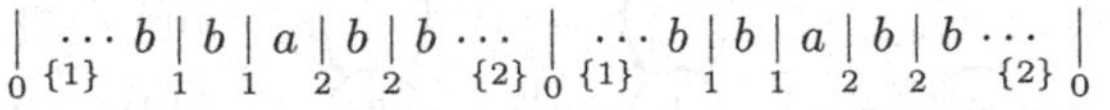

| ··· b | b | a | b | b ··· | ··· b | b | a | b | b ··· |
0 {1} 1 1 2 2 {2} 0 {1} 1 1 2 2 {2} 0

Fig. 3. The word $(b^{-\omega}ab^{\omega})^2$ is accepted

Note that a path labeled by a word x of length J can be considered as a word of length $\hat{J}$ over the alphabet Q. Therefore, a path is said to be *rational* if it is rational as a word. The concatenation of two paths γ and γ' is possible when the last state of γ is also the first state of γ'. In that case, the concatenation of γ and γ' is the path $\gamma\gamma'$ obtained by first concatenating γ and γ' as words and then identifying the last state of γ with the first state of γ'.

4 A Pumping Lemma

In this section, we give a kind of pumping lemma for automata on scattered linear orderings. The well-known pumping lemma for automata on finite words states that an automaton accepting a nonempty set accepts a *small* word. More precisely, if an automaton with n states accepts at least one word, it accepts a word of length smaller than n. A pumping lemma for automata on transfinite words was proved by Wojciechowski [17]. It states that if such an automaton with n states accepts a nonempty set, it accepts a word of length smaller than ω^{n+1}. The following theorem extends this result to automata on scattered linear orderings.

Theorem 2. *Let $\mathcal{A}$ be an automaton on scattered linear orderings and let n be the maximal cardinality of P for any limit transition $P \rightarrow q$ or $q \rightarrow P$. If there is a successful path in $\mathcal{A}$, there is a successful rational path in $\mathcal{A}$ of rank less than n.*

Note first that if a path is rational, its label is also rational but the converse is false. A rational word can be the label of a non rational path. Note also that a path is of rank n if and only if its label is also of rank n. Note also that the integer n in the previous theorem is trivially bounded by the number of states of the automaton.

The assumption that the lengths of the words are scattered is really needed. If general linear orderings are considered, there exists an automaton which accepts exactly words whose length is complete and dense. This automaton does not have a small successful path.

Note finally that the previous theorem is a direct extension of the result of Wojciechowski for automata on transfinite words since an ordinal belongs to the class $V_{n+\frac{1}{2}}$ iff it is smaller than ω^{n+1}.

The previous theorem has the following corollary. The result stated by this corollary is an extension of the well-known fact that any nonempty rational set of infinite words contains an ultimately periodic word.

Corollary 1. *Any nonempty rational set of words on scattered linear orderings contains a rational word.*

5 Accessibility

In this section, we study the accessibility in automata on linear orderings. The main result is that this accessibility can be checked in polynomial time. The key idea is to reduce the accessibility in automata on linear orderings to the accessibility in the usual automata on finite words. From a given automata on linear orderings, we show how to compute in polynomial time an automaton on finite words such that the accessibility in the two automata are equivalent.

The usual automata on finite words have just successor transitions labeled by letters from a finite alphabet. In this section, we use such automata where the label of each transition is a subset of a finite set C. The alphabet the letters are taken from is then the power set of C. In these automata, we modify slightly the definition of the label of a path. The label of path is also a subset of C. It is the union of the labels of all the transitions along the path. Note that the label of the path does depend on the order of the transitions in the path. This definition a very special case of a general notion of an automaton on a monoid. In an automaton on a monoid M, the transitions are labeled by elements of M and the label of a path is the product in M of the labels of the transitions. We here consider the case where the monoid is the power set of C with the union as the inner product. We refer the reader to [6] for a complete introduction to automata on monoids.

We have seen in the previous section that if there is path from p to q in a automaton on scattered linear orderings, there is also a simple path from p to q. This simple path is rational and it is of rank less than the maximal cardinality of a set occurring in a limit transition. The rational paths are constructed by either concatenation or taking the ω power of a cycle followed by an appropriate left limit transition or $-\omega$ power of a cycle preceded by an appropriate right limit transition. By appropriate, we mean that the set of state occurring in the limit transition must be equal to the content of the cycle. The main idea is to construct a new automaton where the iteration of the loop and the limit transition are replaced by a single transition. This cycle followed (or preceded) by a limit transition might be part of another bigger cycle. The content of the smaller cycle is then part of the content of the bigger one and cannot be discarded. The solution to this problem is to keep track of the contents in the labels of the transitions. The letters which label the successors transitions in the initial automaton are meaningless and can be removed. Each transition of the new automaton is labeled by a subset of states.

Let $\mathcal{A}$ be a automaton (Q, A, E, I, F) on linear orderings and let n be the maximal cardinality of P for any limit transition $P \to q$ or $q \to P$. We define a sequence $\mathcal{A}_0, \ldots, \mathcal{A}_n$ of automata as follows. The state set of all these automata is the same state set Q as $\mathcal{A}$ and the sets of initial and final states are also the same as those of $\mathcal{A}$. The transitions of all these automata are labeled by subsets of the set Q of states. Each automaton $\mathcal{A}_k$ is thus equal to $(Q, \mathcal{P}(Q), E_k, I, F)$ where E_k is its set of transitions. The sets $E_0, \ldots, E_n$ of transitions are defined by induction as follows. The set E_0 of transitions of $\mathcal{A}_0$ is given by

$$E_0 = \{(p, \{p, q\}, q) \mid \exists a \in A \ \ (p, a, q) \in E\}\,.$$

The set E_k of transitions of $\mathcal{A}_k$ is then is defined from $\mathcal{A}_{k-1}$ by $E_k = E_{k-1} \cup L_k \cup R_k$ where L_k is given by

$$\begin{aligned} L_k = \{(p, P \cup \{q\}, q) \mid & \exists P \to q \in E \text{ such that } p \in P, \ \ |P| = k \text{ and} \\ & \text{there is a cycle in } \mathcal{A}_{k-1} \text{ of label } P\}. \end{aligned}$$

and R_k is defined analogously for right limit transitions.

The following proposition relates the paths in the automata $\mathcal{A}_k$ with the paths in the automaton $\mathcal{A}$.

Proposition 1. *Let p and q two states of $\mathcal{A}$. There is a path from p to q in $\mathcal{A}_k$ iff there is a rational path of rank less than k in $\mathcal{A}$.*

Proposition 2. *The automaton $\mathcal{A}_n$ can be computed in time $|Q||E|^2$.*

In the construction of the transitions of E_k, it is needed to checked whether for a given subset P of states, there is a cycle in $\mathcal{A}_{k-1}$ whose label is P. The following lemma states that this test can be done in linear time.

Lemma 1. *Given an automaton labeled by subsets of C and let D be a subset of C. It can be checked in linear time whether there is a cycle labeled by D.*

Note that it is crucial that we look for a cycle and not for a path.

Combining the results of Theorem 2 and of Propositions 1 and 2, one gets the following theorem.

Theorem 3. *The accessibility in an automaton (Q, A, E, I, F) on scattered linear orderings can be checked in time $|Q||E|^2$.*

We illustrate the constructions of the automata $\mathcal{A}_0, \ldots, \mathcal{A}_n$ with the following example. Consider the automaton $\mathcal{A}$ pictured in Figure 4. The automaton $\mathcal{A}_0$

which has the same successor transitions as $\mathcal{A}$ is pictured in Figure 5. Since there is a loop in $\mathcal{A}_0$ whose label is $\{1\}$, the limit transition $\{1\} \to 2$ gives a transition $1 \xrightarrow{\{1,2\}} 2$ in $\mathcal{A}_1$ (see Figure 6). Since there is now a loop in $\mathcal{A}_1$ whose label is $\{1,2\}$, the limit transition $0 \to \{1,2\}$ gives two transitions $0 \xrightarrow{\{0,1,2\}} 1$ and $0 \xrightarrow{\{0,1,2\}} 2$ in $\mathcal{A}_2$ (see Figure 7). The final state 1 is now reachable from the initial state 0. This shows that the automaton accepts at least one word. It can be verified that it accepts the word $(ab^\omega)^{-\omega}$.

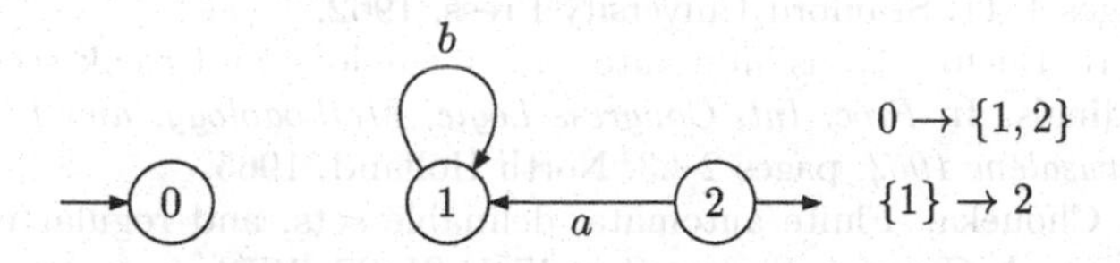

Fig. 4. Automaton $\mathcal{A}$

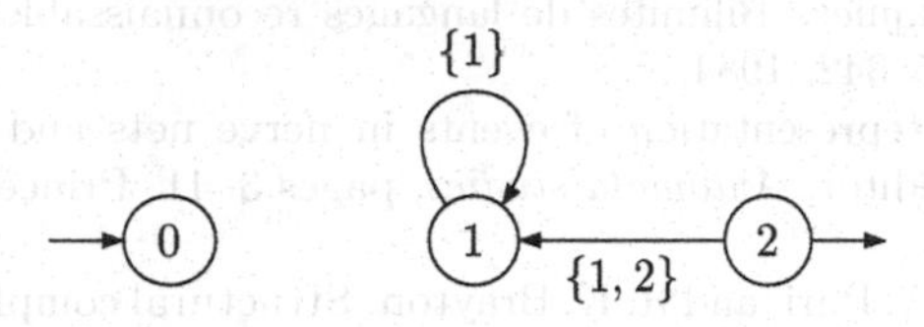

Fig. 5. Automaton $\mathcal{A}_0$

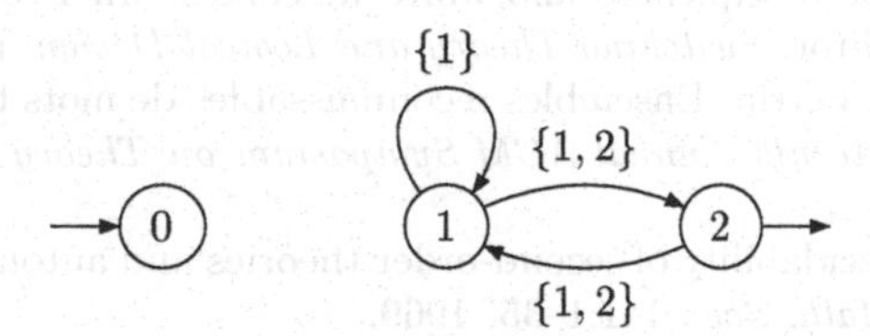

Fig. 6. Automaton $\mathcal{A}_1$

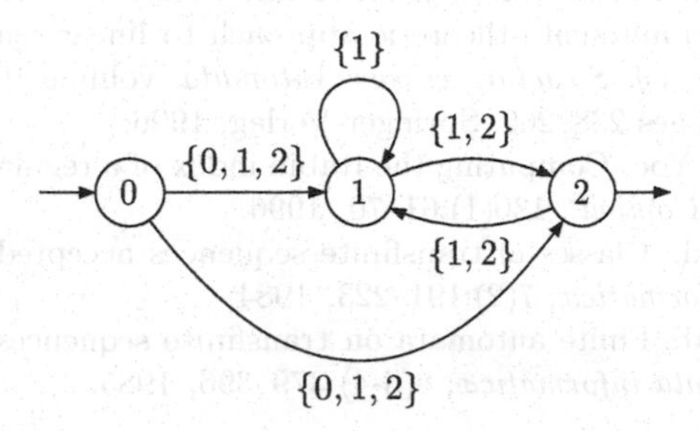

Fig. 7. Automaton $\mathcal{A}_2$

References

[1] V. Bruyère and O. Carton. Automata on linear orderings. In J. Sgall, A. Pultr, and P. Kolman, editors, *MFCS'2001*, volume 2136 of *Lect. Notes in Comput. Sci.*, pages 236–247, 2001. IGM report 2001-12.

[2] J. R. Büchi. Weak second-order arithmetic and finite automata. *Z. Math. Logik und grundl. Math.*, 6:66–92, 1960.

[3] J. R. Büchi. On a decision method in the restricted second-order arithmetic. In *Proc. Int. Congress Logic, Methodology and Philosophy of science, Berkeley 1960*, pages 1–11. Stanford University Press, 1962.

[4] J. R. Büchi. Transfinite automata recursions and weak second order theory of ordinals. In *Proc. Int. Congress Logic, Methodology, and Philosophy of Science, Jerusalem 1964*, pages 2–23. North Holland, 1965.

[5] Y. Choueka. Finite automata, definable sets, and regular expressions over ω^n-tapes. *J. Comput. System Sci.*, 17(1):81–97, 1978.

[6] S. Eilenberg. *Automata, Languages and Machines*, volume A. Academic Press, New York, 1972.

[7] D. Girault-Beauquier. Bilimites de langages reconnaissables. *Theoret. Comput. Sci.*, 33(2–3):335–342, 1984.

[8] S. C. Kleene. Representation of events in nerve nets and finite automata. In C.E. Shannon, editor, *Automata studies*, pages 3–41. Princeton university Press, Princeton, 1956.

[9] S. C. Krishnan, A. Puri, and R. K. Brayton. Structural complexity of ω-languages. In *STACS '95*, volume 900 of *Lect. Notes in Comput. Sci.*, pages 143–156, Berlin, 1995. Springer-Verlag.

[10] D. Muller. Infinite sequences and finite machines. In Proc. of Fourth Annual IEEE Symp., editor, *Switching Theory and Logical Design*, pages 3–16, 1963.

[11] M. Nivat and D. Perrin. Ensembles reconnaissables de mots bi-infinis. In *Proceedings of the Fourteenth Annual ACM Symposium on Theory of Computing*, pages 47–59, 1982.

[12] M. O. Rabin. Decidability of second-order theories and automata on infinite trees. *Trans. Amer. Math. Soc.*, 141:1–35, 1969.

[13] E. Roche and Y. Schabes. *Finite-State Language Processing*, chapter 7. MIT Press, Cambridge, 1997.

[14] J. G. Rosenstein. *Linear ordering.* Academic Press, New York, 1982.

[15] M. Y. Vardi. An automata-theoretic approach to linear temporal logic. In *Logics for Concurrency: Structure versus Automata*, volume 1043 of *Lect. Notes in Comput. Sci.*, pages 238–266. Springer-Verlag, 1996.

[16] T. Wilke and H. Yoo. Computing the Rabin index of a regular language of infinite words. *Inform. Comput.*, 130(1):61–70, 1996.

[17] J. Wojciechowski. Classes of transfinite sequences accepted by finite automata. *Fundamenta informaticæ*, 7(2):191–223, 1984.

[18] J. Wojciechowski. Finite automata on transfinite sequences and regular expressions. *Fundamenta informaticæ*, 8(3-4):379–396, 1985.

On Infinite Terms Having a Decidable Monadic Theory

Didier Caucal

IRISA–CNRS, Campus de Beaulieu, 35042 Rennes, France,
caucal@irisa.fr

Abstract. We study a transformation on terms consisting of applying an inverse deterministic rational mapping followed by an unfolding. Iterating these transformations from the regular terms gives a hierarchy of families of terms having a decidable monadic theory. In particular, the family at level 2 contains the morphic infinite words investigated by Carton and Thomas. We show that this hierarchy coincides with the hierarchy considered by Knapik, Niwiński and Urzyczyn: the families of terms that are solutions of higher order safe schemes. We also show that this hierarchy coincides with the hierarchy defined by Damm, and recently considered by Courcelle and Knapik: the families of terms obtained by iterating applications of first order substitutions to the set of regular terms. Finally, using second order substitutions yields the same terms.

1 Introduction

A general approach to check properties of a finite system is to express these properties by formulas, that can be decided on the generally infinite structure modeling the behaviour of the system. We focus on monadic second order formulas and the systems we consider are the higher order schemes [In76], [Da82] and the deterministic pushdown automata with multi-level stacks [En83].

A simple way to show the decidability of the monadic second order theory of a given graph is to obtain the graph from a finite graph using basic graph transformations that preserve the decidability of the monadic theory. We only use two graph transformations. The first transformation is the unfolding of a graph from a given vertex [CW98]. For instance, the unfoldings of finite graphs are the regular trees (having only a finite number of non isomorphic subtrees) whose each node is of finite out-degree. Therefore, these trees have a decidable monadic theory; it is the case in particular for the complete infinite binary tree [Ra69]. The second graph transformation is given by a finite automaton over labels [Ca96]. Precisely, a rational mapping h associates to each label a a rational language $h(a)$ (over labels and barred labels for moving by inverse arc). And we apply h^{-1} to a graph G to get the graph $h^{-1}(G)$ having an arc $s \xrightarrow{a} t$ when there is a path $s \overset{u}{\Longrightarrow} t$ in G for some word u in $h(a)$. This graph transformation preserves the decidability of the monadic theory: it is a noncopying monadic

K. Diks et al. (Eds): MFSC 2002, LNCS 2420, pp. 165–176, 2002.

second-order definable transduction in the sense of [Co94]. This transformation has a maximality property: starting from the regular trees, we get the same graphs that we can get by monadic definable transductions [Ba98] : the prefix-recognizable graphs [Ca96].

By alternate repetition of unfoldings and inverse rational mappings from the finite trees, we get a hierarchy of tree families and a hierarchy of graph families that have a decidable monadic theory. At level 0, the tree family is by definition the set of finite trees, and the graph family is the set of finite graphs. At level 1, the tree family is the set of regular trees, and the graph family is the set of prefix-recognizable graphs. At higher levels, the tree family and the graph family have never been studied. However for two related hierarchies of (families of finite and infinite) terms, the decidability of the monadic theory has been shown. The first one classifies the solutions of higher order safe schemes [KNU02]. The second one is obtained by iterating first order substitutions starting from regular terms [CK02]. As terms are deterministic trees, we restrict our hierarchy of tree families by using only deterministic rational mappings from terms to terms, and starting from the regular terms. The family at level 2 contains all the morphic infinite words [CT00] (Proposition 3.2). Then, we show that the three hierarchies of term families are equal (Theorems 3.3 and 3.5). In particular, we describe paths in lambda graphs of [KNU02] by giving directly a deterministic rational mapping. And we show that the evaluation of first order substitutions in [CK02] is essentially an inverse deterministic rational mapping followed by an unfolding. Furthermore we show that the evaluation of second order substitutions can be done by applying two inverse deterministic rational mappings each being followed by an unfolding (Proposition 3.4).

2 A Hierarchy of Tree Families

We present two basic graph transformations that preserve the decidability of the monadic theory: the unfolding and the inverse rational mapping. Iterating these transformations gives a hierarchy of graph families, and especially a hierarchy of tree families.

Let $\mathbb{N}$ be the set of nonnegative integers. For any $n \in \mathbb{N}$, we denote $[n] = \{1,\ldots,n\}$ with $[0] = \emptyset$. For any set E, we denote $|E|$ its cardinal and 2^E its powerset. Let L^* be the free monoid generated by any set L of symbols, called *letters*. Any *word* $u \in L^*$ of *length* $|u| \in \mathbb{N}$ is a mapping from $\{1,\ldots,|u|\}$ into L represented by $u(1)\ldots u(|u|) = u$. The word of length 0 is the empty word ε. For any word u, $Occ(u) := \{\ u(i) \mid i \in [|u|]\ \}$ is its set of letters.

Let L be a countable set of symbols for labelling arcs. A simple, oriented and arc labelled *graph* G is a subset of $V{\times}L{\times}V$ where V is an arbitrary set and such that its *label* set

$L_G := \{\ a \in L \mid \exists\ s,t,\ (s,a,t) \in G\ \}$ is finite

but its *vertex* set

$V_G := \{\ s \mid \exists\ a,t,\ (s,a,t) \in G \vee (t,a,s) \in G\ \}$ is finite or countable.

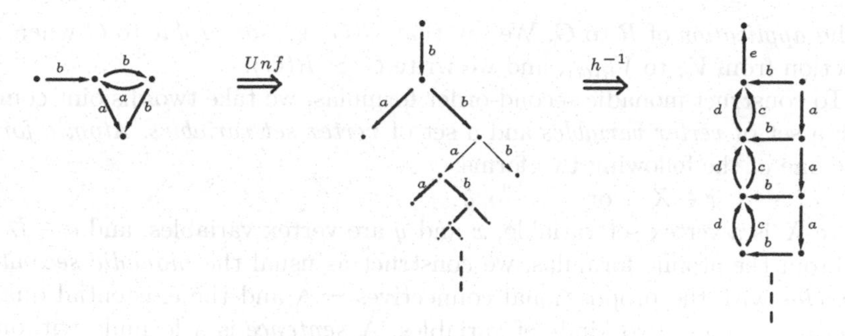

Fig. 2.2. A root unfolding and an inverse finite mapping.

Any (s,a,t) of G is a *labelled arc* of *source* s, of *target* t, with label a, and is identified with the labelled transition $s \xrightarrow[G]{a} t$ or directly $s \xrightarrow{a} t$ if G is understood. For instance, the finite graph $\{r \xrightarrow{b} p\,,\, p \xrightarrow{a} s\,,\, p \xrightarrow{b} q\,,\, q \xrightarrow{a} p\,,\, q \xrightarrow{b} s\}$ has p,q,r,s for vertices, and has a,b for labels, and is represented by the leftmost figure in Figure 2.2. A vertex s is *terminal* if it is source of no arc. We write $s \xrightarrow[G]{P}$ with $P \subseteq L$ if there is $s \xrightarrow[G]{a} t$ for some $a \in P$ and $t \in V_G$. We denote $G_{|W} := \{\, s \xrightarrow[G]{a} t \mid s,t \in W \,\}$ the *restriction* of a graph G to a subset W. Any word $s_0 a_1 s_1 \ldots a_n s_n$ such that $n \geq 0$ and $s_0 \xrightarrow{a_1} s_1 \ldots s_{n-1} \xrightarrow{a_n} s_n$ is a *path* from s_0 to s_n labelled by $w = a_1 \ldots a_n$, and we write $s_0 \overset{w}{\underset{G}{\Longrightarrow}} s_n$ or directly $s_0 \overset{w}{\Longrightarrow} s_n$ if G is understood. We say that a vertex r is a *root* of G if every vertex is accessible from r: $\forall\, s \in V_G,\ \exists\, w,\ r \overset{w}{\Longrightarrow} s$. We denote $Path(G)$ the path language of G, and $L(G,I,F)$ the path labels from a set I to a set F:

$$L(G,I,F) \ := \{\, w \mid \exists\, s \in I, \exists\, t \in F, s \overset{w}{\underset{G}{\Longrightarrow}} t \,\}$$

For instance taking the leftmost graph in Figure 2.2, its path labels from its root to its terminal vertex is $b(ba)^*(a+bb)$. A *trace* of a graph is a path language from and to finite vertex sets. Recall that the traces of finite graphs form the set

$$Rat(L^*) := \{\, L(G,I,F) \mid |G| < \infty \wedge I,F \subseteq V_G \,\}$$

of *rational languages* over L, and we denote $Fin(L^*)$ the family of finite languages over L. A graph is *deterministic* if distinct arcs with the same source have distinct labels: if $r \xrightarrow{a} s$ and $r \xrightarrow{a} t$ then $s = t$. Recall that a graph is a *tree* if it has a root r which is target of no arc, and every vertex $s \neq r$ is target of a unique arc. Any vertex of a tree is also called a *node* and any terminal node is a *leaf.* The *subtree* of a tree G at node s is the restriction $G_{|\{\, t \,\mid\, \exists\, w,\ s \overset{w}{\Longrightarrow} t \}}$ of G to the vertices accessible from s. A *forest* is a graph such that each connected component is a tree.

We will present two known basic graph transformations preserving the graph isomorphism, and also the decidability of the monadic theory. Let us recall the notion of graph isomorphism and the monadic theory of a graph.

Given a binary relation R, we consider the graph

$$R(G) \ := \{\, s' \xrightarrow{a} t' \mid \exists\, s \xrightarrow[G]{a} t,\ s\,R\,s' \wedge t\,R\,t' \,\}$$

of the *application* of R to G. We say that $R(G)$ is *isomorphic* to G when R is a bijection from V_G to $V_{R(G)}$, and we write $G \simeq R(G)$.

To construct monadic second-order formulas, we take two disjoint countable sets: a set of *vertex variables* and a set of *vertex set variables*. *Atomic formulas* have one of the following two forms:

$$x \in X \quad \text{or} \quad x \xrightarrow{a} y$$

where X is a vertex set variable, x and y are vertex variables, and $a \in L$.

From the atomic formulas, we construct as usual the *monadic second-order formulas* with the propositional connectives $\neg, \wedge$ and the existential quantifier $\exists$ acting on these two kinds of variables. A *sentence* is a formula without free variable. The set of monadic second-order sentences $MTh(G)$ satisfied by a graph G forms the *monadic theory* of G. Note that two isomorphic graphs satisfy the same sentences: $MTh(G) = MTh(H)$ if $G \simeq H$. Many articles concern the decidability of the monadic theory for infinite graphs (see among others [Sem84], [MuS85], [Co90], [Th90] and [Th97]).

We present now two known graph transformations: the unfolding and the inverse label mapping. The first transformation is to unfold a graph from all its vertices. The *unfolding* $Unf(G)$ of any graph G is the following forest:

$$Unf(G) \ := \ \{ \ ws \xrightarrow{a} wsat \mid wsat \in Path(G) \ \wedge \ s \xrightarrow[G]{a} t \ \}$$

The unfolding $Unf(G,I)$ of a graph G from a vertex subset I is the restriction of $Unf(G)$ to paths starting from vertices in I *i.e.*

$$Unf(G,I) \ := \ Unf(G)_{|\{ \, su \in Path(G) \, | \, s \in I \, \}}$$

In particular $Unf(G) = Unf(G,V_G)$ and for any vertex $s \in V_G$, $Unf(G,s)$ is a tree (a connected component of $Unf(G)$), called an *unfolding tree* of G. For instance, the unfolding tree from the root of the leftmost graph in Figure 2.2 is given by the middle representation. This tree is deterministic and is a *regular tree*: it has a finite number of non isomorphic subtrees.

Note that for any deterministic graph G, its unfolding $Unf(G,r)$ from any vertex r is isomorphic to the following tree:

$$Tree(G,r) := \{ \ u \xrightarrow{a} ua \mid \exists \ s, \ r \overset{ua}{\underset{G}{\Longrightarrow}} s \ \wedge \ a \in L \ \}$$

where the isomorphism associates to any $s_0a_1s_1 \ldots a_ns_n \in Path(G)$ its label $a_1 \ldots a_n$. The unfolding preserves the decidability of the monadic theory.

Proposition 2.1 [CW98] *Given a graph G and a vertex s, we have*

$$MTh(G) \text{ is decidable} \implies MTh(Unf(G,s)) \text{ is decidable.}$$

The second transformation is the inverse label mapping [Ca96]. To move by inverse arcs, we take a new symbol set $\overline{L} \ := \ \{ \ \overline{a} \mid a \in L \ \}$ in bijection with L. Any transition $s \xrightarrow[G]{\overline{a}} t$ means that $t \xrightarrow[G]{a} s$ is an arc of G. We extend the existence of a path $\overset{w}{\Longrightarrow}$ labelled by a word w in $(L \cup \overline{L})^*$: $s \overset{\varepsilon}{\Longrightarrow} s$ and $s \overset{aw}{\Longrightarrow} t$ if there is r such that $s \xrightarrow{a} r \overset{w}{\Longrightarrow} t$. Given any relation $h \subseteq L{\times}(L \cup \overline{L})^*$ of finite domain *i.e.* a mapping from L into $2^{(L \cup \overline{L})^*}$ such that $Dom(h) := \{ \ a \mid h(a) \neq \emptyset \ \}$ is finite, the *inverse mapping* $h^{-1}(G)$ of any graph G by h is the following graph:

$$h^{-1}(G) \ := \ \{ \ s \xrightarrow{a} t \mid \exists \ w \in h(a), \ s \overset{w}{\underset{G}{\Longrightarrow}} t \ \}$$

and h is a *finite mapping* (resp. *rational mapping*) when $h(a)$ is finite (resp.

rational) for every $a \in L$. For instance, starting from the tree of the middle representation of Figure 2.2 and by applying by inverse the finite mapping defined by $a \mapsto \{\overline{a}aba\}$; $b \mapsto \{\overline{a}aa\}$; $c, d \mapsto \{\overline{a}\,\overline{a}\overline{b}\overline{a}aa\}$; $e \mapsto \{\overline{a}\,\overline{a}\overline{b}\,\overline{b}ba\}$, we get the deterministic graph of the rightmost representation of Figure 2.2. The inverse rational mapping preserves the decidability of the monadic theory.

Proposition 2.3 [Ca96] *Given a graph G and a rational mapping h,*

$$MTh(G) \text{ is decidable } \Longrightarrow MTh(h^{-1}(G)) \text{ is decidable.}$$

Starting from the finite trees and by alternate repetition of inverse rational mappings and unfoldings, we get a hierarchy of families of trees and a hierarchy of families of graphs having a decidable monadic theory. Precisely and for any graph family $\mathcal{F}$, we denote

$[\mathcal{F}] := \{ H \mid \exists\, G \in \mathcal{F},\ G \simeq H \}$ the closure by isomorphism of $\mathcal{F}$

$Unf(\mathcal{F}) := [\{ Unf(G,s) \mid \exists\, G \in \mathcal{F}\ s \in V_G \}]$

the unfolding trees, up to isomorphism, of graphs in $\mathcal{F}$

$Rat^{-1}(\mathcal{F}) := \{ h^{-1}(G) \mid G \in \mathcal{F} \wedge h : L \longrightarrow Rat((L \cup \overline{L})^*) \wedge |Dom(h)| < \infty \}$

the inverse rational mappings of graphs in $\mathcal{F}$.

Taking the set $Tree_0$ of finite trees, we consider the hierarchy $(Tree_n)_{n\geq 0}$ of trees and the hierarchy $(Graph_n)_{n\geq 0}$ of graphs, defined inductively on $n \geq 0$ as follows:

$$Graph_n := Rat^{-1}(Tree_n) \quad \text{and} \quad Tree_{n+1} := Unf(Graph_n)$$

By Proposition 2.1 and 2.3, $\bigcup_{n\geq 0} Graph_n$ is a family of graphs having a decidable monadic theory.

At level 0, by definition $Tree_0$ is the family of finite trees, and $Graph_0$ is the family of finite graphs whose their traces are the rational languages.

At level 1, $Tree_1$ is the family of regular trees of finite degree (each node is node of a finite number of edges). Let us describe the family $Graph_1$.

The family $Graph_1$ is the family of prefix-recognizable graphs [Ca96] : it is the set of prefix transition graphs of labelled recognizable word rewriting systems, or equivalently the set of VR-equational graphs [Ba98]. We get the graphs of $Graph_1$ by ε-closure of the transition graphs of pushdown automata [MuS85] [Ca90] with ε-transitions. An important property is that the inverse rational mappings applied to $Tree_1$ to get $Graph_1 = Rat^{-1}(Tree_1)$ are sufficient to obtain all the graphs which are monadic interpretable on $Tree_1$ (or only on the binary tree) [Ba98]. Note that $Graph_1$ trace the context-free languages, and the trees in $Graph_1$ are all the regular trees. An extension of $Graph_1$ to hypergraphs has been done in [LN01] and [Bl02].

At level $n \geq 2$, we have no general result of $Graph_n$. However we have results on hierarchies of families of deterministic trees, or more exactly of finite and infinite terms [Da82], [KNU01], [KNU02], [CK02], and we will compare these hierarchies with $(Tree_n)_{n\geq 0}$ restricted to terms.

3 A Hierarchy of Term Families

We restrict the previous hierarchy on tree families to a hierarchy of term families $(Term_n)$ obtained from the family of regular terms by iterated inverse determin-

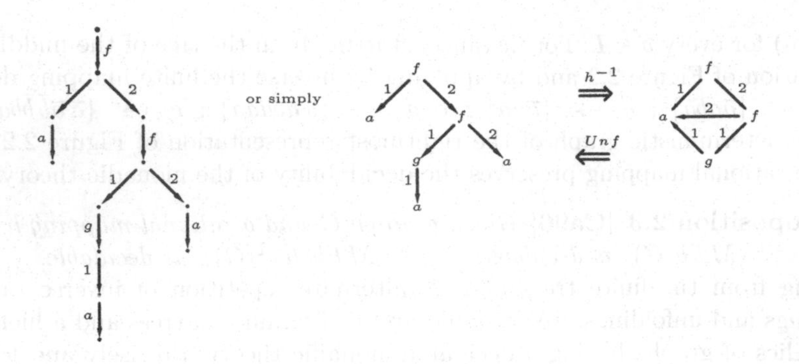

Fig. 3.1. Tree representations and an inverse finite mapping.

istic rational mappings with unfoldings. Any morphic infinite word is in $Term_2$ (Proposition 3.2). We establish that the hierarchy $(Term_n)$ coincides with the hierarchy (Sub_n) of families of terms obtained from the regular terms by iterated first order substitutions (Theorem 3.3). Then, we show that the terms obtained from the regular terms by iterated second order substitutions are the terms of the hierarchy (Proposition 3.4). Finally, we establish that the hierarchy $(Term_n)$ coincides with the hierarchy of families of terms that are solutions of safe schemes (Theorem 3.5).

Let F be a set of symbols called *functions*, graded by a mapping $\varrho : F \longrightarrow \mathbb{N}$ associating to each function f its *arity* $\varrho(f)$, and such that

$$F_n := \{ f \in F \mid \varrho(f) = n \} \quad \text{is countable for every } n \geq 0.$$

The set $T(F)$ of *finite terms* is the smallest subset of F^* such that

$$f \in F \wedge t_1, \ldots, t_{\varrho(f)} \in T(F) \implies f t_1 \ldots t_{\varrho(f)} \in T(F)$$

Particularly the *constant* set $F_0 \subseteq T(F)$. Any finite term, for instance $fafgaa$ with $\varrho(f) = 2, \varrho(g) = 1, \varrho(a) = 0$, is represented by a tree as shown by the leftmost representation of Figure 3.1. The middle representation of Figure 3.1 is simpler and usual. So we have to use vertex labelled graphs.

A coloured graph is a graph with a vertex labelling in a finite subset of F. Precisely, a *coloured graph* $G := \underline{G} \cup c_G$ is the union of a (uncoloured) graph $\underline{G} \subseteq V \times L \times V$ and a *vertex labelling* or *colouring* $c_G \subseteq V \times F$ which is functional: $|c_G \cap \{s\} \times F| \leq 1$ for every $s \in V$, such that its domain $\{ s \mid \exists f, (s,f) \in G \}$ is the *vertex set* V_G of G containing $V_{\underline{G}}$, and such that its image $\{ f \mid \exists s, (s,f) \in G \}$ is finite: each vertex has one colour from a finite set. Any $(s,f) \in G$ is a vertex s labelled f and is also denoted sf. Note that the coloured graph $\{1a\}$ has an empty uncoloured graph and the vertex labelling $1 \mapsto a$.

Any path $s_0 a_1 s_1 \ldots a_n s_n \in Path(\underline{G})$ is now labelled by $w = c(s_0) a_1 c(s_1) \ldots a_n c(s_n)$ and we write $s_0 \overset{w}{\underset{G}{\Longrightarrow}} s_n$ (or directly $s_0 \overset{w}{\Longrightarrow} s_n$ if G is understood) for a path from s_0 to s_n labelled by w. For instance, the path labels $L(G, r, E)$ of the coloured tree G of Figure 3.1 from its root r to its set E of its leaves is the language $\{f1a, f2f1g1a, f2f2a\}$ called the *branch language* of G [Co83].

We extend the unfolding and the inverse mapping to any coloured graph G:

$$Unf(G) := Unf(\underline{G}) \cup \{ wsf \mid ws \in Path(\underline{G}) \wedge sf \in G \}$$

and
$$h^{-1}(G) := \{ s \xrightarrow{a} t \mid \exists\, w \in h(a),\ s \underset{G}{\overset{w}{\Longrightarrow}} t \} \cup \{ s\,h(f) \in G \mid \exists a\ \exists w \in h(a)\ \exists t,\ s \underset{G}{\overset{w}{\Longrightarrow}} t \vee t \underset{G}{\overset{w}{\Longrightarrow}} s \}$$

where h is a mapping from $L \cup F$ into $2^{(L \cup \overline{L} \cup F)^*}$ such that $h(F) \subseteq F$; so the vertices of $h^{-1}(G)$ are the vertices s of $\underline{h^{-1}(G)}$ coloured by $h(f)$ if s is coloured by f in G.

For any coloured graph G and any vertex r, we also consider the unfolding
$$Unf(G,r) := Unf(G)_{|\{ rw \mid rw \in Path(\underline{G}) \}}$$
of G from r, and a canonical representative for $\underline{G}$ deterministic is:
$$Tree(G,r) := \{ u \xrightarrow{a} ua \mid \exists\, s,\ r \underset{G}{\overset{ua}{\Longrightarrow}} s \wedge a \in L \} \cup \{ uf \mid r \underset{G}{\overset{u}{\Longrightarrow}} s \wedge sf \in G \}$$

We also write $Tree(G) := Tree(G,r)$ when r is the unique root of G.

For instance, starting from the tree of Figure 3.1, we apply by inverse the finite mapping defined by $1 \mapsto \{f1a, f1g, g\bar{1}f\bar{2}f1a\}$, $2 \mapsto \{f2f, f\bar{2}f1a\}$ and $x \mapsto x$ for any colour $x \in \{f, g, a\}$, to get the coloured graph of the rightmost representation of figure below. Note that the rightmost graph of Figure 3.1 is a quotient of the tree: its root unfolding is the tree. To represent terms and their quotients, we consider a restriction of coloured graphs.

A *term graph* is a deterministic coloured graph labelled in $L = \mathbb{N} - \{0\}$ such that
$$\{ i \mid \exists\, t,\ s \xrightarrow{i} t \} \ = \ \{1, \ldots, \varrho(c(s))\} \quad \text{for every vertex } s.$$

A general term graph is the graph $\overrightarrow{T(F)}$ of finite terms, defined by
$$\overrightarrow{T(F)} := \{ ft_1 ..t_{\varrho(f)} \xrightarrow{i} t_i \mid f \in F \wedge t_1, \ldots, t_{\varrho(f)} \in T(F) \wedge i \in [\varrho(f)] \} \cup \{ (ft_1 \ldots t_{\varrho(f)})f \mid t_1, \ldots, t_{\varrho(f)} \in T(F) \wedge i \in [\varrho(f)] \}$$
and the *maximal quotient* $Graph(t)$ of any finite term t is
$$Graph(t) \ := \ \overrightarrow{T(F)}_{|\{ s \mid t \Longrightarrow s \}}$$
the restriction of $\overrightarrow{T(F)}$ to its vertices accessible from t (the subterms of t).

For instance $Graph(fafgaa) = \{ fafgaa \xrightarrow{1} a,\ fafgaa \xrightarrow{2} fgaa,\ fgaa \xrightarrow{1} ga, fgaa \xrightarrow{2} a,\ ga \xrightarrow{1} a \}$ with the colouring $c(fafgaa) = c(fgaa) = f$, $c(ga) = g$, $c(a) = a$, and is represented by the rightmost figure in Figure 3.1.

A *term tree* is a term graph which is a tree, and we denote $TermTrees$ the family of term trees. Any finite term t is identified with the isomorphic class of the rooted unfolding of the maximal quotient of t: $Unf(Graph(t), t) = Unf(\overrightarrow{T(F)}, t)$.

More generally, a *term* (finite or infinite) is the isomorphic class $[G]$ of a term tree G whose a standard canonical representative is $Tree(G)$. In particular, the canonical representative of any finite term t is
$$Tree(t) \ := \ Tree(Graph(t)) \ = \ Tree(\overrightarrow{T(F)}, t)$$

For instance $Tree(fafgaa) = \{ \varepsilon \xrightarrow{1} 1, \varepsilon \xrightarrow{2} 2, 2 \xrightarrow{1} 21, 21 \xrightarrow{1} 211, 2 \xrightarrow{2} 22 \}$ with the colouring $c(\varepsilon) = c(2) = f$, $c(21) = g$, $c(1) = c(211) = c(22) = a$.

A well-known fact is that the family of the canonical (representatives of) term trees

$$Terms := \{ Tree(G) \mid G \in TermTrees \}$$

is a complete partial order by taking $\Omega \in F_0$ and by using the following partial order:

$$G \leq_\Omega H \quad \text{if} \quad G - V_G\{\Omega\} \subseteq H \quad \text{for any } G, H \in Terms$$

whose the smallest element is $\{\varepsilon\, \Omega\}$ and such that any increasing chain $G_0 \leq_\Omega \ldots \leq_\Omega G_n \leq_\Omega \ldots$ has a least upper bound

$$sup_{n\geq 0}(G_n) = (\bigcup_{n\geq 0} \underline{G_n}) \cup \{ uf \mid \exists m\, \forall n \geq m,\ uf \in G_n \}$$

For instance and for any $G \in Terms$, we define its *truncation* G_n to the level $n \geq 0$ by

$$G_n := \underline{G}_{|\{u \mid |u| \leq n\}} \cup \{ uf \in G \mid |u| < n \} \cup \{ u\Omega \mid \exists\, f,\ uf \in G \wedge |u| = n \}$$

to obtain an increasing chain $G_0 \leq_\Omega \ldots \leq_\Omega G_n \leq_\Omega \ldots$ of least upper bound G. So any mapping $h\ :\ T(F) \longrightarrow T(F)$ which is monotone

$$Tree(s) \leq_\Omega Tree(t) \implies Tree(h(s)) \leq_\Omega Tree(h(t))$$

is extended into a continuous function $h:\ TermTrees \longrightarrow Terms$ by defining for any $G \in TermTrees$ and taking any $\Omega \in F_0 - Im(c_G)$

$$h(G) := sup_{n\geq 0} Tree(h(t_n)) \quad \text{if no node is labelled by } \Omega$$

where t_n is the unique finite term with $Tree(t_n) = (Tree(G))_n$; otherwise $h(G)$ is undefined.

As we want to produce only terms by inverse rational mappings with unfoldings from terms, we can restrict to any rational mapping h preserving the determinism:

$$G \text{ deterministic} \implies h^{-1}(G) \text{ deterministic}$$

This implication remains true if for each $a \in L$, we restrict $h(a)$ to its minimal prefix subset of $(F(L \cup \overline{L}))^* F$: for any finite automaton recognizing $h(a)$, we determinize it, then we do the synchronized product with the automaton $\{0\}{\times}F{\times}\{1\} \cup \{1\}{\times}(L \cup \overline{L}){\times}\{0\}$ of initial state 0 and of final state 1, and then we remove any arc which is source of a final state. So we may assume that for each $a \in L$, $h(a)$ is recognized by a finite deterministic automaton (A, ι, T) such that each final state (in T) is terminal (source of no arc), and its initial state $\iota \in P := \{ s \in V_A \mid s \xrightarrow[A]{F} \}$ which is disjoint of $\{ s \in V_A \mid s \xrightarrow[A]{L\cup\overline{L}} \}$ and such that

$$s \in P \iff t \in V_A - P \quad \text{for any transition } s \xrightarrow[A]{} t$$

A sufficient condition to have $h^{-1}(G)$ deterministic from a term tree G, is to add the following *determinism condition*:

$$s \xrightarrow[A]{a} \wedge\ s \xrightarrow[A]{b} \ \wedge\ a \in L \implies a = b$$

and in that case, we say that h is a *deterministic rational mapping.*

For any graph family $\mathcal{F}$, we denote

$$DRat^{-1}(\mathcal{F}) := \{ h^{-1}(G) \mid G \in \mathcal{F} \wedge h : L \cup F \longrightarrow Rat((L \cup \overline{L} \cup F)^*) \wedge h(F) \subseteq F \wedge |Dom(h)| < \infty \wedge h \text{ deterministic} \}$$

the inverse deterministic rational mappings of graphs in $\mathcal{F}$. So we restrict the hierarchy $(Tree_n)$ on terms and by using only deterministic rational mappings:

$$Term_0 := \text{the set of regular term trees}$$

$$Term_{n+1} := Unf(DRat^{-1}(Term_n)) \cap TermTrees \ n \geq 0$$

where by commodity with the two next hierarchies, we start to $Tree_1$ restricted to terms. Recall that the terms over F_1 are the infinite words. Particular infinite words having a decidable monadic theory are the infinite *morphic words* [CT00] which are words of the form:

$$\sigma(\tau^\omega(a)) \quad = \quad \sigma(\,a\,u\,\tau(u)\ldots\tau^n(u)\ldots)$$

where σ and τ are morphisms from A^* into itself for some finite $A \subset F_1$ with $a \in A$ and $\tau(a) = au$. The decidability of the monadic theory for these infinite words is also a consequence that they are terms at level 2 of our hierarchy.

Proposition 3.2 *Any morphic word is in* $Term_2$.

We consider the hierarchy of term families defined in [Da82] whose terms have a decidable monadic theory [CK02]. This hierarchy is defined as follows:

$Sub_0 :=$ the set of regular term trees

$Sub_{n+1} := \bigcup\{\ [Subst_{e,u}(Sub_n)] \mid u \in F_0^+ \wedge u(1) \neq \ldots \neq u(|u|) \wedge e \in F_{|u|+1}\ \}$

where for any finite term t, $Subst_{e,u}(t)$ is the finite term without e and is obtained by evaluating the function e as a *first order substitution*: its first argument is the term on which we apply the substitution and its $i+1$-th argument is the term which is substituted to $u(i)$ for each $1 \leq i \leq |u|$. Precisely $Subst_{e,u}(t)$ is defined by induction on the length of any finite term t as follows:

$Subst_{e,u}(ft_1\ldots t_{\varrho(f)}) := f Subst_{e,u}(t_1)\ldots Subst_{e,u}(t_{\varrho(f)}) \quad$ if $f \neq e$

$Subst_{e,u}(et_0t_1\ldots t_{|u|}) := Subst_{e,u}(t_0)[Subst_{e,u}(t_1)/u(1),\ldots, Subst_{e,u}(t_{|u|})/u(|u|)]$

where for any $n \geq 0$, $t, s_1, \ldots, s_n \in T(F)$, $a_1 \neq \ldots \neq a_n \in F_0$, $t[s_1/a_1,\ldots,s_n/a_n]$ is the term obtained by simultaneous replacement in t of a_i by s_i (for $1 \leq i \leq n$), and is defined by induction on the length of t as follows:

$ft_1\ldots t_{\varrho(f)}[s_1/a_1,\ldots,s_n/a_n] := f(t_1[s_1/a_1,\ldots,s_n/a_n])\ldots(t_{\varrho(f)}[s_1/a_1,\ldots,s_n/a_n])$ if $f \notin \{a_1,\ldots,a_n\}$

$a_i[s_1/a_1,\ldots,s_n/a_n] := s_i \quad$ for every $i \in [n]$

Note that the mapping $Subst_{e,u} : T(F) \longrightarrow T(F - \{e\})$ is monotone for $\leq_\Omega$ if $\Omega \notin \{u(1),\ldots,u(|u|)\}$, and we extend by continuity $Subst_{e,u}$ to any term tree $G \in TermTrees$. Finally we extend by union $Subst_{e,u}$ to any set of term trees.

Theorem 3.3 *Starting from the regular terms, the terms generated by n first order substitutions are exactly the terms obtained by applying n inverse deterministic rational mappings each being followed by an unfolding:*

$$Sub_n \;=\; Term_n \quad \textit{for every } n \geq 0\,.$$

We consider now the extension of the first order substitution to the *second order substitution* $Subst_{e,r_1\ldots r_n}$ where $e \in F_{n+1}$ and $r_1,\ldots,r_n$ are elementary terms *i.e.* of the form $f a_1 \ldots a_{\varrho(f)}$ with $f \in F$ and $a_1 \neq \ \ldots\ \neq a_{\varrho(f)} \in F_0$ and such that $r_1(1) \neq \ldots \neq r_n(1)$. This function $Subst_{e,r_1\ldots r_n}$ is first defined by induction on the length of any finite term as follows:

$Subst_{e,r_1\ldots r_n}(ft_1\ldots t_{\varrho(f)}) := f Subst_{e,r_1\ldots r_n}(t_1)\ldots Subst_{e,r_1\ldots r_n}(t_{\varrho(f)})$ if $f \neq e$

and $\quad Subst_{e,r_1\ldots r_n}(et_0t_1\ldots t_n)$

$:= Subst_{e,r_1\ldots r_n}(t_0)[Subst_{e,r_1\ldots r_n}(t_1)/r_1,\ldots, Subst_{e,r_1\ldots r_n}(t_n)/r_n]$

where for any $n \geq 0$, $t, s_1, \ldots, s_n \in T(F)$ and $r_1,\ldots,r_n \in T(F)$ elementary terms with $r_1(1) \neq \ldots \neq r_n(1)$, the term $t[s_1/r_1,\ldots,s_n/r_n]$ is defined by

$$ft_1\ldots t_{\varrho(f)}[s_1/r_1,\ldots,s_n/r_n]$$
$$:= f\,(t_1[s_1/r_1,\ldots,s_n/r_n])\ldots(t_{\varrho(f)}[s_1/r_1,\ldots,s_n/r_n]) \text{ if } f \notin \{r_1(1),\ldots,r_n(1)\}$$
$$:= s_i\,[\,(t_1[s_1/r_1,\ldots,s_n/r_n])/a_1\,,\,\ldots,\,(t_{\varrho(f)}[s_1/r_1,\ldots,s_n/r_n])/a_{\varrho(f)}\,] \quad \text{if } r_i = f\,a_1\ldots a_{\varrho(f)}$$

Note that $Subst_{e,r_1\ldots r_n} : T(F) \longrightarrow T(F-\{e\})$ is monotone for $\leq_\Omega$ if $\Omega \notin Occur(r_1) \cup \ldots \cup Occur(r_n)$, and similarly to the first order substitution, we extend by continuity the second order substitution $Subst_{e,r_1\ldots r_n}$ to any term tree. In particular $Subst_{e,r_1\ldots r_n}$ with $r_1,\ldots,r_n \in F_0$ corresponds to the first order substitution. Any second order substitution preserves the decidability of the monadic theory, and all the terms in $\bigcup_n Term_n$ are also the terms obtained from the regular terms by iterated applications of second order substitutions.

Proposition 3.4 *For any second order substitution $Subst_{e,r_1\ldots r_n}$ and any $q \geq 0$, $G \in Term_q \implies Subst_{e,r_1\ldots r_n}(G) \in Term_{q+2}$ when it is defined.*

We consider the hierarchy of families of terms which are least solutions of higher order recursive schemes [In76] [Da82] and that have a decidable monadic theory when the schemes are safe [KNU02]. By lack of space, it is not possible to reintroduce here the notion of a higher order recursive *scheme*, which is a deterministic grammar between typed terms. The level of a scheme is the maximum level of the left hand side types. The terms considered in [KNU02] are generated by schemes satisfying a safety condition. A scheme P is *safe* if for any rule $fa_1\ldots a_n \longrightarrow t$ there is no subterm s of t with an occurrence a_i in s of type level strictly less than the type level of s. The schemes of level 1 are safe, and have been first considered in [Ni75] and called *recursive program schemes* (see among others [Gu81] and [Co90]).

Theorem 3.5 *For every $n \geq 0$, the terms generated by the safe schemes of level at most n are exactly the terms in $Term_n$ i.e. the terms obtained from the regular terms by applying n inverse deterministic rational mappings each being followed by an unfolding.*

4 Conclusion

Several natural questions arise; we list some of them.

a) Let us begin with a question in [KNU02]. Does any solution of an unsafe scheme can be generated by a safe scheme, and in the negative case, has it a decidable monadic theory?

b) Do we get more terms by allowing non deterministic rational mappings, and applying them not only on terms but on trees *i.e.*

for every $n \geq 0$, $Term_n = Tree_n \cap TermTrees$?

c) Another question is to characterize the family of branch languages of all terms in

$$T := \bigcup_n Term_n\,.$$

Recall that the branch languages for the terms in $Term_1$ is the family of deterministic context-free branch languages [Co83].

d) A last question follows from the fact that the hierarchy ($Term_n$) coincides with the hierarchy of families of terms recognized by the deterministic pushdown automata with multi-level stacks. This last question is the equivalence problem for terms in T:

is the equality of terms in T decidable?

This problem is decidable for terms in $Term_1$ because it is inter-reducible to the equivalence problem of (one level) deterministic pushdown automata, which is decidable [Sen97].

References

[Ba98] K. BARTHELMANN, *When can an equational simple graph be generated by hyperedge replacement*, 23rd MFCS, LNCS 1450, L. Brim, J. Gruska, J. Zlatuska (Eds.), 543–552 (1998).

[Bl02] A. BLUMENSATH, *Axiomatising tree-interpretable structures*, 19th STACS, LNCS 2285, H. Alt, A. Ferreira (Eds.), 596–607 (2002).

[CT00] O. CARTON and W. THOMAS, *The monadic theory of morphic infinite words and generalizations*, 25th MFCS, LNCS 1893, M. Nielsen and B. Rovan (Eds.), 275–284 (2000).

[Ca90] D. CAUCAL, *On the regular structure of prefix rewriting*, 15th CAAP, LNCS 431, A. Arnold (Ed.), 87–102 (1990) [a full version is in Theoretical Computer Science 106, 61–86 (1992)].

[Ca96] D. CAUCAL, *On infinite transition graphs having a decidable monadic theory*, 23rd ICALP, LNCS 1099, F. Meyer auf der Heide, B. Monien (Eds.), 194–205 (1996) [a full version will appear in Theoretical Computer Science].

[CK01] D. CAUCAL and T. KNAPIK, *An internal presentation of regular graphs by prefix-recognizable ones*, Theory of Computing Systems 34-4 (2001).

[Co83] B. COURCELLE, *Fundamental properties of infinite trees*, Theoretical Computer Science 25, 95–169 (1983).

[Co90] B. COURCELLE, *Graph rewriting: an algebraic and logic approach*, Handbook of Theoretical Computer Science Vol. B, J. Leeuwen (Ed.), Elsevier, 193–242 (1990).

[Co90] B. COURCELLE, *Recursive applicative program schemes*, Handbook of Theoretical Computer Science Vol. B, J. Leeuwen (Ed.), Elsevier, 459–492 (1990).

[Co94] B. COURCELLE, *Monadic second-order definable graph transductions: a survey*, Theoretical Computer Science 126, 53–75 (1994).

[CK02] B. COURCELLE and T. KNAPIK, *The evaluation of first-order substitution is monadic second-order compatible*, to appear in Theoretical Computer Science (2002).

[CW98] B. COURCELLE and I. WALUKIEWICZ, *Monadic second-order logic, graph coverings and unfoldings of transition systems*, Annals of Pure and Applied Logic 92, 35–62 (1998).

[Da82] W. DAMM, *The IO and OI hierarchies*, Theoretical Computer Science 20 (2), 95–208 (1982).

[En83] J. ENGELFRIET, *Iterated push-down automata and complexity classes*, 15th STOC, 365–373 (1983).

[Gu81] I. GUESSARIAN, *Algebraic semantics*, LNCS 99, Sringer Verlag, (1981).

[In76] K. INDERMARK, *Schemes with recursion on higher types*, 5th MFCS, LNCS 45, A. Mazurkiewicz (Ed.), 352–358 (1976).

[KNU01] T. Knapik, D. Niwiński and P. Urzyczyn, *Deciding monadic theories of hyperalgebraic trees*, 5th International Conference on Typed Lambda Calculi and Applications, LNCS 2044, Abramsky (Ed.), 253–267 (2001).

[KNU02] T. Knapik, D. Niwiński and P. Urzyczyn, *Higher-order pushdown trees are easy*, 5th FOSSACS, to appear in LNCS, M. Nielsen (Ed.), (2002).

[KV00] O. Kupferman and M. Vardi, *An automata-theoretic approach to reasoning about infinite-state systems*, 12th CAV, LNCS 1855, A. Emerson, P. Sistla (Eds.), 36–52 (2000).

[LN01] S. La Torre and M. Napoli, *Automata-based representations for infinite graphs*, Theoretical Informatics and Applications 35, 311–330 (2001).

[MuS85] D. Muller and P. Schupp, *The theory of ends, pushdown automata, and second-order logic*, Theoretical Computer Science 37, 51–75 (1985).

[Ni75] M. Nivat, *On the interpretation of polyadic recursive schemes*, Symposia Mathematica 15, Academic Press (1975).

[Ra69] M. Rabin, *Decidability of second-order theories and automata on infinite trees*, Transactions of the American Mathematical Society 141, 1–35 (1969).

[Sem84] A. Semenov, *Decidability of monadic theories*, MFCS 84, LNCS 176, W. Brauer (Ed.), 162–175 (1984).

[Sen97] G. Sénizergues, *The equivalence problem for deterministic pushdown automata is decidable*, 24th ICALP, LNCS 1256, P. Degano, R. Gorrieri, A. Marchetti-Spaccamela (Eds.), 671–681 (1997).

[Th90] W. Thomas, *Automata on infinite objects*, Handbook of Theoretical Computer Science Vol. B, J. Leeuwen (Ed.), Elsevier, 135–191 (1990).

[Th97] W. Thomas, *Languages, automata, and logic*, Handbook of Formal Languages, Vol. 3, G. Rozenberg, A. Salomaa (Eds.), Springer, 389–456 (1997).

[Ti86] J. Tiuryn, *Higher-order arrays and stacks in programming: An application of complexity theory to logics of programs*, 12th MFCS, LNCS 233, J. Gruska, B. Rovan, J. Wiedermann (Eds.), 177–198 (1986).

A Chomsky-Like Hierarchy of Infinite Graphs

Didier Caucal[1] and Teodor Knapik[2]

[1] IRISA–CNRS, Campus de Beaulieu,
35042 Rennes, France,
caucal@irisa.fr
[2] ERMIT, Université de la Réunion,
BP 7151, 97715 Saint Denis Messageries Cedex 9, Réunion,
knapik@univ-reunion.fr

Abstract. We consider a strict four–level hierarchy of graphs closely related to the Chomsky hierarchy of formal languages. We provide a uniform presentation of the four families by means of string rewriting.

1 Introduction

Many devices known in computation theory such as automata, machines, transducers or rewrite systems may be used in several ways for encoding infinite graphs. A graph may be associated to a device $\mathcal{D}$ as a transition system that represents its behaviour. In this case we are not interested in any acceptance condition because the device is not used as a recognizer. We simply assimilate it to a reactive system which reads a letter and changes nondeterministically its internal configuration one or more times before reading, in an infinite input stream, the next letter of an *external alphabet* Γ. We call *transition graph* a graph constructed in this way. Yet we have a choice between a graph $\mathfrak{tg}_\Gamma(\mathcal{D})$ that displays all transitions of $\mathcal{D}$, including silent transitions (labelled by τ) or a graph $\mathfrak{tG}_\Gamma(\mathcal{D})$ that reflects only its observable behaviour. In both cases, we may wish the vertices of a graph do not consist of all configurations of $\mathcal{D}$ but belong to a given set C. Indeed, the graphs studied here are *concrete* viz its vertices are named and the words over an *internal alphabet* Σ are used as names. Thus C is a language over Σ which, for decidability related reasons, is assumed to be rational. Transition graphs have been introduced in [15] for pushdown automata, in [11] for linear–bounded automata and in [16, 8] for Turing machines.

Besides transition graphs, infinite graphs may be defined using those machines that are able to accept or generate relations, merely rational transducers and Turing machines. We call these *relation graphs.* In order to obtain edge–labelled graphs, we let each final state of the device correspond to a labelling letter of the external alphabet Γ. For instance, if a single–tape nondeterministic Turing machine $\mathcal{M}$ starts with a word u and stops with a word v in a final state f_a, then the edge $u \xrightarrow{a} v$ belongs to $\mathfrak{rg}_\Gamma(\mathcal{M})$. In a similar way, labelled string–rewriting systems may define relation graphs. Such graphs have an edge $u \xrightarrow{a} v$, whenever a word u may be rewritten into a word v by a rewrite rule which has a label a. Relation graphs have been introduced in [9] for synchronized finite

K. Diks et al. (Eds): MFSC 2002, LNCS 2420, pp. 177–187, 2002.

transducers, in [13] for finite transducers, in [8] for Turing machines and in [5] for labelled string–rewriting systems.

In this paper, we show that several classes of relation and transition graphs have a uniform finite presentation by means of string rewriting. The uniform presentation of this paper has been introduced in [10, 4] and, similarly to the Cayley graph of a group, is based on the following idea: the graph associated to a string–rewriting system $\mathcal{S}$ has an edge $u \xrightarrow{a} v$ whenever au may be rewritten by $\mathcal{S}$ into v, where u and v are irreducible words. However, the vertices of the graph do not consist of all irreducible words but are restricted to a given rational subset of the latter. A slightly different idea introduced in [7] consists in restricting not only the vertices of the graph but also all intermediate rewrite steps to a given rational set. We speak of Cayley–type graphs in the former case and normal–form graphs in the latter case. We show that both provide a uniform presentation of a Chomsky–like hierarchy of infinite graphs.

We are only interested in those graphs which have a rational set of vertices. Since the edges of our graphs are induced by the rewriting relation, we argue that the rationality of this relation is an essential separation criterion for families of graphs presented by string rewriting. This rationality depends on possible overlaps of right–hand sides by left–hand sides of rewrite rules. As shown in [7], only two types of overlaps, namely prefix and right (together with their symmetric versions: suffix and left) lead to rational rewriting relations. We establish that prefix–recognizable (resp. rational) graphs (see survey [18]) correspond to prefix (resp. right) systems. With unrestricted overlaps we get the family of recursively enumerable graphs. We also capture the family of finite graphs located at the bottom of our four–level hierarchy of graphs. The latter is closely related to the Chomsky hierarchy of formal languages via the notion of trace of a graph.

Due to the limitation of the number of pages, proofs are not given in this version but the reader may consult a complete version at
`http://www.univ-reunion.fr/~knapik/publications/mfcs02.ps.gz`

2 Preliminaries

The powerset of a set E is written $\wp(E)$. The set $\{1, \ldots, n\}$ is abbreviated as $[n]$ with $[0] = \varnothing$. We write $\mathcal{D}\text{om}(\mathcal{R})$ (resp. $\mathcal{R}\text{an}(\mathcal{R})$) for the domain (resp. range) of a relation $\mathcal{R}$.

We assume that the reader is familiar with the notions of monoid, word, rational and recognizable subsets of a monoid, regular expression, finite automaton and transducer (see e.g. [3]) as well as pushdown automaton (see e.g. [1]) and Turing machine (see e.g. [12]).

The family of finite (resp. recognizable, rational) subsets of a monoid $\mathcal{M}$ is written $\mathcal{F}\text{in}(\mathcal{M})$ (resp. $\mathcal{R}\text{ec}(\mathcal{M})$, $\mathcal{R}\text{at}(\mathcal{M})$). The free monoid over Σ is written Σ^*, ε stands for the empty word and the length of a word $w \in \Sigma^*$ is denoted by $|w|$.

A *string–rewriting system* (an srs for short) $\mathcal{S}$ is a subset of $\Sigma^* \times \Sigma^*$. An element of $\mathcal{S}$ is called a *rule* and is written $l \to r$; the word l (resp. r) is its left-hand (resp. right-hand) side. In this paper we consider only finite and recognizable srs. (Recall that a recognizable srs is a finite union of relations of the form $L \times R$, where $L, R \in \mathfrak{Rat}(\Sigma^*)$). The *single-step reduction relation* induced by $\mathcal{S}$ on Σ^*, is the binary relation

$$\underset{\mathcal{S}}{\longrightarrow} = \{(xly, xry) \mid x, y \in \Sigma^*,\ l \to r \in \mathcal{S}\} .$$

An $\mathcal{S}$*-redex* of a reduction step $u \underset{\mathcal{S}}{\longrightarrow} v$ is a factor l of u such that $l \to r \in \mathcal{S}$, $u = xly$ and $v = xry$ for some $x, y \in \Sigma^*$. A word u *reduces* into a word v (alternatively v is a *descendant* of u), if $u \overset{*}{\underset{\mathcal{S}}{\longrightarrow}} v$, where $\overset{*}{\underset{\mathcal{S}}{\longrightarrow}}$ is the reflexive–transitive closure of $\underset{\mathcal{S}}{\longrightarrow}$ called the *reduction relation* induced by $\mathcal{S}$ on Σ^*. A word v is *irreducible* w.r.t. $\mathcal{S}$ when v does not belong to $\mathfrak{Dom}(\underset{\mathcal{S}}{\longrightarrow})$. The set $\mathfrak{Irr}(\mathcal{S}) := \Sigma^* \setminus \mathfrak{Dom}(\underset{\mathcal{S}}{\longrightarrow})$ of all irreducible words w.r.t. $\mathcal{S}$ is rational whenever $\mathfrak{Dom}(\mathcal{S})$ is so, since $\mathfrak{Dom}(\underset{\mathcal{S}}{\longrightarrow}) = \Sigma^*(\mathfrak{Dom}(\mathcal{S}))\Sigma^*$.

When the reduction w.r.t. $\mathcal{S} \subseteq \Sigma^* \times \Sigma^*$ is restricted to some specific set of configurations $C \subseteq \Sigma^*$ we speak of *configured rewriting* [7]. Only rational sets of configurations are of interest in this paper. Thus, a *configured system* is a pair $(\mathcal{S}, C)$ where $\mathcal{S} \subseteq \Sigma^* \times \Sigma^*$ and $C \in \mathfrak{Rat}(\Sigma^*)$. We define $\underset{(\mathcal{S},C)}{\longrightarrow} := \underset{\mathcal{S}}{\longrightarrow} \cap\, C \times C$ and the rewriting relation w.r.t. $(\mathcal{S}, C)$ is the reflexive–transitive closure $\overset{*}{\underset{(\mathcal{S},C)}{\longrightarrow}}$ of $\underset{(\mathcal{S},C)}{\longrightarrow}$. We say that $(\mathcal{S}, C)$ is *stable* if for all $u, w \in C$, $v \in \Sigma^*$ such that $u \overset{*}{\underset{\mathcal{S}}{\longrightarrow}} v \overset{*}{\underset{\mathcal{S}}{\longrightarrow}} w$ we have $v \in C$. We have then the following equality $\overset{*}{\underset{(\mathcal{S},C)}{\longrightarrow}} = \overset{*}{\underset{\mathcal{S}}{\longrightarrow}} \cap\, C \times C$. Note that the stability is undecidable in general [7] but the following sufficient condition $\underset{\mathcal{S}}{\longrightarrow}(C) \subseteq C$, where $\underset{\mathcal{S}}{\longrightarrow}(C)$ stands for the image of C under $\underset{\mathcal{S}}{\longrightarrow}$, is decidable whenever $\mathcal{S}$ is finite, recognizable or rational. A configured system $(\mathcal{S}, C)$ is finite (resp. recongnizable) when $\mathcal{S}$ is so. We denote by $\mathfrak{StabCS}$ (resp. $\mathfrak{FinCS}$, $\mathfrak{RecCS}$) the class of stable (resp. finite, recognizable) configured systems.

An essential feature of an srs is the way that left–hand sides overlap right–hand sides. An *LR–overlap of* $\mathcal{S}$ (resp. of $(\mathcal{S}, C)$) is a tuple $(x_1, l_1 \to r_1, y_1, x_2, l_2 \to r_2, y_2)$, where $x_1, y_1, x_2, y_2 \in \Sigma^*$ and $l_1 \to r_1, l_2 \to r_2 \in \mathcal{S}$ are such that $x_1 r_1 y_1 = x_2 l_2 y_2$ (resp. and $x_1 l_1 y_1, x_1 r_1 y_1, x_2 l_2 y_2, x_2 r_2 y_2 \in C$) and at least one of the following holds: $|x_1| \leq |x_2| < |x_1 r_1|$, $|x_2| < |x_1| \leq |x_2 l_2|$. The former inequality corresponds to the following cases:

x_1	r_1	y_1
x_2	l_2	y_2

x_1	r_1	y_1
x_2	l_2	y_2

and the latter corresponds to the following cases:

x_1	r_1	y_1
x_2	l_2	y_2

x_1	r_1	y_1
x_2	l_2	y_2

Seven kinds of LR–overlaps may be distinguished, including the special case when both$x_1 = x_2$ and $r_1 = l_2$ (see [7] for details).

Graphs

By *graph* we always understand here a countable oriented edge–labelled simple graph, viz., a subset of $D \times \Gamma \times D$ where D is a countable set and Γ is a finite alphabet. The family of all such graphs with labels in Γ is written $\mathrm{G}(\Gamma)$ (independently of D). Given a graph $G \subseteq D \times \Gamma \times D$ where D, we write $d \xrightarrow[G]{a} d'$ for an a–labelled edge of G from a vertex $d \in D$ to a vertex $d' \in D$. A *path* of G from d to d' is a sequence of edges $(d_{i-1} \xrightarrow[G]{a_i} d_i)_{i \in [n]}$ such that $d_0 = d$ and $d_n = d'$. The word $a_1 \ldots a_n$ is the *label of the path*. We write $d \overset{w}{\underset{G}{\cdots\cdots\blacktriangleright}} d'$ if there exists a path labelled by $w \in \Gamma^*$ from d to d'. We use the same notation in a more general case when the labels belong to an arbitrary monoid. We also write $d \overset{L}{\underset{G}{\cdots\cdots\blacktriangleright}} d'$, where $L \subseteq \Gamma^*$, to mean that there exists $w \in L$ such that $d \overset{w}{\underset{G}{\cdots\cdots\blacktriangleright}} d'$.

An *inverse* of G is the graph $\overline{G} := \{d' \xrightarrow{\overline{a}} d \mid d \xrightarrow[G]{a} d'\}$, $\overline{G} \subseteq D \times \overline{\Gamma} \times D$ where $\overline{\Gamma}$ is the alphabet of *formal inverses* of Γ. A map $h\colon \Gamma_2 \to \mathfrak{Rat}((\Gamma_1 \cup \overline{\Gamma_1})^*)$, yields the map $h^{-1}\colon \mathrm{G}(\Gamma_1) \to \mathrm{G}(\Gamma_2)$ called *inverse rational mapping*[1] defined as follows $h^{-1}(G) := \{d \xrightarrow{a} d' \mid d \overset{h(a)}{\underset{G \cup \overline{G}}{\cdots\cdots\blacktriangleright}} d'\}$. The image of a family of graphs $\mathcal{G}$ by all inverse rational mappings is written $\mathrm{RatM}^{-1}(\mathcal{G})$.

The *restriction* of $G \subseteq D \times \Gamma \times D$ to $C \subseteq D$, written $G|_C$ is the subgraph of G induced by C, namely $G|_C := G \cap (C \times \Gamma \times C)$. When C is a rational set, we speak of a *rational restriction*. This, of course, makes sense only when the vertices of the graph belong to a monoid $\mathcal{M}$. In this case we also define a multiplication on the left (resp. right) of a graph by an element of the monoid $m \in \mathcal{M}$, viz., $mG := \{mu \xrightarrow{a} mv \mid u \xrightarrow[G]{a} v\}$ (resp. $Gm := \{um \xrightarrow{a} vm \mid u \xrightarrow[G]{a} v\}$).

We denote by $G_1 \equiv G_2$ the existence of an isomorphism between two graphs G_1 and G_2. Given two families of graphs $\mathcal{G}_1$ and $\mathcal{G}_2$, we write $\mathcal{G}_1 \subseteqq \mathcal{G}_2$ if, for every $G_1 \in \mathcal{G}_1$, there exists $G_2 \in \mathcal{G}_2$ such that $G_1 \equiv G_2$. We write $\mathcal{G}_1 \equiv \mathcal{G}_2$ when both $\mathcal{G}_1 \subseteqq \mathcal{G}_2$ and $\mathcal{G}_2 \subseteqq \mathcal{G}_1$.

Given a graph G and two vertices d_1, d_2 of G, we define $\mathrm{L}(G, d_1, d_2) := \{w \mid d_1 \overset{w}{\underset{G}{\cdots\cdots\blacktriangleright}} d_2\}$. The *trace* of a family of graphs $\mathcal{G}$, written $\mathrm{L}(\mathcal{G})$ is defined as follows: $\mathrm{L}(\mathcal{G}) := \{\mathrm{L}(G, d_1, d_2) \mid G \in \mathcal{G};\ d_1, d_2 \text{ are vertices of } G\}$.

3 Graphs Defined by String-Rewriting

Just like a Cayley graph is associated to a group presentation, we define a *Cayley-type graph* $\mathfrak{CG}_\Gamma(\mathcal{S})$ associated to an srs $\mathcal{S} \subseteq \Sigma^* \times \Sigma^*$ as follows:

$$\mathfrak{CG}_\Gamma(\mathcal{S}) := \{u \xrightarrow{a} v \mid a \in \Gamma,\ u, v \in \mathfrak{Irr}(\mathcal{S}),\ au \xrightarrow[\mathcal{S}]{*} v\}$$

[1] An inverse rational mapping $h^{-1}\colon \mathrm{G}(\Gamma_1) \to \mathrm{G}(\Gamma_2)$ is not an inverse of $h\colon \Gamma_2 \to \mathfrak{Rat}((\Gamma_1 \cup \overline{\Gamma_1})^*)$.

where $\Gamma \subseteq \Sigma$. The expressive power is increased when the set of the vertices is restricted to a rational set $C \in \mathfrak{Rat}(\Sigma^*)$:

$$\mathfrak{CG}_\Gamma(\mathcal{S}, C) := \mathfrak{CG}_\Gamma(\mathcal{S})|_C = \{u \xrightarrow{a} v \mid a \in \Gamma,\ u, v \in \mathfrak{Irr}(\mathcal{S}) \cap C,\ au \xrightarrow[\mathcal{S}]{*} v\} .$$

In the case of a configured system $(\mathcal{S}, C)$, we speak of *normal–form graph* $\mathfrak{NG}_\Gamma(\mathcal{S}, C)$ defined as follows:

$$\mathfrak{NG}_\Gamma(\mathcal{S}, C) := \{u \xrightarrow{a} v \mid a \in \Gamma,\ u, v \in \mathfrak{Irr}(\mathcal{S}) \cap C,\ au \xrightarrow[(\mathcal{S},C)]{*} v\} .$$

Note that, by definition, $au \in C$ and every reduction step from au into v is in C which is not the case for Cayley–type graphs (see examples on Fig. 1). In addition $\mathfrak{CG}_\Gamma(\mathcal{S}, C) = \mathfrak{NG}_\Gamma(\mathcal{S}, \Sigma^*)|_C$.

Let us denote by $\mathfrak{Fing}_\Gamma$ the class of finite graphs with edge labels in Γ. As expected, every finite graph may be encoded as both Cayley–type and normal–form graph.

Proposition 3.1. *The following classes of graphs coincide:*

(1) $\mathfrak{Fing}_\Gamma$,
(2) $\{\mathfrak{CG}_\Gamma(\mathcal{S}, C) \mid \mathcal{S} \subseteq \Sigma^* \times \Sigma^*,\ C \in \mathfrak{Fin}(\Sigma^*)\}$,
(3) $\{\mathfrak{NG}_\Gamma(\mathcal{S}, C) \mid \mathcal{S} \subseteq \Sigma^* \times \Sigma^*,\ C \in \mathfrak{Fin}(\Sigma^*)\}$.

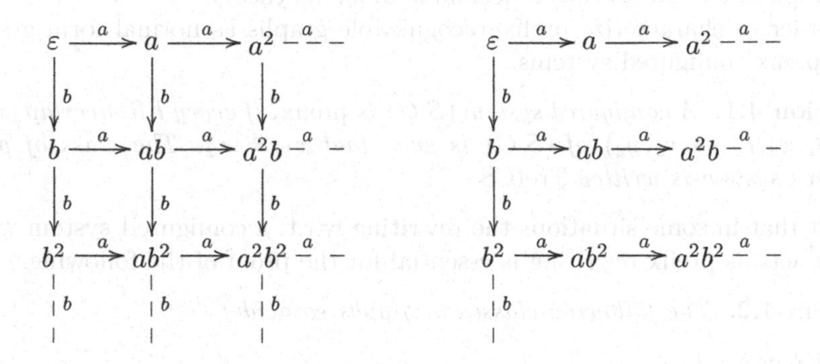

Fig. 1. $\mathfrak{CG}_\Gamma(\mathcal{S}, C)$ and $\mathfrak{NG}_\Gamma(\mathcal{S}, C)$ for $\mathcal{S} = \{ba \to ab\}$ and $C = a^*b^*$

4 Prefix-Recognizable Graphs

A well known restriction in string rewriting is the prefix rewriting [2, 5]. Given an srs $\mathcal{S}$, the *single–step prefix–reduction relation* w.r.t. $\mathcal{S}$, written $\xrightarrow[\mathcal{S}]{}\!\!\!\!\!\rightarrow$ is defined as the right closure of $\mathcal{S}$ by Σ^*: $\xrightarrow[\mathcal{S}]{}\!\!\!\!\!\rightarrow\ := \mathcal{S}\Sigma^* = \{(lx, rx) \mid l \to r \in \mathcal{S},\ x \in \Sigma^*\}$ and the transitive closure of $\xrightarrow[\mathcal{S}]{}\!\!\!\!\!\rightarrow$ is the *prefix–reduction relation* $\xrightarrow[\mathcal{S}]{*}\!\!\!\!\!\rightarrow$. For finite string rewriting systems, the prefix rewriting is equivalent to pushdown automata

[5] provided that the rules of an srs are labelled by Γ. A single–step reduction (resp. prefix–reduction) relation w.r.t. a labelled srs $\mathcal{L} \subseteq \Sigma^* \times \Gamma \times \Sigma^*$ is a graph

$$\mathfrak{gr}_\Gamma(\mathcal{L}) := \underset{\mathcal{L}}{\rightarrow} = \{xly \xrightarrow{a} xry \mid l \xrightarrow{a} r \in \mathcal{L},\ x, y \in \Sigma^*\},$$
$$(\text{resp. } \mathfrak{pgr}_\Gamma(\mathcal{L}) := \underset{\mathcal{L}}{\twoheadrightarrow} = \{ly \xrightarrow{a} ry \mid l \xrightarrow{a} r \in \mathcal{L},\ y \in \Sigma^*\}).$$

In the case of recognizable labelled string rewriting systems, the prefix rewriting together with rational restrictions leads to the family of *prefix–recognizable graphs* introduced in [6]:

$$\mathcal{P}\text{ref}\mathcal{R}\text{ec}\mathfrak{g}_\Gamma := \{ \underset{\mathcal{L}}{\twoheadrightarrow} |_D \mid \mathcal{L} = \bigcup_{i=1}^{k} L_i \times \{a_i\} \times R_i,\ k \in \mathbb{N},\ a_i \in \Gamma,\ L_i, R_i, D \in \mathcal{R}\text{at}(\Sigma^*)\}$$
$$\equiv \{\bigcup_{i=1}^{n} (U_i \xrightarrow{a_i} V_i)W_i \mid n \in \mathbb{N},\ a_i \in \Gamma,\ U_i, V_i, W_i \in \mathcal{R}\text{at}(\Sigma^*)\},$$

where $(U_i \xrightarrow{a_i} V_i)W_i := \{u_i w_i \xrightarrow{a_i} v_i w_i \mid u_i \in U_i,\ v_i \in V_i,\ w_i \in W_i\}$. We shall use both expressions as alternative definitions of this family of graphs (see [6] for the proof of the above). We recall from [6] that $\mathrm{L}(\mathcal{P}\text{ref}\mathcal{R}\text{ec}\mathfrak{g}_\Gamma)$ is exactly the family of context–free languages, $\mathcal{P}\text{ref}\mathcal{R}\text{ec}\mathfrak{g}_\Gamma$ is an effective boolean algebra, the transitive closure of a prefix–recognizable relation is prefix–recognizable and every graph of $\mathcal{P}\text{ref}\mathcal{R}\text{ec}\mathfrak{g}_\Gamma$ has a decidable monadic theory.

In order to characterize prefix–recognizable graphs as normal form graphs, we use prefix configured systems.

Definition 4.1. *A configured system* $(\mathcal{S}, C)$ *is* prefix, *if every LR–overlap* $(x_1, l_1 \rightarrow r_1, y_1, x_2, l_2 \rightarrow r_2, y_2)$ *of* $(\mathcal{S}, C)$ *is such that* $x_1 = x_2$. *The class of prefix configured system is written* $\mathcal{P}\mathrm{refCS}$.

The fact that in some situations the rewriting w.r.t. a configured system which is prefix acts as prefix rewriting is essential for the proof of the following.

Theorem 4.2. *The following classes of graphs coincide:*

(1) $\mathcal{P}\text{ref}\mathcal{R}\text{ec}\mathfrak{g}_\Gamma$,
(2) $\{\mathfrak{NG}_\Gamma(\mathcal{S}, C) \mid (\mathcal{S}, C) \in \mathcal{F}\mathrm{inCS} \cap \mathcal{P}\mathrm{refCS} \cap \mathcal{S}\mathrm{tabCS}\}$,
(3) $\{\mathfrak{NG}_\Gamma(\mathcal{S}, C) \mid (\mathcal{S}, C) \in \mathcal{R}\mathrm{ecCS} \cap \mathcal{P}\mathrm{refCS} \cap \mathcal{S}\mathrm{tabCS}\}$.

We now turn our attention to Cayley–type graphs. In order to characterize prefix–recognizable graphs, it is sufficient to check for overlaps only for descendants of some specific sets.

Definition 4.3. *An srs* $\mathcal{S}$ *is* prefix from $D \subseteq \Sigma^*$, *if every LR–overlap* $(x_1, l_1 \rightarrow r_1, y_1, x_2, l_2 \rightarrow r_2, y_2)$ *of* $\mathcal{S}$ *such that* $x_1 l_1 y_1 \in \overset{*}{\underset{\mathcal{S}}{\rightarrow}}(D)$ *satisfies* $x_1 = x_2$.

For an srs $\mathcal{S}$, the property of being prefix from D guarantees that

$$\overset{*}{\underset{\mathcal{S}}{\rightarrow}} \cap \ ((D \cap \Gamma\mathcal{I}\mathrm{rr}(\mathcal{S})) \times \Sigma^*) = \overset{*}{\underset{\mathcal{S}}{\twoheadrightarrow}} \cap \ ((D \cap \Gamma\mathcal{I}\mathrm{rr}(\mathcal{S})) \times \Sigma^*) \ .$$

Proposition 4.4. *The following problem is decidable:*
instance: $\mathcal{S} \in \mathcal{R}\mathrm{ec}(\Sigma^* \times \Sigma^*)$, $D \in \mathcal{R}\mathrm{at}(\Sigma^*)$,
question: is $\mathcal{S}$ *prefix from* D *?*

For the proof of the above proposition, the following lemma is useful.

Lemma 4.5 ([7]). *For every* $(\mathcal{S}, C) \in \mathcal{R}\mathrm{ecCS} \cap \mathcal{P}\mathrm{refCS} \cap \mathcal{S}\mathrm{tabCS}$, $\xrightarrow[(\mathcal{S},C)]{*}$ *is a rational relation.*

The characterization of prefix–recognizable graphs in terms of Cayley–type graphs is given in the following.

Theorem 4.6. *The following classes of graphs coincide:*
(1) $\mathcal{P}\mathrm{ref}\mathcal{R}\mathrm{ec}\mathfrak{g}_\Gamma$,
(2) $\{\mathfrak{CG}_\Gamma(\mathcal{S}, C) \mid C \in \mathcal{R}\mathrm{at}(\Sigma^*),\ \mathcal{S}$ *is finite and prefix from* $\Gamma C\}$,
(3) $\{\mathfrak{CG}_\Gamma(\mathcal{S}, C) \mid C \in \mathcal{R}\mathrm{at}(\Sigma^*),\ \mathcal{S}$ *is recognizable and prefix from* $\Gamma C\}$.

5 Rational Graphs

A graph $G \subseteq \Sigma^* \times \Gamma \times \Sigma^*$ is *rational* when, for each $a \in \Gamma$, the edge relation $\xrightarrow[G]{a}$ is a rational subset of $\Sigma^* \times \Sigma^*$. Rational graphs and have been introduced in [13]. We denote by $\mathcal{R}\mathrm{at}\mathfrak{g}_\Gamma$ the class of rational graphs with edge labels in Γ. As established in [14], $\mathrm{L}(\mathcal{R}\mathrm{at}\mathfrak{g}_\Gamma)$ is exactly the family of context sensitive languages.

In a more concrete way, rational graphs may be seen as relation graphs of rational transducers with labelled accepting states. To a rational transducer $\mathcal{T} = (Q, \xrightarrow[\mathcal{T}]{}, \iota, (F_a)_{a\in\Gamma})$ where Q is a finite set of states, $\xrightarrow[\mathcal{T}]{} \subseteq Q \times \Sigma^* \times \Sigma^* \times Q$ is a finite transition relation, $\iota \in Q$ is an initial state and $F_a \subseteq Q$ is a set of final states labelled by $a \in \Gamma$, we associate its relation graph:

$$\mathfrak{rg}_\Gamma(\mathcal{T}) := \{u \xrightarrow{a} v \mid a \in \Gamma,\ \exists f \in F_a : \iota \xrightarrow[\mathcal{T}]{(u,v)} f\}$$

We denote by $\mathfrak{rg}_\Gamma\mathrm{RT}$ the family of relation graphs of rational transducers. Of course $\mathfrak{rg}_\Gamma\mathrm{RT} = \mathcal{R}\mathrm{at}\mathfrak{g}_\Gamma$.

In order to characterize rational graphs as normal form graphs, we use right configured systems.

Definition 5.1. *A configured system* $(\mathcal{S}, C)$ *is* right, *if every LR–overlap* $(x_1, l_1 \to r_1, y_1, x_2, l_2 \to r_2, y_2)$ *of* $(\mathcal{S}, C)$ *is such that* $|x_1| \leq |x_2|$ *and* $|x_1 r_1| \leq |x_2 l_2|$*:*

x_1	r_1	y_1
x_2	l_2	y_2

Right configured systems have been studied in [7] and linked to rational relations as follows.

Lemma 5.2 ([7]). *For every* $(\mathcal{S}, C) \in \mathcal{R}\mathrm{ecCS} \cap \mathcal{R}\mathrm{ightCS} \cap \mathcal{S}\mathrm{tabCS}$, $\xrightarrow[(\mathcal{S},C)]{*}$ *is a rational relation.*

The above lemma is useful in proving the following characterization of rational graphs in terms of normal form graphs.

Theorem 5.3. *The following classes of graphs coincide:*

(1) $\mathfrak{Ratg}_\Gamma$,
(2) $\{\mathfrak{NG}_\Gamma(\mathcal{S},C) \mid (\mathcal{S},C) \in \mathfrak{FinCS} \cap \mathfrak{StabCS} \cap \mathfrak{RightCS}\}$,
(3) $\{\mathfrak{NG}_\Gamma(\mathcal{S},C) \mid (\mathcal{S},C) \in \mathfrak{RecCS} \cap \mathfrak{StabCS} \cap \mathfrak{RightCS}\}$,

The second characterization of rational graphs is given in terms of Cayley–type graphs. Again, we have to forbid some overlaps but the restriction needs only to be limited to the context of words that encode the vertices of a graph.

Definition 5.4. *An srs $\mathcal{S}$ is* right from D *if every LR–overlap* $(x_1, l_1 \to r_1, y_1, x_2, l_2 \to r_2, y_2)$ *of $\mathcal{S}$ such that* $x_1 l_1 y_1 \in \xrightarrow[\mathcal{S}]{*}(D)$ *satisfies* $|x_1| \leq |x_2|$ *and* $|x_1 r_1| \leq |x_2 l_2|$*:*

x_1	r_1	y_1
x_2	l_2	y_2

Proposition 5.5. *The following problem is decidable:*
instance: $\mathcal{S} \in \mathfrak{Rec}(\Sigma^* \times \Sigma^*)$, $D \in \mathfrak{Rat}(\Sigma^*)$,
question: is $\mathcal{S}$ right from D ?

We have the following characterization of rational graphs in terms of Cayley–type graphs.

Theorem 5.6. *The following classes of graphs coincide:*

(1) $\mathfrak{Ratg}_\Gamma$
(2) $\{\mathfrak{CG}_\Gamma(\mathcal{S},C) \mid \mathcal{S}$ *is finite and right from* $\Gamma C\}$,
(3) $\{\mathfrak{CG}_\Gamma(\mathcal{S},C) \mid \mathcal{S}$ *is recognizable and right from* $\Gamma C\}$.

6 Recursively Enumerable Graphs

There are at least two possibilities for associating graphs to Turing machines. We may use an off–line Turing machine with a read–only input tape and a work tape. The head of the input tape cannot move backwards. The vertices of the graph are the configurations of the work tape. There is an a–labelled edge from a configuration c_1 to a configuration c_2 if there is a computation of the machine that leads from c_1 to c_2 whereas only one letter a is read on the input tape. We speak of observable transition graph in this case. Of course, as in former sections, it is useful to restrict the vertices of a graph to a rational set C. This definition is introduced in [11] and further discussed in [8].

A second possibility, considered in [8] is based on a relation graph. The accepting states of a single–tape Turing machine $\mathcal{M}$ are labelled by Γ. Formally, $\mathcal{M} = (Q, T, \iota, (F_a)_{a \in \Gamma})$ where Q is a finite set of states, $\iota \in Q$ is an initial state, $F_a \subseteq Q$ is the set of accepting states labelled by $a \in \Gamma$ and T is a finite set of transition rules. The tape alphabet $\Sigma_\Box := \Sigma \cup \{\Box\}$ is extended with the blank $\Box \notin \Sigma$ and transition rules are of the form $pA \to qB\delta$ with $p, q \in Q$, $A, B \in \Sigma_\Box$

and $\delta \in \{+,-\}$. A *single-step computation relation* $\underset{\mathcal{M}}{\longrightarrow} \subseteq \Sigma_{\Box}^* Q \Sigma_{\Box}^* \times \Sigma_{\Box}^* Q \Sigma_{\Box}^*$ of $\mathcal{M}$ is defined as follows

$$
\begin{aligned}
\underset{\mathcal{M}}{\longrightarrow} := &\{(upAv, uBqv) \mid pA \to qB+ \in T,\ u,v \in \Sigma_{\Box}^*\} \cup \\
&\{(up, uBq) \mid p\Box \to qB+ \in T,\ u \in \Sigma_{\Box}^*\} \cup \\
&\{(uCpAv, uqCBv) \mid pA \to qB- \in T,\ C \in \Sigma_{\Box},\ u,v \in \Sigma_{\Box}^*\} \cup \\
&\{(uCp, uqCB) \mid p\Box \to qB- \in T,\ C \in \Sigma_{\Box},\ u \in \Sigma_{\Box}^*\} \cup \\
&\{(pAv, q\Box Bv) \mid pA \to qB- \in T,\ v \in \Sigma_{\Box}^*\} \cup \\
&\{(p, q\Box B) \mid p\Box \to qB- \in T\}
\end{aligned}
$$

The *relation graph* $\mathfrak{tg}_\Gamma(\mathcal{M})$ of $\mathcal{M}$ is defined as follows:

$$\mathfrak{rg}_\Gamma\mathrm{TM} := \{u \xrightarrow{a} \overleftarrow{v}\,\overrightarrow{w} \mid u \in \Sigma^*,\ v,w \in \Sigma_{\Box}^*,\ a \in \Gamma,\ \exists f \in F_a : \iota u \underset{\mathcal{M}}{\overset{*}{\longrightarrow}} vfw\}$$

where $\overleftarrow{v}$ (resp. $\overrightarrow{w}$) stands for the longest suffix (resp. prefix) of v (resp. w) which does not contain $\Box$. The class of relation (resp. transition) graphs of Turing machines is written $\mathfrak{rg}_\Gamma\mathrm{TM}$ (resp. $\mathfrak{t}\mathfrak{G}_\Gamma\mathrm{TM}$). In [8], it is established that $\mathfrak{rg}_\Gamma\mathrm{TM} \equiv \mathfrak{t}\mathfrak{G}_\Gamma\mathrm{TM}$. Thus, the class of *recursively enumerable graphs*, written $\mathrm{RE}\mathfrak{g}_\Gamma$, is the class of graphs of Turing machines: $\mathrm{RE}\mathfrak{g}_\Gamma := \mathfrak{rg}_\Gamma\mathrm{TM} \equiv \mathfrak{t}\mathfrak{G}_\Gamma\mathrm{TM}$. Several other characterizations of this class are given in [8], among which we need the following for the main theorem of this section (i.e. Theorem 6.2).

Theorem 6.1 ([8]). $\mathfrak{rg}_\Gamma\mathrm{TM} \equiv \mathrm{RatM}^{-1}(\mathfrak{rg}_\Gamma\mathrm{RT})$.

Of course, $\mathrm{L}(\mathrm{RE}\mathfrak{g}_\Gamma)$ is the family of recursively enumerable languages.

The following theorem gives characterizations of the class of recursively enumerable graphs in terms of both normal–form graphs and Cayley–type graphs.

Theorem 6.2. *The following families of graphs coincide:*

(1) $\mathrm{RE}\mathfrak{g}_\Gamma$.
(2) $\{\mathfrak{N}\mathfrak{G}_\Gamma(\mathcal{S}, C) \mid \mathcal{S} \in \mathfrak{Fin}(\Sigma^* \times \Sigma^*),\ C \in \mathfrak{Rat}(\Sigma^*)\}$,
(3) $\{\mathfrak{N}\mathfrak{G}_\Gamma(\mathcal{S}, C) \mid \mathcal{S} \in \mathfrak{Rec}(\Sigma^* \times \Sigma^*),\ C \in \mathfrak{Rat}(\Sigma^*)\}$,
(4) $\{\mathfrak{C}\mathfrak{G}_\Gamma(\mathcal{S}, C) \mid \mathcal{S} \in \mathfrak{Fin}(\Sigma^* \times \Sigma^*),\ C \in \mathfrak{Rat}(\Sigma^*)\}$,
(5) $\{\mathfrak{C}\mathfrak{G}_\Gamma(\mathcal{S}, C) \mid \mathcal{S} \in \mathfrak{Rec}(\Sigma^* \times \Sigma^*),\ C \in \mathfrak{Rat}(\Sigma^*)\}$,

To close this section, we point out a similar result of [16, 17] establishing that there is an observational equivalence (up to τ–transitions) between the class of rooted Cayley–type graphs of finite srs $\{\mathfrak{C}\mathfrak{G}_\Gamma(\mathcal{S}, C)_{\mathrm{acc}} \mid \mathcal{S} \in \mathfrak{Fin},\ C \in \mathfrak{Rat}(\Sigma^*)\}$ and the class of rooted transition graphs of Turing machines (with nonobservable transitions τ) $\mathfrak{tg}_\Gamma\mathrm{TM}$.

7 Conclusion

We have considered four families of graphs: finite graphs $\mathfrak{Fin}\mathfrak{g}_\Gamma$, prefix–recognizable graphs $\mathfrak{PrefRec}\mathfrak{g}_\Gamma$, rational graphs $\mathfrak{Rat}\mathfrak{g}_\Gamma$, and recursively enumerable graphs $\mathrm{RE}\mathfrak{g}_\Gamma$. These families form a strict hierarchy analogous to the Chomsky hierarchy of formal languages. Indeed the notion of trace relates each level of the hierarchy of graphs to the Chomsky hierarchy: $L(\mathcal{F}\mathrm{in}\mathfrak{g}_\Gamma)$ is the family of rational languages, $L(\mathcal{P}\mathrm{ref}\mathcal{R}\mathrm{ec}\mathfrak{g}_\Gamma)$ is the family of context–free languages, $L(\mathcal{R}\mathrm{at}\mathfrak{g}_\Gamma)$ is the family of context–sensitive languages and $L(\mathrm{RE}\mathfrak{g}_\Gamma)$ is the family of recursively enumerable languages.

Whereas the Chomsky hierarchy has a uniform presentation in terms of grammars (type 0, 1, 2 and 3), we have shown that string rewriting provides a uniform presentation of a Chomsky–like hierarchy of graphs. We have established that both notions of Cayley–type graph and normal form graph lead to the same Chomsky–like hierarchy. In addition, we have established that in both cases finite string–rewriting systems are as expressive as recognizable ones.

We did not discussed sub-families of the Chomsky–like hierarchy. At least one of them has a known characterization in terms of string rewriting, namely the family of pushdown transition graphs [4].

References

1. J.-M. Autebert, J. Berstel, and L. Boasson. Context–free languages and push–down automata. In G. Rozenberg and A. Salomaa, editors, *Word, Language, Grammar*, volume 1 of *Handbook of Formal Languages*, pages 111–174. Springer–Verlag, 1997.
2. J. R. Bchi. Regular canonical systems. *Archiv fr Math. Logik und Grundlagenforschung*, 6:91–111, 1964.
3. J. Berstel. *Transductions and Context–Free Languages.* B. G. Teubner, Stuttgart, 1979.
4. H. Calbrix and T. Knapik. A string–rewriting characterization of Muller and Schupp's context–free graphs. In V. Arvind and R. Ramanujam, editors, *18th International Conference on Foundations of Software Technology and Theoretical Computer Science*, LNCS 1530, pages 331–342, Chennai, Dec. 1998.
5. D. Caucal. On the regular structure of prefix rewriting. *Theoretical Comput. Sci.*, 106:61–86, 1992.
6. D. Caucal. On infinite transition graphs having a decidable monadic second–order theory. In F. M. auf der Heide and B. Monien, editors, *23th International Colloquium on Automata Languages and Programming*, LNCS 1099, pages 194–205, Paderborn, July 1996. A long version will appear in TCS.
7. D. Caucal. On word–rewriting systems having a rational derivation. In J. Tiuryn, editor, *FoSSaCS '2000*, LNCS 1784, pages 48–62, Berlin, Mar. 2000.
8. D. Caucal. On the transition graphs of Turing machines. In M. Margenstern and Y. Rogozhin, editors, *MCU '2001*, LNCS 2055, pages 177–189, Chisinau, May 2001.
9. B. Khoussainov and A. Nerode. Automatic presentations of structures. In D. Leivant, editor, *Logic and Computational Complexity*, LNCS 960, pages 367–392, Indianapolis, Oct. 1994. Selected papers from the LCC'94 Intl. Workshop.
10. T. Knapik. Spécifications de Thue. Technical Report INF/95/12/02/a, Institut de Recherche en Mathematiques et Informatique Appliquées, 1996.
11. T. Knapik and É. Payet. Synchronized product of linear bounded machines. In G. Ciobanu and G. Paun, editors, *12 th International Symposium on Fundamentals of Computation Theory*, LNCS 1684, pages 362–373, Iaşi, Aug. 1999.

12. A. Mateescu and A. Salomaa. Aspects of classical language theory. In G. Rozenberg and A. Salomaa, editors, *Word, Language, Grammar*, volume 1 of *Handbook of Formal Languages*, pages 175–251. Springer–Verlag, 1997.
13. C. Morvan. On rational graphs. In J. Tiuryn, editor, *FoSSaCS '2000*, LNCS 1784, pages 252–266, Berlin, Mar. 2000.
14. C. Morvan and C. Stirling. Rational graphs trace context-sensitive languages. In A. Pultr, J. Sgall, and P. Kolman, editors, *MFCS '2001*, LNCS 2136, Marianske Lazne, Aug. 2001.
15. D. E. Muller and P. E. Schupp. Pushdown automata, graphs, ends, second–order logic, and reachability problems. In *Proceedings of the 13th Annual ACM Symposium on Theory of Computing*, pages 46–54, Milwaukee, May 1981.
16. É. Payet. *Produit synchronisé pour quelques classes de graphes infinis*. PhD thesis, Université de la Réunion, 2000.
17. É. Payet. Thue specifications, infinite graphs and synchronized product. *Fundamenta Informaticae*, 44(3):265–290, 2000.
18. W. Thomas. A short introduction to infinite automata. In W. Kuich, G. Rozenberg, and A. Salomaa, editors, *Developments in Language Theory, 5th International Conference, DLT 2001*, LNCS 2295, pages 130–144, Vienna, July 2001.

Competitive Analysis of On-line Stream Merging Algorithms

Wun-Tat Chan[1], Tak-Wah Lam[2*]
Hing-Fung Ting[2**], and Prudence W.H. Wong[2,**]

[1] Department of Computing, Hong Kong Polytechnic University, Hong Kong,
cswtchan@comp.polyu.edu.hk
[2] Department of Computer Science, University of Hong Kong, Hong Kong,
{twlam,hfting,whwong}@cs.hku.hk

Abstract. A popular approach to reduce the server bandwidth in a video-on-demand system is to merge the streams initiated at different times. In recent years, a number of on-line algorithms for stream merging have been proposed; the objective is to minimize either the total bandwidth or the maximum bandwidth over all time. The performance of these algorithms was better understood with respect to the first objective. In particular, the connector algorithm [9] is known to be $O(1)$-competitive, and the dyadic algorithm [12] is known to have an almost tight bounds on its average total bandwidth requirement. For minimizing maximum bandwidth, existing results are limited to empirical studies only and no algorithm has been known to be competitive. The main contribution of this paper is the first competitive analysis of the connector and the dyadic algorithms with respect to the maximum bandwidth, showing that both algorithms are 4-competitive. We also give a worst-case analysis of the dyadic algorithm with respect to the total bandwidth, revealing that the dyadic algorithm can be tuned to be 3-competitive.

1 Introduction

This paper gives a competitive analysis of two on-line stream merging algorithms designed for video-on-demand (VOD) systems. A VOD system often receives many requests for a hot video over a short period of time (say, Friday 7 PM to 9 PM). If the video server uses a dedicated video stream to serve each request, the bandwidth requirement for the server is huge. A simple alternative is to batch [1,11,13] the requests so that a single stream is issued to serve requests arriving within some time interval; however, this increases the response time substantially. To reduce the bandwidth requirement without sacrificing the response time, Eager *et al.* [14,15] proposed an approach that allows streams initiated for different requests to merge eventually.

Stream merging is based on the assumption that each client is equipped with extra buffer and can receive data from two streams simultaneously. In its simplest

* This research was supported in part by Hong Kong RGC Grant HKU-7024/01E.
** This research was supported in part by Hong Kong RGC Grant HKU-7103/99E.

K. Diks et al. (Eds): MFSC 2002, LNCS 2420, pp. 188–200, 2002.

setting, stream merging works as follows. A client initially receives data from a dedicated stream X for immediate playback. At a certain time, the client may start to buffer data simultaneously from another stream Y initiated Δ time units earlier (for another client). After listening to both X and Y for another Δ time units, the client no longer needs X as it can play back the next Δ time units using the data in its buffer, while further buffering data from Y simultaneously. Hence, X may terminate now and clients of X all switch to Y. In this case, we say that X *merges* with Y. Note that Y itself may later merge with another earlier stream. The stream merging technique allows several clients to be eventually served using only one full stream, thus saving the server bandwidth.

To support stream merging effectively, we need an on-line algorithm to decide how streams merge with each other. The past few years have witnessed many on-line stream merging algorithms ([6,9,10,12,14,15]). An interesting and important problem is to analyze and compare their strength and weakness. Two measurements, the total bandwidth and the maximum bandwidth over time, are commonly used to characterize the performance of these algorithms [4,6,12,15].

Previous Results: The worst-case performance of existing on-line stream merging algorithms is better understood when the total bandwidth is of concern. Bay-Noy *et al.* [6] showed that simple algorithms like best-fit and nearest-fit are $\Omega(n/\log n)$-competitive where n is the total number of requests (i.e., the total bandwidth of the schedules produced by best-fit and nearest-fit can be $\Omega(n/\log n)$ times that of the optimal off-line schedules). They also proposed the Dynamic Fibonacci tree algorithm whose competitive ratio is bounded by $O(\min\{\log n, \log \frac{1}{2D}\})$, where $D < 1$ is the guaranteed startup delay. More recently, Chan *et al.* [9] have devised a 3-competitive algorithm, which will be referred to as the connector algorithm. For average-case performance, Coffman *et al.* [12] proposed the 2-dyadic algorithm and gave an almost tight bound on its average total bandwidth requirement when the input distribution is Poisson. Regarding maximum bandwidth, no on-line algorithms have been known to have bounded competitive ratios.

Empirical comparison shows that the performance of the connector algorithm, the 2-dyadic algorithm, as well as its variant the ϕ-dyadic algorithm ($\phi \approx 1.618$ is the golden ratio) [4] are quite similar no matter whether the total bandwidth or the maximum bandwidth are used as the measurement. This suggests that with respect to the total bandwidth, the dyadic algorithms might also be $O(1)$-competitive because the connector algorithm is 3-competitive. However, the empirical result does not tell much about the maximum bandwidth requirement of these algorithms against the optimal ones because none of them are known to have bounded competitive ratios (with respect to maximum bandwidth). Nevertheless, there is a general belief that algorithms that perform well under the total bandwidth measurement should also perform well under the maximum bandwidth measurement. Note that in contrast with the case of total bandwidth where a quadratic time optimal off-line algorithm [5] is known, no off-line algorithm is known to compute or approximate schedules with optimal maximum bandwidth.

Our Results: The first result of this paper is on total bandwidth. We study the α-dyadic algorithm (for any $1 < \alpha \leq 2$), which is a generalization of the 2-dyadic and the ϕ-dyadic algorithm. We show that with respect to total bandwidth, the α-dyadic algorithm is 2α-competitive if $1.5 \leq \alpha \leq 2$ and is $2+\frac{1}{2(\alpha-1)}$-competitive if $1 < \alpha \leq 1.5$. It follows that the 2-dyadic is 4-competitive, the ϕ-dyadic is 3.26-competitive, and the 1.5-dyadic algorithm gives the best worst-case performance guarantee with competitive ratio 3. The next result gives the first competitive analysis for the maximum bandwidth measurement. We show that at any time, the bandwidth used by the connector algorithm, and the α-dyadic algorithm with $(1+\sqrt{2})/2 \leq \alpha \leq 2$, are at most four times the optimal ones. Hence, these algorithms are competitive no matter whether the total bandwidth or the maximum bandwidth measure is used. This agrees with the empirical findings mentioned above.

Technically speaking, the results obtained in this paper are based on two novel extensions to an analytical technique, which we call *schedule-sandwiching*, given in our earlier work [10]. Consider any input request sequence I. Let S_I and O_I be an on-line schedule and optimal off-line schedule, respectively, for serving I. The idea of competitive analysis is to obtain an upper bound of the cost of S_I in terms of the cost of O_I. However, due to the different structures of S_I and O_I, it may be difficult to formulate a comparison between the two schedules. The schedule-sandwiching technique carefully constructs an intermediate schedule T_I for I. The most important property of T_I is its regular structure that provides a common ground for the comparison. The cost of T_I may be slightly higher than that of S_I, yet T_I is constructed in such a way that its cost could be upper bounded in terms of the cost of O_I.

The schedule-sandwiching technique has been used to prove that the connector algorithm is 3-competitive [9]. In this case, the intermediate schedule T_I is easy to construct because the connector algorithm designs its on-line schedule S_I based on a pre-defined static schedule for a fully loaded request sequence; it is natural to choose T_I as a restriction of this full schedule with respect to the input I. However, it is more difficult to apply this technique to analyze the α-dyadic algorithm, which does not make reference to any pre-defined static schedule. We have to construct the intermediate schedule T_I based on the α-dyadic algorithm.

To apply the schedule-sandwiching technique to analyze the maximum bandwidth requirement introduces another difficulty. Recall that the total bandwidth of T_I can be higher than that of S_I. If this extra bandwidth usage occurs at the same time, the maximum bandwidth used by T_I will be much larger than that of S_I and this makes T_I a bad reference for comparing S_I and O_I. To ensure that this will not happen, we have to take extra effort in our analysis to consider the exact timing of individual streams. The geometric representation of merging schedules used in studying total bandwidth [9] fails to capture the exact timing (as the absolute time of individual streams is not an important concern before). To overcome the difficult, we introduce a novel notion of time-line in this representation to help analyze the maximum bandwidth.

The organization of the paper is as follows. In Section 2, we review some basic concepts of representing merging schedules. In Section 3, we focus on the total bandwidth measurement and prove a formula on the competitive ratio of the α-dyadic algorithm. In Section 4, we show that under the maximum bandwidth measurement, the connector algorithm is 4-competitive and give a range of α such that the α-dyadic algorithm is also 4-competitive.

Remarks: The simplest setting of stream merging [6,12] assumes that the client receiving bandwidth is twice the playback bandwidth and the client buffer is sufficient to store half of the video. In [9,10], algorithms are proposed to work in the general setting of any buffer size and receiving bandwidth. Our results in this paper can be extended to the general setting as well. To be compatible with previous work which assume the simplest setting and for the sake of easier discussion, we adopt the simplest setting throughout the paper.

Stream merging originates from the model of the pyramid broadcasting scheme proposed by Viswanathan and Imielinski [24], which provides a multicast basis for sharing streams. Various broadcasting schemes [3,18,19,21,23] are introduced afterwards. Some primitive forms of stream merging including patching [7,17,22] and tapping [8] are also studied. Another common technique is piggybacking [2,9,16,20]. With respect to the total bandwidth measurement, it is shown in [9] that the connector algorithm also works in piggybacking systems and the competitive ratio preserves. With respect to the maximum bandwidth measurement, we can show that the connector algorithm is also 4-competitive for piggybacking systems. We omit the details due to space limitation.

2 Preliminaries

We first introduce some defintions. Any stream X can be in two states: *normal* and *exceptional*. When X is in normal state, then its clients receive one unit of video from X in one time unit. When X changes to exceptional state, then X must be coupled with some earlier stream, say Y, and the clients of X receive one unit from X and one unit from Y in one time unit. Furthermore, X will eventually merge with Y, which means that all clients of X will switch to listening to Y and X then terminates. All streams are initially in normal state. A *merging schedule* describes for each stream whether and when it changes to the exceptional state, and with which stream it merges. The literature contains different ways of representing a merging schedule. Below, we review the geometric representation introduced in [9].

Suppose we are serving a video of length ℓ. For simplicity, we assume that $\ell/2$ is an integer. Consider any merging schedule $\mathcal{M}$ for this video. Let X be some stream scheduled by $\mathcal{M}$. Suppose X is initiated at time t_1, changes to exceptional state at time t_2, and terminates at time t_3. We represent X by the following rectilinear line $p(X)$ on a 2-dimensional plane: Starting from the point with coordinate (t_1, t_1), $p(X)$ comprises a right-going horizontal line segment of length $(t_2 - t_1)/2$, followed by a down-going vertical line segment of length $t_3 - t_2$. (See Figure 1 (a) for an example.) The lifespan of X is completely captured by

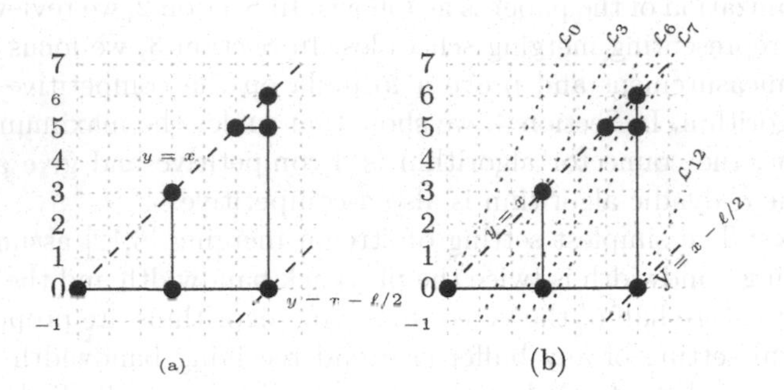

Fig. 1. (a) Geometric representation for a merging schedule with streams initated at time 0, 3, 5, 6. For example, the stream initiated at time 5 is in normal state for two time units, then in exceptional state for another five time units before merging to the first stream. (b) The time-lines $\mathcal{L}_t$ are shown.

the length and shape of $p(X)$. More interestingly, the way $p(X)$ is defined enforces some nice property.

Lemma 1. (see [9]) *Suppose X and Y are two streams scheduled by $\mathcal{M}$. Then X merges with Y if and only if $p(X)$ terminates at some point on $p(Y)$.*

Suppose we put down two nodes at the starting and ending point of each rectilinear lines. Then these lines form a forest in the plane where all the leaves lie on the line $y = x$. Furthermore, it can be shown that all roots must lie on the line $y = x - \ell/2$. Such a forest is called a *connector* in [9], and it is called a *connector tree* if it has only one root. For any set of rectilinear lines S, define $w(S)$, the *weight* of S, to be the sum of the total length of its vertical segments and two times the total length of its horizontal segments. Observe that the weight of a connector is the total running time of the streams, or equivalently, the total bandwidth used by the corresponding schedule. To analyze the total bandwidth of a schedule, it suffices to analyze the weight of the corresponding connector.

To analyze the maximum bandwidth, we introduce the following notion of time in the geometric representation: for any time $t \geq 0$, define $\mathcal{L}_t$, the *time-line* of t, to be the line $y = 2x - t$ on the plane (see Figure 1 (b)). The following lemma gives the real meaning of time-lines.

Lemma 2. *Consider any stream X starting at time t_0 and terminating at time t_1. Then, for any time t within the interval $[t_0, t_1]$, the rectilinear line $p(X)$ intersects the time-line $\mathcal{L}_t$ exactly once. Furthermore, the weight of the portion of $p(X)$ up to the intersection with $\mathcal{L}_t$ is exactly $t - t_0$.*

Proof. By straightforward arithmetic.

For any set of rectilinear lines S, define $c(S, \mathcal{L}_t)$ to be the number of rectilinear lines in S that start at or cross $\mathcal{L}_t$. (Note that we do not count those terminating at $\mathcal{L}_t$.) The following corollary follows directly from Lemma 2.

Corollary 1. *Let $\mathcal{F}$ be the connector representing the schedule $\mathcal{M}$. The number of streams used by $\mathcal{M}$ at any time t is equal to $c(\mathcal{F}, \mathcal{L}_t)$. Thus, the maximum bandwidth used by $\mathcal{M}$ is $\max_t c(\mathcal{F}, \mathcal{L}_t)$.*

Therefore, to estimate the maximum bandwidth of a schedule $\mathcal{M}$, it suffices to analyze $\max_t c(\mathcal{F}, \mathcal{L}_t)$ of the corresponding connector $\mathcal{F}$.

3 Total Bandwidth

This section gives the first competitive analysis for the α-dyadic algorithm with respect to the total bandwidth requirement. Let $C = \langle (t_1, t_1), (t_2, t_2), \ldots, (t_k, t_k) \rangle$ be a sequence of intergal points corresponding to the requests arriving at time $t_1 < t_2 < \cdots < t_k$. Recall that ℓ is the length of the video. As in [6,9], we say that C is *compact* if $t_k - t_1 \leq \ell/2$. Define Δ_C to be the triangle bounded by the points $x = (t_1, t_1), y = (t_1 + \ell/2, t_1)$ and $z = (t_1 + \ell/2, t_1 + \ell/2)$. Let $b(\Delta_C)$ and $r(\Delta_C)$ denote the bottom bounding line $[x, y]$ and the right bounding line $[y, z]$ of Δ_C, respectively. Observe that if C is compact, then all points of C are within Δ_C. Furthermore, as shown in [6], there is a connector tree for C (i.e., with leaves located at the points of C) that lies completely in Δ_C, and the one with the smallest weight is an optimal schedule for C.

Like all existing competitive analysis on stream merging, we will make use of the following results, which enable us to focus on compact sequence. Consider any sequence V of integral points on the line $y = x$. Suppose V can be divided into m maximal and non-overlapping compact sequences $V_1, V_2, \ldots, V_m$. Let $opt(V)$ be an optimal connector for V. Let $\mathcal{G}$ be any on-line algorithm which, given a compact sequence C, produces a connector tree $\mathcal{G}(C)$ for C. The following lemma is given by Bar-Noy and Ladner [6].

Lemma 3. *Suppose there is a number λ such that for any compact sequence C and any connector tree Z for C, $w(\mathcal{G}(C) - b(\Delta_C)) \leq \lambda w(Z - b(\Delta_C))$. Then, $w(\mathcal{G}(V_1)) + w(\mathcal{G}(V_2)) + \cdots + w(\mathcal{G}(V_m)) \leq (\lambda + 2) w(opt(V))$.*

Chan *et al.* [9] showed that the above lemma can be improved as follows.

Lemma 4. *Suppose there is a number λ such that for any compact sequence C and any connector tree Z for C, $w(\mathcal{G}(C) - b(\Delta_C) - r(\Delta_C)) \leq \lambda\, w(Z - b(\Delta_C) - r(\Delta_C))$. Then, $w(\mathcal{G}(V_1)) + w(\mathcal{G}(V_2)) + \cdots + w(\mathcal{G}(V_m)) \leq \max\{\lambda, 3\}\, w(opt(V))$.*

In [9], we show that when $\mathcal{G}$ is the connector algorithm, the condition of Lemma 4 is ture with $\lambda = 3$. It follows that the connector algorithm is 3-competitive. In the rest of this section, we find λ for the α-dyadic algorithm.

3.1 The α-Dyadic Schedule

Suppose $1 < \alpha \leq 2$. The α-dyadic algorithm is a direct generalization of the 2-dyadic algorithm given in [12], and one can refer to the paper for more details. To find a schedule for an input sequence, the α-dyadic algorithm $\mathcal{D}$ divides the sequence into maximal and non-overlapping compact sequences and finds a schedule for each of them. Below, we describe the schedule $\mathcal{D}(C)$ given by $\mathcal{D}$ for the compact sequence C. Recall that any point (t,t) in C corresponds to a request arriving at t. Without loss of generality, we assume that the first request of C arrives at time 0. Note that the stream serving this request must be a full stream (becasue it has no earlier stream to merge) and it is represented by a horizontal line from $(0,0)$ to $(\ell/2,0)$ in $\mathcal{D}(C)$. Since C is compact, all the remaining requests arrives in interval $(0,\ell/2]$, and $\mathcal{D}$ decides a schedule for them recursively as follows.

We explain how $\mathcal{D}$ schedules streams for the set $S \subseteq C$ of requests that arrives in interval $(t_0,t_1]$. We assume there is a request X_0 arrives at t_0, and $\mathcal{D}$ has scheduled a stream for it. Let X be the earliest request in S that arrives after $t_0+(t_1-t_0)/\alpha$, and Y be the last request in S. Suppose X and Y arrive at t_X and t_Y, respectively. To decide a schedule for S, $\mathcal{D}$ schedules a stream for X that runs in normal state for $2(t_Y-t_X)$ time units and then in exceptional state for (t_X-t_0) time units before merging with the stream for X_0, or equivalently, $\mathcal{D}$ adds a horizontal line from (t_X,t_X) to (t_Y,t_X) followed by a vertical line from (t_Y,t_X) to (t_Y,t_0) in $\mathcal{D}(C)$. For those requests arrive in $(t_0,\lfloor t_0+(t_1-t_0)/\alpha\rfloor]$, as well as those in $(t_X,t_1]$, $\mathcal{D}$ handles them recursively (there is no request in interval $(\lfloor t_0+(t_1-t_0)/\alpha\rfloor,t_X)$).

Note that by scheduling the stream for X in such way, we can proved inductively (on the levels of recursion) that all streams initiated after X can merge with the one for X eventually, and the stream for X can correctly merge with the stream for X_0 (see [12]). Furthermore, we remark that the schedule $\mathcal{D}(C)$ can be decided in an on-line manner; the key observations are (1) the earliest time that the stream for X can merge with the stream for X_0 is $t_X+(t_X-t_0)\geq 2t_0+2(t_1-t_0)/\alpha-t_0\geq t_1$ (as $\alpha\leq 2$), i.e., after the last request in S arrives; and (2) even if the stream for X is already in exceptional state when the last request arrives, it can return to normal state as if it had never changed to exceptional state simply by discarding the extra data it received from some other stream.

3.2 Analysis of the α-Dyadic Schedule

We now apply the schedule-sandwiching technique to analyze the total bandwidth of the schedule $\mathcal{D}(C)$ described in the previous section. To give an intermediate connector tree that is a good reference for comparing $\mathcal{D}(C)$ and the optimal one, we first introduce the connector tree $\mathcal{D}^*$, whose construction is very similar to that of $\mathcal{D}(C)$, except that when it recursively schedules streams in some interval $(t_0,t_1]$, it always assumes that there is a point (t_1,t_1) in C. The formal definition of $\mathcal{D}^*$ is given below.

Recall that the first request in C arrives at time 0. We add a horizontal line from $(0,0)$ to $(\ell/2,0)$ in $\mathcal{D}^*$ for this request. The remaining requests, which arrive in time interval $(0,\ell/2]$, are scheduled recursively as follows. Suppose $\mathcal{D}^*$ is scheduling streams for the set S_0 of requests arriving in the interval $(t_0,t_1]$. Let $S=S_0\cup\{(t_1,t_1)\}$, and let t_X be the earliest time a request in S arrived after

$t_0 + (t_1 - t_0)/\alpha$. For this request, we add to $\mathcal{D}^*$ a horizontal line from (t_X, t_X) to (t_1, t_X) and then a vertical line from (t_1, t_X) to (t_1, t_0). If S has requests in $(t_0, \lfloor t_0 + (t_1 - t_0)/\alpha \rfloor]$, we handle them recursively, and similarly for $(t_X, t_1]$.

Note that $\mathcal{D}^*$ may contain leaves outside C. To restrict our attention back to C, we define the restriction of a tree as follows. For any tree T and any subset L of leaves of T, define $T\|_L$, the *restriction* of T on L, to be the subtree comprises all points in L, as well as all edges and nodes along the lines of T from some point in L to the root. Our intermediate connector tree for the schedule-sandwiching analysis is just $\mathcal{D}^*\|_C$. Note that $\mathcal{D}(C)$ and $\mathcal{D}^*\|_C$ are very similar. However, from their constructions given in this and the previous section, one may find out that the weight of $\mathcal{D}(C)$ is often smaller than that of $\mathcal{D}^*\|_C$. (For example, in $\mathcal{D}(C)$, the horizontal line from (t_X, t_X) has length $t_Y - t_X$, while the one in $\mathcal{D}^*\|_C$ has length $t_1 - t_X$, and $t_1 \geq t_Y$.)

Recall that to apply Lemma 4, we need to find a number λ such that for any connector tree T for C, we have

$$w(\mathcal{D}(C) - b(\Delta_C) - r(\Delta_C)) \leq \lambda\, w(T - b(\Delta_C) - r(\Delta_C)). \tag{1}$$

We will find such λ and derive (1) through a sequence of inequalities involving $\mathcal{D}^*\|_C$. The following lemma gives the first such inequality.

Lemma 5. $w(\mathcal{D}(C) - b(\Delta_C) - r(\Delta_C)) \leq w(\mathcal{D}^*\|_C - b(\Delta_C) - r(\Delta_C))$

Proof. By construction. Details will be given in full paper.

To relate $w(\mathcal{D}^*\|_C - b(\Delta_C) - r(\Delta_C))$ and $w(T - b(\Delta_C) - r(\Delta_C))$, we give below a conservative transformation of T to some subtree Z of $\mathcal{D}^*$ that covers all C. We will prove that our transformation guarantees $w(Z - b(\Delta_C) - r(\Delta_C)) \leq \lambda\, w(T - b(\Delta_C) - r(\Delta_C))$ where

$$\lambda = \begin{cases} 2\alpha & \text{if } 1.5 \leq \alpha \leq 2, \\ \frac{1}{2(\alpha-1)} + 2 & \text{if } 1 < \alpha < 1.5. \end{cases}$$

Note that Z must contain $\mathcal{D}^*\|_C$ as a subtree, and thus we have $w(\mathcal{D}^*\|_C - b(\Delta_C) - r(\Delta_C)) \leq w(Z - b(\Delta_C) - r(\Delta_C)) \leq \lambda w(T - b(\Delta_C) - r(\Delta_C))$. Together with Lemmas 4 and 5, we have the following theorem.

Theorem 1. *The α-dyadic algorithm $\mathcal{D}$ is λ-competitive.*

We transform T to a subtree of $\mathcal{D}^*$ recursively. To consider the first step, let $K = \ell/2$ and $g = \lfloor K/\alpha \rfloor$. Let (s, s) be the lowest point in C such that $s > g$. Consider the five points $p = (g, g)$, $q = (g, 0)$, $r = (K, 0)$, $u = (K, s)$, and $v = (s, s)$. By construction, $\mathcal{D}^*$ does not have any line segments in the rectangular region $\mathcal{H}$ bounded by these five points, while the four line segments $[p, q]$, $[q, r]$, $[r, u]$ and $[u, v]$ are in $\mathcal{D}^*$. If T has some lines within $\mathcal{H}$, we remove all of them, and then add the necessary line segments on $[p, q]$, $[q, r]$, $[r, u]$ and $[u, v]$ to restore the tree structure. We repeat this process on the left and right subtrees of the root of the resulting tree recursively.

Let Z be the resulting tree of the above transform. The following lemma relate Z to T.

Lemma 6. $w(Z - b(\Delta_C) - r(\Delta_C)) \le \lambda\, w(T - b(\Delta_C) - r(\Delta_C))$.

Proof. W.L.O.G., suppose T is the connector tree for C such that $w(T - b(\Delta_C) - r(\Delta_C))$ is the smallest. Consider the first step of the transformation. We claim that the total weight of the lines added is at most λ times that of the lines removed. To prove the claim, we first observe that if T has some lines within $\mathcal{H}$, T must contain a path crossing $\mathcal{H}$ vertically or horizontally, i.e., the path passes through a single column or a single row of $\mathcal{H}$. This can be proved similarly as in [10]. Suppose T contains a path crossing $\mathcal{H}$ vertically (the other case can be proved symmetrically). Then, the weight of the lines removed $\mathcal{E}_r$ is at least $w([r, u]) + w(H_{\mathtt{in}})$, where $H_{\mathtt{in}}$ is the set of horizontal lines of T that are within $\mathcal{H}$; and the weight of the lines added $\mathcal{E}_a$ is at most $w([r, u]) + w([p, q]) + w([u, v]) + w(H_{\mathtt{in}})$. Note that the value of α affects the length of the bounding lines of $\mathcal{H}$. We can show that $\mathcal{E}_r - \mathcal{E}_a \le (\lambda - 1)w([r, u])$, implying that the total weight of the lines added is at most λ times the total weight of the lines removed. By comparing the total weights of the lines removed and added, we conclude that for the final transformed tree Z, $w(Z - b(\Delta_C) - r(\Delta_C)) \le \lambda\, w(T - b(\Delta_C) - r(\Delta_C))$.

4 Maximum Bandwidth

In this section we give the first formal analysis on the maximum bandwidth measurement. We show that with respect to this measurement, the connector algorithm is 4-competitive, and the α-dyadic algorithms are also 4-competitive for many values of α. As in the case of total bandwidth, we first show in Section 4.1 that we can focus on compact sequences. Then we review the connector algorithm in Section 4.2. Finally, in Section 4.3 we give the competitive analysis.

4.1 Compact Point Sequence

Recall that $c(S, \mathcal{L}_t)$ is the number of rectilinear lines in S that start at or cross the timeline $\mathcal{L}_t$. Let V be a sequence of integral points on $y = x$, and $opt(V)$ be a connector for V with the smallest $\max_t c(opt(V), \mathcal{L}_t)$. Suppose V can be divided into m maximal and non-overlapping compact sequences $V_1, V_2, \ldots, V_m$. Let $\mathcal{G}$ be an on-line algorithm that, given any compact sequence C, returns a connector tree $\mathcal{G}(C)$ for C. The following lemma suggests that we can focus our analysis on compact sequence.

Lemma 7. *Let $t > 0$ be any integer. Suppose there is a $\beta > 0$ such that for any compact sequence C and any connector tree Z for C, $c(\mathcal{G}(C), \mathcal{L}_t) \le \beta\, c(Z, \mathcal{L}_t)$. Then, $c(\mathcal{G}(V_1), \mathcal{L}_t) + c(\mathcal{G}(V_2), \mathcal{L}_t) + \cdots + c(\mathcal{G}(V_m), \mathcal{L}_t) \le \beta\,(c(opt(V), \mathcal{L}_t) + 4)$*

Proof. For each $1 \le i \le m$, let O_i be the set of rectilinear lines in $opt(V)$ starting at some points in V_i. Let $O_i \cap \Delta_{V_i}$ denote the portion of the lines in O_i that lies within Δ_{V_i}. Observe that $T_i = (O_i \cap \Delta_{V_i}) \cup b(\Delta_{V_i}) \cup r(\Delta_{V_i})$ is a connector tree for V_i, and thus $c(\mathcal{G}(V_i), \mathcal{L}_t) \le \beta c(T_i, \mathcal{L}_t)$. Furthermore, $c(T_i, \mathcal{L}_t) \le c(O_i, \mathcal{L}_t) + 2$. We claim that there are at most two T_i, say T_{i_1} and T_{i_2}, such that $c(T_i, \mathcal{L}_t) \ne 0$.

Then, $\sum_{1\leq i\leq m} c(\mathcal{G}(V_i), \mathcal{L}_t) \leq \beta \sum_{1\leq i\leq m} c(T_i, \mathcal{L}_t) = \beta(c(T_{i_1}, \mathcal{L}_t) + c(T_{i_2}, \mathcal{L}_t)) \leq \beta(c(O_{i_1}, \mathcal{L}_t) + 2 + c(O_{i_2}, \mathcal{L}_t) + 2) \leq \beta(c(opt(V), \mathcal{L}_t) + 4)$, and the lemma follows.

To prove the claim, consider the smallest i such that $c(T_i, \mathcal{L}_t) \neq 0$. Let (s_i, s_i) be the first request point in T_i. Note that all streams scheduled by T_i terminate before $s_i + \ell$ and it follows that $t \leq s_i + \ell$. On the other hand, for any $j \geq i+2$, all requests scheduled by T_j arrive after $s_i + \ell$ (becasue $V_1, V_2, \ldots, V_m$ are maximal non-overlapping compact sequences). Thus $\mathcal{L}_t$ will not cross any such T_j, and it can only cross at most two connector trees, namely T_i and T_{i+1}.

4.2 The Connector Algorithm

This section gives a brief description on the connector algorithm $\mathcal{A}$ (for details, see [9]). Given an input sequence, $\mathcal{A}$ divides it into maximal and non-overlapping compact point sequences and then constructs a connector tree for each of them. Consider the compact sequence $C = \langle(t_1, t_1), (t_2, t_2), \ldots, (t_n, t_n)\rangle$ where $t_1 < t_2 < \cdots < t_n$. Let C_i be the first i points of C. Without loss of generality, suppose $t_1 = 0$. $\mathcal{A}$ constructs the connector tree $\mathcal{A}(C)$ for C by incrementally constructing the connector trees $\mathcal{A}(C_i)$ for C_i. All these trees lie on the triangle bounded by $a = (0,0), b = (K, 0)$ and $c = (K, K)$ where $K = \ell/2$. Note that $(t_1, t_1) = a$ and $\mathcal{A}(C_1)$ is simply the horizontal line segment $[(0,0), (K, K)]$. $\mathcal{A}$ will construct $\mathcal{A}(C_i)$ from $\mathcal{A}(C_{i-1})$, and the process needs the following tree $\mathcal{R}$.

Let $g(n) = 1 + \lfloor 2n/3 \rfloor$. The root of $\mathcal{R}$ is at the point $b = (K, 0)$. All the $K+1$ integral points on the line segment $[a, c]$ are leaves of $\mathcal{R}$. The root has two subtrees; its left subtree has its root at the point $(g(K) - 1, 0)$ and covers all the bottom $g(K)$ leaves along $[a, c]$, its right subtree has its root at $(K, g(K))$ and covers the rest $K + 1 - g(K)$ leaves. Both subtrees are constructed recursively until only one leaf remains.

Upon receiving (t_i, t_i), $\mathcal{A}$ constructs $\mathcal{A}(C_i)$ by augmenting $\mathcal{A}(C_{i-1})$ with the following line σ_i. Let (x_i, y_i) be the least common ancestor of (t_{i-1}, t_{i-1}) and (t_i, t_i) in $\mathcal{R}$. The new line σ_i comprises a horizontal line segment from (t_i, t_i) to the point (x_i, t_i), and a vertical line segment from (x_i, t_i) down to the highest point that is also a point on $\mathcal{A}(C_{i-1})$.

4.3 Analysis of the Connector and the α-Dyadic Algorithms

In this section we analyze the bandwidth used at any time by the connector algorithm $\mathcal{A}$ and the α-dyadic algorithm $\mathcal{D}$. The main step is again to transform an optimal connector to an intermediate one with more regular structure. Recall the compact request sequence C in Section 4.2. The intermediate connector tree for analyzing $\mathcal{A}$ is $\mathcal{R}\|_C$. For $\mathcal{D}$, the intermediate tree is $\mathcal{D}^*\|_C$ (see Section 3.2). The following two lemmas relate the bandwidth usage among these connectors.

Lemma 8. *At any time t, we have $c(\mathcal{R}\|_C, \mathcal{L}_t) \geq c(\mathcal{A}(C), \mathcal{L}_t)$ for the connector algorithm $\mathcal{A}$, and $c(\mathcal{D}^*\|_C, \mathcal{L}_t) \geq c(\mathcal{D}(C), \mathcal{L}_t)$ for the α-dyadic algorithm $\mathcal{D}$.*

Proof. Recall that $C_i = \langle(t_1, t_1), (t_2, t_2), \ldots (t_i, t_i)\rangle$ is the first i points of C. To prove $c(\mathcal{R}\|_C, \mathcal{L}_t) \geq c(\mathcal{A}(C), \mathcal{L}_t)$, it suffices to prove $c(\mathcal{R}\|_{C_i}, \mathcal{L}_t) \geq c(\mathcal{A}(C_i), \mathcal{L}_t)$

by induction on i. Obviously, it is true for $i = 1$. Suppose it is true for $i - 1$. Let (x_i, y_i) be the least common ancestor of (t_{i-1}, t_{i-1}) and (t_i, t_i) in $\mathcal{R}$. Note that $\mathcal{R}\|_{C_i}$ is constructed by adding to $\mathcal{R}\|_{C_{i-1}}$ the rectilinear line π_i from (t_i, t_i) to (x_i, y_i) in $\mathcal{R}$, and thus $c(\mathcal{R}\|_{C_i}, \mathcal{L}_t) = c(\mathcal{R}\|_{C_{i-1}}, \mathcal{L}_t) + c(\pi_i, \mathcal{L}_t)$. On the other hand, recall that we construct $\mathcal{A}(C_i)$ by adding to $\mathcal{A}(C_{i-1})$ the line σ_i from (t_{i-1}, t_{i-1}) to (x_i, t_i) and then down to the highest point that is also a point on $\mathcal{A}(C_{i-1})$ and thus $c(\mathcal{A}(C_i), \mathcal{L}_t) = c(\mathcal{A}(C_{i-1}), \mathcal{L}_t) + c(\sigma_i, \mathcal{L}_t)$.

Let ρ be the rectilinear line consisting of the horizontal line from (t_i, t_i) to (x_i, t_i) and the vertical line from (x_i, t_i) to (x_i, y_i). Since the starting and ending points of π_i and ρ are the same, both of them cross the same set of time-lines and thus $c(\pi_i, \mathcal{L}_t) = c(\rho, \mathcal{L}_t)$. On the other hand, $\sigma_i \subseteq \rho$ and the set of time-lines crossed by σ_i is a subset of those crossed by ρ, implying that $c(\sigma_i, \mathcal{L}_t) \leq c(\rho, \mathcal{L}_t)$. Therefore, $c(\sigma_i, \mathcal{L}_t) \leq c(\pi_i, \mathcal{L}_t)$, and the first inequality follows. We can prove the second inequality similarly.

The following lemma shows that the cost of the intermediate connector trees are not too large.

Lemma 9. *Let T be any connector tree for the sequence $C' = C \cup \{(\ell/2, \ell/2)\}$. Then, at any time t, we have $c(\mathcal{R}\|_{C'}, \mathcal{L}_t) \leq 4\,c(T, \mathcal{L}_t)$, and for any α-dyadic algorithm $\mathcal{D}$ with $(1+\sqrt{2})/2 \leq \alpha \leq 2$, we have $c(\mathcal{D}^*\|_{C'}, \mathcal{L}_t) \leq 4\,c(T, \mathcal{L}_t)$.*

Proof. We apply the transformation described in Section 3.2 to prove the lemma. For example, to show $c(\mathcal{R}\|_{C'}, \mathcal{L}_t) \leq 4\,c(T, \mathcal{L}_t)$, we transform T into another tree Z that is a subtree of $\mathcal{R}$ covering C' as a subset of its leaves. Note that $c(\mathcal{R}\|_{C'}, \mathcal{L}_t) \leq c(Z, \mathcal{L}_t)$ since every subtree of R that covers C' must contain $\mathcal{R}\|_{C'}$ as a subtree. On the other hand, we show that $c(Z, \mathcal{L}_t) \leq 4\,c(T, \mathcal{L}_t)$. Thus, we have $c(\mathcal{R}\|_{C'}, \mathcal{L}_t) \leq 4\,c(T, \mathcal{L}_t)$. Details of the proof will be given in the full paper.

Theorem 2. *(1) The connector algorithm is 4-competitive. (2) The α-dyadic algorithm is 4-competitive when $(1+\sqrt{2})/2 \leq \alpha \leq 2$.*

Proof. (1) By Lemmas 8 and 9, we have $c(\mathcal{A}(C), \mathcal{L}_t) \leq 4\,c(T, \mathcal{L}_t)$. Together with Lemma 7, for any point sequence V, $c(\mathcal{A}(V), \mathcal{L}_t) \leq 4\,(c(\mathit{opt}(V), \mathcal{L}_t) + 4)$. Therefore, $\max_t c(\mathcal{A}(V), \mathcal{L}_t) \leq 4\,(\max_t c(\mathit{opt}(V), \mathcal{L}_t) + 4)$. (2) Suppose $(1+\sqrt{2})/2 \leq \alpha \leq 2$. By Lemmas 8 and 9, we have $c(\mathcal{D}(C), \mathcal{L}_t) \leq 4\,c(T, \mathcal{L}_t)$. Together with Lemma 7, for any point sequence V, $c(\mathcal{D}(V), \mathcal{L}_t) \leq 4\,(c(\mathit{opt}(V), \mathcal{L}_t) + 4)$. Therefore, $\max_t c(\mathcal{D}(V), \mathcal{L}_t) \leq 4\,(\max_t c(\mathit{opt}(V), \mathcal{L}_t) + 4)$. Thus, the theorem follows.

References

[1] C. C. Aggarwal, J. L. Wolf, and P. S. Yu. On optimal batching policies for video-on-demand storage servers. In *Proc. International Conference on Multimedia Computing and Systems*, pages 253–258, 1996.

[2] C. C. Aggarwal, J. L. Wolf, and P. S. Yu. On optimal piggyback merging policies for video-on-demand systems. In *Proc. ACM Sigmetrics*, pages 200–209, 1996.

[3] C. C. Aggarwal, J. L. Wolf, and P. S. Yu. A permutation-based pyramid broadcasting scheme for video-on-demand systems. In *Proc. International Conference on Multimedia Computing and Systems*, pages 118–126, 1996.

[4] A. Bar-Noy, J. Goshi, R. E. Ladner, and K. Tam. Comparison of stream merging algorithms for media-on-demand. In *Proc. Multimedia Computing and Networking*, pages 115–129, 2002.

[5] A. Bar-Noy and R. E. Ladner. Efficient algorithms for optimal stream merging algorithms for media-on-demand. Manuscript, 2000.

[6] A. Bar-Noy and R. E. Ladner. Competitive on-line stream merging algorithms for media-on-demand. In *Proc. 12th ACM-SIAM Symposium on Discrete Algorithms*, pages 364–373, 2001.

[7] Y. Cai, K. A. Hua, and K. Vu. Optimizing patching performance. In *Proc. Multimedia Computing and Networking*, pages 204–215, 1999.

[8] S. W. Carter and D. D. E. Long. Improving bandwidth efficiency of video-on-demand. *Computer Networks*, 31(1-2):99–111, 1999.

[9] W. T. Chan, T. W. Lam, H. F. Ting, and W. H. Wong. A unified analysis of hot video schedulers. In *Proc. 34th ACM Symposium on Theory of Computing*, 2002. To appear.

[10] W. T. Chan, T. W. Lam, H. F. Ting, and W. H. Wong. On-line stream merging in a general setting. *Theoretical Computer Science*. To appear.

[11] T. Chiueh and C. Lu. A periodic broadcasting approach to video-on-demand service. In *Proc. Multimedia Computing and Networking*, pages 162–169, 1995.

[12] E. Coffman, P. Jelenkovi, and P. Momcilovic. The dyadic stream merging algorithm. *Journal of Algorithms*. To Appear.

[13] A. Dan, D. Sitaram, and P. Shahabuddin. Dynamic batching policies for an on-demand video server. *Multimedia Systems*, 4(3):112–121, 1996.

[14] D. Eager, M. Vernon, and J. Zahorjan. Optimal and efficient merging schedules for video-on-demand servers. In *Proc. 7th ACM Multimedia*, pages 199–202, 1999.

[15] D. Eager, M. Vernon, and J. Zahorjan. Minimizing bandwidth requirements for on-demand data delivery. *IEEE Transactions on Knowledge and Data Engineering*, 13(5):742–757, 2001.

[16] L. Golubchik, J. C. S. Lui, and R. R. Muntz. Adaptive piggybacking: A novel technique for data sharing in video-on-demand storage servers. *Multimedia Systems*, 4(3):140–155, 1996.

[17] K. A. Hua, Y. Cai, and S. Sheu. Patching: A multicast technique for true video-on-demand services. In *Proc. 6th ACM Multimedia*, pages 191–200, 1998.

[18] K. A. Hua and S. Sheu. Skyscraper broadcasting: A new broadcasting scheme for metropolitan video-on-demand systems. In *Proc. SIGCOMM*, pages 89–100, 1997.

[19] L.-S. Juhn and L.-M. Tseng. Harmonic broadcasting for video-on-demand service. *IEEE Transactions on Broadcasting*, 43(3):268–271, 1997.

[20] S. W. Lau, J. C. S. Lui, and L. Golubchik. Merging video streams in a multimedia storage server: Complexity and heuristics. *Multimedia Systems*, 6(1):29–42, 1998.

[21] J.-F. Paris, S. W. Carter, and D. D. E. Long. Efficient broadcasting protocols for video on demand. In *Proc. 6th International Symposium on Modeling, Analysis and Simulation of Computer and Telecommunication Systems*, pages 127–132, 1998.

[22] S. Sen, L. Gao, J. Rexford, and D. Towsley. Optimal patching schemes for efficient multimedia streaming. In *Proc. 9th International Workshop on Network and Operating Systems Support for Digital Audio and Video*, pages 44–55, 1999.

[23] Y.-C. Tseng, C.-M. Hsieh, M.-H. Yang, W.-H. Liao, and J.-P. Sheu. Data broadcasting and seamless channel transition for highly-demanded videos. In *Proc. 19th Annual Conference IEEE Computer and Communication Societies*, pages 727–736, 2000.
[24] S. Viswanathan and T. Imielinski. Metropolitan area video-on-demand service using pyramid broadcasting. *Multimedia Systems*, 4(3):197–208, 1996.

Coloring k-Colorable Semirandom Graphs in Polynomial Expected Time via Semidefinite Programming

Amin Coja-Oghlan*

Humboldt-Universität zu Berlin, Institut für Informatik – Forschergruppe Algorithmen, Struktur, Zufall, Unter den Linden 6, 10099 Berlin, Germany,
coja@informatik.hu-berlin.de

Abstract. The analysis of algorithms on semirandom graph instances intermediates smoothly between the analysis in the worst case and the average case. The aim of this paper is to present an algorithm for finding a large independent set in a semirandom graph in polynomial expected time, thereby extending the work of Feige and Kilian [4]. In order to preprocess the input graph, the algorithm makes use of SDP-based techniques. The analysis of the algorithm shows that not only is the expected running time polynomial, but even arbitrary moments of the running time are polynomial in the number of vertices of the input graph. The algorithm for the independent set problem yields an algorithm for k-coloring semirandom k-colorable graphs, and for the latter algorithm a similar result concerning its running time holds, as also for the independent set algorithm. The results on both problems are essentially best-possible, by the hardness results obtained in [4].

1 Introduction

The analysis of algorithms on semirandom graph instances intermediates smoothly between the analysis in the worst case and the (so-called) average case analysis, i.e., the consideration of algorithms on random graphs. In semirandom models, an instance is made up of a random share and a worst case part "added by the adversary". The smaller the random share is, the more the semirandom model resembles the worst case, and the harder the according algorithmic problem becomes. Since algorithms that make use of special properties of random graphs extensively will (in general) not be useful for semirandom instances, one may conclude that algorithms that work properly in the semirandom case are comparatively robust.

The first problem studied in this paper is closely related to the maximum independent set problem (find an independent set S in a graph $G = (V, E)$ such that $\#S = \alpha(G)$), which is NP-hard (cf. [5]). In fact, in [6] even a strong non-approximability result is given. The problem remains hard even if it is known

* Research supported by the Deutsche Forschungsgemeinschaft (grant DFG FOR 413/1-1)

K. Diks et al. (Eds): MFSC 2002, LNCS 2420, pp. 201–211, 2002.

that $\alpha(G) \geq \alpha n$ for some constant $0 < \alpha < 1$ independent of $n = \#V$, and we are required to exhibit some independent set $S \subset V$ of size $\#S \geq \alpha n$.

The following semirandom model of the last problem, which we may call the "hidden independent set problem", will be studied in this paper. First, the adversary picks a set $S \subset V$ of cardinality αn. Then, a random bipartite graph G_0 is constructed by including every $V \setminus S$-S-edge with probability $p = p(n)$ independently of all other edges. Finally, the adversary constructs a graph $G \supset G_0$ by adding some $V \setminus S$-V-edges. We emphasize the fact that the adversary's decisions are *not* random decisions. Note that the adversary is allowed to work local maxima into $V \setminus S$. The above model has been introduced in [4].

We shall present an algorithm that in the case $p \gg \ln n/n$ on input G within polynomial expected time finds an independent set of size $\geq \alpha n$, where probability is taken over the choice of the random bipartite graph G_0. In fact, we will prove that even arbitrary moments of the running time of the algorithm are polynomial in the input length. Moreover, by a hardness result given in [4], in the case $p \ll (\ln n)/n$ there is no algorithm that on input G with high probability finds an independent set of size $\geq \alpha n$, unless BPP $\supset$ NP.

The second problem studied in this paper is k-coloring, i.e. given a graph $G = (V, E)$ such that $\chi(G) \leq k$, it is required to exhibit a proper k-coloring of G. Although 2-coloring is easy, it is hard to 4-color a 3-colorable graph (cf. [8]). Semirandom instances of the k-coloring problem are created as follows. First the adversary partitions the vertices $V = \{1, \ldots, n\}$ into k different classes $V_1, \ldots, V_k$ of size n/k each. Then, a random k-partite graph G_0 is constructed by letting each edge $\{v, w\}$, $v \in V_i$, $w \in V_j$, $i \neq j$, be present in G_0 with probability $p = p(n)$ independently of all others. Finally, the adversary may add further edges $\{v, w\}$, $v \in V_i$, $w \in V_j$, $i \neq j$, thereby completing the instance G. The above model has been introduced in [2].

Applying our algorithm for the independent set problem, we obtain an algorithm that within polynomial expected time k-colors semirandom k-colorable graphs, provided $p \gg \ln(n)/n$. Indeed, we will see that aribtrary moments of the running time are polynomials. As in the case of the independent set problem, a hardness result given in [4] states that there is no polynomial time algorithm that in the case $p \ll \ln(n)/n$ k-colors semirandom k-colorable graphs, unless BPP $\supset$ NP. Therefore, with respect to the number of random edges needed our algorithm is essentially best possible.

Related work has been done by Blum and Spencer [2], by Subramanian [10], and by Feige and Kilian [4]. In [2] and [10] the k-coloring problem is studied in the case $p \geq n^{\varepsilon - 1}$, where $\varepsilon > 0$ does not depend on n. In [4] a polynomial time algorithm is presented that in the case $p \geq (1 + \varepsilon)\ln(n)/(\alpha n)$ with high probability finds a hidden independent set of size αn. Moreover, it is shown that in the case $p \leq (1 - \varepsilon)\ln(n)/(\alpha n)$ no such algorithm exists, unless NP $\subset$ BPP. By applying their algorithm for the independent set problem to the k-coloring problem, Feige and Kilian obtain a polynomial time algorithm that in the case $p \geq (1+\varepsilon)k\ln(n)/n$ k-colors semirandom k-colorable graphs. Further, it is shown that in the case $p \leq (1 - \varepsilon)\ln(n)/n$ no such algorithm exists, unless NP $\subset$ BPP.

In this paper we extend the techniques established in [4] in order to obtain algorithms with a polynomial expected running time. Treating the independent set problem first (in sections 1–3), we shall encounter the k-coloring problem indirectly (in section 4).

2 A Semi-random Model for the Independent Set Problem

We shall state a semirandom model that generates instances of the hidden independent set problem. Let $\alpha \in (0;1)$ be a constant, $V = \{1,\ldots,n\}$, $V_1 = \{1,\ldots,\lfloor(1-\alpha)n\rfloor\}$, and $V_2 = V \setminus V_1$. Let $p = p(n) \in (0;1)$ be a sequence of probabilities. As a convenience we can write $p = \frac{\omega}{\alpha n}$. Then, an instance $G = (V,E)$ can be constructed as follows:

1. We let $G_0 = (V, E_0)$ be a random bipartite graph obtained by deciding independently with probability p for each edge $\{v,w\}$, $v \in V_1$, $w \in V_2$, whether $\{v,w\}$ is in E_0.
2. Then, the adversary adds some edges $\{v,w\}$, $v \in V_1$, $w \in V_1 \cup V_2$, in order to obtain G_1.
3. Finally, the application of a permutation $\sigma \in S_n$ chosen by the adversary to the vertices of G_1 yields G.

A graph constructed according to the above experiment is denoted by $\mathcal{G}_{n,\alpha,p}$. We shall consider $\mathcal{G}_{n,\alpha,p}$ as a random variable, where probability is taken with respect to the choice of G_0. By S we shall denote the image of V_2 under σ. The algorithmic problem associated with the semirandom model $\mathcal{G}_{n,\alpha,p}$ is clearly to find S or another independent set of G of size $\geq \alpha n$. We will arrive at the following threshold result about the complexity of the hidden independent set problem.

Theorem 1. *Let α be fixed. Assume that $p = \omega/(\alpha n)$, $\omega \gg \ln n$. Then there is an algorithm that within expected polynomial time finds an independent set of size $\geq \alpha n$ in $\mathcal{G}_{n,\alpha,p}$.*

If $p = (1-\varepsilon)\ln(n)/(\alpha n)$, $\varepsilon > 0$, then there is no polynomial time algorithm that with high probability finds an independent set of size $\geq \alpha n$ in $\mathcal{G}_{n,\alpha,p}$, unless BPP $\supset$ NP.

Proof. The first part will follow from the analysis of our algorithm `Find` below; the second part of the assertion is a hardness result from [4]. □

One may ask whether picking the permutation σ randomly (under the uniform distribution on S_n) results in an easier computational problem. It is, however, not hard to prove that both problems are equivalent.

Thus, the only random object in the construction of $\mathcal{G}_{n,\alpha,p}$ is G_0. Consequently, our algorithmic approach to finding sufficently large independent sets in $\mathcal{G}_{n,\alpha,p}$ will have to make substantial use of "typical" properties of G_0. In order to design an algorithm with expected polynomial running time, we shall extend the consideration given in [4] by taking into account such graphs that lack typical properties to a certain degree. Thus, on the one hand, we have to present algorithmic techniques that even work if G_0 looks pathological, and, on the other hand, we need to estimate the probability that G_0 lacks some typical properties. An estimate of this kind is the following. Let us assume $p = \omega/n$, $\omega \gg \ln n$.

Proposition 2. *Let $c \in (0,1)$ be fixed. Let $d \in \{1, 2, \ldots\}$ be arbitrary but constant. Let $1 \le \eta \le c\alpha n$. Then the probability that there exists some subset $X \subset V_1$, $\#X \le \frac{c\alpha n}{d}$, such that $\#N_{G_0}(X) \le d \cdot \#X - \eta$ is bounded from above by $\exp\left(-\frac{(1-c)\omega\eta}{4d}\right)$, provided $\omega \gg \ln n$. (By $N_{G_0}(X)$ we denote the set of all vertices that in G_0 are adjacent to some vertex in X).*

Proof. Let $X \subset V_1$ be a subset of cardinality k, $1 \le k \le c\alpha n/d$, and let $Y \subset V_2$ be of cardinality dk. There are at most $\binom{n}{k}\binom{\alpha n}{dk}$ possible choices of X and Y. The probability that $N_{G_0}(X) \subset Y$ is $\le (1-p)^{k(\alpha n - dk)}$. We have

$$\binom{n}{k}\binom{\alpha n}{dk}(1-p)^{k(\alpha n-dk)} \le \left(\left(\frac{en}{k}\right)\left(\frac{e\alpha n}{dk}\right)^d \exp(-(1-c)\omega)\right)^k.$$

Furthermore, because $\omega \gg \ln n$, we conclude

$$\ln\left\{\left(\frac{en}{k}\right)\left(\frac{e\alpha n}{dk}\right)^d \exp(-(1-c)\omega)\right\} \le (d+2)\ln(n) - (1-c)\omega \le -\frac{(1-c)\omega}{2}.$$

Thus the probability that there exist sets X and Y as above is at most $e^{-(1-c)k\omega/2}$. We may assume that $k = \#X \ge \eta/d$. Hence, the estimate

$$\sum_{k=\eta/d}^{c\alpha n/d} \exp(-(1-c)k\omega/2) \le \sum_{k=\eta/d}^{\infty} \exp(-(1-c)k\omega/2) \le \exp\left(-\frac{(1-c)\eta\omega}{4d}\right)$$

proves the proposition. □

The following lemma is a consequence of Hall's theorem.

Lemma 3. *Suppose that the condition stated in Proposition 2 holds. If $X \subset V_1$ satisfies $\#X \le c\alpha n/d$, then in G_0 there exists a d-fold matching M from X to V_2 with defect $\le \eta$; i.e. M is a set of V_1-V_2-edges such that at most η vertices in X are incident with $< d$ edges in M and no vertex in V_2 is incident with more than one edge in M.*

Proof. Let V' consist of d copies $v^{(1)}, \ldots, v^{(d)}$ of each verrtex $v \in X$, the vertices V_2, and additional vertices $d_1, \ldots, d_\eta \notin V$. Let E' consist of all edges $\{v^{(i)}, w\}$ such that $\{v, w\} \in G_0$, and all edges $\{v^{(i)}, d_j\}$ ($v \in X$, $i = 1, \ldots, d$, $w \in V_2$, $j = 1, \ldots, \eta$). Then $H = (V', E')$ is a bipartite graph that satisfies the assumption of Hall's theorem. Hence, in H there exists a complete matching M' from X to V_2. After contracting the d copies $v^{(1)}, \ldots, v^{(d)}$ for all $v \in X$, there are at most η vertices $v \in X$ that are connected with one of the d_i's. The remaining $\ge \#X - \eta$ vertices are connected with d vertices in V_2, and every vertex in V_2 is used at most once. □

Lemma 4. *Let $\delta > 0$ be fixed. Then, the probability that the number of vertices $v \in V_1$ that in G_0 have less than $\omega/10$ neighbours exceeds δn is $\ll e^{-\gamma n}$, where $\gamma \gg 1$.*

Proof. Let X be the number of vertices $v \in V_1$ that in G_0 have less than $\omega/10$ neighbours. Then X has binomial distribution with parameters $(1-\alpha)n$ and

$$p_0 = P(d_{G_0}(v) < \omega/10) \leq \exp(-\omega/1000).$$

Thus, $\lambda = \mathrm{E}(X) \leq n \exp(-\omega/1000)$. Let $\varphi(x) = (1+x)\ln(1+x) - x$, $x \geq -1$. Then

$$\lambda\varphi(\delta n/\lambda) \geq \frac{\delta n}{2} \ln\left(1 + \frac{\delta n}{\lambda}\right) \gg n,$$

and the estimate

$$P(X \geq 2\delta n) \leq P(X \geq \lambda + \delta n) \leq \exp(-\lambda\varphi(\delta n/\lambda)).$$

proves our assertion. □

Proposition 5. *Let $\omega' > 0$ be a constant. Then with probability $\geq 1 - e^{-\gamma n}$, $\gamma \gg 1$, there exists some subgraph $G_0' \subset G_0$ that has the following properties.*

1. *Each vertex $v \in V_1$ has $\leq \omega'$ neighbours.*
2. *Let $X \subset V_1, Y \subset V_2$ be subsets of cardinality $\#X = \#Y = \frac{3n \ln \omega'}{\omega'}$. Then, there is an X-Y-edge.*

Proof. Let us consider the following experiment producing a random bipartite graph $G_{\omega'}$. Every vertex $v \in V_1$ chooses precisely ω' neighbours in V_2 uniformly at random and independently of all others. Let $X \subset V_1$ and $Y \subset V_2$ be subsets of cardinality $K = 2n\ln(\omega')/\omega'$. The probability that there is no X-Y-edge is at most

$$\left(\binom{\alpha n - \#Y}{\omega'} \Big/ \binom{\alpha n}{\omega'}\right)^K \leq \omega'^{-2K}.$$

Hence, the probability that there exists sets X and Y as above such that there is no X-Y-edge is at most

$$\binom{\alpha n}{K}\binom{n}{K}\omega'^{-2K} \leq \left(\frac{e}{\ln \omega'}\right)^{2K} \leq \exp\left(-\frac{4n\ln(\omega')}{\omega'}\right), \tag{1}$$

provided $\omega' \geq 30$.

Let $\delta < \ln(\omega')/\omega'$. By Lemma 4, with probability $1 - \exp(-\gamma n)$, $\gamma \gg 1$, all up to at most δn vertices $v \in V_1$ have chosen $\Omega(\omega)$ random neighbours in V_2. Thus, G_0 consists of $\Omega(\omega)$ independently chosen random bipartite graphs $G_{\omega'}$. By (1), the probability that for each of these $\Omega(\omega)$ random bipartite graphs there exists sets X and Y as above such that there is no X-Y-edge is at most $\exp(-\Omega(\omega n))$. □

3 Preprocessing the Input via Semidefinite Programming

Our algorithm makes use of an SDP based partitioning algorithm that is implicit in [1]. The input of this polynomial time algorithm, `Vectors`, consists of a graph $G = (V, E)$, and a number $K \in \{1, 2, \ldots\}$. The output satisfies the following conditions.

- If $\alpha(G) \geq K$, then `Vectors` outputs a map $\varphi : V_0 \to S^{\#V-1}$, where $V_0 \subset V$ is a subset of cardinality $\#V_0 = K/2$. For any edge $\{v, w\} \in E$ connecting two vertices $v, w \in V_0$ we have $(\varphi(v)|\varphi(w)) < -\frac{K}{2n}$. (By $S^{\#V-1}$ we denote the $\#V - 1$-dimensional sphere.)
- If $\alpha(G) < K$, then `Vectors` either produces a map φ as mentioned or outputs "fail".

We want to use the map φ produced by `Vectors` in order to partition the vertex set of the input graph. Therefore, we need another procedure `Round` that rounds the output of `Vectors` to discrete values. We either let `Round` be the randomized rounding via vector projections from [7] or the deterministic rounding from [9].

Algorithm 6. `Filter`(G, α, ε)
Input: A graph $G = (V, E)$, where $V = \{1, \ldots, n\}$; a number $\alpha \in (0; 1)$; $\varepsilon > 0$.
Output: A set $\mathcal{M}$ of subsets of V.

1. Put $\mathcal{M} = \emptyset$. Let D be a sufficiently large number depending only on ε and α but not on n. Start with $l = 0$, $G_0 = G$, and $V_0 = V$. If `Vectors`$(G_l, \lfloor \frac{\varepsilon \alpha n}{3(D+1)} \rfloor)$ outputs "fail", go to 2. Otherwise, let $\varphi_{l+1} : V_{l+1} \to S^{n-1}$ be the output of `Vectors`. Put $l = l + 1$, $G_l = G[V \setminus (V_1 \cup \cdots \cup V_l)]$, and iterate.
2. (a) Put $t = \omega_0^\eta$ for some suitable $\eta \in (0; 1)$ and some large constant $\omega_0 \in \{1, 2, 3, \ldots\}$ depending only on α and ε but not on n. For $i = 1, \ldots, l$, let $(V_{i1}, \ldots, V_{it}) =$ `Round`$(G[V_i], \varphi_i, t)$.
   (b) For any i, j, compute a maximal matching M_{ij} in $G[V_{ij}]$. Put $I_{ij} = V_{ij} \setminus V(M_{ij})$, where $V(M_{ij})$ denotes the set of all vertices incident with an edge in M_{ij}.
3. Let $\mathcal{P}$ be the set of all subsets of $\{(i, j)|\ i, j \text{ as above}\}$. Put $\mathcal{M} = \mathcal{M} \cup \{\bigcup_{(i,j) \in I} I_{ij} |\ I \in \mathcal{P}\}$.
4. If `Round` is randomized, run 2 and 3 n^2 times, each time with independent coin tosses. Output $\mathcal{M}$.

Similar preprocessing techniques have been used in [4]. A detailed analysis of `Filter` can be found in my technical report [3]; the following proposition states the result of the analysis.

Proposition 7. *Let $G = \mathcal{G}_{n,\alpha,p}$ be the input graph of `Filter`, $p = \omega/n$, $\omega \gg \ln n$. Then, with probability $\geq 1 - e^{-\gamma n}$, $\gamma \gg 1$, taken over the choice of G_0 and the coin tosses of `Filter` (if `Round` is randomized), the output $\mathcal{M}$ of `Filter` enjoys the following properties.*

1. *There exists some $I \in \mathcal{M}$ such that $\#I \cap S \geq (\alpha - 5\varepsilon)n$ and $\#I \setminus S \leq \varepsilon \#I$.*
2. *$\#\mathcal{M} = O(n^2)$.*

4 Finding Large Independent Sets

Algorithm 8. `Find`(G, α)
Input: A graph $G = (V, E)$, $V = \{1, \ldots, n\}$, a number $\alpha \in (0; 1)$.
Output: A set $S'' \subset V$.

1. Run `Filter`$(G, \alpha, \varepsilon = \frac{\alpha^2}{1000})$. Let $\mathcal{M}$ be the output of `Filter`.
2. For each $\eta = 0, 1, \ldots, n/2$, run 3. Then go to 6.
3. For any $I \in \mathcal{M}$, run 4.
4. (a) Construct the following network N: The vertices of N are
   - s and t,
   - s_v for each $v \in I$,
   - t_w for each $w \in V$.

   The arcs of N are
   - (s, s_v) for $v \in I$,
   - (t_w, t) for $w \in V$,
   - (s_v, t_w) for $\{v, w\} \in E$, $v \in I$.

   The capacity c is given by
   - $c(s, s_v) = \lceil \frac{\alpha}{\varepsilon} \rceil$ for $v \in I$,
   - $c(t_w, t) = 7$ for $w \in V$,
   - $c(s_v, t_w) = 1$ for $\{v, w\} \in E$, $v \in I$.

   (b) Compute a maximum integer flow f in N and put
   $$L = \{v \in I | \ f(s, s_v) = c(s, s_v)\}$$
   and $S'_1 = I \setminus L$.
   (c) For each subset $D' \subset S'_1$, $\#D' \leq \frac{8\alpha\eta}{\varepsilon}$, run step 5 with $S' = S'_1 \setminus D'$.
5. (a) For any $D \subset V$, $\#D \leq \eta$, run (b).
   (b) For $\tau = 0, \ldots, \log_2(n)$, run (c)–(e).
   (c) If S' is an independent set of size $\geq \alpha n$, terminate with output S'. Otherwise, try to find some $Y \subset V' \setminus S'$, $\#Y \leq 8\eta$, such that $S' \cup Y$ is an independent set of size $\geq \alpha n$. If this is the case, terminate with output $S' \cup Y$.
   (d) Compute $V' = \{v \in V | \ N(v) \cap S' = \emptyset\} \setminus D$. Construct the following network N: The vertices of N are
   - s and t,
   - s_v for each $v \in V' \setminus S'$,
   - t_w for each $w \in V'$.

   The arcs of N are
   - (s, s_v) for $v \in V' \setminus S'$,
   - (t_w, t) for $w \in V'$,
   - (s_v, t_w) for $\{v, w\} \in E$, $v \in V' \setminus S'$, $w \in V'$.

   The capacity c satisfies
   - $c(s, s_v) = 12$ for $v \in V' \setminus S'$,
   - $c(t_w, t) = 5$ for $w \in V'$,
   - $c(s_v, t_w) = 1$ for $\{v, w\} \in E$, $v \in V' \setminus S'$, $w \in V'$.

(e) Compute a maximum integer flow f in N, and put

$$L = \{v \in V' \setminus S' | \ f(s, s_v) = c(s, s_v)\}.$$

Let $S' = V' \setminus L$.

6. Find a maximum independent set in G by enumerating all subsets of V.

As for the analysis of Find, we shall assume that the graph G_0 has the following property: For some fixed $c \in (0;1)$ sufficiently close to 1, and some fixed but sufficiently large $d \in \{1, 2, 3, \ldots\}$, any subset $U \subset V_1$, $\#U \leq can/d$, has at least $d\#U - \eta$ neighbours in G_0 for some $\eta \in \{0, 1, \ldots, n/2\}$ independent of U. Moreover, we shall assume that there exists $I \in \mathcal{M}$ satisfying

$$\#I \cap S \geq (\alpha - 5\varepsilon)n, \ \#I \setminus S < \varepsilon\#I. \tag{2}$$

If there is no such η or no such I, then our algorithm will surely find a sufficiently large independent set by running step 6. Furthermore, let us assume that step 3 has picked a set I satisfying (2).

Under the assumptions made, the set S_1' in 4b admits a set $D' \subset V_1$, $\#D' \leq 6\alpha\eta/\varepsilon$, such that $S' = S_1' \setminus D' \subset S$ and

$$\#S' \geq \left(\alpha - \frac{17\varepsilon}{\alpha}\right) n.$$

Thus the algorithm will finally run step 5 with the "correct" S'. Then, we observe that one of the following conditions is satisfied: Either $\#S \setminus S' \leq 8\eta$ or with $\tilde{V} = \{v \in V | \ N(v) \cap S = \emptyset\}$ we have $\#\tilde{V} \leq (\#S \setminus S')/4$. If the first statement holds, 5c run with $D = \emptyset$ will find a sufficiently large independent set and terminate.

In the second case, we have

$$\#\tilde{V} \setminus S \leq \frac{\#S \setminus S'}{4} < \frac{\#S}{16}.$$

Therefore, there exists a 12-fold matching M^* from $\tilde{V} \setminus S$ with defect $\leq \eta$. Let D^* be the set of unmatched vertices $v \in \tilde{V} \setminus S$. Finally, 5a picks $D = D^*$. Then $V' = \tilde{V} \setminus D^*$, and M^* is a complete 12-fold matching from $V' \setminus S$ to S. Hence,

$$\#L \cap (S \setminus S') \leq \frac{\#S \setminus S'}{4}$$

and $L \supset V' \setminus S$.

As for the running time of Find, we note that step 1 runs in polynomial time. If η is the smallest integer satisfying our assumption above, then the number of computations in steps 3–5 is a polynomial plus

$$O\left(\left(\eta \cdot \binom{n}{\eta}\right)^{3\gamma}\right) \leq O\left(\exp(3\gamma\eta \ln n)\right),$$

where $\gamma > 0$ is a sufficiently large constant. Conversely, we have to estimate the probability that there exists some $T \subset V_1$, $\#T \leq c\alpha n/d$, that in G_0 has at most $d\#T - \eta$ neighbours. By Proposition 2, this probability is at most $\exp(-\Omega(\omega\eta))$. If we assume $\omega \gg \ln n$, we obtain

$$\eta \ln n \ll \omega\eta.$$

Finally, if $\mathcal{M}$ does not contain a set I such that

$$\#I \cap S \geq (\alpha - 5\varepsilon)n, \; \#I \setminus S < \varepsilon\#I$$

then either G_0 violates the property stated in Proposition 5 or, if `Round` is randomized, we have by chance chosen bad random vectors. The probability that one of these two events occurs is $O(\exp(-\gamma n))$, $\gamma \gg 1$.

In summary, not only is the expected running time a polynomial with respect to n, but so too are arbitrarily high moments of the running time.

Proposition 9. *Let $G = \mathcal{G}_{n,\alpha,p}$, $p = \omega/(\alpha n)$, $\omega \gg \ln n$. Then for each fixed $\alpha \in (0;1)$,* `Find`*(G, α) determines an independent set of size $\geq \alpha n$ within expected polynomial time with respect to n.*

5 Coloring k-Colorable Graphs

We consider the following semirandom model of k-colorable graphs (cf. [4]). Let $V = \{1, \ldots, n\}$. Let V_1 consist of the vertices $1, \ldots, n/k$, let V_2 consist of $1 + n/k, \ldots, 2n/k$ and so on up to V_k. For any two vertices $v \in V_i$, $w \in V_j$, $i \neq j$, the edge $\{v, w\}$ is present with probability $p = p(n)$, independently of all other edges. Then, the adversary adds further edges $\{v, w\}$, $v \in V_i$, $w \in V_j$, $i \neq j$ and finally permutes the vertices V in order to obtain $G = \mathcal{G}_{n,k,p}$. By $\tilde{V}_1, \ldots, \tilde{V}_k$ we denote the images of $V_1, \ldots, V_k$ under the permutation chosen by the adversary. The analysis of the coloring algorithm given below leads to the following result.

Theorem 10. *Suppose $p = \omega/n$, $\omega \gg \ln n$. Then there exists an algorithm that k-colors $\mathcal{G}_{n,p,k}$ within expected polynomial time.*

If $p = (1-\varepsilon)\ln(n)/n$, then there is no polynomial time algorithm that k-colors $\mathcal{G}_{n,p,k}$ with probability $1 - o(1)$, unless BPP $\supset$ NP.

Let us assume $p = \omega/n$, $\omega \gg \ln n$. In the case $k = 3$, the algorithm `Find` immediately gives an algorithm that 3-colors $G = \mathcal{G}_{n,k,p}$ within expected polynomial time. We just need to adjust 5c as follows: Whenever the algorithm discovers an independent set $\tilde{S}$ of size $\geq n/3$, we check whether $G - \tilde{S}$ is bipartite. If we can achieve a proper 2-coloring of $G - \tilde{S}$, then we have successfully 3-colored G. Otherwise, `Find` has not yet discovered the "right" independent set, and we proceed. This approach works because a careful analysis of `Find` shows that `Find` eventually discovers *one of the hidden independent sets $\tilde{V}_1, \ldots, \tilde{V}_k$.*

Because the algorithm only needs to recover one of the hidden sets $\tilde{V}_1, \ldots, \tilde{V}_k$, 3-coloring is especially simple. In order to present an algorithm for 4-coloring, it is necessary to adapt the algorithm `Find` and obtain a new algorithm denoted by $\mathcal{F}$ as follows.

1. In addition to G and α, $\mathcal{F}$ requires a number $\eta_{\max}$ and a parameter z at the input. In step 2, we let $\mathcal{F}$ try $\eta = 0, \ldots, \eta_{\max}$ instead of $\eta = 0, \ldots, n/2$.
2. The parameter z represents the state of the internal counters that $\mathcal{F}$ uses to enumerate D and D' in 4c and 5a and for 5c. $\mathcal{F}$ starts its computation with the next state following z. If $z = \Box$, then $\mathcal{F}$ starts with the initial state for all counters.
3. Whenever $\mathcal{F}$ recovers a sufficiently large independent set $\tilde{S}$ (in 5c), $\mathcal{F}$ outputs $\tilde{S}$ and the current state z of its counters. Thus, if we run $\mathcal{F}$ on input z, then $\mathcal{F}$ resumes its computation as if $\tilde{S}$ had not been a sufficiently large independent set.
4. Step 6 is replaced by: Terminate with output "fail".

We present an algorithm for 4-coloring. This algorithm can be easily adapted for k-coloring. In order to keep matters simple, we assume that `Round` works deterministically.

Algorithm 11. `4-Color`(G)
Input: A 4-colorable graph $G = (V, E)$, where $V = \{1, \ldots, n\}$.
Output: A 4-coloring of G.

1. Put $\eta_{\max} = 0$, $z_1 = \Box$.
2. (a) If $\mathcal{F}(G, 1/4, \eta_{\max}, z_1)$ fails, put $\eta_{\max} = \eta_{\max} + 1$, $z_1 = \Box$ and iterate 2.
   (b) If $\mathcal{F}$ outputs an independent set S_1 and a state z_1, then let G' be the graph obtained from G by turning S_1 into a clique and go to 3.
   (c) If $\eta_{\max} > n/2$, then color G by enumerating all subsets of V and output a proper 4-coloring.
3. (a) Let $z_2 = \Box$.
   (b) If $\mathcal{F}(G', 1/4, \eta_{\max}, z_2)$ fails, iterate 2a.
   (c) Otherwise let z_2, S_2 be the output of $\mathcal{F}(G', 1/4, \eta_{\max}, z_2)$. Let $G'' = G - S_1 - S_2$.
   (d) If 2-coloring G'' is successful, then output the resulting proper 4-coloring of G. Otherwise iterate 3b.

The analysis of the above algorithm implies the first part of Theorem 10. The hardness result is again from [4].

References

1. Alon, N., Kahale, N.: Approximating the independence number via the ϑ-function. Math. Programming **80** (1998) 253–264.
2. Blum, A., Spencer, J.: Coloring random and semirandom k-colorable graphs. J. of Algorithms **19(2)** (1995) 203–234
3. Coja-Oghlan, A.: Zum Färben k-färbbarer semizufälliger Graphen in erwarteter Polynomzeit mittels Semidefiniter Programmierung. Technical Report **141**, Fachbereich Mathematik der Universität Hamburg (2002)
4. Feige, U., Kilian, J.: Heuristics for semirandom graph problems. preprint (2000)
5. Garey, M.R., Johnson, D.S.: Computers and intractability. W.H. Freeman and Company 1979

6. Håstad, J.: Clique is hard to approximate within $n^{1-\varepsilon}$. Proc. 37th Annual Symp. on Foundations of Computer Science (1996) 627–636
7. Karger, D., Motwani, R., Sudan, M.: Approximate graph coloring by semidefinite programming. Proc. of the 35th. IEEE Symp. on Foundations of Computer Science (1994) 2–13
8. Khanna, S., Linial, N., Safra, S.: On the hardness of approximating the chromatic number, Proc. 2nd Israeli Symp. on Theory and Computing Systems (1992) 250–260
9. Mahajan, S., Ramesh, H.: Derandomizing semidefinite programming based approximation algorithms. Proc. 36th IEEE Symp. on Foundations of Computer Science (1995) 162–169
10. Subramanian, C.: Minimum coloring random and semirandom graphs in polynomial expected time. Proc. 36th Annual Symp. on Foundations of Computer Science (1995) 463–472

On Word Equations in One Variable

Robert Dąbrowski and Wojtek Plandowski

Institute of Informatics, University of Warsaw,
Banacha 2, 02-097 Warszawa, Poland,
{kulisty,wojtekpl}@mimuw.edu.pl.

Abstract. For a word equation E of length n in one variable x occurring $\#_x$ times in E a resolution algorithm of $O(n + \#_x \log n)$ time complexity is presented here. This is the best result known and for the equations that feature $\#_x < \frac{n}{\log n}$ it yields time complexity of $O(n)$ which is optimal. Additionally, we prove that the set of solutions of one-variable word equations is either of the form F where F is a set of $O(\log n)$ words or of the form $F \cup (uv)^+u$ where F is a set of $O(\log n)$ words and u, v are some words such that uv is a primitive word.

1 Introduction

One of the most famous (and complicated) algortihms existing in the litterature is Makanin's algorithm [11]. The algorithm takes as an input a word equation and decides whether or not the equation has a solution. It has been improved several times. The best version works in EXPSPACE [8]. The current version (including the proof of correctness) occupies 50 pages [5]. Recently, new algorithms were found [13,15]. The first one works nondeterministically in polynomial time with respect to the length of the input equation and the logarithm of the length of the minimal solution. The best upper bound for the length of the minimal solution is double exponential [14] so with this bound the algorithm works in NEXPTIME. The algorithm in [15] works in PSPACE. The algorithms solving the problem of satisfiability of general word equations cannot be called efficient. We cannot even expect efficiency since the problem is NP-hard [1,10].

However, if we concentrate on some classes of word equations, then the polynomial time algorithms exist. For two-variable word equations there are two polynomial time algorithms [2,9]. The best one works in $O(n^6)$ time (n is the length of the input equation) so although the algorithm is polynomial it cannot be called efficient. An algorithm which can be called efficient is known for equations in one variable [7]. It works in $O(n \log n)$ time. There is also an efficient $O(n \log^2 n)$ time algorithm for special equations with two variables [12].

We deepen the analysis in [7] and obtain an algorithm working in $O(n + \#_x \log n)$ time where $\#_x$ is the number of occurences of the variable in the input equation. The algorithm consists of two phases: in the first one it finds $O(\log n)$ candidates for a solution, in the second one it verifies in $O(\#_x)$ time if a candidate is a solution. The second part assumes that the alphabet Σ is of constant size or that it is a set of numbers $1..|\Sigma|$.

K. Diks et al. (Eds): MFSC 2002, LNCS 2420, pp. 212–220, 2002.

As a side effect of our considerations we prove that the set of solutions of one-variable word equations is either of the form F where F is a set of $O(\log n)$ words or of the form $F \cup (uv)^{+}u$ where F is a set of $O(\log n)$ words and u, v are some words such that uv is primitive.

2 Preliminaries

A *word equation* over alphabet Σ in one variable $x \notin \Sigma$ is a pair $E = (L, R)$ of words $L, R \in (\Sigma \cup \{x\})^{\star}$ and is usually written as $E : L = R$. A solution of E is a homomorphism $\varphi : (\Sigma \cup \{x\})^{\star} \to \Sigma^{\star}$ leaving the letters of Σ invariant (i.e. $\varphi(a) = a$ for all $a \in \Sigma$) and such that $\varphi(L) = \varphi(R)$. The solution is uniquely identified by a mapping of x into $\Sigma^{\star}$ for which the homomorphism is a canonical extension. By *Sol*(E) we denote the *solution set* of equation E. The length of an equation E is denoted by $|E| = |LR|$.

The *prefix, proper prefix, suffix* and *proper suffix* relations on words are denoted respectively by $\sqsubseteq$,$\sqsubset$, $\sqsupseteq$ and $\sqsupset$. For a word B, denote by B^{ω}, an infinite word which is an infinite repetition of the word B. We start by recalling basic facts. Words $A, B \in \Sigma^{\star}$ are *conjugate* iff there exist two words $u, v \in \Sigma^{\star}$ such that $A = uv$ and $B = vu$. A non-empty word $w \in \Sigma^{\star}$ is *primitive* iff it has exactly $|w|$ distinct conjugations or, equivalently, if it is not a power of a different word. For arbitrary words A and w we define $A/w = \max\{k : w^k \sqsubseteq A\}$. Every non-empty word $A \in \Sigma^{\star}$ is a power of a unique primitive word $w \in \Sigma^{\star}$ called its *primitive root*, that is $A = w^{A/w}$.

Proposition 1 ([4]). *For an arbitrary word A let $u, v, w \sqsubseteq A$ be distinct primitive words such that $u^2 \sqsubset v^2 \sqsubset w^2 \sqsubseteq A$. Then $|u| + |v| \le |w|$.*

By *Rpp*(A) we shall denote the sequence of all *repetitive primitive prefixes* of A, that is all primitive $w \sqsubseteq A$ such that $w^2 \sqsubseteq A$. As an immediate consequence of Proposition 1 we have

Proposition 2 ([4]). *Let Rpp*$(A) = \{w_1, w_2, \ldots, w_k\}$ *and* $w_1 \sqsubset w_2 \sqsubset \ldots \sqsubset w_k$. *Then* $|w_{i+2}| \ge 2|w_i|$. *Moreover* $k \le 2\log|A|$ *and* $\sum_{i=1}^{k} |w_i| \le 2|A|$.

For given u, v we define $(uv)^{\star}u = \{(uv)^k u : k \ge 0\}$ and $(uv)^{+}u = \{(uv)^k u : k \ge 1\}$. Word A is *periodic* with *period* uv iff $A \in (uv)^{+}u$.

Proposition 3 (Periodicity lemma). *Let p, q be periods of given word A. If $|p| + |q| \le |A| + 1$ then $\gcd(p, q)$ is also a period of A.*

In a few places we will refer to algorithm in [7]. This algorithm consists of two phases. In the first one it finds $O(\log n)$ pairs of words (u_i, v_i) such that *Sol*$(E) \subseteq \bigcup_i (v_i u_i)^{\star} v_i$. In the second one it finds *Sol*$(E) \cap (v_i u_i)^{\star} v_i$ for all i. This phase uses a procedure finding, for given u, v the set *Sol*$(E) \cap (vu)^{\star}v$ in $O(n)$ time.

A word equation with one variable is of the form

$$E : A_0 x A_1 x \ldots\ x A_s = B_0 x B_1 x \ldots x B_t$$

where A_i, $B_i \in \Sigma^\star$. If the equation E has a solution, then, by cancelling the same leftmost and rightmost symbols from the left-hand side and right-hand side of E, we can reduce the equation to the form

$$E : A_0 x A_1 x \ldots\ x A_s = x B_1 x \ldots x B_t$$

where A_0 is not the empty word and either A_s is a nonempty word and B_t the empty word or vice versa. This form of equation implies that any solution x is a prefix of the word A_0^ω. If $s \neq t$, then the length argument reduces the number of possible solutions of E to at most one. The verification whether this word is a solution can be done in $O(|E|)$ time [7]. If $s = t = 1$, then E is of the form $A_0 x = x B_1$ which is well known in combinatorics of words. This equation has a solution only if there are two words p, q such that pq is a primitive word and $A_0 = (pq)^i$ and $B_1 = (qp)^i$ for some i and the set of all solutions is the set $(pq)^\star p$. In our further analysis we assume $s = t \geq 2$.

Denote by $\#_x$ the number of occurrences of the variable x in E. Then $\#_x = s + t = 2s$.

3 Solutions

Lemma 1. *Let E be an equation and let u, v be two words such that uv is primitive. Then*

$$Sol(E) \cap (uv)^+ u = \begin{cases} \emptyset & (0) \\ \{(uv)^k u\} \text{ for certain } k \geq 1 & (1) \\ (uv)^+ u & (\infty) \end{cases}$$

Proof. It has been shown in [7] that the problem of computing all integers $n \geq 1$ such that $(uv)^n u \in Sol(E)$ can be reduced to solving a Diophantine equation in n. Since the equation may only have: (0) no solutions at all; (1) exactly one solution for certain $k \geq 1$; and (∞) infinitely many solutions for all $\{k : k \geq 1\}$, this completes the proof. □

For simplicity we call, for given u, $v \in \Sigma^*$, $k \geq 0$, $(uv)^k u$ a *finite solution* and, for given u, $v \in \Sigma^*$, $(uv)^+ u$ an *infinite solution*. We show next that there is at most one infinite solution and it can be found in linear time.

Theorem 1. *For given equation E there exists at most one infinite solution and it can be found in time $O(|E|)$.*

Proof. The proof falls naturally into two parts.

(1) Assume $|A_0| \leq |B_1|$. Denote by B' the prefix of B_1 of length $|B'| = |A_0|$. Fix any $X \in Sol(E)$. Since $A_0 X = X B'$, A_0 has to conjugate B'. Therefore the primitive roots of A_0 and B' also have to conjugate each other and thus be respectively equal to uv and vu. Therefore $Sol(E) \subseteq (uv)^\star u$ and we can find $Sol(E)$ in time $O(|E|)$ running the algorithm described in [7]. Since $Sol(E) \subseteq (uv)^\star u$, it follows from Lemma 1 that there can be at most one infinite solution.

(2) Assume $|A_0| > |B_1|$. Denote by A' the prefix of A_0 of length $|A'| = |A_0| - |B_1|$. Fix any $X \in Sol(E)$ of length $|X| \geq |A'|$. Since $A_0X = XB_1A'$, A_0 has to conjugate B_1A'. Analogously to the previous case we conclude that there is at most one infinite solution and that it can be found in time $O(|E|)$.

□

We follow to show that for the set of finite solutions it is possible to find in linear time its logaritmically bounded superset.

Theorem 2. *Let S be the set of finite solutions of equation E. It is possible to find in time $O(|E|)$ a set $\widehat{S}$ of* finite solution candidates *of size $|\widehat{S}| = O(\log|E|)$ and such that $S \subseteq \widehat{S}$.*

Proof. An immediate conclusion from the proof of Theorem 1 is that for the equations that feature $|A_0| \leq |B_1|$ all the solutions can be detected in time $O(|E|)$, and for the equations that feature $|A_0| > |B_1|$ the solutions not shorter than $|A_0| - |B_1|$ can be detected in time $O(|E|)$. In both cases the respective solution sets contribute to the magnitude of $\widehat{S}$ in only a constant factor. Thus without loss of generality it suffices to focus on the equations that feature $|A_0| > |B_1|$ and search candidates for solutions shorter than $|A_0| - |B_1|$. Let $|A_0| > |B_1|$. Fix $X \in S$ of length $|X| < |A_0| - |B_1|$. Then $XB_1X \sqsubset A_0X \sqsubset A_0A_0$ and consequently B_1XB_1X is a proper prefix of $B_1A_0A_0$. Let w denote the primitive root of B_1X. Therefore $w^2 \sqsubset B_1A_0A_0$ and our aim is to search for finite solution candidates with respect to the elements of $Rpp(B_1A_0A_0)$. Then $B_1X = w^k$ for some $k \geq 1$. Hence there is a unique pair of words (u, v) such that u is not the empty word and there are numbers i, j such that $w = vu$ and $B_1 = (vu)^iv$ and $X = (uv)^ju$, see Fig. 1. We say that (u, v) is the unique *generator* for w. All $u = (uv)^0u$ we put into the set $\widehat{S}$. For each $w \in Rpp(B_1A_0A_0)$ we will put into $\widehat{S}$ at most one word of the form $(uv)^ju$, $j > 0$, where (u, v) is the generator for w. In the following we may assume that $X = (uv)^ju$ and $j > 0$ and our aim is to find unique j such that $X = (uv)^ju$ can be a solution of E.

Following Knuth, Morris and Pratt [3] let

$$P[j] = \max\{0 \leq k < j : B_1A_0A_0[1 \ldots k] \sqsupset B_1A_0A_0[1 \ldots j]\}.$$

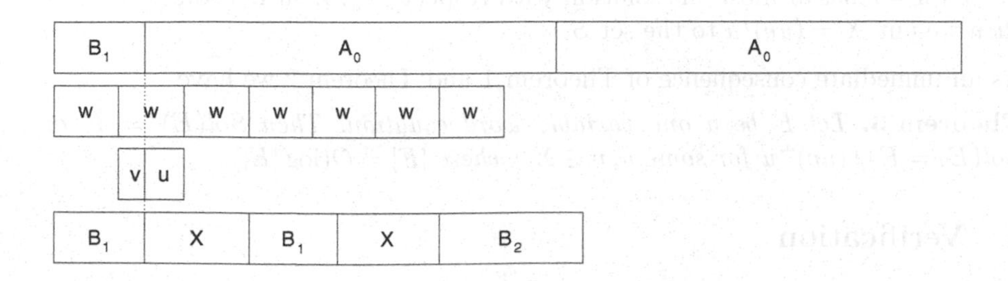

Fig. 1. The pair (u, v) is a generator for w.

The *failure table* P can be calculated in time $O(|B_1A_0A_0|)$ [3]. The property of P that $j - P[j]$ is the length of the shortest period of $B_1A_0A_0[1..j]$ allows us to calculate $Rpp(B_1A_0A_0)$ in total time $O(|B_1A_0A_0|)$. More precisely

$$w \in Rpp(B_1A_0A_0) \iff 2|w| - P[2|w|] = |w|.$$

Fix $w \in Rpp(|B_1A_0A_0|)$. To find its respective solution candidate we analyze the sequence of consecutive occurrences of w in both sides of the equation. First we calculate the number of consecutive occurrences of w in $B_1A_0A_0$. Since $ww \sqsubset B_1A_0A_0$, it follows from the periodicity lemma and primitivity of w that there is at most one w for which the sequence of its consecutive occurences does not end inside $B_1A_0A_0$ and such w can be found in time $O(|B_1A_0A_0|)$. To find its respective solution candidate we run in time $O(|E|)$ the algorithm described in [7]. Therefore without loss of generality we may assume the sequence ends inside $B_1A_0A_0$. We can use the failure table P to find for each w the maximum number of its consecutive occurrences in $B_1A_0A_0$ in total time $O(|B_1A_0A_0|)$. Consider a sequence of w starting in the right hand side of equation E. All such w that their sequence ends inside B_1 can be discarded, since they could never be primitive roots of B_1X. Thus let (u, v) be the unique generator for $w = vu$. Since B_1 must equal $(vu)^iv$ for some $i \geq 0$, we use failure table for B_1 to determine for all w in total time $O(|B_1|)$ which of the cases takes place here.

Case $i \geq 2$. Then it follows from the periodicity lemma and primitivity of w that there is at most one such w. To find a candidate corresponding to w we use the algorithm in [7].

Case $i = 1$. Then there are at most two such w. Indeed, for the set

$$Rpp(B_1A_0A_0) = \{w_1, \ldots, w_k\}$$

of repetitive primitive prefixes we have $2|w_i| \leq |w_{i+2}|$. Since $|w| \leq |B_1| < |ww|$, at most two consecutive words w_i, w_{i+1} meet the conditions. Therefore for a constant number of words w we simply run the algorithm described in [7].

Case $i = 0$. Then $B_1 = v$. Now, let t be the first index such that $B_t \neq B_1$. if $B_t = (vu)^jv$, $j \geq 1$, then again for a constant number of words w we run the algorithm described in [7]. Otherwise $B_t \neq (vu)^jv$ for any $j \geq 0$ and for each w we find the place its sequence ends inside B_tuv (note that B_tuv is a prefix of B_tX) in total time $O(|B_t|)$ using a failure table for B_tA_0. Suppose that $s = B_tuv/w$ and $r = B_1A_0A_0/w$. Then $(B_1X)^tB_tX/w = B_1A_0A_0/w$ so $jt + t + s = r$ has at most one solution with respect to j. If such a solution exists then we put $X = (uv)^ju$ to the set $\widehat{S}$. □

As an immediate consequence of Theorem 1 and Theorem 2 we have

Theorem 3. *Let E be a one variable word equation. Then $Sol(E) = F$ or $Sol(E) = F \cup (uv)^+u$ for some $u, v \in \Sigma^*$ where $|F| = O(\log |E|)$.*

4 Verification

Throughout this section we assume that the alphabet Σ is of constant size or $\Sigma = 1..|\Sigma|$.

Lemma 2. *Given a finite set Π of words of total length $n = \sum_{u \in \Pi} |u|$ and over an alphabet Σ, after an $O(n)$-time preprocessing it is possible to answer in time $O(1)$ if for given a, b being some prefixes of words in Π it is true, that $a \sqsubseteq b$.*

Proof. The preprocessing phase goes as follows. First we build a *trie* tree to represent all the words in set Π. Additionally we keep a link from every letter in every word in Π to its respective node in the tree. Then we traverse the *trie* tree from the root in *prefix* order and assign the nodes consecutive numbers. Then we traverse the *trie* tree in *postfix* order and assign every node the range of numbers in the subtree rooted at the node, including the node's own number. Thus every node is additionally assigned its subtree's minimum number and maximum number. This can be done in time $O(n)$. We follow now the links from the last letters of a and b to check if the respective nodes are in a child-parent relation in *trie* tree, that is if one of the nodes is within the range of the other node. This can be done in time $O(1)$, which completes the proof. □

For given word u by a *prefix table* we mean the array $Pref[1..|u|]$ such that $Pref[j]$ is the length of longest word starting at j in u which is a prefix of u. Formally, $Pref[j] = \max\{0 < k \leq |u| - j + 1 : u[j \ldots j + k - 1] \sqsubseteq u\} \cup \{0\}$. Note that any prefix of u starting at position j in u is a prefix of $u[1..Pref[j] - 1]$.

Now, we generalize the notion of a prefix table to a set of words Π.

Definition 1. *For given set Π of words by* prefix table *we call:*

$$Pref[u, j] = (v, l)$$

where

$$l = \max\{0 < k \leq |u| - j + 1 : u[j \ldots j + k - 1] \sqsubseteq v\} \cup \{0\}$$

and

$$\forall w \in \Pi \; l \geq \max\{0 < k \leq |u| - j + 1 : u[j \ldots j + k - 1] \sqsubseteq w\} \cup \{0\}$$

for any $u \in \Pi$, $1 \leq j \leq |u|$.

Lemma 3. *Given a finite set Π of words of total length $n = \sum_{u \in \Pi} |u|$ and over an alphabet Σ, it is possible to construct the prefix table for Π in time $O(n)$.*

Proof. Let G be the Aho-Corasick automaton for Π, that is a deterministic finite automaton accepting the set of all words containing a word from Π as a suffix. Since Σ is of constant size or a set $1..|\Sigma|$, it is possible to construct G in time $O(n)$. For details on the Aho-Corasick algorithm see [3]. We recall that the Aho-Corasick automaton is a trie for the set of words Π with additional edges $s \rightarrow t$ labeled by symbols $b \in \Sigma$ such that tb is the longest proper suffix of s which is a prefix of a word in Π. If the automaton goes through such an edge we say that it *backtracks*.

Fix word $u \in \Pi$. We shall fill the table $Pref$ for u iteratively. Clearly $Pref[u, 1] = (u, |u|)$. We run the Aho-Corasick automaton on the word u starting from position $i = 2$ in u. If G goes down the trie for Π then we do nothing. Assume that G backtracks at position $j + 1$ going from state v to the

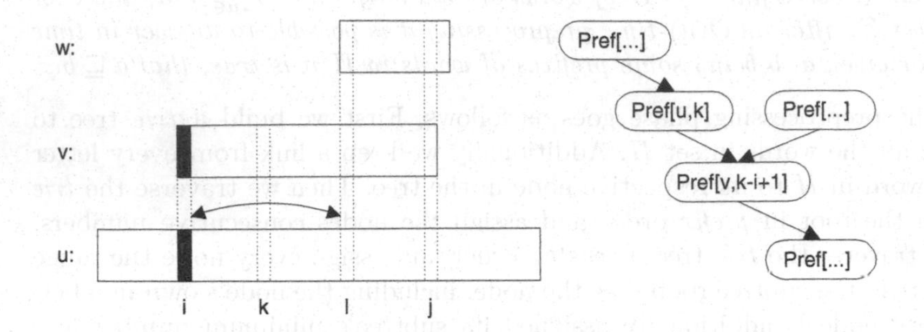

Fig. 2. On the left: The Aho Corasick algorithm backtracked at position $j+1$ from a v to w. On the right: the dependency graph.

state w, see Fig. 2. It follows from the properties of Aho-Corasick automaton, that $Pref[u,i] = (v, j-i+1)$. Now we have to fill the missing values for $Pref[u,i+1] \dots Pref[u,l-1]$.

We start with an observation, that for any $k \in (i,l)$ we have

$$Pref[u,k] = \min\{Pref[v,k-i+1], j-k\}. \tag{1}$$

Since naive recurrent computation could yield multiple calculation of certain prefix table values, we resolve the problem dynamically. Instead of filling those positions we say that $Pref[u,k]$ *depends* on $Pref[v,k-i+1]$ and build *dependency* graph, see Fig. 2. The dependency graph is a forest of trees since the outdegree of each node in the graph is at most one and there are no cycles in the graph. Indeed, every edge leads to a pair with a smaller second coordinate. Using the dependency graph we compute the values in the trees in a root-to-leaves order using equality (1). It is possible, since every tree mustbe rooted at a node that has already been calculated. Therefore the total computation takes time $O(n)$. □

Theorem 4. *Given an equation $E : L = R$ over an alphabet Σ, after an $O(|E|)$-time preprocessing, it is possible to answer in time $O(\#_x)$ if given candidate $\widehat{X} \sqsubseteq A_0$ is a solution of E.*

Proof. Let Π be the set of *equation words*, that is $\Pi = \{A_0, \dots, A_s, B_1, \dots, B_s\}$. First we follow the proof of Lemma 2 and build a *trie* tree to represent the equation words Π. Next we build the *prefix* table for the equation words Π. Since $\widehat{X} \sqsubset A_0$, $\widehat{X}$ is represented both in the *trie* tree and the *prefix* table. It follows from Lemmas 2 and 3 that the preprocessing can be done in time $O(|E|)$. Let homomorphism φ be the canonical extension to domain $\Sigma \cup \{x\}$ of a mapping $x \to \widehat{X}$. We slice both $\varphi(L)$ and $\varphi(R)$ at the positions where any of the words $\Pi \cup \{\widehat{X}\}$ begin/end, see Fig. 3. Thus we get $\Theta(\#_x)$ pairs of slices. We

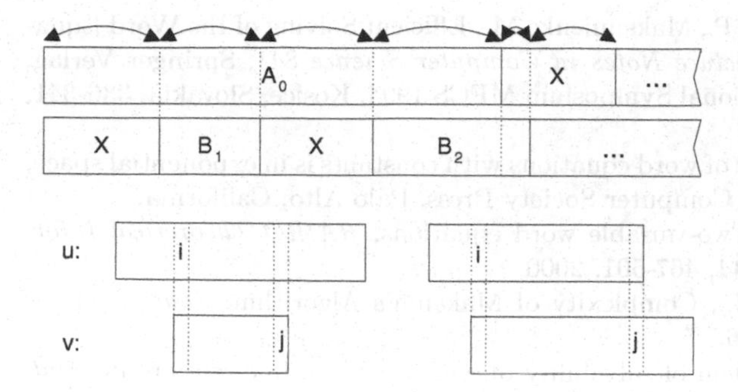

Fig. 3. Top: Slices correspond to arches. Down: Two possible situations of occuring a slice between positions i in u and j in v. In the first one $j = |v|$

shall compare the respective slices iteratively from left to right. In any case if we get a mismatch, we end with a "not a solution" answer. It suffices to prove a single comparison takes $O(1)$ steps. It is easy to note that the equation words the slices origin from can be in at most two distinct configurations, see Fig. 3. Let u be the equation word that does not start in given slice or any of the words otherwise. Let v be the other equation word. We use the prefix table for Π and find $Pref[u, i] = (w, k)$. It follows from Lemma 3 that it can be done in $O(1)$ time. Now we check if $v[1..j] \sqsubseteq w[1..k]$. It follows from Lemma 2 that it can be done in $O(1)$ time. If the comparison results true, then we cancel the slice and iterate the algorithm. Since every iteration takes $O(1)$ time and the number of iterations is equal to $O(\#_x)$, verification time for one solution candidate is $O(\#_x)$. This completes the proof. □

As an immediate consequence of Theorem 4 we obtain

Theorem 5. *Let Σ be of constant size or $\Sigma = 1..|\Sigma|$. There is an algorithm finding all solutions of a word equation with one variable which works in $O(|E| + \#_x \log |E|)$ time.*

References

1. Angluin D., Finding pattern common to a set of string, *in Proc. STOC'79*, 130-141, 1979.
2. Charatonik W., Pacholski L., Word equations in two variables, *Proc. IWWERT'91*, LNCS 677, 43-57, 1991.
3. Crochemore M., Rytter W., Text Algorithms, Oxford University Press, 1994.
4. Crochemore M., Rytter W., Periodic suffixes of strings, *in:* Acts of Sequences, 1991.
5. Diekert V., Makanin's algorithm, a chapter in a book on combinatorics of words, 1998, personal comunication (Volker.Diekert@informatik.uni-stuttgart.de).
6. Jaffar J., Minimal and complete word unification, *Journal of the ACM* **37**(1), 47-85, 1990.

7. Eyono Obono S., Goralcik P., Maksimienko M., Efficient Solving of the Word Equations in One Variable, *Lecture Notes in Computer Science* 841, Springer Verlag, Proc. of the 19th International Symposium MFCS 1994, Kosice, Slovakia, 336-341, Springer, 1994.
8. Gutierrez C., Satisfiability of word equations with constants is in exponential space, in: *Proc. FOCS'98*, IEEE Computer Society Press, Palo Alto, California.
9. Ilie L., Plandowski W., Two-variable word equations, *RAIRO Theoretical Informatics and Applications* **34**, 467-501, 2000.
10. Koscielski A., Pacholski L., Complexity of Makanin's Algorithm, *Journal of the ACM* **43**(4), 670-684, 1996.
11. Makanin G. S., The problem of solvability of equations in a free semigroup, *Mat. Sb.*, **103**(2), 147-236. In Russian; English translation in: *Math. USSR Sbornik*, **32**, 129-198, 1977.
12. Neraud J., Equations in words: an algorithmic contribution, *Bull. Belg. Math. Soc.*, **1**, 253-283, 1994.
13. Plandowski W., Rytter W., Application of Lempel-Ziv encodings to the solution of word equations, in: *Proc. ICALP'98*, LNCS 1443, 731-742, 1998.
14. Plandowski W., Satisfiability of word equations with constants is in NEXPTIME, *Proc. STOC'99*, 721-725, ACM Press, 1999.
15. Plandowski W., Satisfiability of word equations with constants is in PSPACE, *Proc. FOCS'99*, 495-500, IEEE Computer Society Press, 1999.

Autoreducibility of Random Sets: A Sharp Bound on the Density of Guessed Bits

Todd Ebert[1] and Wolfgang Merkle[2]

[1] California State University, Long Beach,
Department of Computer Engineering and Computer Science,
1250 Bellflower Blvd., Long Beach, Ca 90840, U.S.A.,
ebert@cecs.csulb.edu

[2] Ruprecht-Karls-Universität Heidelberg,
Mathematisches Institut,
Im Neuenheimer Feld 294, D–69120 Heidelberg, Germany,
merkle@math.uni-heidelberg.de

Abstract. A set $A \subseteq \{0,1\}^*$ is called i.o. Turing-autoreducible if A is reducible to itself via an oracle Turing machine that never queries its oracle at the current input, outputs either $A(x)$ or a don't-know symbol on any given input x, and outputs $A(x)$ for infinitely many x. If in addition the oracle Turing machine terminates on all inputs and oracles, A is called i.o. truth-table autoreducible. Ebert and Vollmer obtained the somewhat counterintuitive result that every Martin-Löf random set is i.o. truth-table-autoreducible and investigated the question of how dense the set of guessed bits can be when i.o. autoreducing a random set. We show that rec-random sets are never i.o. truth-table-autoreducible such that the set of guessed bits has strictly positive constant density in the limit, and that a similar assertion holds for Martin-Löf random sets and i.o. Turing-autoreducibility. On the other hand, our main result asserts that for any computable function r that goes non-ascendingly to zero, any rec-random set is i.o. truth-table-autoreducible such that the set of guessed bits has density bounded from below by $r(m)$.

1 Introduction

For the moment, call a set A *autoreducible* if there is an effective procedure that on input x computes $A(x)$ while having access to the values $A(y)$ for $y \neq x$. Intuitively speaking, for an autoreducible set the information on $A(x)$ can be effectively recovered from the remainder of the set. For example, for any set X the set $\{x0 : x \in X\} \cup \{x1 : x \in X\}$ is autoreducible. Autoreducibility was first studied by Trakhtenbrot [20], see Wagner and Wechsung [22, pp. 483ff]. Since then, the concept of autoreducibility has received some attention in the literature. For example, Buhrman et al. [5] emphasize the role of time-bounded versions of autoreducibility as a tool for proving separation results in complexity theory.

Now consider the question of whether an effective random set, say, a Martin-Löf random set, can be autoreducible. By definition, the bits of an autoreducible

K. Diks et al. (Eds): MFSC 2002, LNCS 2420, pp. 221–233, 2002.

set depend on each other in an effective way. This suggests that by exploiting the dependencies, we might come up with an effective betting strategy that succeeds on this set. Indeed, Martin-Löf-random sets are never autoreducible and, similarly, rec-random sets are not autoreducible by reductions that are confined to non-adaptive queries. The two latter assertions can be obtained as corollaries to Theorem 7 below.

Furthermore, we might ask whether for a random set R it is at least possible to recover some of the values $R(x)$ from the values $R(y)$ with $y \neq x$. Trivially, this is possible for finitely many places x, so consider the case of infinitely many x. For the moment, call a set A *i.o. autoreducible* if there is an effective procedure that for infinitely many x computes $A(x)$ while having access to the values $A(y)$ for $y \neq x$, whereas for all other inputs the procedure signals "don't-know". For example, any set that has an infinite computable subset is i.o. autoreducible, and, in particular, any infinite recursively enumerable set is i.o. autoreducible. On the other hand, by a standard diagonalization argument one can easily construct a set that is not i.o. autoreducible.

For an i.o. autoreducible set R there are infinitely many places x where the value of $R(x)$ depends in an effective way on the remainder of the set R. Thus at first glance one may be tempted to assume that effective random sets are not i.o. autoreducible. So the following result of Ebert and Vollmer [7,8] seems counterintuitive. *Every* Martin-Löf random set is i.o. autoreducible. Moreover, this result has a finite version known as the *hat problem* which is described in section 5 and has received substantial attention outside of academia [16].

In addition to the issue of simply i.o. autoreducing a random set, Ebert and Vollmer [8] consider the question of how dense the set of guessed bits can be when i.o. autoreducing a random set. A set is i.o. autoreducible with density $r(m)$ if it is i.o. autoreducible such that for all m, at least a fraction of $r(m)$ of the first m bits of the set is guessed. In Theorem 7 it is shown that rec-random sets are never i.o. truth-table-autoreducible with strictly positive constant density (i.e., with density $r(m) = \varepsilon m$ for some $\varepsilon > 0$), and that a similar assertion holds with respect to Martin-Löf random sets and i.o. Turing-autoreducibility. On the other hand, Theorem 11, our main result, asserts that for any computable function r that goes non-ascendingly to 0, any rec-random set is i.o. truth-table-autoreducible with density $r(m)$. So we obtain essentially matching bounds on the density of guessed bits for i.o. truth-table-autoreductions of rec-random sets. These results improve on the work of Ebert and Vollmer and, in particular, our results close the gap between the previously shown bounds, which amount to constant density (for i.o. truth-table autoreductions of Martin-Löf random sets only) and to density worse than $1/\sqrt{m}$. Furthermore, our results on constant bounds are shown by standard probabilistic arguments, which are simpler than the proofs used previously.

1.1 Notation

We use standard notation which is elaborated further in the references [2,4,11].

We consider words over the binary alphabet $\{0,1\}$, which are ordered by the usual length-lexicographical ordering and the $(i+1)$st word in this ordering is denoted by s_i, hence for example s_0 is the empty word λ. Occasionally, we identify words with natural numbers via the mapping $i \mapsto s_i$.

The terms *set* and *class* refer to sets of words and to sets of sets of words, respectively. For any set A, we write $A(x)$ for the value of the characteristic function of A at x and we identify A with the infinite binary sequence $A(0)A(1)\ldots$.

An *assignment* is a (total) function from some subset of the natural numbers to $\{0,1\}$. An assignment is *finite* iff its domain is finite. An assignment with domain $\{0,\ldots,n-1\}$ is identified in the natural way with a word of length n. The restriction of an assignment β to a set I is denoted by $\beta|I$, thus, in particular, for any set X, the assignment $X|I$ has domain I and agrees there with X.

The class of all sets is referred to as *Cantor space* and is denoted by $\{0,1\}^\infty$. The class of all sets that have a word x as common prefix is called the *cylinder generated by* x and is denoted by $x\{0,1\}^\infty$. For a set W, let $W\{0,1\}^\infty$ be the union of all the cylinders $x\{0,1\}^\infty$ where the word x is in W. We write Prob[.] for probability measures and $\mathbf{E}$[.] for expected values; unless stated otherwise, all probabilities refer to the uniform measure on Cantor space.

2 Random Sets

In this section, we briefly review effective random sets and related concepts. For more details please refer to the references [1,2,11].

Imagine a player that successively places bets on the individual bits of the characteristic sequence of an unknown set A. The betting proceeds in rounds $i = 1, 2, \ldots$. During round i, the player receives as input the length $i-1$ prefix of A and then, first, decides whether to bet on the ith bit being 0 or 1 and, second, determines the stake that shall be bet. The stake might be any fraction between 0 and 1 of the capital accumulated so far, i.e., in particular, the player is not allowed to incur debts. Formally, a player can be identified with a *betting strategy* $b : \{0,1\}^* \rightarrow [-1,1]$ where on input w the absolute value of $b(w)$ is the fraction of the current capital that shall be at stake and the the bet is placed on the next bit being 0 or 1 depending on whether $b(w)$ is negative or non-negative.

The player starts with positive, finite capital. At the end of each round, in case of a correct guess, the capital is increased by that round's stake and, otherwise, is decreased by the same amount. So given a betting strategy b, we can inductively compute the corresponding *payoff function* d_b by applying the equations

$$d_b(w0) = d_b(w) - b(w) \cdot d_b(w), \qquad\qquad d_b(w1) = d_b(w) + b(w) \cdot d_b(w) \;\; .$$

Intuitively speaking, the payoff $d_b(w)$ is the capital the player accumulates till the end of round $|w|$ by betting on a set that has the word w as a prefix.

Definition 1. *A betting strategy* b succeeds *on a set* A *if the corresponding payoff function* d_b *is unbounded on the prefixes of* A, *i.e., if*

$$\limsup_{m \in \omega} d_b(A|\{0,\ldots,m\}) = \infty.$$

A betting strategy is computable if it is confined to rational values and there is a Turing machine that on input w outputs an appropriate finite representation of $b(w)$. In the context of recursion theory, usually computable betting strategies are considered [1,17,19,23], while in connection with complexity classes one considers betting strategies that in addition are computable within appropriate resource-bounds [2,10,11,13].

Definition 2. *A set is* rec-random *if no computable betting strategy succeeds on it.*

Besides rec-random sets, we consider Martin-Löf-random sets [12]. For the sake of brevity, we only state their definition and in passing note that the class of Martin-Löf-random sets is properly contained in the class of rec-random sets, where both classes have uniform measure 1. Please see Li and Vitányi [9] for more details. Let $\mathrm{W}_0, \mathrm{W}_1, \ldots$ be the standard enumeration of the recursively enumerable sets [18].

Definition 3. *A class $\mathcal{N}$ is called a* Martin-Löf null class *iff there exists a computable function $g \colon \mathbb{N} \to \mathbb{N}$ such that for all i*

$$\mathcal{N} \subseteq \mathrm{W}_{g(i)}\{0,1\}^\infty \quad \text{and} \quad \mathrm{Prob}[\mathrm{W}_{g(i)}\{0,1\}^\infty] < \frac{1}{2^i} .$$

A set is Martin-Löf-random *is it is not contained in any Martin-Löf null class.*

We conclude this section by stating a well-known result for the construction of betting strategies, and leave its proof as an exercise (hint: for each round, distribute the capital uniformly with respect to the elements of Θ that are consistent with the information currently known).

Remark 4. Let I be a finite set and let Θ be a non-empty subset of all assignments on I. Then there is a betting strategy b that, by betting on places in I, increases its capital by a factor of $\alpha = 2^{|I|}/|\Theta|$ for all sets B where the restriction of B to I is in Θ.

3 Autoreducibility

In this section, we give formal definitions of the concepts of autoreducibility of Turing and truth-table type. Recall the concept of an *oracle Turing machine* [4], which is a Turing machine that during its computation has access to a set X, the oracle. In case an oracle Turing machine M with oracle X terminates on input x, let $M(X,x)$ denote the computed value; otherwise, $M(X,x)$ is undefined. But rather than use standard oracle machines M whose defined outputs $M(X,x)$ belong to $\{0,1\}$, we also allow the machines to output a special "don't-know-symbol" $\perp$, which has the intended meaning of signaling that the correct value is not known. We emphasize however that the results of this paper still hold when using the standard model, in that rather than output "don't-know", the Turing machine would simply query the oracle about the value of x, and output that value. This of course would require a slightly different definition of i.o. autoreducibility, which was originally used by Ebert [7].

Definition 5. *Let M be an oracle Turing machine (with output in $\{0, 1, \bot\}$) and let $Q(M, A, x)$ be the set of query words occurring during the computation of M on input x and with oracle A. Then M* autoreduces *a set A* on *a set E if and only if, for all inputs x, $x \notin Q(M, A, x)$ and either $A(x) = M(A, x)$ or $M(A, x) = \bot$, with the case $A(x) = M(A, x)$ occurring for all x in E.*

A set A is i.o. Turing autoreducible *iff it is autoreduced by some oracle Turing machine on an infinite set. A set A is* Turing autoreducible *iff it is autoreduced by some oracle Turing machine on the set of all words.*

The definitions of *i.o. truth-table autoreducibility* and *truth-table autoreducibility* are the same, except that M is now required to be total, i.e., for all oracles X and for all inputs x, the computation of $M(X, x)$ eventually terminates. Such machines can be represented by a pair of computable functions $h(x)$ and $g(x)$ where, for input x, $h(x)$ computes the query words and $g(x)$ computes the machine output as a function of how the queries about those words are answered. From here on *T-autoreducibility* and *tt-autoreducibility* stand for Turing and truth-table autoreducibility respectively.

In addition to i.o. autoreducing a random set, Ebert and Vollmer [8] considered the density of the set of guessed bits when i.o. autoreducing a random set.

Definition 6. *For all $m > 0$, the* density $\rho(E, m)$ *of a set E up to m is defined by*

$$\rho(E, m) = \frac{|E \cap \{s_0, \ldots, s_{m-1}\}|}{m}.$$

An oracle Turing machine M i.o. T-autoreduces a set X with density $r(m)$ *if M i.o. T-autoreduces X on a set E such that $\rho(E, m) \geq r(m)$ for all $m > 0$ (i.e., the density of the set of words x such that M guesses $X(x)$ is always at least $r(m)$).*

A set A is called i.o. T-autoreducible with density $r(m)$ *if there is some oracle Turing machine that i.o. T-autoreduces A with density $r(m)$. Concepts like* i.o. tt-autoreducible with density $r(m)$ *are defined in the same manner.*

Relative to the above definition, Ebert and Vollmer [8] proved that no Martin-Löf random set is i.o. tt-autoreducible with strictly positive constant density (i.e., with density $r(m) = \varepsilon m$ for some constant $\varepsilon > 0$), and that every Martin-Löf random set can be i.o. tt-autoreduced with density bounded above by $1/\sqrt{m}$.

We improve on these results in two respects. First, in Theorem 7, we show that no rec-random set is i.o. tt-autoreducible with positive constant density and that Martin-Löf random sets are not i.o. T-autoreducible with such density. Second, in Theorem 11, we close the gap between constant density and density $1/\sqrt{m}$ by showing that any rec-random set can be i.o. tt-autoreduced with density $1/g(m)$ for any unbounded and non-decreasing computable function g. Equivalently, we can achieve density $r(m)$ for any computable rational-valued function r that goes non-ascendingly to 0. Hence, when tt-autoreducing a rec-random set, the fraction of guessed bits cannot be constant in the limit but can be chosen to decrease as slowly as any computable function that goes non-ascendingly to 0.

4 Autoreducing Random Sets: Negative Results on Density

Ebert and Vollmer [8] show that no Martin-Löf-random set is i.o. tt-autoreducible with positive constant density. Theorem 7 extends this statement to the larger class of rec-random sets, while for Martin-Löf-random sets the result is strengthened to i.o. T-autoreducibility. Theorem 7 is demonstrated by probabilistic arguments, which are more simple than previous methods.

Theorem 7. (a) *No rec-random set is i.o. tt-autoreducible with strictly positive constant density.*
(b) *No Martin-Löf-random set is i.o. T-autoreducible with strictly positive constant density.*

Proof. The proofs of (a) and (b) rely on Claims 1 and 2 below. For the sake of brevity, we only demonstrate (a). Assertion (b) then follows by a similar but slightly more involved argument. Moreover, the proof of Claim 1 is left as an easy exercise.

Fix an oracle Turing machine M and a rational $\varepsilon_0 > 0$. We want to show that if M i.o. autoreduces a set with density ε_0, then this set cannot be random, i.e., an appropriate betting strategy succeeds on this set. We argue that for any w, the fraction of sets in the cylinder $w\{0,1\}^\infty$ that are i.o. autoreduced by M with density ε_0 is bounded by a constant $\delta < 1$. Let $\varepsilon = \varepsilon_0/2$ and for any m let $i(m) = \lceil \frac{m}{\varepsilon} \rceil$ and $I(m) = \{s_m, \ldots, s_{m+i(m)-1}\}$; that is, in the case where $m = |w|$, the interval $I(m)$ contains the first $i(m)$ words not in the domain of w. For any set X and any finite set I, let

$$\begin{aligned} \text{correct}(X,I) &= |\{x \in I : M(X,x) = X(x)\}|, \\ \text{incorrect}(X,I) &= |\{x \in I : M(X,x) = 1 - X(x)\}| . \end{aligned}$$

That is, the set of all inputs x in I such that $M(X,x)$ is defined and differs from $\perp$ is partitioned into the inputs of correct and incorrect guesses. In the remainder of this proof and with M and ε understood from the context, we say a set X is consistent with a word w if

(i) X is an extension of w,
(ii) for all x in $I(m)$, $M(X,x)$ exists and is computed without querying the oracle at place x,
(iii) $\text{incorrect}(X, I(m)) = 0$,
(iv) $\text{correct}(X, I(m)) \geq \varepsilon |I(m)|$.

Claim 1. *If M i.o. T-autoreduces a set with density ε_0, then this set is consistent with any prefix of itself.*

For any word w, let the random set X_w be an extension of w where the bits of X_w that are not already determined by w are obtained by independent tosses of a fair coin. Let $\delta = 1/(1+\varepsilon)$ and observe that $\delta < 1$ due to $\varepsilon > 0$.

Claim 2. *For any word w, the probability that X_w is consistent with w is at most δ.*

Proof. Fix w of length m. We claim that

$$\mathbf{E}[\text{correct}(X_w, I(m))] \quad = \quad \mathbf{E}[\text{incorrect}(X_w, I(m))] \,. \tag{1}$$

For proof, observe that the values $X_w(y)$ with y in $I(m)$ are stochastically independent, and hence by (ii) the same holds for $X_w(x)$ and $M(X_w, x)$. Furthermore, since $X_w(x)$ is chosen uniformly from $\{0, 1\}$, it follows that the expected number of correct and incorrect answers at x are equal. Thus, Equation 1 follows by linearity of expectation.

We conclude the proof of Claim 2 by distinguishing two cases.

Case I: $\mathbf{E}[\text{correct}(X_w, I(m))] \leq \delta\varepsilon|I(m)|$.

The random variable $\text{correct}(X_w, I(m))$ is non-negative, hence the case assumption implies that the probability that $\text{correct}(X_w, I(m))$ is at least $\varepsilon|I(m)|$ is at most δ. Also, the probability that X_w is consistent with w is at most δ by condition (iv) in the definition of consistency.

Case II: $\mathbf{E}[\text{correct}(X_w, I(m))] > \delta\varepsilon|I(m)|$.

By (1), we also have $\mathbf{E}[\text{incorrect}(X_w, I(m))] > \delta\varepsilon|I(m)|$. The latter implies that the probability that $\text{incorrect}(X_w, I(m))$ is strictly larger than 0 is at least $\delta\varepsilon$ since by definition, the random variable $\text{incorrect}(X_w, I(m))$ is bounded by $|I(m)|$. By condition (iii), the probability that X_w is consistent is then at most

$$1 - \delta\varepsilon = 1 - \frac{\varepsilon}{1+\varepsilon} = \frac{1}{1+\varepsilon} = \delta \,. \qquad \square$$

Proof of (a). We can assume that M is in fact a tt-reduction, say, $M = (g, h)$. Fix A such that M i.o. tt-autoreduces A with density $r(m) = \varepsilon_0$. It suffices to show that A is not rec-random, and this is done by constructing a computable betting strategy that succeeds on A.

Partition the set of words into consecutive intervals $J_0, J_1, \ldots$ where the cardinality of J_i is denoted by l_i. The J_i are defined via specifying the l_i inductively. For all i, let $s_i = l_0 + l_1 + \ldots + l_i$. Let $l_0 = 1$ and for all $i \geq 0$ choose l_{i+1} so large that J_{i+1} contains $I(s_i)$ as well as the query sets $h(x)$ for all x in $I(s_i)$. This way for any word w of length s_i, the consistency with w of any set X that extends w depends only on the restriction of X to the interval J_{i+1}. Call an assignment on J_{i+1} consistent with such a word w if the assignment is the restriction of a set that is consistent with w. Then for any given word w of length s_i, the following holds with respect to the assignments on J_{i+1}: it can be decided effectively whether a given assignment is consistent with w; if w is a prefix of A, then the assignment obtained by restricting A is consistent; the consistent assignments comprise a fraction of at most δ of all assignments.

These assertions hold by definition of consistency and by Claims 1 and 2, respectively. Now consider the betting strategy that for each interval J_i, uses the capital accumulated up to the first element of the interval in order to bet according to Remark 4 against all assignments on this interval that are not consistent with the already seen length s_i prefix of the unknown set. If the unknown set is indeed A, then by the preceding discussion on each interval, b increases its capital at least by the constant factor $1/\delta > 1$. Furthermore, the betting strategy b is computable since consistency can be decided effectively. □

As a corollary to Theorem 7, we have that no rec-random set is truth-table autoreducible, and that no Martin-Löf random set is Turing autoreducible. These corollaries are complemented by the following results of Merkle and Mihailović [14]. There are rec-random sets that are weak truth-table autoreducibility. There are Martin-Löf random sets that are selfreducible with respect to the reducibility *being recursively enumerable in*, where selfreducible means autoreducible while asking only queries strictly smaller than the current input.

A set is p-random if it cannot be succeeded upon by a betting strategy that is computable in polynomial time [2,11]. Apparently, the arguments used in this section do not carry over to show that p-random sets are not T- or tt-autoreducible via oracle machines that run in polynomial time. The latter statements have been considered by Buhrman et al. [6]. They argue that these questions are closely related to major open questions in complexity theory.

5 Autoreducing Random Sets: Positive Results on Density

Ebert and Vollmer [8] prove that any Martin-Löf random sequence is i.o. tt-autoreducible and that in fact a certain density can be achieved. These results rely on the use of appropriate error-correcting codes, and our main result Theorem 11 is shown by similar methods. First we review perfect one-error-correcting codes and their relation to i.o. autoreducibility.

Definition 8. *Any subset of* $\{0,1\}^n$ *is called a* code *of (*codeword) length n. *The* unit sphere *around a word* w *is the set of all words in* $\{0,1\}^n$ *that differ from* w *by at most one position. A code* C *of length* n *is called* perfect one-error-correcting *if the set* $\{0,1\}^n$ *is the disjoint union of the unit spheres around the words in* C.

For any perfect one-error-correcting code of length n, it is necessary that the unit-sphere volume of $(n+1)$ divides 2^n. This condition is also sufficient [21].

Fact 9. Let $n = 2^k - 1$ for integer $k > 0$. Then there is a perfect one-error-correcting code C_n of length n. Furthermore, the codes C_n can be chosen so that $\bigcup_{\{n:n=2^k-1\}} C_n$ is decidable in polynomial time. For example, such codes are given by the well-known family of binary Hamming codes [21].

The hat problem and its solution [7,8,16] (see also Aspnes et al. [3] for related problems) described in Remark 10 show neatly how error-correcting codes can be used when autoreducing random sets.

Remark 10. In the hat problem with n players, a binary sequence of n bits, the true word, is chosen by independent tosses of a fair coin. Player i is assigned bit i (or, equivalently, is assigned one of two possible hat colors according to bit i). Afterwards, each player may cast a guess on its own bit under the following conditions. The players know the bits of all other player but not their own bit. Once the game begins, the players are neither allowed to communicate nor do they know whether the other players have already guessed. The team wins if and only if there is no incorrect guess and at least one correct guess.

We now provide a strategy for the hat problem with n players where n has the form $2^k - 1$. Fix a perfect one-error-correcting code C_n of length n and call the unit spheres with center in C_n the designated spheres. Each player behaves according to the following rule, where the consistent words of player i are the true word and the unique word that differs from the true word at position i. In case both consistent words are in the same designated sphere, guess according to the assumption that the true word is the (unique) consistent word that differs from the center of this sphere. Otherwise, do not guess.

Under this strategy, all players guess incorrectly and the team looses in case the true word is the center of a designated sphere. Otherwise, i.e., if the true word differs from the center of some (uniquely determined) sphere at position i, the team wins since player i guesses correctly while no other players guess. Hence the team wins with probability $1 - 1/(n+1)$, the fraction of non-center words.

The solution of the hat problem described in Remark 10 is used in the proof of Theorem 11.

Theorem 11. *Let $g : \mathbb{N} \to \mathbb{N}$ be any computable function that is unbounded and non-decreasing. Then every rec-random set is i.o. tt-autoreducible with density $r(m) = 1/g(m)$.*

An equivalent formulation of Theorem 11 is the following. For any computable, rational-valued function r that goes non-ascendingly to 0, every rec-random set is i.o. tt-autoreducible with density $r(m)$. This formulation shows that the bound given by Theorem 11 essentially matches the negative result on constant bounds in Theorem 7.

Proof. Fix any rec-random set R. We have to show that R is i.o. tt-autoreducible with density $r(m)$ by some oracle Turing machine M. For every $k > 0$, let $l_k = 2^k - 1$. For further use, fix perfect one-error-correcting codes C_k of length l_k such that given x and k, we can decide effectively whether x is in $\bigcup \mathrm{C}_k$. The function g is unbounded, thus we can define a computable sequence $t_0 < t_1 < \dots$ where $g(t_k) \geq l_{k+1}$ and hence $r(t_k) \leq \frac{1}{l_{k+1}}$. For every $k > 0$, partition the set of words into consecutive intervals $J_k^1, J_k^2, \dots$ of length l_k. Let c_k be minimum such that the t_kth word s_{t_k} is in $J_k^{c_k}$ and let $H_k = J_k^2 \cup \dots \cup J_k^{c_k}$.

We construct M from oracle machines $M_1, M_2, \dots$, to which we refer as modules. Intuitively, module k is meant to ensure density $1/l_k$ in the interval between s_0 and s_{t_k}. Before defining the modules, we describe their properties and how they are combined to form M.

Module k never queries any place outside the set $J_k^1 \cup H_k$ and outputs $\bot$ on all inputs that are not in H_k. Furthermore, on oracle R, module k never makes a wrong guess and guesses exactly one bit in each of the intervals J_k^2 through $J_k^{c_k}$. In addition, we will ensure that the M_k are uniformly effective in the sense that there is an oracle machine M_0 such that the values $M_k(X, x)$ and $M_0(X, \langle x, k\rangle)$ either are both undefined or are both defined and equal (where $\langle .,. \rangle$ denotes the usual pairing function [18]).

Then M is obtained by running the modules in parallel as follows. For input x, if x is equal to s_0 or s_1, then M outputs $R(x)$. Otherwise, M determines the finitely many k such that $x \in H_k$ and simulates the corresponding modules with input x and the current oracle. If any of these modules outputs a value different from $\bot$, then M outputs the value output by the least such module; otherwise, M outputs $\bot$. From the properties stated so far, we can already prove that M works as required.

Claim 1. *The oracle Turing machine M i.o tt-autoreduces the set R with density $r(m)$.*

Proof. From the module assumptions, it is immediate that M is computable, queries its oracle non-adaptively, and never queries the oracle at the current input. Moreover, on oracle R, the modules always guess correctly, hence so does M.

Let E be the set of inputs x such that $M(R, x)$ differs from $\bot$. Fix k and assume that m is in J_k^1 through $J_k^{c_k}$. Then we have $\rho(E, m) \geq 1/l_k$. For $k = 1$, this follows because the intervals have length $l_1 = 1$ and thus, by assumption, module 1 and hence M guess all bits in these intervals. For $k > 1$, M guesses the first two bits of the first interval, while module k and hence M guess at least one bit in the remaining intervals, where the intervals have length l_k. So if m is in interval J_k^j, up to m there are at most jl_k words and at least j guesses, so the density up to m is at least $1/l_k$. Hence $\rho(E, m) \geq 1/l_k$.

In order to prove $\rho(E, m) \geq r(m)$, fix any m and choose k such that $t_{k-1} \leq m < t_k$. Then we have

$$\rho(E, m) \geq \frac{1}{l_k} \geq r(t_{k-1}) \geq r(m) , \tag{2}$$

where the inequalities follow because m is in J_k^1 through $J_k^{c_k}$, by choice of t_{k-1}, and since r is non-ascending. □

In order to construct modules that have the required properties, let w_k^j be the restriction of R to J_k^j, i.e., $R = w_k^1 w_k^2 \ldots$, where $|w_k^j| = l_k$, and let $w_k = w_k^1 \oplus \ldots \oplus w_k^{c_k}$, where $\oplus$ is bit-wise exclusive-or. The idea underlying the construction is as follows. The code words in C_k comprise such a small fraction of all words of length l_k that in case infinitely many words w_k were in C_k, there would be a computable betting strategy that succeeds on R. But R is assumed to be rec-random, hence w_k is not in C_k for almost all k. So in order to guess bits of w_k and then also of w_k^1 through $w_k^{c_k}$, we can apply the solution to the hat problem described in Remark 10.

Claim 2. *For almost all k, w_k is not in C_k.*

Proof. We bet on the bits of an unknown set X. Similar to the definition of w_k, let v_k^j be the restriction of X to the interval J_k^j and let $v_k = v_k^1 \oplus \ldots \oplus v_k^{c_k}$. Recall that C_k contains a fraction of $a_k = 1/2^k$ of all length l_k words and observe that the mapping $o_k : u \mapsto v_k^1 \oplus \ldots \oplus v_k^{c_k-1} \oplus u$ is a bijection. Thus o_k maps a fraction of a_k of all length l_k words to C_k. When betting on the places in $J_k^{c_k}$, we have already seen v_k^1 through $v_k^{c_k-1}$. Under the assumption that v_k is in C_k, we can exclude all but a fraction of a_k of the possible assignments to $J_k^{c_k}$ and hence, by betting in favor of these assignments we can increase our capital by a factor of $1/a_k$ in case the assumption is true.

Now consider the following betting strategy. For every k, a portion of a_k of the initial capital 1 is exclusively used for bets on the interval $J_k^{c_k}$. On each such interval, the bets are in favor of the a_k-fraction of assignments that o_k maps to C_k. Then the capital a_k increases to 1 for all k such that v_k is in C_k and consequently the strategy succeeds on any set such that the latter occurs for infinitely many k. But no computable betting strategy can succeed on the rec-random sequence R, hence Claim 2 follows. We leave it to the reader to show that this strategy is indeed computable. Observe in this connection that each word is contained in at most finitely many intervals of the form $J_k^{c_k}$ and consequently at most finitely many of the strategies related to these intervals might act in parallel. □

Now construct the modules. By Claim 1, fix k_0 such that w_k is not in C_k for all $k > k_0$. First assume $k \leq k_0$. Consider the least elements of the intervals J_k^2 through $J_k^{c_k}$ and for all these x, hard-wire $R(x)$ into module k. On all these x, module k outputs $R(x)$, while the module outputs $\perp$ for all other inputs.

Next assume $k > k_0$. On inputs that are not in H_k, module k simply outputs $\perp$. Now consider an input x in H_k and suppose that x is the i_0th element of interval $J_k^{j_0}$, where then $2 \leq j_0 \leq c_k$. For the given oracle X, define v_k and the v_k^j as in the proof of Claim 2, i.e., v_k^j is the restriction of X to J_k^j and v_k is the bit-wise exclusive-or of the v_k^j. Then on input x, module k queries its oracle at the remaining words in $J_k^1 \cup H_k$, i.e., the module obtains all bits of the words $v_k^j = v_k^j(1)v_k^j(2)\ldots v_k^j(l_k)$, $j = 1, \ldots, c_k$, except for the bit $X(x) = v_k^{j_0}(i_0)$. In order to guess this bit, the module tries to guess bit $v_k(i_0)$. From the latter and from the already known bits $v_k^j(i_0)$ for $j \neq j_0$, the bit $v_k^{j_0}(i_0)$ can be computed easily since $v_k(i_0)$ is the exclusive-or of the $v_k^j(i_0)$.

In order to guess $v_k(i_0)$, module k mimics the solution of the hat problem with l_k players that is given in Remark 10. More precisely, the module computes the remaining bits of v_k from the $v_k^j(i)$ and obtains two consistent words u_0 and u_1. In case for some r, the word u_r is in C_k, the module guesses $u_{1-r}(i_0)$ while the module abstains from guessing, otherwise. By assumption on k, on oracle R the word v_k is not in C_k, hence the discussion in Remark 10 shows that module k never guesses incorrectly and guesses exactly one bit in each of the intervals J_k^2 through $J_k^{c_k}$ (in fact, for some i, in each interval bit i is guessed). □

Acknowledgments

We acknowledge helpful hints from Klaus Ambos-Spies and Heribert Vollmer.

References

1. K. Ambos-Spies and A. Kučera. Randomness in computability theory. In P. A. Cholak et al. (eds.), *Computability theory and its applications. Current trends and open problems.* Proceedings of a 1999 AMS-IMS-SIAM joint summer research conference, Boulder, USA, AMS Contemporary Mathematics 257:1–14, 2000.
2. K. Ambos-Spies and E. Mayordomo. Resource-bounded measure and randomness. In A. Sorbi (ed.), *Complexity, logic, and recursion theory*, p. 1-47. Dekker, New York, 1997.
3. J. Aspnes, R. Beigel, M. Furst, and S. Rudich. *The expressive power of voting polynomials*, Combinatorica 14(2), p. 135-148, 1994.
4. J. L. Balcázar, J. Díaz, and J. Gabarró. *Structural Complexity*, volumes I and II. Springer, 1995 and 1990.
5. H. Buhrman, L. Fortnow, D. van Melkebeek, and L. Torenvliet, *Separating complexity classes using autoreducibility.* SIAM Journal on Computing 29:1497-1520, 2000.
6. H. Buhrman, D. van Melkebeek, K. W. Regan, D. Sivakumar, M. Strauss, *A generalization of resource-bounded measure, with application to the BPP vs. EXP problem.* SIAM Journal on Computing 30(2):576-601, 2000.
7. T. Ebert. *Applications of Recursive Operators to Randomness and Complexity.* Ph.D. Thesis, University of California at Santa Barbara, 1998.
8. T. Ebert, H. Vollmer. *On the Autoreducibility of Random Sequences.* In: M. Nielsen and B. Rovan (eds.), Mathematical Foundations of Computer Science 2000, Lecture Notes in Computer Science 1893: 333-342, Springer, 2000.
9. M. Li and P. Vitányi. *An Introduction to Kolmogorov Complexity and Its Applications*, second edition. Springer, 1997.
10. J. H. Lutz. Almost everywhere high nonuniform complexity. *Journal of Computer and System Sciences*, 44:220–258, 1992.
11. J. H. Lutz. The quantitative structure of exponential time. In L. A. Hemaspaandra and A. L. Selman (eds.), *Complexity Theory Retrospective* II, p. 225–260, Springer, 1997.
12. P. Martin-Löf. The definition of random sequences. *Inform. and Control* 9(6):602–619, 1966.
13. E. Mayordomo. *Contributions to the Study of Resource-Bounded Measure.* Doctoral dissertation, Universitat Politècnica de Catalunya, Barcelona, Spain, 1994.
14. W. Merkle and N. Mihailović. *On the construction of effective random sets.* In: Mathematical Foundations of Computer Science 2002, this volume.
15. P. Odifreddi. *Classical Recursion Theory.* North-Holland, Amsterdam, 1989.
16. S. Robinson. Why mathematicians now care about their hat color. *NY Times*, April 10, 2001.
17. C.-P. Schnorr. A unified approach to the definition of random sequences. *Mathematical Systems Theory*, 5:246–258, 1971.
18. R. I. Soare. *Recursively Enumerable Sets and Degrees.* Springer, 1987.
19. S. A. Terwijn. *Computability and Measure.* Doctoral dissertation, Universiteit van Amsterdam, Amsterdam, Netherlands, 1998.

20. B. A. Trakhtenbrot, *On Autoreducibility.* Soviet Math. Doklady, 11:814-817, 1970.
21. J.H. van Lint. *Introduction to coding theory*, third edition. Springer, 1999.
22. K. Wagner and G. Wechsung. *Computational Complexity.* Deutscher Verlag der Wissenschaften, Berlin, 1986.
23. Y. Wang. *Randomness and Complexity.* Doctoral dissertation, Universität Heidelberg, Mathematische Fakultät, INF 288, Heidelberg, Germany, 1996.

Two-Way Finite State Transducers with Nested Pebbles

Joost Engelfriet and Sebastian Maneth

LIACS, Leiden University, P.O.Box 9512, 2300 RA Leiden, The Netherlands,
{engelfri,maneth}@liacs.nl

Abstract. Two-way finite state transducers are considered with a fixed number of pebbles, of which the life times are nested. In the deterministic case, the transductions computed by such pebble transducers are closed under composition, and they can be realized by the composition of one-pebble transducers. The ranges of the k-pebble transducers form a hierarchy with respect to k, their finiteness problem is decidable, and they can be generated by compositions of k macro tree transducers. Related results hold in the nondeterministic case.

1 Introduction

Two-way finite automata with a pebble recognize regular languages only [3]. With more than one pebble they are equivalent to the multihead finite automata, which recognize all logarithmic space languages, see, e.g., [17,24,26]. Recently, it has been shown that restricting the pebbles to have nested life times, the two-way pebble automata still recognize regular languages only [15] and this also holds when generalized to trees [8]. Even more recently, a corresponding tree transducer model was proposed as a general model for XML-based query languages: the so-called pebble tree transducer [20,28] (related models are considered in, e.g., [1,21,22]). Here we study some theoretical properties of the pebble tree transducer of [20,28] restricted to monadic trees, i.e., to strings: the two-way finite state transducer with nested pebbles (pebble transducer, for short).

For a string w, $|w|$ denotes its length, and $w(i)$ denotes its ith element. The empty string is denoted λ. For binary relations R and R', $R \circ R' = \{(a,c) \mid \exists b : (a,b) \in R, (b,c) \in R'\}$. For classes $\mathcal{R}$ and $\mathcal{R}'$ of binary relations, $\mathcal{R} \circ \mathcal{R}' = \{R \circ R' \mid R \in \mathcal{R}, R' \in \mathcal{R}'\}$, and $\mathcal{R}^{k+1} = \mathcal{R}^k \circ \mathcal{R}$.

A two-way finite state transducer (also called two-way generalized sequential machine, or 2gsm) is a finite state automaton with a two-way input tape, surrounded by the endmarkers $\triangleleft$ and $\triangleright$, and a one-way output tape. A *k-pebble transducer* is a two-way finite state transducer that additionally carries k pebbles, each with a unique name, say, $1, \ldots, k$. Initially there are no pebbles on the input tape, but during its computation the transducer can drop pebbles on the squares of the input tape, lift them, drop them again, etc. In one step of its computation, the transducer can determine which pebbles are lying on the current square of the input tape, lift one of these pebbles, or drop a new pebble.

K. Diks et al. (Eds): MFSC 2002, LNCS 2420, pp. 234–244, 2002.

However, the life times of the pebbles should be nested. This means that at each moment of time, pebbles 1 to ℓ (for some $0 \leq \ell \leq k$) are on the input tape, and at such a moment the only pebble that can be dropped on the current square is pebble $\ell + 1$, whereas the only pebble that can be lifted from the current square is pebble ℓ. In the case of one pebble this is, of course, no restriction.

Formally, a k-pebble transducer is a system $M = (\Sigma, \Delta, Q, q_0, F, \delta)$ where Σ and Δ are the input and output alphabet, respectively, Q is the finite set of states, $q_0 \in Q$ is the initial state, $F \subseteq Q$ is the set of final states, and δ is the finite set of transitions. Each transition is of the form $(q, \sigma, b) \mapsto (q', \varphi, w)$, with $q, q' \in Q$, $\sigma \in \Sigma \cup \{\lhd, \rhd\}$, $b \in \{0,1\}^k$, $\varphi \in \{\text{left}, \text{right}, \text{drop}, \text{lift}\}$, and $w \in \Delta^*$. Intuitively, such a transition means that if M is in state q, σ is the symbol on the current square, and, for every $1 \leq m \leq k$, pebble m is on the current square iff $b(m) = 1$, then M can go into state q', output the string w, and execute the instruction φ, i.e., move the reading head one square to the left or right, drop pebble $\ell + 1$ (the "next" pebble), or lift pebble ℓ (the "last" pebble); note that these instructions are undefined if, respectively, $\sigma = \lhd$, $\sigma = \rhd$, all k pebbles are on the input tape, or pebble ℓ is not on the current square.

For a given input string $u \in \Sigma^*$, the input tape of M contains the string $\lhd u \rhd$. The squares of this tape are numbered $0, 1, \ldots, |u|, |u| + 1$. Accordingly, a *configuration* on u is a triple (q, i, π) with $q \in Q$, $i \in \{0, \ldots, |u|+1\}$ representing the position of the reading head, and $\pi \in \{\#, 0, \ldots, |u| + 1\}^k$, where $\pi(m) = i$ means that pebble m is on square i, and $\pi(m) = \#$ that it is not on the input tape. The one step computation relation $\vdash_u$ is defined on 4-tuples (q, i, π, v) where (q, i, π) is a configuration and $v \in \Delta^*$ is the content of the output tape, in the obvious way. The *transduction computed by* M, denoted τ_M, is $\{(u, v) \in \Sigma^* \times \Delta^* \mid (q_0, 0, \#^k, \lambda) \vdash_u^* (q, i, \pi, v)$ for some configuration (q, i, π) with $q \in F\}$.

A k-pebble transducer M is *deterministic* if it does not have two transitions with the same left-hand side. In that case, τ_M is a partial function from Σ^* to Δ^*. The class of transductions computed by k-pebble transducers is denoted by PT_k, and by DPT_k for the deterministic transducers. The union of these classes over all $k \in \mathbb{N}$ is denoted PT and DPT, respectively.

In database terminology (cf. [20,28]), the formalism of deterministic pebble transducers can be viewed as a query implementation language. From this point of view, each (deterministic) pebble transducer M computes a query, viz. the query τ_M. An input string u is the current contents of a database (semi-structured as a string), and the output string v, with $(u, v) \in \tau_M$, is the result of the query. As an easy example of such a query, let R be some regular expression; the query asks for the sequence of all substrings s of u that satisfy the regular expression, i.e., with $s \in L(R)$, the regular language corresponding to R. This query can be computed by a 2-pebble transducer M that systematically investigates all substrings s of u by dropping pebble 1 on the first symbol of s and pebble 2 on its last symbol, and then simulating a finite automaton for $L(R)$ on s. If s is in $L(R)$, M outputs $\$s$ (where \$ is a marker). Note that the life times of the pebbles are nested: pebble 1 is moved one square to the right when pebble 2 is lifted after having visited all squares to the right of pebble 1.

For a query language it is desirable that it is closed under composition. In fact, a query is often based on a user view of the database, which itself can be considered as a query. Closure under composition then guarantees that such a query can be transformed into a direct query, which can then be implemented in the usual way. In the next section we show that the query language of pebble transducers is indeed closed under composition. We also show that each query can be decomposed into 1-pebble queries, which means that the user who understands views and 1-pebble queries, need not be bothered with the more complex k-pebble queries. In Section 3 we prove that (even for nondeterministic queries) $k+1$ pebbles are more powerful than k pebbles, in the sense that the $(k+1)$-pebble transducer can define more queries than the k-pebble transducer.

Finally, in Section 4, we investigate the views corresponding to queries defined by deterministic pebble transducers. A deterministic k-pebble transducer M defines the view $\{\tau_M(u) \mid u \in \Sigma^*\}$: the set of all possible results of the query τ_M, i.e., the range of its transduction. An important issue in XML-based query languages is type checking of views [20,23,28]: given a view V and a type T, decide whether $V \subseteq T$ (or, find such a T: type inference). Here, a type is usually a regular tree language (as defined by, e.g., a DTD). The main result of [20] is that type checking is decidable for pebble tree transducers. We show that, for a view V of a pebble transducer, it is decidable whether $V - T$ is finite, i.e., whether all documents in V have type T with the exception of finitely many ("almost always" type checking). This is based on the result that each such view can be generated by a composition of macro tree transducers, a tree transformation system known from tree language theory [12,13]. Using this same result we show that with $k+1$ pebbles more views can be defined than with k pebbles, thus strengthening the result for queries in Section 3 (in the deterministic case).

2 Decomposition and Composition

In this section we show that DPT is closed under composition, and that, for every k, $\mathrm{DPT}_k \subseteq \mathrm{DPT}_1^k$, i.e., every deterministic k-pebble transduction can be decomposed into k deterministic 1-pebble transductions. This implies that $\mathrm{DPT} = \mathrm{DPT}_1^*$, i.e., the pebble queries are the compositions of the 1-pebble queries. For the nondeterministic transductions we show that $\mathrm{DPT} \circ \mathrm{PT} \subseteq \mathrm{PT}$, and that every nondeterministic k-pebble transduction can be decomposed into k deterministic 1-pebble transductions and one nondeterministic 0-pebble transduction. This implies that $\mathrm{PT} = \mathrm{DPT} \circ \mathrm{PT}_0$. Note that $\mathrm{PT}_0 = \mathrm{2NGSM}$, the class of nondeterministic two-way finite state transductions. We start with our decomposition result, which follows by induction from the next lemma.

Lemma 1. *For every $(k+1)$-pebble transducer M, a deterministic 1-pebble transducer M_1 and a k-pebble transducer M' can be constructed such that $\tau_M = \tau_{M_1} \circ \tau_{M'}$. If M is deterministic, then M' is also deterministic.*

Proof. Let b and e be two new symbols. The deterministic 1-pebble transducer M_1 translates each input string $u = a_1 a_2 \cdots a_n$ of M into the string $\mathrm{peb}(u) =$

$\bar{b}a_1a_2\cdots a_n e\, b\bar{a}_1a_2\cdots a_n e\cdots ba_1a_2\cdots \bar{a}_n e\, ba_1a_2\cdots a_n\bar{e}$, i.e., the concatenation of all copies of bue in which one position is barred (intuitively, this is a possible position of pebble 1 of M). It should be clear how to program M_1: it puts its pebble consecutively on squares 0 to $n+1$ of the input tape $\triangleleft u \triangleright$, and for each such pebble position it copies the input tape to the output, barring the symbol on the pebbled square (and translating $\triangleleft$ and $\triangleright$ into b and e, respectively).

The k-pebble transducer M', on input peb(u), simulates the computation of M on input u. It uses the specific form of peb(u) to simulate pebble 1 of M, and it uses pebble m to simulate pebble $m+1$ of M, for $1 \leq m \leq k$. When M has no pebbles on its input tape, M' simulates M on the string of barred symbols of peb(u): $\bar{b}\bar{a}_1\cdots\bar{a}_n\bar{e}$. In other words, if M moves left, then M' moves to the first barred symbol to its left (and similarly if M moves right); initially, M' moves one square to the right (to the square containing $\bar{b}$). When M drops pebble 1, say on a_i, M' from then on simulates M on the substring $ba_1\cdots\bar{a}_i\cdots a_n e$ of peb(u), interpreting the barred symbol as the position of pebble 1 of M, and using its k pebbles for pebbles 2 to $k+1$ of M. □

Theorem 1. *For every* $k \geq 1$, $\mathrm{DPT}_k \subseteq \mathrm{DPT}_1^k$ *and* $\mathrm{PT}_k \subseteq \mathrm{DPT}_1^k \circ \mathrm{PT}_0$.

This result can be used to give a simple proof of the known fact that the domains of our transducers, i.e., the languages accepted by two-way k-pebble automata (with nested pebbles) are regular. For $k=0$ (no pebbles) this is the classical result of [27,25]. For $k=1$, i.e., for two-way 1-pebble automata, it is the, also classical, result of [3]. For arbitrary k it was first proved (for a slightly restricted case) in [15] and then in [8] for the more general case of trees (see also [20]). The alternative proof is as follows: Since $\mathrm{PT}_k \subseteq \mathrm{PT}_1^{k+1}$, it suffices to show that the regular languages are closed under inverse PT_1 transductions. Let M be a 1-pebble transducer, and R a regular language. By a standard product construction, a 1-pebble transducer M' can be constructed such that $\tau_{M'} = \{(u,v) \in \tau_M \mid v \in R\}$. Hence, the domain of M' is $\tau_M^{-1}(R)$. By the result of [3] this domain is regular. The languages accepted by two-way *alternating* k-pebble automata are also regular [20]. Theorem 1 (which clearly also holds for alternating pebble automata) again provides a simple proof, now also based on the case $k=0$ in [4] (for the case $k=1$ see [16,2]).

Theorem 2. DPT *is closed under composition, and* $\mathrm{DPT} \circ \mathrm{PT} \subseteq \mathrm{PT}$.

Proof. Let M_1 be a deterministic k_1-pebble transducer, and M_2 a k_2-pebble transducer. We will describe a pebble transducer M such that $\tau_M = \tau_{M_1} \circ \tau_{M_2}$. Clearly, we may assume that M_1 produces exactly one output symbol at each computation step, and that M_1 produces output $\triangleleft v \triangleright$ rather than v.

The idea of the construction is the same as the one for the closure under composition of multi-head finite state transductions, i.e., log-space reductions (see [19]): Each square of the output tape of M_1 (which is the input tape of M_2) is represented by the configuration of M_1 that produces the square in the next computation step. Note that since M_1 is deterministic, that square is unique: there are no repeating configurations in M_1's computation (M first checks whether the input is in the, regular, domain of M_1). To simulate M_2 on this output tape, M has to keep track of (at most) k_2+1 such configurations, one for each pebble of M_2 and one for M_2's reading head. The states of these configurations are kept in the finite control of M, and for each configuration M uses a group of (at most) k_1+1 pebbles, one for each pebble of M_1 and one for M_1's reading head. The ordering of these pebbles is as follows: first the group of pebbles corresponding to pebble 1 of M_2, ordered in the same way as the pebbles of M_1, with the reading head pebble last, then the group of pebbles corresponding to pebble 2 of M_2, similarly ordered, etcetera, and finally the group of pebbles corresponding to the reading head of M_2. Additionally, M needs an auxiliary group of k_1+1 pebbles, to simulate a computation of M_1, as explained below.

Initially, M drops one pebble on the first square of the input tape, and starts the simulation of M_2. During this simulation, if M_2 moves one square to the right, then M simulates one step of M_1, starting in the configuration c_r corresponding to the reading head of M_2. If M_2 moves one square to the left, then M computes the configuration of M_1 previous to c_r. It does that by simulating the computation of M_1 from the initial configuration, using the auxiliary group of pebbles, and checking for each computed configuration whether the next step will give c_r. If so, M lifts all auxiliary pebbles and applies the reverse of the transition used in that step to c_r (where 'left' is the reverse of 'right', and 'drop' is the reverse of 'lift'). If M_2 drops a pebble then M "copies" c_r (by dropping another pebble at each pebble of c_r), and if M_2 lifts a pebble then the configuration corresponding to this pebble is the same as c_r and hence M "removes" c_r (by lifting the pebbles of c_r). Finally, if M_2 checks whether its current square contains a certain pebble, then M checks whether the configuration corresponding to this pebble is the same as c_r. It is straightforward to show that the life times of M's pebbles are nested. □

It would be interesting to know whether composition can be realized with less than $(\ell+2)(k+1)$ pebbles. We note here that the deterministic 0-pebble transductions (i.e., the deterministic two-way finite state transductions) are closed under composition, see [5]; thus, 0 pebbles are used rather than 2. We conjecture that in general $(k_1+1)(k_2+1)-1$ pebbles suffice; by the discussion preceding Theorem 3 in the next section, this would be optimal.

3 Hierarchy of Transductions

It is easy to see that the deterministic transductions form a hierarchy, i.e., that $\mathrm{DPT}_k \subset \mathrm{DPT}_{k+1}$ for every k (where $\subset$ denotes proper inclusion). We will show that the same holds for the nondeterministic transductions, with the same counter-examples, because for every k-pebble transducer that computes a function there is an equivalent deterministic k-pebble transducer. Hence $\mathrm{PT}_k \subset \mathrm{PT}_{k+1}$, and the same holds for the pebble tree transducers of [20].

Let M_k be a deterministic k-pebble transducer that computes the function n^{k+1}, i.e., with $\tau_{M_k} = \{(a^n, a^{n^{k+1}}) \mid n \in \mathbb{N}\}$. Such a transducer is easy to construct: It can use its pebbles to count to n^k (where a pebble at square i, $1 \leq i \leq n$, represents the digit $i-1$ in the n-ary number system), as is well known for multi-head automata. Obviously, during this counting, the life times of the pebbles are nested. At each counting step, M_k copies its input string a^n to the output. It should be clear that τ_{M_k} cannot be computed by any deterministic $(k-1)$-pebble transducer: on input a^n, such a transducer has $O(n^k)$ configurations ($k-1$ pebbles plus the reading head), and since its computation does not repeat configurations, the length of the output string is also $O(n^k)$. We now show that τ_{M_k} can neither be computed by a *non*deterministic $(k-1)$-pebble transducer. Let $\mathcal{F}$ be the class of all functions.

Theorem 3. *For every $k \in \mathbb{N}$, $\mathrm{PT}_k \cap \mathcal{F} = \mathrm{DPT}_k$.*

Proof. Let $M = (\Sigma, \Delta, Q, q_0, F, \delta)$ be a k-pebble transducer such that τ_M is a function, i.e., for every u there is at most one v such that $(u, v) \in \tau_M$. We will construct a deterministic k-pebble transducer M'' with $\tau_{M''} = \tau_M$. For $k = 0$ see Theorem 4.9 of [7] (or Theorem 22 of [9]). The idea is that M'' simulates M, and whenever M has a nondeterministic choice, M'' finds out which of these choices leads to an accepting computation and then executes one of those. To find out whether a configuration leads to acceptance, we use the fact that the domain of a two-way k-pebble automaton is regular (as discussed after Theorem 1).

For $q \in Q$, let A_q be the deterministic finite automaton that accepts all strings $(a_0, b_0)(a_1, b_1) \cdots (a_{n+1}, b_{n+1})$ such that (1) $a_0 = \triangleleft$, $a_{n+1} = \triangleright$, $a_1, \ldots, a_n \in \Sigma$, $b_i \in \{0,1\}^{k+1}$, for every $m \leq k$ there is at most one i such that $b_i(m) = 1$, there is exactly one i such that $b_i(k+1) = 1$, and (2) M has at least one accepting computation when started in configuration (q, i, π), where $b_i(k+1) = 1$ and $\pi(m) = j$ iff $b_j(m) = 1$. This language is indeed regular, because there is a two-way k-pebble automaton accepting it (the automaton first goes into the required configuration, by inspecting the input string, and then simulates M).

It is now easy to construct a deterministic $(k+1)$-pebble (rather than k-pebble) transducer M' that is equivalent to M: Whenever M has a choice between several next steps, M' checks for each next configuration whether or not M has an accepting computation, and then takes one of the successful next steps. To check successfulness of a configuration, M' drops an additional pebble on the current square and then simulates A_q (for the current state q of M) on the current input tape, to check whether the (coding of the) configuration is accepted by A_q; note that the pebbles are viewed as part of the input of A_q.

To show that successfulness of a configuration can be checked without having an additional pebble, we use Lemma 3 of [18] (see also [5,9,24]): any kind of two-way automaton can, in its finite control, keep track of the state of a deterministic finite automaton A (i.e., the state in which A arrives after processing the part of the input tape to the left of the reading head of the two-way automaton). In particular, M'' behaves in the same way as M', but additionally keeps track of the state of each automaton A_q, $q \in Q$. Moreover, since the above result also holds for deterministic finite automata that process the input tape from right to left, M'' can keep track, for each A_q, of the set of states of A_q that lead to acceptance by A_q of the part of the input tape to the right of the reading head (consider the deterministic finite automaton obtained by applying the standard subset construction to the reversal of A_q). This information allows M'' to check at each moment whether or not the (coding of the) current configuration is accepted by A_q. Note that this depends on the fact that pebbles are always dropped or lifted at the position of the reading head. □

4 Hierarchy of Output Languages

The *output language* of a k-pebble transducer M, denoted by out(M), is the range of τ_M, i.e., $\text{out}(M) = \{v \mid (u, v) \in \tau_M \text{ for some } u\}$. In database terminology, as discussed in the Introduction, out(M) is the view corresponding to query τ_M. By OUT(DPT$_k$) we denote the class of all output languages of deterministic k-pebble transducers. In this section we show that these languages form a proper hierarchy, i.e., OUT(DPT$_k$) is a proper subclass of OUT(DPT$_{k+1}$) for all k. This strengthens the hierarchy of deterministic transductions in Section 3. We do not know whether the output languages of nondeterministic k-pebble transducers form a proper hierarchy. To prove properness of the above hierarchy we will use old and recent results from the theory of tree transducers, in particular macro tree transducers. For a survey on macro tree transducers, see Chapter 4 of the book [13]; to define them here, we need some standard terminology on trees.

A ranked alphabet is an alphabet Σ together with a mapping $\text{rank}_\Sigma : \Sigma \to \mathbb{N}$. For $k \in \mathbb{N}$, $\Sigma^{(k)}$ denotes $\{\sigma \in \Sigma \mid \text{rank}_\Sigma(\sigma) = k\}$. The set of trees over Σ, denoted T_Σ, is the smallest set of strings $T \subseteq \Sigma^*$ such that $\sigma T^k \subseteq T$ for every $\sigma \in \Sigma^{(k)}$, $k \geq 0$. Instead of $\sigma t_1 \cdots t_k$ we will write $\sigma(t_1, \ldots, t_k)$. For an alphabet Ω, $T_\Sigma(\Omega)$ denotes the set $T_{\Sigma \cup \Omega}$, where every symbol of Ω has rank 0. The yield of a tree t is denoted yt. For a tree language $L \subseteq T_\Sigma$, $yL = \{yt \mid t \in L\}$; for a class $\mathcal{L}$ of tree languages, $y\mathcal{L} = \{yL \mid L \in \mathcal{L}\}$: a class of (string) languages. We assume the reader to be familiar with elementary tree language theory [14]. The class of regular tree languages is denoted REGT.

A *macro tree transducer* (mtt, for short) is a term rewriting system, specified as a tuple $M = (Q, \Sigma, \Delta, q_0, R)$ where Q is the ranked alphabet of *states* (of rank ≥ 1), Σ and Δ are ranked alphabets of *input* and *output symbols*, respectively, $q_0 \in Q^{(1)}$ is the *initial state*, and R is the finite set of *rewrite rules*. For every $q \in Q^{(m+1)}$ and $\sigma \in \Sigma^{(k)}$, R contains exactly one rule $q(\sigma(x_1, \ldots, x_k), y_1, \ldots, y_m) \to t$ with $t \in T_{Q \cup \Delta}(\{x_1, \ldots, x_k, y_1, \ldots, y_m\})$ and every state occurring in t has one of the variables $x_1, \ldots, x_k$ as its first argument. The *transduction computed by* M, denoted τ_M, is the total function $\{(s, t) \in T_\Sigma \times T_\Delta \mid q_0(s) \Rightarrow_M^* t\}$, where $\Rightarrow_M$ denotes a rewriting step by one of the rules, as usual. The class of transductions computed by mtt's is denoted D$_t$MTT (where D$_t$ stands for "total deterministic"). A *top-down tree transducer* is an mtt of which all states have rank 1; the corresponding class of transductions is denoted D$_t$T.

Intuitively, a macro tree transducer breaks down the input tree in the first argument of each state, and builds up (or, accumulates) the output tree in the other arguments of each state. For these purposes it uses the variables $x_1, \ldots, x_k$ and $y_1, \ldots, y_m$, respectively.

Example 1. Here we present a macro tree transducer $M = (Q, \Sigma, \Delta, q_0, R)$ such that, for every input tree $s \in T_\Sigma$, $y\tau_M(s) = \text{peb}'(ys)$ where $\text{peb}'(a_1a_2 \cdots a_n) = b\bar{a}_1a_2 \cdots a_n e\, ba_1\bar{a}_2 \cdots a_n e \cdots ba_1a_2 \cdots \bar{a}_n e$. The peb′-function is a slight variation of the peb-function in the proof of Lemma 1. The mtt M has states q_0 of rank 1 and q of rank 3. Let us assume that the input symbols of M all have rank 0 or 2. The output alphabet Δ contains δ of rank 2 and γ of rank 3, and $\Delta^{(0)} = \Sigma^{(0)} \cup \{\bar{a} \mid a \in \Sigma^{(0)}\} \cup \{b, e\}$. The rules of M are as follows, for every $a \in \Sigma^{(0)}$ and $\sigma \in \Sigma^{(2)}$:

$q_0(a) \to \gamma(b, \bar{a}, e)$
$q_0(\sigma(x_1, x_2)) \to \delta(q(x_1, b, \delta(x_2, e)), q(x_2, \delta(b, x_1), e))$
$q(a, y_1, y_2) \to \gamma(y_1, \bar{a}, y_2)$
$q(\sigma(x_1, x_2), y_1, y_2) \to \delta(q(x_1, y_1, \delta(x_2, y_2)), q(x_2, \delta(y_1, x_1), y_2))$.

Note that, as required, each occurrence of q in a right-hand side of a rule has x_1 or x_2 as first argument. To understand how M works, note that if s' is a subtree of s with yield $a_i \cdots a_j$, then q translates s' into a tree with yield $ba_1 \cdots \bar{a}_i \cdots a_n e \cdots \cdots ba_1 \cdots \bar{a}_j \cdots a_n e$, provided its two accumulating arguments y_1 and y_2 contain trees with yield $ba_1 \cdots a_{i-1}$ and $a_{j+1} \cdots a_n e$, respectively. It is easy to generalize this construction to an arbitrary input alphabet Σ.

For $k \geq 1$, we denote by $\text{D}_\text{t}\text{MTT}^k(\text{REGT})$ the class of images of the regular tree languages under the $\text{D}_\text{t}\text{MTT}^k$ transductions, i.e., $\{\tau(R) \mid \tau \in \text{D}_\text{t}\text{MTT}^k, R \in \text{REGT}\}$. Thus, the class of (string) languages obtained by taking the yields of these tree languages is denoted $y\text{D}_\text{t}\text{MTT}^k(\text{REGT})$. Intuitively, they are the languages generated by the composition of k macro tree transductions from the regular tree languages. For convenience we will abuse our notation for $k = 0$ and denote the class $y\text{D}_\text{t}\text{T}(\text{REGT})$ also by $y\text{D}_\text{t}\text{MTT}^0(\text{REGT})$.

We will show that $\text{OUT}(\text{DPT}_k)$ is included in $y\text{D}_\text{t}\text{MTT}^k(\text{REGT})$. In the proof we need the following closure property. Note that $\text{DPT}_0 = \text{2DGSM}$, the class of deterministic two-way finite state transductions.

Lemma 2. *For $k \geq 1$, $y\text{D}_\text{t}\text{MTT}^k(\text{REGT})$ is closed under DPT_0 transductions.*

Proof. This can be shown by a series of results on tree transducers. First of all, the class $\text{D}_\text{t}\text{MTT}^k(\text{REGT})$ is closed under intersection with regular tree languages: $\tau(R_1) \cap R_2 = \tau(R_1 \cap \tau^{-1}(R_2))$, and REGT is closed under intersection and under inverse macro tree transductions (Theorem 7.4(1) of [12]). It now follows from Lemma 5.22 and Theorem 4.12 of [12] and from Lemma 11 of [10] that $\text{D}_\text{t}\text{MTT}^k(\text{REGT})$ is closed under DT^R transductions (the class of tree transductions computed by, not necessarily total, deterministic top-down tree transducers with regular look-ahead; for a definition see, e.g., Section 18 of [14]). The final, essential, result to be used is the following direct consequence of Theorem 5.5 of [11] (where DPT_0 is denoted DCS): if $\mathcal{L}$ is a class of tree languages closed under DT^R transductions, then $y\mathcal{L}$ is closed under DPT_0 transductions. □

Theorem 4. *For every $k \in \mathbb{N}$, $\text{OUT}(\text{DPT}_k) \subseteq y\text{D}_\text{t}\text{MTT}^k(\text{REGT})$.*

Proof. For $k = 0$, it follows from Corollary 4.11 and Lemma 3.2.3 of [11] and from Lemma 5.22 of [12] that $\text{OUT}(\text{DPT}_0) \subseteq y\text{D}_\text{t}\text{T}(\text{REGT})$. Now let $k \geq 1$, and let M be a deterministic k-pebble transducer. We may assume that τ_M is undefined on input λ. Consider the peb-function in the proof of Lemma 1. By that proof, there is a deterministic 0-pebble transducer M' such that $\tau_M = \text{peb}^k \circ \tau_{M'}$. Hence $\text{out}(M) = \tau_{M'}(\text{peb}^k(\Sigma^+))$, where Σ is the input alphabet of M. Let Σ_r be a ranked alphabet such that $yT_{\Sigma_\text{r}} = \Sigma^+$. It is easy to construct an mtt N such that $y\tau_N(s) = \text{peb}(ys)$ for every input tree s: in Example 1, replace the right-hand side of the first rule by $\gamma(\gamma(\bar{b}, a, e), \gamma(b, \bar{a}, e), \gamma(b, a, \bar{e}))$ and replace the right-hand side t of the second rule by $\gamma(\beta(\bar{b}, x_1, x_2, e), t, \beta(b, x_1, x_2, \bar{e}))$, where β has rank 4. Hence, since $T_{\Sigma_\text{r}} \in \text{REGT}$, $\text{peb}^k(\Sigma^+) = \text{peb}^k(yT_{\Sigma_\text{r}}) \in y\text{D}_\text{t}\text{MTT}^k(\text{REGT})$. Since $\tau_{M'}$ is a DPT_0 transduction, Lemma 2 now implies that $\text{out}(M) \in y\text{D}_\text{t}\text{MTT}^k(\text{REGT})$. □

Since all these results are effective, and the finiteness problem for languages in $y\text{D}_\text{t}\text{MTT}^k(\text{REGT})$ is decidable [6], Theorem 4 implies that the finiteness problem for output languages of deterministic pebble transducers is decidable. Hence, since $\text{OUT}(\text{DPT}_k)$ is closed under intersection with regular languages (by an obvious product construction), it is decidable for a deterministic pebble transducer M and a regular language T whether or not $\text{out}(M) - T$ is finite: "almost always" type checking, cf. the Introduction.

It has recently been proved [10] that the language classes $y\text{D}_\text{t}\text{MTT}^k(\text{REGT})$ form a proper hierarchy with respect to k. We will use the main tool of that paper to show that the counter-examples to this hierarchy can already be found in $\text{OUT}(\text{DPT}_k)$, which implies that also the language classes $\text{OUT}(\text{DPT}_k)$ form a proper hierarchy.

Let Δ and Ω be disjoint alphabets. A string $w_1 a_1 w_2 a_2 \cdots a_{l-1} w_l a_l w_{l+1}$ with $l \geq 0$, $a_i \in \Delta$, and $w_i \in \Omega^*$, is called a d-*string* for the string $a_1 \cdots a_l \in \Delta^*$ if all strings $w_2, \ldots, w_l$ are distinct (i.e., all Ω-strings between Δ-symbols are different). For an arbitrary string $u \in (\Delta \cup \Omega)^*$ we denote by $\text{res}_\Delta(u)$ the projection of u on Δ^*, i.e., res_Δ is the homomorphism that erases all symbols of Ω and is the identity on the symbols of Δ. Now let $L \subseteq \Delta^*$ and $L' \subseteq (\Delta \cup \Omega)^*$. We say that L' is d-*complete* for L if $\text{res}_\Delta(L') = L$ and L' contains at least one d-string for every string of L. We now state Theorem 18 of [10].

Lemma 3. *Let Δ and Ω be disjoint alphabets and let $L \subseteq \Delta^*$ and $L' \subseteq (\Delta \cup \Omega)^*$ be languages such that L' is* d-*complete for L.*
For every $k \in \mathbb{N}$, if $L' \in y\text{D}_\text{t}\text{MTT}^{k+1}(\text{REGT})$, then $L \in y\text{D}_\text{t}\text{MTT}^k(\text{REGT})$.

To use this result, we construct for every deterministic k-pebble transducer M a deterministic $(k+1)$-pebble transducer $\text{conf}(M)$ such that $\text{out}(\text{conf}(M))$ is d-complete for $\text{out}(M)$. We may assume that M produces at most one output symbol at each computation step. The idea is to insert just before each output symbol a coding of the configuration of M at the moment the output symbol is produced. Thus, on input u, $\text{conf}(M)$ simulates M, but whenever M produces an output symbol (in Δ), $\text{conf}(M)$ first drops an extra pebble on the current square i and outputs a coding of the current configuration (q, i, π) of M (coded as a string over an alphabet Ω disjoint with the output alphabet Δ). More precisely, $\text{conf}(M)$ outputs the symbol q and then copies the input to the output, including a coding of the pebbles: each symbol σ of the input tape, on some square j, is output as the symbol (σ, b) where $b \in \{0,1\}^{k+1}$ with $b(m) = 1$ iff pebble m is on square j. Clearly, since all configurations of M during its computation are distinct, out(conf(M)) is d-complete for out(M).

Theorem 5. *For every* $k \in \mathbb{N}$, $\mathrm{OUT}(\mathrm{DPT}_k) \subset \mathrm{OUT}(\mathrm{DPT}_{k+1})$.

Proof. Let M_1 be a deterministic 1-pebble transducer that translates a^n into $(a^n b)^n$; M_1 counts to n with its pebble, and at each counting step copies the input to the output, with an additional b. Thus, $\mathrm{out}(M_1) = \{(a^n b)^n \mid n \in \mathbb{N}\}$ is in $\mathrm{OUT}(\mathrm{DPT}_1)$. Since $\mathrm{OUT}(\mathrm{DPT}_0) \subseteq y\mathrm{D_tMTT}^0(\mathrm{REGT})$ by Theorem 4, and $\{(a^n b)^n \mid n \in \mathbb{N}\} \notin y\mathrm{D_tMTT}^0(\mathrm{REGT})$ by Theorem 3.16 of [7], $\mathrm{out}(M_1)$ is in $\mathrm{OUT}(\mathrm{DPT}_1) - \mathrm{OUT}(\mathrm{DPT}_0)$. It follows from the above property of 'conf' and from Lemma 3 that if out(M) is in $\mathrm{OUT}(\mathrm{DPT}_k) - y\mathrm{D_tMTT}^{k-1}(\mathrm{REGT})$, then out(conf($M$)) is in $\mathrm{OUT}(\mathrm{DPT}_{k+1}) - y\mathrm{D_tMTT}^{k}(\mathrm{REGT})$. Thus, $\mathrm{out}(\mathrm{conf}^k(M_1))$ is in $\mathrm{OUT}(\mathrm{DPT}_{k+1})$, but not in $y\mathrm{D_tMTT}^{k}(\mathrm{REGT})$, and hence, by Theorem 4, not in $\mathrm{OUT}(\mathrm{DPT}_k)$. □

The proof also shows that Theorem 4 is optimal, i.e., $\mathrm{OUT}(\mathrm{DPT}_k)$ is not included in $y\mathrm{D_tMTT}^{k-1}(\mathrm{REGT})$. In other words, languages that show properness of the macro tree transducer hierarchy $y\mathrm{D_tMTT}^{k}(\mathrm{REGT})$ (see [10]) can already be found among the output languages of deterministic pebble transducers.

Acknowledgement

We are grateful to Frank Neven for drawing our attention to the relation between automata theory and query languages, and in particular to the pebble tree transducer model of [20,28]. We thank the referees for constructive comments.

References

1. G. J. Bex, S. Maneth, F. Neven; A formal model for an expressive fragment of XSLT, Information Systems 27 (2002), 21–39
2. J. C. Birget; Two-way automata and length-preserving homomorphisms, Math. Syst. Theory 29 (1996), 191–226
3. M. Blum, C. Hewitt; Automata on a 2-dimensional tape, in: Proc. 8th IEEE Symp. on Switching and Automata Theory, 155–160, 1967
4. A. K. Chandra, D. C. Kozen, L. J. Stockmeyer; Alternation, J. of the ACM 28 (1981), 114–133
5. M. P. Chytil, V. Jákl; Serial composition of 2-way finite-state transducers and simple programs on strings, in: Proc. 4th ICALP (A. Salomaa, M. Steinby, eds.), Lect. Notes in Comput. Sci. 52, Springer-Verlag, 135–147, 1977
6. F. Drewes, J. Engelfriet; Decidability of finiteness of ranges of tree transductions, Inform. and Comput. 145 (1998), 1–50
7. J. Engelfriet; Three hierarchies of transducers, Math. Systems Theory 15 (1982), 95–125
8. J. Engelfriet, H.J. Hoogeboom; Tree-walking pebble automata, in: *Jewels are Forever* (J. Karhumäki et al., eds.), Springer-Verlag, 72–83, 1999

9. J. Engelfriet, H.J. Hoogeboom; MSO definable string transductions and two-way finite state transducers, ACM Trans. on Computational Logic 2 (2001), 216-254
10. J. Engelfriet, S. Maneth; Output string languages of compositions of deterministic macro tree transducers, Technical Report 01-04, Leiden University, 2001. To appear in J. of Comp. Syst. Sci.
11. J. Engelfriet, G. Rozenberg, G. Slutzki; Tree transducers, L systems, and two-way machines, J. of Comp. Syst. Sci. 20 (1980), 150–202
12. J. Engelfriet, H. Vogler; Macro tree transducers, J. of Comp. Syst. Sci. 31 (1985), 71–146
13. Z. Fülöp, H. Vogler; *Syntax-Directed Semantics - Formal Models based on Tree Transducers*, EATCS Monographs on Theoretical Computer Science (W. Brauer, G. Rozenberg, A. Salomaa, eds.), Springer-Verlag, 1998
14. F. Gécseg, M. Steinby; Tree languages, in: *Handbook of Formal Languages, Volume 3* (G. Rozenberg, A. Salomaa, eds.), Chapter 1, Springer-Verlag, 1997
15. N. Globerman, D. Harel; Complexity results for two-way and multi-pebble automata and their logics, Theor. Comput. Sci. 169 (1996), 161–184
16. P. Goralčik, A. Goralčiková, V. Koubek; Alternation with a pebble, Inform. Proc. Letters 38 (1991), 7–13
17. J. Hartmanis; On non-determinancy in simple computing devices, Acta Informatica 1 (1972), 336–344
18. J. E. Hopcroft, J. D. Ullman; An approach to a unified theory of automata, The Bell System Technical Journal 46 (1967), 1793–1829. Also in: IEEE Record of the 8th Annual Symposium on Switching and Automata Theory, 140–147, 1967
19. N. D. Jones; Space-bounded reducibility among combinatorial problems, J. of Comp. Syst. Sci. 11 (1975), 68–85
20. T. Milo, D. Suciu, V. Vianu; Typechecking for XML transformers, in: Proc. 19th ACM Symposium on Principles of Database Systems (PODS 2000), 11–22, 2000. To appear in J. of Comp. Syst. Sci.
21. F. Neven, J. Van den Bussche; Expressiveness of structured document query languages based on attribute grammars, in: Proc. 17th ACM Symposium on Principles of Database Systems (PODS 1998), 11–17, 1998. To appear in J. of the ACM
22. F. Neven, T. Schwentick, V. Vianu; Towards regular languages over infinite alphabets, in: Proc. MFCS 2001 (J. Sgall, A. Pultr, P. Kolman, eds.), Lect. Notes in Comput. Sci. 2136, Springer-Verlag, 560–572, 2001
23. Y. Papakonstantinou, V. Vianu; DTD inference for views of XML data; Proc. 19th ACM Symposium on Principles of Database Systems (PODS 2000), 35–46, 2000
24. H. Petersen; The equivalence of pebbles and sensing heads for finite automata, in: Proc. FCT'97 (B. S. Chlebus, L. Czaja, eds.), Lect. Notes in Comput. Sci. 1279, Springer-Verlag, 400–410, 1997
25. M.O. Rabin, D. Scott; Finite automata and their decision problems, IBM J. Res. Devel. 3 (1959), 115–125
26. R. W. Ritchie, F. N. Springsteel; Language recognition by marking automata, Inf. and Control 20 (1972), 313–330
27. J.C. Shepherdson; The reduction of two-way automata to one-way automata, IBM J. Res. Devel. 3 (1959), 198–200
28. V. Vianu; A Web Odyssey: from Codd to XML, in: Proc. 20th ACM Symposium on Principles of Database Systems (PODS 2001), 1–15, 2001

Optimal Non-preemptive Semi-online Scheduling on Two Related Machines

Leah Epstein[1,*] and Lene M. Favrholdt[2,**]

[1] School of Computer Science,
The Interdisciplinary Center, Herzliya, Israel,
lea@idc.ac.il.

[2] Department of Mathematics and Computer Science,
University of Southern Denmark, Odense, Denmark,
lenem@imada.sdu.dk

Abstract. We consider the following non-preemptive semi-online scheduling problem. Jobs with non-increasing sizes arrive one by one to be scheduled on two uniformly related machines, with the goal of minimizing the makespan. We analyze both the optimal overall competitive ratio, and the optimal competitive ratio as a function of the speed ratio ($q \geq 1$) between the two machines. We show that the greedy algorithm LPT has optimal competitive ratio $\frac{1}{4}(1+\sqrt{17}) \approx 1.28$ overall, but does not have optimal competitive ratio for every value of q. We determine the intervals of q where LPT is an algorithm of optimal competitive ratio, and design different algorithms of optimal competitive ratio for the intervals where it fails to be the best algorithm. As a result, we give a tight analysis of the competitive ratio for every speed ratio.

1 Introduction

The Problem. In this paper we study non-preemptive semi-online scheduling on two uniformly related machines. In the model of uniformly related machines, each machine has a *speed* and each job has a *size* which is the time it takes to complete it on a machine with unit speed. The jobs arrive one by one in order of non-increasing sizes. Each job must be assigned to one of the machines without any knowledge of future jobs (except for a bound on their size that follows from the size of the current job). Since the jobs are known to have non-increasing sizes, the problem cannot be seen as on-line but semi-online. We study the non-preemptive case, where it is not allowed to split a job in more parts and run the various parts on different machines. The goal is to minimize the *makespan*, i.e., the latest completion time of any job.

The processing time of a job on a given machine is also called the *load* of the job on that machine. The load of a machine is the sum of the loads of the jobs assigned to it. Thus, the makespan is the maximum load of any machine.

---

* Research supported in part by the Israel Science Foundation, (grant no. 250/01)

** Supported in part by the Danish Natural Science Research Council (SNF) and in part by the Future and Emerging Technologies program of the EU under contract number IST-1999-14186 (ALCOM-FT).

K. Diks et al. (Eds): MFSC 2002, LNCS 2420, pp. 245–256, 2002.

Since we study the case of two machines, the important parameter is the speed ratio $q \geq 1$ between the two machines. Without loss of generality, we assume that the faster machine has speed 1, and the other machine has speed $\frac{1}{q}$. We denote the faster machine by M_1 and the other machine by M_q.

Preliminaries. The quality of a semi-online algorithm, similarly to on-line algorithms, is measured by the *competitive ratio* which is the worst case ratio of the cost (the makespan, in this paper) of the semi-online algorithm to the cost of an optimal off-line algorithm which knows the whole sequence in advance.

The semi-online algorithm under consideration as well as its makespan is denoted by SONL. Similarly, the optimal off-line algorithm as well as its makespan is denoted by OPT. Thus, the competitive ratio of an algorithm SONL is

$$C = \inf\{c \mid \text{SONL} \leq c \cdot \text{OPT}, \text{ for any input sequence}\}.$$

For any $c \geq C$, SONL is said to be *c-competitive.*

For the first k jobs of an input sequence, we denote the makespan of the semi-online algorithm by SONL_k and the makespan of the optimal off-line algorithm by OPT_k. The jobs in a sequence are denoted $J_1, J_2, \ldots, J_\ell$ and their sizes are denoted $p_1, p_2, \ldots, p_\ell$.

The greedy algorithm LPT (Longest Processing Time first) was originally designed by Graham [6] for off-line scheduling on identical machines. It sorts the jobs by non-increasing sizes and schedules them one by one on the least loaded machine. This algorithm also works for the semi-online version where the jobs arrive in order of non-increasing sizes. The natural extension for uniformly related machines is as follows:

Algorithm LPT: Assign each arriving job J (of size p) to the machine that would finish it first. Formally, for each machine i let L_i be its load before the arrival of J. The job J is assigned to the fastest machine i for which $L_i + \frac{p}{s_i}$ is minimized.

Previous Work. All previous study of this problem on non-identical machines involves a study of the LPT algorithm. For two machines, Mireault, Orlin and Vohra [7] give a complete analysis of LPT as a function of the speed ratio. They show that the interval $q \geq 1$ is partitioned into nine intervals, and introduce a function which gives the competitive ratio in each interval (they consider the off-line problem, so they do not use the term competitive ratio). Some properties of LPT were already shown earlier. Graham [6] shows that the exact approximation ratio of LPT is $\frac{7}{6}$ for two identical machines. Seiden, Sgall and Woeginger [8] show that this is tight, i.e., LPT has the best possible competitive ratio for the problem. For two related machines, [5] shows that for any speed ratio, the performance ratio of LPT is at most $\frac{1}{4}(1+\sqrt{17}) \approx 1.28$.

For m identical machines, Graham [6] shows that the exact approximation ratio of LPT is $\frac{4}{3} - \frac{1}{3m}$. For three machines, [8] gives a general lower bound of $\frac{1}{6}(1+\sqrt{37}) \approx 1.18$. For a general setting of m related machines, Friesen [4] shows that the overall approximation ratio of LPT is between 1.52 and $\frac{5}{3}$. Dobson [2]

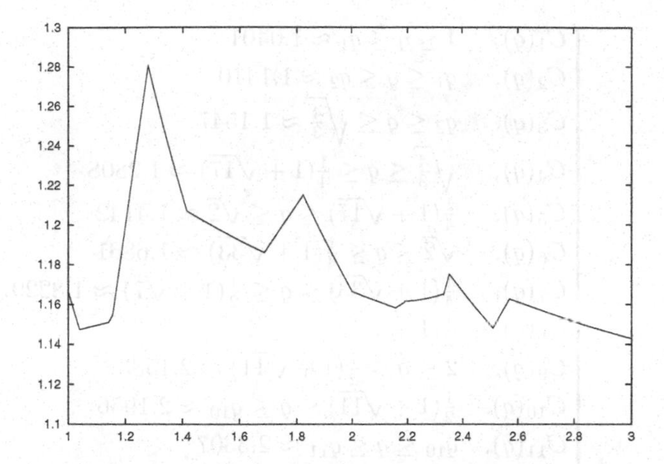

Fig. 1. The competitive ratio as a function of q

claims to improve the upper bound to $\frac{19}{12} \approx 1.58$. Unfortunately, his proof does not seem to be complete.

Our Results. In this paper we give the exact competitive ratio as a function of the speed ratio q for semi-online non-preemptive scheduling on two related machines with non-increasing job sizes (see Figure 1). The function involves 15 distinct intervals, as defined in Theorem 1.

In some of those intervals, we give general lower bounds which match the upper bounds in [7]. In those cases, LPT is an optimal semi-online algorithm. In the other intervals, we design new algorithms and prove matching general lower bounds. We show that, in terms of overall competitive ratio, $\frac{1}{4}(1+\sqrt{17})$ is the optimal competitive ratio achieved at $q = \frac{1}{4}(1+\sqrt{17})$ by LPT.

It is interesting to examine our results in the view of the results for on-line scheduling on two related machines. Unlike our problem, for that problem, LPT is optimal for all $q \geq 1$. The overall competitive ratio is ϕ ($\phi \approx 1.618$ is the golden ratio). For $q \leq \phi$ the competitive ratio is $1 + \frac{q}{q+1}$ and the competitive ratio for $q \geq \phi$ is $1 + \frac{1}{q}$. There are only two distinct intervals and the worst competitive ratio is achieved at ϕ. Surprisingly, for both problems, the highest competitive ratio is equal to the value of q for which it is achieved. The upper bounds, as well as the overall lower bound are given in [1], the other lower bounds are given in [3].

2 The Function

Theorem 1. *The optimal competitive ratio for semi-online scheduling on two related machines (with speed ratio $q \geq 1$) is described by the following function (see Figure 1).*

$$C(q) = \begin{cases} C_1(q), & 1 \le q \le q_1 \approx 1.0401 \\ C_2(q), & q_1 \le q \le q_2 \approx 1.1410 \\ C_3(q), & q_2 \le q \le \sqrt{\frac{4}{3}} \approx 1.1547 \\ C_4(q), & \sqrt{\frac{4}{3}} \le q \le \frac{1}{4}(1+\sqrt{17}) \approx 1.2808 \\ C_5(q), & \frac{1}{4}(1+\sqrt{17}) \le q \le \sqrt{2} \approx 1.4142 \\ C_6(q), & \sqrt{2} \le q \le \frac{1}{4}(1+\sqrt{33}) \approx 1.6861 \\ C_7(q), & \frac{1}{4}(1+\sqrt{33}) \le q \le \frac{1}{2}(1+\sqrt{7}) \approx 1.8229 \\ C_8(q), & \frac{1}{2}(1+\sqrt{7}) \le q \le 2 \\ C_9(q), & 2 \le q \le \frac{1}{2}(1+\sqrt{11}) \approx 2.1583 \\ C_{10}(q), & \frac{1}{2}(1+\sqrt{11}) \le q \le q_{10} \approx 2.1956 \\ C_{11}(q), & q_{10} \le q \le q_{11} \approx 2.3307 \\ C_{12}(q), & q_{11} \le q \le \frac{1}{4}(3+\sqrt{41}) \approx 2.3508 \\ C_{13}(q), & \frac{1}{4}(3+\sqrt{41}) \le q \le q_{13} \approx 2.5111 \\ C_{14}(q), & q_{13} \le q \le q_{14} \approx 2.5704 \\ C_{15}(q), & q \ge q_{14}, \end{cases}$$

$$C_1(q) = \frac{2}{3} + \frac{1}{2q}, \quad C_2(q) = 1 + \frac{1}{2}\left(4q^2 + 4q - 1 - \sqrt{(4q^2+4q-1)^2 - 4q^2}\right),$$

$$C_3(q) = \frac{6q+4}{3q+6}, \quad C_4(q) = q, \quad C_5(q) = \frac{q+2}{2q}, \quad C_6(q) = \frac{2q+3}{2q+2},$$

$$C_7(q) = \frac{2q+1}{q+2}, \quad C_8(q) = \frac{2q+3}{3q}, \quad C_9(q) = \frac{2q+3}{2q+2}, \quad C_{10}(q) = \frac{3q+2}{2q+3},$$

$$C_{11}(q) = \frac{q^2+3+\sqrt{q^4-6q^2+24q+9}}{6q}, \quad C_{12}(q) = \frac{q}{2}, \quad C_{13}(q) = \frac{3}{4} + \frac{1}{q},$$

$$C_{14}(q) = 1 + \frac{q^2+2q-2-\sqrt{q^4+8q+4}}{2q+4}, \quad C_{15}(q) = \frac{2q+2}{2q+1},$$

q_1 *is the largest real root of* $84q^4 - 24q^3 - 80q^2 + 6q + 9$,
q_2 *is the largest real root of* $27q^4 + 48q^3 - 54q^2 - 48q + 8$,
q_{10} *is the smallest real root of* $3q^4 - 9q^3 - 8q^2 + 21q + 18$,
q_{11} *is the largest real root of* $q^3 - 2q - 8$,
q_{13} *is the largest real root of* $20q^4 - 39q^3 - 46q^2 + 32q + 32$,
q_{14} *is the largest real root of* $4q^5 + 2q^4 - 24q^3 - 23q^2 + 6q + 8$.

The proof of the upper bound is given in Section 5 and the proof of the lower bound in Section 6.

3 Analysis of the Overall Competitive Ratio

In this section we show that LPT is optimal if we are only interested in the overall competitive ratio. That is, the maximum competitive ratio achieved by any set of two uniformly related machines. One way to determine the overall competitive ratio would be to analyze each interval separately, and take the maximum. However, we use previous results to do it in a simpler way.

Theorem 2. *The overall optimal competitive ratio for non-preemptive scheduling on two related machines is* $\frac{1}{4}(1+\sqrt{17})$.

Proof. We use a result given already in [5]. That paper shows that LPT on two related machines has a competitive ratio of at most $\frac{1}{4}(1+\sqrt{17})$. Later in this paper we show that for $q=\frac{1}{4}(1+\sqrt{17})$, the competitive ratio of any semi-online algorithm is at least $\frac{1}{4}(1+\sqrt{17})$, which completes the proof. □

The conclusion is that LPT is an optimal semi-online algorithm in the sense that its overall competitive ratio is best possible. However, analyzing the competitive ratio as a function of the speed ratio q gives the surprising result that LPT is not optimal for all q.

4 Properties and Assumptions

In this section we describe a few facts and assumptions used in the upper bound analysis.

We assume without loss of generality that OPT = 1. Thus, the ratio $\frac{\text{SONL}}{\text{OPT}} =$ SONL. For a given input sequence, we denote the total size of jobs by P and the size of the last job by p_ℓ. Note that $P \le 1+\frac{1}{q} = \frac{q+1}{q}$ always holds, since the total size of jobs scheduled by OPT is at most 1 on M_1 and at most $\frac{1}{q}$ on M_q.

In the upper bound analysis, we consider worst case sequences of each possible length separately. Thus, we can assume that the makespan is determined by the last (and hence the smallest) job of the sequence, since otherwise it is handled by the analysis of shorter sequences.

Lemma 1. *For a given input sequence and a given semi-online algorithm, assume that J_ℓ is scheduled according to the LPT rule and that $SONL > SONL_{\ell-1}$. Let $P_1^{\ell-1}$ and $P_q^{\ell-1}$ be the total size of jobs assigned to M_1 and M_q, respectively, just before the arrival of J_ℓ. Then,*

$$SONL = \min\{P_1^{\ell-1} + p_\ell,\ q(P_q^{\ell-1} + p_\ell)\}.$$

Proof. LPT schedules J_ℓ on a machine such that the resulting load on that machine is minimized. By the assumption that J_ℓ determines the makespan, the final makespan is equal to the load on the machine running J_ℓ. □

The following lemma appears in [7]. For completeness, we prove it here as well.

Lemma 2. *For a given input sequence and a given semi-online algorithm SONL, assume that the last job J_ℓ is scheduled according to the LPT rule and that $SONL > SONL_{\ell-1}$. Then $SONL \le 1 + \frac{q}{q+1}p_\ell$.*

Proof. Let $P_q^{\ell-1}$ and $P_1^{\ell-1}$ denote the total size of jobs assigned to the slow and the fast machine, respectively, just before the last job is assigned. Note that $P = P_q^{\ell-1} + P_1^{\ell-1} + p_\ell$. Since the last job is assigned by the LPT rule, and by the assumption that the last job determines the makespan, the final makespan is $\min\{q(P_q^{\ell-1} + p_\ell), P_1^{\ell-1} + p_\ell\}$ (using Lemma 1). This is bounded by the convex combination

$$\begin{aligned}\frac{1}{q+1}q(P_q^{\ell-1} + p_\ell) + \frac{q}{q+1}(P_1^{\ell-1} + p_\ell) &= \frac{q}{q+1}(P_q^{\ell-1} + P_1^{\ell-1} + 2p_\ell) \\ &= \frac{q}{q+1}(P + p_\ell) \leq 1 + \frac{q}{q+1}p_\ell .\end{aligned}$$

□

We will sometimes use Lemma 2 in the following form.

Corollary 1. *For a given input sequence and a given semi-online algorithm SONL, assume that the last job J_ℓ is scheduled according to the LPT rule and that SONL > $SONL_{\ell-1}$.*

$$\text{If } SONL > C, \text{ then } p_\ell > (C-1)\frac{q+1}{q} .$$

Lemma 3. *For a given input sequence and a given semi-online algorithm, assume that J_ℓ is scheduled according to the LPT rule and that SONL > $SONL_{\ell-1}$. Let $k \in \mathbb{N}$.*

If $SONL > 1 + \frac{1}{(k+1)(q+1)}$ *then OPT runs at most k jobs on M_q. If* $SONL > 1 + \frac{q}{(k+1)(q+1)}$ *then OPT runs at most k jobs on M_1.*

The proof is given in the full paper.

5 New Algorithms

In this section we present algorithms of optimal competitive ratio, for intervals where LPT is not optimal. We first mention all intervals where LPT is an optimal algorithm. In [7] the exact performance ratio of LPT is given. In all intervals where the lower bound in Section 6 matches the upper bound in [7], clearly LPT has optimal competitive ratio.

The names we use for the intervals are taken from the definition of the function. Hence, we deal with intervals 1–15. The intervals where LPT is optimal are as follows: The first interval is the point $q = 1$. For $q = 1$, it is known [6] that the competitive ratio of LPT is $\frac{7}{6}$ and that this is the best possible competitive ratio for any semi-online algorithm [8]. However, for $q = 1 + \varepsilon$, for small $\varepsilon > 0$, this paper shows that LPT is not an optimal semi-online algorithm.

The other intervals where LPT is optimal are $\frac{1}{6}(1 + \sqrt{37}) \leq q \leq q_{\mathrm{LPT}}$ and $q \geq q_{14} \approx 2.57$, where $q_{\mathrm{LPT}} \approx 2.04$ is the largest real root of $4q^3 - 4q^2 - 10q + 3$.

This leaves the following intervals to deal with. Intervals 1–4 (not including $q = 1$ in interval 1, and interval 4 only up to $\frac{1}{6}(1 + \sqrt{37})$) and intervals 9–14 (interval 9 starting only at q_{LPT}). We design three new algorithms *Slow-LPT,*

Balanced-LPT and Opposite-LPT. Slow-LPT has optimal competitive ratio in the interval $1 < q < \frac{1}{6}(1+\sqrt{37})$. Balanced-LPT has optimal competitive ratio in the intervals $q_{\mathrm{LPT}} < q \leq q_{10}$ and $\frac{1}{4}(3+\sqrt{41}) \leq q < q_{14}$, and Opposite-LPT has optimal competitive ratio in the interval $q_{10} \leq q \leq \frac{1}{4}(3+\sqrt{41})$.

As can be seen in the next section, the most difficult sequences for the algorithms are quite short sequences (up to six jobs). For longer sequences, the last job is relatively small. Thus, as shown by Lemma 2, the algorithm benefits from the non-increasing order in this case. As may be seen from the definitions of the three new algorithms, the most difficult decision is either the decision for the second job, or the first and the third jobs.

5.1 The First Four Intervals

We design an algorithm Slow-LPT which has optimal competitive ratio in the interval $1 < q < \frac{1}{6}(1+\sqrt{37}) \approx 1.18$. Intuitively, the reason why LPT fails in this interval is that the slow machine is not much slower than the faster one. Since the fast machine does not dominate the slow machine so easily, it often makes sense to use the slow machine first, and keep the fast machine free for future jobs. The algorithm is actually optimal in the interval $1 \leq q \leq \frac{1}{4}(1+\sqrt{17}) \approx 1.28$, giving an alternative algorithm with optimal competitive ratio in the interval $\frac{1}{6}(1+\sqrt{37}) \leq q \leq \frac{1}{4}(1+\sqrt{17})$.

Algorithm Slow-LPT

Assign J_1 to M_q. Assign J_2 to M_1.
If $q(p_1+p_3) \leq C(q)(p_2+p_3)$, assign J_3 to M_q, and otherwise to M_1.
Assign the rest of the jobs by the LPT rule.

The algorithm Slow-LPT is analyzed in the full paper.

5.2 Intervals 9 and 10, 13 and 14

In intervals 9–14, only sequences of five jobs and less can be slightly problematic, unlike the intervals 1–4, where sequences of six jobs had to be considered. Both algorithms for intervals 9–14 have a special rule for the second job only. In this section, we consider the algorithm Balanced-LPT, which tries to assign the second job to M_q.

Algorithm Balanced-LPT

Assign J_1 to M_1.
If $qp_2 > C(q)(p_1+p_2)$, assign J_2 to M_1, and otherwise to M_q.
Assign the rest of the jobs by the LPT rule.

Intervals 9 and 10. Due to space restrictions we analyze Balanced-LPT only in intervals 9 and 10. Recall that $C_9(q) = \frac{2q+3}{2q+2}$ and $C_{10}(q) = \frac{3q+2}{2q+3}$, and that intervals 9 and 10 correspond to the interval $2 \leq q \leq q_{10} \approx 2.20$. In interval 9,

$C_{10}(q) \le C_9(q)$, and in interval 10, $C_9(q) \le C_{10}(q)$. Thus, in intervals 9 and 10, $C(q) = \max\{C_9(q), C_{10}(q)\}$.

By Lemma 3, for any sequence such that SONL $> C(q)$, OPT runs at most one job on M_q, since $C(q) \ge C_9(q) = \frac{2q+3}{2q+2} = 1 + \frac{1}{2(q+1)}$. Similarly, OPT runs at most 4 jobs on M_1, since $\frac{1}{2(q+1)} \ge \frac{q}{5(q+1)}$ for $q \le 2.5$. This means that we need only consider sequences of at most five jobs. If an optimal off-line algorithm does not run any jobs on M_q, Balanced-LPT will not break the ratio, so we will assume that OPT schedules exactly one job on M_q and at most four on M_1.

In intervals 9 and 10, Balanced-LPT always assigns J_2 to M_q, since $C(q)(p_1 + p_2) \ge C_9(q)(p_1 + p_2) \ge 2C_9(q)p_2 \ge qp_2$ for $q \le \frac{1}{2}(1 + \sqrt{13}) \approx 2.30$. This shows that sequences of at most two jobs cannot break the ratio.

By Lemma 1, if there are at least three jobs, SONL $\le P - p_2$. If OPT does not run J_1 on M_q, we get OPT $\ge P - p_2$. This leaves only the case where OPT runs J_1 on M_q and all other jobs on M_1 to consider.

Three Jobs. By the assumption that OPT runs J_1 on M_q and all other jobs on M_1, OPT $= \max\{qp_1, p_2 + p_3\}$. By Lemma 1, SONL $\le p_1 + p_3 \le 2p_1 \le qp_1 \le$ OPT, since Balanced-LPT runs J_2 on M_q.

Four Jobs. Since OPT runs J_1 on M_q and J_2, J_3, and J_4 on M_1, $p_1 \le \frac{1}{q}$ and $p_2 + p_3 + p_4 \le 1$. Combining the latter inequality with $p_2 \ge p_3 \ge p_4$ gives $p_3 + p_4 \le \frac{2}{3}$. Thus, using Lemma 1, we have

$$\text{SONL} \le p_1 + p_3 + p_4 \le \frac{1}{q} + \frac{2}{3} = \frac{2q+3}{3q} \le \frac{2q+3}{2q+2} = C_9(q), \quad \text{for } q \ge 2.$$

Five Jobs. If Balanced-LPT assigns at least one of the jobs J_3 and J_4 to M_q, SONL $\le p_1 + p_3 + p_5 = P - p_2 - p_4 \le 1 + \frac{1}{q} - 2p_5$. By Corollary 1, SONL $> C_9(q)$ implies

$$\text{SONL} < 1 + \frac{1}{q} - 2(C_9(q) - 1)\frac{q+1}{q} = 1 + \frac{1}{q} - \frac{2(q+1)}{q(2q+2)} = 1,$$

which is a contradiction.

Otherwise, SONL $\le q(p_2 + p_5)$. Since OPT runs J_2, J_3, J_4, and J_5 on the fast machine, $p_3 + p_4 + p_5 \le 1 - p_2$. Thus, $p_5 \le \frac{1}{3}(1 - p_2)$, and SONL $\le q(p_2 + p_5) \le \frac{q}{3}(1 + 2p_2)$. Furthermore, SONL $\le P - p_2 \le 1 + \frac{1}{q} - p_2$, so $p_2 \le 1 + \frac{1}{q} -$ SONL. Thus, if SONL $> C_{10}(q)$, we get $p_2 < 1 + \frac{1}{q} - \frac{3q+2}{2q+3} = \frac{1}{q} - \frac{q-1}{2q+3}$, and

$$\text{SONL} \le \frac{q}{3}(1 + 2p_2) < \frac{q}{3}\left(1 + \frac{2}{q} - \frac{2q-2}{2q+3}\right) = \frac{3q+2}{2q+3} = C_{10}(q),$$

which is a contradiction.

Note that since the analysis is valid for all of interval 9, this means that Balanced-LPT has optimal competitive ratio for $2 \le q \le q_{\text{LPT}} \approx 2.04$, as well as LPT.

The analysis of Balanced-LPT in intervals 13 and 14 is given in the full paper.

5.3 Intervals 11 and 12

When assigning the second job, the following algorithm tries to do the opposite of LPT. If $qp_2 < p_1 + p_2$, LPT puts J_2 on M_q, so Opposite-LPT puts J_2 on M_1, unless $p_1 + p_2 > C(q)qp_2$. Similarly, if $qp_2 \geq p_1 + p_2$, Opposite-LPT puts J_2 on M_q, unless $qp_2 > C(q)(p_1 + p_2)$.

Algorithm Opposite-LPT

Assign J_1 to M_1.
Assign J_2 to M_1 if one of the following holds:

$$qp_2 < p_1 + p_2 \leq C(q)\, qp_2 \quad \textit{or} \quad qp_2 > C(q)(p_1 + p_2).$$

Otherwise, assign J_2 to M_q.
Assign the rest of the jobs by the LPT rule.

The analysis of Opposite-LPT is given in the full paper.

6 Matching Lower Bounds

In this section we present job sequences that prove the lower bounds matching the upper bounds of Section 5 or, in the intervals where LPT is optimal, the bounds of LPT as given in [7].

For all of the intervals 4–15, q is at least as large as the competitive ratio (the competitive ratio is bounded by $\frac{1}{4}(1+\sqrt{17}) \approx 1.28$), so any algorithm scheduling the first job on M_q has a competitive ratio of at least $C(q)$. Thus, when proving lower bounds for these intervals, it can be assumed that the algorithm schedules the first job on M_1. In the intervals 1–3 we take care of both possibilities (the first job may be assigned to either machine). In all sequences, unless otherwise mentioned, jobs are scaled so that if the sequence is completed, OPT = 1. All sequences have between three and six jobs, most of them have exactly five jobs.

Interval 1. ($1 \leq q \leq q_1 \approx 1.04$): $C_1(q) = \frac{2}{3} + \frac{1}{2q}$. The sequence consists of five jobs with sizes

$$p_1 = p_2 = \frac{1}{2q}, \quad p_3 = p_4 = p_5 = \frac{1}{3}.$$

The schedule of OPT_2 is achieved by running one job on each machine. This gives $\mathrm{OPT}_2 = \frac{1}{2}$. If SONL schedules both jobs on M_1, $\mathrm{SONL}_2 = \frac{1}{q} = \frac{2}{q}\,\mathrm{OPT}_2 > C_1(q)\,\mathrm{OPT}_2$. Thus, assume that SONL schedules one of the first two jobs on M_q.

For the complete sequence, OPT runs the first two jobs on M_q and the other jobs on M_1. If SONL assigns two of the last three jobs to M_1, $\mathrm{SONL} \geq \frac{1}{2q} + \frac{2}{3}$, and if SONL assigns two of the last three jobs to M_q, $\mathrm{SONL} \geq \frac{1}{2} + \frac{2}{3}q \geq \frac{1}{2q} + \frac{2}{3}$.

In the next two intervals, we build the sequence assuming that J_1 is assigned to the slow machine by the algorithm, since algorithms that put the first job on M_1, are not $C(q)$-competitive in these two intervals. The sequence proving this is based on a sequence from [7]. The definition of the sequence (as a function of p_1) is

$$p_2 = \frac{3+2q-2q^2}{2q^2+q}p_1, \;\; p_3 = p_4 = p_5 = \frac{q+1}{2q+1}p_1.$$

Note that for this sequence, the makespan of OPT is not necessarily 1.

For $1 \le q \le 1+\sqrt{7}$, $0 \le \frac{3+2q-2q^2}{2q^2+q} \le 1$, so the sequence is well-defined and $p_2 \le p_1$. Furthermore, since $q \le \frac{1}{6}(1+\sqrt{37}) \approx 1.18$, $p_3 \le p_2$.

Since $\frac{3+2q-2q^2}{2q^2+q} \le \frac{1}{q}$ for $q \ge 1$, $\mathrm{OPT}_2 = p_1$. If SONL puts J_2 on M_1,

$$\frac{\mathrm{SONL}_2}{\mathrm{OPT}_2} = \frac{p_1+p_2}{p_1} = \frac{3q+3}{2q^2+q} > \frac{3q+3}{2q^2+2q} = \frac{3}{2q} \ge \frac{3}{2}\sqrt{\frac{3}{4}} > 1.299 > C(q).$$

Otherwise, $\mathrm{SONL} \ge q(p_2+2p_3) = p_1+2p_3 = \frac{4q+3}{2q+1}p_1$. The optimal makespan is, however, $\frac{3q+3}{2q+1}p_1$, achieved by scheduling the first two jobs on M_q and the last three on M_1. This gives a ratio of $\frac{4q+3}{3q+3}$, which is larger than $C(q)$ in intervals 2 and 3.

Interval 2. ($q_1 \le q \le q_2 \approx 1.14$):
$C_2(q) = 1+\frac{1}{2}\left(4q^2+4q-1-\sqrt{(4q^2+4q-1)^2-4q^2}\right)$. The sequence consists of five jobs with sizes

$$p_1 = \frac{1}{q} - \frac{2q+1}{q+1}p_5, \;\; p_2 = \frac{2q+1}{q+1}p_5, \;\; p_3 = 1-2p_5,$$

$$p_4 = p_5 = \frac{q+1}{2q}\left(4q^2+4q-1-\sqrt{(4q^2+4q-1)^2-4q^2}\right).$$

The first job is larger than the second job, since $\frac{1}{q} > \frac{4q+2}{q+1}p_5$ for $q < \frac{1}{4}(1+\sqrt{13}) \approx 1.15$. This is equivalent to $p_5 < \frac{q+1}{4q^2+2q}$ for $q < \frac{1}{4}(1+\sqrt{13})$. Since $\frac{q+1}{4q^2+2q} \le \frac{1}{3}$ for $q \ge 1$, this also implies $p_3 \ge p_4$. Finally, $p_2 \ge p_3$, since $p_5 \ge \frac{q+1}{4q+3}$ for $q \le \frac{1}{4}(1+\sqrt{13})$.

If the algorithm assigns J_2 to M_q, $\mathrm{SONL}_2 = 1$. However, we show that $\mathrm{OPT}_2 \le \frac{2}{3}$. Clearly, $\mathrm{OPT}_2 \le \max\{p_1, qp_2\}$. By the proof that $p_1 \ge p_2$, $p_2 \le \frac{1}{2q}$. Thus, $qp_2 \le \frac{1}{2}$. Furthermore, $p_3 \le \frac{1}{2}$, and $p_5 \ge \frac{1}{2}(1-p_3) \ge \frac{1}{4}$. Now,

$$p_1 = \frac{1}{q} - \frac{2q+1}{q+1}p_5 \le \frac{1}{q} - \frac{2q+1}{4q+4} \text{ for all } q \ge 1.$$

Thus, to be $C(q)$-competitive, the algorithm must put J_2 on M_1.

If the algorithm assigns the third job to M_q,

$$\begin{aligned}\mathrm{SONL}_3 \ge (p_1+p_3)q &= 1 - \frac{2q^2+q}{q+1}p_5 + q - \frac{2q^2+2q}{q+1}p_5 \\ &= q+1-\frac{4q^2+3q}{q+1}p_5.\end{aligned}$$

On the other hand,

$$\mathrm{OPT}_3 \le \max\{qp_1, p_2+p_3\} = \max\{1-\frac{2q^2+1}{q+1}p_5, 1-\frac{1}{q+1}p_5\} = 1-\frac{1}{q+1}p_5.$$

Thus, in this case $\mathrm{SONL}_3 \le C(q)\mathrm{OPT}_3$, if and only if $q+1-\frac{4q^2+3q}{q+1}p_5 \le C(q)(1-\frac{1}{q+1}p_5)$. Note that $p_5 = \frac{1}{q}(C(q)-1)(q+1)$. Substituting this in the inequality, we get $1+\frac{1}{2}(4q^2+4q-1-\sqrt{(4q^2+4q-1)^2-4q^2}) \le C(q) \le 1+\frac{1}{2}(4q^2+4q-1+\sqrt{(4q^2+4q-1)^2-4q^2}) = C_2(q)$. Thus, if the algorithm is better than $C_2(q)$-competitive in this interval, it assigns the third job to M_1.

Now, consider the whole sequence. OPT runs J_1 and J_2 on M_q and the other jobs on M_1. If SONL runs J_4 and J_5 on M_q,

$$\mathrm{SONL} \ge (p_1+2p_5)q = (\frac{1}{q}-\frac{2q+1}{q+1}p_5+\frac{2q+2}{q+1}p_5)q = 1+\frac{q}{q+1}p_5 = C_2(q).$$

Otherwise,

$$\mathrm{SONL} \ge (p_2+p_3+p_5) = \frac{2q+1}{q+1}p_5+1-\frac{q+1}{q+1}p_5 = 1+\frac{q}{q+1}p_5 = C_2(q).$$

Interval 3. $(q_2 \le q \le \sqrt{4/3} \approx 1.15)$: $C_3(q) = \frac{6q+4}{3q+6}$. This is the only case of a lower bound sequence that consists of six jobs. Let $\beta = \frac{4-3q^2}{3q(q+2)}$. In this interval, $\beta > 0$. The sequence is

$$p_1 = 1-2p_5,\ \ p_2 = p_3 = p_4 = \frac{1}{3q},\ \ p_5 = p_6 = \frac{1}{3q}-\beta.$$

The first job is larger than the second job, because $p_1 = 1-2p_5 > 1-2p_2 > p_2$, since $p_2 < \frac{1}{3}$.

Recall that we can assume that the algorithm schedules J_1 on M_q. We now show that we can also assume that among the following three jobs, exactly one is scheduled on M_q. First note that

$$\begin{aligned}\mathrm{OPT}_4 &\le \max\{p_1+p_2, q(p_3+p_4)\} = \max\{1-2(\frac{1}{3q}-\beta)+\frac{1}{3q},\frac{2}{3}\}\\ &= \max\{1-\frac{1}{3q}+2\beta,\frac{2}{3}\} = 1-\frac{1}{3q}+2\beta\\ &= \frac{3q(q+2)-(q+2)+2(4-3q^2)}{3q(q+2)} = \frac{6+5q-3q^2}{3q(q+2)}.\end{aligned}$$

Now, if SONL puts all three jobs J_2, J_3, J_4 on M_1, $\mathrm{SONL}_4 \ge \frac{1}{q} = \frac{6+5q-3q^2}{3q+6}\,\mathrm{OPT}_4$, which is larger than $C_3(q)\,\mathrm{OPT}_4$ for $q < \frac{1}{6}(5+\sqrt{97}) \approx 2.47$. If SONL puts two of the jobs J_2, J_3, and J_4 on M_q,

$$\mathrm{SONL}_4 \ge q(p_1+p_3+p_4) = q\left(1-2\left(\frac{1}{3q}-\beta\right)+\frac{2}{3q}\right) = q(1+2\beta) > 1,$$

yielding an even worse ratio.

If J_5 and J_6 are both scheduled on M_1,

$$\mathrm{SONL} \ge p_3+p_4+p_5+p_6 = \frac{2}{3q}+\frac{2}{3q}-2\beta = \frac{4(q+2)-2(4-3q^2)}{3q(q+2)} = \frac{6q+4}{3q+6}.$$

Otherwise,

$$\begin{aligned}\text{SONL} &\geq q(p_1 + p_4 + p_5) = q(1 - 2p_5 + p_4 + p_5) = q(1 + p_4 - p_5) \\ &= q(1 + \beta) = \frac{6q + 4}{3q + 6} .\end{aligned}$$

For intervals 4–9, we use sequences given in [7] as negative examples for LPT. We show that those sequences are in fact lower bound sequences for any semi-online algorithm. In intervals 4 and 9, the proof holds for the entire interval, even though LPT is not optimal in the complete interval. Due to space restrictions, we give the proofs for all other intervals only in the full paper.

7 Conclusion

We have given a complete analysis of deterministic semi-online algorithms for two related machines and non-increasing job sizes. It is left as an open problem to analyze the behavior of randomized algorithms for two machines. For a general setting of m machines, it should be difficult to give a complete analysis depending on the speeds. However, it is intriguing to close the open question: what is the best overall competitive ratio for m machines?

References

1. Y. Cho and S. Sahni. Bounds for List Schedules on Uniform Processors. *SIAM Journal on Computing*, 9(1):91–103, 1980.
2. G. Dobson. Scheduling Independent Tasks on Uniform Processors. *SIAM Journal on Computing*, 13(4):705–716, 1984.
3. L. Epstein, J. Noga, S. S. Seiden, J. Sgall, and G. J. Woeginger. Randomized Online Scheduling on Two Uniform Machines. *Journal of Scheduling*, 4(2):71–92, 2001.
4. D. K. Friesen. Tighter Bounds for LPT Scheduling on Uniform Processors. *SIAM Journal on Computing*, 16(3):554–560, 1987.
5. T. Gonzalez, O. H. Ibarra, and S. Sahni. Bounds for LPT Schedules on Uniform Processors. *SIAM Journal on Computing*, 6(1):155–166, 1977.
6. R. L. Graham. Bounds on Multiprocessing Timing Anomalies. *SIAM J. Appl. Math*, 17:416–429, 1969.
7. P. Mireault, J. B. Orlin, and R. V. Vohra. A Parametric Worst Case Analysis of the LPT Heuristic for Two Uniform Machines. *Operations Research*, 45:116–125, 1997.
8. S. Seiden, J. Sgall, and G. J. Woeginger. Semi-Online Scheduling with Decreasing Job Sizes. *Operations Research Letters*, 27(5):215–221, 2000.

More on Weighted Servers or FIFO is Better than LRU

Leah Epstein[1,*], Csanád Imreh[2,**], and Rob van Stee[3,***]

[1] School of Computer Science, The Interdisciplinary Center,
P.O.B 167, 46150 Herzliya, Israel,
lea@idc.ac.il

[2] Max-Planck Institut für Informatik,
Stuhlsatzenhausweg 85, Saarbrücken, Germany,
on leave from University of Szeged, Hungary,
imreh@mpi-sb.mpg.de

[3] Institut für Informatik, Albert-Ludwigs-Universität,
Georges-Köhler-Allee, 79110 Freiburg, Germany,
vanstee@informatik.uni-freiburg.de

Abstract. We consider a generalized 2-server problem on the uniform space in which servers have different costs. Previous work focused on the case where the ratio between these costs was very large. We give results for varying ratios. For ratios below 2.2, we present an optimal algorithm which is trackless. We present a general lower bound for trackless algorithms depending on the cost ratio, proving that our algorithm is the optimal trackless algorithm up to a constant factor for any cost ratio. The results are extended for the case where we have two sets of servers with different costs.

1 Introduction

The weighted k-server problem was introduced by Fiat and Ricklin [8]. In this problem, we are given a metric space $M = (A, d)$ with k mobile servers, and each server s_i has a weight $x_i > 0$. Here A is a set of points and d is a distance function (metric). At each step, a request $r \in M$ is issued that has to be served by one of the servers by moving to r. The cost for server s_i to serve request r is $d(r, s_i) \cdot x_i$. This problem is an on-line problem: each time that there is a request, it needs to be served before the next request becomes known. We denote the cost of an algorithm ALG on a request sequence σ by ALG(σ). We denote an optimal off-line algorithm that knows all the input in advance by OPT. The goal of an

* Research supported in part by the Israel Science Foundation (grant no. 250/01).
** Research supported by Future and Emerging Technologies programme of the EU under contract number IST-1999-14186 (ALCOM-FT) and by the Hungarian National Foundation for Scientific Research, Grant TO30074.
*** Work supported by the Deutsche Forschungsgemeinschaft, Project AL 464/3-1, and by the European Community, Projects APPOL and APPOL II.

K. Diks et al. (Eds): MFSC 2002, LNCS 2420, pp. 257–268, 2002.

on-line algorithm $\mathcal{A}$ is to minimize its competitive ratio $\mathcal{R}(\mathcal{A})$, which is defined as the smallest value $\mathcal{R}$ that satisfies

$$\mathcal{A}(\sigma) \leq \mathcal{R} \cdot \text{OPT}(\sigma) + c,$$

for any request sequence σ and some constant c (independent of σ).

In [8] a doubly exponential upper bound in k is given for uniform spaces. Furthermore, for the special case where only two weights are allowed, a $k^{O(k)}$ competitive algorithm is presented. They also show that the competitive ratio is at least $k^{\Omega(k)}$ in any space with at least $k+1$ points. In [11], a simple upper bound of kw_{avg}/w_{min} is proven for the general case, where w_{min} is the minimal and w_{avg} is the average weight of the servers.

The special case of two servers and uniform spaces was investigated in [7]. There a 5-competitive version of the Work Function Algorithm and matching lower bound, and a 5-competitive memoryless randomized algorithm with matching lower bound are presented.

All previous work (except the simple result of [11]) focuses on the asymptotic case where the ratio between the weights of the servers tends to ∞. We consider instead the case of smaller ratios and obtain the surprising result that for the weighted 2-server problem in a uniform space, an algorithm that uses both its servers equally is optimal as long as the ratio between the weights is at most 2.2.

We also consider the more general case, where we have κ servers with speed 1 and κ servers with speed x. The total number of servers is $k = 2\kappa$. Since we only investigate uniform spaces, the problem can also be seen as a caching problem where we have two caches of size κ: the cheap cache C, and the expensive cache E. This type of cache is called *a two-level cache* [1,5]. For this reason, we borrow some terminology from caching theory. We formulate our problem as follows. The algorithm has to serve a sequence of requests for pages. If the requested page is not in either of the caches, then we have to put it into one of the caches, evicting a page from the chosen cache if it is full. This event is called a fault. The set of possible pages is called the slow memory. Moving a page into C has cost 1, and moving a page into E has a cost of $x \geq 1$.

All of the previous algorithms for the weighted server problem store information about most of the appearing request points: the algorithm SAMPLE of [8] has a counter of the points, in the work function algorithm of [7] this information is coded in the work function. This yields that these algorithms might have an extremely large space requirement if there is a large number of different requested points. For the original k-server problem, a class of algorithms called trackless is introduced in [3] to avoid this problem. In the case of trackless algorithms, for each request point the algorithm is only given as input the distances of the current server positions to the request point. The algorithm may memorize such distance values, but it is restricted from explicitly storing any points of the metric space.

For the special case of uniform spaces, the trackless property changes into the rule that the algorithm is not allowed to use bookmarks in the slow memory to distinguish between the pages. This means that in the case of a fault, the decision as to which cache is used for the requested page and which page is removed from this cache must be independent of the requested page itself. This restriction seems to be very strong at first look, but we must note that for paging the optimal deterministic marking algorithms are trackless [9], and even for the more general web caching problem the optimal algorithm is trackless [13]. Randomized trackless algorithms for the paging problem are investigated in [2]. It is shown that it is necessary to use bookmarks to reach the optimal $\log k$ competitive ratio.

Our Results: The results we show in this paper are as follows:

For $\kappa = 1$ we introduce a trackless algorithm which is based on the well-known paging algorithm FIFO. As mentioned before this algorithm has optimal competitive ratio for $x \le 2.2$. Specifically it has competitive ratio $\max\{2, 3(1+x)/4\}$. We also analyze a modified version of the other well-known paging algorithm LRU, and we obtain that in the weighted case FIFO is better than LRU, which has the competitive ratio $1 + x$. This is in sharp contrast to the intuition from previous literature [6] and practice. This surprising result hints that even in practice it might be the case that two-level caches (of relatively small size) would work more efficiently using FIFO rather than using the standard LRU. A third algorithm we study is an adaptation of the widely known algorithm BALANCE. We show that BALANCE performs even worse than LRU for infinitely many values of x (its competitive ratio is $2x$ for even values of x). All the above results hold for $\kappa = 1$, however the adaptations of FIFO and LRU are defined not only for $\kappa = 1$ but also for general values of κ.

Next we move on to trackless algorithms and $\kappa \ge 1$. We show that for such algorithms the competitive ratio must grow (at least) linearly as a function of κx (it is at least $\kappa(1+x)/2$ for $\kappa \ge 2$ and at least $x/2+1$ for $\kappa = 1$). For the sake of completeness we also show a simple upper bound of $\kappa(1+x)$ for general κ (which is a special case of the upper bound from [11]), pointing out that all marking algorithms (including our version of LRU) are optimal trackless algorithms up to a constant factor 2 (in terms of competitive ratio). The proof also holds for our version of FIFO.

2 Marking Algorithms

To begin with, we give a simple proof for the competitiveness of marking algorithms, which is a special case of the bound from [11].

We first partition the request sequence into phases in the following well-known way: each phase $i \ge 1$ is the maximal subsequence of the request sequence that contains requests to at most k distinct pages, and that starts with the first request after phase $i-1$ ends. Phase 1 starts with the first request of the sequence.

The marking algorithms, which are defined in [9], unmark all the pages in the caches at the beginning of each phase. When a page from one of the caches is requested we mark it. If there is a fault, then we evict some unmarked page (determined by the algorithm), load the requested page and mark it. From the definition of the phase partitioning, we can see that a phase is ended when we have a fault and every page in the caches is marked. We unmark the pages and the new phase is started.

Theorem 1. *For any marking algorithm $\mathcal{A}$, $\mathcal{R}(\mathcal{A}) \leq \kappa(1+x)$. Moreover, we also have $\mathcal{R}(\text{FIFO}) \leq \kappa(1+x)$.*

Proof. Every marking algorithm has a cost of at most $\kappa(1+x)$ per phase, whereas OPT has a cost of at least 1 per phase. This also holds for FIFO. □

3 Two Servers

It is possible to adapt the well-known paging algorithms FIFO and LRU for the current problem by considering the two caches as one big cache and ignoring the difference in costs. We denote these adaptations also by FIFO and LRU. We begin by proving the following theorem.

Theorem 2. *For $\kappa = 1$,*

$$\mathcal{R}(\text{FIFO}) = \max\left(2, \frac{3}{4}(1+x)\right).$$

Definition 1. *A* relevant *request sequence is a request sequence on which* FIFO *faults on every request.*

We can map any request sequence σ onto a relevant request sequence σ' by removing all requests on which FIFO does not fault. This does not affect the cost of FIFO or its decisions: a request in σ' is put into the cheap cache by FIFO if and only if FIFO puts the corresponding request in σ in the cheap cache. Hence $\text{FIFO}(\sigma') = \text{FIFO}(\sigma)$. Moreover, $\text{OPT}(\sigma') \leq \text{OPT}(\sigma)$. Therefore we only need to consider relevant request sequences to determine the competitive ratio of FIFO. Note that for a relevant request sequence, a phase always consists of two consecutive requests. W.l.o.g. we can assume that OPT only faults on a page when it is requested.

Lemma 1. *Any three consecutive requests in a relevant request sequence are to distinct pages.*

Proof. If this were not the case, FIFO would not fault on every request.

Lemma 2. *Suppose $\{a_i\}_{i=1}^{\ell}$ is a relevant sequence. For any $1 \leq j \leq \ell - 2$,* OPT *faults on at least one of the requests a_{j+1} and a_{j+2}.*

Proof. We use Lemma 1. If OPT does not fault on a_{j+1}, then it has a_j and a_{j+1} in its cache after the request for a_j. Thus OPT must fault on a_{j+2}, since that is a different page. □

Definition 2. *A relevant interval is a subsequence $\{a_i\}_{i=j_1}^{j_2}$ of a request sequence with the following property:* OPT *faults on $a_j \iff (j - j_1) \bmod 2 = 0$.*

By Lemma 2 we can partition any relevant sequence into relevant intervals. Note that all relevant intervals end with a fault by OPT, and have odd length.

Lemma 3. *For $f \geq 1$, OPT has at least f expensive faults on a relevant interval of length at least $4f + 1$.*

Proof. Consider a relevant interval of length at least 5 and partition it into f subintervals of length 4, and one final subinterval of length at most 3. Consider a subinterval of length 4. It has the form

$$\mathbf{a}b\,\mathbf{c}d$$

where OPT faults only on a and c. Since OPT does not fault on d, that page must be in OPT's cache at that point. By Lemma 1, d cannot be b or c. If also $d \neq a$, then after OPT loads a, d must also already be in the cache. But then OPT would fault on b, since $b \neq a$ by Lemma 1. Hence $d = a$. Since a remains in the cache when OPT serves c, either a or c must be loaded into the expensive cache by OPT.

Therefore OPT has at least one expensive fault per subinterval. □

Lemma 4. *For any relevant request sequence σ of length ℓ, we have*

$$\text{OPT}(\sigma) \geq \min\left(\frac{2}{3}, \frac{1+x}{4}\right) \cdot \ell.$$

Proof. We partition σ into relevant intervals. Consider any relevant interval I and denote its length by $\ell(I)$. We show that for each such interval, $\text{OPT}(I) \geq \min(2/3, (1+x)/4) \cdot \ell(I)$.

If $\ell(I) = 1$, $\text{OPT}(I) \geq 1 \geq 2/3$. If $\ell(I) = 3$, $\text{OPT}(I) \geq 2 = \frac{2}{3} \cdot 3$.

Suppose $\ell(I) \geq 5$, and write $f = \lfloor \ell(I)/4 \rfloor$. We use Lemma 3.

If $\ell(I) = 4f + 1$ then OPT has $2f + 1$ faults, among which at least f are expensive. This is a cost of $(f + 1 + fx)/(4f + 1)$ per request.

If $\ell(I) = 4f + 3$ then OPT has $2f + 2$ faults, among which at least f are expensive. This is a cost of $(f + 2 + fx)/(4f + 3)$ per request.

Both of these are at least $(1 + x)/4$ or at least $2/3$. □

Lemma 5. *For $\kappa = 1$, $\mathcal{R}(\text{FIFO}) \leq \max(2, 3(1+x)/4)$.*

Proof. Consider a relevant request sequence σ. We partition σ into phases, and compare the cost of FIFO to the *average* cost of OPT per phase. By Lemma 4, we have that $\text{OPT}(p) \geq \min(4/3, (1+x)/2)$ for any phase p on average. We also have $\text{FIFO}(p) = 1 + x$ exactly. Therefore, $\mathcal{R}(\text{FIFO}) \leq \max(2, 3(1+x)/4)$. □

We now show a general lower bound for FIFO, that holds if both the cheap cache and the expensive cache have some size $\kappa \geq 1$.

Lemma 6. *For $\kappa \geq 1$,*

$$\mathcal{R}(\text{FIFO}) \geq \max\left(2\kappa, \frac{2\kappa+1}{2\kappa+2}\kappa(1+x)\right).$$

Proof. Consider the following request sequence:

$$(1\ 2\ 3\ \ldots\ 2\kappa+1)^N.$$

We define two different offline strategies to serve this sequence, and use one or the other depending on x.

For $x \geq (2\kappa+3)/(2\kappa+1)$, OPT has $1,\ldots,\kappa$ in the cheap cache at the start, and $\kappa+1,\ldots,2\kappa$ in the expensive cache. OPT only faults on pages $1,\ldots,\kappa$ and $2\kappa+1$, each time evicting the page from this set which will be requested the furthest in the future (LFD [4]). In $2\kappa+1$ phases, there are $2\kappa(2\kappa+1)$ requests in total and $2\kappa(\kappa+1)$ requests to pages $1,\ldots,\kappa$ and $2\kappa+1$. OPT has a fault once every κ requests to these pages, so it has $2(\kappa+1)$ faults in $2\kappa+1$ phases.

FIFO pays $\kappa(1+x)$ per phase, so by letting N grow without bound we find

$$\mathcal{R}(\text{FIFO}) \geq \frac{2\kappa+1}{2\kappa+2}\kappa(1+x)\ .$$

For $x < (2\kappa+3)/(2\kappa+1)$, OPT uses its entire cache in a round-robin fashion. Thus OPT pays on average $(1+x)/2$ per FIFO phase, so

$$\mathcal{R}(\text{FIFO}) \geq 2\kappa. \qquad \square$$

Proof of Theorem 2. This follows from Lemma 6 for the case $\kappa = 1$ and from Lemma 5. □

FIFO cannot have optimal competitive ratio for every x since according to [7] there exists a constant competitive algorithm whereas the competitive ratio of FIFO grows linearly with x. However for relatively small values of x, FIFO has optimal competitive ratio.

Theorem 3. FIFO *is an optimal on-line algorithm for the weighted server problem and* $\kappa = 1$ *in the interval* $1 \leq x \leq \sqrt{249}/6 - 1/2 \approx 2.1299556$.

Proof. We prove a lower bound that matches the upper bound of FIFO from Theorem 2. This even holds for a 3-point space.

We construct a sequence of requests, which consists of requests for at most three different pages: 0, 1 and 2. Consequently there are 6 different possible configurations of the cache. Two basic ingredients for the proof are similar to the lower bound for the k-server problem [10]. Many new ingredients are added to the analysis, that becomes more involved for the related case.

As in [10], the sequence of requests is constructed in such a way that the on-line algorithm has a fault in every request. That is, in each step a request is added for the only page that is not present in either of the two caches of the on-line algorithm. In order to give a lower bound on the general (not strict) competitive ratio, we build a long enough sequence. Note that the on-line cost of such a sequence of N requests is at least N. Another similarity to [10] is a comparison of the on-line algorithm to several off-line algorithms. We keep five off-line algorithms that all process the complete sequence along with the on-line algorithm, and consider their average cost. This is a lower bound for the optimal off-line cost.

A sequence of requests that is constructed in this way can be transformed into a sequence of costs that the on-line algorithm pays in every step. We define a cost sequence as a sequence of 1's and x's, where 1 indicates a fault of the cheap cache whereas x indicates a fault of the expensive cache. Given a starting configuration of the on-line algorithm (without loss of generality $C = \{0\}$ and $E = \{1\}$) and a cost sequence, it is easy to recall the request sequence. We define a set of pattern sequences S which consists of a finite number of cost sequences of bounded length s. The patterns in S form a prefix code. In other words, there is a unique way to partition any cost sequence into members of S (a remainder of length at most $s-1$ might be left over). We call the points in the cost sequence where a member of S ends 'breakpoints'. The starting point of the complete sequence is also considered a breakpoint.

Besides having to serve the request sequence, a second requirement from the off-line algorithms is that at every breakpoint, each of the six algorithms (that is, the five off-line algorithms and the on-line algorithm) have a different configuration. This means that at every breakpoint, exactly one algorithm has each possible configuration. Note that we do not require a certain order among the configurations of the off-line algorithms.

For each of the patterns $p \in S$, we compute the cost of the on-line algorithm on p and compute the cost of the five off-line algorithms to serve all requests and end up in a valid configuration (i.e. so that all six final configurations are again all different). Note that this gives room to the off-line algorithms to choose which configuration each of them ends up in, and there are $5! = 120$ possibilities.

As we need to show a lower bound of $\mathcal{R}$, we compare the on-line cost on p, denoted by $\text{ONL}(p)$, to the total cost of the off-line algorithms $\text{OFF}(p)$, and show $5 \cdot \text{ONL}(p) \geq \mathcal{R} \cdot \text{OFF}(p)$. This implies that for the total off-line cost we have

$$\text{OFF}(\sigma) \leq \frac{5}{\mathcal{R}} \cdot \text{ONL}(\sigma) + 5(1+x) + 5(s-1)x.$$

The value $5(1+x)$ is the setup cost of the five off-line algorithms to reach the correct starting configurations, and $5(s-1)x$ is an upper bound on the cost of serving the 'remainder' of the request sequence which is not a pattern in S. As OPT is an optimal off-line algorithm, its cost is at most the average cost of the five off-line algorithms and so $\text{OPT}(\sigma) \leq 1/\mathcal{R} \cdot \text{ONL}(\sigma) + 1 + sx$. We get that

$$\text{ONL}(\sigma) \geq \mathcal{R} \cdot \text{OPT}(\sigma) - 1 - sx.$$

It is left to show the sets S that imply the lower bound. We show a pattern set S_1 that proves a lower bound of 2 for every value of x in the interval $[1, 5/3]$, and a pattern set S_2 which proves the lower bound $3(1+x)/4$ for every value of x in the interval $[5/3, 2.1299556]$. See Table 1. To check that those patterns are indeed valid (give the correct lower bound on the competitive ratio) it is only required to solve a linear or a quadratic equation for each pattern p. It is also easy to see that both pattern sets form a prefix code. The cost of the on-line algorithm follows directly from the pattern whereas the off-line costs require a short proof. The details of that are omitted. □

Even though all the data was carefully verified, the origin of the pattern sets is a computer program we used. The program performs an exhaustive search on all 120 permutations and finds the cheapest cost of the off-line algorithms for each possible pattern. The result given here is an output of the program when it checked all patterns up to a length of 11. Using this program for patterns of length 13, we also found that FIFO is optimal for $x \leq 2.206$. An extension of the program for patterns of length at most 15 improves the bound by a negligible amount.

Table 1. Left are the patterns for $x \in [1, 5/3]$, right for $x \in [5/3, 2.1299556]$.

Pattern p	ONL(p)	OFF(p)
111	3	7
$xx1$	$2x+1$	$2x+5$
$11x$	$x+2$	$x+6$
xxx	$3x$	$2x+5$
$x111$	$x+3$	$x+9$
$x1x1$	$2x+2$	$5x+5$
$1x1x$	$2x+2$	$5x+5$
$1xx1$	$2x+2$	$2x+8$
$x11x$	$2x+2$	$2x+8$
$x1xx$	$3x+1$	$4x+6$
$1xxx$	$3x+1$	$2x+8$
$1x111$	$x+4$	$3x+8$
$1x11x11$	$2x+5$	$4x+13$
$1x11xx1$	$3x+4$	$4x+13$
$1x11x1x$	$3x+4$	$5x+12$
$1x11xxx$	$4x+3$	$3x+14$

Pattern p	ONL(p)	OFF(p)
x	x	$x+2$
$11x$	$x+2$	$x+6$
$1xx$	$2x+1$	9
1111	4	8
$111x$	$x+3$	$x+7$
$1x11x$	$2x+3$	15
$1x1xx$	$3x+2$	$x+14$
$1x1111$	$x+5$	$x+13$
$1x111x$	$2x+4$	$2x+12$
$1x1x1x$	$3x+3$	20
$1x1x11x$	$3x+4$	$x+20$
$1x1x1111$	$2x+6$	$5x+11$
$1x1x111x$	$3x+5$	$3x+17$

Interestingly, it is possible to show that LRU is strictly worse than FIFO, in contrast to [6]. We have the following result.

Lemma 7. $\mathcal{R}(\text{LRU}) = 1 + x$.

Proof. The upper bound follows from Theorem 1.

Denote the starting caches of LRU by $C = \{0\}$ and $E = \{1\}$. Without loss of generality, we may assume LRU starts by using the cheap cache. We use the following sequence: $(202101)^N$. For one iteration of this sequence (six requests), LRU pays $2 + 2x$. On the other hand, an offline algorithm can keep 0 in E at all times and only pay 2 per iteration. By letting N grow without bound, we get the desired result. □

Note that the lower bound from Lemma 6 also holds for LRU, so for large κ both algorithms have a competitive ratio that tends to $\kappa(1 + x)$. We end this section with a short analysis of two other natural algorithms. When x is large, it is tempting to use an algorithm which uses only the cheap cache, i.e. each time there is a page fault, the algorithm replaces the page in C by the requested page. This algorithm is not competitive at all. Consider the on-line cache in the beginning of the sequence, and denote $C = \{a\}$ and $E = \{b\}$. Let c and d be pages such that all four pages a, b, c, d are distinct. The sequence simply alternates between pages c and d. Let $2N$ be the length of the sequence. Then the cost of the algorithm is $2N$, whereas OPT can initialize its caches by $C = \{c\}$ and $E = \{d\}$ and pay only $1 + x$. As N grows, the competitive ratio grows without bound.

Another natural option is to apply the BALANCE algorithm, trying to balance the costs of the two caches. As BALANCE performs as well as LRU and FIFO for $x = 1$ (competitive ratio 2 for paging), it is interesting to see how it performs for other values of x. The modified definition of BALANCE for $x > 1$ is as follows:

Keep a counter for each place in the cache (B_C and B_E). On a page fault, if $B_C + 1 \leq B_E + x$, replace the page in C and increase B_C by 1; otherwise replace the page in E and increase B_E by x.

Lemma 8. *The competitive ratio of* BALANCE *is at most* $2x$. *This is tight for infinitely many values of* x.

Proof. We again reduce a request sequence σ to a relevant request sequence and cut it into parts. Each part contains some number of cheap faults followed by one expensive fault, except for the last part which might not contain an expensive fault. Let N be the number of parts that contain an expensive fault. The cost of BALANCE consists of two amounts; cost for cheap faults and cost for expensive faults. These are stored in B_C and B_E. It is easy to show that after a page is put into E (i.e. at the start of a new part), we have $B_C \leq B_E < B_C + 1$. Hence the cost of BALANCE is $B_E + B_C \leq 2B_E = 2Nx$. To give a bound on OPT, note that there are $N - 1$ subsequences where BALANCE has a cheap fault followed (immediately) by an expensive fault followed by a cheap fault (of the next part). These 3 faults must be on 3 distinct pages and hence OPT must have at least one fault, or $N - 1$ faults in total. Consequently, the competitive ratio approaches $2x$ as the length of the sequence grows.

To show tightness we consider (integer) even values of x, say $x = 2y, y \in \mathbb{N}$. Let $C = \{0\}$ and $E = \{1\}$ the initial configuration of BALANCE. The sequence consists of N parts, each of which is the sequence $(20)^y 2(10)^y 1$. BALANCE has a fault on every request and pays $4x$ for each part. However, an offline algorithm can keep 0 in E at all times and only have two faults per part. This gives the competitive ratio $2x$. □

4 Lower Bounds for Trackless Algorithms

In this part we prove a lower bound for trackless algorithms which is linear in κx. First we prove the lower bound for the case $\kappa = 1$, and then for general κ.

Theorem 4. *For $x \geq 2$, any trackless algorithm for $\kappa = 1$ has a competitive ratio of at least $x/2 + 1$.*

Note that for $x < 2$, we can use the general lower bounds of the previous section. Specifically Theorem 3 gives a lower bound of 2 for the complete interval $1 \leq x \leq 2$.

Proof. We construct a sequence in pieces, and bound the ratio for the pieces. We have a distinguished page denoted by a, this page is kept in E by OFF, and is never placed into E by ONL. The configurations of the caches are called inverse if the caches of OFF are in the configuration $E = \{a\}$, $C = \{y\}$ and the caches of ONL are in $E = \{y\}$, $C = \{a\}$ for some page y. Each piece starts and ends when the algorithms are in an inverse configuration.

The first inverse configuration can be reached as follows. Denote the page that ONL has in cache C by a and the page that it has in E by b. Then, if this is not already the case, OFF loads a into E and b into C. Thus OFF has a startup-cost of at most $1 + x$. We are now ready to start the first piece.

Suppose the caches are ONL : $C = \{a\}, E = \{b\}$; OFF : $C = \{b\}, E = \{a\}$. The next request is for c. We use a case analysis.

Case 1: c is placed by ONL into E. OFF loads c into C and the piece ends. The configuration is inverse and the cost ratio during this piece is $x \geq x/2 + 1$ for $x \geq 2$.

Case 2: c is placed by ONL into C.

Case 2.1: The next page is placed by ONL into E. Then this request is for the page d. Denote by ℓ_E the number of pages which are placed into E by ONL after this request. These ℓ_E requests are alternating between b and d, and the $\ell_E + 1$-st request is for a. (If ONL never puts another page into C, then this is the last piece. By extending it arbitrarily long, the ratio for this piece tends to $x \geq x/2+1$.) After the ℓ_E+1-st request we are again in an inverse configuration. During this piece the cost of OFF is $\ell_E + 2$, and the cost of ONL is $(\ell_E + 1)x + 2$. Therefore the ratio is at least $x/2 + 1$.

Case 2.2: The next page is placed by ONL into C. Then this is a request for a.

Case 2.2.1: The next page is placed into E. Then we request d, and the piece is ended. OFF has cost 2, and ONL has a cost of $x + 2$.

Case 2.2.2: The next page is placed into C. Then we request c, and we are at the same configuration as we were after the first request for c (the start of case 2). During this loop OFF has cost 0, and ONL has cost 2. We can therefore ignore this loop and continue as at the start of case 2. If ONL never puts another page in E, then this is the last piece. By extending it arbitrarily long, the ratio for this piece tends to ∞.

Since we considered all the possible cases, and the ratio of the costs is at least $x/2 + 1$ in all cases, we are done. □

Theorem 5. *Any trackless algorithm for $\kappa \geq 2$ has a competitive ratio of at least $\kappa(x+1)/2$.*

Proof. Consider an arbitrary trackless online algorithm. Consider two sets of pages, $A = \{a_1, \ldots, a_{\kappa+1}\}$ and $B = \{b_1, \ldots, b_{\kappa+1}\}$. The sequence is constructed in such a way that the online algorithm always has pages from A in C, and pages from B in E. If the algorithm will place the next page into C, then this request is to the page from A which is not currently in C. Otherwise, the request is to the page from B which is not currently in E.

Consider a long sequence of requests produced by this rule, denote by p the number of online faults in C, and by q the number of faults in E. Then ONL has a cost of $p + qx$.

To estimate the offline cost for the request sequence, we consider two offline algorithms. The first algorithm OFF_1 uses E for $A_1, \ldots, A_\kappa$ and one memory cell from C for $A_{\kappa+1}$. Furthermore it uses the other $\kappa - 1$ cells (called active cells) for serving the requests for B, always evicting the page from B which will be requested the furthest in the future (LFD [4]). The other offline algorithm called OFF_2 works in the same way, but with the roles of A and B interchanged. Thus OFF_2 has the pages from A continuously in its cache.

Consider first the cost of OFF_1. It has at most $\kappa x + 1$ cost on the requests from the set A, and a starting cost of $\kappa - 1$, placing the first $\kappa - 1$ requests for the pages from set B. We can bound the remaining cost in a similar way as it is done in [12]. We partition the sequence of requests to pages in B into subsequences of length κ called κ-parts. Consider a κ-part. Suppose there is a request for a page from B, which is not contained in the $\kappa - 1$ active cells of OFF_1. Consider the first such request in the κ-part. Denote the set of active cells after servicing this request by C', the evicted page by b_i, and the remaining element of $B \setminus (C' \cup \{b_i\})$ by b_j. If during the rest of this κ-part there is a request for b_j, then there will be no other fault in the κ-part: by the LFD rule there is no further request for b_i or for the page which was evicted when b_j was placed into the cache. On the other hand, if there is no request for b_j, then there is no fault during the next $\kappa - 2$ requests (there is no request for b_i).

This yields that OFF_1 has at most 2 faults during a κ-part. Therefore we showed that the cost of OFF_1 is at most $\kappa(x+1) + \lceil 2q/\kappa \rceil$. Similarly, OFF_2 has a cost of at most $\kappa(x+1) + \lceil 2p/\kappa \rceil$. Therefore, we obtained that the optimal offline cost is at most $\kappa(x+1) + \lceil 2\min(p,q)/\kappa \rceil$. By considering arbitrarily long sequences, where we have $p + q \to \infty$, we find that the competitive ratio of any on-line algorithm is at least

$$\frac{p + qx}{2\min(p,q)/\kappa} \geq \frac{(1+x)\kappa}{2},$$

which ends the proof. □

Corollary 1. *Any trackless algorithm has a competitive ratio at least half that of any marking algorithm.*

References

1. A. Aggarwal, B. Alpern, A. K. Chandra, and M. Snir. A model for hierarchical memory. In *Proceedings of the 19th ACM Symposium on the Theory of Computing*, pages 305–313. ACM, 1987.
2. W. W. Bein, R. Fleischer, L. L. Larmore, Limited bookmark randomized online algorithms for the paging problem, *Information Processing letters*, 76: 155–162, 2000.
3. W. W. Bein, L. L. Larmore, Trackless online algorithms for the server problem, *Information Processing letters*, 74 no 1-2: 73–79, 2000.
4. L. Belady, A study of replacement algorithms for virtual storage computers, *IBM Systems Journal* 5: 78–101, 1966
5. Marek Chrobak and John Noga. Competitive algorithms for multilevel caching and relaxed list update (extended abstract). In *Proceedings of the Ninth ACM-SIAM Symp. on Discrete Algorithms*, pages 87–96. ACM/SIAM, 1998.
6. M. Chrobak and J. Noga. LRU is better than FIFO. *Algorithmica*, 23:180–185, 1999.
7. M. Chrobak, J. Sgall, The weighted 2-server Problem, in *Proc. of STACS 2000, LNCS 1770*, Springer-Verlag Berlin, 593–604, 2000.
8. A. Fiat, M. Ricklin, Competitive algorithms for the weighted server problem, *Theoretical Computer Science*, 130: 85–99, 1994.
9. A. Karlin, M. Manasse, L. Rudolph, D. Sleator, Competitive snoopy caching, *Algorithmica*, 3(1): 79–119, 1988.
10. M. Manasse, L. A. McGeoch, and D. Sleator. Competitive algorithms for server problems. *Journal of Algorithms*, 11:208–230, 1990.
11. L. Newberg, The K-server problem with distinguishable servers, Master's Thesis, Univ. of California at Berkeley, 1991.
12. D. Sleator, R. E. Tarjan, Amortized efficiency of list update and paging rules, *Communications of the ACM*, 28:202–208, 1985.
13. N. Young, Online file caching, *Proc. 9th Annual ACM-SIAM Symp. on Discrete Algorithms*, 82–86, 1998 (to appear in *Algorithmica*)

On Maximizing the Throughput of Multiprocessor Tasks*

Aleksei V. Fishkin[1] and Guochuan Zhang[1,2]

[1] Institut für Informatik und Praktische Mathematik,
Christian-Albrechts-Universität zu Kiel,
Olshausenstrasse 40, 24 098 Kiel, Germany,
{avf,gzh}@informatik.uni-kiel.de
[2] Department of Mathematics,
Zhejiang University,
Hangzhou 310027, China,
zgc@math.zju.edu.cn

Abstract. We consider the problem of scheduling n independent multiprocessor tasks with due dates and unit processing times, where the objective is to compute a schedule maximizing the throughput. We derive the complexity results and present several approximation algorithms. For the parallel variant of the problem, we introduce the first-fit increasing algorithm and the latest-fit increasing algorithm, and prove that their worst-case ratios are 2 and $2 - 1/m$, respectively ($m \geq 2$ is the number of processors). Then we propose a revised algorithm with worst-case ratio bounded by $3/2 - 1/(2m - 2)$ (m is even) and $3/2 - 1/(2m)$ (m is odd). For the dedicated variant, we present a simple greedy algorithm. We show that its worst-case ratio is bounded by $\sqrt{m} + 1$. We straighten this result by showing that the problem (even for a common due date $D = 1$) cannot be approximated within a factor of $m^{1/2-\varepsilon}$ for any $\varepsilon > 0$, unless $NP = ZPP$.

Keywords. Multiprocessor task, throughput, complexity, approximation algorithm.

1 Introduction

In the traditional theory of scheduling, each task is processed by only one processor at a time. However, due to the rapid development of parallel computer systems, new theoretical approaches have emerged to model scheduling on parallel architectures. One of these is scheduling multiprocessor tasks [5,9].

In this paper we address the following multiprocessor scheduling problem. We are given a set of n tasks $T = \{1, \dots, n\}$ and a set of m processors $M = \{1, 2, \dots, m\}$. Each task $j \in T$ has a unit processing time $p_j = 1$ and integer due date d_j. Each processor can work on at most one task at a time and a task can (or may need to be) processed simultaneously by several processors. Here we

* Supported by Alexander von Humboldt Foundation.

K. Diks et al. (Eds): MFSC 2002, LNCS 2420, pp. 269–279, 2002.

assume that all tasks are available at time zero and the objective is to maximize the *throughput* $\sum \bar{U}_j$, where $\bar{U}_j = 0$ if task j completes after d_j, and $\bar{U}_j = 1$ otherwise.

We deal with two basic variants of this problem. In the *dedicated* variant, for each task $j \in T$ there is given a prespecified set $fix_j \subseteq M$ which indicates the task must be processed by all the processors of fix_j. In the *parallel* variant, the multiprocessor architecture is disregarded and for each task $j \in T$ there is given a prespecified number $size_j \in M$ which indicates that the task can be processed by any subset of processors of the cardinality equal to $size_j$.

We call fix_j the *type* and $size_j$ the *size* of job $j \in T$, and use D to denote the largest due date $\max_j d_j$. We say that tasks have a *common* due date if $d_j = D$ for all tasks j. We call a task j *early* if it meets its due dates d_j, and *lost* otherwise. To refer to the several variants of the above scheduling problem, we use the standard notation scheme by Graham at al. [9]. We write $P|\, fix_j, p_j = 1|\sum \bar{U}_j$ and $P|\, size_j, p_j = 1|\sum \bar{U}_j$ for the dedicated and parallel variants of our problem, respectively.

Known and New Results: Both of the variants of the multiprocessor scheduling problem have been studied. However, the previous research has mainly focused on the objectives of minimizing *the makespan* $C_{\max}$ and *the sum of completion times* $\sum C_j$. As a rule, scheduling unit multiprocessor tasks in these cases is strongly NP-hard [14,11], and recently there have been proposed a number of different approximation algorithms, e.g. [1,3,6,7,14,16]. In classical scheduling theory, there are several results known for the objective of minimizing the (weighted) number of late tasks, e.g. [13,15,2]. Up to our knowledge, no results are known for the multiprocessor tasks scheduling problem with the throughput objective of maximizing the number of early tasks.

In this paper, focusing on the throughput objective, we give the complexity results and present several approximation algorithms, for both parallel and dedicated variants of the problem. The quality of an approximation algorithm A is measured by its *worst-case ratio* defined as

$$R_A = \sup_T \{N_{\mathrm{OPT}}(T)/N_A(T)\},$$

where $N_A(T)$ denotes the number of early tasks in the schedule produced by the approximation algorithm A, and $N_{\mathrm{OPT}}(T)$ denotes the number of early tasks in an optimal schedule for a task set T.

In the first part of the paper we consider the parallel variant of the problem. We prove that it is strongly NP-hard and present a number of different approximation algorithms. We start with two simple greedy algorithms, namely, FFIS and LFIS. We prove that the worst-case ratio of FFIS is 2, and the worst-case ratio of LFIS is $2 - 1/m$, respectively. Then, by refining the algorithm LFIS we get an improved algorithm HA with the worst-case ratio at most $3/2 - 1/(2m)$ (m is odd) and $3/2 - 1/(2m-2)$ (m is even).

In the second part of the paper we consider the dedicated variant. Each dedicated task requires a subset of processors. Hence, two tasks that share a processor cannot be processed at the same time. Then, if all tasks have a common due date, we can adopt the complexity result for MAXIMUM CLIQUE [10]. Accordingly, we prove that our problem cannot be approximated within $m^{1/2-\varepsilon}$ unless $NP = ZPP$, where m is the number of processors and $\varepsilon > 0$ is any given small number. On the other hand, we are able to show that the worst-case ratio of a greedy algorithm does not exceed $\sqrt{m}+1$.

Interestingly, there are a number of different relations to some well-known combinatorial problems. Just beyond the relation to MAXIMUM CLIQUE, we can find that BIN PACKING and MULTIPLE KNAPSACK correspond to the parallel variant of our problem. We discuss this in successive sections.

The paper is organized as follows: Section 2 presents the results on the parallel model, and Section 3 on the dedicated model. Conclusions are given in Section 4.

2 Scheduling Parallel Tasks

We are given a set of n tasks $T = \{1, \ldots, n\}$ and a set of m processors $M = \{1, 2, \ldots, m\}$. Each task $j \in T$ has a unit processing time $p_j = 1$, an integral due date d_j, and requires $size_j$ processors. The goal is to maximize the *throughput* $\sum \bar{U}_j$, i.e. the number of *early* tasks j that meet their due dates d_j.

Theorem 2.1. *Problem $P|\,size_j, p_j = 1|\sum \bar{U}_j$ is strongly* NP*-hard.*

Proof. Recall 3-PARTITION [8]:
INSTANCE: Set A of $3k$ elements, a bound $B \in \mathbb{Z}^+$, and a size $s(a) \in \mathbb{Z}^+$ for each $a \in A$ such that $B/4 < s(a) < B/2$ and such that $\sum_{a\in B} s(a) = kB$.
QUESTION: Can A be partitioned into k disjoint sets $A_1, A_2, \ldots, A_k$ such that, for $1 \le i \le k$, $\sum_{a\in A_i} s(a) = B$?

We transform 3-PARTITION to our problem. First, we take a set of B processors. Next, we replace each $a \in A$ by a single task j_a with the unit processing time, size $s(a)$ and due date $D = k$. (There are $3k$ tasks and all of them have a common due date k.) Clearly, this instance can be constructed in polynomial time from the 3-PARTITION instance, and 3-PARTITION is *YES* if and only if all the tasks meet the common due date. Since 3-PARTITION is strongly NP-complete [8], our problem is strongly NP-hard. □

We then concentrate on efficient approximation algorithms.

FIRST FIT INCREASING SIZE (FFIS):
Reindex the tasks j of T in non-decreasing order of sizes $size_j$. Select the tasks one by one and assign them as early as possible. If a task j cannot be assigned to meet its due date d_j, it gets lost (will not be processed).

One can see that if all tasks have a common due date, then the problem is just a version of BIN PACKING studied by Coffman, Leung & Ting in 1978 [4]. Later, Kellerer [12] proved that there is a PTAS for the MULTIPLE KNAPSACK problem, that yields a PTAS for $P|\,size_j, p_j = 1, d_j = D|\sum \bar{U}_j$ as well. From this point of view, our problem is solved completely. We thus turn to the general case that tasks have individual due dates.

We need some definitions. A task j is *large* if its $size_j$ is greater than 1/2, and *small* otherwise. Let $0 = \bar{d}_0 < \bar{d}_1 < \cdots < \bar{d}_L = D$ be all distinct due dates, where $D = \max_j \bar{d}_j$. Then, we define time slots $I_t = (t-1, t]$, $t = 1, \ldots, D$. Consider an arbitrary algorithm A scheduling tasks on time slots. We write $m(I_t)$ and N_t to denote the number of processors occupied and the number of tasks scheduled in time slot I_t, $t = 1, \ldots, D$. We say that I_t is *closed* if A meets the first task for which there is no room in I_t, and *open* if it is not closed yet. In addition, we partition $(\bar{d}_0, \bar{d}_L] = \cup_{t=1}^{D} I_t$ into *blocks* $B(1), \ldots, B(k)$:

- the first block $B(1) = (\bar{d}_0, \bar{d}_{i_1}]$ with the smallest $\bar{d}_{i_1}$ such that all tasks j in $B(1)$ have due dates $d_j \le \bar{d}_{i_1}$,
- each further block $B(s)$ is the smallest interval $(\bar{d}_{i_{s-1}}, \bar{d}_{i_s}]$ in which all tasks j have due dates $\bar{d}_{i_{s-1}} < d_j \le \bar{d}_{i_s}$.

(Notice that it can happen that there is only one block or there are L blocks.) Accordingly, we use $lost(s)$ and $sch(s)$ to denote the set of the lost tasks and the set of early tasks with due dates in block $B(s)$.

Theorem 2.2. *For $P|\, size_j, p_j = 1 | \sum \bar{U}_j$, the worst-case ratio $R_{\text{FFIS}} = 2$.*

Proof. We first prove that $R_{\text{FFIS}} \le 2$. Consider an optimal schedule with N_{OPT} tasks and the FFIS schedule with N_{FFIS} tasks. Remove from the optimal schedule all the tasks involved in the FFIS schedule. Let ℓ_t be the number of the left tasks in time slot I_t. We prove that $\sum_{t=1}^{D} \ell_t$ is at most $N_{\text{FFIS}} = \sum_{t=1}^{D} N_t$. In this case we get $N_{\text{FFIS}} \ge N_{\text{OPT}}/2$.

Recall that all the ℓ_t left tasks in the optimal schedule are lost in the FFIS schedule. Hence, in each time slot I_t the left ℓ_t tasks of the optimal schedule are not smaller in size than those of the FFIS schedule. Since FFIS schedules the tasks by non-decreasing order of sizes, the number of scheduled tasks N_t cannot be less than ℓ_t. Thus we have $N_t \ge \ell_t$ for all $t = 1, \ldots, D$.

The bound is tight. Consider two tasks: task a with $size_a = 1$, $d_a = 2$, and task b with $size_b = m$, $d_b = 1$. In an optimal schedule both of the tasks meet their due dates, but FFIS loses task b. □

LATEST FIT INCREASING SIZE (LFIS):
Reindex the tasks j of T in non-decreasing order of sizes $size_j$. Select the tasks one by one and assign them as late as possible. If a task j cannot be assigned to meet its due date d_j, it gets lost (will not be processed).

Lemma 2.3. *For $P|\, size_j, p_j = 1 | \sum \bar{U}_j$, the worst-case ratio $R_{\text{LFIS}} \le 2 - 1/m$.*

Proof. Take the following instance: There are m *small* tasks, each task j ($j = 1, \ldots, m$) has size $size_j = 1$ and due date $d_j = j$; and $m - 1$ *large* tasks, each task j ($j = m+1, \ldots, 2m$) has size $size_j = m$, but all of them have a common due date $D = m$. All tasks are early in the optimal schedule. At the same time, LFIS schedules the small tasks, but loses all the large tasks. □

Theorem 2.4. *For* $P| size_j, p_j = 1| \sum \bar{U}_j$*, the worst-case ratio* $R_{\text{LFIS}} = 2 - 1/m$.

Proof. We show that $R_{\text{LFIS}} \leq 2 - 1/m$, and then we complete by Lemma 2.3. We prove by a contradiction. Assume that $R_{\text{LFIS}} > 2 - 1/m$. Accordingly, let $T_{\min}$ be the minimum task set, in terms of the number of tasks, such that $N_{\text{OPT}}(T_{\min})/N_{\text{LFIS}}\ (T_{\min}) > 2 - 1/m$. For all task sets T with $|T| < |T_{\min}|$, it follows $N_{\text{OPT}}(T)/N_{\text{LFIS}}\ (T) \leq 2 - 1/m$.

Assume that LFIS runs on $T_{\min}$. Then, we can claim the following.

Lemma 2.5. *There are no open time slots.*

Proof. Assume that a time slot I_t is open, where $\bar{d}_{i_s} < t \leq \bar{d}_{i_{s+1}}$. Then, tasks j with due dates $d_j > \bar{d}_{i_s}$ are early, and removing all these tasks from $T_{\min}$, we get a smaller set. □

Lemma 2.6. *For each block* $B(s)$*, the tasks of* $lost(s)$ *are not smaller in size than the tasks of* $sch(s)$*, and none of task in* $lost(s)$ *fits in any of the time slots of block* $B(s)$.

Proof. Let block $B(s) = (\bar{d}_{i_{s-1}}, \bar{d}_{i_s}]$ and let j be the first lost task in $lost(s)$ (This task j has the smallest size among the tasks of $lost(s)$). Then, its due date $\bar{d}_{i_{s-1}} < d_j \leq \bar{d}_{i_s}$, and its $size_j$ is at least the size of the tasks of $sch(s)$ with due dates at most d_j. Since j is lost by LFIS, $size_j + m(t) > m$ holds for each time slot I_t, $t = \bar{d}_{i_{s-1}}+1, \ldots, d_j$. If $d_j = \bar{d}_{i_s}$, we have done. Consider the case $d_j < \bar{d}_{i_s}$. There can be several distinct due dates in $(d_j, \bar{d}_{i_s}]$, and we have to handle them separately. By the definition of a block, there is a task in $sch(s)$, say j_1, which starts before d_j and has due date $d_{j_1} > d_j$ (If there is no such task, block $B(s)$ becomes $(\bar{d}_{i_{s-1}}, d_j]$). Then, $size_j \geq size_{j_1}$ and $size_j + m(t) \geq size_{j_1} + m(t) > m$ for each time slot I_t, $t = d_j + 1, \ldots, d_{j_1}$ (Due to LFIS $size_j$ is at least the size of any task in $(d_j, d_{j_1}]$). Next, consider the reduced interval $[d_{j_1} + 1, \bar{d}_{i_s})$ and find a task, say j_2, which starts before d_{j_1} and has due date $d_{j_2} > d_{j_1}$. Then, $size_j \geq size_{j_1} \geq size_{j_2}$ and $size_j + m(t) > m$ for each time slot I_t, $t = d_{j_1} + 1, \ldots, d_{j_2}$ (Due to LFIS $size_{j_2}$ is at least the size of the tasks in $(d_{j_1}, d_{j_2}]$). We continue this process until the reduced interval is empty. In the end, we get a sequence of at most $\bar{d}_{i_s} - d_j$ tasks $j_1, j_2, j_3 \ldots$ in $sch(s)$ that "restrict" job j from being in $(d_j, \bar{d}_{i_s}]$, and "bound" its size. We combine all facts and complete the proof. □

Analogously we can prove the following lemma.

Lemma 2.7. *For any two blocks* $B(s)$ *and* $B(s')$ *(with* $s' < s$*) the tasks of* $lost(s)$ *are not smaller in size than the tasks of* $sch(s')$*, and no task of* $lost(s)$ *fits in any of the time slots of* $B(s)$. □

Now recall that the number of time slots is D. In the LFIS schedule of $T_{\min}$, each time slot contains at least one task (Lemma 2.5). Hence $N_{\text{LFIS}}\ (T_{\min}) \geq D$. For an optimal schedule of $T_{\min}$, let h be the extra number of early tasks ($N_{\text{OPT}}(T_{\min}) = N_{\text{LFIS}}\ (T_{\min}) + h$), and let S^* be the total size of all early tasks ($S^* \leq mD$). We want to find a bound on h.

Consider the following. Take the LFIS schedule and try to assign h extra tasks that are lost within blocks $B(1), \ldots, B(k)$. Consider the first block $B(1)$ and the lost tasks of $lost(1)$. Within the h tasks we select the ones in $lost(1)$. Since none of these tasks fits in any of the time slots of $B(1)$ (Lemma 2.6), we put one task per time slot extra and assign all of them in $B(1)$. Next, we take block $B(2)$ and the remainder of the h tasks in $lost(2)$. None of these tasks can be scheduled in $B(2)$ (Lemma 2.7), but we assign all of them in a similar manner in the time slots of $B(2)$. We continue until all of the h tasks are assigned.

By the above construction, the total size of the h extra tasks and the total size of the h time slots to which these tasks are assigned is at least $h(m+1)$. Since each other time slot contains at least one task, the total leftover size is at least $D-h$. However, the total size $h(m+1)+(D-h) \leq S^* = mD$ and we have $h \leq D(m-1)/m$. From another side, $N_{\rm OPT}(T_{\min})/N_{\rm LFIS}\ (T_{\min}) > 2-1/m$ gives $(N_{\rm LFIS}\ (T_{\min})+h)/N_{\rm LFIS}\ (T_{\min}) > 2-1/m$ and $h > D(m-1)/m$ (Here $N_{\rm LFIS}\ (T_{\min}) \geq D$). It is a contradiction. □

Here we give a result which will be used later.

Lemma 2.8. *If all early tasks are small,* LFIS *schedules at least* $\frac{2m-2}{3m-4}N_{\rm OPT}$ *tasks.*

Proof. See the proof of Theorem 2.4. The difference is that each time slot of the LFIS schedule contains at least two tasks. Then $h(m+1)+2(D-h) \leq mD$ and it follows $h \leq D(m-2)/(m-1)$. Therefore, $N_{\rm OPT}(T_{\min})/N_{\rm LFIS}\ (T_{\min}) \leq (T_{\min}+h)/N_{\rm LFIS}\ (T_{\min}) \leq 1+(m-2)/(2m-2) = 3/2+1/(2m-2)$. □

Notice that both FFIS and LFIS attach importance to the task sizes. In some sense, FFIS "groups" small tasks together, whereas LFIS "spreads" them (see the above "bad" examples). Can we do something better?

It seems that *Earliest Due Date (*EDD) rule – *schedule tasks in non-decreasing order of their due dates* – cannot help. For example, take k large tasks with $size_j = m$ and $d_j = k$ ($j = 1, \ldots, k$), and $m(k+1)$ small tasks with $size_j = 1$ and $d_j = k+1$ ($j = k+1, \ldots, (m+1)(k+1)$). Then, $N_{\rm OPT} = m(k+1)$, but EDD schedules only k large tasks and m small tasks. Thus, as $k \to \infty$, the ratio tends to m. However, we can combine all our ideas together.

HYBRID ALGORITHM (HA):

1. Divide the tasks of T into small ones and large ones.
2. Schedule the set of small tasks by LFIS. If there are no time slots open, go to Step 5.
3. Start from the first open time slot and go further taking the tasks in a slot and indexing them from the bottom of the slot. Then, reschedule the indexed tasks in a first-fit manner.
4. If there is a time slot which contains a single small task, say j_s, put this task j_s into the set of large tasks.
5. Schedule the set of large tasks by EDD.

Notice that we break ties in a favor of smaller size.

Lemma 2.9. *For $P|\,size_j, p_j = 1|\sum \bar{U}_j$, the worst-case ratio*

$$R_{\text{HA}} \geq \begin{cases} 3/2 - 1/(2k-2) & \text{if } m = 3k, \\ 3/2 - 1/(2k-1) & \text{if } m = 3k+1, \\ 3/2 - 1/(2k) & \text{if } m = 3k+2. \end{cases}$$

Proof. Consider the following instance. There are $3n$ tasks – for each $i = 1, \dots, n$ there are three tasks a_i, b_i, c_i with sizes x_i, y_i, z_i and due date i, respectively. Let $x_i, y_i \leq z_i$; $x_i \geq x_{i+1}$; $y_i \geq y_{i+1}$; $z_i < z_{i+1}$, and $x_i + y_i + z_i = m + 1$, $x_{i+1} + y_{i+1} + z_i = m$ (Below we specify the values of n, x_i, y_i and z_i with respect to the number of processors m). Then, HA schedules a_i and b_i in time slots I_i and lose all tasks c_i. However, one can schedule a_i, b_i and c_{i+1} in time slots I_i for $i = 1, n-1$, and schedule c_n in I_n. The number of tasks is $2n$ and $3n-2$, respectively. We specify the precise values as follows: (1) if $m = 3k$, define $n = 2k-2$, $x_i + y_i = 2k - i$ and $z_i = k + i + 1$, for $i = 1, \dots, 2k-2$; (2) if $m = 3k+1$, define $n = 2k-1$, $x_i + y_i = 2k - i + 1$ and $z_i = k + i + 1$, for $i = 1, \dots, 2k-1$; (3) if $m = 3k+2$ define $n = 2k$, $x_i + y_i = 2k - i + 2$ and $z_i = k + i + 1$, for $i = 1, \dots, 2k$. □

Theorem 2.10. *For $P|\,size_j, p_j = 1|\sum \bar{U}_j$, the worst-case ratio $R_{\text{HA}} \leq \frac{3m-4}{2m-2}$ when m is even, and $R_{\text{HA}} \leq \frac{3m-1}{2m}$ when m is odd.*

Proof. First consider the case when there are no open time slots after Step 2 of HA. Let N_S and N_L be the number of small and large tasks scheduled, respectively. Then $N_{\text{HA}} = N_S + N_L$. If I_k is the last time slot containing small tasks, then there is no small task j with due date $d_j > k$. Thus, at most $D - k$ (large) tasks can be additionally scheduled after time k. Assume that there are h tasks more in the optimum. Following the same line of ideas as in the proof of Lemma 2.8, we get $h(m+1) + 2(k-h) \leq mk$. Hence $h \leq k(m-2)/(m-1)$ and

$$\begin{aligned} N_{\text{OPT}}/N_{\text{HA}} &= (N_S + N_L + h)/(N_S + N_L) \leq 1 + h/(2k) \\ &\leq (3m-4)/(2m-2). \end{aligned}$$

Now consider the case when there is an open time slot after Step 2 of HA. Let I_t be this time slot. We put the tasks into three groups: (S1) small tasks completed before I_t, i.e. in the closed time slots; (S2) small tasks rescheduled at or after I_t except the small task j_s (if any) from Step 4; (L) large tasks scheduled and the small task j_s (if any) from Step 4. The tasks of (S1) have due dates smaller than t, and we can use Theorem 2.8. The tasks of (S2) have due dates at least t, and all of the small tasks with due dates at least t are scheduled. Let k_2 be the number of time slots occupied by the tasks of (S2). Each of these time slots contains at least two tasks. Let N_{S1}, N_{S2} and N_L be the number of tasks in (S1), (S2) and (L), respectively. Then, $N_{\text{HA}} = N_{S1} + N_{S2} + N_L$.

We consider the following three cases: **a)** task j_s shares a time slot with a large task **b)** task j_s stays alone; **c)** there is no task j_s.

We start with the last case. Take an optimal schedule. Assume that h_1 more tasks are assigned before time slot I_t, and h_2 more tasks are assigned at of after I_t. Then, $N_{\mathrm{OPT}} \leq N_{\mathrm{HA}} + h_1 + h_2$. As in the above analysis, we get $h_1 \leq (t-1)(m-2)/(m-1)$ (There are $t-1$ time slots). From $N_{S1} \geq 2(t-1)$ we have $(N_{S1}+h_1)/N_{S1} \leq 3/2 - 1/(2m-2)$. It is not hard to prove that $(N_{S2}+N_L+h_2)/(N_{S2}+N_L) \leq 3/2 - 1/(2m-2)$ when m is even. Suppose that it does not hold. Then $h_2 > (m-2)(N_{S2}+N_L)/(2m-2)$. On the other hand, $h_2(m+1) + (N_L + k_2 - h_2)(m/2+1) \leq m(N_L + k_2)$, or simplifying, $h_2 \leq (m-2)(N_L+k_2)/m$. Thus, we have $N_{S2} < (m-2)N_L$ and $h_2 > N_{S2}/2$. Since $N_{S2} \geq 2k_2$, we get $h_2 > k_2$. It is impossible. Hence, when m is even, $(N_{S2}+N_L+h_2)/(N_{S2}+N_L) \leq 3/2 - 1/(2m-2)$ and

$$\begin{aligned} N_{\mathrm{OPT}}/N_{\mathrm{HA}} &= (N_{S1}+h_1+N_{S2}+N_L+h_2)/(N_{S1}+N_{S2}+N_L) \\ &\leq (3m-4)/(2m-2). \end{aligned}$$

When m is odd, similarly we can prove that $(N_{S2}+N_L+h_2)/(N_{S2}+N_L) \leq 3/2 - 1/(2m)$ and thus

$$\begin{aligned} N_{\mathrm{OPT}}/N_{\mathrm{HA}} &= (N_{S1}+h_1+N_{S2}+N_L+h_2)/(N_{S1}+N_{S2}+N_L) \\ &\leq (3m-1)/(2m). \end{aligned}$$

Finally, in case **a)** we regard the time slot with j_s as one of the slots constructed by the small tasks of (S2), and in case **b)** we can regard the time slot with j_s as one of the slots constructed by the large tasks of (L). The above analysis remains valid in both cases. □

3 Scheduling Dedicated Tasks

We are given a set of n tasks $T = \{1,\dots,n\}$ and a set of m processors $M = \{1,2,\dots,m\}$. Each task $j \in T$ has a unit processing time $p_j = 1$, an integer due date d_j, and requires the processors of fix_j. The goal is to maximize the *throughput* $\sum \bar{U}_j$, i.e. the number of *early* tasks j that meet their due dates d_j.

Consider problem $P|\,fix_j, p_j = 1, d_j = 1|\sum \bar{U}_j$ where all tasks have a common due date $D = 1$. Then, there is a relationship between this problem and MAXIMUM INDEPENDENT SET: Given a graph $G(V,E)$ on n vertices, find a maximum independent set of vertices, i.e. a subset $V' \subseteq V$ such that no two vertices in V' are joined by an edge in E and $|V|$ is maximum.

Suppose we have an instance of our problem. For a graph $G = (V,E)$, a vertex v_j in V corresponds to a task j in T, and two vertices v_j and $v_{j'}$ are joined by an edge e in E iff $fix_j \cap fix_{j'} \neq \emptyset$. Then, scheduling maximum number of the tasks of T in $[0,1]$ is equivalent to finding a maximum independent set in G.

Suppose we have an instance of MAXIMUM INDEPENDENT SET. For a task set T, a task j_v in T corresponds to a vertex v in V, a processor i_e in M corresponds to an edge e in E, and $i_e \in fix_{j_v}$ iff $v \in e$. Then, finding a maximum independent set is equivalent to finding a maximum subset of T that can be scheduled in $[0,1]$.

The number of processors is no more than $n(n-1)/2$, where n is the number of vertices (tasks). If our problem can be approximated within $O(f(m))$ (m is the number of processors), then MAXIMUM INDEPENDENT SET can be approximated within $O(f(n^2))$. In [10], it was proved by Hastad that MAXIMUM INDEPENDENT SET can not be approximated within $n^{1-\varepsilon}$ unless NP=ZPP. Hence, we get the following theorem immediately.

Theorem 3.1. *$P|\,fix_j, p_j = 1|\sum \bar{U}_j$ cannot be approximated within $m^{1/2-\varepsilon}$ for any given small $\varepsilon > 0$, unless NP=ZPP.*

> FIRST FIT INCREASING TYPE (FFIT):
> Select tasks j in T by non-decreasing order of $|fix_j|$, and schedule them in a first-fit manner before D. If a task can not be scheduled before time D, it gets lost (will not be processed).

Theorem 3.2. *For $P|\,fix_j, p_j = 1, d_j = D|\sum \bar{U}_j$, the worst-case ratio $\sqrt{m} \leq R_{\mathrm{FFIT}} \leq \sqrt{m} + 1$.*

Proof. Note that one task can block at most $\min\{|fix_j|, m/|fix_j|\}$ tasks. It can be observed from the following facts: task j occupies $|fix_j|$ processors, and thus at most $|fix_j|$ tasks can be blocked; for each blocked task ℓ, $|fix_\ell| \geq |fix_j|$ and thus at most $m/|fix_j|$ tasks can be blocked.

Let L_t and L_t^* be the set of tasks scheduled in time slot $I_t = [t-1, t]$ ($t = 1, \ldots, D = \max_j d_j$) of the FFIT schedule and an optimal schedule, respectively. Let $lost_t^*$ be the tasks in L_t^*, which are lost in the FFIT schedule. We prove that $|lost_t^*| \leq \sqrt{m}|L_t|$. Let $j = 1, \ldots, k$ be tasks in L_t such that $|fix_1| \leq |fix_2| \leq \cdots \leq |fix_k|$. Clearly, any task in $lost_t^*$ is blocked by some task in L_t. Let B_1^* be the tasks in $lost_t^*$ blocked by task 1, and B_j^* be the tasks in $lost_t^* \setminus (B_1^* \cup \cdots \cup B_{j-1}^*)$ blocked by task j, for $j = 2, \ldots, k$. Then, for any task ℓ_j in B_j^*, $|fix_{\ell_j}| \geq |fix_j|$ holds, and we have $|B_j^*| \leq \sqrt{m}$. Therefore, $|lost_j^*| \leq \sqrt{m}|L_j|$, and hence $N_{\mathrm{OPT}} \leq (\sqrt{m}+1)N_{\mathrm{FFIT}}$.

The following simple instance shows that $R_{\mathrm{HA}} \geq \sqrt{m}$. Given $\sqrt{m}+1$ tasks j, each of which has $|fix_j| = \sqrt{m}$ and $d_j = 1$. The last $\sqrt{m}$ tasks are compatible with each other, but are incompatible with the first one. Then, $N_{\mathrm{OPT}} = \sqrt{m}$ and $N_{\mathrm{FFIT}} = 1$. □

We extend the above result to the general case.

> LATEST FIT INCREASING TYPE (LFIT):
> Select tasks j in T by non-decreasing order of $|fix_j|$, and schedule each task j in a latest-fit manner before d_j. If a task j can not be scheduled before time d_j, it gets lost (will not be processed).

Theorem 3.3. *For $P|\,fix_j, p_j = 1|\sum \bar{U}_j$, the worst-case ratio $\sqrt{m} \leq R_{\mathrm{LFIT}} \leq \sqrt{m} + 1$.*

Proof. The proof is similar to the one of Theorem 3.2. □

4 Conclusions

In this paper we have considered the scheduling problem to maximize the number of early multiprocessor tasks on both dedicated processors and parallel processors. For the parallel model, several heuristics have been proposed and analyzed. The best algorithm we have obtained has a worst-case ratio no more than 3/2. For the dedicated model, no polynomial-time algorithms can have a worst-case ratio $m^{1/2-\varepsilon}$ for any $\varepsilon > 0$, while we have shown that a greedy algorithm has a worst-case ratio at most $\sqrt{m}+1$. Although multiprocessor task scheduling has been studied extensively, the objective of maximizing throughput is somehow new. Our work raises the following questions: For the parallel variant, is there a PTAS or is it APX-Hard? How is the approximability for the general case that the processing times of tasks are non-identical?

Another interesting question is designing on-line algorithms for the problem.

Acknowledgment

We would like to thank Klaus Jansen for valuable discussions.

References

1. F. Afrati, E. Bampis, A.V. Fishkin, K. Jansen, and C. Kenyon. Scheduling to minimize the average completion time of dedicated tasks. In *Proceedings 20th Conference on Foundations of Software Technology and Theoretical Computer Science*, LNCS 1974, pages 454-464. Springer Verlag, 2000.
2. P. Brucker. *Scheduling Algorithms.* Springer Verlag, 1998.
3. X. Cai, C.-Y. Lee, and C.-L. Li. Minimizing total completion time in two-processor task systems with prespecified processor allocation. *Naval Research Logistics*, 45:231-242, 1998.
4. E.G. Coffman, J.Y-T. Leung, and D.W. Ting. Bin packing: maximizing the number of pieces packed. *Acta Informatica*, 9:263-271, 1978.
5. M. Drozdowski. Scheduling multiprocessor tasks - an overview. *European Journal of Operational Research*, 94:215-230, 1996.
6. A. Feldmann, J. Sgall, and S.-H. Teng. Dynamic scheduling on parallel machines. *Theoretical Computer Science*, 130:49-72, 1994.
7. A.V. Fishkin and K. Jansen, and L. Porkolab. On minimizing average weighted completion time of multiprocessor tasks with release dates. In *Proceedings 28th International Colloquium on Automata, Languages and Programming*, LNCS 2076, pages 875-886. Springer Verlag, 2001.
8. M.R. Garey and D.S. Johnson. *Computers and intractability: A guide to the theory of NP-completeness.* Freeman, San Francisco, CA, 1979.
9. R.L. Graham, E.L. Lawler, J.K. Lenstra, and A.H.G. Rinnooy Kan. Optimization and approximation in deterministic scheduling: a survey. *Annals of Discrete Mathematics*, 5:287-326, 1979.
10. J. Hastad. Clique is hard to approximate within $n^{1-\varepsilon}$. *Acta Mathematica*, 182:105-142, 1999.

11. J.A. Hoogeveen, S.L. Van de Velde, and B. Veltman. Complexity of scheduling multiprocessor tasks with prespecified processor allocations. *Discrete Applied Mathematics*, 55:259-272, 1994.
12. H. Kellerer. A polynomial time approximation scheme for the multiple knapsack problem. *RANDOM-APPROX*, pages 51-62, 1999.
13. E.L. Lawler. Sequencing to minimize the weighted number of of tardy jobs. *RAIRO Recherche opérationnele*, S10:27-33, 1976.
14. E.L. Lloyd. Concurrent task systems. *Operations Research*, 29:189-201, 1981.
15. C.L. Monma. Linear-time algorithms for scheduling on parallel processors. *Operation Research*, 37:116-124, 1982.
16. J. Turek, W. Ludwig, J. Wolf, and P. Yu. Scheduling parallel tasks to minimize average response times. In *Proceedings 5th ACM-SIAM Symposium on Discrete Algorithms*, pages 112-121, 1994.

Some Results on Random Unsatisfiable k-Sat Instances and Approximation Algorithms Applied to Random Structures

Andreas Goerdt and Tomasz Jurdziński*

Technische Universität Chemnitz,
Fakultät für Informatik,
Professur Theoretische Informatik,
goerdt@informatik.tu-chemnitz.de,
tju@informatik.tu-chemnitz.de

Abstract. It is known that random k-Sat instances with at least cn random clauses where $c = c_k$ is a suitable constant are unsatisfiable with high probability. These results are obtained by estimating the expected number of satisfying assignments and thus do not provide us with an efficient algorithm. Concerning efficient algorithms it is only known that formulas with $n^{\varepsilon} \cdot n^{k/2}$ clauses with k literals over n underlying variables can be efficiently certified as unsatisfiable. The present paper is the result of trying to lower the preceding bound. We obtain better bounds for some specialized satisfiability problems.

Introduction

The investigation of random k-Sat instances is a current topic of Theoretical Computer Science, see [9] an the literature cited there. This is in part due to the interesting threshold behaviour in that there exist $c_k = c_k(n)$ such that random k-Sat instances with asymptotically less than $c_k n$ random clauses are satisfiable with high probability (whp) whereas for more than $c_k n$ random clauses we have unsatisfiability with high probability (i.e. with probability tending to 1 when n the number of variables goes to infinity).

We are interested in "efficient certification" of unsatisfiability of a random k-Sat instance. That is we look for an efficient, deterministic algorithm which, given a propositional formula as input, either gives an inconclusive answer or states that the input formula is unsatisfiable. We require the algorithm to be *correct* in that the unsatisfiable answer implies that the input formula is really unsatisfiable. Assume we are given a family of probability spaces of random inputs which are unsatisfiable with high probability. "Efficient certification" requires in addition that an unsatisfiability algorithm as above is *complete* for the probability space in question. That is it answers "unsatisfiable" with high probability for inputs from this space. Note that this does not mean that it answers

* Supported by DFG, grant GO 493/1-1. On leave from Wrocław University, Institute of Computer Science, Wrocław, Poland.

K. Diks et al. (Eds): MFSC 2002, LNCS 2420, pp. 280–291, 2002.

unsatisfiable on all unsatisfiable inputs, but only on most of them. We always use the notion of efficient certification in this sense, also for other properties than unsatisfiability.

From [9] and [12] it is essentially known that for random k-Sat instances with $n^{\varepsilon} \cdot n^{k/2}$ k-clauses we can efficiently certify unsatisfiability. Note that probabilistically we know much more: Random k-Sat instances with a linear number of clauses are unsatisfiable with high probability. It is thus an obvious program to lower the bound of $n^{\varepsilon} \cdot n^{k/2}$. Apparently two recent papers are motivated by this problem, [8] and [3]. However, a better bound as the one above is by now not known. Note that it is rather the rule than the exception that we know probalistically much more than what we can efficiently certify.

We look at the following specialized satisfiability problems: For l-Out-Of-k-Sat we require that *at least* l literals per clause must be *true* in order that the formula is satisfied. The ususal satsfiability problem has $l = 1$. For $l = 2$ and $k = 5$ we can certify unsatisfiability for $n^{\varepsilon} \cdot n^{3/2}$ random 5-clauses. This is much better than the satisfiability bound for $n^{\varepsilon} \cdot n^{5/2}$ clauses which was obtained for 1-Out-Of-5-Sat, i.e. 5-Sat [12]. We think that it is only a technical matter to extend these results to other values of l and k. To obtain our result we make some observations concerning the efficient certification of "discrepancy properties", see Definition 1.1, of random 3-uniform hypergraphs. These observations may be of independent interest.

Moreover, applying classical approximation algorithms known from the literature we can efficiently certify unsatisfiability in the Not-All-Equal-3-Sat sense for a linear number of random clauses. The same holds for non-k-colorability of random graphs with linear number of edges. These *explicit* applications seem to be a conceptual novelty.

Section 1 serves to introduce the techniques of the present paper in a clear and simple context. Section 2 contains our results concerning 3-uniform hypergraphs. In section 3 we look into the l-Out-Of-k-Sat problem. And in the final section we deal with the Not-All-Equal-3-Sat problem and k-colorability.

1 Certifying Discrepancy Properties of Random Graphs

This section is mainly based on [7] and serves as a motivation for the next section in which discrepancy problems for hypergraphs are treated. In particular we refer to pages 71 ff of [7]. We consider undirected graphs $G = (V, E)$ as usual where $E \subseteq \{\{x, y\} \mid x, y \in V, x \neq y\}$. For $x \in V$ we denote by d_x the degree of x. For $X, Y \subseteq V$ we write $E(X, Y) = \{(x, y) \mid \{x, y\} \in E, x \in X \text{ and } y \in Y\}$, and $e(X, Y) = |E(X, Y)|$. Note that $E(X, Y)$ consists of *ordered* pairs. Let $E(X) = \{\{x, y\} \mid \{x, y\} \in E, x, y \in X\}$ be the set of *edges* of G inside of X and $e(X) = |E(X)|$.

Let $e = e(V)$ denote the number of all edges of the graph $G(V, E)$. The edge density of $G = (V, E)$ where $V = \{1, \ldots, n\}$ is $\rho = e/\binom{n}{2}$ ($\approx 2e/n^2$). When picking a set $\{x, y\}$ of two vertices $x \neq y$ uniformly at random ρ is the probability that $\{x, y\} \in E$. Picking a random set $X \subseteq V$ with $|X| = \alpha n$

the expectation of $e(X)$ is by linearity of expectation $\rho \cdot \binom{\alpha n}{2}$ ($\approx e \cdot \alpha^2$). For $Y = V \backslash X$ similar considerations apply.

Definition 1.1 (Discrepancy Notions). *(a) The discrepancy of $X \subseteq V$ with $|X| = \alpha n$ in the graph G is $Disc_G(X) = |e(X) - \rho \cdot \binom{\alpha n}{2}|$*
(b) The discrepancy of X and $Y \subseteq V$ where $Y = V \backslash X$ is $Disc_G(X, Y) = |e(X, Y) - \rho \cdot \alpha n \cdot (1 - \alpha)n|$.
(c) Generally we are interested in asymptotic considerations: For all n we have graphs $G = (V, E)$ where $V = \{1, \dots, n\}$ and we think of n as going to infinity. For $1 \geq \alpha \geq 0$ and α constant, we say G is of low α-discrepancy iff for all X, Y as above

$$e(X) = e \cdot \alpha^2 \cdot (1 + o(1)) \text{ and } e(X, Y) = 2 \cdot e \cdot \alpha \cdot (1 - \alpha) \cdot (1 + o(1)).$$

For α constant low α-discrepancy means that the discrepancy is asymptotically negligible with respect to the expectation of $e(X)$ and $e(X, Y)$.

The probability space $G_{n,p}$ is the space of random graphs over the set of vertices $V = \{1, \dots, n\}$ in which each possible edge is thrown in with probability p. Usually our graphs are moderately sparse in that $p = n^{\varepsilon}/n$ with $\varepsilon > 0$ being a constant.

Lemma 1.2. *Let $1 \geq \alpha \geq 0$ be fixed. The subsequent statements hold with high probability for a random G from $G_{n,p}$ with $p = n^{\varepsilon}/n$.*
(a) For all vertices x of G we have for the degree of x that $d_x = n^{\varepsilon} \cdot (1 + o(1))$.
(b) G is of low α-discrepancy.

Proof. (sketch) The proof of part (a) is based on the fact that the degree of the vertex is a random variable following binomial distribution with parameters $n - 1$ and p. The rest follows by application of Chernoff Bounds. For part (b), observe that for any $X \subseteq V$ with $|X| = \alpha n$, $e(X)$ is a binomially distributed random variable with parameters $\binom{\alpha n}{2}$ and p. □

Given graphs G on the set of vertices $V = \{1, \dots, n\}$ for each n and $d = d(n)$ we say that G is *almost d-regular* if for *all* $x \in V$ the degree $d_x = d(1 + o(1))$.

Picking a suitable function for the $o(1)$-terms, for example $1/\log_2 n$ should do, Lemma 1.2(a) can trivially be efficiently checked. As far as Lemma 1.2 (b) is concerned this is not so obvious as the property of having low α-discrepancy is co-NP-complete (see a remark on page 71 in [7]). For regular graphs, the Eigenvalues of the adjacency matrix are used to obtain efficient bounds on the discrepancy, see [1] page 119 ff. The Linear Algebra required can be found in [17]. In the regular case we have that the all-1-'s-vector is an Eigenvector of the adjacency matrix whose Eigenvalue is the degree of the graph considered. In "almost regular" case this does not hold any more and we have to resort to the Laplacian matrix instead of the adjacency matrix, as in [7]. For $G = (V, E)$ with $V = \{1, \dots, n\}$ we define the matrix $L_G(x, y)$ for $x, y \in V$ by $L_G(x, y) = 1$

if $x = y$ and $d_x \geq 1$ and $L_G(x,y) = 0$ otherwise. The matrix $H_G(x,y)$ for $x, y \in V$ is defined by $H_G(x,y) = 1/\sqrt{d_x \cdot d_y}$ if $\{x,y\} \in E$ and 0 otherwise. The Laplacian matrix of G is $\mathcal{L}_\mathcal{G} = L_G - H_G$. Throughout we assume that for all x the degree satisfies $d_x \geq 1$, so L_G is the identity matrix. For $v = (\sqrt{d_1}, \ldots, \sqrt{d_n})^{tr}$ (tr denoting the transposed column vector) it is easy to check that $H_G \cdot v = v$. That is v is an Eigenvector with Eigenvalue 1 of H_G. As H_G is symmetric and real valued it is known that H_G has n real Eigenvalues ρ_i which can be ordered as $\rho_0 = 1 \geq \rho_1 \geq \rho_2 \geq \ldots \geq \rho_{n-1}$. The Laplacian $\mathcal{L}_\mathcal{G}$ has the same Eigenvectors as H_G and the Eigenvalues are $\lambda_i = 1 - \rho_i$. Let $\bar{\lambda}$ be defined by

$$\bar{\lambda} = Max\{|1 - \lambda_i| \mid i \neq 0\} = Max\{|\rho_i| \mid i \neq 0\}.$$

For $G = (V,E)$ and $X \subseteq V$ the volume of X is $Vol\,X = \sum_{x \in X} d_x$. For $G = (V, E)$ with $V = \{1, \ldots n\}$ being almost d-regular we have for all $X \subseteq V$ that $Vol\,X = d \cdot |X| \cdot (1 + o(1))$. Under the premise of almost d-regularity the results on page 72 , 73 of [7] read as

Lemma 1.3. *[7] For $1 \geq \alpha \geq 0$ constant and for all $X \subseteq V$ with $|X| = \alpha n$ and $Y = V \backslash Y$ we have:*
(a) $|e(X) - (1/2) \cdot n \cdot d \cdot \alpha^2 \cdot (1 + o(1))| \leq \bar{\lambda} \cdot (1/2) \cdot n \cdot d \cdot \alpha \cdot (1 + o(1))$.
(b) $|e(X, Y) - n \cdot d \cdot \alpha \cdot (1 - \alpha) \cdot (1 + o(1))| \leq \bar{\lambda} \cdot n \cdot d \cdot \sqrt{\alpha(1 - \alpha)} \cdot (1 + o(1))$. □

We use this fact only for α constant and $\bar{\lambda} = o(1)$. In this case it reads $e(X) = (1/2) \cdot n \cdot d \cdot \alpha^2 \cdot (1 + o(1))$ and $e(X, Y) = n \cdot d \cdot \alpha \cdot (1 - \alpha) \cdot (1 + o(1))$.

We need to consider the usual 0-1-adjacency matrix fo a random graph G from $G_{n,p}$ where $p = n^\varepsilon / n$. This adjacency matrix is symmetric and has n real Eigenvalues. It is known (see e.g. proof in [9]) that the second largest in absolute value Eigenvalue of such an adjacency matrix is $o(n^\varepsilon)$ with high probability. As the non-zero entries of the matrix H_G are $-1/\sqrt{d_x \cdot d_y} = -1/n^\varepsilon \cdot (1 + o(1))$ this implies that $\bar{\lambda} = o(1)$ with high probability. The way to estimate the second largest Eigenvalue (the trace method) is the same for both matrices, see [9] for the details.

Now we can efficiently certify in the sense of the introduction that a random graph G from $G_{n,p}$ has low α-discrepancy: First check almost d-regularity for $d = n^\varepsilon$. Second approximate $\bar{\lambda}$ and check the property that $\bar{\lambda} = o(1)$. If both checks are positive, which they are with high probability, we know that the graph at hand has low α-discrepancy.

2 Certifying Discrepancy Properties of Random Uniform Hypergraphs

We are interested in 3-uniform hypergraphs $H = (V, E)$ where $V = \{1, \ldots, n\}$ is a standard set of n vertices and $E \subseteq V \times V \times V$ is a set of *ordered* 3-tuples. For $X, Y, Z \subseteq V$ we let $E(X, Y, Z) = E \cap (X \times Y \times Z)$, thus $E(X, X, X)$ is the set of edges inside of X. Let $e(X, Y, Z) = |E(X, Y, Z)|$. We say an edge is of *type* (X, Y, Z) if it belongs to $X \times Y \times Z$. For $0 \leq \alpha \leq 1$ we say as in the case of graphs that the hypergraph $H = (V, E)$ with $V = \{1, \ldots, n\}$ and $e = |E|$ has *low α-discrepancy* if for all $X \subseteq V$ with $X = \alpha n$ and $Y = V \backslash X$ we have that $e(X, X, X) = \alpha^3 \cdot e \cdot (1 + o(1))$ and $e(X, X, Y) = \alpha^2 \cdot (1 - \alpha) \cdot e \cdot (1 + o(1))$ and analogously for the remaining 6 possibilities of placing X and Y into the 3 positions. Let $d = d(n)$ be given. The hypergraph H is almost d-regular iff for each $x \in V$ the number of occurrences of x at each of the 3 positions of edges is asymptotically equal to d, that is $d(1 + o(1))$, and the number of edges which contain a particular pair of vertices is only a fraction of $o(1)$ of all edges.

Throwing each possible edge with probability p we get the probabiltiy space $H_{n,p}$ of random 3-uniform hypergraphs over n vertices. For $X \subseteq V$ with $|X| = \alpha n$ the random variable $e(X, X, X)$ follows the binomial distribution with parameters $(\alpha n)^3$ and p. As in Lemma 1.2 we get that $e(X, X, X)$ is with high probability concentrated at $\alpha^3 \cdot n^{1+\varepsilon}$ provided that $p = n^\varepsilon / n^2$. The analogous fact applies to all types of edges with X and Y. As in the case of graphs this applies to *all* sets X and the random hypergraph has low α-discrepancy with high probability.

As we have no obvious notion of adjacency matrix, let alone Laplacian matrix, for hypergraphs, discrepancy properties of random hypergraphs cannot be as easily certified as in the case of graphs in the end of the preceding section. In spite of the fact that there are some attempts to generalize spectral graph theory to hypergraphs, results relevant to our context are not known to the best of the authors' knowledge, see [10] and [5]. Results closest to our concern are obtained in [6]. Unfortunately the results are only useful for k-uniform hypergraphs with $\geq n^{k-1}$ edges. This is too dense to be interesting to us. The related theory of quasi-random hypergraphs is by now only developed for $\Omega(n^k)$ k-uniform edges, see the references in [6].

In order to deal with sparser hypergraphs we define some graphs associated with a hypergraph. The product graph defined below looks to the authors like a new concept in the present context. A similar graph is only used in [9].

Definition 2.1. *Let $H = (V, E)$ be a 3-uniform hypergraph.*
(a) The projection graph on the first and second coordinate is the graph $G = (V, F)$ where F contains all edges $\{x, y\}$ with $x \neq y$ such that there is an z with $(x, y, z) \in E$ (or $(y, x, z) \in E$). Such projection graphs may be defined for all pairs of two coordinates.
(b) The product graph of H with respect to the first coordinate is the graph $G = (W, F)$ where $W = V \times V$ and F is defined as follows: For $x_1, x_2, y_1, y_2 \in V$ with $(x_1, x_2) \neq (y_1, y_2)$ we have $\{(x_1, x_2)\ ,\ (y_1, y_2)\} \in F$ iff there exists an $u \in V$ such that $(u, x_1, y_1) \in E$ and $(u, x_2, y_2) \in E$. Product graphs can be defined similarly with respect to other coordinates. □

In case that H is almost d-regular we have that the product graph has asymptotically $n \cdot \binom{d}{2} = 1/2 \cdot d^2 n(1 + o(1))$ edges. We are mainly interested in the case when the underlying hypergraph is a random hypergraph. In this case we have

Corollary 2.2. *Let $H = (V, E)$ be a random 3-uniform hypergraph from $H_{n,p}$ where $p = n^{\varepsilon}/n^{3/2}$ and $\varepsilon > 0$ is a constant.*
(a) The projection graphs are random graphs from $G_{n,p'}$ with $p' = 2n^{(1/2)+\varepsilon}/n + o(1)$.
(b) With $d = 2n^{2\varepsilon}$ the product graph is almost d-regular with high probability. Moreover, with high probability the product graph has $\bar{\lambda} = o(1)$. □

We omit the proof of the above corollary here, the proof of part (b) exploits trace method, see [9] for details (however, some slight modification is required, because in [9] adjacency matrix is considered, not the Laplacian).

The following observation is one of our contributions to the area of discrepancy.

Theorem 2.3. *Let α with $0 \leq \alpha \leq 1$ be a fixed constant and let $d = d(n)$ going to infinity be given. Let $V = \{1, \dots, n\}$ and let $H = (V, E)$ be a 3-uniform hypergraph having the following properties:*

- *H is almost d-regular.*
- *The projection graphs of H have low α-discrepancy.*
- *The product graph of H with respect to the first coordinate has low α^2-discrepancy.*

Then H itself has low α-discrepancy.

One can prove the above theorem by contradiction. Assuming that there exists a constant $\varepsilon > 0$ such that in H the asymptotic fraction of edges of type (X, X, X) is $\alpha^3 + \varepsilon$, we obtain contradiction to the assumption that the product graph of H (with respect to the first coordinate) has low α^2 discrepancy. Due to limited space, details are omitted.

Now we can efficiently certify, as in the case of graphs, that a random 3-uniform hypergraph H from $H_{n,p}$ where $p = n^{\varepsilon}/n^{3/2}$ has low α-discrepancy: First check d-regularity with $d = n^{(1/2)+\varepsilon}$. Note that for ε constant and $< 1/2$ simple expectation calculations together with Markov's inequality show that the number of edges inducing double occurrences of the same two vertices is only an asymptotically negligible fraction of all edges with high probability. Next we approximate $\bar{\lambda}$'s of the projection and product graphs and certify that they are $o(1)$. Both checks must be positive with high probability as follows from Corollary 2.2 and the remarks at the end of the preceding section. The preceding theorem and Lemma 1.3 imply low α-discrepancy of H.

3 Applications to Random Unsatisfiable k-Sat Instances

Let Var_n be a standard set of n propositional variables, let Lit_n be the set of literals over Var_n, as usual. The set of k-clauses over Var_n, $\text{Clause}_{n,k}$ simply is the k-fold cartesian product of Lit_n (other common notions of clauses would not change our results). We have that $|\text{Clause}_{n,k}| = (2n)^k$. Clauses that contain only non-negated variables (negated variables, resp.) are called *all-positive* clauses (*all-negative* clauses respectively). The probability space of random k-Sat instances $\mathrm{Form}_{n,k,p}$ is obtained by throwing each k-clause independently with probability p.

In [8] the following Lemma 3.1 is proved for random 3-Sat instances with a linear number of clauses using approximation algorithms based on semidefinite programming. The subsequent Lemma 3.1 is weaker in that we assume that the number of clauses is superlinear as opposed to [8]. However our proof method is different than Feige's. Despite being of independent interest due to its conceptual simplicity its purpose is to introduce the technique of the subsequent 4-Sat result.

Lemma 3.1. *We consider the space* $\mathrm{Form}_{n,3,p}$ *where* $p = n^{\varepsilon}/n^2$, $\varepsilon > 0$ *being a constant. Let* a *be an arbitrary truth value assignment with* F_a *being the set of variables assigned* $false$ *by* a *and* T_a *the set of variables assigned true. For a random* F *we can efficiently certify the following properties:*
(a) If a *satisfies* F *then* $|F_a| = (n/2)(1 + o(1))$ *and* $T_a = (n/2)(1 + o(1))$.
(b) If a *satisfies* F *then the asymptotic fraction of all-positive clauses of each of the following types,* $(F_a, F_a, T_a), (F_a, T_a, F_a), (T_a, F_a, F_a), (T_a, T_a, T_a)$ *among the all-positive clauses is* $1/4$. *Analogous statements hold for the remaining types of clauses.*

For the notion of "type" see the beginning of the preceding section. Note that (a) means that we can efficiently certify that no assignment which assigns $true$ or $false$ to $n/2 \cdot (1 + \delta)$ variables with $\delta > 0$ a constant satisfies a random F. Note that (b) implies that the number of positive clauses with exactly 2 literals from T_a is negligible when compared to the number of all clauses, provided a satisfies F.

Proof. (a) Assume that the assignment a assigns $false$ to $(n/2)(1+2\delta)$ of the n underlying propositional variables and δ is a constant. We look at the 3-uniform hypergraph of all-positive clauses of F. Let G_1 be the projection graph on the first and second coordinate. From Corollary 2.2 and Lemma 1.2 we know that G_1 is of low $(1/2)+\delta$-discrepancy with high probability. Therefore by Definition 1.1 (c) the asymptotic fraction of $E(F_a)$ among all edges of G_1 is $((1/2) + \delta)^2$. As the positive clauses are almost d-regular for a suitable d we know that the asymptotic fraction of all-positive clauses of type $(F_a, F_a, -)$ is $(1/4) + \delta + \delta^2$. Now as the assignment a satisfies F by assumption we have that the last variable of the clauses must be from the set T_a and we get a fraction of $(1/4) + \delta + \delta^2$ clauses of type (F_a, F_a, T_a). By looking at the projection graph on the first and third coordinate we get the same asymptotic fraction of clauses of type (F_a, T_a, F_a) Now we look at the projection graph on the second and third coordinate. With high probability this graph is of low $(1/2) + \delta$-discrepancy, too. We get that the asymptotic fraction of the number of edges with one vertex from T_a and the other one from F_a is $2 \cdot ((1/2) - \delta) \cdot (1/2 + \delta) = 1/2 - 2 \cdot \delta^2$. However almost d-regularity of the hypergraph and the consideration above implies that this fraction must be asymptotically $(1/2) + 2\delta + 2\delta^2$. As low discrepancy and almost d-regularity properties can be efficiently certified the claim holds. For the other case one should look at the all-negative clauses.

(b) We use the same principles as in (a). We can assume that $|T_a| = (n/2)(1 + o(1))$ and $|F_a| = (n/2)(1+o(1))$. By low 1/2-discrepancy of the projection graphs and d-regularity of the formula for a suitable $d = d(n)$ we know that the asymptotic fraction of all-positive clauses of each of the types, $(F_a, F_a, -), (F_a, -, F_a), (-, F_a, F_a), (T_a, T_a, -), (T_a, -, T_a)$, and $(-, T_a, T_a)$ is 1/4. The clauses corresponding to the first 3 types must have a literal from T_a in the free position because otherwise the assignment would not satisfy the formula. Thus 3/4 of all clauses are already spent and it must be that asymptotically 1/4 of all clauses is of type (T_a, T_a, T_a). □

Now we come to clause size $k = 4$. When considering the space $\text{Form}_{n,4,p}$ with $p = n^\varepsilon/n^2$ the number of clauses is with high probability $> n^{2+\varepsilon}$. Splitting each clause into two halves of two literals each we get a graph with n^2 vertices and $> n^{2+\varepsilon}$ random edges. Discrepancy certifications as in [12] certify the unsatisfiability of a random formula. Not much is known for sparser random 4-Sat instances and this simple trick does not work any longer because the degree of most of the n^2 vertices of the graph would be 0. However an analogue to Lemma 3.1 can be proved for sparser random 4-Sat instances. And this implies that we can get more when we consider the l-Out-Of-k-Sat problem which is a natural generalization of the k-Sat problem in that we require that at least l literals of each clause must be *true* in order for the formula to be satisfied, see Corollary 3.3

Theorem 3.2. *Let a be an assignment of the n propositional variables with true and false. Let T_a be the set of variables set to true and F_a be the set of variables set to false. For a random formula F from $\text{Form}_{n,4,p}$ where $p = n^\varepsilon/n^{5/2}$ we can efficiently certify the following properties, provided the assignment a satisfies F:*
(a) $|T_a|$ and $|F_a|$ is asymptotically $n/2$.
(b) Among the all-positive clauses of F the asymptotic fraction of clauses of each of the following types is 1/8. The types are all possibilities of having an odd number of variables from T_a:
$(F_a, F_a, F_a, T_a), \cdots, (T_a, F_a, F_a, F_a), (T_a, T_a, T_a, F_a), \cdots, (F_a, T_a, T_a, T_a)$.
The analogous statement holds for the remaining kinds of clauses. □

We omit the proof of the above theorem, it is based on the ideas from Lemma 3.1.

Now let us return to the l-Out-Of-k-Sat problem. For $l = 2$ and $k = 4$ the 3-Sat result [9] implies certification of unsatisfiability for $n^{3/2+\varepsilon}$ many clauses. Surprisingly the same bound holds for 2-Out-Of-5-Sat, using the preceding considerations.

Corollary 3.3. *For F from $\text{Form}_{n,5,p}$ where $p = n^\varepsilon/n^{7/2}$ we can efficiently certify that F is unsatisfiable in the 2-Out-Of-5-Sat sense.*

Proof. We first look at the projection on the first 4 coordinates. This projection must be satisfied in order to make the given 5-Sat instance *true* in the 2-Out-Of-5-Sat sense. The projection is dense enough to apply Theorem 3.2. Therefore we have an asymptotic fraction of 1/8 of each of the following types among the all-positive clauses (we omit the index "$_a$" for simplicity):

$(F, F, F, T, -), (F, F, T, F, -), (F, T, F, F, -), (T, F, F, F, -)$. In order to make at least 2 literals *true* per clause the unspecified literal must be from the set of literals set to *true*, T. Looking at the projection hypergraph on the last 3 coordinates we see that we have too many edges of type (F, F, T) namely asymptotically a fraction of $1/4$ where low $1/2$-discrepancy would allow only $1/8$. Efficient certification follows from our previous considerations. □

We have not formally checked the details but the authors think that this result can be generalized in a natural way to arbitrary k and l.

4 Approximation Algorithms Applied to Random Instances

Not-All-Equal-3-Sat Problem. A 3-Sat instance is satisfied by the assignment a in the Not-All-Equal-3-Sat sense iff each clause contains at least 1 literal which is *true* under a and at least 1 literal which is *false*. In the Not-All-Equal context we modify our set of clauses $\text{Clause}_{n,k}$ in that we do not allow double occurences of literals inside clauses or x and $\neg x$ both in one clause. The Not-all-Equal-3-Sat problem is well known to be $\mathcal{NP}$-complete.

For random instances, there exists threshold values $c = c(n)$ such that for any $\epsilon > 0$ formulas with at most $(1 - \epsilon) \cdot c \cdot n$ clauses are satisfiable whereas formulas with at least $(1 + \epsilon) \cdot c \cdot n$ are unsatisfiable with high probability [2] (it is shown in [2] that $1.514 \leq c \leq 2.215$).

We show how to apply an approximation algorithm from [14] in order to get the efficient certification of unsatisfiability in the Not-All-Equal sense for a linear number of random 3-clauses.

Lemma 4.1. *Let $\varepsilon > 0$ be a small constant. Let F be a random $F \in Form_{n,3,p}$ where $p = c/n^2$ and c is a sufficiently large constant. With high probability we have that for all assignments a, a fraction of at most $7/8 + \varepsilon$ of the all-positive clauses is satisfied in the Not-All-Equal-3-Sat sense.* □

We need the following result from [14]:

Theorem 4.2. *[14] There exists polynomial time deterministic algorithm which for* each *3-Sat instance F finds a truth assignment that satisfies at least $0.878 \cdot Opt(F)$ clauses of F in the Not-All-Equal sense. Here $Opt(F)$ is the maximum over all truth assignments of the number of clauses of F which can be satisfied in the Not-All-Equal sense.*

Based on this fact we consider the following algorithm:
The input to the algorithm is an instance F of the Not-All-Equal-3-Sat problem. Let P be the set of all-positive clauses of F. Run the algorithm from Theorem 4.2 on the set of clauses P. Let S be the set of clauses satisfied in the Not-All-Equal sense by the assignment found by the algorithm. If $|S| < 0.876 \cdot |P|$ then the algorithm answers that F is unsatisfiable, in the Not-All-Equal sense. Otherwise the algorithm fails in that it gives an inconclusive answer.

The algorithm is correct in that it never finds that a fomula is unsatisfiable which is satisfiable in the Not-All-Equal sense. This follows from the fact that if the algorithm answers "unsatisfiable" then, by Theorem 4.2, the maximal number of clauses satisfied by F is $|S|/0.878 < |S|/0.876 < |P|$. Completeness in the sense that for random F the algorithm determines unsatisfiability with high probability follows with Lemma 4.1. From this lemma we know that for all assignments the fraction of all-positive clauses satisfied in the Not-All-Equal sense is with high probability at most $7/8 + \varepsilon$ with $\varepsilon > 0$ a small constant provided we pick c sufficiently large. As $7/8 = 0.875$ the algorithm must answer unsatisfiable with high probability provided we pick ε small enough.

Using simple reduction we can show that the above algorithm may be also applied for certification of unsatisfiability in Half-4-Sat sense ([3]) for random formulas with linear number of clauses, improving the certification algorithm proposed in [3] (which needs $n \log n$ clauses).

Certification of Non-k-colorability. Now, we concentrate on a graph-theoretic $\mathcal{NP}$-complete problem, k-colorability. A graph $G = (V, E)$ is called k-colorable if there exists a coloring $f : V \to \{1, \ldots, k\}$ so that no edge of G is monochromatic under f. Similarly to satisfiability problems, there exist threshold values $c_k = c_k(n)$ such that random graphs from $G_{n,p}$ are non-k-colorable with high probability if $p \geq c_k/n$. Krivelevich [15] proposed an algorithm for deciding k-colorability whose expected running time is polynomial in the probability space $G_{n,p}$ if $p \geq c/n$ where $c = c(k)$ is a sufficiently large constant. The value p is chosen such that the graph is non-k-colorable with high probability (thus the algorithm efficiently certifies non-k-colorability). Here, we propose a direct application of a known approximation algorithm for some related problem which gives a polynomial time algorithm that certifies non-k-colorability with high probability for distribution $G_{n,p}$, where $p \geq c/n$ for relatively small values of c. For clarity of presentation, we mainly concentrate on 3-colorability.

Let a k-partition of the graph G be the partition of the set of vertices of G into k subsets. Let crossing edges of a k-partition be edges whose ends belong to different subsets of the partition. We need the following result from [11, 16]:

Theorem 4.3. *[11, 16] There exists polynomial time deterministic algorithm which for each graph $G = (V, E)$ finds a k-partition of G such that the number of crossing edges for this partition is at least $d_k Opt(G)$ for $d_k > 1 - \frac{1}{k}$ and d_k going to $1 - \frac{1}{k} + 2\frac{\ln k}{k^2}$ when k goes to infinity; in particular $d_3 \geq 0.8002$. Here $Opt(G)$ is the maximum number of croosing edges over all k-partitions.*

Our result is based on the following probabilistic observation.

Lemma 4.4. *Let G be a random graph, $G \in G_{n,p}$ where $p = c/n$ and $c \geq 248$. With high probability we have that for all 3-partitions of G the number of crossing edges is not bigger than 0.8 times the number of all edges of G.* □

The proof this lemma is based on the fact that the expectation of the maximal number of crossing edges (over all 3-partitions) is smaller than $0.8|E|$, namely this is $\leq \frac{2}{3}|E|$ whp. Thus, by Chernoff Bounds, the probability of any linear deviation from the expectation is exponentially small. The rest is obtained by choosing appropriate constants.

Based on these facts we consider the following algorithm: The input to the algorithm is a graph $G(V, E)$ chosen according to $G_{n,p}$. Run the algorithm from Theorem 4.3 on G. Let q be the number of crossing edges in the partition found by the algorithm. If $q < d_k|E|$ then the algorithm answers that G is not k-colorable. Otherwise the algorithm fails, i.e. it gives an inconclusive answer.

Note that k-colorability of the graph $G(V, E)$ induces k-partition of G such that all edges in E are crossing edges. Thus, if the approximate solution obtained by the algorithm from [11] contains less than $d_k|E|$ crossing edges then each k-partition of G contains less than $|E|$ edges and G is not k-colorable. Completeness in the sense that for random graph G from $G_{n,p}$ with $p \geq c/n$ the algorithm determines non-k-colorability whp follows from the fact that the maximal number of crossing edges (over all k-partitions) is with high probability a fraction $1 - 1/k + \varepsilon$ (for each $\varepsilon > 0$) of all edges, smaller than the approximation ratio of the algorithm from Theorem 4.3. In particular for $k = 3$, we know from Lemma 4.4 that the maximal number of crossing edges in 3-partition is whp not bigger than 0.8 times the number of all edges of the graph, while the approximation ratio of the algorithm is bigger than 0.8.

Conclusions

For the Not-All-Equal-3-Sat (k-colorability, resp.) problem we have shown how the direct application of an approximation algorithm yields an algorithm certifying the unsatisfiabiltiy (non-k-colorability, resp.) of a random instance. This application is possible because the approximation ratio of the algorithm for the naturally associated maximization problem is larger than the fraction of the number of clauses a maximal solution has with high probability in a random instance. It may be worthwhile to look for other problems on random instances if approximation algorithms can be applied in this sense. However, often we have bounds on the approximation ratio making this impossible (see e.g. bounds for 3-Sat and Not-All-Equal-k-Sat where $k \geq 4$ in [13]).

Lately, Feige [8] turns, what we do for Not-All-Equal-3-SAT, on its head: Assuming that the certification of unsatisfiability is hard for random 3-Sat instances with a sufficiently large linear number of clauses interesting bounds on approximation ratios for several problems are obtained. The investigation uses the observation as stated in Lemma 3.1 as a starting point. It is an obvious program to try to obtain bounds on approximation ratios from intractability assumptions about random k-Sat instances with larger k. Theorem 3.2 might be a starting point here. Lemma 3.1 and Theorem 3.2 can be seen as an efficient reduction of unsatisfibility to solving equations over the 2-element field (up to an asymptotically but not algorithmically negligible rest). In [4] approximation algorithms for the maximum number of satisfiable equations over a finite field are considered. Can we apply these algorithms in the present context?

As the Laplacian matrix allows easy certification of low discrepancy properties of random graphs it seems worthwhile to think about calculating the $\bar{\lambda}$ of random graphs with a linear number of edges. The case of edge probability $p = n^{\varepsilon}/n$ considered by us is relatively simple because of almost regularity. For sparser random graphs almost regularity does not hold. To show Lemma 3.1 for a linear numer of clauses Feige resorts to the well known approximation algorithms for maximum cut by Goemans and Williamson. Using the conjecture that $\bar{\lambda}$ is small for random graphs with a linear number of edges should give a different (simpler?) proof of this result.

References

1. Noga Alon, Joel Spencer. *The Probabilistic Method.* John Wiley and Sons 1992.
2. D. Achlioptas, A. Chtcherba, G. Istrate, C. Moore, *The Phase Transition in NAE-SAT and 1-in-k SAT.* In Proceedings SODA 2001, SIAM, 721-722.
3. E. Ben-Sasson, Y. Bilu, *A Gap in Average Proof Complexity,* Electronic Colloquium on Computational Complexity (ECCC),003, 2002.
4. P. Berman, M. Karpinski, *Approximating Minimum Unsatisfiability of Linear Equations*, Electronic Colloquium on Computational Complexity, 025, 2001.
5. M. Bolla. *Spectra, Euclidean representations and clusterings of hypergraphs*, Discrete Mathematics 117, 1993, 19-39.
6. Fan R. K. Chung. *The Laplacian of a Hypergraph.* In Expanding Graphs. DIMACS Series in Discrete Mathematics and Theoretical Computer Science 10, 1993, American Mathematical society, 21-36.
7. Fan R. K. Chung. *Spectral Graph Theory.* Conference Board of the Mathematical Sciences,Regional Conferencec Series in Mathematics 92, America Mathematical Society, 1997.
8. U. Feige, *Relations between Average Case Complexity and Approximation Complexity,* In Proceedings STOC 2002, ACM, to appear.
9. J. Friedman, A. Goerdt, *Recognizing more unsatisfiable random 3SAT instances,* Proc. of ICALP 2001, LNCS 2076, 310-321.
10. J. Friedman, A. Wigderson, *On the second Eigenvalue of hypergraphs*, Combinatorica 15, 1995, 43-65.
11. A. M. Frieze, M. Jerrum, *Improved Approximation Algorithms for MAX k-CUT and MAX BISECTION*, Algorithmica 18(1), 1997, 67–81.
12. A. Goerdt, M. Krivelevich. *Efficient recognition of unsatisfiable random k-SAT instances by spectral methods*, Proc. of STACS 2001, LNCS 2010, 294-304.
13. J. Hastad. *Some optimal inapproximability results.* Journal of the ACM, 48(4), 2001, 798-859.
14. V. Kann, J. Lagergren, A. Panconesi, *Approximability of maximum splitting of k-Sets and some other Apx-complete problems*, IPL 58(3), 1996, 105-110.
15. M. Krivelevich, *Deciding k-colorability in expected polynomial time*, Information Processing Letters, 81(1), 2002, 1–6.
16. S. Mahajan, H. Ramesh, *Derandomizing Approximation Algorithms Based on Semidefinite Programming*, SIAM J. Comput. 28(5), 1999, 1641–1663.
17. Gilbert Strang. *Linear Algebra and its Applications.* Harcourt Brace Jovanovich Publishers, San Diego 1988.

Evolutive Tandem Repeats Using Hamming Distance

Richard Groult[1,*], Martine Léonard[1], and Laurent Mouchard[2,**]

[1] LIFAR – ABISS, Faculté des Sciences, 76821 Mont Saint Aignan Cedex, France
[2] UMR 6037 – ABISS, Faculté des Sciences, 76821 Mont Saint Aignan Cedex, France
and
Dept. Computer Science, King's College London, London WC2R 2LS, England,
{Richard.Groult,Martine.Leonard,Laurent.Mouchard}@univ-rouen.fr

Abstract. *In this paper, we present an algorithm for detecting a "new" type of approximate repeat in texts, named evolutive tandem repeat. An evolutive tandem repeat consists in the concatenation of a series of copies, where every copy might slightly differ from its predecessor. We are presenting in this paper a new* $O(\ell.|w|^2)$ *algorithm for computing evolutive tandem repeats in a word* w, ℓ *being the length of a copy. This algorithm relies on the use of equivalence classes and graphs and has been implemented in LEDA.*

1 Introduction

Since 1995 more and more whole-genome projects are providing us with large DNA sequences [FL+95,BWO$^+$96]. Genomic sequences usually contain numerous repeats. These repeats can be classified according to the length of a consensus motif (from a few letters to several hundreds), the number of copies (from a few to several thousands), the distance between two consecutive occurrences of a motif to name a few. Repetitions are now used as a main tool for DNA fingerprinting, crime investigation, several disease diagnoses [ADD$^+$01,HMD00].

In this paper, we introduce *evolutive tandem repeats with jumps.* An evolutive tandem repeat with jumps consists in a series of copies having the following properties: each copy is very similar to its predecessor and its successor (for a given distance) and the copies are almost contiguous. An approximate tandem repeat makes use of a consensus model, every copy participating to this repeat being very similar to this overall model. An evolutive tandem repeat has no need for a consensus model, the first and the last copies might be completely different but everytime we are considering two successive copies participating to the repeat, they are very similar to each other.

Some of these repeats have been observed in the human genome sequences but unfortunately, the lack of appropriate algorithm and software prevented

* supported by a French Ministry of Research grant
** partially supported by Programme inter-EPST Bio-informatique and by GenoGRID (ACI GRID)

K. Diks et al. (Eds): MFSC 2002, LNCS 2420, pp. 292–304, 2002.

us from detecting these repeats in a more systematic way. We previously developped an algorithm that search for evolutive tandem repeats in musical sequences [IKMV00] but this algorithm mostly relies on an efficient representation of music scores and therefore can't be used directly for biological sequences. Numerous algorithms searching for various kinds of repeats have been developped [Ben98,KS99,LL00,SM98] but none of these algorithms are able to locate evolutive tandem repeats, as far as we know.

In this article, we present an algorithm that finds all maximal evolutive tandem repeats in a word w in $O(\ell|w|^2)$-time where ℓ is the length of the copies. It uses the Hamming distance, that means that the copy process does not alter the length of the copies, but only modify their content.

The article is organized as follows: in section 2, we present some basic definitions for words and the Hamming distance. In section 3, we introduce repeats named evolutive tandem repeats. In section 4, we present the equivalence classes technique together with the two graphs we will use. In section 5, we propose an improvement and phrase perspectives and finally in section 6, we conclude.

2 Preliminaries

Let Σ be an alphabet and Σ^* its free monoid. A *word* (resp. *non empty word*) over Σ is an element of Σ^* (resp. Σ^+). The letter of a word w occurring at position i is denoted by w_i. The *length* $|w|$ of a word w is the number of letters of w, i.e. $w = w_1 \cdots w_{|w|}$. We denote by $u.v$ the concatenation of two words u and v. We denote by $w[i,j] = w_i w_{i+1} \cdots w_{j-1} w_j$ for $1 \le i \le j \le |w|$ the *factor* of w starting at position i and ending at position j. The concatenation of n copies of u is denoted by u^n. There exist several distances we can use for genomic sequences, we will consider in this article the *Hamming distance*, denoted by d_H: the Hamming distance between two words of equal length is the number of positions at which their corresponding letters differ. For u, v two words of Σ^* of length ℓ, $d_H(u,v) = \text{Card}\{i \in \{1, \dots, \ell\} \mid u_i \neq v_i\}$.

Definition 2.1 (Exact Tandem Repeat). *An* exact tandem repeat *is a tuple* $(v, m, \ell, n, (p_i)_{1 \le i \le n})$ *where v is a word (the repeat by itself), m is a word of length ℓ namely the* model, *n is the number of* copies *of m in v, p_i are the starting positions of the copies* $c_i = v[p_i, p_i + \ell - 1]$ *with* $p_1 = 1, p_n + \ell - 1 = |v|$ *and* $p_1 < p_2 < \ldots < p_n$, *and* $p_{i+1} = p_i + \ell$, $\forall i \in \{1, \dots, n-1\}$ *(it means that the copies are contiguous).*

Example 2.2. $(\underline{aca}\overline{aca}\underline{aca}\overline{aca}, aca, 3, 4, (1, 4, 7, 10))$ is an exact tandem repeat.

Remark 2.3. Note that $p_i = (i-1) \times \ell + 1$ and that $v = c_1 \ldots c_n = m^n$.

There exist algorithms, such as [Cro81], in $O(|w| \log |w|)$-time, that find all exact tandem repeats in a word w with an additional constraint: m is not itself an exact tandem repeat.

This kind of repeat is *exact*, all the copies being identical. In biological sequences such repeats do not appear very often: the copy process is rarely exact, the copies remain similar but not identical. In order to consider similar copies instead of exact copies, we have to define *approximate tandem repeats* using the Hamming distance to identify similar copies. First we explain what similar means by defining the neighborhood of a word, then we define approximate tandem repeats.

Definition 2.4 (Neighborhood). *Given a word m of length ℓ and an integer ε. The* neighborhood $\mathcal{N}(m,\varepsilon) = \{u \in \Sigma^* \text{ of length } \ell \mid d_H(u,m) \leq \varepsilon\}$ *is the set of all words of length ℓ having at most ε mismatches with m.*

In what follows, m is the *model*, a word of length ℓ over Σ and ε is the *error rate*, an integer we will use for approximate repeats.

Definition 2.5 (Approximate Tandem Repeat with Jumps). *An* approximate tandem repeat with jumps *(a.t.r. for short) is a tuple $(v, m, \varepsilon, j, \ell, n, (p_i)_{1\leq i\leq n})$ where v is a word, j is the maximal length of a jump, n is the number of copies, p_i are the starting positions of the copies $c_i = v[p_i, p_i + \ell - 1]$, $p_1 = 1, p_n + \ell - 1 = |v|$, $|p_{i+1} - (p_i + \ell)| \leq j$, $\forall i \in \{1, \ldots, n-1\}$ and $c_i \in \mathcal{N}(m,\varepsilon)$, $\forall i \in \{1, \ldots, n\}$.*

Example 2.6.
$(\underline{aag}\overline{aag}t\underline{ca\overline{c}}\overline{aa}\underline{cag}, cag, 1, 1, 3, 5, (1, 4, 8, 10, 13))$ is an a.t.r with jumps.

The parameter j represents the maximal length of a jump:

$$\begin{cases} p_{i+1} - (p_i + \ell) > 0 & \text{there is a gap between } c_i \text{ and } c_{i+1}, \\ p_{i+1} - (p_i + \ell) = 0 & c_i \text{ and } c_{i+1} \text{ are contiguous,} \\ p_{i+1} - (p_i + \ell) < 0 & \text{there is an overlap between } c_i \text{ and } c_{i+1}. \end{cases}$$

To make a long story short, an a.t.r. with jumps can be depicted as a series of copies of equal length ℓ, every copy being in a neighborhood of the model and the distance between the starting positions of two contiguous copies being "close to ℓ".

3 Global vs Local: The Evolutive Repeats

Each copy participating to an a.t.r. belongs to a neighborhood $\mathcal{N}(m,\varepsilon)$. The error rate ε, which is supposed to be "small", prevents two copies from being too far from each other in terms of Hamming distance. It means that, if we are looking for an a.t.r. in a text, we are observing this text and the a.t.r. from a *global* viewpoint.

Let w be a word containing the two a.t.r. $(v, m, \varepsilon, j, \ell, n, (p_i)_{1\leq i\leq n})$ and $(v', m', \varepsilon', j', \ell, n', (p'_i)_{1\leq i\leq n'})$ such that v (resp. v') occurs in w at position $p(v)$ (resp. $p(v')$) i.e. $v = w[p(v), p(v)+|v|-1]$ (resp. $v' = w[p(v'), p(v')+|v'|-1]$). We denote $d = p(v') - (p(v)+|v|)$. We suppose that we have the following properties:

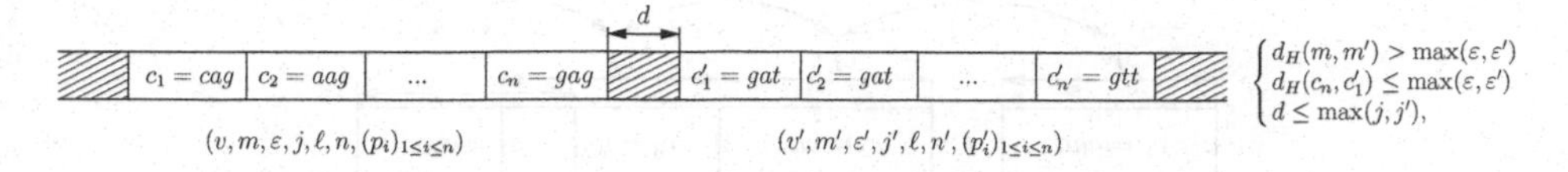

Fig. 1. Merging two contiguous approximate tandem repeats

$d_H(m,m') > \max(\varepsilon,\varepsilon')$, $d_H(c_n,c'_1) \leq \max(\varepsilon,\varepsilon')$ and $d \leq \max(j,j')$, as depicted in Fig. 1.

We can merge these two a.t.r. to obtain a longer a.t.r. but in counterpart we have to allow a larger error rate, that is:

$$(v'', m, \varepsilon'', j'', \ell, n+n', ((p_i)_{1\leq i\leq n}, (|v|+d+p'_i)_{1\leq i\leq n'}))$$

where:
$v'' = w[p(v), p(v') + |v'| - 1]$, $\varepsilon'' = d_H(m,m') + \max(\varepsilon,\varepsilon')$ and $j'' = \max(j,j')$. Increasing the error rate is not necessarily what the user is ready to pay for a longer a.t.r., we might imagine that he already chose a maximal error rate and that no increase is permitted.

Consider now that the copies may evolve during the copy process: two successive copies differ by at most ε errors. The main idea we are developing in this section is that the degradation might be propagated gradually, i.e. there is no model, but each factor that participates to this new kind of repeat is an exact or approximate copy of its predecessor and its successor whenever they exist.

3.1 Definitions

Definition 3.1 (Evolutive Tandem Repeat with Jumps). *An* evolutive tandem repeat with jumps *(e.t.r. for short) is a tuple* $(v, c_1, \varepsilon, j, \ell, n, (p_i)_{1\leq i\leq n})$ *where* v *is a word,* j *is the maximal jump,* n *is the number of copies,* p_i *are the starting positions of the copies* $c_i = v[p_i, p_i+\ell-1]$ *and* $p_1 = 1$, $p_n + \ell - 1 = |v|$, $|p_{i+1} - (p_i+\ell)| \leq j$, $\forall i \in \{1,\ldots,n-1\}$ *and* $d_H(c_i, c_{i+1}) \leq \varepsilon$, $\forall i \in \{1,\ldots,n-1\}$.

Example 3.2. $(\underline{aaa}t\overline{aac}ag\overline{cgc}, aaa, 1, 1, 3, 4, (1,5,8,10))$ is an e.t.r. with jumps: $p_1 = 1$, $p_2 = 5$ (gap), $p_3 = 8$ and $p_4 = 10$ (overlap) corresponding to $c_1 = aaa$, $c_2 = aac$, $c_3 = agc$ and $c_4 = cgc$ (c.f Fig. 2).

Definition 3.3 (Maximal e.t.r.). *Let* w *be a word and* v *a factor of* w*. An e.t.r* $(v, c_1, \varepsilon, j, \ell, n, (p_i)_{1\leq i\leq n})$ *is* maximal *in* w *if there exists no factor* v' *of* w *and no e.t.r* $(v', c'_1, \varepsilon, j, \ell, n', (p'_i)_{1\leq i\leq n'})$ *such that:*
$n' > n$ *and there exists* $i_0 \in \{1,\ldots,n'\}$ *such that* $p_k = p'_{i_0+k-1}, \forall k \in \{1,\ldots,n\}$.

Example 3.4. Let a word $w = aaaagacgaggcgg$ and $\ell = 3$. The e.t.r. $etr_1 = (aagacgagg, aag, 1, 1, 3, 3, (1,4,7))$ is not maximal in w since the repeat $etr_2 = (aagacgaggcgg, aag, 1, 1, 3, 4, (1,4,7,10))$ contains more copies. In this case, we say that etr_2 "contains" etr_1 and remark that etr_2 is a maximal e.t.r. in w.

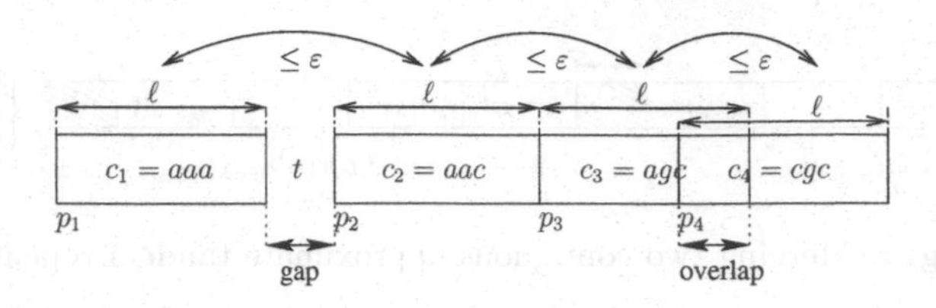

Fig. 2. Example of an evolutive tandem repeat with jumps

3.2 The Real Evolutive Tandem Repeats

An e.t.r. $(v, c_1, \varepsilon, j, \ell, n, (p_i)_{1\leq i\leq n})$ is said to be a *real e.t.r.* when there exists no $k \in \{1, \ldots, n\}$ such that $(v, c_k, \varepsilon, j, \ell, n, (p_i)_{1\leq i\leq n})$ is an a.t.r..

For example, $(agcaccaacgac, agc, 1, 0, 3, 4, (1, 4, 7, 10))$ is not a real e.t.r. because $(agcaccaacgac, aac, 1, 0, 3, 4, (1, 4, 7, 10))$ is an a.t.r..

Lemma 3.5. *Let* $r = (v, c_1, \varepsilon, \ell, n, (p_i)_{1\leq i\leq n})$ *be an e.t.r.:*

- *if there exist* i *and* $i' \in \{1, \ldots, n\}$ *such that* $d_H(c_i, c_{i'}) > 2\varepsilon$ *then* r *is a real e.t.r. (1),*
- *if there exists* $k \in \{1, \ldots, n\}$ *such that* $\forall i \in \{1, \ldots, n\}$, $d_H(c_k, c_i) \leq \varepsilon$ *then* r *is also an a.t.r. (2).*

Proof.
(1) Let $i, i' \in \{1, \ldots, n\}$ such that $d_H(c_i, c_{i'}) > 2\varepsilon$.
Then, $\forall m$ a word of length ℓ, $d_H(c_i, m) + d_H(c_{i'}, m) \geq d_H(c_i, c_{i'}) > 2\varepsilon$, so $d_H(c_i, m) > \varepsilon$ or $d_H(c_{i'}, m) > \varepsilon$ (because $a + b \geq c \Rightarrow a \geq c/2$ or $b \geq c/2$), then we can't find any model m, so r is a real e.t.r..
(2) Let $k \in \{1, \ldots, n\}$ such that $\forall i \in \{1, \ldots, n\}$, $d_H(c_k, c_i) \leq \varepsilon$.
Then $(v, c_k, \varepsilon, \ell, n, (p_i)_{1\leq i\leq n})$ is an a.t.r.. □

Note that there is no real e.t.r. such that $1 \leq n \leq 3$ since $(v, c_1, \varepsilon, j, \ell, 2, (p_1, p_2))$ and $(v, c_2, \varepsilon, j, \ell, 3, (p_1, p_2, p_3))$ are a.t.r..

4 Algorithms

In this section, we present an algorithm that finds all maximal e.t.r. in a word w for a given integer ℓ. This research is carried out in 4 steps.

First, we compute a series of equivalence relations $(E_k)_{k\in\{1,\ldots,\ell\}}$ in ascendant order. Each equivalent class of E_k is the set of starting positions of a factor of length k in w. Then, using E_ℓ-classes, we build a non-oriented graph, named the E_ℓ-class graph. Using this graph, we obtain a series of factors such that two consecutive elements in the series are slightly different. Then, we find the factors which are almost contiguous in w by constructing an oriented graph, named the E_ℓ-position graph. Finally, we look for all the longest paths in the position graph to find maximal e.t.r..

4.1 Equivalence Classes of Positions

Given a word w, we determine the positions of all factors of length ℓ in w by using a series of equivalence relations.

Definition 4.1 (Equivalence Relation E_ℓ). *Let w be a word and ℓ an integer not greater than $|w|$. Two positions i and i' are ℓ-equivalent, denoted by $i \; E_\ell \; i'$, if the factors of length ℓ occurring at positions i and i' are identical, that is:*

$$\forall i, i' \in \{1, ..., |w| - \ell + 1\}, i \; E_\ell \; i' \Leftrightarrow w[i, i + \ell - 1] = w[i', i' + \ell - 1].$$

E_ℓ is an equivalence relation. We will denote the set of E_ℓ-classes in w by $\mathcal{P}_\ell(w) = \{(C_i^\ell)_{1 \le i \le p}\}$, where p is the number of E_ℓ-classes (we will omit w in what follows). This is the set of equivalence classes associated with E_ℓ and we will denote by $f_i^\ell = w[i, i + \ell - 1]$ the factor associated with C_i^ℓ.

Crochemore [Cro81] depicts an algorithm that builds all E_ℓ-classes of a given word w in $O(|w| \log |w|)$-time. He uses this partitioning method to locate all exact tandem repeats in a string. This technique has been used for various kinds of repeats [IKMV00,IM99,IMP96,SEVS95].

Finally, we know all factors of length ℓ in w and their starting positions. We now have to compute the Hamming distance for every possible pair of factors $(f_i^\ell, f_{i'}^\ell)$ for $i \neq i'$ and represent all these distances in a convenient way.

4.2 E_ℓ-Class Graph

While searching for e.t.r., we have to be sure that two contiguous copies are similar. We must, therefore, compute the Hamming distance for every possible pair of factors. This can be done in $O(\ell p^2)$-time where p is the number of E_ℓ-classes in $\mathcal{P}_\ell$. We, therefore, extend the Hamming distance by defining "$d_H(C_i^\ell, C_{i'}^\ell) = d_H(f_i^\ell, f_{i'}^\ell)$" (Hamming distance between classes) and represent these distances by constructing a non-oriented graph namely the *E_ℓ-class graph*. In what follows we denote by (i, i', d) an edge labeled d between nodes i and i'.

Definition 4.2 (E_ℓ-Class Graph). *Let w be a word, ℓ an integer and p the number of E_ℓ-classes denoted by $(C_i^\ell)_{1 \le i \le p}$ corresponding to factors $(f_i^\ell)_{1 \le i \le p}$. The E_ℓ-class graph corresponding to E_ℓ and w is the non-oriented graph $EG_\ell(w) = (N, E)$ such that $N = \{1, \cdots, p\}$ and $E = \{(i, i', d_H(f_i^\ell, f_{i'}^\ell))$ for $(i, i') \in N \times N\}$.*

In what follows, we will restrict this graph by considering only edges such that $d_H(f_i^\ell, f_{i'}^\ell) \le \varepsilon$. This restricted graph is denoted by $EG_\ell(w, \varepsilon)$ or EG_ℓ since w and ε are both constant.

Example 4.3. $\varepsilon = 1$ (c.f. Fig. 5):

```
    1 2 3 4 5 6 7 8 9 10 11 12
w=  a c a t a c a a c a  c  a
```

index i	1	2	3	4	5	6	7
classes C_i^3	{1,5,8,10}	{2}	{3}	{4}	{6}	{7}	{9}
factors f_i^3	*aca*	*cat*	*ata*	*tac*	*caa*	*aac*	*cac*

Construct E_ℓ-classes graph(ε, $\mathcal{P}_\ell$)

1 $N \leftarrow \{1, \ldots, |\mathcal{P}_\ell|\}$
2 $E \leftarrow \emptyset$
3 **for each** class $C_i^\ell \in \mathcal{P}_\ell$ **do**
4 **for each** class $C_{i'}^\ell \in \mathcal{P}_\ell$, $i' > i$ **do**
5 ▷ Add edges when the distance is smaller than ε
6 **if** $d_H(C_i^\ell, C_{i'}^\ell) \leq \varepsilon$ **then**
7 $E \leftarrow E \cup \{(i, i', d_H(C_i^\ell, C_{i'}^\ell))\}$
8 **return** (N, E)

Fig. 3. Algorithm of the construction of the E_ℓ-class graph

Time and space complexities (c.f. algorithm Fig. 3) are both quadratic in p since we compute the Hamming distance in $O(\ell)$-time for two factors of length ℓ and we have to consider $\frac{p\times(p-1)}{2}$ pairs of factors leading to $O(\ell p^2)$-time complexity and $O(p^2)$-space complexity.

4.3 ℓ-Position Graph

We now determine the factors which are almost contiguous in the sequence by building an oriented graph (namely *ℓ-position graph*) associated with E_ℓ. Its nodes are labeled with the positions $\{1, \ldots, |w| - \ell + 1\}$ of all factors of length ℓ and there exists an edge between two nodes if the Hamming distance between their associated factors is not greater than a given ε. The ℓ-position graph is computed from the E_ℓ-class graph. In what follows we denote by (i, i', d) an edge labeled d from the node i to the node i'.

Definition 4.4 (ℓ-Position Graph). *Let w be a word and ε, j integers. The* ℓ-position graph *corresponding to w, ε and j is the oriented graph $PG_\ell(w, \varepsilon, j) = (N, E)$ where $N = \{1, ..., |w| - \ell + 1\}$ and $E = \{(i, i', i' - (i + \ell))$ for $(i, i') \in N \times N$, $i < i'$such that $|i' - (i + \ell)| \leq j$, $d_H(w[i, i + \ell - 1], w[i', i' + \ell - 1] \leq \varepsilon\}$.*

To increase readability we will denote by EG_ℓ for $EG_\ell(w, \varepsilon)$ and PG_ℓ for $PG_\ell(w, \varepsilon, j)$, w, ε and j being constant.

Remark 4.5. The ℓ-position graph is acyclic since an edge between two nodes i and i' where $i < i'$ is oriented from i to i'.

Example 4.6. (c.f. Fig.6).

Consider i and i' two integers such that $0 < i < i' \leq |w| - \ell + 1$. Let denote by d the distance $i' - (i + \ell)$ between the two fragments starting at position i and i'. Two cases are possible:

- i and i' belong to the same E_ℓ-class C_k^ℓ, i.e. $f_k^\ell = w[i, i+\ell-1] = w[i', i'+\ell-1]$. The Hamming distance between $w[i, i + \ell - 1]$ and $w[i', i' + \ell - 1]$ equals zero and therefore is smaller than ε. So, if $|d|$ is smaller than j, we add an edge from i to i' labeled d in PG_ℓ (we have an exact tandem repeat with jumps, c.f. algorithm Fig. 4 lines 3-10).

CONSTRUCTION ℓ-POSITION GRAPH(ε, ℓ, j, $\mathcal{P}_\ell$, EG_ℓ)

1 $N \leftarrow \emptyset$
2 $E \leftarrow \emptyset$
3 $\triangleright$ Create edges corresponding to exact tandem repeats with jumps
4 **for each** class $C_k^\ell \in \mathcal{P}_\ell$ **do**
5 $\quad\triangleright$ Create an edge between two positions in C_k^ℓ which are almost contiguous
6 $\quad$**for each** pair of positions $(i, i') \in C_k^\ell \times C_k^\ell$ such that $i < i'$ **do**
7 $\quad\quad d \leftarrow i' - (i + \ell)$
8 $\quad\quad$**if** $|d| \leq j$ **then**
9 $\quad\quad\quad N \leftarrow N \cup \{i, i'\}$
10 $\quad\quad\quad E \leftarrow E \cup \{(i, i', d)\}$
11 $\triangleright$ Create edges corresponding to approximate tandem repeats with jumps
12 **for each** class $C_k^\ell \in \mathcal{P}_\ell$ **do**
13 $\quad$**for each** class $C_{k'}^\ell \in \mathcal{P}_\ell$ such that $k < k'$ and $d_H(C_k, C_{k'}) \leq \varepsilon$ **do**
14 $\quad\quad\triangleright$ nodes labeled k and k' are adjacent in EG_ℓ
15 $\quad\quad$**for each** pair of positions $(i, i') \in C_k^\ell \times C_{k'}^\ell$, such that $i < i'$ **do**
16 $\quad\quad\quad d \leftarrow i' - (i + \ell)$
17 $\quad\quad\quad$**if** $|d| \leq j$ **then**
18 $\quad\quad\quad\quad N \leftarrow N \cup \{i, i'\}$
19 $\quad\quad\quad\quad E \leftarrow E \cup \{(i, i', d)\}$
20 **return** (N, E)

Fig. 4. Construction of the ℓ-position graph

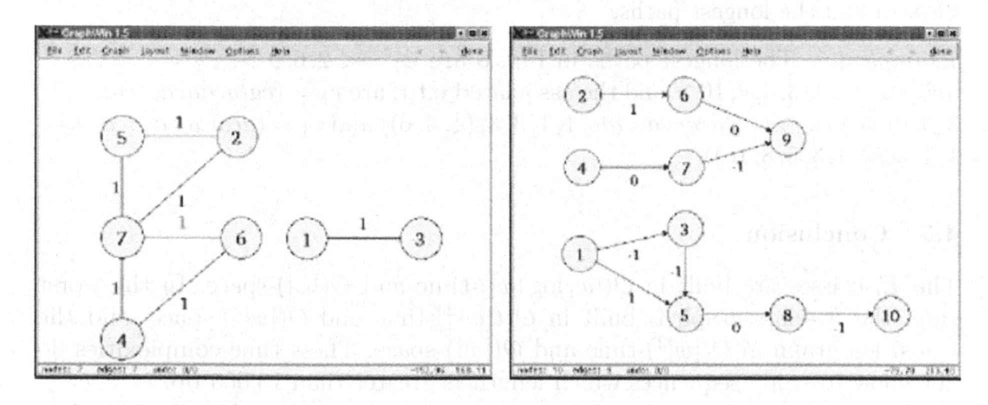

Fig. 5. Equivalence graph

Fig. 6. position graph

– i and i' belong to two different E_ℓ-classes C_k^ℓ and $C_{k'}^\ell$, then f_k^ℓ is different from $f_{k'}^\ell$. Recall that the Hamming distance between f_k^ℓ and $f_{k'}^\ell$ is smaller than ε if and only if there exists an edge between the two nodes labeled k and k' in EG_ℓ. So if k and k' are connected in EG_ℓ and if $|d|$ is smaller than j (corresponding to a.t.r.) then we add an edge from i to i' labeled d in PG_ℓ (c.f. algorithm Fig. 4).

This algorithm consider all the pairs (i, i') such that $i < i'$ and $w[i, i+\ell-1] = w[i', i'+\ell-1]$ (lines 3-10) or $0 < d_H(w[i, i+\ell-1], w[i', i'+\ell-1]) \leq \varepsilon$ (lines 11-19). In the worst case, all the $|w| - \ell + 1$ factors of length ℓ are similar according to the Hamming distance and therefore we have to consider $\frac{(|w|-\ell+1)\times(|w|-\ell)}{2}$ pairs (i, i') with $i < i'$.

Finally, the worst-case time complexity is $O(|w|^2)$. The ℓ-position graph contains at most $|w| - \ell + 1$ nodes and one node has at most $(2j + 1)$ edges, so the space complexity is $O(|w|)$.

At the end of this step, there exists an edge (i, i', d) between two nodes in PG_ℓ if and only if the factors occurring at positions i and i' are slightly different and are almost contiguous in the sequence.

4.4 Longest Path

We are looking for all the maximal e.t.r. appearing in a word w.
Let $r = (v, c_1, \varepsilon, j, \ell, n, (p_i)_{1 \leq i \leq n})$ be an e.t.r. in w. Let $(s_i)_{i \in \{1,\ldots,n\}}$ be the starting positions of the copies $(c_i)_{i \in \{1,\ldots,n\}}$ of r, then $s_i = p_i + s_1 - 1$. Since r is an e.t.r., we have for all i in $\{1, \ldots, n-1\}$, $|p_{i+1} - (p_i + \ell)| \leq j$ and $d_H(c_i, c_{i+1}) \leq \varepsilon$. Therefore, there exists an edge between s_i and s_{i+1} in PG_ℓ and then a path $\psi = < s_1, s_2, \ldots, s_n >$ in PG_ℓ. If r is a maximal e.t.r. we cannot extend r with another copy and ψ is a longest path in PG_ℓ. Such a path must start with a node without any in-edge and finish with a node without any out-edge. Since the graph is acyclic, this starting node exists and we use an in-depth method visit to find the longest paths.

Example 4.7. The longest paths in Fig. 6 are $\psi_1 = < 2, 6, 9 >$, $\psi_2 = < 4, 7, 9 >$ and $\psi_3 = < 1, 3, 5, 8, 10 >$ and the associated e.t.r. are $r_1 = (catacaacac, cat, 1, 1, 3, 3, (1, 5, 8))$, $r_2 = (tacaacac, tac, 1, 1, 3, 3, (1, 4, 6))$ and $r_3 = (acatacaacaca, aca, 1, 1, 3, 5, (1, 3, 5, 8, 10))$.

4.5 Conclusion

The E_ℓ-classes are built in $O(|w| \log |w|)$-time and $O(|w|)$-space. In the worst case, the E_ℓ-class graph is built in $O(\ell|w|^2)$-time and $O(|w|^2)$-space, and the ℓ-position graph in $O(|w|^2)$-time and $O(|w|)$-space. These time complexities do not allow to study sequences which length is greater than 13,000 bp.

In the next section we propose some improvements that decrease the time complexity.

5 Improvement: Progressive Computation of the Distances

To reduce the time complexity of the algorithm building the E_ℓ-class graph, the idea we presented in sections 4.1 and 4.2 might be combined, i.e the computation of distances and the construction of the E_ℓ-class graph will be carry out simultaneously.

Remark 5.1. Given two words $u = u_1 \cdots u_r$ and $v = v_1 \cdots v_r$, we have

$$d_H(u, v) = d_H(u[1, r-1], v[1, r-1]) + \begin{cases} 1 \text{ if } u_r \neq v_r, \\ 0 \text{ otherwise.} \end{cases}$$

The algorithm we use to build E_ℓ-classes refines at each step a class C_i^ℓ by splitting it into the classes $C_{i_1}^{\ell+1}, C_{i_2}^{\ell+1}, \cdots, C_{i_r}^{\ell+1}$, named *child classes.* We store the information that the *parent*-class of $C_{i_1}^{\ell+1}, C_{i_2}^{\ell+1}, \cdots$ and $C_{i_r}^{\ell+1}$ is C_i^ℓ. Therefore, according to the previous remark, we compute the distance between two classes $C_{i_k}^{\ell+1}$ and $C_{i'_{k'}}^{\ell+1}$ in $\mathcal{P}_{\ell+1}$ knowing their *parent*-classes. We denote by p_i^ℓ a position in the class C_i^ℓ and $f_i^\ell = w[p_i^\ell, p_i^\ell + \ell - 1]$:

$$d_H(C_{i_k}^{\ell+1}, C_{i'_{k'}}^{\ell+1}) = d_H(parent(C_{i_k}^{\ell+1}), parent(C_{i'_{k'}}^{\ell+1})) + \begin{cases} 1 \text{ if } w[p_{i_k}^{\ell+1} + \ell] \neq w[p_{i'_{k'}}^{\ell+1} + \ell], \\ 0 \text{ otherwise.} \end{cases} \tag{1}$$

Let EG_ℓ be the E_ℓ-class graph obtained at step ℓ of the construction of the equivalent classes. At step $(\ell+1)$, we compute new classes and store the indexes of the *parent* classes of each created class. We build the new E_ℓ-class graph $EG_{\ell+1}$, whose nodes are classes of $\mathcal{P}_{\ell+1}$. We will see now for the edges.

Lemma 5.2. *Let C_i^ℓ be a class of $\mathcal{P}_\ell$ and $C_{i_1}^{\ell+1}, \ldots, C_{i_r}^{\ell+1}$ its child classes in $\mathcal{P}_{\ell+1}$ then $d_H(C_{i_k}^{\ell+1}, C_{i_{k'}}^{\ell+1}) = 1$ for all k, k' in $\{1, \ldots, r\}$, $k \neq k'$.*

Proof. Two child classes of a same parent class only differ on their last position. □

Now, consider $C_{i_k}^{\ell+1}$ and $C_{i'_{k'}}^{\ell+1}$ two classes of $\mathcal{P}_{\ell+1}$. Let denote $d_p = d_H(parent(C_{i_k}^{\ell+1}),$
$parent(C_{i'_{k'}}^{\ell+1}))$ and $d_c = d_H(C_{i_k}^{\ell+1}, C_{i'_{k'}}^{\ell+1})$. Two cases are possible :

- if $parent(C_{i_k}^{\ell+1}) = parent(C_{i'_{k'}}^{\ell+1})$ then (c.f. Lemma 5.2) an edge $(i_k, i'_{k'}, 1)$ is created in $EG_{\ell+1}$ (c.f. algorithm Fig. 7 line 4-5);
- if $parent(C_{i_k}^{\ell+1}) \neq parent(C_{i'_{k'}}^{\ell+1})$, two cases are possible:
  - $d_p > \varepsilon$: then no edge is created (c.f. Eq.(1));
  - $0 < d_p \leq \varepsilon$: there exists an edge (i, i', d_p) in EG_ℓ. An edge $(i_k, i'_{k'}, d_c)$ is created in $EG_{\ell+1}$ if and only if d_c is not greater than ε (c.f. algorithm Fig. 7 lines 6-13).

  So we compute the distance between two child classes only if their parents are adjacent in EG_ℓ.

Lemma 5.3. *Let Σ be an alphabet, v a word over Σ and ε an integer.*

$$|\mathcal{N}(v, \varepsilon)| = \sum_{e=0}^{e=\varepsilon} (|\Sigma| - 1)^e \binom{|v|}{e}.$$

Proof. Let $e \leq \varepsilon$ and u be a word such that $d_H(u, v) = e$. Since there are $\binom{|v|}{e}$ ways of choosing e positions in a word v, we obtain $\binom{|v|}{e}$ different sets of error positions. For a single error position i, there exist exactly $|\Sigma| - 1$ possible letters that differ from $v[i]$. Therefore for each set of error positions, there exist

```
CONSTRUCT E_{ℓ+1}-CLASS GRAPH(w, ε, EG_ℓ, P_{ℓ+1})
1  N ← {1,...,|P_{ℓ+1}|}
2  E ← ∅
3  for each node C_i^ℓ of EG_ℓ do
4      for each pair of child classes (C_{i_k}^{ℓ+1}, C_{i_{k'}}^{ℓ+1}) of C_i^ℓ such that i_k < i'_k do
5          E ← E ∪ {(i_k, i_{k'}, 1)}
6      for each adjacent class C_{i'}^ℓ of C_i^ℓ such that i < i' do
7          Let (i, i', d) be the edge between i and i' in EG_ℓ
8          for each child class C_{i_k}^{ℓ+1} of C_i^ℓ do
9              for each child class C_{i'_{k'}}^{ℓ+1} of C_{i'}^ℓ such that i_k < i'_{k'} do
10                 if w[p_{i_k}^{ℓ+1} + ℓ] ≠ w[p_{i'_{k'}}^{ℓ+1} + ℓ] then
11                     d ← d + 1
12                 if d ≤ ε then
13                     E ← E ∪ {(i_k, i'_{k'}, d)}
14 return (N, E)
```

Fig. 7. Computation of the new E_ℓ-class graph

$(|\Sigma|-1)^e$ words u such that $d_H(u,v)=e$. Then there are $(|\Sigma|-1)^e\binom{|v|}{e}$ words that exactly differ with v at e positions. Finally, there are $\sum_{e=0}^{e=\varepsilon}(|\Sigma|-1)^e\binom{|v|}{e}$ words in $|\mathcal{N}(v,\varepsilon)|$. □

Let C_i^ℓ be an E_ℓ-class and f_i^ℓ the associated factor. The child classes of C_i^ℓ correspond to factors $f_i^\ell a$, $a\in\Sigma$, then C_i^ℓ has at most $|\Sigma|$ child classes.

First, for each class C_i^ℓ, we add an edge for all pairs of its child classes (lines 4-5). So we have to consider at most $\mathcal{P}_\ell \times \frac{|\Sigma|\times(|\Sigma|-1)}{2}$ pairs.

Next, we consider all pairs $(C_i^\ell, C_{i'}^\ell)$ of adjacent E_ℓ-classes. For each class, there exist at most $\sum_{e=0}^{e=\varepsilon}(|\Sigma|-1)^e\binom{|w|}{e}$ adjacent classes (c.f. Lemma 5.3). Moreover, we showed that an E_ℓ-class has at most $|\Sigma|$ child classes, so the time complexity of lines 6-13 is at most $\mathcal{P}_\ell \times [\sum_{e=0}^{e=\varepsilon}(|\Sigma|-1)^e\binom{|w|}{e}]\times|\Sigma|\times|\Sigma|$.

Then the time complexity of this algorithm is at worst

$$\mathcal{P}_\ell \times \frac{|\Sigma|\times(|\Sigma|-1)}{2} + \mathcal{P}_\ell \times [\sum_{e=0}^{e=\varepsilon}(|\Sigma|-1)^e\binom{|w|}{e}]\times|\Sigma|\times|\Sigma|.$$

We will see it more precisely. Since there exist at most $|\Sigma|^\ell$ different words of length ℓ but at most $|w|-\ell+1$ factors of length ℓ of w, then $|\mathcal{P}_\ell|\le\min(|\Sigma|^\ell, |w|-\ell+1) = min_1$ and, given an E_ℓ-class, the number of adjacent classes is lower than $\min(min_1-1, \sum_{e=0}^{e=\varepsilon}(|\Sigma|-1)^e\binom{|w|}{e}) = min_2$. Finally, the time complexity is at worst

$$min_1 \times \frac{|\Sigma|\times(|\Sigma|-1)}{2} + min_1 \times min_2 \times |\Sigma|\times|\Sigma|.$$

6 Conclusion

We have presented an algorithm for detecting approximate tandem repeats and more generally evolutive tandem repeats with jumps using the Hamming distance. Even if using Hamming distance seems restrictive, we already found a few promising repetitions that were not detected with other repeat detection programs. A `C++` program using the `LEDA` library has been implemented. Its overall time and space complexities prevent it from being used for whole chromosomes on a conventional machine but a very straighforward parallel algorithm can handle chromosomes by splitting these large sequences into smaller ones. We are currently improving the parallel version on a Linux cluster.

References

[ADD+01] S.A. Ahrendt, P.A. Decker, K. Doffek, B. Wang, L. Xu, M.J. Demeure, J. Jen, and D. Sidransky. Microsatellite instability at selected tetranucleotide repeats is associated with p53 mutations in non-small cell lung cancer. *Cancer Research*, 60(9):2488–2491, 2001.

[Ben98] G. Benson. An algorithm for finding tandem repeats of unspecified pattern size. In S. Istrail, P. Pevzner, and M. Waterman, editors, *Proceedings of the 2nd Annual International Conference on Computational Molecular Biology (RECOMB-98)*, pages 20–29, New York, March 22–25 1998. ACM Press.

[BWO+96] C.J. Bult *et al.*. Complete genome sequence of the methanogenic archaeon, Methanococcus jannaschii. *Science*, 273:1058–1073, 1996.

[Cro81] M. Crochemore. An optimal algorithm for computing the repetitions in a word. *Inf. Process. Lett.*, 12(5):244–250, 1981.

[FL+95] Fleischmann R.D. *et al.* Whole-genome random sequencing and assembly of Haemophilus influenzae *Science*, 269:496–512, 1995.

[HMD00] K. Heinimann, H. Muller, and Z Dobbie. Microsatellite instability in colorectal cancer. *New England Journal of Medicine*, 342(21):1607–1608, 2000.

[IKMV00] C. S. Iliopoulos, M. Kumar, L. Mouchard, and S. Venkatesh. Motif evolution in polyphonic musical sequences. In L. Brankovic and J. Ryan, editors, *Proceedings of the 11th Australasian Workshop On Combinatorial Algorithms*, pages 53–66, Hunter Valley, Australia, 2000.

[IM99] C. S. Iliopoulos and L. Mouchard. An $O(n \log n)$ algorithm for computing all maximal quasiperiodicities in strings. In C. S. Calude and M. J. Dinneen, editors, *Combinatorics, Computation and Logic. Proceedings of DMTCS'99 and CATS'99*, Lecture Notes in Computer Science, pages 262–272, Auckland, New-Zealand, 1999. Springer-Verlag, Singapore.

[IMP96] C. S. Iliopoulos, D. W. G. Moore, and K. Park. Covering a string. *Algorithmica*, 16:289–297, September 1996.

[KS99] S. Kurtz and C. Schleiermacher. Reputer - fast computation of maximal repeats in complete genomes. *Bioinformatics*, 15(5), 1999.

[LL00] A. Lefebvre and T. Lecroq. Computing repeated factors with a factor oracle. In L. Brankovic and J. Ryan, editors, *Proceedings of the 11th Australasian Workshop On Combinatorial Algorithms*, pages 145–158, Hunter Valley, Australia, 2000.

[SEVS95] M.-F. Sagot, V. Escalier, A. Viari, and H. Soldano. Searching for repeated words in a text allowing for mismatches and gaps. In R. Baeza-Yates and U. Manber, editors, *Proceedings of the 2nd South American Workshop on String Processing*, pages 101–116, Valparaíso, Chile, 1995.

[SM98] M. Sagot and E. W. Myers. Identifying satellites in nucleic acid sequences. In S. Istrail, P. Pevzner, and M. Waterman, editors, *Proceedings of the 2nd Annual International Conference on Computational Molecular Biology (RECOMB-98)*, pages 234–242, New York, March 22–25 1998. ACM Press.

Subgraph Isomorphism, log-Bounded Fragmentation and Graphs of (Locally) Bounded Treewidth*

MohammadTaghi Hajiaghayi[1] and Naomi Nishimura[2]

[1] Laboratory for Computer Science, Massachusetts Institute of Technology, Cambridge, MA, U.S.A.,
hajiagha@theory.lcs.mit.edu

[2] School of Computer Science, University of Waterloo, Waterloo, Ontario, Canada,
nishi@uwaterloo.ca

Abstract. The subgraph isomorphism problem, that of finding a copy of one graph in another, has proved to be intractable except when certain restrictions are placed on the inputs. In this paper, we introduce a new property for graphs (a generalization on bounded degree) and extend the known classes of inputs for which polynomial-time subgraph isomorphism algorithms are attainable. In particular, if the removal of any set of at most k vertices from an n-vertex graph results in $O(k \log n)$ connected components, we say that the graph is a log-*bounded fragmentation* graph. We present a polynomial-time algorithm for finding a subgraph of H isomorphic to a graph G when G is a log-bounded fragmentation graph and H has bounded treewidth; these results are extended to handle graphs of *locally bounded treewidth* (a generalization of treewidth) when G is a log-bounded fragmentation graph and has constant diameter.

1 Introduction

In this paper we consider the subgraph isomorphism problem, in which we search for a subgraph of host graph H isomorphic to the source (or pattern) graph G. This problem arises in application areas ranging from text processing to biology and chemistry.

Since the subgraph isomorphism problem is NP-complete [13], algorithms have been developed for restricted classes of inputs; specific classes of graphs for which there exist polynomial-time algorithms include trees, two-connected outerplanar graphs, and two-connected series-parallel graphs, all graphs of bounded treewidth. However, the problem remains NP-complete for unrestricted graphs of bounded treewidth, also known as partial k-trees [21].

---

* This research was done while the first author was a graduate student in Computer Science department of the University of Waterloo. The research was supported by the Natural Sciences and Engineering Research Council of Canada and Communication and Information Technology Ontario.

K. Diks et al. (Eds): MFSC 2002, LNCS 2420, pp. 305–318, 2002.

Graphs of bounded treewidth are known for their good algorithmic properties; in particular, any problem that can be expressed in the language of extended monadic second order logic (EMSOL) can be solved using a general dynamic-programming approach [1]. Although the subgraph isomorphism problem can be expressed in EMSOL when the source graph is fixed, additional restrictions are needed to solve the two-input problem in polynomial time. Matoušek and Thomas [18] presented an $O(n^{k+2})$-time algorithm for bounded degree partial k-trees, and an $O(n^{k+4.5})$-time algorithm for k-connected partial k-trees, later improved to $O(n^{k+2})$ [9].

Theorem 1. *(Theorem 5.14 [18]) Suppose graph G is connected, the maximum degree of G is bounded by a constant, and H is a partial k-tree. There are $O(|V(G)|^{k+1} \cdot |V(H)|)$-time algorithms which solve isomorphism and its subgraph and induced subgraph versions.* □

Gupta and Nishimura [15] provided a new approach which generalized to other embeddings [15], finding the largest common subgraph [8], and finding a maximum packing (the maximum number of disjoint copies of a source graph in a host graph) [10].

The significance of these results stems from the intractability of more general problems. Subgraph isomorphism is NP-complete when the source graph is a tree, and the host graph is a partial 2-tree which has at most one node of degree greater than three [18], when the source and host graphs each are partial k-trees with all but k nodes of degree at most $k+2$ [14], or when the source and host graphs are both partial k-trees but the source graph is not k-connected [14].

In this paper, we extend the bounded degree result of Matoušek and Thomas to handle a more general property. A graph is a log-*bounded fragmentation* graph if the removal of at most k vertices results in $O(k \log n)$ connected components, where n is the number of vertices of the graph. The results obtained for these graphs are extended to graphs of locally bounded treewidth, a generalization of treewidth in which there are constant bounds on the treewidth of neighbourhoods in the graph, though not necessarily in the overall graph.

We begin the paper with an explanation of terminology (Section 2) and an overview of Arnborg and Proskurowski's general dynamic-programming approach to problems on graphs of bounded treewidth (Section 3). In Section 4, we demonstrate how the result of Matoušek and Thomas can be extended to graphs with the log-bounded fragmentation property, further generalizing the approach to handle graphs of locally bounded treewidth in Section 5. Finally, we conclude with a list of open problems and potential extensions for future work in Section 6. Where necessary, full proofs are omitted due to space limitations.

2 Preliminaries

We assume the reader is familiar with general concepts of graph theory such as directed graphs, trees and planar graphs. The reader is referred to standard references for appropriate background [7].

In this paper, all graphs are finite and simple, and are undirected unless indicated otherwise. A graph G is represented by $G = (V, E)$, where V (or $V(G)$) is the set of vertices and E (or $E(G)$) is the set of edges; $n = |V|$. We denote an undirected (directed) edge e between u and v by $\{u, v\}$ ((u, v)). The maximum and minimum degrees of G are denoted by $\Delta(G)$ and $\delta(G)$, respectively. The distance between vertices u and v is the length of the shortest path from u to v. We define the *r-neighborhood* of a set $S \subseteq V(G)$, denoted by $N_G^r(S)$, to be the set of vertices at distance at most r from at least one vertex of S; if $S = \{v\}$ we simply use the notation $N_G^r(v)$. The *diameter* of G, denoted by $diam(G)$, is the maximum distance between any pair of vertices of G. Two disjoint sets S and S' of vertices of undirected (directed) graph G are *adjacent* if and only if there are $u \in S$ and $v \in S'$ such that $\{u, v\} \in E(G)$ ($(u, v) \in E(G)$ or $(v, u) \in E(G)$).

A graph $G' = (V', E')$ is a *subgraph of* G if $V' \subseteq V$ and $E' \subseteq E$. A graph $G' = (V', E')$ is an *induced subgraph of* G, denoted by $G[V']$, if $V' \subseteq V$ and E' contains all edges of E which have both end vertices in V'. G is a *supergraph* of G' if G' is a (not necessarily induced) subgraph of G.

This paper is devoted to solving special cases of the subgraph isomorphism problem, where an *isomorphism* ϕ from (directed) graph G into (directed) graph H' is a one-to-one mapping between vertices of G and H' such that for each pair $u, v \in V(G)$, $\{u, v\} \in E(G)$ ($(u, v) \in E(G)$) if and only if $\{\phi(u), \phi(v)\} \in E(H')$ ($(\phi(u), \phi(v)) \in E(H')$). For a set $S \subseteq V(G)$, we define $\phi(S) = \bigcup_{v \in S} \phi(v)$. A (directed) graph G is *isomorphic* to a (directed) graph H if and only if there is an isomorphism ϕ from G into H such that $\phi(G) = V(H)$. A graph G is *subgraph isomorphic* to H if there is a subgraph H' of H which is isomorphic to G, and is *induced subgraph isomorphic* to H if there exists an induced subgraph H' of H isomorphic to G.

The set of components of a graph G is represented by $\mathcal{C}(G)$, where each element of $\mathcal{C}(G)$ is a connected graph. For a set $\mathcal{D} \subseteq \mathcal{C}(G)$, we denote the set of all vertices in components of $\mathcal{D}$ by $V(\mathcal{D})$ and the set of all such edges by $E(\mathcal{D})$. The graph resulting from the removal from G of a set S of vertices and all adjacent edges is denoted by $G[V - S]$, as it is the induced subgraph on $V - S$. A set S is called a *separator* if $|\mathcal{C}(G[V - S])| > 1$. For $k > 0$, graph G is called *k-connected* if every separator has size at least k.

The notion of treewidth was introduced by Robertson and Seymour [19] and plays an important role in their fundamental work on graph minors. To define this notion, first we consider the representation of a graph as a tree, which is the basis of our algorithms.

Definition 1. *[19] A* tree decomposition *of a graph $G = (V, E)$, denoted by $TD(G)$, is a pair (χ, T) in which $T = (I, F)$ is a tree and $\chi = \{\chi_i | i \in I\}$ is a family of subsets of $V(G)$ such that: $\bigcup_{i \in I} \chi_i = V$; for each edge $e = \{u, v\} \in E$ there exists an $i \in I$ such that both u and v belong to χ_i; and for all $v \in V$, the set of nodes $\{i \in I | v \in \chi_i\}$ forms a connected subtree of T.*

To distinguish between vertices of the original graph G and vertices of T in $TD(G)$, we call vertices of T *nodes* and their corresponding χ_i's *bags*. We define the *terminal subgraph* $G_{[z]}$ for a node z of $TD(G)$ to be the subgraph of G induced over vertices of χ_z and bags of descendants of z in $TD(G)$. The maximum size of a bag in $TD(G)$ minus one is called the *width* of the tree decomposition. The *treewidth* of a graph G ($tw(G)$) is the minimum width of any tree decomposition of G. A graph G is a *partial k-tree* if G has treewidth at most k (van Leeuwen [17]).

The *bounded fragmentation* property plays an important role in our results.

Definition 2. *A graph G is a $(k, g(k,n))$-*bounded fragmentation *graph if $|\mathcal{C}(G[V-S])| \leq |g(k,n)|$ for every $S \subseteq V(G)$ of size at most k, where g is a function of k and n. A graph G is a* totally $g(k,n)$-bounded fragmentation *graph if it is a $(k, g(k,n))$-bounded fragmentation graph for all $0 \leq k \leq n$. A graph G is a k-*log-bounded fragmentation *graph (or just* log-bounded fragmentation *graph if it is clear from context) if G is a $(k, O(k \log n))$-bounded fragmentation graph. Finally a graph G is a* totally log-bounded fragmentation *graph if it is a k-log-bounded fragmentation graph for all $0 \leq k \leq n$.*

The lemmas below demonstrate that bounded degree graphs form a strict subset of log-bounded fragmentation graphs.

Lemma 1. *Connected graphs with maximum degree $O(\log n)$ are totally* log-*bounded fragmentation graphs.* □

Lemma 2. *Suppose the vertices of a graph G can be partitioned into $O(\log n)$ subsets such that for each set S, there exists a path in the graph that contains exactly the vertices in S. Then G is a totally* log-*bounded fragmentation graph.* □

Example 1. Consider a Hamiltonian graph F_n which is constructed from a path of length n by connecting one of its vertices to all its non-neighbors. Since vertices of every Hamiltonian graph can be covered by one path, F_n is a totally $(k+1)$-bounded fragmentation graph.

Bounded fragmentation can be viewed as measure of the reliability of a network, as it demonstrates the number of failures that can be tolerated while still supporting communication among $\Omega(n)$ of the remaining nodes. The reader is referred to Hajiaghayi's thesis [16] for further discussion of this concept, other applications of bounded fragmentation graphs, and other examples, such as graphs with maximum independent set or maximum matching of constant size, planar 3-connected graphs or graphs with certain minimum degree.

3 General Dynamic-Programming Approach for Graphs of Bounded Treewidth

In this section we briefly outline one of the general approaches used to derive polynomial-time algorithms for NP-complete problems restricted to graphs of bounded treewidth, namely the dynamic-programming approach of *computing tables of characterizations of partial solutions* [3], as formalized by Bodlaender [5]. First, we obtain a *nice tree decomposition* [6] of G, $TD(G) = (\chi, T)$, such that T is a rooted binary tree whose nodes have the following properties: a *leaf* node has no children; a *separator* node has a single child, and its bag is a subset of its child's bag; and a *join* node has two children, and its bag is the union of its children's bags. Since for any partial k-tree it is possible in linear time to construct a tree decomposition of width k [4] and transform it into an $O(k \cdot |V(G)|)$-node nice tree decomposition of the same width [6], throughout this paper we assume the existence of nice tree decompositions of graphs.

We then use dynamic programming to work bottom-up through the nice tree decomposition, where at each node z we use information computed at its children to determine properties of the terminal subgraph $G_{[z]}$. Using Bodlaender's terminology, to determine the *solution* to the problem, we make use of *partial solutions*, namely restrictions of solutions to $G_{[z]}$. Each partial solution is characterized by an *extension*, which relates a partial solution to a solution, and a *characteristic*, a property required for a partial solution to be able to be extended to a solution. The algorithm proceeds by determining the *full set of characteristics* for z (the set of all characteristics of partial solutions on $G_{[z]}$) using the full sets of characteristics for its children. Finally, the solution to the original problem is determined from the full set of characteristics for the root of $TD(G)$.

4 Subgraph Isomorphism for log-Bounded Fragmentation Graphs

We make use of the approach discussed in Section 3 to extend Theorem 1 to bounded fragmentation graphs; Corollary 1 follows directly from Theorem 2, which we prove below.

Theorem 2. *Suppose G is a $(k+1, g(k+1,n))$-bounded fragmentation graph and H is a partial k-tree for $k \geq 2$. There are $O(g(k+1,n) \cdot 2^{2g(k+1,n)}|V(G)|^{k+1} \cdot |V(H)|)$-time algorithms which solve isomorphism and subgraph and induced subgraph versions of this problem.*

Corollary 1. *There are polynomial-time algorithms for testing graph isomorphism and its subgraph and induced subgraph versions when graph H has bounded treewidth and graph G is a* log*-bounded fragmentation graph.* □

It is worth noting that Corollary 1 implies several new results. Theorem 1 guarantees polynomial running times when the source graph has constant degree; we can extend the result to provide the same guarantee when the source graph has $O(\log n)$ degree, since by Lemma 1 such a graph is a totally log-bounded fragmentation graph. Our algorithms also apply to graphs coverable by $O(\log n)$ paths (Lemma 2), which are not necessarily handled by previous results.

To prove Theorem 2, we first use dynamic programming (Section 3) to solve the induced subgraph isomorphism problem and then extend the algorithm to handle the subgraph isomorphism problem (we note that the graph isomorphism problem follows from the induced subgraph isomorphism problem for $|V(G)| =$

$|V(H)|$). Using Bodlaender's terminology, the solution for the induced subgraph isomorphism problem is an isomorphism ϕ from G into an induced subgraph H' of H. To define a partial solution, we consider the possible structure of an isomorphism ϕ restricted to an induced subgraph of $H_{[z]}$ for a node z of a nice tree decomposition $TD(H)$. Intuitively, this restriction maps a subgraph G' of G into $H_{[z]}$; the vertices of G' are those vertices of G whose images are in $V(H_{[z]})$, and the edges of G' are edges of G images of whose end-vertices are adjacent in $H_{[z]}$. This mapping is an isomorphism φ from G' into an induced subgraph of $H_{[z]}$ such that $\varphi(v) = \phi(v)$ for $v \in G'$, and we call it a *partial isomorphism*. Each partial solution will be a partial isomorphism.

To understand the nature of the partial solutions, we observe that each possible graph G' consists of vertices in a separator S in $V(G)$ along with some of the connected components resulting from the removal of S from G. More formally, we let S consist of vertices in $V(G)$ mapped to χ_z by the isomorphism ϕ from G into H, or $S = \{v \in V(G) | \phi(v) \in \chi_z\}$. Furthermore, consider the connected components resulting from the removal of S, $\mathcal{C}(G[V-S]) = \{C_1, \cdots, C_h\}$. We now observe that since each graph isomorphic to a connected graph is connected, each $\phi(V(C_i))$, $1 \leq i \leq h$, is a connected subgraph of H. As χ_z is a separator for H, and $\phi(V(C_i))$ is a connected subgraph of H which does not intersect χ_z, each $\phi(V(C_i))$ is completely inside of $H_{[z]}$ or completely outside of $H_{[z]}$. Let $\mathcal{D} = \{D_1, \cdots, D_l\}$ be those components of $\mathcal{C}(G[V-S])$ whose images are completely inside $H_{[z]}$. The set of vertices that map to $H_{[z]}$, $V(G')$, is thus the union of S and vertex sets of a number of components of $\mathcal{C}(G[V-S])$. To see that G' is an induced graph on these vertices, we observe that since $\phi(S \cup V(\mathcal{D}))$ is a subset of $V(H_{[z]})$ and ϕ is an isomorphism form G into H, for all $u, v \in S \cup V(\mathcal{D})$ such that $\{u, v\} \in E(G)$, $\varphi(u) = \phi(u)$ and $\varphi(v) = \phi(v)$ are adjacent in $H_{[z]}$.

To complete the specification of our algorithm, we must define the characteristic of a partial solution with respect to a node z of $TD(H)$, that is, the essence of a partial solution which determines how a partial solution can be extended to a solution. Intuitively, a partial solution is defined by G' (itself defined by S and $\mathcal{D}$, as discussed in the previous paragraph) as well as the way vertices in S are mapped. Using properties in upcoming Definition 4 as labels, we specify the sets S and $\mathcal{D}$, which uniquely determine edges of G' (Property [Pc]). For the remainder of the characteristic, we observe that information is essential if it has an impact on way the mapping can be extended, that is, the way that vertices outside of G' can be mapped. In particular, we argue that it is only the images of S that we need to record; we store the mapping $\psi(v) = \varphi(v)$ for $v \in S$ (Property [Pb]). If instead a vertex v is in $V(\mathcal{D})$, then we observe that $\varphi(v) = \phi(v)$ is in $V(H_{[z]}) - \chi_z$ and (since S is a separator) all neighbors of v in G are in $S \cup V(\mathcal{D})$. Since no neighbour of v is mapped outside of $H_{[z]}$, whether or not the mapping can be extended does not depend on the exactly where in $H_{[z]}$ the vertex v is mapped. Finally, we note that since $\phi(v)$ is an isomorphism and $\varphi(v)$ is obtained from $\phi(v)$, $\varphi(u) \neq \varphi(v)$ for all $u, v \in S \cup V(\mathcal{D})$ (Property [Pa]).

Combining all the necessary information, we form an *iso-triple* $(S, \mathcal{D}, \psi)$, defined formally below. We consider all possible such triples, not all of which can be extended to partial isomorphisms. Not each triple $(S, \mathcal{D}, \psi)$ is necessarily a characteristic; we call a partial isomorphism φ from which a characteristic $(S, \mathcal{D}, \psi)$ is obtained an *extension* of this characteristic. We note that the characteristic $(S, \mathcal{D}, \psi)$ might be obtained from several partial isomorphisms and thus have several extensions.

Formalizing the notions in the previous paragraph, we now define the terminology related to characteristics.

Definition 3. *An* iso-triple ξ of G into H relative to a node z of $TD(H)$ *is a triple* $(S, \mathcal{D}, \psi)$ *where* $S \subseteq V(G)$, $\mathcal{D} \subseteq \mathcal{C}(G[V - S])$, *and* ψ *is a one-to-one mapping from* S *into* χ_z.

Definition 4. *An* extension φ of an iso-triple $\xi = (S, \mathcal{D}, \psi)$ relative to a node z of $TD(H)$ *is a mapping* φ *from* $S \cup V(\mathcal{D})$ *into* $H_{[z]}$ *with these properties: [Pa]* $\varphi(u) \neq \varphi(v)$ *for all* $u, v \in S \cup V(\mathcal{D})$; *[Pb]* $\varphi(v) = \psi(v)$ *for* $v \in S$ *and* $\varphi(v) \notin \chi_z$ *for* $v \in V(\mathcal{D})$; *and [Pc] for each* $u, v \in S \cup V(\mathcal{D})$, $\{u, v\} \in E(G)$ *if and only if* $\{\varphi(u), \varphi(v)\} \in E(H)$.

Definition 5. *A* characteristic of a partial isomorphism (CPI) ξ of G into H relative to a node z of $TD(H)$ *is an iso-triple* $(S, \mathcal{D}, \psi)$ *which has an extension* φ *(not necessarily unique).*

To use the general dynamic-programming approach, we must specify how to determine the full set of characteristics, namely all CPIs relative to a node z of $TD(H)$. This set can be represented by a *full set array* indexed by all iso-triples, where the element in the array for iso-triple $(S, \mathcal{D}, \psi)$ is **true** if and only if the iso-triple is a CPI. In order to guarantee a polynomial-time algorithm, we need to show that the size of a full set is polynomial (Lemma 3), that the solution to the original problem can be derived from the full set computed for the root (Lemma 4), and that the full set of a node can be built from the full sets of its children (if they exist).

Lemma 3. *The number of all iso-triples and the number of CPIs relative to a node* z *of* $TD(H)$ *is in* $O((k+1)! \cdot 2^{g(k+1,n)} \cdot |V(G)|^{k+1})$. □

Lemma 4. *There exists an induced subgraph isomorphism* ϕ *from* G *into* H *if and only if there exists a CPI* $(S, \mathcal{D}, \psi)$ *of* G *into* H *relative to the root of* $TD(H)$ *such that* $\mathcal{D} = \mathcal{C}(G[V - S])$, *where* S *can be empty.* □

We now show how full set arrays can be determined for leaf nodes, separator nodes, and join nodes of $TD(H)$. For a leaf z, since $H_{[z]} = H[\chi_z]$ and hence $\mathcal{D}$ is empty, ψ is equal to its extension φ. Thus for each iso-triple ξ relative to a leaf, we can use brute force to check whether or not ψ is an extension of ξ (further details appear in Algorithm Induced below). Determining a CPI for a separator node requires checking conditions of a CPI of its child to ensure that the two CPIs are *separator-consistent* (Lemma 5) and determining a CPI for a join node requires checking conditions of CPIs of its children, this time to sure that they are *join-consistent* (Lemma 6). The subroutines **SeparatorFS** and **JoinFS** of Algorithm Induced, presented below, demonstrate how the conditions can be checked.

Lemma 5. *There exists a CPI* $\xi = (S, \mathcal{D}, \psi)$ *relative to a separator node* z *of* $TD(H)$ *if and only if there exists a CPI* $\xi' = (S', \mathcal{D}', \psi')$ *relative to the child* z' *of* z *such that* ξ *and* ξ' *are* separator-consistent, *that is, they satisfy the following conditions: 1.* $S = \{v \in S' | \psi'(v) \in \chi_z\}$*; 2.* $\mathcal{D}' = \{D' \in \mathcal{C}(G[V - S']) | D'$ *is a subgraph of some* $D \in \mathcal{D}\}$*; and 3.* $\psi(v) = \psi'(v)$ *for* $v \in S$. □

Lemma 6. *There exists a CPI* $\xi = (S, \mathcal{D}, \psi)$ *relative to a join node* z *of* $TD(H)$ *ifand only if there exist CPIs* $\xi_i = (S_i, \mathcal{D}_i, \psi_i)$ *relative to the children* z_1 *and* z_2 *of* z *such that* ξ *is* join-consistent *with* ξ_1 *and* ξ_2*, that is, if the following conditions are satisfied: 1.* $S_i = \{v \in S | \psi(v) \in \chi_{z_i}\}$ *for* $i = 1, 2$*; 2. the components of* $\mathcal{D}$ *are partitioned into* $\mathcal{D}_1$ *and* $\mathcal{D}_2$*; 3.* $\psi_i(v) = \psi(v)$ *for* $v \in S_i$ *for* $i = 1, 2$*; and 4. for* $u, v \in S$, $\{u, v\} \in E(G)$ *if and only if* $\{\psi(u), \psi(v)\} \in E(H)$. □

Our induced subgraph isomorphism algorithm, Algorithm Induced, consists of preprocessing steps, followed by computation of CPIs for all nodes in $TD(H)$. More specifically, we fix an ordering on the vertices of G and then index each component in $\mathcal{C}(G[V - S])$ by its lowest numbered vertex; information relating these components are stored in in *Subs* (smaller components in a larger component) and *Sup* (a larger component containing a smaller component) in lines 3–10. Next, all iso-triples are stored in the set *Iso* (lines 11–14); in order to be able to create one set for all nodes, we observe that S and $\mathcal{D}$ can be used for multiple nodes, and that we can store ψ as a one-to-one mapping from S into set $\{1, 2, \cdots, k+1\}$ (line 13), where for each node z of $TD(H)$ the vertices of χ_z are assigned to numbers in the set by an arbitrary ordering.

To fill in the full set arrays, we observe that for a leaf it suffices to check whether or not ψ is an extension of ψ (lines 19–22). For the remaining types of nodes, we use procedures **SeparatorFS** and **JoinFS**, specified below. Finally, we check whether or not there exists a CPI relative to the root of $TD(H)$ (lines 27–28) satisfying the condition in Lemma 4.

Algorithm Induced: testing induced subgraph isomorphism
Input: G : a $(k+1, g(k+1, n))$-bounded fragmentation graph
$TD(H)$: a nice tree decomposition of a partial k-tree H
Output: **true** if G is isomorphic to an induced subgraph of H, **false** otherwise
Variables:
$Subs[S, S', C']$: specifies components of $G[V - S]$ which are contained in $C' \in \mathcal{C}(G[V - S'])$, where $S' \subset S$
$Sup[S, S', C]$: specifies a component of $G[V - S']$ which contains $C \in \mathcal{C}(G[V - S])$, where $S' \subset S$
$FSA[z, \xi]$: **true** if ξ is in the full set of node z, **false** otherwise
Iso : A set containing all iso-triples

1 **if** $|V(G)| > |V(H)|$ **return false**;
2 **let** $\alpha_1, \alpha_2, \cdots, \alpha_{|TD(H)|}$ be nodes of $TD(H)$
in reverse breadth-first search order;
3 **for** each set S of at most $k+1$ vertices of G
4 find $\mathcal{C}(G[V - S])$

5   **for** each set $S' \subset S$
6   find $\mathcal{C}(G[V - S'])$
7   **for** each component $C'_i \in \mathcal{C}(G[V - S'])$
8   **let** $Subs[S, S', C'_i] \leftarrow \{C_{j_1}, \cdots C_{j_h}\}$
                    such that $V(C_{j_l}) \subseteq V(C'_i)$, $1 \leq l \leq h$
9   **for** each component $C_i \in \mathcal{C}(G[V - S])$
10   **let** $Sup[S, S', C_i] \leftarrow C'_j$ such that $V(C_i) \subseteq V(C'_j)$
11 **for** each $S \subseteq V(G)$ of size at most $k+1$
12   **for** each $\mathcal{D} \subseteq \mathcal{C}(G[V - S])$
13   **for** each one-to-one mapping ψ from S into set $\{1, 2, \cdots, k+1\}$
14   **let** $Iso \leftarrow Iso \cup \{\xi = (S, \mathcal{D}, \psi)\}$
15 **for** each node α_i of $TD(H)$, i from 1 to $|TD(H)|$
16   **for** each iso-triple $\xi = (S, \mathcal{D}, \psi)$ in Iso
17         $FSA[\alpha_i, \xi] \leftarrow$ **false**;
18   **if** α_i is a *leaf* node
19   **for** each iso-triple ξ in Iso
20   **let** $\varphi \leftarrow \psi$;
21   **if** $\mathcal{D} = \emptyset$ **and**
            for each $u, v \in S$, $\{u, v\} \in E(G)$
                    if and only if $\{\varphi(u), \varphi(v)\} \in E(H)$
22   **let** $FSA[\alpha_i, \xi] \leftarrow$ **true**;
23   **else if** α_i is a *separator* node
24   **SeparatorFS**(α_i);
25   **else if** α_i is a *join* node
26   **JoinFS**(α_i);
27 **for** each iso-triple $\xi = (S, \mathcal{D}, \psi)$ in Iso
28   **if** $FSA[root, \xi] =$ **true and** $\mathcal{D} = \mathcal{C}(G[V - S])$ **return true**;
29 **return false**;

In procedure **SeparatorFS**, we find all CPIs of a separator node z from CPIs of its child z'. For each CPI ξ' relative to z' (line S2), we construct an iso-triple ξ relative to z which is separator-consistent with ξ' and thus by Lemma 5 is a CPI relative to z (lines S3–S11). The three conditions of separator-consistency are satisfied by our construction of S on line S3 (Condition 1), ψ on line S4 (Condition 3), and $\mathcal{D}$ on lines S5–S11 (Condition 2). We set $\mathcal{D}$ to be the set of all supergraphs of components $C' \in \mathcal{D}'$ (line S7), and then to ensure that no member of $\mathcal{D}$ is a supergraph of components in $\mathcal{C}(G[V - S']) - \mathcal{D}'$, we form the set $\mathcal{D}''$ of all components of $\mathcal{C}(G[V - S'])$ which are contained in components $D \in \mathcal{D}$ (line S10) and check that $\mathcal{D}'$ and $\mathcal{D}''$ are equal (line S11).

SeparatorFS(z)
Input: z : a separator node of $TD(H)$
S1 **let** $z' \leftarrow$ the child of z;
S2 **for** each iso-triple $\xi' = (S', \mathcal{D}', \psi')$ in Iso such that $FSA[z', \xi']$ **is true**
S3   **let** $S \leftarrow \{v \in S' | \psi'(v) \in \chi_z\}$;

```
S4        let ψ(v) ← ψ'(v) for v ∈ S;
S5        𝒟 ← ∅;
S6        for each C' ∈ 𝒟'
S7              let 𝒟 ← 𝒟 ∪ {Sup[S', S, C']};
S8        𝒟'' ← ∅;
S9        for each C ∈ 𝒟
S10             let 𝒟'' ← 𝒟'' ∪ Subs[S', S, C];
S11       if 𝒟'' = 𝒟'
S12             let ξ ← (S, 𝒟, ψ);
S13             let FSA[z, ξ] ← true;
```

To find all CPIs of a join node z from CPIs of its children z_1 and z_2 in procedure **JoinFS**, we check each possible iso-triple with respect to the conditions for join-consistency defined in Lemma 6. In particular, we construct all possible iso-triples ξ_1 and ξ_2 relative to its children; ξ is a CPI if both ξ_1 and ξ_2 are CPIs (as checked on line J12) and join-consistent with ξ. The conditions of join-consistency are satisfied by our constructions of S_1 and S_2 on line J3 (Condition 1) and ψ_1 and ψ_2 on line J4 (Condition 3) as well as our check for edges on line J5 (Condition 4) and partitions of components in lines J6–J9 (Condition 2). To check Condition 2, after partitioning elements of $\mathcal{D}$ into $\mathcal{D}_1$ and $\mathcal{D}_2$ (line J6) we need to check that each element $D \in \mathcal{D}$, which is in $\mathcal{D}_i$, is a component of $G[V - S_i]$. Since all outgoing edges from $V(D)$ go into S, it is sufficient to check on line J9 whether there is any edge between $V(D)$ and $S - S_i$, and if so, reject the possible partition, as then the component $D' \in \mathcal{C}(G[V - S_i])$ which is a supergraph of D has at least one vertex in $S - S_i$.

JoinFS(z)

```
Input:   z   : a join node of TD(H)
J1  let z_i ← ith child of z, i = 1,2;
J2  for each iso-triple ξ in Iso
J3        let S_i ← {v ∈ S | ψ(v) ∈ χ_{z_i}}, i = 1,2;
J4        let ψ_i(v) ← ψ(v) for v ∈ S_i, i = 1,2;
J5        if for all u,v ∈ S, {u,v} ∈ E(G) if and only if {ψ(u), ψ(v)} ∈ E(H)
J6              for each partition of elements of 𝒟 into 𝒟_1 and 𝒟_2
J7                    let bool ← true;
J8                    for each D ∈ 𝒟, which is in 𝒟_i or just for each D ∈ 𝒟_i?
J9                          if V(Sup[S, S_i, D]) ∩ S − S_i ≠ ∅ let bool ← false;
J10                   if bool = true
J11                         let ξ_i ← (S_i, 𝒟_i, ψ_i), i = 1,2;
J12                         if FSA[z_1, ξ_1] = FSA[z_2, ξ_2] =true
J13                               let FSA[z, ξ] ← true;
```

To prove the correctness of the algorithm and obtaining the running time, first we define an invariant and present two lemmas for separator and join nodes.

Definition 6. *We say that the FSA array relative to a node z of $TD(H)$ is in* correct form, *if for each ξ relative to z, $FSA[z, \xi] =$* **true** *if and only if ξ is a CPI relative to z.*

Lemma 7. *Given the FSA array of the child z' of a separator node z in correct form, Procedure* **SeparatorFS** *fills in the FSA array relative to z in correct form in $O(g(k+1,n) \cdot 2^{g(k+1,n)}|V(G)|^{k+1})$ time.* □

Lemma 8. *Given the FSA arrays of children z_1 and z_2 of a join node z in correct form, Procedure* **JoinFS** *fills in the FSA array relative to z in correct form in $O(g(k+1,n) \cdot 2^{2g(k+1,n)}|V(G)|^{k+1})$ time.* □

The correctness and running time, as stated in Theorem 3, follow from Lemmas 3, 4, 7, and 8.

Theorem 3. *The above algorithm solves the induced subgraph isomorphism problem in $O(g(k+1,n) \cdot 2^{2g(k+1,n)}|V(G)|^{k+1} \cdot |V(H)|)$ time.* □

By making the use of Theorem 3 and the results below, it is possible to obtain a proof of Theorem 2.

Definition 7. *The* subdivision *of an edge $\{u, w\}$ is the operation of deleting this edge and adding a new vertex v and two new edges $\{u, v\}$ and $\{v, w\}$. The graph obtained from graph G by subdivisions of all its edges is denoted by G^*.*

Lemma 9. *If G is a partial k-tree and $k \geq 2$, then G^* is a partial k-tree.* □

Lemma 10. *If G is a $(k, g(k,n))$-bounded fragmentation graph and k is a constant, G^* is a $(k, O(g(k,n)))$-bounded fragmentation graph.* □

Lemma 11. *(Lemma 1.4 [18]) Let G and H be two graphs. G is subgraph isomorphic to H if and only if G^* is induced subgraph isomorphic to H^*.* □

We can further generalize Theorem 2 to directed graphs by considering directions of edges in consistency checking, (we replace edge $\{u, v\}$ by (u, v) in lines 21 and J5). Hence if G is a directed graph with a $(k+1, g(k+1,n))$-bounded fragmentation underlying graph and H is a directed graph, whose underlying graph is a partial k-tree, then there are $O(g(k+1,n) \cdot 2^{2g(k+1,n)}|V(G)|^{k+1} \cdot |V(H)|)$-time algorithms which solve isomorphism, subgraph isomorphism and induced subgraph isomorphism.

It is worth mentioning that all current algorithms for constructing a tree decomposition of a graph of treewidth at most k have a hidden constant factor that is at least exponential in k, and hence they are impractical in general. However, in the cases $k = 2, 3, 4$, practical linear-time algorithms exist [2,18,20] and for these cases, the polynomial-time algorithms presented in this section could be used in practice.

5 Extension to Graphs of Locally Bounded Treewidth

It is possible to extend our polynomial-time algorithm for testing subgraph isomorphism to graphs of locally bounded treewidth, defined by Eppstein [12] as a generalization of the notion of treewidth.

Definition 8. *The* local treewidth *of a graph G is the function* $\text{ltw}^G : \mathbb{N} \to \mathbb{N}$ *that associates with every* $r \in \mathbb{N}$ *the maximum treewidth of an r-neighborhood in G. We set* $\text{ltw}^G(r) = \max_{v \in V(G)}\{\text{tw}(G[N_G^r(v)])\}$*, and we say that a graph class* $\mathcal{C}$ *has* bounded local treewidth (or locally bounded treewidth) *when there is a function* $f : \mathbb{N} \to \mathbb{N}$ *such that for all* $G \in \mathcal{C}$ *and* $r \in \mathbb{N}$, $\text{ltw}^G(r) \leq f(r)$.

Theorem 4. *Subgraph and induced subgraph isomorphism can be solved in* $O((ltw(diam(G)) \log |V(G)|) \cdot 2^{O(ltw(diam(G)) \log |V(G)|)} |V(G)|^{ltw(diam(G))+1} \cdot |V(H)|)$ *time when graph H is of minor-closed family of graphs of locally bounded treewidth and graph G is a totally* log*-bounded fragmentation graph and has constant diameter.* □

The bounded diameter of the source graph is crucial to the result, as without this condition, the problem is NP-complete.

Theorem 5. *Let graph H have locally bounded treewidth with* $\Delta(H) = 3$ *and graph G have bounded treewidth with* $\Delta(G) = 2$*. The subgraph isomorphism problem for the source graph G and the host graph H is NP-complete.* □

One interesting consequence of Theorem 4 is that it duplicates Eppstein's result on testing subgraph isomorphism for fixed patterns [11,12], restricted to graphs of locally bounded treewidth.

Corollary 2. *For a fixed pattern G and a graph H of minor-closed family of graphs of locally bounded treewidth, subgraph isomorphism and induced subgraph isomorphism can be tested in* $O(|V(H)|)$ *time.* □

Using this result, Eppstein also showed that other problems such as finding diameter (if we know the graph has bounded diameter), h-clustering for constant h and finding girth (if we know the graph has bounded girth) can be tested in $O(n)$ time. The reader is referred to the original papers for details.

6 Conclusions and Future Work

In this paper, we presented a polynomial-time algorithm for the subgraph isomorphism when the source graph is log-bounded-fragmentation and the host graph has bounded treewidth. In addition, we demonstrated how this algorithm can be generalized to graphs of locally bounded treewidth.

Possible extensions of these results include generalizations of techniques and results to bounded fragmentation graphs and graphs of locally bounded treewidth. Using a different techique, Gupta and Nishimura proved that when both the pattern graph G and the source graph H are k-connected partial k-trees, the subgraph isomorphism problem has a polynomial-time solution [15]; their techniques apply to an embedding that generalizes subgraph isomorphism. It might be possible to generalize their approach for H of locally bounded treewidth, including an algorithm for the minor containment problem under certain restrictions.

Subgraph isomorphism can be generalized to finding the largest common subgraph of two graphs. Brandenburg [8] showed that if two graphs G and H are k-connected partial k-trees, the problem of finding the largest common k-connected subgraph can be solved in polynomial time. In addition, using techniques of Brandenburg [8] and Gupta and Nishimura [14], one can observe that finding the largest common $(k-1)$-connected subgraph is NP-complete. As yet, the complexity of the problem is unknown for bounded fragmentation graphs.

Acknowledgments

We would like to thank Professor Jim Geelen of the University of Waterloo for his thoughtful comments.

References

1. Stefan Arnborg, Jens Lagergren, and Detlef Seese. Problems easy for tree-decomposable graphs (extended abstract). In *Proc., 15th Internat. Colloquium on Automata, Languages, and Programming*, pages 38–51, 1998.
2. Stefan Arnborg and Andrzej Proskurowski. Characterization and recognition of partial 3-trees. *SIAM J. Algebraic Discrete Methods*, 7(2):305–314, 1986.
3. Stefan Arnborg and Andrzej Proskurowski. Linear time algorithms for NP-hard problems restricted to partial k-trees. *Discrete Appl. Math.*, 23(1):11–24, 1989.
4. Hans L. Bodlaender. A linear-time algorithm for finding tree-decompositions of small treewidth. *SIAM J. Comput.*, 25(6):1305–1317, 1996.
5. Hans L. Bodlaender. Treewidth: algorithmic techniques and results. In *Proc., 22nd Internat. Symposium on Mathematical Foundations of Computer Science*, pages 29–36, 1997.
6. Hans L. Bodlaender. A partial k-arboretum of graphs with bounded treewidth. *Theoret. Comput. Sci.*, 209(1-2):1–45, 1998.
7. John A. Bondy and U. S. R. Murty. *Graph Theory with Applications*. American Elsevier Publishing Co., Inc., New York, 1976.
8. Franz J. Brandenburg. Pattern matching problems in graphs. Manuscript, 2001.
9. Anders Dessmark, Andrzej Lingas, and Andrzej Proskurowski. Faster algorithms for subgraph isomorphism of k-connected partial k-trees. *Algorithmica*, 27(3-4):337–347, 2000.
10. Anders Dessmark, Andrzej Lingas, and Andrzej Proskurowski. Maximum packing for k-connected partial k-trees in polynomial time. *Theoret. Comput. Sci.*, 236(1-2):179–191, 2000.
11. David Eppstein. Subgraph isomorphism in planar graphs and related problems. *Journal of Graph Algorithms and Applications*, 3(3):27 pp. (electronic), 1999.
12. David Eppstein. Diameter and treewidth in minor-closed graph families. *Algorithmica*, 27(3-4):275–291, 2000.

13. Michael R. Garey and David S. Johnson. *Computers and Intractability: A Guide to the Theory of NP-completeness.* W. H. Freeman and Co., San Francisco, Calif., 1979.
14. Arvind Gupta and Naomi Nishimura. The complexity of subgraph isomorphism for classes of partial k-trees. *Theoret. Comput. Sci.*, 164(1-2):287–298, 1996.
15. Arvind Gupta and Naomi Nishmura. Sequential and parallel algorithms for embedding problems on classes of partial k-trees. In *Proc., 4th Scandinavian Workshop on Algorithm Theory*, pages 172–182, 1994.
16. MohammadTaghi Hajiaghayi. Algorithms for Graphs of (Locally) Bounded Tree-width. Master's thesis, University of Waterloo, September 2001. `http://citeseer.nj.nec.com/hajiaghayi01algorithms.html`.
17. Jan van Leeuwen. Graph algorithms. In *Handbook of Theoretical Computer Science, Vol. A*, pages 525–631. Elsevier, Amsterdam, 1990.
18. Jiří Matoušek and Robin Thomas. On the complexity of finding iso- and other morphisms for partial k-trees. *Discrete Math.*, 108(1-3):343–364, 1992.
19. Neil Robertson and Paul D. Seymour. Graph minors II. Algorithmic aspects of tree-width. *J. Algorithms*, 7(3):309–322, 1986.
20. Daniel P. Sanders. On linear recognition of tree-width at most four. *SIAM J. Discrete Math.*, 9(1):101–117, 1996.
21. Maciej M. Sysło. The subgraph isomorphism problem for outerplanar graphs. *Theoret. Comput. Sci.*, 17(1):91–97, 1982.

Computing Partial Information out of Intractable One – The First Digit of 2^n at Base 3 as an Example*

Mika Hirvensalo and Juhani Karhumäki

[1] Department of Mathematics, University of Turku,
FIN-20014, Turku, Finland.
[2] Turku Centre for Computer Science, Lemminkäisenkatu 14 A, 4th floor,
FIN-20520, Turku, Finland,
{mikhirve,karhumak}@cs.utu.fi

Abstract. The problem of finding 2^n when n is given is mathematically trivial, but computationally *intractable*. We show how to compute the most significant digit of the ternary representation of 2^n in polynomial time.

1 Introduction

Exponent functions are mathematically, in many aspects, among the simplest ones. For example, their derivatives are particularly easy to compute. On the other hand, from algorithmic point of view they are terrible. Although the value of an exponent function, say 2^n for example, is very easy to compute in theory, it is out of the capacity of modern computers in practice. More precisely, the value of such a function cannot be computed by an algorithm running in polynomial time with respect to the size of the input, i.e. the length of n. This, indeed, follows since the output is exponential in terms of the length of n. The same applies for many other numbers, for example, for the well-known Fibonacci numbers.

Having this situation a natural question is: Can we compute some partial information about an exponent function in polynomial time, for example its length at a given base? This is the problem area we pose here. We concentrate to a very simple function: for a given n consider the ternary representation of 2^n, i.e., the function $n \mapsto f(n) = (2^n)_3$. We believe this already captures difficulties encountered in our questions.

Let $f(n)$ be defined as above. What can be computed easily is the last, or even last k, for a fixed k, digits of $f(n)$. Indeed, such values form an ultimately periodic sequence of numbers (cf. Figure 1), so that the threshold and the period can be computed in constant time. Hence the problem reduces to the computation $n \mapsto n \pmod{p}$, where p is the precomputed period. Therefore it can be done in linear time.

On the other hand, to compute even the length of $f(n)$ in polynomial time is not obvious. We shall show, that this can be done, and moreover, that the first

* Supported by the Academy of Finland under grant 44087.

K. Diks et al. (Eds): MFSC 2002, LNCS 2420, pp. 319–327, 2002.

n	$(2^n)_\mathbf{3}$	$\lvert(2^n)_\mathbf{3}\rvert$	$\lfloor 0.6n \rfloor + 1$
1	2	1	1
2	11	2	2
3	22	2	2
4	121	3	3
5	1012	4	4
6	2101	4	4
7	11202	5	5
8	100111	6	5(!)
9	200222	6	6
10	1101221	7	7
11	2210212	7	7
12	12121201	8	8
13	102020102	9	8(!)

Fig. 1. Ternary representations of 2^n for $n = 1, 2, \ldots, 13$.

digit of $f(n)$ can also be computed in polynomial time. The algorithm itself is not complicated, nor difficult to analyze. However, to prove its correctness seems to require deep results on transcendental numbers.

To compute a fixed number of first digits of $f(n)$ does no seem to be essentially more difficult question. The situation, however, changes if we ask to compute the "middle" digit of $f(n)$. We do not know how to do it in polynomial time. A more general question is: given $n \in \mathbb{N}$ and $i \leq \lfloor \log_3 f(n) \rfloor + 1$, can we compute the ith digit of $f(n)$ in polynomial time?

Our problems can be seen as examples of interesting general algorithmic questions. How much partial information about intractable algorithmic problem can be computed in practice, that is in polynomial time?

2 Preliminaries

In this section we introduce some definitions and notations used in the following sections.

For any real number x, notation $\lfloor x \rfloor$ stands for the largest integer M satisfying $M \leq x$, and $\log_d x$ stands for the d-ary logarithm of x. We also use the standard notation $\ln x$ for the natural logarithm of x.

Let $d > 1$ be an integer. Each natural number M admits a unique representation as

$$M = a_0 d^{\ell-1} + a_1 d^{\ell-2} + \ldots + a_{\ell-2} d^1 + a_{\ell-1}, \tag{1}$$

where $a_i \in \{0, 1, \ldots, d-1\}$ and $a_0 \neq 0$. The string $a_0 a_1 \ldots a_{\ell-1}$ is called the *d-ary representation of* M, and denoted by $M_\mathbf{d}$. The *length* of d-ary representation of M is defined to be the length of string $M_\mathbf{d}$ and denoted by $|M_\mathbf{d}|$. Notice that we use boldface subscripts to separate between the numbers and their representations.

Especially, $|M|$ stands for the *absolute value* of M whereas $|M_\mathbf{d}|$ stands for the length of the string representing M.

Equation (1) gives inequalities $M \geq d^{\ell-1}$ and

$$M \leq (d-1)(d^{\ell-1} + d^{\ell-2} + \ldots + d + 1) = d^\ell - 1 < d^\ell.$$

Combining these two estimates we see that $d^{\ell-1} \leq M < d^\ell$, or, equivalently, $\ell - 1 \leq \log_d M < \ell$, which is to say that $|M_\mathbf{d}| = \ell = \lfloor \log_d M \rfloor + 1$. In the case $d = 3$ we say that $M_\mathbf{3}$ is the *ternary* representation of M.

Let $q = \frac{a}{b}$ be a nonzero rational number such that $\gcd(a, b) = 1$. The *height* of q is defined as $H(q) = \max\{|a|, |b|\}$, and the *size* of q as $S(q) = |a_\mathbf{d}| + |b_\mathbf{d}|$, where d is some fixed base where both a and b are represented.

As defined above, the size of a rational number $q = \frac{a}{b}$ depends on the number system on which a and b are represented. On the other hand, since $\log_d M < |M_\mathbf{d}| \leq \log_d M + 1$ holds for any integer M, and because $\log_d M = (\ln d)^{-1} \ln M$, we see that the length of M is $\Theta(\ln M)$ in any representation system. Here and hereafter, we exclude the unary representations.

3 The Algorithms

In this section we represent the algorithms for computing $|(2^n)_\mathbf{3}|$ and the first symbol of $(2^n)_\mathbf{3}$, and analyze their complexities. In the remaining sections we prove that the algorithms represented work correctly. We begin with some observations.

Without loss of generality we can assume that the input n is given in binary representation, so the *size* of input is $|n_\mathbf{2}|$, which, as seen in the previous section, is $\Theta(\ln n)$. We can therefore state our problem rigorously as follows: given an input $n_\mathbf{2} \in \{0, 1\}^*$, compute $(2^n)_\mathbf{3} \in \{0, 1, 2\}^*$.

As seen in the previous sections, the size of the input is $\Theta(\ln n)$, whereas the output size is $\Theta(n)$, which is exponential in the input size. It follows that the problem is intractable by necessity.

On the other hand, computing $n_\mathbf{2} \mapsto |(2^n)_\mathbf{3}|$ is a very different problem. Now the output should be the length of the ternary representation of 2^n, or, from the algorithmic point of view, a *word* which represents the length. Again, without loss of generality, we can require the output in binary, which means that the output size would be

$$\begin{aligned} ||(2^n)_\mathbf{3}|_\mathbf{2}| &= \lfloor \log_2 |(2^n)_\mathbf{3}| \rfloor + 1 \leq \log_2(\lfloor n \log_3 2 \rfloor + 1) + 1 \\ &\leq \log_2(n \log_3 2 + n) + 1 = \log_2 n + \log_2(\log_3 2 + 1) + 1, \end{aligned}$$

in particular, it is in $O(\ln n)$.

Equation

$$|(2^n)_\mathbf{3}| = \lfloor n \log_3 2 \rfloor + 1 \tag{2}$$

gives a good starting point for computing the length, but the straightforward utilization of (2) contains at least two problematic features.

First, knowing n and $\log_3 2$ precisely enough allows us to compute their product, but it must be noted that we should be able to compute $\log_3 2 \approx 0.6$ at least up to precision $\frac{1}{n}$, since for larger imprecisions the outcome would be incorrect. This is illustrated in the rightmost column of Figure 1. This problem is easy to handle, and the beginning of the next section is devoted to this.

The second, and more severe problem is, that an approximation for $\log_3 2$, does not directly offer any tools to compute $\lfloor n \log_3 2 \rfloor$, no matter how precise the approximation is! To see this, let β_n, $n = 1, 2, 3, \ldots$ be a sequence of irrational numbers, and b_n, $n = 1, 2, 3, \ldots$, be a sequence of their very precise rational approximations, $|b_n - \beta_n| \ll 1$ for each n. Let us take some n, and assume, for instance, that $b_n < \beta_n$. If the interval (b_n, β_n) happens to contain an integer M, then $\lfloor b_n \rfloor = M - 1$, whereas the correct value $\lfloor \beta_n \rfloor$ is equal to M. In other words, if we do not have apriori knowledge on the distance between β_n and the nearest integer M, we cannot certainly find the value $\lfloor \beta_n \rfloor$ by using only an approximation b_n of β_n. In the next section, we use deep a result of Alan Baker to solve this problem.

When computing the most significant (leftmost) digit of $(2^n)_3$, we note that an estimation similar to that one used in the previous section shows that if $(2^n)_3 \in 1\{0,1,2\}^*$, then

$$3^{\ell-1} \leq 2^n < 2 \cdot 3^{\ell-1},$$

whereas $(2^n)_3 \in 2\{0,1,2\}^*$ implies

$$2 \cdot 3^{\ell-1} \leq 2^n < 3^{\ell},$$

where $\ell = |(2^n)_3|$. Thus, to recover the most significant digit of $(2^n)_3$ is to decide whether or not the inequality

$$2 \cdot 3^{\ell-1} \leq 2^n \iff 3^{\ell-1} \leq 2^{n-1} \tag{3}$$

holds. Inequality (3) is equivalent to

$$\ell - 1 \leq (n-1) \log_3 2, \tag{4}$$

and (4) will be used for finding the first digit of $(2^n)_3$.

Now we are ready for the algorithms. The constant C occurring in the algorithms will be explained in the next section.

Algorithm 1.
Input: A natural number n in binary representation.
Output: $|(2^n)_3|$ in binary representation.

1. Compute a rational approximation q of $\log_3 2$ so precise that
$$|q - \log_3 2| \leq \frac{1}{\ln 3} \frac{1}{n^{C+1}}.$$
2. Compute qn.
3. Compute $\lfloor qn \rfloor$.
4. Output $\lfloor qn \rfloor + 1$.

In the next section we show that step 1 can be accomplished in time $O(\ln^3 n$ $(\ln\ln n)^2)$, and, moreover, the approximating q has size $O(\ln n \ln\ln n)$. Especially, the numerator of $q = \frac{a}{b}$ has size $O(\ln n \ln\ln n)$, and hence the multiplication of q by n in step 2 can be done in time $O((\ln n \ln\ln n)^2)$, and the resulting number $qn = \frac{an}{b}$ has size $O(\ln n \ln\ln n + \ln n) = O(\ln n \ln\ln n)$. For the third step, an ordinary division of an by b is enough to reveal $\lfloor qn \rfloor$, and because of the numerator and the nominator sizes, it can be done in time $O((\ln n \ln\ln n)^2)$. As verified earlier, the outcoming number has size $O(\ln n)$, which implies that the last computation included in the fourth step can be performed in time $O(\ln n)$.

As a conclusion: Computation $\mathbf{n_2} \mapsto |(2^n)_\mathbf{3}|_\mathbf{2}$ can be performed in time $O(\ln^3 n(\ln\ln n)^2)$, or, to put it into other format, in time $O(|\mathbf{n_2}|^3 \ln^2 |\mathbf{n_2}|)$.

Algorithm 2.
Input: A natural number n in binary representation.
Output: The leftmost digit of string $(2^n)_\mathbf{3}$.

1. Compute $\ell = |(2^n)_\mathbf{3}|$ by using Algorithm 1.
2. Compute a rational approximation q of $\log_3 2$ so precise that

$$|q - \log_3 2| \le \frac{1}{\ln 3} \frac{1}{(n-1)^{C+1}}.$$

3. Compute numbers $\ell - 1$ and $(n-1)q$.
4. Decide, whether $\ell - 1 \le (n-1)q$.
5. If $\ell - 1 \le (n-1)q$, output 2, otherwise output 1.

The complexity analysis of Algorithm 2 is similar to that of the Algorithm 1, and the outcoming complexity is $O(|\mathbf{n_2}|^3 \ln^2 |\mathbf{n_2}|)$.

4 Validity of the Algorithms

Approximating $\log_3 2$

It is a well-known fact that the series

$$\ln(1+x) = x - \frac{1}{2}x^2 + \frac{1}{3}x^3 - \frac{1}{4}x^4 + \ldots$$

converges if $|x| < 1$. By substituting $x = -y$ we see that

$$\ln \frac{1}{1-y} = -\ln(1-y) = y + \frac{1}{2}y^2 + \frac{1}{3}y^3 + \frac{1}{4}y^4 + \ldots,$$

and, a further substitution $y = \frac{\alpha-1}{\alpha}$ shows us that

$$\ln \alpha = \sum_{i=1}^{\infty} \frac{1}{i} \left(\frac{\alpha - 1}{\alpha} \right)^i \tag{5}$$

converges for any fixed $\alpha > 1$. Hereafter we assume that $\alpha > 1$ is a fixed integer.

The remainder of (5) can be easily estimated as

$$\left|\sum_{i=M+1}^{\infty}\frac{1}{i}\left(\frac{\alpha-1}{\alpha}\right)^i\right|$$
$$<\left(\frac{\alpha-1}{\alpha}\right)^{M+1}\sum_{i=0}^{\infty}\left(\frac{\alpha-1}{\alpha}\right)^i$$
$$=\alpha\left(\frac{\alpha-1}{\alpha}\right)^{M+1}.$$

Thus, by choosing

$$M > 1 + \frac{\ln\frac{\alpha}{\epsilon}}{\ln\frac{\alpha}{\alpha-1}} \tag{6}$$

the remainder is less than ϵ.

In the case $\epsilon = \frac{1}{n^T}$ we see that in order to approximate $\ln\alpha$ within precision $\frac{1}{n^T}$, it is sufficient to take

$$M = 1 + \frac{T\ln n + \ln\alpha}{\ln\frac{\alpha}{\alpha-1}} = \Theta(T\ln n)$$

first summands of the series (5). Notice that all the summands in the series are positive, which implies that an approximation obtained by taking M first terms is always smaller than the actual value of the logarithm.

Let us choose $M = \Theta(T\ln n)$. Then, by writing

$$\sum_{i=1}^{M}\frac{1}{i}\left(\frac{\alpha-1}{\alpha}\right)^i = \frac{1}{M!\alpha^M}\sum_{i=1}^{M}\frac{M!}{i}\alpha^{M-i}(\alpha-1)^i \tag{7}$$

we see that the nominator of (7) is at most $M!\alpha^M$, and its length is of order

$$\ln(M!\alpha^M) = M\ln M + O(M) + M\ln\alpha = O(M\ln M)$$
$$= O(T\ln n\ln(T\ln n)) = O(\ln n\ln\ln n).$$

An estimate for the numerator can be found in the similar way: First, a trivial estimate shows that the numerator is at most $M\cdot M!\alpha^M(\alpha-1)^M$, whose length is of order

$$\ln M + M\ln M + O(M) + M\ln\alpha + M\ln(\alpha-1) = O(M\ln M)$$
$$= O(\ln n\ln\ln n)$$

similarly as in the case of the nominator.

Number $M!$, or, to be precise, its representation, can be computed in time $O(M(\ln M!)^2) = O(M^3\ln^2 M) = O(\ln^3 n(\ln\ln n)^2)$. On the other hand, number α^M can be found in time $O(M\cdot(M\ln\alpha)^2) = O(M^3) = O(\ln^3 n)$, and the product $M!\alpha^M$ in time $O(\ln^2 M!) = O((\ln n\ln\ln n)^2)$. It follows that the nominator of (7) can be computed in time $O(\ln^3 n(\ln\ln n)^2)$. A similar estimate can be found also for the numerator.

As a conclusion, we get the following theorem:

Theorem 1. *Let $\alpha > 1$ and T be fixed natural numbers. It is possible to compute a rational number $q < \ln \alpha$ such that $\alpha - q < \frac{1}{n^T}$ in time $O(\ln^3 n (\ln \ln n)^2)$. Moreover, the size of q is $O(\ln n \ln \ln n)$.*

Now, a rational approximation for $\log_3 2 = \frac{\ln 2}{\ln 3}$ can be found by finding rational approximations r and s for both $\ln 2$ and $\ln 3$, respectively. Assume that $\ln 2 = r + \epsilon_1$ and $\ln 3 = s + \epsilon_2$, where $0 < \epsilon_1, \epsilon_2 \leq \epsilon$. If we further assume that s and r are so good approximations that $1 \leq s < \ln 3$ and $r > 0$, we can estimate their ratio as

$$\left|\frac{\ln 2}{\ln 3} - \frac{r}{s}\right| = \frac{|s \ln 2 - r \ln 3|}{s \ln 3} \leq \frac{|\epsilon_1 s - \epsilon_2 r|}{\ln 3}.$$

If $\epsilon_1 s - \epsilon_2 r \geq 0$, then $|\epsilon_1 s - \epsilon_2 r| = \epsilon_1 s - \epsilon_2 r < \epsilon_1 s < \epsilon \ln 3$, whereas $\epsilon_1 s - \epsilon_2 r < 0$ implies $|\epsilon_1 s - \epsilon_2 r| = \epsilon_2 r - \epsilon_1 s \leq \epsilon_2 r < \epsilon \ln 2$. In both cases,

$$\left|\frac{\ln 2}{\ln 3} - \frac{r}{s}\right| \leq \frac{|\epsilon_1 s - \epsilon_2 r|}{\ln 3} < \frac{\epsilon \ln 3}{\ln 3} = \epsilon.$$

The above inequality states that if r and s are approximations of $\ln 2$ and $\ln 3$ at most ϵ apart from the actual values, then their ratio $\frac{r}{s}$ is an approximation of $\log_3 2$ at most ϵ apart from the real value. Moreover, it is clear that $\frac{r}{s}$ has size $O(\ln n \ln \ln n)$, if both r and s have.

Baker's Result

The extra information we use for computing $\ell = \lfloor n \log_3 2 \rfloor + 1$ is provided in the following theorem, the proof can be found in [1].

Theorem 2 (A. Baker, 1966). *Let $\alpha_1, \ldots, \alpha_k$ be non-zero algebraic numbers with degrees at most d and heights at most A. Further, let $\beta_0, \ldots, \beta_k$ be algebraic numbers with degrees at most d and heights at most $B \geq 2$. Then for*

$$\Lambda = \beta_0 + \beta_1 \ln \alpha_1 + \ldots + \beta_k \ln \alpha_k$$

we have either $\Lambda = 0$ or $|\Lambda| > B^{-C}$, where C is an effectively computable number depending only on k, d, A, and the original determinations of the logarithms.

Remark 1. In the case when $\beta_0 = 0$ and $\beta_1, \ldots, \beta_k$ are rational integers, it has been shown that the theorem holds with $C = C'(\ln A)^k \ln(\ln A)^k$ and C' depends only on k and d [1].

Now we choose $k = 2$, $\beta_0 = 0$, $\beta_1 = m \in \mathbb{N}$, $\beta_2 = -n \in -\mathbb{N}$, $\alpha_1 = 3$, and $\alpha_2 = 2$. Then we can take $d = 1$, $A = 3$, and $B = \max\{|m|, |n|\}$. It is easy to show that $m \ln 3 - n \ln 2 \neq 0$, so we have the following theorem as a special case of Theorem 2.

Theorem 3. *There exists a constant C such that*

$$|m \ln 3 - n \ln 2| > \frac{1}{n^C},$$

whenever $m, n \in \mathbb{N}$ and $m \leq n$.

Remark 2. In fact, the previous theorem would already follow from a result of Gelfond, see [2]. It can be also shown that the above theorem holds with $C = 13.3$ [3].

The above theorem directly implies that

$$|m - n\log_3 2| > \frac{1}{\ln 3}\frac{1}{n^C} \tag{8}$$

for any natural numbers $m < n$. Important points here are that C is independent of n and m, and can be computed.

Theorem 4. *Let C be the constant mentioned in (8). If $q \in \mathbb{Q}$ is an approximation of $\log_3 2$ so precise that $|q - \log_3 2| \leq \frac{1}{\ln 3}\frac{1}{n^{C+1}}$, then $\lfloor n\log_3 2\rfloor = \lfloor nq\rfloor$.*

Proof. Let $M = \lfloor nq\rfloor$. The assumption on q verifies that $M + 1 \leq n$, so the premise of Theorem 3 is satisfied. Now $M \leq nq < M + 1$, and we must show that also $M \leq n\log_3 2 < M + 1$. For that purpose, we first assume, for the sake of contradiction, that $n\log_3 2 < M$. But then we would have

$$|M - n\log_3 2| = M - n\log_3 2 \leq nq - n\log_3 2,$$

which is at most $\frac{1}{\ln 3}\frac{1}{n^C}$ by the assumption. In virtue of inequality (8) this is a contradiction, and hence $M \leq n\log_3 2$. On the other hand, if $n\log_3 2 \geq M + 1$, then

$$|M + 1 - n\log_3 2| = n\log_3 2 - (M + 1) < n\log_3 2 - nq \leq \frac{1}{\ln 3}\frac{1}{n^C},$$

which is absurd by inequality (8). Thus $M \leq n\log_3 2 < M + 1$, which shows that $\lfloor n\log_3 2\rfloor = M = \lfloor qn\rfloor$.

Theorem 4 directly implies the correctness of Algorithm 1. We emphasize again that the required approximation q can be computed in time $O(|n_2|^3 \ln^2 |n_2|)$. For Algorithm 2, we state the following theorem.

Theorem 5. *Let $q \in \mathbb{Q}$ be an approximation of $\log_3 2$ so precise that*

$$|q - \log_3 2| \leq \frac{1}{\ln 3}\frac{1}{(n-1)^{C+1}}, \tag{9}$$

Then $\ell - 1 \leq (n-1)q$ if and only if $\ell - 1 \leq (n-1)\log_3 2$.

Proof. Assume first that $\ell - 1 \leq (n-1)\log_3 2$, but $\ell - 1 > (n-1)q$. Then

$$\begin{aligned}|(n-1)\log_3 2 - (\ell - 1)| &= (n-1)\log_3 2 - (\ell - 1)\\ &< (n-1)\log_3 2 - (n-1)q.\end{aligned}$$

By (9), the latest expression is at most $\frac{1}{\ln 3}\frac{1}{(n-1)^C}$, which contradicts inequality (8). Similarly, assuming $\ell - 1 \leq (n-1)q$ and $\ell - 1 > (n-1)\log_3 2$ leads into contradiction.

The above theorem verifies the correctness of Algorithm 2.

5 Conclusion

We have shown how to compute $|(2^n)_{\mathbf{3}}|$ and the first symbol of $(2^n)_{\mathbf{3}}$ in polynomial time with respect to $|n_{\mathbf{2}}|$. We believe that our approach, together with more general results of [1], allows also to compute k first symbols of $(2^n)_{\mathbf{3}}$ for any fixed k, or more generally, k first symbols of $(p)_{\mathbf{q}}$ for $p, q \geq 2$ and $p \neq q$.

On the other hand we do not see how to compute the middle symbol of $(2^n)_{\mathbf{3}}$ in polynomial time. This motivates the following problem.

Problem. Does there exist a polynomial time algorithm computing, for a given n and $i \leq \lfloor n \log_3 2 \rfloor + 1$, the ith symbol of string $(2^n)_{\mathbf{3}}$ in polynomial time with respect to $|n_{\mathbf{2}}| = \Theta(\ln n)$?

Other problems are to consider our questions for some other exponentially growing sequences of numbers, such as the Fibonacci sequence.

Acknowledgement

We wish to thank Tapani Matala-aho for providing the reference to the work of Georges Rhin [3].

References

1. A. Baker: *Transcendental Number Theory.* Cambridge University Press (1975).
2. A. Gelfond: *Sur les approximations des nombres transcendants par des nombres algébriques.* Dokl. Acad. Sci. URSS 2, 177–182, (in russian and french) (1935).
3. G. Rhin: *Approximants de Padé et mesures effectives d'irrationalité.* Séminaire de Théorie des Nombres, Paris 1985–86, 155–164, Progr. Math., 71, Birkhäuser Boston, Boston, MA (1987).

Algorithms for Computing Small NFAs

Lucian Ilie*,** and Sheng Yu***

Department of Computer Science, University of Western Ontario,
N6A 5B7, London, Ontario, Canada,
ilie|syu@csd.uwo.ca

Abstract. We give new methods for constructing small nondeterministic finite automata (NFA) from regular expressions or from other NFAs. Given an arbitrary NFA, we compute the largest right-invariant equivalence on the states and then merge the states in the same class to obtain a smaller automaton. When applying this method to position automata, we get a way to convert regular expressions into NFAs which can be arbitrarily smaller than the position, partial derivative, and follow automata. In most cases, it is smaller than all NFAs obtained by similar constructions.

Keywords: regular expressions, nondeterministic finite automata, algorithms, positions, partial derivatives, follow relations, quotients, invariant equivalences

1 Introduction

The importance of regular expressions for applications is well known. They describe lexical tokens for syntactic specifications and textual patterns in text manipulation systems. Regular expressions have become the basis of standard utilities such as scanner generators (lex), editors (emacs, vi), or programming languages (perl, awk), see [ASU86,Fr98]. The implementation of the regular expressions is done using finite automata. As the deterministic finite automata (DFA) obtained from regular expressions can be exponentially larger in size and minimal nondeterministic ones (NFA) are hard to compute (the problem is PSPACE-complete, see [Yu97]), other methods need to be used to find small automata. One classical solution, due to Thompson, is to construct NFAs with ε-transitions (εNFA) which have linear size. Another solution is to construct an NFA which is not minimal but "reasonably" small; a well-known construction in this case is the position automaton, due to Glushkov [Gl61] and McNaughton-Yamada [McNYa60]; the size of their NFA is between linear and quadratic and can be computed by the algorithm of Brügemann-Klein [BrK93] in quadratic time.

* corresponding author

** Research partially supported by NSERC grant R3143A01.

*** Research partially supported by NSERC grant OGP0041630.

K. Diks et al. (Eds): MFSC 2002, LNCS 2420, pp. 328–340, 2002.

There are several improvements of these constructions, of which we mention the most important ones. For εNFAs, Sippu and Soisalon-Soininen [SiSo88] gave another construction which builds smaller εNFAs than Thompson's. For NFAs, we mention several constructions. Antimirov [An96] constructed an NFA based on partial derivatives. Champarnaud and Ziadi [ChZi01a,ChZi01b] improved very much Antimirov's $\mathcal{O}(n^5)$ algorithm for the construction of such NFA; their algorithm runs in quadratic time. They proved also that the partial derivative automaton is a quotient of the position automaton and so it is always smaller. Hromkovič et al. [HSW01] gave the construction with the best worst case so far; their NFA has size at most $\mathcal{O}(n(\log n)^2)$ and, by the algorithm of Hagenah and Muscholl [HaMu00], it can be computed in time $\mathcal{O}(n(\log n)^2)$. However, the number of states is artificially increased to reduce the number of transitions.

It is unexpected that this very important problem, of computing small NFAs from regular expressions, did not receive more attention. Also, the more general problem of reducing the size of arbitrary NFAs was even less investigated. We address these problems and give new methods to construct small finite automata from regular expressions as well as new methods to reduce the size of arbitrary NFAs.

Our basic idea resembles the minimization algorithm for DFAs where indistinguishable states are first computed and then merged; see, e.g., [HoUl79]. The idea cannot be applied directly to NFAs and we give a discussion on what is the best way to do it. It is unexpected that this old idea was not used for reducing NFAs. This is probably because in the case of DFAs a very nice mathematical object is obtained – the (unique) minimal DFA – while for NFAs this is hopeless. However, an interesting mathematical object can also be obtained in the case of NFAs. We show how to compute the largest right-invariant equivalence on states and then merge states according to this equivalence.

The above idea can be easily employed to construct small NFAs from regular expressions. Simply take any NFA obtained from a regular expression and reduce it according to the algorithm. The automata we use as a start are either the position automaton or quotients of it. We compare our NFAs to the position and partial derivative automata, as well as with the new follow automata introduced in [IlYu02].

Follow automata are a new type of automata obtained from regular expressions and they seem to have several remarkable properties. The idea is to build first εNFAs which are always smaller than the ones of Thompson [Th68] or Sippu and Soisalon-Soininen [SiSo88]. Their size is at most $\frac{3}{2}|\alpha| + \frac{5}{2}$ which is very close to the optimal because of the lower bound $\frac{4}{3}|\alpha| + \frac{5}{2}$. Then, elimination of ε-transitions from these εNFAs is used. Although this method of constructing NFAs has, apparently, nothing to do with positions, it turns out, unexpectedly, that the NFA it produces is a quotient of the position automaton with respect to the equivalence given by the follow relation; therefore giving the name of follow automaton. The follow automaton is always smaller and faster to compute than the position automaton. It uses optimally the information from the positions of the regular expression and thus it cannot be improved this way. It is in general uncomparable with Antimirov's partial derivative automaton.

The NFA we obtain from a regular expression by state merging is always smaller than the partial derivative automaton and follow automaton since both are quotients of the position automaton with respect to right-invariant equivalences. Moreover, it can be arbitrarily smaller! Even if the size in the worst case is quadratic, on most examples it is smaller than all known constructions and seems to be a good candidate for the best method to reduce the size of NFAs.

Due to limited space, all proofs are omitted.

2 Basic Notions

We recall here the basic definitions we need throughout the paper. For further details we refer to [HoUl79] or [Yu97].

Let A be an alphabet and A^* the set of all words over A; ε denotes the empty word and the length of a word w is denoted $|w|$. A *language* over A is a subset of A^*. A *regular expression* over A is $\emptyset$, ε, or $a \in A$, or is obtained from these applying the following rules finitely many times: for two regular expressions α and β, the *union*, $\alpha + \beta$, the *catenation*, $\alpha \cdot \beta$, and the *star*, α^* are regular expressions. For a regular expression α, the language denoted by it is $L(\alpha)$, and $\varepsilon(\alpha)$ is ε if $\varepsilon \in L(\alpha)$ and $\emptyset$ otherwise. Also, $|\alpha|$ and $|\alpha|_A$ denote the size of α and the number of occurrences of letters from A in α, respectively.

A *finite automaton* is a quintuple $M = (Q, A, q_0, \delta, F)$, where Q is the set of states, $q_0 \in Q$ is the initial state, $F \subseteq Q$ is the set of final states, and $\delta \subseteq Q \times (A \cup \{\varepsilon\}) \times Q$ is the transition mapping; we shall denote, for $p \in Q, a \in A \cup \{\varepsilon\}$, $\delta(p, a) = \{q \in Q \mid (p, a, q) \in \delta\}$. The automaton M is called *deterministic* (DFA) if $\delta : Q \times A \to Q$ is a (partial) function, *nondeterministic* (NFA) if $\delta \subseteq Q \times A \times Q$, and *nondeterministic with ε-transitions* (εNFA) if there are no restrictions on δ. The *language* recognized by M is denoted $L(M)$. The *size* of a finite automaton M is $|M| = |Q| + |\delta|$, that is, we count both states and transitions.

Let $\equiv$ be an equivalence relation over Q. For $q \in Q$, $[q]_\equiv$ denotes the equivalence class of q w.r.t. $\equiv$ and, for $S \subseteq Q$, $S/_\equiv$ denotes the quotient set $S/_\equiv = \{[q]_\equiv \mid q \in S\}$. We say that $\equiv$ is *right invariant* w.r.t. M iff (i) $\equiv \subseteq (Q - F)^2 \cup F^2$ (that is, final and non-final states are not $\equiv$-equivalent) and (ii) for any $p, q \in Q$ and any $a \in A$, if $p \equiv q$, then $\delta(p, a)/_\equiv = \delta(q, a)/_\equiv$. If $\equiv$ is right invariant, the *quotient automaton* $M/_\equiv$ is constructed by $M/_\equiv = (Q/_\equiv, A, [q_0]_\equiv, \delta_\equiv, F/_\equiv)$ where $\delta_\equiv = \{([p]_\equiv, a, [q]_\equiv) \mid (p, a, q) \in \delta\}$. Notice that $Q/_\equiv = (Q - F)/_\equiv \cup F/_\equiv$, so we do not merge final with non-final states. Also, we have that $L(M/_\equiv) = L(M)$.

3 Position Automata

We recall in this section the well-known construction of the position automaton, discovered independently by Glushkov [Gl61] and McNaughton and Yamada [McNYa60].

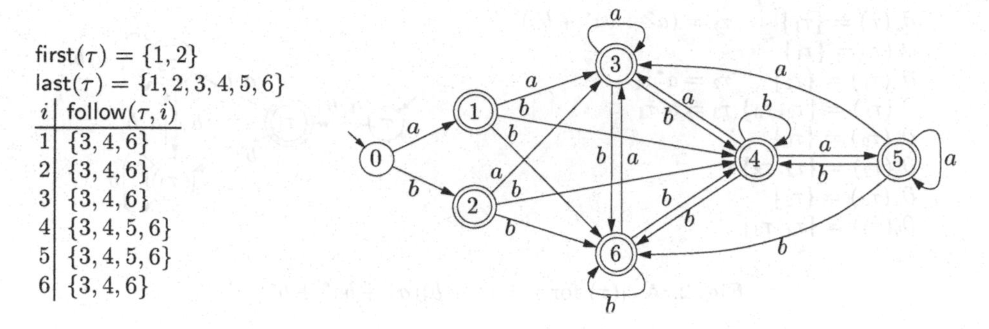

$\mathsf{first}(\tau) = \{1, 2\}$

$\mathsf{last}(\tau) = \{1, 2, 3, 4, 5, 6\}$

i	$\mathsf{follow}(\tau, i)$
1	$\{3, 4, 6\}$
2	$\{3, 4, 6\}$
3	$\{3, 4, 6\}$
4	$\{3, 4, 5, 6\}$
5	$\{3, 4, 5, 6\}$
6	$\{3, 4, 6\}$

Fig. 1. $\mathbf{A}_{\text{pos}}(\tau)$ for $\tau = (a + b)(a^* + ba^* + b^*)^*$

Let α be a regular expression. Put $\mathsf{pos}(\alpha) = \{1, 2, \ldots, |\alpha|_A\}$ and $\mathsf{pos}_0(\alpha) = \mathsf{pos}(\alpha) \cup \{0\}$. All letters in α are made different by marking each letter with its position in α; denote the obtained expression $\overline{\alpha} \in \overline{A}^*$, where $\overline{A} = \{a_i \mid a \in A, 1 \leq i \leq |\alpha|_A\}$. For instance, if $\alpha = a(baa + b^*)$, then $\overline{\alpha} = a_1(b_2 a_3 a_4 + b_5^*)$. Notice that $\mathsf{pos}(\alpha) = \mathsf{pos}(\overline{\alpha})$. The same notation will also be used for removing indices, that is, for unmarked expressions α, the operator $\bar{\cdot}$ adds indices, while for marked expressions $\overline{\alpha}$ the same operator, $\bar{\cdot}$, removes the indices: $\overline{\overline{\alpha}} = \alpha$. We extend the notation for arbitrary structures, like automata, in the obvious way. It will be clear from the context whether $\bar{\cdot}$ adds or removes indices.

Three mappings first, last, and follow are then defined as follows (see, e.g., [BrK93]). For any regular expression α and any $i \in \mathsf{pos}(\alpha)$, we have:

$$\begin{aligned} \mathsf{first}(\alpha) &= \{i \mid a_i w \in L(\overline{\alpha})\}, \\ \mathsf{last}(\alpha) &= \{i \mid w a_i \in L(\overline{\alpha})\}, \\ \mathsf{follow}(\alpha, i) &= \{j \mid u a_i a_j v \in L(\overline{\alpha})\}. \end{aligned} \tag{1}$$

For reasons that are made clear in the next section, we extend $\mathsf{follow}(\alpha, 0) = \mathsf{first}(\alpha)$. Also, let $\mathsf{last}_0(\alpha)$ stand for $\mathsf{last}(\alpha)$ if $\varepsilon(\alpha) = \emptyset$ and $\mathsf{last}(\alpha) \cup \{0\}$ otherwise.

The *position automaton* for α is $\mathbf{A}_{\text{pos}}(\alpha) = (\mathsf{pos}_0(\alpha), A, \delta_{\text{pos}}, 0, \mathsf{last}_0(\alpha))$ with $\delta_{\text{pos}} = \{(i, a, j) \mid j \in \mathsf{follow}(\alpha, i), a = \overline{a_j}\}$. As shown by Glushkov [Gl61] and McNaughton and Yamada [McNYa60], $L(\mathbf{A}_{\text{pos}}(\alpha)) = L(\alpha)$.

Example 1. Consider the regular expression $\tau = (a + b)(a^* + ba^* + b^*)^*$. The marked version of τ is $\overline{\tau} = (a_1 + b_2)(a_3^* + b_4 a_5^* + b_6^*)^*$. The values of the mappings first, last, and follow for τ and the corresponding position automaton $\mathbf{A}_{\text{pos}}(\tau)$ are given in Fig. 1.

4 Partial Derivative Automata

Another construction we recall is the partial derivative automaton, introduced by Antimirov [An96]. If the set of all partial derivatives of α (see [An96]) is $\mathrm{PD}(\alpha) = \{\partial_w(\alpha) \mid w \in A^*\}$, then the *partial derivative automaton* is $\mathbf{A}_{\text{pd}}(\alpha) =$

$\partial_a(\tau) = \{\tau_1\}$ $\quad \tau_1 = (a^* + ba^* + b^*)^*$
$\partial_b(\tau) = \{\tau_1\}$
$\partial_a(\tau_1) = \{\tau_2\}$ $\quad \tau_2 = a^*\tau_1$
$\partial_b(\tau_1) = \{\tau_2, \tau_3\}$ $\quad \tau_3 = b^*\tau_1$
$\partial_a(\tau_2) = \{\tau_2\}$
$\partial_b(\tau_2) = \{\tau_2, \tau_3\}$
$\partial_a(\tau_3) = \{\tau_2\}$
$\partial_b(\tau_3) = \{\tau_2, \tau_3\}$

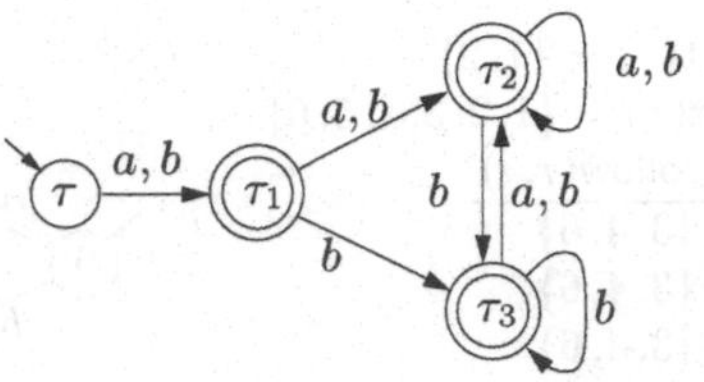

Fig. 2. $\mathbf{A}_{\mathrm{pd}}(\tau)$ for $\tau = (a + b)(a^* + ba^* + b^*)^*$

$(\mathrm{PD}(\alpha), A, \delta_{\mathrm{pd}}, \alpha, \{q \in \mathrm{PD}(\alpha) \mid \varepsilon(q) = \varepsilon\})$, where $\delta_{\mathrm{pd}}(q, a) = \partial_a(q)$, for any $q \in \mathrm{PD}(\alpha), a \in A$. Antimirov proved that $L(\mathbf{A}_{\mathrm{pd}}(\alpha)) = L(\alpha)$.

Example 2. Consider the regular expression τ from Example 1. The partial derivatives of τ are computed in Fig. 2 where also its partial derivative automaton $\mathbf{A}_{\mathrm{pd}}(\tau)$ is shown.

Champarnaud and Ziadi [ChZi01a,ChZi01b] proved that the partial derivative automaton is a quotient of the position automaton and gave also an algorithm to compute the partial derivative automaton in quadratic time. It can be seen that the automaton in Fig. 2 is a quotient of $\mathbf{A}_{\mathrm{pos}}(\tau)$ by comparing with Fig. 1.

5 Follow Automata

A new type of automata obtained from regular expressions, follow automata, introduced in [IlYu02], is the topic of this section.

We start with an algorithm for constructing εNFAs from regular expressions. The regular expression is assumed *reduced* such that 'many' redundant $\emptyset$'s, ε's, and $*$'s are eliminated. The precise way to do such reductions is omitted.

Algorithm 3. Given a regular expression α, the algorithm constructs an εNFA for α inductively, following the structure of α, and is shown in Fig. 3.

The steps should be clear from the Fig. 3 but we bring improvements at each step:

(a) after catenation (Fig. 3(v)): denote the state common to the two automata by p; (a1) if there is a single transition outgoing from p which, in addition, is labelled ε, say $p \xrightarrow{\varepsilon} q$, then the transition is removed and p and q are merged; otherwise (a2) if there is a single transition incoming to p, say $q \xrightarrow{\varepsilon} p$, then the transition is removed and p and q are merged;

(b) after iteration (Fig. 3(vi)), denote the middle state by p. If there is a cycle containing p such that all its transitions are labelled by ε, then all transitions in the cycle are removed and all states in the cycle are merged.

After the end of all steps in Fig. 3 (and the corresponding improvements (a) and (b)), we do some further improvements:

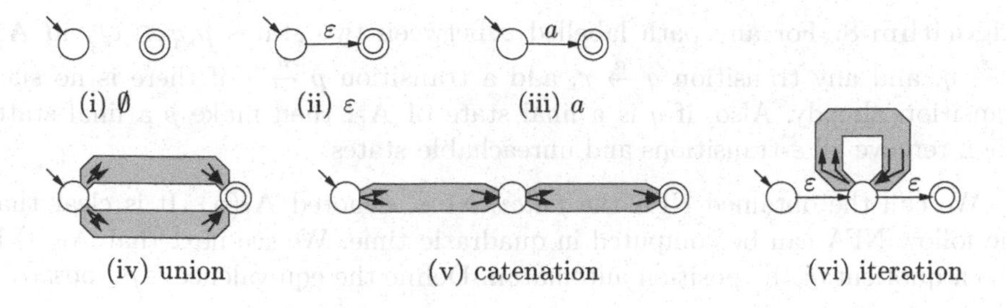

Fig. 3. The construction of $\mathbf{A}_{\mathrm{f}}^{\varepsilon}$

(c) if there is only one transition leaving the initial state and is labelled ε, say $q_0 \stackrel{\varepsilon}{\to} p$, then the transition is removed and q_0 and p merged;
(d) in case of multiple transitions, that is, transitions with the same source, target, and label, only one transition is kept, the others are removed.

We call the obtained nondeterministic finite automaton with ε-transitions the *follow* εNFA (the reason for this name will be clear later) and denote it $\mathbf{A}_{\mathrm{f}}^{\varepsilon}(\alpha) = (Q_f^{\varepsilon}, A, 0_f, \delta_f^{\varepsilon}, q_f)$. The next theorem says that this εNFA is always smaller than the ones obtained by the constructions of Thompson [Th68] and Sippu and Soisalon-Soininen [SiSo88]. We give also an example showing that it can be much smaller. Notice that in the example we do not use the improvements (a)-(d) at all since we want to emphasize the superiority of the core of our construction. It is easy to construct expressions, using (a)-(d), for which our construction gives an arbitrarily smaller automaton.

Theorem 4. For any regular expression α, the size of $\mathbf{A}_{\mathrm{f}}^{\varepsilon}(\alpha)$ is smaller than or equal to the size of the εNFAs obtained from α using the constructions of Thompson and Sippu and Soisalon-Soininen.

Example 5. For the regular expression $\alpha = a_1 + a_2 + \cdots + a_n$, $\mathbf{A}_{\mathrm{f}}^{\varepsilon}(\alpha)$ has size $n+2$ (2 states, n transitions), Thomson's has size $9n-6$ ($4n-2$ states, $5n-4$ transitions), and Sippu and Soisalon-Soininen's has size $5n-2$ ($2n$ states, $3n-2$ transitions).

The next theorem gives the upper bound on the size of our $\mathbf{A}_{\mathrm{f}}^{\varepsilon}$. The upper bound in Theorem 6(iii) is very close to the lower bound in Lemma 7 and hence very close to optimal.

Theorem 6. For any reduced regular expression α we have:
(i) $L(\mathbf{A}_{\mathrm{f}}^{\varepsilon}(\alpha)) = L(\alpha)$,
(ii) Algorithm 3 computes $\mathbf{A}_{\mathrm{f}}^{\varepsilon}(\alpha)$ in time $\mathcal{O}(|\alpha|)$, and
(iii) $|\,\mathbf{A}_{\mathrm{f}}^{\varepsilon}(\alpha)| \le \frac{3}{2}|\alpha| + \frac{5}{2}$.

Lemma 7. Let $\alpha_n = (a_1^* + a_2^*)(a_3^* + a_4^*) \cdots (a_{2n-1}^* + a_{2n}^*)$. Every εNFA accepting $L(\alpha_n)$ has size at least $8n - 1 = \frac{4}{3}|\alpha_n| + \frac{1}{3}$.

We describe next the way ε-transitions are eliminated from $\mathbf{A}_{\mathrm{f}}^{\varepsilon}$.

Algorithm 8. For any path labelled ε between two states $p, q \in Q_f^\varepsilon$ in $\mathbf{A}_\mathsf{f}^\varepsilon$, $p \overset{\varepsilon}{\rightsquigarrow} q$, and any transition $q \overset{a}{\to} r$, add a transition $p \overset{a}{\to} r$ if there is no such transition already. Also, if q is a final state of $\mathbf{A}_\mathsf{f}^\varepsilon$, then make p a final state. Then remove all ε-transitions and unreachable states.

We call the obtained NFA the *follow* NFA, denoted $\mathbf{A}_\mathsf{f}(\alpha)$. It is clear that the follow NFA can be computed in quadratic time. We see next that $\mathbf{A}_\mathsf{f}(\alpha)$ is also a quotient of the position automaton. Define the equivalence $\equiv_f \subseteq \mathsf{pos}_0(\alpha)^2$ by

$$i \equiv_f j \quad \text{iff} \quad \begin{array}{l}\text{(i) both } i, j \text{ or none belong to } \mathsf{last}(\alpha) \text{ and} \\ \text{(ii) } \mathsf{follow}(\alpha, i) = \mathsf{follow}(\alpha, j)\end{array}$$

Notice that we restrict the equivalence so that we do not make equivalence between a final and a non-final state in $\mathbf{A}_{\mathsf{pos}}(\alpha)$. Also, $\equiv_f$ is a right-invariant equivalence, so we can make the quotient.

Theorem 9. $\mathbf{A}_\mathsf{f}(\alpha) \simeq \mathbf{A}_{\mathsf{pos}}(\alpha)/_{\equiv_f}$.

We notice that the restriction we imposed on $\equiv_f$ so that final and non-final states in $\mathsf{last}_0(\alpha)$ cannot be $\equiv_f$-equivalent is essential, as shown by the expression $\alpha = (a^*b)^*$. Here $\mathsf{follow}(\alpha, i) = \{1, 2\}$, for any $0 \le i \le 2$. However, merging all three states of $\mathbf{A}_{\mathsf{pos}}(\alpha)$ is an error as the resulting automaton would accept the language $(a + b)^*$.

A first consequence of Theorem 9 is that the follow NFA is always smaller than or equal to the position automaton. As we shall see later, it can be much smaller.

Example 10. We give an example of an application of Theorem 9. For the same regular expression $\tau = (a+b)(a^* + ba^* + b^*)^*$ from Example 1, we build in Fig. 4 the $\mathbf{A}_\mathsf{f}^\varepsilon(\tau)$ and then give the equivalence classes of $\equiv_f$ and the automaton $\mathbf{A}_\mathsf{f}(\tau)$.

We finally notice that the follow automaton uses the whole information which comes from the positions of α. To see this, we consider the follow automaton for a marked expression, $\mathbf{A}_\mathsf{f}(\overline{\alpha})$, which is a deterministic automaton. Let the minimal automaton equivalent to it be $\mathsf{min}(\mathbf{A}_\mathsf{f}(\overline{\alpha}))$. Then $\overline{\mathsf{min}(\mathbf{A}_\mathsf{f}(\overline{\alpha}))}$ is an NFA accepting $L(\alpha)$ which can be computed in time $\mathcal{O}(|\alpha|^2 \log |\alpha|)$ using the minimization algorithm of Hopcroft [Ho71]. This is, in fact, another way of using positions to compute NFAs for regular expressions. However, it is interesting to

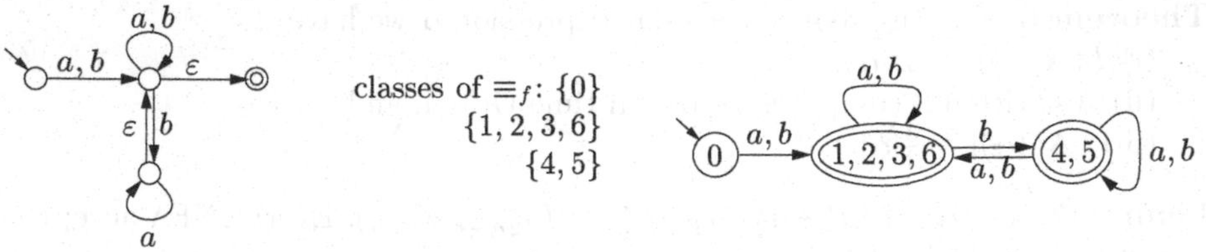

Fig. 4. $\mathbf{A}_\mathsf{f}^\varepsilon(\tau)$ and $\mathbf{A}_\mathsf{f}(\tau) = \mathbf{A}_{\mathsf{pos}}(\tau)/_{\equiv_f}$ for $\tau = (a+b)(a^* + ba^* + b^*)^*$

notice that $\overline{\mathsf{min}(\mathbf{A}_\mathsf{f}(\overline{\alpha}))}$ brings no improvement over $\mathbf{A}_\mathsf{f}(\alpha)$ since it can be shown that $\overline{\mathsf{min}(\mathbf{A}_\mathsf{f}(\overline{\alpha}))} \simeq \mathbf{A}_\mathsf{f}(\alpha)$.

6 Basic Idea for Reducing Arbitrary NFAs

We consider now the more general problem of constructing small NFAs when starting from arbitrary NFAs. Indeed, the results here can be applied to the problem of constructing small NFAs from regular expressions, simply by constructing first any NFA from a given regular expression and then reduce its size.

The basic idea comes from the algorithm for DFA minimization, where indistinguishable states are merged together, thus reducing the size of the automaton. We say that two states p and q are *distinguishable* if there is a word w which can lead the automaton from p to a final state but cannot lead from q to a final state. In the deterministic case, we start by distinguishing between final and non-final states. Then, two states that have a transition by the same letter to some distinguishable states become distinguishable. With nondeterminism, we have sets of states and it is no longer clear how we should decide that two states are distinguishable. In fact, this seems very difficult to decide. What we shall do is find a superset, such that we are sure we do not merge state which we should not. There can always be states which could be merged but detecting those is too expensive.

Therefore, our goal is to compute an equivalence relation (denoted later in this section $\equiv_R$) such that any two states $\equiv_R$-equivalent are indistinguishable but the converse need not be true. To avoid confusions, we shall call such states *equivalent.* So, equivalent states will be indistinguishable but non-equivalent states will not necessarily be distinguishable. As in the deterministic case, we compute the complement of this equivalence (denoted by $\not\equiv_R$); that is, we compute which states are not equivalent.

The starting point is that final states are not equivalent to non-final states; this is the only reasonable way to start with. Then, we discuss how to use current information about non-equivalent states in order to compute further non-equivalent pairs. Assume p and q are two states about which we do not know whether they are equivalent or not and a is a letter. (We shall complete the automaton such that both $\delta(p,a)$ and $\delta(q,a)$ are non-empty.) Two candidates for the condition implying non-equivalence of p and q come first to mind:

(i) any state in $\delta(p,a)$ is not equivalent to any state in $\delta(q,a)$,

(ii) there is a state in $\delta(p,a)$ and one in $\delta(q,a)$ which are not equivalent.

We see next why neither of (i) and (ii) is good. The condition in (i) is too strong as it turns out that we mark as non-equivalent too few pairs and then merge states which must not be merged. For example, in the first NFA in Fig. 5, the states 1 and 5 will be merged and the resulting automaton accepts a strictly bigger language. On the other hand, the condition (ii) is too weak and we obtain a too small reduction of the automaton. For instance, in the second NFA in Fig. 5, the states 1 and 5 are not merged although they clearly should.

The right condition is in-between (i) and (ii):

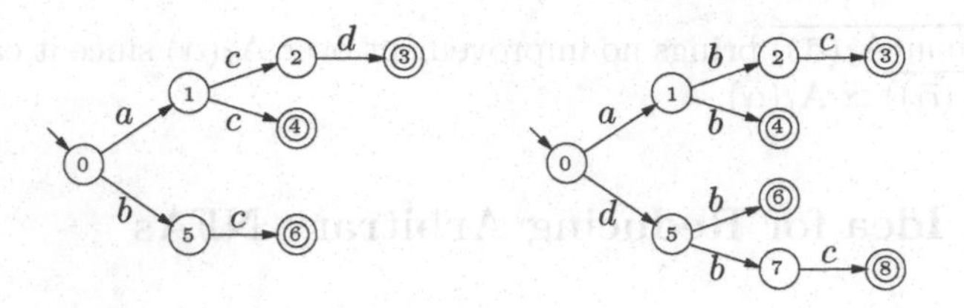

Fig. 5. (i) and (ii) are not good conditions

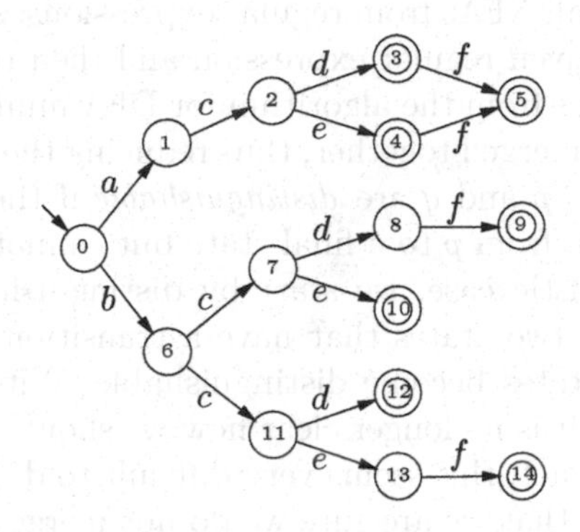

Fig. 6. $\mathbf{A}_{\mathrm{R}}$ is the best we can hope

(iii) there is a state in $\delta(p,a)$ which is not equivalent to any state in $\delta(q,a)$.

We shall see that (iii) is precisely what we need to merge as many states as one can hope. Before that, we see an example showing that finding indistinguishable states is hopeless. In Fig. 6, we have that 3 and 8 are non-equivalent (as being final and non-final, resp.) and also 4 and 13 are non-equivalent. Therefore, no matter how we find further non-equivalent states, 2 and 7 will become non-equivalent, since $\delta(2,d) = \{3\}$, $\delta(7,d) = \{8\}$. Similarly, 2 and 11 will be non-equivalent. From this, it follows that 1 and 6 will be non-equivalent. But, clearly, 1 and 6 are indistinguishable.

7 Computing Inequivalent States

We show next an algorithm to compute the equivalent states. Notice that we consider automata with arbitrarily many starting states.

Algorithm 11. Given an NFA $M = (Q, A, I, \delta, F)$, the algorithm computes a relation $\equiv_R$ on states; $p \equiv_R q$ means p and q are equivalent and therefore can (and will) be merged.
1. Completion of the automaton M: add a new (non-initial and non-final) state, denoted $\emptyset$, obtaining the new set of states $Q^{\emptyset} = Q \cup \{\emptyset\}$; add then all transitions which do not exist – the new transition function is $\delta^{\emptyset} = \delta \cup \{(p,a,\emptyset) \mid \delta(p,a) = \emptyset\} \cup \{(\emptyset,a,\emptyset) \mid a \in A\}$.

2. Start with $\not\equiv_R = \emptyset$; we compute the complement $\not\equiv_R$ of $\equiv_R$ (non-equivalent states)
3. **for** any $(p,q) \in (Q^\emptyset - F) \times F$ **do**
   $\not\equiv_R \leftarrow \not\equiv_R \cup \{(p,q),(q,p)\}$ (do not merge final with non-final)
4. **while** $\exists p,q \in Q^\emptyset, \exists a \in A, \exists r \in \delta^\emptyset(p,a), \forall s \in \delta^\emptyset(q,a), r \not\equiv_R s$ **do** (cond. (iii))
   choose one such pair (p,q) and $\not\equiv_R \leftarrow \not\equiv_R \cup \{(p,q),(q,p)\}$
5. **return** $\equiv_R = ((Q^\emptyset \times Q^\emptyset) - \not\equiv_R) - \{(\emptyset,\emptyset)\}$.

Notice that no pair in $\equiv_R$ contains the state $\emptyset$, so $\equiv_R$ is over Q. (We assume M has no useless states.) The next theorem shows that $\equiv_R$ is as good as we can hope.

Theorem 12. $\equiv_R$ is the largest equivalence over Q which is right-invariant w.r.t. M.

The automaton $M/_{\equiv_R}$ is the one we look for. It follows from Theorem 12 that it recognizes the same language as M does. Also, it can be very small as seen in the examples in the next sections.

8 Merging the States of $\mathbf{A}_{pos}$

We apply now the construction in the previous section to NFAs obtained from regular expressions. Denote $\mathbf{A}_\mathrm{R}(\alpha) = \mathbf{A}_\mathrm{pos}(\alpha)/_{\equiv_R}$. The following result is a corollary of Theorem 12 and the properties of $\mathbf{A}_\mathrm{pd}$ and $\mathbf{A}_\mathrm{f}$.

Theorem 13. $\mathbf{A}_\mathrm{R}(\alpha)$ is always smaller than or equal to either of $\mathbf{A}_\mathrm{f}(\alpha)$ and $\mathbf{A}_\mathrm{pd}(\alpha)$.

As seen in Example 18, $\mathbf{A}_\mathrm{R}(\alpha)$ can be arbitrarily smaller! Notice also that we need not compute $\mathbf{A}_\mathrm{pos}(\alpha)$ first; $\mathbf{A}_\mathrm{R}(\alpha)$ can be obtained from either of $\mathbf{A}_\mathrm{f}(\alpha)$ or $\mathbf{A}_\mathrm{pd}(\alpha)$.

Example 14. We show here the automaton $\mathbf{A}_\mathrm{R}(\tau)$ for the same expression $\tau = (a+b)(a^* + ba^* + b^*)^*$ from Example 1. Here, $\not\equiv_R$ will distinguish only final and nonfinal states. Therefore, we have the classes of $\equiv_R$ in Fig. 7 where also the corresponding automaton is shown.

classes of $\equiv_R$: $\{0\}$
$\{1,2,3,4,5,6\}$

Fig. 7. $\mathbf{A}_\mathrm{R}(\tau)$ for $\tau = (a+b)(a^* + ba^* + b^*)^*$

9 Examples

We give in this section several examples to compare the above constructions. We consider the automaton $\mathbf{A}_{\mathrm{R}}(\alpha)$ only in the last example where we show it can be arbitrarily smaller than the others.

Example 15. Consider $\alpha_1 = (a_1 + \varepsilon)^*$ and define inductively, for all $i \geq 1$, $\alpha_{i+1} = (\alpha_i + \beta_i)^*$, where β_i is obtained from α_i by replacing each a_j by $a_{j+|\alpha_i|_A}$. For instance, $\alpha_3 = (((a_1+\varepsilon)^* + (a_2+\varepsilon)^*)^* + ((a_3+\varepsilon)^* + (a_4+\varepsilon)^*)^*)^*$. We have $|\,\mathbf{A}_{\mathrm{pos}}(\alpha_n)| = |\,\mathbf{A}_{\mathrm{pd}}(\alpha_n)| = \Theta(|\alpha_n|^2)$ and $|\,\mathbf{A}_{\mathrm{f}}(\alpha_n)| = \Theta(|\alpha_n|)$.

Example 16. Consider the regular expression $\alpha_n = a_1(b_1+\cdots+b_n)^* + a_2(b_1+\cdots+b_n)^* + \cdots + a_n(b_1+\cdots+b_n)^*$. We have $|\,\mathbf{A}_{\mathrm{pos}}(\alpha_n)| = \Theta(|\alpha_n|^{3/2})$, $|\,\mathbf{A}_{\mathrm{f}}(\alpha_n)| = \Theta(|\alpha_n|)$, and $|\,\mathbf{A}_{\mathrm{pd}}(\alpha_n)| = \Theta(|\alpha_n|^{1/2})$.

Example 17. Consider the regular expression (generalization of one for identifiers in programming languages) $\alpha_n = (a_1 + a_2 + \cdots + a_n)(a_1 + a_2 + \cdots + a_n + b_1 + b_2 + \cdots + b_m)^*$. We have $|\,\mathbf{A}_{\mathrm{f}}(\alpha_n)| = |\,\mathbf{A}_{\mathrm{pd}}(\alpha_n)| = \Theta(|\alpha_n|)$ and $|\,\mathbf{A}_{\mathrm{pos}}(\alpha_n)| = \Theta(|\alpha_n|^2)$.

Example 18. Consider the expression $\alpha_n = (a+b)^*(a+b)^* \cdots (a+b)^*$. We have $|\,\mathbf{A}_{\mathrm{pos}}(\alpha_n)| = \Theta(|\alpha_n|^2)$ and $|\,\mathbf{A}_{\mathrm{f}}(\alpha_n)| = |\,\mathbf{A}_{\mathrm{pd}}(\alpha_n)| = \Theta(|\alpha_n|^2)$, while $\mathbf{A}_{\mathrm{R}}(\alpha_n)$ has a fixed size (independent of n), $|\,\mathbf{A}_{\mathrm{R}}(\alpha_n)| = 3$.

10 Reducing NFAs in Both Directions

When reducing NFAs according to $\equiv_R$, we considered only the distinguishability of the states to the right, that is, for two states p and q, we considered the words that led from p or q to final states. The same thing can be done symmetrically to the left, considering words that lead from the initial state(s) to p or q. We do not have to do really anything new. This symmetric reduction is the same as before but for the *reversed* automaton, say M^R, where all transitions are reversed and initial and final states interchanged. Denote the equivalence obtained as before but for M^R by $\equiv_L$; $\equiv_L$ is the largest left-invariant equivalence w.r.t. M. Combining the $\equiv_R$ and $\equiv_L$ is a very powerful method of reducing the size of NFAs. For instance, consider the regular expression $\alpha_n = \bigoplus_{1 \leq i_j \leq 2} a_{1i_1} a_{2i_2} \cdots a_{ni_n}$ (this is $\bigodot_{i=1}^{n}(a_{i1} + a_{i2})$ after opening the parentheses). For $n = 3$, α_3 looks like $ace+acf+ade+adf+bce+bcf+bde+bdf$. We have: $|\alpha_n| = 2n2^n - 1$, $|\,\mathbf{A}_{\mathrm{pos}}(\alpha_n)| = \Theta(|\alpha_n|)$, $|\,\mathbf{A}_{\mathrm{pd}}(\alpha_n)| = \Theta(\frac{|\alpha_n|}{\log|\alpha_n|})$, $|\,\mathbf{A}_{\mathrm{f}}(\alpha_n)| = \Theta(|\alpha_n|)$, and $\mathbf{A}_{\mathrm{pos}}(\alpha_n)$ can be reduced using $\equiv_R$ and $\equiv_L$ to an automaton of size $3n + 1 = \Theta(\log|\alpha_n|)$, that is, exponentially smaller.

11 Conclusions and Further Research

We gave a general method to reduce the size of arbitrary NFAs. When applying the construction to regular expressions, we obtain NFAs which can be arbitrarily smaller (!) than position, partial derivative, or follow automata. Although the best worst case is still the one given by Hromkovič et al. [HSW01], our construction seems to perform better on most examples. The worst case seems to be not very relevant. Notice that we have not compared our automata with those of Chang and Paige [ChPa97] since we do not work with compressed automata.

Several important problems remain to be investigated further.

Problem 1. Improve the running time of Algorithm 11 (which runs clearly in low polynomial time).

Problem 2. We have seen that using $\equiv_R$ and $\equiv_L$ together can result in much smaller NFAs. Find a way to combine the two optimally, that is, to obtain the highest reduction of the size. As seen from the position automaton for the regular expression $\alpha = ac + ad + bc$, the optimal way to do it may not be unique.

Problem 3. If combining optimally $\equiv_R$ and $\equiv_L$ turns out to be hard, find an algorithm which uses both but which is also fast.

References

[ASU86] Aho, A., Sethi, R., Ullman, J., *Compilers: Principles, Techniques, and Tools*, Addison-Wesley, MA, 1988.

[An96] Antimirov, V., Partial derivatives of regular expressions and finite automaton constructions, *Theoret. Comput. Sci.* **155** (1996) 291–319.

[BeSe86] Berry, G, Sethi, R., From regular expressions to deterministic automata, *Theoret. Comput. Sci.* **48** (1986) 117–126.

[BrK93] Brüggemann-Klein, A., Regular expressions into finite automata, *Theoret. Comput. Sci.* **120** (1993) 197–213.

[Br64] Brzozowski, J., Derivatives of regular expressions, *J. ACM* **11** (1964) 481–494.

[ChZi01a] Champarnaud, J.-M., Ziadi, D., New finite automaton constructions based on canonical derivatives, in: S. Yu, A. Paun, eds., *Proc. of CIAA 2000*, Lecture Notes in Comput. Sci. 2088, Springer-Verlag, Berlin, 2001, 94–104.

[ChZi01b] Champarnaud, J.-M., Ziadi, D., Computing the equation automaton of a regular expression in $\mathcal{O}(s^2)$ space and time, in: A. Amir, G. Landau, *Proc. of 12th CPM*, Lecture Notes in Comput. Sci. 2089, Springer-Verlag, Berlin, 2001, 157–168.

[ChPa97] Chang, C.-H., Paige, R., From regular expressions to DFA's using compressed NFA's, *Theoret. Comput. Sci* **178** (1997) 1–36.

[Fr98] Friedl, J., *Mastering Regular Expressions*, O'Reilly, 1998.

[Gl61] Glushkov, V.M., The abstract theory of automata, *Russian Math. Surveys* **16** (1961) 1–53.

[HaMu00] Hagenah, C., Muscholl, A., Computing ϵ-free NFA from regular expressions in $O(n\log^2(n))$ time, *Theor. Inform. Appl.* **34** (4) (2000) 257–277.

[Ho71] Hopcroft, J., An $n\log n$ algorithm for minimizing states in a finite automaton, *Proc. Internat. Sympos. Theory of machines and computations*, Technion, Haifa, 1971, Academic Press, New York, 1971, 189–196.

[HoUl79] Hopcroft, J.E., Ullman, J.D., *Introduction to Automata Theory, Languages, and Computation*, Addison-Wesley, Reading, Mass., 1979.

[HSW01] Hromkovic, J., Seibert, S., Wilke, T., Translating regular expressions into small ϵ-free nondeterministic finite automata, *J. Comput. System Sci.* **62** (4) (2001) 565–588.

[IlYu02] Ilie, L., Yu, S., Constructing NFAs by optimal use of positions in regular expressions, in: A. Apostolico, M. Takeda, *Proc. of 13th CPM*, Lecture Notes in Comput. Sci., Springer-Verlag, Berlin, 2002, to appear.

[McNYa60] McNaughton, R., Yamada, H., Regular expressions and state graphs for automata, *IEEE Trans. on Electronic Computers* **9** (1) (1960) 39–47.

[SiSo88] Sippu, S., Soisalon Soininen, E., *Parsing Theory: I Languages and Parsing*, EATCS Monographs on Theoretical Computer Science, Vol. 15, Springer-Verlag, New York, 1988.

[Th68] Thompson, K., Regular expression search algorithm, *Comm. ACM* **11** (6) (1968) 419–422.

[Yu97] Yu, S., Regular Languages, in: G. Rozenberg, A. Salomaa, *Handbook of Formal Languages, Vol. I*, Springer-Verlag, Berlin, 1997, 41–110.

Space-Economical Construction of Index Structures for All Suffixes of a String

Shunsuke Inenaga[1], Ayumi Shinohara[1,2], Masayuki Takeda[1,2], Hideo Bannai[3], and Setsuo Arikawa[1]

[1] Department of Informatics, Kyushu University 33, Fukuoka 812-8581, Japan,
[2] PRESTO, Japan Science and Technology Corporation (JST),
{s-ine,ayumi,takeda,arikawa}@i.kyushu-u.ac.jp
[3] Human Genome Center, University of Tokyo, Tokyo 108-8639, Japan,
bannai@ims.u-tokyo.ac.jp

Abstract. The *minimum all-suffixes directed acyclic word graph* (*MASDAWG*) of a string w has $|w|+1$ initial nodes, where the dag induced by all reachable nodes from the k-th initial node conforms with the DAWG of the k-th suffix of w. A new space-economical algorithm for the construction of $MASDAWG(w)$ is presented. The algorithm reads a given string w from right to left, and constructs $MASDAWG(w)$ *without suffix links*. It performs in time linear in the output size. Furthermore, we introduce the *minimum all-suffixes compact DAWG* (*MASCDAWG*). CDAWGs are known to be more space-economical than DAWGs, and thus $MASCDAWG(w)$ requires smaller space than $MASDAWG(w)$. We present an on-line (right-to-left) algorithm to build $MASCDAWG(w)$ without suffix links, whose running time is also linear in its size.

1 Introduction

Pattern matching on strings is one of the most fundamental and important problems in Theoretical Computer Science. When a pattern is flexible and a text is fixed, the problem can be solved in time proportional to the length of the pattern by using a suitable *index structure*.

An example of widely explored patterns is the *variable-length-don't-care pattern* (*VLDC-pattern*) which includes a symbol $\star$, a *wildcard* matching any string. Formally, when Σ is an alphabet, a VLDC-pattern is an element of set $(\Sigma \cup \{\star\})^*$. For example, $a\star ab\star$ is a VLDC-pattern, where $a, b \in \Sigma$. VLDC-patterns are sometimes called *regular patterns* as in [11]. The language of a VLDC-pattern (or a regular pattern) is the set of strings obtained by replacing $\star$'s in the pattern by arbitrary strings. This language corresponds to a class of the *pattern languages* proposed in [1].

The smallest automaton to recognize all VLDC-patterns matching a given text string was introduced in [8]. It is essentially the same structure as the minimum dag representing all substrings of every suffix of a string, which is called the *minimum all-suffixes directed acyclic word graph* (*MASDAWG*). The MASDAWG for a string w is the minimization of the DAWGs for all suffixes of

K. Diks et al. (Eds): MFSC 2002, LNCS 2420, pp. 341–352, 2002.

w. It has $|w|+1$ initial nodes, in which the dag induced by all reachable nodes from the k-th initial node conforms with the DAWG of the k-th suffix of w. Some applications of MASDAWGs were presented in [8].

The size of the DAWG for a string w is $O(|w|)$ [2]. This implies that the total size of the DAWGs of all suffixes of w is $O(|w|^2)$. Hence, the MASDAWG for w can be constructed in $O(|w|^2)$ time by minimizing the DAWGs [10]. On the other hand, it has been proven that the size of the MASDAWG of w is $\Theta(|w|^2)$ [8]. The direct construction of MASDAWGs that avoids the creation of redundant nodes and edges is therefore important, considering the reduction of space requirements. The first algorithm to directly build the MASDAWG of a string was given in [8]. It performs in *on-line* manner, that is, it processes a given string from left to right, a character by a character, and converts the MASDAWG of w to the MASDAWG of wa.

The algorithm of [8] can efficiently construct MASDAWGs by means of *suffix links*, kinds of failure transitions, like most linear-time algorithms constructing index structures (e.g., see [13,9,12,2,3,5,7,4,6]). On the other hand, it is also the fact that the memory space required by suffix links is non-ignorable. Moreover, for each node, the algorithm additionally requires to keep the length of the longest string that reaches to the node, in the construction phase. These values are unnecessary in order to examine whether a given pattern occurs or not in the specified suffix. In this paper, we present a new algorithm to construct MASDAWGs *without suffix links nor length information*, which thus permits us to save memory space. The algorithm is best understood as one constructing MASDAWGs in 'right-to-left' on-line manner. Namely, it builds the MASDAWG of aw by adding some nodes and edges to the MASDAWG of w.

Furthermore, we aim to reduce the space requirement by *compacting* the structure itself. We focus on the *compact DAWG* (*CDAWG*) whose space requirement is strictly smaller than that of the DAWG, both theoretically and practically [3,5]. Its all-suffixes version, named the *minimum all-suffixes CDAWG* (*MASCDAWG*), is introduced in this paper. We also present an on-line (right-to-left) algorithm to construct the MASCDAWG in linear time with respect to its size, without using suffix links nor length information.

2 Minimum All-Suffixes Directed Acyclic Word Graphs

Strings x, y, and z are said to be a *prefix*, *factor*, and *suffix* of string $w = xyz$, respectively. The sets of prefixes, factors, and suffixes of a string w are denoted by *Prefix*(w), *Factor*(w), and *Suffix*(w), respectively. The empty string is denoted by ε, that is, $|\varepsilon| = 0$. Let $\Sigma^+ = \Sigma^* - \{\varepsilon\}$. The factor of a string w that begins at position i and ends at position j is denoted by $w[i:j]$ for $1 \leq i \leq j \leq |w|$. For convenience, let $w[i:j] = \varepsilon$ for $j < i$. Let $w[i:] = w[i:|w|]$ for $1 \leq i \leq |w|+1$. Assume S is a subset of Σ^*. For any string $u \in \Sigma^*$, $u^{-1}S = \{x \mid ux \in S\}$.

Let $w \in \Sigma^*$. We define an equivalence relation $\equiv_w$ on Σ^* by

$$x \equiv_w y \Leftrightarrow x^{-1}\mathit{Suffix}(w) = y^{-1}\mathit{Suffix}(w).$$

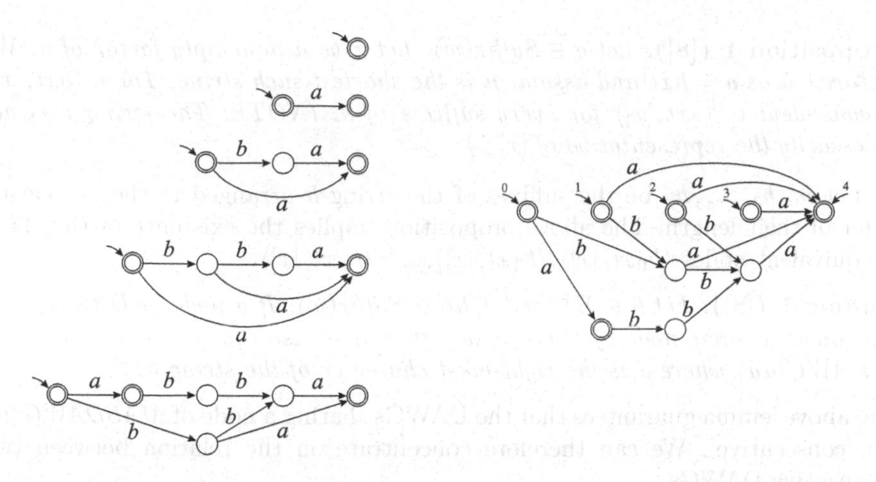

Fig. 1. The naive *ASDAWG*(w) is shown on the left, where $w = abba$. *MASDAWG*(w) is displayed on the right.

Let $[x]_w$ denote the equivalence class of a string $x \in \Sigma^*$ under $\equiv_w$. The longest element in the equivalence class $[x]_w$ for $x \in \mathit{Factor}(w)$ is called its *representative*.

Definition 1. *DAWG*(w) *is the dag* (V, E) *such that*

$$V = \{[x]_w \mid x \in \mathit{Factor}(w)\},$$
$$E = \{([x]_w, a, [xa]_w) \mid x, xa \in \mathit{Factor}(w), a \in \Sigma\}.$$

Definition 2. *ASDAWG*(w) *is a kind of dag with* $|w| + 1$ *initial nodes, designated by* $0, 1, \ldots, |w|$, *in which the subgraph consisting of the nodes reachable from the k-th initial node and their out-going edges is* *DAWG*($w[k+1:]$).

The simple collection of *DAWG*($w[1:]$), *DAWG*($w[2:]$),..., *DAWG*($w[n]$), *DAWG*($w[n+1:]$) ($n = |w|$) is an example of *ASDAWG*(w), referred to as the *naive* *ASDAWG*(w). The number of nodes of the naive *ASDAWG*(w) is $O(|w|^2)$. By minimizing the naive *ASDAWG*(w), we can obtain the *minimum* *ASDAWG*(w), which is denoted by *MASDAWG*(w). The naive *ASDAWG*($abba$) and *MASDAWG*($abba$) are shown in Fig. 1. The minimization is performed based on the equivalence relation defined as follows. Each node of *ASDAWG*(w) is represented by a pair $\langle u, [x]_u \rangle$ with $u \in \mathit{Suffix}(w)$ and $x \in \mathit{Factor}(u)$. The equivalence relation, denoted by $\sim_w$, is defined by

$$\langle u, [x]_u \rangle \sim_w \langle v, [y]_v \rangle \Leftrightarrow x^{-1}\mathit{Suffix}(u) = y^{-1}\mathit{Suffix}(v).$$

A node of *MASDAWG*(w) corresponds to an equivalence class under $\sim_w$. We write $\langle u, [x]_u \rangle$ simply as $\langle u, [x] \rangle$ in case no confusion occurs.

Theorem 1 ([8]). *When* $|\Sigma| \geq 2$, *the number of nodes of* *MASDAWG*(w) *for a string* w *is* $\Theta(|w|^2)$. *It is* $\Theta(|w|)$ *for a unary alphabet.*

Proposition 1 ([8]). *Let $u \in \mathit{Suffix}(w)$. Let x be a nonempty factor of u. We factorize u as $u = hxt$ and assume h is the shortest such string. Then, $\langle hxt, [x] \rangle$ is equivalent to $\langle sxt, [x] \rangle$ for every suffix s of h. (NOTE: The string x is not necessarily the representative of $[x]_u$.)*

Let $h_0, h_1, \ldots, h_r$ be the suffixes of the string h arranged in the decreasing order of their length. The above proposition implies the existence of the chain of equivalent nodes $\langle h_0xt, [x] \rangle, \langle h_1xt, [x] \rangle, \ldots, \langle h_rxt, [x] \rangle$.

Lemma 1 ([8]). *Let $h \in \Sigma^+$ and $u, hu \in \mathit{Suffix}(w)$. If a node of DAWG($u$) is equivalent to some node of DAWG(hu), then it is also equivalent to some node of DAWG(au) where a is the right-most character of the string h.*

The above lemma guarantees that the DAWGs sharing a node of *MASDAWG*(w) are 'consecutive'. We can therefore concentrate on the relation between two consecutive DAWGs.

From now on, we consider what happens when constructing *MASDAWG*(au) from *MASDAWG*(u). Due to Lemma 1, we only investigate the relationship between *DAWG*(au) and *DAWG*(u).

Lemma 2. *Let $a \in \Sigma$ and $u \in \Sigma^*$. For any string $x \in \mathit{Factor}(u) - \mathit{Prefix}(au)$, it holds that $\langle au, [x] \rangle \sim_{au} \langle u, [x] \rangle$.*

Proof. $x^{-1}\mathit{Suffix}(au) = x^{-1}(\{au\} \cup \mathit{Suffix}(u)) = x^{-1}\{au\} \cup x^{-1}\mathit{Suffix}(u) = x^{-1}\mathit{Suffix}(u)$, because $x^{-1}\{au\} = \emptyset$ for $x \notin \mathit{Prefix}(au)$. □

The above lemma implies that we have only to care about the prefixes of au in order to construct *MASDAWG*(au) from *MASDAWG*(u). We need not modify nor change the structure of *MASDAWG*(u): it is kept static.

Lemma 3. *Let $a \in \Sigma$ and $u \in \Sigma^*$. For any $x \in \mathit{Prefix}(u)$ and $y \in \Sigma^*$, if $\langle au, [ax] \rangle \sim_{au} \langle u, [y] \rangle$ then $[x]_u = [y]_u$.*

Proof. Since $x \in \mathit{Prefix}(u)$, there exists $s \in \Sigma^*$ such that $u = xs$. By the assumption, $(ax)^{-1}\mathit{Suffix}(au) = y^{-1}\mathit{Suffix}(u)$. Since s is included in the left set, s is also included in the right set, i.e. $s \in y^{-1}\mathit{Suffix}(u)$, which implies $ys \in \mathit{Suffix}(xs)$, thus $y \in \mathit{Suffix}(x)$. We have two cases according to $x \in \mathit{Prefix}(au)$.

(Case 1) When $x \in \mathit{Prefix}(au)$. Since $x \in \mathit{Prefix}(axs)$, $x = a^i$ and $y = a^j$ for some integers $j \leq i$. Suppose $j < i$, and let $k = i - j > 0$. Then $a^ks \in y^{-1}\mathit{Suffix}(u)$ while $a^ks \notin (ax)^{-1}\mathit{Suffix}(au)$, that contradicts with the assumption that $(ax)^{-1}\mathit{Suffix}(au) = y^{-1}\mathit{Suffix}(u)$. Thus $j = i$, which yields $y = x = a^i$.

(Case 2) When $x \notin \mathit{Prefix}(au)$.

$$
\begin{aligned}
y^{-1}\mathit{Suffix}(u) &= (ax)^{-1}\mathit{Suffix}(au) && \text{by the assumption} \\
&\subseteq x^{-1}\mathit{Suffix}(au) && \text{since } x \in \mathit{Suffix}(ax) \\
&= x^{-1}\mathit{Suffix}(u) && \text{since } x \notin \mathit{Prefix}(au) \\
&\subseteq y^{-1}\mathit{Suffix}(u) && \text{since } y \in \mathit{Suffix}(x)
\end{aligned}
$$

Thus we have $x^{-1}\mathit{Suffix}(u) = y^{-1}\mathit{Suffix}(u)$, that is, $[x]_u = [y]_u$. □

The path in $MASDAWG(u)$ spelling out u is called its 'backbone'. The above lemma shows that if a node $\langle au, [ax]\rangle$ on the 'backbone' of $MASDAWG(au)$ is equivalent to a node of $MASDAWG(u)$, the node $\langle au, [ax]\rangle$ is also on the 'backbone' of $MASDAWG(u)$. This fact is crucial in order that our algorithm, which will be given in the sequel, performs in time linear in the size of $MASDAWG(u)$.

For the prefixes of string au, we have the following lemma.

Lemma 4. *Let $a \in \Sigma$ and $u \in \Sigma^*$. Let $ax \in Prefix(au)$ be the shortest string which satisfies $\langle au, [ax]\rangle \sim_{au} \langle u, [x]\rangle$. Then for any longer prefix $axv \in Prefix(au)$, it holds that $\langle au, [axv]\rangle \sim_{au} \langle u, [xv]\rangle$.*

Proof. Since $\langle au, [ax]\rangle \sim_{au} \langle u, [x]\rangle$, $(ax)^{-1}Suffix(au) = x^{-1}Suffix(u)$. Thus, $(axv)^{-1}Suffix(au) = v^{-1}((ax)^{-1}Suffix(au)) = v^{-1}(x^{-1}Suffix(u)) = (xv)^{-1}Suffix(u)$. □

Remark that the node $\langle u, [xv]\rangle$ already exists in $MASDAWG(u)$, since $xv \in Prefix(u)$. Thus the above lemma guarantees that all nodes we have to newly create in $MASDAWG(au)$ are $\langle au, [t]\rangle$ for strings $t \in Prefix(z)$, where z is the longest prefix of au which does *not* satisfy $\langle au, [ax]\rangle \sim_{au} \langle u, [x]\rangle$. Now the next question is how to *efficiently* check whether $\langle au, [ax]\rangle \sim_{au} \langle u, [x]\rangle$ or not for each $x \in Prefix(u)$. Our idea is to count the cardinality of the set $x^{-1}Suffix(u)$.

Lemma 5. *Let $a \in \Sigma$ and $u \in \Sigma^*$. For any $x \in Factor(u)$, $\langle au, [ax]\rangle \sim_{au} \langle u, [x]\rangle$ if and only if $|(ax)^{-1}Suffix(au)| = |x^{-1}Suffix(u)|$.*

Proof. We first show that $(ax)^{-1}Suffix(au) \subseteq x^{-1}Suffix(u)$. Let us choose $s \in (ax)^{-1}Suffix(au)$ arbitrarily. Then $axs \in Suffix(au) = \{au\} \cup Suffix(u)$. If $axs = au$, then $xs = u$. Otherwise, $axs \in Suffix(u)$. Since xs is a suffix of axs, we know that xs is also a suffix of u. In both cases, we have $xs \in Suffix(u)$, which implies that $s \in x^{-1}Suffix(u)$. Thus $(ax)^{-1}Suffix(au) \subseteq x^{-1}Suffix(u)$. It yields that $(ax)^{-1}Suffix(au) = x^{-1}Suffix(u)$ if and only if $|(ax)^{-1}Suffix(au)| = |x^{-1}Suffix(u)|$. By the definition of $\sim_{au}$, we have proved the lemma. □

We associate each node $\langle u, [x]\rangle$ with the cardinality of the set, $|x^{-1}Suffix(u)|$, denoted by $\#\langle u, [x]\rangle$. Note that $\#\langle u, [u]\rangle = 1$ since $u^{-1}Suffix(u) = \{\varepsilon\}$, and that $\#\langle u, [\varepsilon]\rangle = |u| + 1$ since $\varepsilon^{-1}Suffix(u) = Suffix(u)$.

Lemma 6. *Let $a \in \Sigma$ and $u \in \Sigma^*$. For any $x \in Prefix(u)$, $\#\langle au, [ax]\rangle = \#\langle u, [ax]\rangle + 1$.*

Proof. Since $x \in Prefix(u)$, $\#\langle au, [ax]\rangle = |(ax)^{-1}Suffix(au)| = |(ax)^{-1}(\{au\} \cup Suffix(u))| = |(ax)^{-1}\{au\} \cup (ax)^{-1}Suffix(u))| = \#\langle u, [ax]\rangle + 1$. □

The whole algorithm is shown in Fig. 2. Since the algorithm manipulates an input string w from right to left, we number the characters in w as $w = w_n w_{n-1} \ldots w_1$. An edge is represented by a triple (r, w_i, s), where s, r are nodes and w_i is the character for the label of the edge.

Theorem 2. *For any string $w \in \Sigma^*$, our algorithm constructs $MASDAWG(w)$ in time linear in its size.*

The on-line (right-to-left) construction of $MASDAWG(w)$ where $w = abaa\$$ is displayed in Fig. 3.

```
Algorithm Construction of MASDAWG(w = w_n w_{n-1} ... w_1).
1  create new nodes s_0;
2  #(s_0) := 1;   #(nil) := 0;
3  initNode[0] := s_0; node := s_0;
4  for i := 1 to n do
5      s := FIND(node, w_i);
6      target := NEWTARGETNODE(s, i - 1, node);
7      newNode := create a new node with copying all out-going edges of node;
8      add or overwrite edge (newNode, w_i, target);
9      #(newNode) := i;
10     initNode[i] = newNode;
11     node = newNode;

function NEWTARGETNODE(Node s, int j, Node backbone) : Node
1  nextNumSuf := #(s) + 1;
2  if nextNumSuf = #(backbone) then return backbone;    /* redirection */
3  nextBackbone := FIND(backbone, w_j);
4  newNode := create a new node with copying all out-going edges of s;
5  s := FIND(s, w_j);
6  target := NEWTARGETNODE(s, j - 1, nextBackbone);
7  add or overwrite edge (newNode, w_j, target);
8  #(newNode) := nextNumSuf;
9  return newNode;

function FIND(Node s, char c) : Node
1  if s has the c-edge then
2      let (s, c, r) be the c-edge from s;
3      return r;
4  else return nil;
```

Fig. 2. The algorithm to construct $MASDAWG(w)$.

3 Minimum All-Suffixes Compact Directed Acyclic Word Graphs

To achieve a more space-economical index structure for all suffixes of a string, we turn our attention to a compact directed acyclic word graph (CDAWG) and consider its all-suffixes version.

Assume S is a subset of Σ^*. For any string $u \in \Sigma^*$, $Su^{-1} = \{x \mid xu \in S\}$. Let $w \in \Sigma^*$. We define an equivalence relation $\equiv'_w$ on Σ^* by

$$x \equiv'_w y \Leftrightarrow Prefix(w)x^{-1} = Prefix(w)y^{-1}.$$

Let $[x]'_w$ denote the equivalence class of a string $x \in \Sigma^*$ under $\equiv'_w$. The longest element in the equivalence class $[x]'_w$ for $x \in Factor(w)$ is also called its *representative*, and is denoted by $\overset{w}{\overrightarrow{x}}$. For any string $x \in Factor(w)$, there uniquely exists string $\alpha \in \Sigma^*$ such that $\overset{w}{\overrightarrow{x}} = x\alpha$.

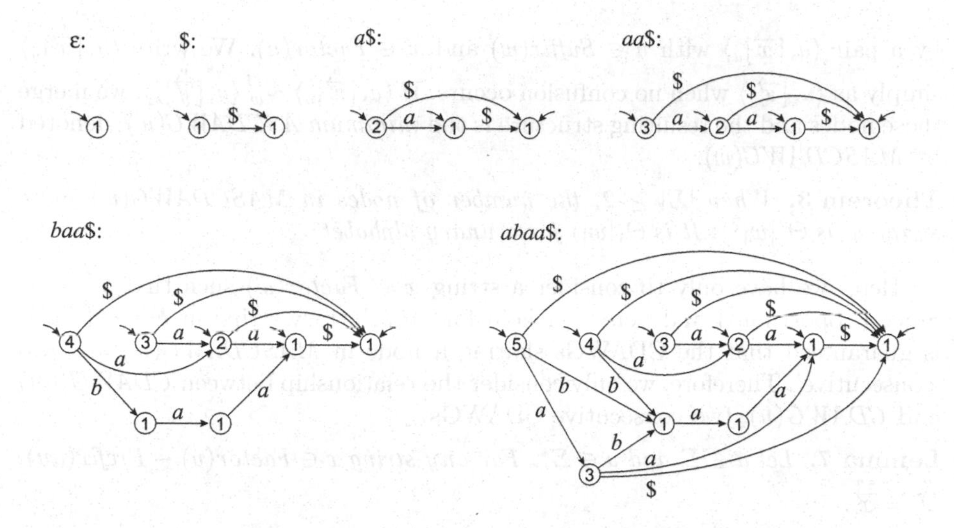

Fig. 3. Construction of $MASDAWG(abaa\$)$. Each node is marked by $\#\langle u, [x]\rangle$ where $u = abaa\$$ and $x \in Factor(u)$.

Proposition 2. *Let $x \in Factor(w)$. Assume $\overset{w}{\overrightarrow{x}} \notin Suffix(w)$. Then, x occurs in w at least twice.*

Proof. For a contradiction, assume x occurs in w only once. Then, we have $|Prefix(w)x^{-1}| = 1$. Let $w = hxy$. Since x occurs in w only once, $|Prefix(w)x^{-1}|$ $= |Prefix(w)(xy)^{-1}|$. Thus $x \equiv'_w xy$ and $\overset{w}{\overrightarrow{x}} = xy$. However, $xy \in Suffix(w)$, a contradiction. Consequently, x appears in w at least twice. □

Definition 3. *$CDAWG(w)$ is the dag (V, E) such that*

$V = \{[\overset{w}{\overrightarrow{x}}]_w \mid x \in Factor(w)\}$,

$E = \{([\overset{w}{\overrightarrow{x}}]_w, a\beta, [\overset{w}{\overrightarrow{xa}}]_w) \mid x, xa \in Factor(w),\ a \in \Sigma,\ \beta \in \Sigma^*,\ \overset{w}{\overrightarrow{xa}} = xa\beta,\ \overset{w}{\overrightarrow{x}} \neq \overset{w}{\overrightarrow{xa}}\}$.

The following corollary derives from Lemma 2.

Corollary 1. *Assume that w terminates with a unique symbol \$. Then, for any string $x \in Factor(w) - Suffix(w)$, node $[\overset{w}{\overrightarrow{x}}]_w$ is of out-degree more than one.*

Namely, $CDAWG(w)$ is the compaction of $DAWG(w)$ where any nodes of out-degree one are removed and their edges are modified accordingly.

Definition 4. *$ASCDAWG(w)$ is a kind of dag with $|w| + 1$ initial nodes, designated by $0, 1, \ldots, |w|$, in which the subgraph consisting of the nodes reachable from the k-th initial node and their out-going edges is $CDAWG(w[k+1:])$.*

We now introduce the minimized version of $ASCDAWG(w)$, which is well defined similarly to MASDAWGs. Each node of $ASCDAWG(w)$ can be represented by a pair $\langle u, [\overset{u}{\overrightarrow{x}}]_u \rangle$ with $u \in \mathit{Suffix}(w)$ and $x \in \mathit{Factor}(u)$. We write $\langle u, [\overset{u}{\overrightarrow{x}}]_u \rangle$ simply as $\langle u, [\overset{u}{\overrightarrow{x}}] \rangle$ when no confusion occurs. If $\langle u, [\overset{u}{\overrightarrow{x}}]_u \rangle \sim_w \langle v, [\overset{v}{\overrightarrow{y}}]_v \rangle$, we merge these nodes and the resulting structure is the *minimum* $\mathit{ASCDAWG}(w)$, denoted by $\mathit{MASCDAWG}(w)$.

Theorem 3. *When $|\Sigma| \geq 2$, the number of nodes in $\mathit{MASCDAWG}(w)$ for a string w is $\Theta(|w|^2)$. It is $\Theta(|w|)$ for a unary alphabet.*

Here, we have only to consider a string $x \in \mathit{Factor}(w)$ such that $\overset{w}{\overrightarrow{x}} = x$. Since Proposition 1 and Lemma 1 hold for an arbitrary string in $\mathit{Factor}(w)$, it is guaranteed that the CDAWGs sharing a node in $\mathit{MASCDAWG}(w)$ are also 'consecutive'. Therefore, we only consider the relationship between $\mathit{CDAWG}(au)$ and $\mathit{CDAWG}(u)$, two consecutive CDAWGs.

Lemma 7. *Let $a \in \Sigma$ and $u \in \Sigma^*$. For any string $x \in \mathit{Factor}(u) - \mathit{Prefix}(au)$, $\overset{u}{\overrightarrow{x}} = \overset{au}{\overrightarrow{x}}$.*

Proof. Since $x \notin \mathit{Prefix}(au)$, there is no new occurrence of x in au. It implies that $a(\mathit{Prefix}(u)x^{-1}) = \mathit{Prefix}(au)x^{-1}$. Thus we have $[x]'_u = [x]'_{au}$. Consequently, $\overset{u}{\overrightarrow{x}} = \overset{au}{\overrightarrow{x}}$. □

The above lemma ensures that any implicit node of $\mathit{CDAWG}(u)$ does not become explicit in $\mathit{CDAWG}(au)$ if it is not associated with a prefix of au. It follows from this lemma and Lemma 2 that we do not need to modify nor change the structure of $\mathit{MASCDAWG}(u)$ when constructing $\mathit{MASCDAWG}(au)$.

Lemma 8. *Let $a \in \Sigma$ and $u \in \Sigma^*$. For any $x, z \in \mathit{Factor}(u)$, if $\overset{au}{\overrightarrow{ax}} = az$ then $\overset{u}{\overrightarrow{z}} = z$.*

Proof. Suppose contrarily that $\overset{u}{\overrightarrow{z}} \neq z$. That means there exists $y \in \Sigma^*$ such that $\mathit{Prefix}(u)y^{-1} = \mathit{Prefix}(u)z^{-1}$ and $|y| > |z|$. Then $\mathit{Prefix}(au)(ay)^{-1} = (\mathit{Prefix}(au)y^{-1})a^{-1} = (a(\mathit{Prefix}(u)y^{-1}))a^{-1} = (a(\mathit{Prefix}(u)z^{-1}))a^{-1} = \mathit{Prefix}(au)(az)^{-1} = \mathit{Prefix}(au)(ax)^{-1}$. Thus $ay \equiv'_{au} ax$ and $|ay| > |az|$. It contradicts the assumption $\overset{au}{\overrightarrow{ax}} = az$. □

Lemma 9. *Let $a \in \Sigma$ and $u \in \Sigma^*$. For any $x \in \mathit{Prefix}(u)$ and $y \in \Sigma^*$ satisfying $\langle au, [\overset{au}{\overrightarrow{ax}}]_{au} \rangle \sim_{au} \langle u, [\overset{u}{\overrightarrow{y}}]_u \rangle$, there exists $z \in \mathit{Prefix}(u)$ such that $[\overset{u}{\overrightarrow{z}}]_u = [\overset{u}{\overrightarrow{y}}]_u$.*

Proof. Let z be the string with $\overset{au}{\overrightarrow{ax}} = az$. Then we have $\overset{u}{\overrightarrow{z}} = z$ by Lemma 8. Moreover, $z \in \mathit{Prefix}(u)$ since $x \in \mathit{Prefix}(u)$. Since $\langle au, [az]_{au} \rangle = \langle au, [\overset{au}{\overrightarrow{ax}}]_{au} \rangle \sim_{au} \langle u, [\overset{u}{\overrightarrow{y}}]_u \rangle$, we have $[z]_u = [\overset{u}{\overrightarrow{y}}]_u$ by Lemma 3. Thus $[\overset{u}{\overrightarrow{z}}]_u = [\overset{u}{\overrightarrow{y}}]_u$. □

Lemma 9 shows that if node $\langle au, [\overset{au}{\overrightarrow{ax}}]_{au} \rangle$ on the 'backbone' of $\mathit{MASCDAWG}(au)$ is equivalent to a node of $\mathit{MASCDAWG}(u)$, the node $\langle au, [\overset{au}{\overrightarrow{ax}}]_{au} \rangle$ is also on the 'backbone' of $\mathit{MASCDAWG}(u)$. It corresponds to Lemma 3.

We have the following lemma which corresponds to Lemma 4.

Lemma 10. *Let $a \in \Sigma$ and $u \in \Sigma^*$. Let $ax \in Prefix(au)$. Let $\overset{au}{\overrightarrow{ax}}$ be the shortest string for which there exists $z \in Prefix(u)$ such that $\langle au, [\overset{au}{\overrightarrow{ax}}]_{au} \rangle \sim_{au} \langle u, [\overset{u}{\overrightarrow{z}}]_u \rangle$. Let $\overset{au}{\overrightarrow{ax}} = ay$. Then for any longer prefix $ayv \in Prefix(au)$, there exists $s \in Prefix(u)$ such that $\langle au, [\overset{au}{\overrightarrow{ayv}}]_{au} \rangle \sim_{au} \langle u, [\overset{u}{\overrightarrow{s}}]_u \rangle$.*

Proof. Let $\overset{au}{\overrightarrow{ayv}} = as$. By Lemma 8, $\overset{u}{\overrightarrow{s}} = s$. Since $yv \in Prefix(u)$, $s \in Prefix(u)$. Let $\overset{u}{\overrightarrow{z}} = t$. By the assumption $\langle au, [\overset{au}{\overrightarrow{ax}}]_{au} \rangle \sim_{au} \langle u, [\overset{u}{\overrightarrow{z}}]_u \rangle$, we have $\langle au, [ay] \rangle \sim_{au} \langle u, [t] \rangle$. Since $y \in Prefix(u)$, $\langle au, [ay] \rangle \sim_{au} \langle u, [y] \rangle$ by Lemma 3. Note that $y \in Prefix(s)$. Hence we have $\langle au, [as] \rangle \sim_{au} \langle u, [s] \rangle$ by Lemma 4. Because $as = \overset{au}{\overrightarrow{ayv}}$ and $s = \overset{u}{\overrightarrow{s}}$, it holds that $\langle au, [\overset{au}{\overrightarrow{ayv}}]_{au} \rangle \sim_{au} \langle u, [\overset{u}{\overrightarrow{s}}]_u \rangle$. □

We remark that the equivalence $\langle au, [\overset{au}{\overrightarrow{ax}}]_{au} \rangle \sim_{au} \langle u, [\overset{u}{\overrightarrow{z}}]_u \rangle$ can also be examined by checking the cardinalities of the corresponding sets, as is the case of MASDAWGs. Hereby we have shown that *MASCDAWG(w)* can be constructed in a similar way to *MASDAWG(w)*. The only thing not clarified yet is whether or not *MASCDAWG(w)* can be built in time linear in its size. We establish the following lemmas to support the linearity.

Lemma 11. *Let $a \in \Sigma$ and $w \in \Sigma^*$. For any $x, z \in Factor(w)$, if $\overset{w}{\overrightarrow{ax}} = az$ then $\overset{w}{\overrightarrow{z}} = z$.*

Proof. For a contradiction, assume $\overset{w}{\overrightarrow{z}} \neq z$. Then there exists $y \in \Sigma^*$ such that $Prefix(w)y^{-1} = Prefix(w)z^{-1}$ and $|y| > |z|$. Then $Prefix(w)(ay)^{-1} = (Prefix(w)y^{-1})a^{-1} = (Prefix(w)z^{-1})a^{-1} = Prefix(w)(az)^{-1}$. Thus $ay \equiv'_{au} az$ and $|ay| > |az|$. It contradicts the assumption $\overset{w}{\overrightarrow{ax}} = az$. □

Note that the statement of the above lemma slightly differs from that of Lemma 8.

Lemma 12. *Let $a, b \in \Sigma$ and $w \in \Sigma^*$. Let $x, y \in Factor(w)$ such that $\overset{w}{\overrightarrow{xb}} = xby \neq w$. If $axb \in Factor(w)$, then $axby \in Factor(w)$, and $\overset{w}{\overrightarrow{axby'}} = \overset{w}{\overrightarrow{axby}}$ for any $y' \in Prefix(y)$.*

Proof. Since $axb \in Factor(w)$ and $xby \neq w$, there always exists $z \in \Sigma^*$ such that $\overset{w}{\overrightarrow{axb}} = axbz \in Factor(w)$. By Lemma 11, $\overset{w}{\overrightarrow{xbz}} = xbz$. Since $\overset{w}{\overrightarrow{xb}} = xby$, $y \in Prefix(z)$. Because $axbz \in Factor(w)$, $axby \in Factor(w)$. For any $y' \in Prefix(y)$, $axbz \equiv'_w axby'$ since $\overset{w}{\overrightarrow{axb}} = axbz$. Therefore $\overset{w}{\overrightarrow{abxy'}} = abxz = \overset{w}{\overrightarrow{abxy}}$. □

Suppose $\overset{w}{\overrightarrow{x}} = x$. If we in advance know node $[\overset{w}{\overrightarrow{x}}]_w$ has an out-going edge labeled with by, we can avoid to scan the whole string xby in traversing the path $axby$ from the initial node of *CDAWG(w)*. Moreover, it is guaranteed that the path by from the (explicit or implicit) node for ax consists of one edge: no explicit node is contained in the path. This is a key to achieve an algorithm that constructs *MASCDAWG*(w) in linear time with respect to its size.

The whole algorithm is shown in Fig. 4. Here we also read an input string w from right to left, and thus w is written as $w = w_n w_{n-1} \dots w_1$. The label

```
Algorithm Construction of MASCDAWG(w = w_n w_{n-1} ... w_1).
1   create new nodes s_0,s_1,s_2;
2   #(s_0) := 1; #(s_1) := 1; #(s_2) := 2; #(nil) := 0;
3   endpos(s_0) := 0; endpos(s_1) := 1; endpos(s_2) := 2; endpos(nil) := 0;
4   add edges (s_1, 1, s_0), (s_2, 1, s_0), (s_2, 2, s_0);
5   initNode[0] := s_0; initNode[1] := s_1; initNode[2] := s_2; node := s_2;
6   for i := 3 to n do
7       (s, k, p, r) := CANONIZE(FASTFIND(node, i, 1));
8       target := NEWTARGETNODE((s, k, p, r), i - 1, node);
9       newNode := create a new node with copying all out-going edges of node;
10      add or overwrite edge (newNode, i, target);
11      #(newNode) := i;   endpos(newNode) := i;
12      initNode[i] = newNode;
13      node = newNode;

function NEWTARGETNODE(refQuartet (s, k, p, r), int j, Node backbone) : Node
1   nextNumSuf := #(r) + 1;
2   if nextNumSuf = #(backbone) then return backbone;   /* redirection */
3   let (backbone, ℓ, nextBackbone) be the w_j-edge from backbone;
4   m := ℓ - endpos(nextBackbone);   /* length of this edge */
5   if k = p then   /* explicit node */
6       newNode := create a new node with copying all out-going edges of s;
7       (s, k, p, r) := CANONIZE(FASTFIND(s, j, m));
8       target := NEWTARGETNODE((s, k, p, r), j - m, nextBackbone);
9       add or overwrite edge (newNode, j, target);
10      #(newNode) := nextNumSuf;   endpos(newNode) := j;
11      return newNode;
12  else if w_p = w_j then /* implicit and next characters are the same */
13      (s, k, p, r) := CANONIZE(s, k, p - m, r);   /* skip m characters */
14      return NEWTARGETNODE((s, k, p, r), j - m, nextBackbone);
15  else   /* implicit and next characters are different */
16      newNode := create a new node;   /* edge split */
17      add new edges (newNode, p, r) and (newNode, j, s_0);
18      #(newNode) := nextNumSuf;   endpos(newNode) := j;
19      return newNode;

function FASTFIND(Node s, int i, int length) : refQuartet
/* compute the position from s along the string w_i w_{i-1} ... w_{i-length+1} */
/* remark that the first character w_i is only compared */
1   if s has the w_i-edge then
2       let (s, ℓ, r) be the w_i-edge from s;
3       return (s, ℓ, ℓ - length, r);
4   else return (s, i, i - length, nil);

function CANONIZE( refQuartet (s, k, p, r) ) : refQuartet
/* when the referenced position is an explicit node, canonize the expression */
1   if k > p and p = endpos(r) then return (r, p, p, r);
2   else return (s, k, p, r);
```

Fig. 4. The algorithm to construct *MASCDAWG*(w).

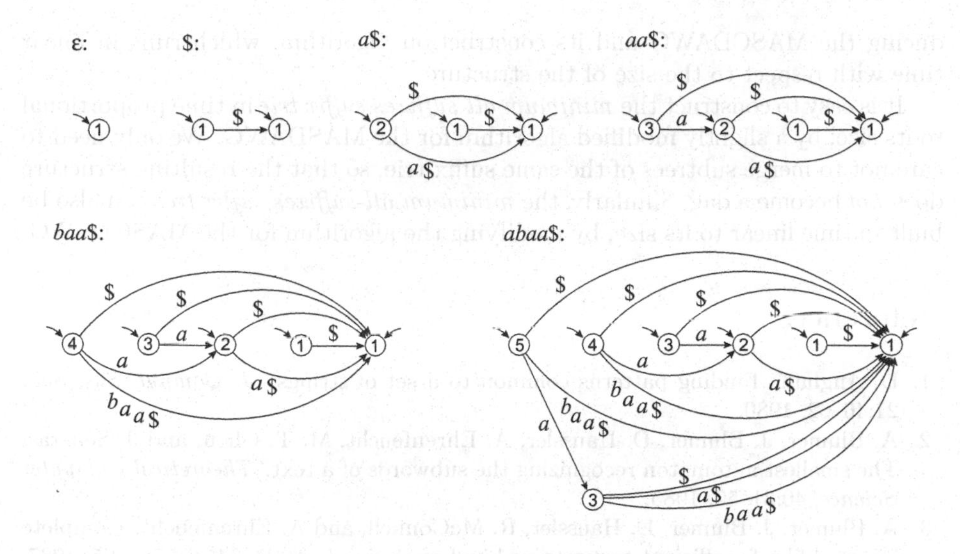

Fig. 5. Construction of *MASCDAWG(abaa$)*.

$w_i w_{i-1} \dots w_j$ of each edge can be represented by a pair of the beginning position i and the ending position $j - 1$. ($i > j - 1$) If the string corresponding to the label appears in w more than once, we represent it by the leftmost occurrence. This way we can assign *endpos*(s) to a node s, where *endpos*(s) indicates the ending position of every in-coming edge of s. Thereby, we represent each edge by a triple (r, i, s), where r, s are explicit nodes. An implicit node corresponding to some factor x of w can be represented by a triple (r, k, p), where r is an explicit parent node of the implicit node. Assuming the representative of the equivalence class associated with r is y, $x = yu$ where $u = w_k w_{k-1} \dots w_p$. The quartet (r, k, p, s) is called the *reference quartet*, where s is the closest explicit child node of r reachable via the w_k-edge from r. When $|p - k|$ is minimum, the quartet (r, k, p, s) is called the *canonical reference quartet*.

Theorem 4. *For any string $w \in \Sigma^*$, our algorithm constructs MASCDAWG(w) in time linear in its size.*

The on-line (right-to-left) construction of *MASCDAWG*(w) where $w = abaa\$$ is displayed in Fig. 5.

4 Concluding Remarks

We proposed a new space-economical algorithm to construct MASDAWGs without suffix links, running in time linear in the output size. As shown in [8], there are several important applications for MASDAWGs. Therefore, reducing memory space needed in the construction of MASDAWGs is considerably significant. We have also accomplished further reduction of the space requirement, by introducing the MASCDAWG and its construction algorithm, which runs in linear time with respect to the size of the structure.

It is easy to construct the *minimum all-suffixes suffix trie* in time proportional to its size, by a slightly modified algorithm for the MASDAWG. We only need to care not to merge subtrees of the same suffix trie, so that the resulting structure does *not* become a dag. Similarly, the *minimum all-suffixes suffix tree* can also be built in time linear to its size, by modifying the algorithm for the MASCDAWG.

References

1. D. Angluin. Finding patterns common to a set of strings. *J. Comput. Sys. Sci.*, 21:46–62, 1980.
2. A. Blumer, J. Blumer, D. Haussler, A. Ehrenfeucht, M. T. Chen, and J. Seiferas. The smallest automaton recognizing the subwords of a text. *Theoretical Computer Science*, 40:31–55, 1985.
3. A. Blumer, J. Blumer, D. Haussler, R. McConnell, and A. Ehrenfeucht. Complete inverted files for efficient text retrieval and analysis. *J. ACM*, 34(3):578–595, 1987.
4. M. Crochemore and W. Rytter. *Text Algorithms.* Oxford University Press, New York, 1994.
5. M. Crochemore and R. Vérin. On compact directed acyclic word graphs. In J. Mycielski, G. Rozenberg, and A. Salomaa, editors, *Structures in Logic and Computer Science*, volume 1261 of *Lecture Notes in Computer Science*, pages 192–211. Springer-Verlag, 1997.
6. D. Gusfield. *Algorithms on Strings, Trees, and Sequences.* Cambridge University Press, New York, 1997.
7. S. Inenaga, H. Hoshino, A. Shinohara, M. Takeda, S. Arikawa, G. Mauri, and G. Pavesi. On-line construction of compact directed acyclic word graphs. In A. Amir and G. M. Landau, editors, *Proc. 12th Annual Symposium on Combinatorial Pattern Matching (CPM'01)*, volume 2089 of *Lecture Notes in Computer Science*, pages 169–180. Springer-Verlag, 2001.
8. S. Inenaga, M. Takeda, A. Shinohara, H. Hoshino, and S. Arikawa. The minimum dawg for all suffixes of a string and its applications. In *Proc. 13th Annual Symposium on Combinatorial Pattern Matching (CPM'02)*, 2002. (to appear).
9. E. M. McCreight. A space-economical suffix tree construction algorithm. *J. ACM*, 23(2):262–272, 1976.
10. D. Revuz. Minimization of acyclic deterministic automata in linear time. *Theoretical Computer Science*, 92(1):181–189, 1992.
11. T. Shinohara. Polynomial-time inference of pattern languages and its applications. In *Proc. 7th IBM Symp. Math. Found. Comp. Sci.*, pages 191–209, 1982.
12. E. Ukkonen. On-line construction of suffix trees. *Algorithmica*, 14(3):249–260, 1995.
13. P. Weiner. Linear pattern matching algorithms. In *Proc. 14th Annual Symposium on Switching and Automata Theory*, pages 1–11, 1973.

An Explicit Lower Bound of $5n - o(n)$ for Boolean Circuits

Kazuo Iwama and Hiroki Morizumi

School of Informatics, Kyoto University, Kyoto 606-8501, Japan,
{iwama, morizumi}@kuis.kyoto-u.ac.jp

Abstract. The current best lower bound of $4.5n - o(n)$ for an explicit family of Boolean circuits [3] is improved to $5n - o(n)$ using the same family of Boolean function.

1 Introduction

The main purpose of this paper is to give a better lower bound for Boolean circuits which compute an explicit family of Boolean functions. This problem is well-known as one of the most fundamental and challenging problems in complexity theory, but its research progress has not been so dramatic: In 1974 Schnorr first gave such a lower bound of $2n$ in [5]. Then Paul proved a $2.5n$-lower bound for a different family of Boolean functions [4]. Stockmeyer gave the same $2.5n$ bound for a larger family of functions [6]. Blum improved this bound to $2.75n$ for the same family of functions [1] and then gave a $3n$-lower bound for a modified family of functions [2] in 1984. This $3n$-bound is still the best result for circuits over the base B_2 where one can use any two-input gate. All these results are based on the so-called "gate-elimination" approach.

In 1991, Zwick gave a lower bound of $4n - O(1)$ for Boolean circuits over the base U_2 that eliminates the XOR and its complement from B_2 [8], which had been the best record for a decade. Recently this bound was improved to $4.5n - o(n)$ by Lachish and Raz [3], where they invented a new family of Boolean functions called *Strongly-Two-Dependent* functions. Suppose that a function f is Strongly-Two-Dependent. Then we can prove that any circuit C computing f does not have the following structure: There exist two input-nodes x_i and x_j such that both have degree two and both are connected to the same gate (i.e., x_i is connected to gates v_l and v_m and x_j also to v_l and v_m). This structure was an important bottleneck in the proof of [8]; removing it plays a key role in the improvement. (After [3], Wegener mentioned that such a function is already known and called *weighted sum* due to Savicky and Zak, see pp. 137-138 of [7].)

In this paper, we further improve this $4.5n$-lower bound to roughly $5n$, using the same family of Boolean functions. Our basic ideas will be given in Sec. 3, but the improvement is mainly due to the redesign of case analysis. As with [3] and [8], our case analysis is still rather complicated, but a lot of efforts have been made to simplify it, which, we believe, has improved the readability of the paper greatly and will certainly give a significant support to the new idea of Lachish and Raz.

K. Diks et al. (Eds): MFSC 2002, LNCS 2420, pp. 353–364, 2002.

2 Boolean Circuits over U_2

Almost all notations are the same as [3]. Define the basis U_2 to be the set of all Boolean functions $f : \{0,1\}^2 \to \{0,1\}$ of the sort

$$f(x,y) = ((x \oplus a) \wedge (y \oplus b)) \oplus c,$$

where $a, b, c \in \{0,1\}$. Our Boolean circuits in this paper are those over the basis U_2, which are directed acyclic graphs with nodes of in-degree 0 or 2. Nodes of in-degree 0 are called *input-nodes*, and each one of them is labeled by a variable in $\{x_1, \cdots, x_n\}$ or a constant 0 or 1. Input-nodes labeled by a constant are called *constant-nodes.* Nodes of in-degree 2 are called *gate-nodes*, and each one of them has two *input* and an *output*, and is labeled by a function in U_2. There is a single specific node called the *output-node.* For nodes u and v, $u \to v$ means that the output of the node u is directly connected to one of the v's inputs. $u \xrightarrow{*} v$ means that there is a path from u to v. For a Boolean circuit C, $OUT_C(v)$ denotes the set of gate-nodes, u, such that $v \to u$. Also $IN_C(v)$ denotes the set of input-nodes, u, such that $u \xrightarrow{*} v$.

Let $X = \{x_1, \cdots, x_n\}$ be the set of input-variables. Given an assignment $\sigma \in \{0,1\}^n$ to the variables in X, we denote by $C(\sigma)$ the value of the output of the circuit C on the assignment $x_i = \sigma_i, 1 \le i \le n$. Similarly, for any node v in the circuit C, we denote by $C_v(\sigma)$ the value of the output of the gate-node v on the assignment $x_i = \sigma_i$. We say that two Boolean circuits C_1 and C_2 are equivalent ($C_1 \equiv C_2$) if they compute the same function. Wlog, we can assume that for every input-variable x_i, there is only one input-node labeled by x_i.

The *size* of a circuit C is the number of gate-nodes in it. We denote this number by $Size(C)$. The circuit complexity of a Boolean function $F : \{0,1\}^n \to \{0,1\}$ is the minimal size of a Boolean circuit that computes F. We denote this number by $Size(F)$. The depth of a node v in a Boolean circuit C is the length of the longest path from v to the output-node, denoted by $Depth_C(v)$. The depth of a circuit C, $Depth(C)$, is the maximal depth of a node v in the circuit. The *degree* of a node v in a Boolean circuit C, denoted by $Degree_C(v)$, is the node's out-degree. We denote by $Degeneracy(C)$ the number of input-variables that have degree one in C. Let x be an input-variable that has degree one in C. Then a node v is called *defect* if $x \to v$. Otherwise, v is called *non-defect* or *ND.* For our lower bound proof, we use the following measure (see the next section for its purpose):

$$SD(C) = Size(C) - Degeneracy(C).$$

Recall that each gate-node v, having inputs x and y, has the functionality defined by $f(x,y) = ((x \oplus a) \wedge (y \oplus b)) \oplus c$. If we assign value a to x then the value of its output is fixed regardless of another input y. In this case, we say that fixing $x = a$ *blocks* the gate-node v or simply x *blocks* v. Similarly for y.

A *restriction* θ is a mapping from a set of n variables to $\{0, 1, \star\}$. We apply a restriction θ to a Boolean function $F : \{0,1\}^n \to \{0,1\}$ in the following way: For any variable x_i that is mapped by θ to a constant $a_i \in \{0,1\}$, we assign a_i to x_i. We leave all the other variables untouched. We refer to the resulting Boolean function by $F|_\theta$. We use the similar notation, $C|_\theta$, for a Boolean circuit C.

3 Basic Ideas

Suppose that S is a family of Boolean functions satisfying the following condition: Let $f \in S$ depend on n variables. Then for any restriction θ which maps a single variable to 0 or 1, the $(n-1)$-variable Boolean function $f|_\theta$ is also in S. For example, let S_3 be the family of the symmetric functions f_i ($i = 0, 1, 2$) such that $f_i = 1$ if the number of 1's in the assignment is $3m + i$ (m is an integer). Then S_3 satisfies the above condition.

Now we briefly explain how the gate-elimination works. Consider a circuit C of n variables which computes a function in S_3. We look at a gate v_1 whose two inputs are both input-nodes, say x_1 and x_2. (Such a gate v_1 must exist.) Suppose that x_1 has three links to gates, v_1, v_2 and v_3. See Fig. 1. (In general, each figure illustrates input-nodes denoted by x_i and gate-nodes denoted by v_i. A node is sometimes accompanied by an integer such as 3 and 2+ to show its outdegree. 2+ means that the outdegree is two or more.) We now fix the value of x_1 so that it will block the gate v_1. Then one can see that each of the other three gates is also blocked or becomes a "through gate", i.e., its output is equal to the remaining input or its negation. In any case, we can remove all those four gates without changing the function. (If the through gate becomes a negation we need to change the gate-type of the next gate.) Thus we can remove four gates by fixing one input variable. Since the new function after this fixing operation is also in S_3, one can repeat this argument. If we can remove at least four gates in each iteration, we can prove a lower bound of $4n$.

Of course, we cannot be so optimistic; for example, both x_1 and x_2 might be degree two (see Fig. 2.). Then the same argument as above can only claim a lower bound of $3n$. One can notice, however, if we fix the value of x_1, the other x_2 becomes degree one in the new circuit. The great idea of [8] is to include the number of such input-gates x, those x whose degree is one, into the new measure SD for the "amortized size" of the circuit C, i.e., $SD(C) = Size(C) - Degeneracy(C)$. Now $SD(C)$ decreases four (three in *Size* and one in *Degeneracy*) in the case of Fig. 2. $Degeneracy(C)$ is usually small for natural Boolean functions, which means decreasing SD has almost the same effect as decreasing the number of actual gates.

Thus an increased *Degeneracy* contributes to a better lower bound. One should notice, however, that *Degeneracy* can *decrease* in some case. For example, suppose that the gate v_4, in Fig 1 is defect, i.e., it receives a link from

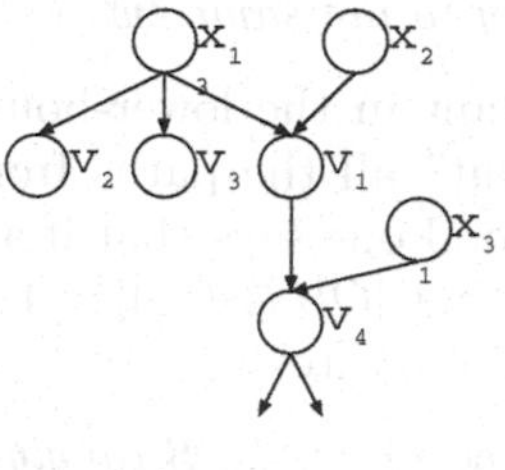

Fig. 1. Gate-elimination

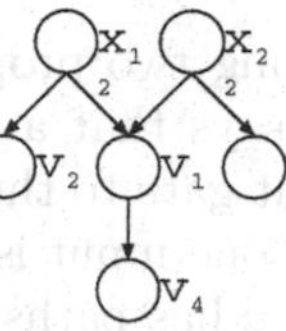

Fig. 2. Small-degree Case

input-node, say x_3, whose degree is one. Then, if $|OUT_C(v_4)| \geq 2$ and if v_4 becomes a through-gate, then the degree of x_3 becomes two or more in the new circuit, namely, *Degeneracy* decreases by one. Thus fixing x_1 actually decreases the number of gates by four, but if we use the new measure SD then it may decrease only by three. In [3], the authors overcome this difficulty by changing the definition of SD slightly, i.e., $SD'(C) = Size(C) - 0.5 \cdot Degeneracy(C)$. Now one can see that for both cases of Figs 1 and 2, we can prove a lower bound of $3.5n$, i.e., we lose $0.5n$ for Fig. 2, and gain $0.5n$ for Fig. 1.

In the present paper, we use the original definition of SD, i.e., $SD(C) = Size(C) - Degeneracy(C)$. Therefore, if *Degeneracy* increases by fixing an input, then it is desirable for us. The opposite case obviously becomes harder. What we do is to make a careful case-analysis to show that all involved gates are ND. See, e.g., the proof of Lemma 5.4, for more details.

4 Strongly Two Dependent Boolean Functions

For our lower-bound proof, we use the same family of Boolean function as [3], which is summarized as follows (slightly but not essentially different from [3]): Let $F : \{0,1\}^n \rightarrow \{0,1\}$ be a Boolean function and $F[i,j,a,b]$, $1 \leq i < j \leq n$, $a, b \in \{0,1\}$, be a Boolean function $F|_{\theta[i,j,a,b]}$ where $\theta[i,j,a,b]$ is a restriction that maps x_i and x_j to a and b, respectively. Then F is called *Two-Dependent* if for any i and j, $1 \leq i < j \leq n$, $F[i,j,0,0], F[i,j,0,1], F[i,j,1,0]$ and $F[i,j,1,1]$ are all different functions. (For example, if F is a symmetric function then it is not Two-Dependent since $F[i,j,0,1] = F[i,j,1,0]$.) Let $X_m \subseteq \{x_1, \cdots x_n\}$ be a set of m variables, and θ_m be a restriction which maps X_m to {0,1}. Then F is called *(n,k)-Strongly-Two-Dependent* if $F|_{\theta_m}$ is always Two-Dependent for any $0 \leq m \leq n-k$, any X_m and any θ_m. ($F|_{\theta_m}$ is obviously $(n-m,k)$-Strongly-Two-Dependent.) It is proved in [3] that there exists an (n,k)-Strongly-Two-Dependent Boolean function for any (sufficiently large) integer n and $k = O(\log n)$ and one can construct in polynomial time.

As mentioned before, Two-Dependent functions have the following property:

Proposition 4.1. [3] *Let $F : \{0,1\}^n \rightarrow \{0,1\}$ be a Two-Dependent Boolean function over the set of variables $X = \{x_1, \cdots, x_n\}$. Let C be a Boolean circuit that computes F. Then, the following is never satisfied in C: There exist two input variables x_i, x_j such that $OUT_C(x_i) = OUT_C(x_j)$ and $|OUT_C(x_i)| = |OUT_C(x_j)| = 2$ (i.e., x_i, x_j are connected directly to the same two gate-nodes).*

The following two properties are also important in the lower-bound proof. The first one says that a restriction does not "cut" all the paths from a non-restricted input-gate to the final output. The second one says that if a gate v is defect, i.e., its one input is connected to x_i such that $|OUT_C(x_i)| = 1$, then the other input of v has paths from many different input-gates.

Proposition 4.2. [3] *Let $F : \{0,1\}^n \rightarrow \{0,1\}$ be an (n,k)-Strongly-Two-Dependent Boolean function over the set of variables $X = \{x_1, \cdots, x_n\}$. Let C be* *a Boolean circuit that computes F. Then, the following is never satisfied in C: There exist an input-variable x_i, and a set of less than $n-k$ other input-variables X' and a restriction θ that maps each input-variables in X' to a constant in $\{0,1\}$, such that, in $C|_\theta$ every path that connects x_i to the output-node contains a gate-node that computes a constant function.*

Proposition 4.3. [3] *Let $F : \{0,1\}^n \to \{0,1\}$ be an (n,k)-Strongly-Two-Dependent Boolean function and let C be a Boolean circuit that computes F. Let v be a gate-node in C and let v' be the node such that $v' \to v$. Assume that $x_i \to v$ for an input-variable x_i such that $Degree_C(x_i) = 1$ (,i.e., the node v is defect.) Then, if the node v' computes a non constant function, then $|IN_C(v)| \geq n - k$.*

For the gate-elimination, it is convenient if the circuit does not include *degenerate* gates, i.e., those that do not contribute to the computation process of the Boolean circuit. The following propositions give methods of removing such gates without any harm.

Proposition 4.4. [3] *Let C be a Boolean circuit. Assume that C contains one of the following degenerate cases: (1) A gate-node v such that for some constant $a \in \{0,1\}$ and any assignment $\sigma \in \{0.1\}^n$, we have $C_v(\sigma) = a$. (2) A gate-node v such that a constant node is connected directly to v. (3) A gate-node v which is not the output of the circuit such that $Degree_C(v) = 0$. (4) A gate-node v such that its two inputs are connected to the same gate. Then, there exists a Boolean circuit $C \equiv C'$ such that $Size(C) \geq Size(C') + 1$*

Proposition 4.5. *Let $F : \{0,1\}^n \to \{0,1\}$ be an (n,k)-Strongly-Two-Dependent Boolean function and let C be a Boolean circuit that computes F. There exists a Boolean circuit $C' \equiv C$ such that $SD(C) \geq SD(C')$ and $Degeneracy(C') \leq k$ and C' does not contain any of the following degenerate cases: (1) Any of the degenerate cases of Proposition 4.4. (2) An input-variable x_i such that $|OUT_{C'}(x_i)| \geq 2$ and for $u, v \in OUT_{C'}(x_i), u \to v$.*

Proof. Only the following differs from the proof in [3]: Suppose that a gate-node v is removed to avoid a case in (1) to (4) of Proposition 4.4. The argument in [3] says that v's two inputs might be connected to input-nodes whose degree is both one. Then *Degeneracy* decreases only by two and SD increases by one (because of the factor 0.5), but that is canceled since *Size* decreases by one due to the removal of v. Recall that our new definition of SD has the factor 1.0 to *Degeneracy*, which could decreases *Degeneracy* by two in the above situation. Fortunately, however, it never happens in our family of circuits that v's both inputs are connected to degree-one input-nodes by Proposition 4.3. ■

5 The Lower Bound

5.1 The Lower Bound

As with [3], our main task is to prove the following Lemma 5.1. Then our main theorem, Theorem 5.2, is almost its corollary.

Lemma 5.1. *Let $F : \{0,1\}^n \to \{0,1\}$ be an (n,k)-Strongly-Two-Dependent Boolean function and assume that $n-k \geq k+4$ and $n-k \geq 5$. Let C be a Boolean circuit that computes F. Then, there exists a set of one or two input-variables X' (i.e., $|X'| \leq 2$) and a constant $c_i \in \{0,1\}$ for each $x_i \in X'$ such that for the restriction θ that maps each variable $x_i \in X'$ to c_i, the following is satisfied: There exists a Boolean circuit $C' \equiv C|_\theta$ such that*

$$SD(C) \geq SD(C') + 5 \cdot |X'|.$$

Theorem 5.2. *Let $F : \{0,1\}^n \to \{0,1\}$ be an (n,k)-Strongly-Two-Dependent Boolean function such that $k = o(n)$. Then,*

$$Size(F) \geq 5n - o(n)$$

Proof. Let C a Boolean circuit that computes F. We generate a sequence of Boolean circuit $C_0, ..C_l$ by iteratively applying Lemma 5.1 to C. (Note that this is possible by the definition of Strongly-Two-Dependent). More formally, we have $C_0 = C$ and C_{i+1} is obtained from C_i by applying Lemma 5.1. We stop when the number of remaining input-variables is smaller than $2k+4$ or $k+5$. By Lemma 5.1, $SD(C) \geq SD(C_l) + 5n - o(n)$. By Proposition 4.5, we can assume that $Degeneracy(C) \leq k$. Therefore, $Size(C) \geq 5n - o(n)$, which immediately implies the theorem. ■

5.2 Preliminaries for the Proof of Lemma 5.1

In this and the next sections (5.2 and 5.3), we always treat Boolean circuits which compute (n,k)-Strongly-Two-Dependent Boolean functions such that $n-k \geq k+4$ and $n-k \geq 5$, which is often omitted to mention. Also, we always assume that the circuits do not include degenerate cases described in Proposition 4.5. Those nodes can be removed without increasing SD as mentioned in its proof. Furthermore we can always assume that the number of degenerate variables is at most k by Proposition 4.5. Our argument in the rest of the paper has the standard structure, which is explained in the proof of our first lemma:

Lemma 5.3. *Suppose that there is an input-variable x_i, such that, (i) $OUT_C(x_i) = \{v_1, v_2, v_3\}$ and (ii) $OUT_C(v_1) \cup OUT_C(v_2) \cup OUT_C(v_3)$ includes at least three ND gate-nodes. Then SD decreases by at least five by fixing x_i appropriately.*

Proof. Since $OUT_C(v_1) \cup OUT_C(v_2) \cup OUT_C(v_3)$ includes at least three ND gate-nodes, considering the following two cases is enough:

Case A: $OUT_C(v_1)$ or $OUT_C(v_2)$ or $OUT_C(v_3)$ includes at least two different ND gate-nodes: wlog, we can assume that $OUT_C(v_1)$ includes such gate-nodes. (see Fig. 3 (i).) One can see that, by fixing x_1 appropriately, we can block v_1, which allows us to remove v_4 and v_5, too. v_2 and v_3 can also be removed. Note that the gate-nodes v_1 to v_5 are all different by Proposition 4.5 and v_4 to v_5 are ND gates by the assumption of the lemma. v_1 to v_3 are also ND by Proposition 4.3. Hence removing v_1 to v_5 does not decrease $Degeneracy(C)$.

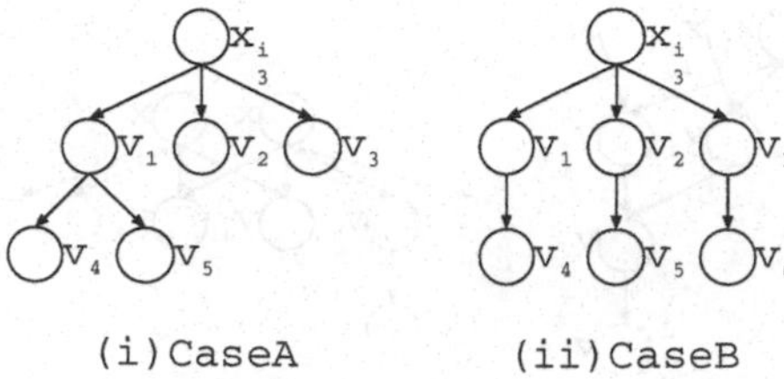

Fig. 3. Lemma 5.3

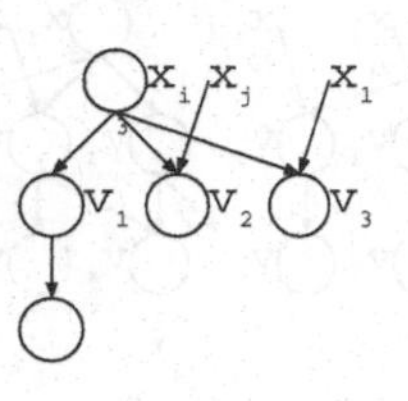

Fig. 4. Lemma 5.4

(*Degeneracy*(C) may increase, but that is not important for us since *increasing* Degeneracy forces SD to decrease.) To summarize all these situations, we write as follows (when a gate is removed since its output is fixed (e.g., by being blocked), we say that the gate is "killed"):

Fix x_i s.t. v_1 blocked $\Rightarrow$ Killed: v_1, Removed: v_1, v_2, v_3, v_4, v_5.
Nondefect: v_1, v_2, v_3 (by Proposition 4.3) v_4, v_5 (by (ii)) $\Rightarrow$ *Degeneracy*: ± 0.

Case B: Each of $OUT_C(v_1)$, $OUT_C(v_2)$ and $OUT_C(v_3)$ includes at least one ND gate-node, v_4, v_5 and v_6, respectively, which are all different (see Fig. 3 (ii)). One can see that, by fixing x_1 appropriately, we can block at least two of v_1, v_2 and v_3 regardless of their gate-types. wlog, we assume that v_1 and v_2 are blocked, which allows us to remove v_4 and v_5, too. v_3 can also be removed. Note that the gate-nodes v_1 to v_5 are all different by Proposition 4.5 and v_4 to v_5 are ND gates by the assumption of the lemma. v_1 to v_3 are also ND by Proposition 4.3. To summarize:

Fix x_i s.t. v_1, v_2 blocked $\Rightarrow$ Killed: v_1, v_2, Removed: v_1, v_2, v_3, v_4, v_5.
Nondefect: v_1, v_2, v_3 (by Proposition 4.3) v_4, v_5 (by (ii)) $\Rightarrow$ *Degeneracy*: ± 0.

∎

Lemma 5.4. *Suppose that there is an input-variable x_i, such that, (i) $OUT_C(x_i) = \{v_1, v_2, v_3\}$, (ii) $OUT_C(v_1)$ includes at least one ND gate-node and (iii) $IN_C(v_2) = \{x_i, x_j\}$ and $IN_C(v_3) = \{x_i, x_l\}$ where x_j and x_l are both input-variables such that $i \neq j$, $i \neq l$. Then, SD decreases by at least five by fixing x_i appropriately.*

Proof. See Fig. 4. Three main cases, A, B and C exists:

Case A: $OUT_C(v_1) \cap OUT_C(v_2) = \emptyset$ and $OUT_C(v_1) \cap OUT_C(v_3) = \emptyset$ and $OUT_C(v_2) \cap OUT_C(v_3) = \emptyset$: See Fig. 5 (i). v_4 is an ND gate-node guaranteed by (ii) above. v_4, v_5 and v_6 are all different gate-nodes by the condition of the case. v_5 and v_6 are also ND by Proposition 4.3. (By fixing x_i and x_j, we can block v_5. Similarly for v_6.) Thus, we can apply Lemma 5.3.

Case B: $OUT_C(v_1) \cap OUT_C(v_2) \neq \emptyset$ or $OUT_C(v_1) \cap OUT_C(v_3) \neq \emptyset$: wlog, assume that $OUT_C(v_1) \cap OUT_C(v_2) \neq \emptyset$. See Fig. 5 (ii). Two subcases exist:

Case B.1: Suppose that we can fix x_i such that it blocks v_1, v_2: There is at least one ND gate-node, say v_6, in $OUT_C(v_1) \cup OUT_C(v_2) \cup OUT_C(v_4)$ other than v_4 for the following reason: Suppose that all gate-nodes (except v_4) in $OUT_C(v_1) \cup OUT_C(v_2) \cup OUT_C(v_4)$ are defect. Then by setting appropriate values to the input-nodes connected to these defect nodes and by setting x_l to block v_3, all the paths from x_i are blocked. Since the number of defect gate-nodes is at most k, this fact contradicts Proposition 4.2. Note that v_6 is obviously

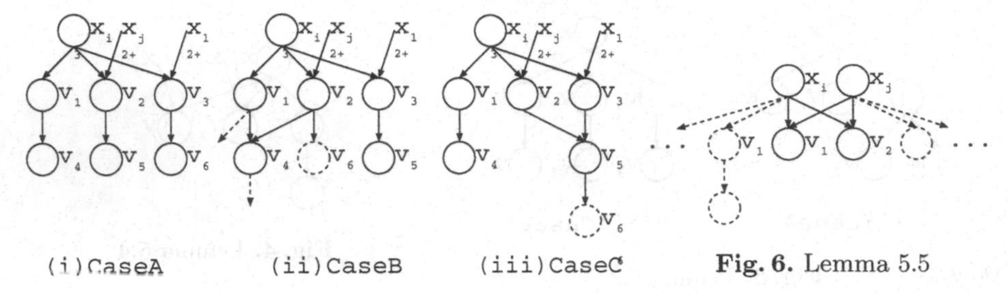

Fig. 6. Lemma 5.5

Fig. 5. Main cases of Lemma 5.4

different from v_1, v_2 or v_4 and it is also different from v_3 whose two inputs are both input-nodes. To summarize:

Fix x_i s.t. v_1, v_2 blocked $\Rightarrow$ Killed: $v_1, v_2 \rightarrow v_4$, Removed: v_1, v_2, v_3, v_4, v_6.
Nondefect: v_1, v_2, v_3 (by Proposition 4.3) v_4 (obvious) v_6 (mentioned above)
$\Rightarrow$ *Degeneracy*: ± 0.

Remark. The above argument breaks if v_4 is the output gate since the paths from x_1 *to the output gate* can no longer be blocked. However, v_4 cannot be the output gate since it is killed only by fixing a few input nodes. In the following we often omit mentioning this fact in similar situations.

Case B.2: We can fix x_i such that it blocks v_1, v_3 or v_2, v_3: This case is easy and details are omitted.

Case C: $OUT_C(v_2) \cap OUT_C(v_3) \neq \emptyset$: v_4 is an ND gate-node guaranteed by (ii) above. See Fig. 5 (iii).

Case C.1: Suppose that we can fix x_i such that it blocks v_2 and v_3: If $OUT_C(v_5)$ does not includes v_1, $OUT_C(v_5)$ must include a gate-nodes, say v_6, that is different from v_1, v_2, v_3 or v_5 and is ND (by Proposition 4.3 since $IN_C(v_5) = 3$). Such a case is proved like Case B.1. If $OUT_C(v_5)$ includes v_1, then we fix x_i such that it blocks v_2 and v_3, which kills v_5 and then kills v_1 also. To summarize:

Fix x_i s.t. v_2, v_3 blocked $\Rightarrow$ Killed: $v_2, v_3 \rightarrow v_5 \rightarrow v_1$, Removed: v_1, v_2, v_3, v_4, v_5.
Nondefect: v_1, v_5 (obvious) v_2, v_3 (by Proposition 4.3) v_4 (assumption)
$\Rightarrow$ *Degeneracy*: ± 0.

Case C.2: We can fix x_i such that it blocks v_1, v_2 or v_1, v_3: Details are omitted. ■

Lemma 5.5. *Suppose that there are two input-variables x_i, x_j, such that $OUT_C(x_i) \supseteq \{v_1, v_2\}$ and $OUT_C(x_j) \supseteq \{v_1, v_2\}$ and $OUT_C(x_i) \cup OUT_C(x_j) \neq \{v_1, v_2\}$. Then, $OUT_C(x_i) \cup OUT_C(x_j)$ includes at least one gate-nodes v_l such that v_l is different from v_1, v_2, and $OUT_C(v_l)$ includes at least one ND gate-node.*

Proof. See Fig. 6. Suppose that there are no such v_l. Then all gate-nodes, say u, except v_1 and v_2 in $OUT_C(x_i) \cup OUT_C(x_j)$ (if any) are connected to defect nodes. Those defect gate-nodes are blocked by their corresponding inputs, by which we can remove all such u's. Thus, by setting at most k input variable, the circuit is converted to C' such that (i) C' is still Strongly-Two-Dependent by the definition of Strongly-Two-Dependent and (ii) $OUT_C(x_i) = OUT_C(x_j) = \{v_1, v_2\}$. But this contradicts Proposition 4.1. ■

5.3 Proof of Lemma 5.1

Let $F : \{0,1\}^n \rightarrow \{0,1\}$ be an (n,k)-Strongly-Two-Dependent Boolean function and assume that $n - k \geq k + 4$ and $n - k \geq 5$. Let C be a Boolean circuit that computes F. Let v_1 be a gate-node such that $Depth(v_1) = Depth(C) - 1$ (we can always find such v_1). The nodes that are connected to v_1 are both input-variables, say x_1 and x_2. By Proposition 4.3, $Degree_C(x_1) \geq 2, Degree_C(x_2) \geq 2$. Four main cases exist:

Case 1: $Degree_C(x_1) \geq 4$ or $Degree_C(x_2) \geq 4$: Omitted.

Case 2: $Degree_C(x_1) = 3$ and $Degree_C(x_2) = 3$: Three subcases exist:

Case 2.1: $|OUT_C(x_1) \cap OUT_C(x_2)| = 3$: See Fig. 7 (i). v_4 is ND by Proposition 4.3. Thus, by Lemma 5.4, SD decreases by at least five by fixing x_1 appropriately.

Case 2.2: $|OUT_C(x_1) \cap OUT_C(x_2)| = 2$: See Fig. 7 (ii). By Lemma 5.5, $OUT_C(v_3)$ or $OUT_C(v_4)$ includes an ND gate-node. wlog, assume that $OUT_C(v_3)$ includes an ND gate-node, say v_5. Thus, SD decreases by at least five by fixing x_1 by Lemma 5.4.

Case 2.3: $|OUT_C(x_1) \cap OUT_C(x_2)| = 1$: See Fig. 7 (iii). Let v_6 be a gate-node in $OUT_C(v_1)$. wlog, we can assume that $Depth_C(v_6) = Depth(C) - 2$. By the condition of Case 2.3, v_1 through v_5 are all different. By Proposition 4.5, v_6 is different from v_2 through v_5. Thus, v_1 through v_6 are all different. Four subcases exist: Let w be a node ($\neq v_1$) such that $w \rightarrow v_6$.

Case 2.3.1: $Degree_C(v_1) \geq 2$: Omitted.

Case 2.3.2: $Degree_C(v_1) = 1$ and the node w is equal to v_2, v_3, v_4 or v_5: wlog, assume that $w = v_2$. See Fig. 8. Since $Depth_C(v_6) = Depth(C) - 2$, the nodes that are connected to v_2 are both input-variables. By setting x_2 to block v_1 and x_i to block v_2, all the paths from x_1 except the path through v_3 are blocked. By this fact and Proposition 4.2, $OUT_C(v_3)$ includes at least one ND gate-node. Thus, by Lemma 5.4, SD decreases by at least five by fixing x_1.

Case 2.3.3: $Degree_C(v_1) = 1$ and w is not equal to v_2, v_3, v_4 or v_5, and w is not an input-node (Case 2.3.4 is the case that w is an input-node): Let w be v_7 and see Fig. 9. Since v_7 is obviously different from v_1 or v_6, v_1 through v_7 are all different. Note that $Depth_C(v_7) = Depth(C) - 1$ since $Depth_C(v_6) = Depth(C) - 2$, which means the nodes connected to v_7 are both input-variables. By Proposition 4.3, the degree of these two input-variables are two or more. In

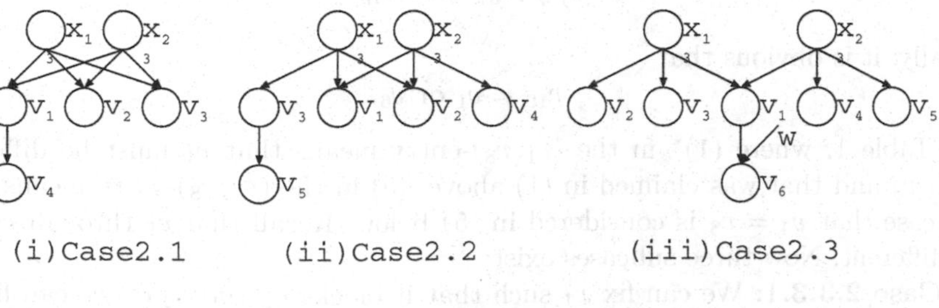

Fig. 7. Subcases of Case 2

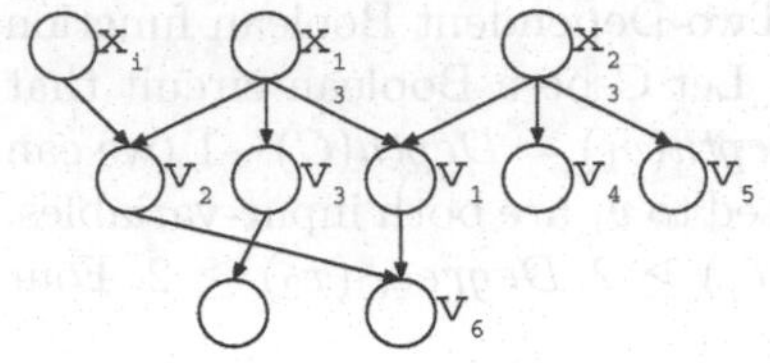

Fig. 8. Case 2.3.2

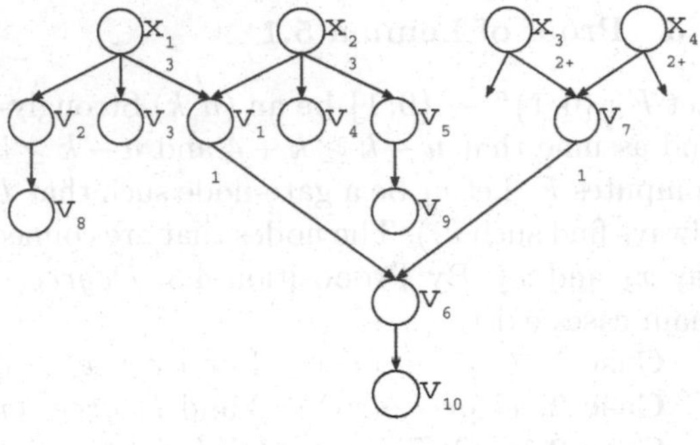

Fig. 9. Case 2.3.3

the following, we only prove the case that $Degree_C(v_7) = 1$. If $Degree_C(v_7) \geq 2$, then we can apply Case 1, 2.1, 2.2, 2.3.1, 3 or 4.

Suppose that there are two ND gate-nodes in $OUT_C(v_2) \cup OUT_C(v_3)$. Since these gates are obviously different from v_6, $OUT_C(v_1) \cup OUT_C(v_2) \cup OUT_C(v_3)$ includes three ND gate-nodes. Thus we can apply
Lemma 5.3. Otherwise, suppose that $OUT_C(v_2) \cup OUT_C(v_3)$ includes only defect nodes. Then we can block all the paths from x_1 by setting the input-nodes corresponding to those defect nodes and x_2 (to block v_1), which contradicts Proposition 4.2. Thus, from now on we can assume that $OUT_C(v_2) \cup OUT_C(v_3)$ includes exactly one ND gate-node, say v_8. Suppose that $OUT_C(v_8)$ includes no ND gate-nodes. Then the similar contradiction to Proposition 4.2 happens. Thus, $OUT_C(v_8)$ includes one or more ND gate-node. wlog, we can assume that v_8 is in $OUT_C(v_2)$. Similarly for v_9. Also, let v_{10} be a gate-nodes in $OUT_C(v_6)$. Now all gates are illustrated in Fig. 9.

Since v_8 is different from v_3 by Proposition 4.5 (and others are obvious),

$$v_8 \neq v_1, v_2, v_3 \text{ or } v_6. \tag{1}$$

Similarly

$$v_9 \neq v_1, v_4, v_5 \text{ or } v_6. \tag{2}$$

Since two inputs of v_7 are both input-nodes,

$$v_7 \neq v_8, v_9 \text{ or } v_{10}. \tag{3}$$

Finally, it is obvious that

$$v_{10} \neq v_1 \text{ or } v_6. \tag{4}$$

See Table 1, where (1)* in the (v_1, v_8)-entry means that v_1 must be different from v_8 and that was claimed in (1) above. (5) in the (v_4, v_8)-entry means that the case that $v_4 = v_8$ is considered in (5) below. Recall that v_1 through v_7 are all different. Now three subcases exist:

Case 2.3.3.1: We can fix x_1 such that it blocks v_1 and v_2 or we can fix x_2 such that it blocks v_1 and v_5: This case is easy and omitted.

Table 1. Case 2.3.3

	v_1	v_2	v_3	v_4	v_5	v_6	v_7	v_8	v_9	v_{10}
v_8	(1)*	(1)*	(1)*	(5)	(5)	(1)*	(3)*	-	-	-
v_9	(2)*	(5)	(5)	(2)*	(2)*	(2)*	(3)*	(7)	-	-
v_{10}	(4)*	(6)	(6)	(6)	(6)	(4)*	(3)*	(6)	(6)	-

Case 2.3.3.2: If v_1 is blocked, then its output blocks v_6: If v_6 is killed, then v_7 can be removed since $Degree_C(v_7) = 1$. Therefore:

Fix x_1 s.t. v_1 blocked $\Rightarrow$ Killed: $v_1 \to v_6$, Removed: v_1, v_2, v_3, v_6, v_7.
Nondefect: v_1, v_6, v_7 (obvious) v_2, v_3 (by Proposition 4.3)
$\Rightarrow$ *Degeneracy*: ± 0.

Case 2.3.3.3: Neither Case 2.3.3.1 nor Case 2.3.3.2 applies: Further subcases exist:

Case 2.3.3.3.1: v_8 is equal to v_4 or v_5, or v_9 is equal to v_2 or v_3 (denoted by (5) in Table 1): Assume that v_8 is equal to v_4. We can block v_4 $(= v_8)$ by x_2 and can kill v_6 by x_3 and x_4 (through v_7). Since we are now assuming the case that $OUT_C(v_2) \cup OUT_C(v_3)$ does not include ND gate-nodes other than v_8, this fact contradicts Proposition 4.2 (all the paths from x_1 can be blocked). Similarly for the case that $v_5 = v_8, v_2 = v_9$ and $v_3 = v_9$.

Case 2.3.3.3.2: v_{10} is equal to v_2, v_3, v_4, v_5, v_8 or v_9 (denoted by (6) in Table 1): Assume that v_{10} is equal to v_2. We can block v_1 by x_2, and we can kill v_2 $(= v_{10})$ by x_3 and x_4 since we are now assuming that if v_1 is blocked v_6 becomes a through-gate. Hence, $OUT_C(v_3)$ must include an ND gate, say u, by Proposition 4.2. Recall that we are now assuming that $OUT_C(v_2) \cup OUT_C(v_3)$ has only one ND gate-node (the other cases were already discussed). Hence u must be v_8, namely, both v_2 and v_3 are connected to v_8. This is exactly the same when $v_3 = v_{10}$. Let u_1 be an ND gate-node in $OUT_C(v_8)$ which must exist by Proposition 4.2. Now, if we can fix x_1 such that it blocks v_1 and v_3, then:

Fix x_1 s.t. v_1, v_3 blocked $\Rightarrow$ Killed: v_1, v_3, Removed: v_1, v_2, v_3, v_6, v_8.
Nondefect: v_1, v_6, v_8 (obvious) v_2, v_3 (by Proposition 4.3)
$\Rightarrow$ *Degeneracy*: ± 0.

if we can fix x_1 such that it blocks v_2 and v_3, then:

Fix x_1 s.t. v_2, v_3 blocked $\Rightarrow$ Killed: $v_2, v_3 \to v_8$, Removed: v_1, v_2, v_3, v_8, u_1.
Nondefect: v_1, v_8 (obvious) v_2, v_3 (by Proposition 4.3) u_1 (above)
$\Rightarrow$ *Degeneracy*: ± 0.

Similarly for the case that $v_4 = v_{10}$ and $v_5 = v_{10}$.

Assume that v_{10} is equal to v_8. We can block v_1 by x_2, and we can kill v_8 $(= v_{10})$ by x_3, x_4 since we are now assuming that if v_1 is blocked v_6 becomes a through-gate. Since we are now assuming the case that $OUT_C(v_2) \cup OUT_C(v_3)$ includes only one ND gate-node $(= v_8)$, this fact contradicts Proposition 4.2 (all the paths from x_1 can be blocked). Similarly for the case that $v_9 = v_{10}$.

Case 2.3.3.3.3: Now one can see that what remains to be considered is the case that $v_8 = v_9$ and the case that all the gates are different. Suppose that v_1 through v_{10} are all different: We block v_2 by x_1 and v_5 by x_2. This assignment kills v_6 (Reason: Recall that we cannot block v_1 and v_2 or v_1 and v_5 at the same time. Hence the current value of neither x_1 nor x_2 blocks v_1. Since we are now assuming that if v_1 is blocked, then its output, say z, does not block v_6, the current output of v_1 must be $\overline{z}$ (otherwise v_1's output would be constant), which does block v_6). To summarize:

Fix x_1 s.t. v_2 blocked and Fix x_2 s.t. v_5 blocked $\Rightarrow$

Killed: $v_1, v_2, v_5 \to v_6$, Removed: $v_1, v_2, v_3, v_4, v_5, v_6, v_7, v_8, v_9, v_{10}$.

Nondefect: v_1, v_6, v_7 (obvious) $v_2, v_3, v_4, v_5, v_{10}$ (by Proposition 4.3)

v_8, v_9 (above) $\Rightarrow$ *Degeneracy*: ± 0.

Case 2.3.3.3.4: v_8 is equal to v_9 (denoted by (7) in Table 1): We can assume that all the other gate-nodes are different. Recall that $OUT_C(v_8)$ includes at least one ND gate-node, say, u_2. One can easily see that we can remove this new gate by the same assignment as Case 2.3.3.3.3.(The output values of its two parent nodes are both fixed.) If u_2 is only such ND gate node and is equal to v_3, then we can again claim that $OUT_C(v_3)$ includes a new ND gate-node, say, u_3, which is removed by the same assignment. We can continue this argument for the cases that $u_2 = v_4, u_2 = v_{10}, u_3 = v_4$ and so on.

We omit the remaining cases for space limitation.

Case 2.3.4: $Degree_C(v_1) = 1$ and w is an input-node: The idea for Case 2.3.3 might help.

Case 3: $Degree_C(x_1) = 3$ and $Degree_C(x_2) = 2$ or $Degree_C(x_1) = 2$ and $Degree_C(x_2) = 3$: This case is not so diffecult.

Case 4: $Degree_C(x_1) = 2$ and $Degree_C(x_2) = 2$: There are several subcases, some of which are similar to Case 2.3.

Acknowledgment

We are truly grateful to Ran Raz who carefully read the manuscript and gave us several comments about the proof.

References

1. N. Blum. A 2.75n-lower bound on the network complexity of boolean functions. *Tech. Rept.* A 81/05, *Universität* des Saarlandes, 1981.
2. N. Blum. A Boolean function requiring 3n network size. *Theoret. Comput. Sci.*, 28, pp. 337-345, 1984.
3. O. Lachish and R. Raz. Explicit lower bound of $4.5n - o(n)$ for Boolean circuits. *Proc. STOC'01*, pp. 399-408, 2001.
4. W. Paul. A 2.5n-lower bound on the combinational complexity of boolean functions. *SIAM J. Comput.* 6, pp. 427-443, 1977.
5. C. Schnorr. Zwei lineare untere Schranken für die Komplexität Boolescher Funktionen. *Computing* 13, pp. 155-171, 1974.
6. L. Stockmeyer. On the combinational complexity of certain symmetric Boolean functions. *Math. System Theory* 10, pp. 323-336, 1977.
7. I. Wegener. Branching programs and binary decision diagrams. *SIAM Monographs on Discrete Mathematics and Applications*, 1999.
8. U. Zwick. A 4n lower bound on the combinatorial complexity of certain symmetric Boolean functions over the basis of unate dyadic Boolean functions. *SIAM J. Comput.* 20, pp. 499-505, 1991.

Computational Complexity in the Hyperbolic Plane

Chuzo Iwamoto*, Takeshi Andou, Kenichi Morita, and Katsunobu Imai

Hiroshima University, Graduate School of Engineering,
Higashi-Hiroshima, 739-8527 Japan,
chuzo@hiroshima-u.ac.jp

Abstract. This paper presents simulation and separation results on the computational complexity of cellular automata (CA) in the hyperbolic plane. It is shown that every $t(n)$-time nondeterministic hyperbolic CA can be simulated by an $O(t^3(n))$-time deterministic hyperbolic CA. It is also shown that for any computable functions $t_1(n)$ and $t_2(n)$ such that $\lim_{n\to\infty} \frac{(t_1(n))^3}{t_2(n)} = 0$, $t_2(n)$-time hyperbolic CA are strictly more powerful than $t_1(n)$-time hyperbolic CA. This time hierarchy holds for both deterministic and nondeterministic cases. As for the space hierarchy, hyperbolic CA of space $s(n) + \varepsilon(n)$ are strictly more powerful than those of space $s(n)$ if $\varepsilon(n)$ is a function not bounded by $O(1)$.

Keywords. cellular automata; hyperbolic geometry; complexity

1 Introduction

One of the simplest models for parallel recognition of languages is a cellular automaton (CA). A CA is an array of identical finite-state automata, called cells, which are uniformly interconnected. Naturally, a CA is defined as a tiling of identical regular polygons in the Euclidean plane such that no two polygons overlap each other except their shared border edge.

In the Euclidean plane, there are only three types of tilings of identical polygons, i.e., triangular CA, square CA, and hexagonal CA. On the other hand, the situation is quite different in the hyperbolic plane; a CA can be defined as a tiling of identical regular right-angled n-gons, where $n \geq 5$ is an arbitrary integer. In the hyperbolic plane, given a line l and a point A not on l, there is more than one line through A that is parallel to l, although there is exactly one such line l in the Euclidean plane.

In [8], a CA was defined as a tiling of identical regular right-angled pentagons in the hyperbolic plane (hyperbolic CA), and it was shown that NP-complete problems can be solved in polynomial time by hyperbolic CA. At a recent workshop, Iwamoto et al. [5] reported that this result can be improved as follows: The class of languages accepted by polynomial-time hyperbolic CA is PSPACE.

* Corresponding author. This research was supported in part by Scientific Research Grant, Ministry of Education, Japan.

K. Diks et al. (Eds): MFSC 2002, LNCS 2420, pp. 365–374, 2002.

Namely, time complexity on hyperbolic CA is polynomially equivalent to space complexity on Turing machines.

In this paper, we present simulation algorithms and separation results on the computational complexity of CA in the hyperbolic plane. It is shown that every $t(n)$-time nondeterministic hyperbolic CA can be simulated by an $O(t^3(n))$-time deterministic hyperbolic CA. It is also shown that for any computable functions $t_1(n)$ and $t_2(n)$ such that $\lim_{n\to\infty} \frac{(t_1(n))^3}{t_2(n)} = 0$, there is a language which can be accepted by a $t_2(n)$-time deterministic hyperbolic CA, but not by any $t_1(n)$-time nondeterministic hyperbolic CA. Hence, a polynomial increase of time strictly enlarges the class of languages accepted by both deterministic and nondeterministic CA in the hyperbolic plane. On the other hand, it is open whether the linear-speedup theorem holds for hyperbolic CA. As for space complexity, we prove that the linear-compression theorem does not hold for deterministic hyperbolic CA. It is shown that there is a language which can be accepted by an $(s(n)+\varepsilon(n))$-space hyperbolic CA, but not by any $s(n)$-space hyperbolic CA, where $\varepsilon(n)$ is an arbitrary computable function not bounded by $O(1)$. Here, an $s(n)$-space hyperbolic CA can use cells within distance $s(n)$ from the "root" cell.

It is known that hyperbolic CA have the same power as the CA implemented on an *extended tree*, where each node has a horizontal edge with the next node in the same level. Strictly speaking, we will show that every $t(n)$-time *nondeterministic* hyperbolic CA can be simulated by an $O(t(n))^3$-time deterministic *binary-tree* CA (with no horizontal edges). Thus, neither nondeterminism nor additional horizontal edges increase the computational power of binary-tree CA if polynomial-time complexity is considered. On the other hand, additional horizontal edges play crucial roles in the proof of the space-hierarchy theorem for hyperbolic CA.

As for hierarchy results in the Euclidean space, there is a huge amount of literature. It is known [1] that $s_2(n)$-space TMs are more powerful than $s_1(n)$-space TMs if $s_2(n) \neq O(s_1(n))$. If the number of tape symbols is fixed, then $(s(n) + \log s(n) + (2+\epsilon)\log\log s(n))$-space TMs are more powerful than $s(n)$-space TMs for any $\epsilon > 0$ [3]. For time complexity, it is known that $t_2(n)\log t_1(n)$-time TMs are more powerful than $t_1(n)$-time TMs if $t_2(n) \neq O(t_1(n))$ [2]. The gap $\log t_1(n)$ is replaced by $t_1(n)$ if one-tape TMs are considered. On the other hand, the gap $\log t_1(n)$ can be removed if the model is CA; namely, $t_2(n)$-time CA are more powerful than $t_1(n)$-time CA [4]. If one-tape off-line computations are considered, $(t_2(n)\frac{\log t_1(n)}{\log n})$-time TMs are more powerful than $t_1(n)$-time TMs [6].

In the following section, we give definitions of hyperbolic CA. Main theorems are summarized in Section 3. The proofs are given in Sections 4, 5, and 6.

2 Cellular Automata in the Hyperbolic Plane

2.1 Tiling of the Hyperbolic Plane

Definitions and notations of CA in the hyperbolic plane are mostly from [7]. In Poincaré's unit disk model (see Fig. 1(a)), the hyperbolic plane is the set

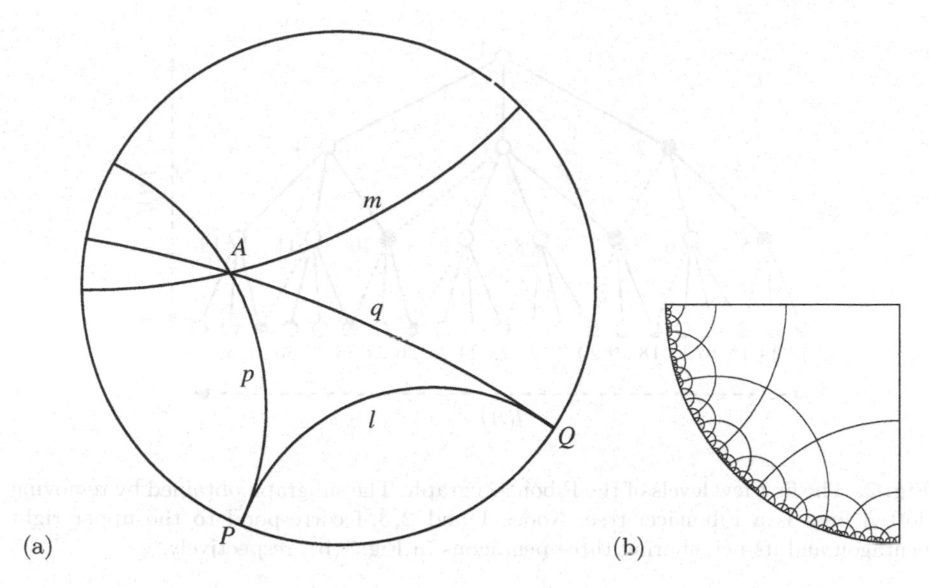

Fig. 1. (a) p and q are parallel to l, with points at infinity P and Q, respectively. (b) Tiling of regular right-angled pentagons in the south-western quarter.

of points lying in the open unit disk of the Euclidean plane. The lines of the hyperbolic plane, called *h-lines*, are either circular arcs which are orthogonal to the unit circle or diametral segments. Consider points on the unit circle as points at infinity.

In the Euclidean plane, given a line l, and a point A not lying on l, there is exactly one line passing through A not intersecting l. On the other hand, in the hyperbolic plane, there are infinitely many h-lines passing through A not intersecting l (see Fig. 1(a)). In the Euclidean plane, two lines are parallel if they do not intersect. If the points at infinity are added to the Euclidean plane, parallel lines are characterized as the lines passing through the same point at infinity. In the hyperbolic plane, h-lines sharing a common point at infinity are called *parallel*. For more details on the hyperbolic plane, the reader is referred to [8].

Fig. 1(b) illustrates a tiling of identical regular right-angled pentagons in the south-western quarter of the hyperbolic plane, constructed in [8]. Note that all pentagons are of the same shape and size, although they seem to be warped out of shape because of the projection of the hyperbolic plane onto the Euclidean plane. An array of pentagons gives rise to cellular automata in the hyperbolic plane (hyperbolic cellular automata, HCA). Two cells can exchange information if their pentagons share a common edge.

2.2 Fibonacci Trees

If we assign a point to each pentagon of the quarter and connect two points of which the corresponding pentagons share a common side, then we obtain the

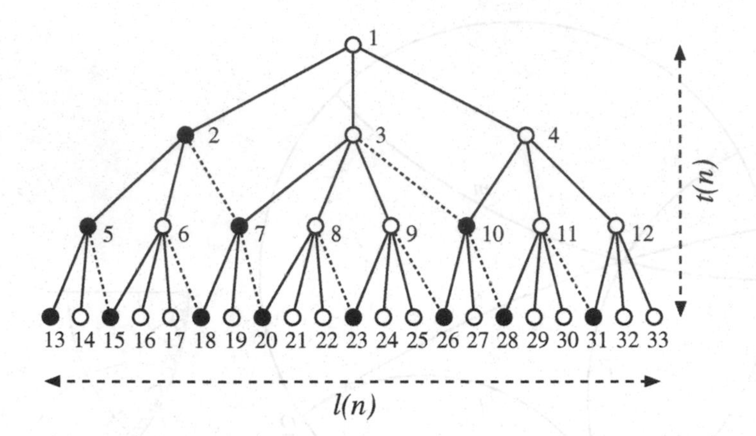

Fig. 2. The first few levels of the Fibonacci graph. The subgraph obtained by removing dotted lines is a Fibonacci tree. Nodes 1 and 2, 3, 4 correspond to the upper right pentagon and its neighboring three pentagons in Fig. 1(b), respectively.

dual graph of the tiling in the south western quarter of the hyperbolic plane. We call it a *Fibonacci graph* F (see Fig. 2).

As suggested by the figure, it is easy to transform a Fibonacci graph into a *Fibonacci tree* T by erasing the edges represented by dotted lines. The tree T is an infinite tree where each node has either 2 or 3 successors; we call it a *black* or *white* node, respectively (see Fig. 2). The root is a white node. The first successor of *each* node is a black node. The other successor(s) is a (are) white node(s). Let f_i denote the Fibonacci numbers, i.e., $f_0 = 0$, $f_1 = 1$, and $f_{i+2} = f_{i+1} + f_i$. In the ith level, the numbers of black and white nodes are f_{2i} and f_{2i+1}, respectively.

The Fibonacci graph F results from T by adding dotted edges. For example, an edge is added from node 3 to node 10, where 10 is the first successor of 4 which is the right of 3. From each node except the rightmost one, an edge is introduced in this rule. If four Fibonacci trees corresponding to the four quarters of the hyperbolic plane were put together, then in the resulting graph each node would have degree 5. A CA embedded in the Poincaré unit disk can be simulated by a CA embedded in the south-western quarter of the disk; the CA on the quarter can be obtained by folding the disk twice. Therefore, we can assume that every node has five neighbors.

2.3 Hyperbolic CA

If identical finite automata are put on the nodes of a Fibonacci graph, then the resulting system is a hyperbolic cellular automaton (HCA). We denote the set of states of a cell by Q and the local transition function by $\delta : Q \times Q^5 \to 2^Q$. In a Euclidean CA, each cell can distinguish between its northern, eastern, southern, and western neighbors. Similarly in a HCA each cell can distinguish between the five directions to its neighbors, provided that it knows which one is its father.

It follows from results in [7] that there is a HCA which achieves the following: When started with the configuration where the root is in a designated state and all other cells are in a quiescent state, after i steps of the HCA each node which has distance i from the root knows the following information: (i) whether it is a black node or a white node, (ii) whether it is a node along the leftmost or rightmost branch, and if so, which neighbors are "missing", (iii) which of its neighbors is its parent in the pure tree T, (iv) which of its neighbors is the "additional" parent in F, if one exists, (v) which of its neighbors is the "additional" successor in F, if one exists, and (vi) which of its regular successors is the left one, which is the right one, and which is the middle one, if one exists. Therefore, we will assume that such information is always presented in each cell from the beginning of the computation.

At step $t = 0$, the input string $a_1 a_2 \cdots a_n$ is given in the first n cells of the leftmost path of the Fibonacci tree. Computations of a nondeterministic HCA given some input are described as a computation tree: All nodes are configurations, the root is the initial configuration of the HCA for the given input, and the children of a configuration C are exactly those configurations which can be reached from C in one step allowed by the transition function. The configuration is said to be *accepting* (resp. *rejecting*) if the root cell 1 is in an accepting (resp. rejecting) state. Leaves of the computation tree are final configurations, they may be accepting or rejecting. Certain paths in the computation tree may be infinite. A nondeterministic HCA is said to be *deterministic* if each configuration has at most one child.

A HCA is said to be $t(n)$*-time bounded* if, for every accepted input of length n, there is an accepting configuration whose distance from the root configuration is at most $t(n)$ in the computation tree. A deterministic HCA is said to be $s(n)$*-space bounded* if all non-quiescent cells are within distance $s(n)$ from the root cell 1 in any configuration during the computation.

3 Main Results

Let $bin(n)$ denote the binary representation of the value n. A function $t(n)$ is said to be *computable* if, for each n, there is a Turing machine (TM) which, given the value n encoded in unary, generates $bin(t(n))$ in time $O(t(n))$ on all inputs of length n. All functions computable by TMs are also computable by CA. Furthermore, it is known [4] that if a function $t(n)$ is computable by an $O(t(n) - n)$-time TM (to which value n is given as a binary string of length $\lceil \log n \rceil$), then there is a one-dimensional CA which makes exactly $t(n)$ moves.

Now we are ready to present our main results.

Theorem 1. *Suppose that $t(n)$ is an arbitrary computable function. For any $t(n)$-time nondeterministic HCA H, there is an $O(t^3(n))$-time deterministic HCA which can simulate H.*

The proof of this theorem is given in Section 4. The second theorem is a separation result for classes of languages accepted by HCA.

Theorem 2. *Suppose that $t_1(n)$ and $t_2(n)$ are computable functions such that $\lim_{n\to\infty} \frac{(t_1(n))^3}{t_2(n)} = 0$. Then, there is a language which can be accepted by a $t_2(n)$-time deterministic HCA, but not by any $t_1(n)$-time nondeterministic HCA.*

The proof is given in Section 5. From Theorem 2, one can see that a polynomial increase of time strictly enlarges the class of languages accepted by deterministic and nondeterministic CA in the hyperbolic plane. On the other hand, it is open whether the linear-speedup theorem holds for hyperbolic CA. As for space complexity, the following theorem implies that the linear-compression theorem does not hold for HCA.

Theorem 3. *Suppose that $s_1(n)$ and $s_2(n)$ are computable functions such that $s_2(n) - s_1(n) \neq O(1)$. Then, there is a language which can be accepted by an $s_2(n)$-space deterministic HCA, but not by any $s_1(n)$-space deterministic HCA.*

We show this theorem in Section 6.

4 Deterministic Simulation of Nondeterministic HCA

In this section, we prove Theorem 1. Let M_1 be an arbitrary $t(n)$-time nondeterministic hyperbolic CA (NHCA) on the Fibonacci graph of depth $t(n)$. We construct an $O(t^3(n))$-time deterministic hyperbolic CA (DHCA), say, M_2, which simulates M_1. In order to simulate NHCA M_1, DHCA M_2 decides whether there is an accepting computation of M_1. Let Q be the set of M_1's cell states. First of all, M_1 computes the value of $t(n)$.

4.1 Basic Properties of NHCA

Before describing details, we observe basic properties of $t(n)$-time NHCA. Consider a Fibonacci graph of depth $t(n)$ (see Fig. 2). Nodes on the leftmost path are $f_1, f_3, f_5, \ldots, f_{2t(n)+1}$, leaves are $f_{2t(n)+1}, f_{2t(n)+1}+1, \ldots, f_{2t(n)+3}-1$, and the number of leaves is $f_{2t(n)+2}$. (In Fig. 2, they are $1, 2, 5, 13$; $13, 14, \ldots, 33$; and 21, respectively.) During a $t(n)$-step computation, the state of each cell changes $t(n)$ times. Thus, the computation in a single cell is represented by a sequence of length $t(n)+1$ in Q^*, called a *state-sequence*. The state-sequence of each cell must result from state-sequences of its five neighbors according to M_1's transition function δ (see, e.g., cell 3 and its neighbors $1, 7, 8, 9, 10$ in Fig. 2).

Consider a path from the root to a leaf in the Fibonacci tree (see, e.g., path 1-3-9-23 in Fig. 2). The computations on the path are represented by $t(n)+1$ state-sequences; the concatenation of them is represented by a string of length $(t(n)+1)^2$. The computations on the path depend on cells adjacent to the path. (In Fig. 2, state-sequences of cells $1, 3, 9, 23$ depend on state-sequences of cells $2, 3, 4$; $1, 7, 8, 9, 10$; $3, 23, 24, 25, 26$; $8, 9$, respectively. Note that paths are considered in the Fibonacci *tree*, but neighbors are considered in the Fibonacci *graph.*) Hence, computations in cells on each path and their neighbors are represented by a binary string, say, *path*, of length $\tau(n) = 6c(t(n)+1)^2$, where c is some constant depending on $|Q|$. We denote by $path(i)$ the binary string for the path from the root to leaf node $f_{2t(n)+1} + i$, where $0 \le i \le l(n)$ and $l(n) = f_{2t(n)+2} - 1$. (In Fig. 2, $f_{2t(n)+1} = 13$ and $0 \le i \le 20$.)

4.2 Deterministic Algorithm Using Binary Tree Structure

A binary tree can be constructed in the hyperbolic plane. (See Fig. 2. Use white nodes reachable from the root cell). We first "guess" three binary strings $path(0)$, $path(\lceil l(n)/2 \rceil)$, and $path(l(n))$ *deterministically.* Namely, M_2 generates all the different binary patterns of length $3\tau(n)$. In Fig. 3, regard the left and right branches as 0 and 1, respectively. Then, M_2 constructs a pair of sons at every node of depth $3\tau(n)$, and go to the second $3\tau(n)$ levels (see the next paragraph). Also, M_2 verifies whether there are no inconsistencies in each binary string of length $3\tau(n)$ according to M_1's transition function.

In the second $3\tau(n)$ levels, each pair of sons generate a pair of paths of length $2\tau(n)$. Each path is followed by a binary *tree* of depth $\tau(n)$. The left path is used for storing both $path(0)$ and $path(\lceil l(n)/2 \rceil)$, and the right path is for $path(\lceil l(n)/2 \rceil)$ and $path(l(n))$. These strings was moved from the first $3\tau(n)$ levels to the second $3\tau(n)$ levels, which can be done in $O(\tau(n))$ time by making use of the firing squad synchronization algorithm [9]. In the pair of binary trees of height $\tau(n)$, we guess $path(\lceil l(n)/4 \rceil)$ and $path(\lceil 3l(n)/4 \rceil)$, respectively.

Similarly, in the third $3\tau(n)$ levels, we guess $path(\lceil l(n)/8 \rceil)$, $path(\lceil 3l(n)/8 \rceil)$, $path(\lceil 5l(n)/8 \rceil)$, and $path(\lceil 7l(n)/8 \rceil)$.

After repeating the above procedure $\log l(n)$ times, the current $3\tau(n)$ levels contain $path(i), path(i')$, and $path(i'')$, where $i' = i+1$ or $i' = i$, and $i'' = i'+1$ or $i'' = i'$. Now, M_2 verifies whether there are no inconsistencies in the relation among strings $path(i), path(i')$, and $path(i'')$. The results are transmitted from

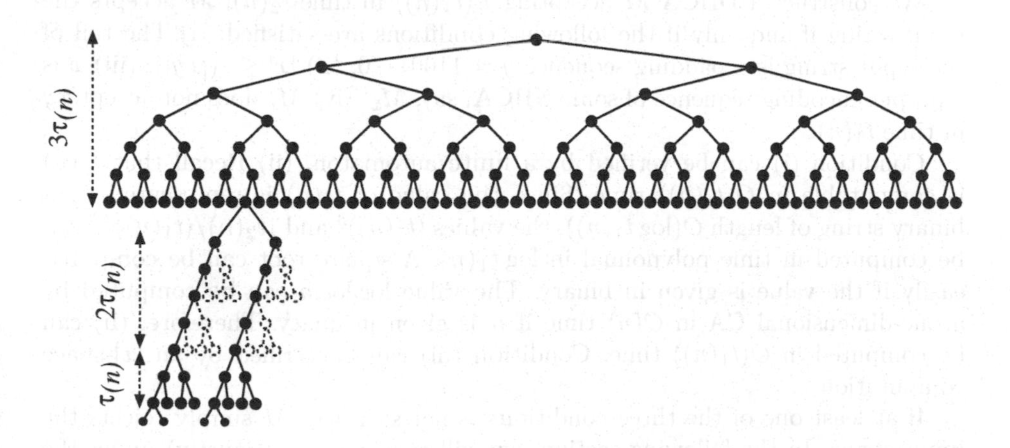

Fig. 3. Binary tree.

leaves toward the root according to the following strategy: There is at least one proper guess among $2^{\tau(n)}$ ones such that both sons are followed by proper guesses. Namely, each branch for the two sons computes an AND function, and each binary tree of height $\tau(n)$ computes a $2^{\tau(n)}$-bit OR function.

Each $3\tau(n)$ levels can be constructed in $O(\tau(n))$ time, and the procedure for $3\tau(n)$ levels was repeated $\log l(n)$ times. Since $\tau(n) = 6c(t(n)+1)^2$ and $l(n) = f_{2t(n)} - 1 = 2^{O(t(n))}$, the time complexity is bounded by $O(t^3(n))$. This completes the proof.

5 Time Hierarchy of Hyperbolic Cellular Automata

In this section, we prove Theorem 2. The proof is by diagonalization. We construct a $t_2(n)$-time DHCA, say, M, accepting a language $L(t_1(n))$ which cannot be accepted by any $t_1(n)$-time nondeterministic hyperbolic CA (NHCA). Here, $t_1(n)$ and $t_2(n)$ are computable functions such that $\lim_{n\to\infty} \frac{(t_1(n))^3}{t_2(n)} = 0$.

5.1 Language Not Accepted by $t_1(n)$-Time NHCA

All languages in this section are over $\{0,1\}$. It is known that any CA can be encoded into a binary string, say, x, by standard encoding. It can be checked in linear time whether the given string encodes a proper CA. Let $\varepsilon(n) = \min\{\lfloor\sqrt{\lfloor t_2(n)/(t_1(n))^3\rfloor}\rfloor, \log\log n\}$. Note that $\varepsilon(n) \neq O(1)$. (The reason why we define such a function is given later.) Let M_x denote the NHCA whose encoding sequence is x. If x is not a proper encoding sequence, we regard M_x as a NHCA accepting $\emptyset$.

Let $L(t_1(n)) = \{xy \mid M_x$ does not accept xy within time $t_1(|xy|)$, $y = 1100\cdots0$, 11 does not appear in x as a substring, and $|x| \le \varepsilon(|xy|)\}$. Here, y is a padding sequence, and its prefix 11 indicates the boundary. It is obvious that any $t_1(n)$-time NHCA cannot accept $L(t_1(n))$.

We construct a DHCA M accepting $L(t_1(n))$ in time $t_2(n)$. M accepts the input string if and only if the following conditions are satisfied: (i) The tail of the input string is a padding sequence $y = 1100\cdots0$. (ii) $|x| \le \varepsilon(|xy|)$. (iii) x is a proper encoding sequence of some NHCA, say, M_x. (iv) M_x does not accept xy in time $t_1(n)$.

Condition (i) can be verified by a finite automaton. (ii) Recall that $t_1(n)$ is computable in $O(t_1(n))$ time. Since the value of $t_1(n)$ is represented by a binary string of length $O(\log t_1(n))$, the values $(t_1(n))^3$ and $\lfloor t_2(n)/(t_1(n))^3\rfloor$ can be computed in time polynomial in $\log t_1(n)$. A square root can be computed easily if the value is given in binary. The value $\log\log n$ can be computed by a one-dimensional CA in $O(n)$ time if n is given in unary. Therefore, (ii) can be computed in $O(t_1(n))$ time. Condition (iii) can be verified by an $|x|$-space computation.

If at least one of the three conditions is not satisfied, M simply rejects the input string. In the following section, we will consider condition (iv) under the assumption that conditions (i), (ii), and (iii) are satisfied.

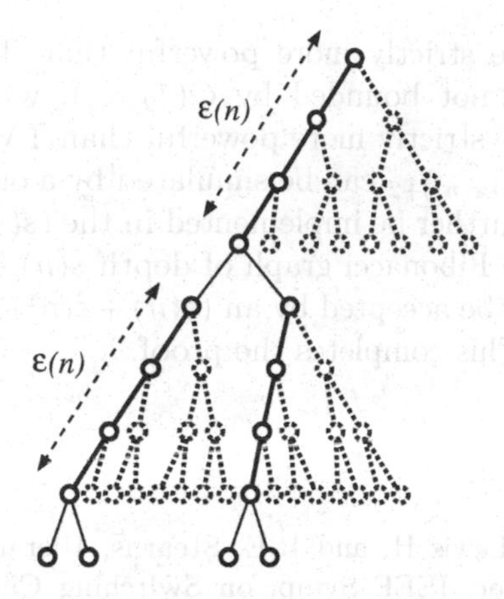

Fig. 4. Small binary trees of height $\varepsilon(n)$ connected in the binary-tree structure.

5.2 DHCA Accepting the Language $L(t_1(n))$

In order to verify condition (iv), M decides whether there is an accepting computation of M_x on input xy as follows.

Recall that a binary tree can be constructed in the hyperbolic plane. In the binary tree, we construct *small binary trees* of height $\varepsilon(n)$ whose leftmost leaf has two small binary trees (see Fig. 4). Namely, we construct a binary tree of small binary trees. We store encoding sequence x of NHCA M_x into the leftmost path of every small binary tree. Therefore, the root node of every small binary tree can always access the transition function of M_x.

In the following, we regard each small binary tree in Fig. 4 as a single node in Fig. 3. Now we can perform the same procedure as in Section 4.2. If M finds that there is a $t_1(n)$-step accepting computation on M_x, then M rejects x; otherwise, M accepts x.

Each $3\tau(n)$ levels can be constructed in $O(\tau(n)\varepsilon(n))$ time, and the procedure for $3\tau(n)$ levels was repeated $\log l(n)$ times. Since $\tau(n) = 6c(t_1(n)+1)^2$ and $l(n) = f_{2t_1(n)} - 1 = 2^{O(t_1(n))}$, the time complexity is bounded by $O((t_1(n))^3\varepsilon(n))$. Since $\varepsilon(n) \leq \sqrt{t_2(n)/(t_1(n))^3}$, we obtain $(t_1(n))^3 \leq t_2(n)/(\varepsilon(n))^2$. Therefore, the complexity $O((t_1(n))^3\varepsilon(n))$ is written as $O(t_2(n)/\varepsilon(n))$, which is less than $t_2(n)$. This completes the proof.

6 Space Hierarchy of Hyperbolic Cellular Automata

In this section, we show that for any computable functions $s(n)$ and $\varepsilon(n) \neq O(1)$, $(s(n)+\varepsilon(n))$-depth HCA are strictly more powerful than $s(n)$-depth HCA.

As was mentioned in Section 4.1, a HCA has $f_{2s(n)+3} - 1$ cells in the first $s(n)+1$ levels. (In Fig. 2, there are $f_9 - 1$ $(= 33)$ nodes in the Fibonacci graph of depth 3.) Thus, every $s(n)$-depth HCA can be simulated by an $O(f_{2s(n)})$-space TM because $f_{2s(n)+3} = O(f_{2s(n)})$. It is known [1] that if $s_2(n) \neq O(s_1(n))$,

TMs of space $s_2(n)$ are strictly more powerful than TMs of space $O(s_1(n))$. Since $f_{2(s(n)+\varepsilon(n))+2}$ is not bounded by $O(f_{2s(n)})$, we can say that TMs of space $f_{2(s(n)+\varepsilon(n))+2}$ are strictly more powerful than TMs of space $O(f_{2s(n)})$.

TMs of space $f_{2(s(n)+\varepsilon(n))+2}$ can be simulated by a one-dimensional CA with the same space. It can further be implemented in the $(s(n)+\varepsilon(n))$th and $(s(n)+\varepsilon(n)-1)$st levels of the Fibonacci graph of depth $s(n)+\varepsilon(n)$. Therefore, there is a language which can be accepted by an $(s(n)+\varepsilon(n))$-depth HCA, but not by any $s(n)$-depth HCA. This completes the proof.

References

1. J. Hartmanis, P.M. Lewis II, and R.E. Stearns, Hierarchies of memory limited computations, in: Proc. IEEE Symp. on Switching Circuit Theory and Logical Design, 1965, 179–190.
2. J. Hartmanis and R.E. Stearns, On the computational complexity of algorithms, Trans. Amer. Math. Soc., **117** (1965) 285–306.
3. K. Iwama and C. Iwamoto, Improved time and space hierarchies of one-tape off-line TMs, in: Proc. Int'l Symp. on Mathematical Foundations of Computer Science (LNCS1450), 1998, 580–588.
4. C. Iwamoto, T. Hatsuyama, K. Morita, and K. Imai, Constructible functions in cellular automata and their applications to hierarchy results, Theor. Comput. Sci. **270** 1/2 (2002) 797–809. (Preliminary version: Proc. 12th Int'l Symp. on Fundamentals of Computation Theory (LNCS1684), 1999, 316–326.)
5. C. Iwamoto, M. Margenstern, K. Morita, and T. Worsch, P = NP in the space of CA of the hyperbolic plane, Presentation at the 7th Int'l Workshop on Cellular Automata, Automata 2001 (IFIP WG 1.5), Hyères , France, September, 2001.
6. K. Loryś, New time hierarchy results for deterministic TMs, in: Proc. Int'l Symp. on Theoretical Aspects of Computer Science (LNCS577), 1992, 329–336.
7. M. Margenstern, New tools for cellular automata in the hyperbolic plane, Journal of Universal Computer Science, **6** 12 (2000) 1226–1252.
8. M. Margenstern and K. Morita, NP problems are tractable in the space of cellular automata in the hyperbolic plane, Technical report, Publications of the I.U.T. of Metz, 38p. 1998. (Journal version: Theor. Comput. Sci. **259** (2001) 99–128.)
9. J. Mazoyer, A 6-state minimal time solution to the firing squad synchronization problem, Theoret. Comput. Sci., **50** (1987) 183–238.

On a Mereological System for Relational Software Specifications*

Ryszard Janicki

Department of Computing and Software, McMaster University,
Hamilton, ON, L8S 4K1 Canada,
janicki@mcmaster.ca

Abstract. The concept of being a *part of* is defined and analysed in an algebraic framework and then apply to the algebra of relations. A motivation that comes from Software Engineering is discussed.

1 Introduction

Main motivation for this work was provided by an attempt to define a formal semantics for *tabular expressions* [10,11,13]. Tabular expressions (Parnas et al. [1,14,18]) are means to represent the complex relations that are used to specify and document software systems. The technique is quite popular in software industry. It was first developed in work for the U.S. Navy and applied to the A-7E aircraft [2]. The ideas were picked up by Grumman, the U.S. Air Force, Bell Laboratories and many others. The tabular notation have also been applied in Canada by Ontario Hydro in Darlington Nuclear Plant (see [14,18,19]). However a formal semantics of tabular expressions has only recently been developed [10,11,13]. When software engineers discuss a specification using *tabular expressions*, the statements like "this is *a part of* a bigger relation", "this relation is composed of the following *parts*", etc., can be heard very often.

To illustrate what *tabular expressions* are, we will analyze one *very* simple example.

Consider the following relation $G \subseteq IN \times OUT$, where $IN = Reals \times Reals$, $OUT = Reals \times Reals \times Reals$, x_1, x_2 are the variables over IN, y_1, y_2, y_3 are variables over OUT, and

$$(x_1, x_2)G(y_1, y_2, y_3) \iff \begin{cases} \left.\begin{array}{l} y_1 = x_1 + x_2 \wedge y_2 x_1 - x_2 = y_2^2 \\ \wedge\, y_3 + x_1 x_2 = |y_3|^3 \end{array}\right\} \text{if } x_2 \leq 0 \\ \left.\begin{array}{l} y_1 = x_1 - x_2 \wedge x_1 + x_2 + x_2 y_2 = |y_2| \\ \wedge\, y_3 = x_1 \end{array}\right\} \text{if } x_2 > 0 \end{cases}$$

The relation G is more readable defined by a *tabular expression* in Figure 1. This kind of tabular expression is called a *vector table* and its intuitive meaning is practically self-explained. It reads that if $x_2 \leq 0$ then $y_1 = x_1 + x_2$, y_2 must satisfy $y_2 x_1 - x_2 = y_2^2$, y_3 must satisfy $y_3 + x_1 x_2 = |y_3|^3$, and similarly for $x_2 > 0$.

* Partially supported by NSERC of Canada and CITO of Ontario Grants.

K. Diks et al. (Eds): MFSC 2002, LNCS 2420, pp. 375–386, 2002.

	$x_2 \leq 0$	$x_2 > 0$
$y_1 =$	$x_1 + x_2$	$x_1 - x_2$
$y_2 \vert$	$y_2 x_1 - x_2 = y_2^2$	$x_1 + x_2 y_2 = \vert y_2 \vert$
$y_3 \vert$	$y_3 + x_1 x_2 = \vert y_3 \vert^3$	$y_3 = x_1$

Fig. 1. The relation G defined by a *vector table*. The symbol "=" after y_1 indicates that the value of y_1 is a function of other variables, the symbol "|" after y_2 and y_3 indicates that the relationship between y_i, $i = 2, 3$, is relational and not functional.

The relation G is a composition of its "atomic" parts, i.e. $G = \bigcirc G_{i,j}$, where $G_{i,j}$, $i = 1, 2$, $j = 1, 2, 3$, are relations defined by expressions in single cells. For instance $G_{1,3} \subseteq IN_{1,3} \times OUT_{1,3}$, where $IN_{1,3} = Reals \times (-\infty, 0\rangle$, $OUT_{1,3} = Reals$, and $(x_1, x_2)G_{1,3}y_3 \iff y_3 + x_1x_2 = |y_3|^3$. The relation $G_{1,1}$ is a function $G_{1,1} : IN_{1,1} \to OUT_{1,1}$, with $IN_{1,1} = IN_{1,3}$ and $OUT_{1,1} = Reals$, and $(x_1, x_2)G_{1,1}y_1 \iff y_1 = G_{1,1}(x_1, x_2) = x_1 + x_2$. The relations $G_{i,1}$ are functions, which is indicated by the symbol "=" after variable y_1 in the left header. The symbol "|" after y_2 and y_3 indicates that $G_{i,2}$ and $G_{i,3}$ are relations with y_2 and y_3 as respective range variables.

Of course every $G_{i,j}$ is a part of G, every subset of G is a part of G, every relation defined by a tabular expression derived from the tabular expression that defines G by removing any number of rows and columns, is also a part of G. There are many types of tabular expressions, we gave a simple example of only one of them. For all tabular expressions, the global relation/function is defined as a composition of its parts, for each type of tabular expressions, an appropriate composition operation is different. The concept of "part of" is essential for specification techniques based on tabular expressions. For more details the reader is referred to [1,10,13,14].

One of the biggest advantages of tabular expression technique is the ability to define a relation R that describes the properties of the system specified, as an easy to understand composition[1] of the relations R_α, $\alpha \in I$, where R_α is a *part of* R.

The problem is that the standard algebra of relations lacks the formal concept of being a *part of*. The concept of subset is not enough, for instance if $A \subseteq B$ and $D = B \times C$, then A is not a subset of D, but according to standard intuition it is a *part of* D.

Attempts to formalize the concept of "part of" go back to S. Leśniewski (1916-1937,[22]), and H. Leonard, N. Goodman (1930-1950,[7,17]), however thay have never become very popular from the application view point. Leśniewski's systems are different from the standard set theory based on Zermello-Freankl ax-

[1] The word "composition" here means "the act of putting together" (*Oxford English Dictionary*, 1990), not "the" composition of relations that is usually denoted by ";" or "∘" ([3,16]. In this sense "∪" is a composition.

ioms [16], which makes their straightforward application quite difficult[2]. Leonard and Goodman Calculus of Individuals was defined in terms of the standard set theory, but it was too much influenced by "spacial reasoning" (see [7]), and as a result of this it is too rigid for many applications. It resembles theories based on the concept of a lattice (which was motivated by the algebra of sets, which is "spacial" indeed!), but many posets defined by "part of" like relations that may occur in practice are not lattices. Nevertheless our approach can be regarded as a refinement of Leonard-Goodman approach (see Final Comments for more details).

The basic difference between this paper and the models mentioned above is that we *do not* start with an axiomatic definition of a "part of" relation. We assume that the complex objects can be built from the more primitive ones by using "constructor" operations, and the less complex objects can be derived from the more complex ones by using "destructor" operations. Our mereological system is kind of an abstract algebra, and the relation "part of" can be derived from the set of "constructor" and "destructor" operations.

Since posets generated by "part of" relations are not lattices in many cases, the concept of a *grid*, an extension of a lattice, is introduced and analysed. We believe that any poset generated by a "part of" relation should be a grid. Grids are a genaralisation of weak lattices introduced and analysed in [12]. From the application to tabular expression viewpoint, Section 4 seems to be the most important, even though it is very sketchy due to a lack of space.

This paper is a continuation of the approach initialized in [12].

2 Grids

For the sake of self-completeness we shall start with rather well known definitions. Let $(X, \preceq)$ be a partial order (poset) and let $A \subseteq X$. An element $a \in X$ is called *upper bound*) (*lower bound*) of A iff $\forall x \in A.\ x \preceq a$ ($\forall x \in A.\ a \preceq x$). Let $\mathrm{ub} A$ ($\mathrm{lb} A$) denote the set of all *upper bounds* (*lower bounds*) of A. A poset $(X, \preceq)$ is called *bounded* if there are $\top, \bot \in X$ such that $\{\top\} = \mathrm{ub}\, X$ and $\{\bot\} = \mathrm{lb}\, X$. The element $\top$ is called the *top* of X, and the element $\bot$ is called *bottom* of X. An element $a \in A$ is a *minimal* (*maximal*) element of A iff $\forall x \in A.\ \neg(x \prec a)$ ($\forall x \in A.\ \neg(a \prec x)$). The set of all *minimal* (*maximal*) elements of A will be denoted by $\min A$ ($\max A$). An element $a \in X$ is called the *least upper bound* (*supremum*) of A, denoted $\sup A$, iff $a \in \mathrm{ub}\, A$ and $\forall x \in \mathrm{ub}\, A.\ a \preceq x$. An element $a \in X$ is called the *greatest lower bound* (*infimum*) of A, denoted $\inf A$, iff $a \in \mathrm{lb}\, A$ and $\forall x \in \mathrm{lb}\, A.\ x \preceq a$. A poset $(X, \preceq)$ is called a *lattice* iff for all $a, b \in X$ both $\sup\{a, b\}$ and $\inf\{a, b\}$ do exist.

The set $\min A$ ($\max A$) is *complete* iff $\forall x \in A.\ \exists a \in \min A.\ a \preceq x$ ($\forall x \in A.\ \exists a \in \max A.\ x \preceq a$).

[2] Tarski's cylindric algebras were partially motivated by Leśniewski's work ("a circle is a part of a cylinder [8]"), even though no formal result of Leśniewski was ever used. A very recent application of Leśniewski's ideas to approximate reasoning can be found in [20,21].

Let cmin A = min A (cmax A = max A) if min A is complete (max A is complete), and undefined otherwise.

The complete minimal/maximal elements were introduced in [12], the remaining definitions are well known ([3,16]).

For every set $A \subseteq X$, we define cminub = cmin ub A, cmaxlb = cmax lb A.

For every $k \geq 0$, we define cminub$^k A$ as: cminub$^0 A = A$ and cminub$^{k+1} A$ = cminub (cminub$^k A$), and similarly for cmaxlb$^k A$.

Directly from the definitions one may prove the following results.

Corollary 1. *For every* $a \in X$, *we have:* ub $\{a\}$ = lb $\{a\}$ = cmin $\{a\}$ = cmax $\{a\}$ = cminub $\{a\}$ = cmaxlb $\{a\}$ = $\{a\}$, *and* $\sup\{a\} = \inf\{a\} = a$. ■

Lemma 1. *For every* $A \subseteq X$

1. $|\text{cminub}^j A| = 1 \Rightarrow |\text{cminub}^{j+1} A| = 1$.
2. $|\text{cmaxlb}^j A| = 1 \Rightarrow |\text{cmaxlb}^{j+1} A| = 1$. ■

For every $A \subseteq X$, let $d_t(A)$ ($d_b(A)$), *degree of the top of A* (*degree of the bottom of A*), be the smallest k such that $|\text{cminub}^k A| = 1$ ($|\text{cmaxlb}^k A| = 1$).

For every $A \subseteq X$, let us define top A, *the top of A*, as

$$\text{top } A = a \iff \{a\} = \text{cminub}^{d_t(A)} A.$$

Similarly, we define bot A, *the bottom of A*, as

$$\text{bot } A = a \iff \{a\} = \text{cmaxlb}^{d_b(A)} A.$$

Of course it may happen that neither top A, nor bot A exists.

We will write $a = \text{top}^k A$ ($a = \text{bot}^k A$) if $a = \text{top } A$ and $d_t(A) = k$ ($a = \text{bot } A$ and $d_b(A) = k$).

Lemma 2. 1. *If* sup A *exists then* sup A = top$^1 A$, *and if* inf A *exists then* inf A = bot$^1 A$.

2. *If* wsup A (winf A)*exists then* wsup A = top$^2 A$ (winf A = bot$^2 A$). *Both* wsup A *and* winf A *were introduced in [12] as extensions of* sup A *and* inf A. ■

A poset $(X, \preceq)$ is called a *grid* if for each $a, b \in X$, top$\{a, b\}$ and bot$\{a, b\}$ exist.

A poset $(X, \preceq)$ is called a *grid of degree* (k, n) if it is a grid and for every finite $A \subseteq X$, $d_t(A) \leq k$, $d_b(A) \leq n$.

A grid $(X, \preceq)$ is *regular* iff $\forall A \subseteq X. \forall B \subseteq A.$ top $B \preceq$ top A and bot $A \preceq$ bot B.

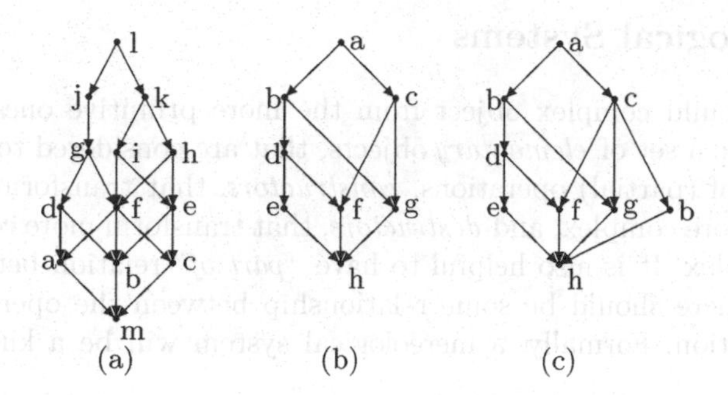

Fig. 2. (a) The poset is a *grid of degree* $(3,3)$ and for example: $g = \text{top}^2\{a,b\}$, $k = \text{top}^3\{b,c\}$, $\text{cminub}\,\{b,c\} = \{e,f\}$, $\text{cminub}^2\{b,c\} = \text{cminub}\,\{e,f\} = \{i,h\}$, $\text{cminub}^3\{b,c\} = \text{cminub}^2\{e,f\} = \text{cminub}\,\{i,h\} = \{k\}$, $\text{cminub}\,\{a,b\} = \{d,f\}$, $\text{cminub}^2\{a,b\} = \text{cminub}\,\{d,f\} = \{g\}$. The grid is not regular since $\text{top}\,\{a,b,c\} = f \preceq g = \text{top}\,\{a,b\}$.
(b) The poset is a *grid of degree* $(2,2)$. We have $e \sqcup f = d$, $d \sqcup g = b$, so $(e \sqcup f) \sqcup g = b$, but $f \sqcup g = a$, $e \sqcup a = a$, so $e \sqcup (f \sqcup g) = a$, i.e. $(e \sqcup f) \sqcup g \neq e \sqcup (f \sqcup g)$. Note that $\text{top}^2\{e,f,g\} = b$. By reversing arrows the same example works for "$\sqcap$".
(c) An example of an *atomistic grid* of degree $(2,2)$.

Corollary 2. *1. A lattice is a regular grid of degree* $(1,1)$.
2. A weak lattice of [12] is a grid of degree $(2,2)$. ∎

Figure 2(a) illustrates the concepts introduced above.

A grid is called *infimum complete* (*supremum complete*) iff for every $A \subseteq X$, bot A (top A) do exist. Immediately from the definitions we conclude that if top X exists than $\text{top}\,X = \sup X = \top$, and dually, if bot X exists, then $\text{bot}\,X = \inf X = \bot$. A grid is *complete* if it is infimum complete and supremum complete.

Let $a, b \in X$, and let $A \subseteq X$. We define:

$$a \sqcup b = \text{top}\,\{a,b\},\;\; a \sqcap b = \text{bot}\,\{a,b\},\;\; \sqcup_{a\in A}\, a = \text{top}\,A,\;\; \sqcap_{a\in A}\, a = \text{bot}\,A.$$

Unfortunately it might happen that $a \sqcup (b \sqcup c) \neq (a \sqcup b) \sqcup c$, or $a \sqcap (b \sqcap c) \neq (a \sqcap b) \sqcap c$, as in the example in Figure 2(b).

Let $(X, \preceq)$ be an infimum complete grid. The minimal elements of $X \setminus \{\bot\}$ are called *atoms*.

Let $atoms_{\preceq}(X)$ denote the set of all atoms of $(X, \preceq)$.

Let $\mathcal{A}(X)$ be the least set satisfying the following properties:

- every atom of X belongs to $\mathcal{A}(X)$,
- for every $A \subseteq \mathcal{A}(X)$, top $A \in \mathcal{A}(X)$.

An infimum complete grid is called *atomistic* iff $X \setminus \{\bot\} = \mathcal{A}(X)$.

Neither the grid from Figure 2(a) nor from Figure 2(b) is atomistic. For example for Figure 2(b), $\mathcal{A}(X) = \{a,b,d,e,f,g\}$, so $c \notin \mathcal{A}(X)$. An example of atomistic grid which is not a lattice is given on Figure 2(c).

3 Mereological Systems

How can we build complex object from the more primitive ones? Usually we assume to have a set of *elementary* objects, that are considered to be "atomic", and two sets of (partial) operations, *constructors*, that transform less complex objects into more complex, and *destructors*, that transform more complex object into less complex. It is also helpful to have *"part of"* relation between objects. And clearly there should be some relationship between the operators and the "part of" relation. Formally, a mereological system will be a kind of abstract algebra.

By a *mereological system* we mean a tuple

$$MS = (X, E, \perp, \Theta, \Delta, \eta),$$

where:

- X is a set of *elements*,
- $E \subseteq X$ is a set of *elementary* elements,
- $\perp \in X \setminus E$ is an *empty* element,
- Θ is the set of *constructors*, each $\theta \in \Theta$ is a partial function $\theta : X^k \to X$,
- Δ is the set of *destructors*, each $\delta \in \Delta$ is a partial function $\delta : X^k \to X$,
- $\eta : X \setminus \{\perp\} \to 2^E$ is a total function interpreted as the *elementary elements assignment* function.

For every set of functions $\mathcal{F}$, and every set A, let $A^{\mathcal{F}}$ denote the smallest set containing A and closed under $\mathcal{F}$.

We will assume that each element is either elementary, or empty, or it can be constructed from the elementary elements by using the constructors and destructors. It can be formally expressed as:

$$E^{\Theta \cup \Delta} = X. \tag{1}$$

Since the elemets of E are "atomic", so they cannot be decomposed any further, which can be formally expressed as:

$$E^{\Delta} \subseteq E \cup \{\perp\}. \tag{2}$$

Let $\sqsubseteq$ be a relation on X interpreted as a "part of". What properties should it satisfy? Since operations from Θ are used to construct more complex elements, one property $\sqsubseteq$ should satisfy is:

$$\forall \theta \in \Theta.\ \theta(a_1, \ldots, a_k) = b \ \Rightarrow \ (\forall i = 1, \ldots, n.\ a_i \sqsubseteq b.$$

The very similar reasoning can be applied to "destructors", with one exception. If $\delta \in \Delta$ and the arity of δ is bigger than 1, i.e. $\delta : X^k \to X$, and $k > 1$, then the first argument of δ is treated differently then the remaining $k - 1$. If $\delta(a_1, a_2, \ldots, a_k) = b$, then we say that b was obtained from a_1 with "help" of $a_2, \ldots, a_k$. In other words we assume that:

$$\forall \delta \in \Delta.\ \delta(a_1, a_2, \ldots, a_k) = b \ \Rightarrow \ b \sqsubseteq a_1).$$

All the above lead us to the following concepts. Let $\dot{\sqsubseteq} \subseteq X \times X$ be the following relation:

$$a \dot{\sqsubseteq} b \iff (\exists \theta \in \Theta . \exists a_1, \ldots, a_k \in X. \ \exists i \in \{1, ..., k\}. \ a = a_i \wedge b = \theta(a_1, \ldots, a_k)) \vee$$
$$(\exists \delta \in \Delta . \exists b_2, \ldots, b_k \in X. \ a = \delta(b, b_2, ..., b_k)).$$

Define the relation $\sqsubseteq \subseteq X \times X$ as: $\sqsubseteq = \dot{\sqsubseteq}^*$, i.e. $\sqsubseteq$ is a reflexive and transitive closure of $\dot{\sqsubseteq}$.

The relation $\sqsubseteq$ is fully defined by MS, so, if necessary, it will be denoted by $\sqsubseteq_{MS}$.

We require the following property from $\sqsubseteq$:

$$(X, \sqsubseteq) \text{ is an infimum complete grid with } \bot = \inf X. \quad (3)$$

The function η that to each element $a \in X$ associates the set of elemetary elements the element a is constructed from, should satisfy the following properties.

$$\begin{array}{l} \forall e \in E. \ \eta(e) = \{c\}, \\ \forall \theta \in \Theta. \ \eta(\theta(a_1, \ldots, a_k)) = \eta(a_1) \cup \ldots \cup \eta(a_k), \\ \forall \delta \in \Delta. \ \eta(\delta(a_1, \ldots, a_k)) \subseteq \eta(a_1). \end{array} \quad (4)$$

For every $A \subseteq X$, let $\eta(A) = \bigcup_{a \in A} \eta(a)$, $X{\downarrow_A} = \{x \mid x \in X \wedge \eta(x) \subseteq \eta(A)\}$, $\sqsubseteq{\downarrow_A} = \{(x,y) \mid x \sqsubseteq y \wedge x, y \in A\}$. One may prove that for every $A \subseteq X$, $(X{\downarrow_A}, \sqsubseteq{\downarrow_A})$, is a poset.

We Will Use the Following Convention: $\overline{\text{ub}}\, A$ denotes the set of all upper bounds of A in $(X{\downarrow_A}, \sqsubseteq{\downarrow_A})$, $\overline{\text{top}}\, A$ denotes the top of A in $(X{\downarrow_A}, \sqsubseteq{\downarrow_A})$, $a \overline{\sqcup} b = \overline{\text{top}}\, \{a, b\}$, $a \overline{\sqcap} b = \overline{\text{bot}}\, \{a, b\}$, etc.

We will call $\overline{\text{ub}}\, A$ the *local* set of upper bounds of A, $\overline{\text{top}}\, A$ the *local* top of A, etc.

Lemma 3. *1.* $E = atoms_{\sqsubseteq}(X)$
2. $\forall a, b \in X. \ a \sqsubseteq b \ \Rightarrow \ \eta(a) \subseteq \eta(b)$
3. $\forall A \subseteq X. \sup A = \overline{\sup}\, A$, $\inf A = \overline{\inf}\, A$, $\overline{\text{top}}\, A \sqsubseteq \text{top}\, A$, $\text{bot}\, A \sqsubseteq \overline{\text{bot}}\, A$ ■

We can now define our final requirement any mereological system must satisfy:

$$\text{For every } A \subseteq X, \ (X{\downarrow_A}, \sqsubseteq{\downarrow_A}) \text{ is an infimum complete regular grid.} \quad (5)$$

Note that the condition (5) implies the condition (3).

We will say that a tuple $MS = (X, E, \bot, \Theta, \Delta, \eta)$, is a mereological systems if the conditions (1),(2),(4),(5) are satisfied.

A mereological system MS is *atomistic* (*complete*) if for every $A \subseteq X$, $(X{\downarrow_A}, \sqsubseteq{\downarrow_A})$ is atomistic (*complete*).

A mereological system $MS = (X, E, \bot, \Theta, \Delta, \eta)$ is *constructive* if $E^{\Theta} = X \setminus \{\bot\}$.

Corollary 3. *If MS is constructive then the mapping η is entirely defined by the elements of Θ, so MS can be defined as a tuple $(X, E, \bot, \Theta, \Delta)$.* ■

For every mereological system $MS = (X, E, \Theta, \Delta, \eta)$, (X, E) is called a *domain* of MS. Let $\mathcal{MS}$ be a family of mereological systems with a common domain. We will say that a mereological system MS_U is *universal* for $\mathcal{MS}$ if for every $MS \in \mathcal{MS}$, $a \sqsubseteq_{MS} b \;\Rightarrow\; a \sqsubseteq_{MS_U} b$.

Example 1. For every set A, let $\hat{A} = \{\{a\} \mid a \in A\}$ be the set of all singletons generated by A, i.e. if $A = \{a, b\}$, then $\hat{A} = \{\{a\}, \{b\}\}$.

Let $E = \hat{D_1} \cup \hat{D_2}$, where $D_1 = \{a, b\}$, $D_2 = \{1, 2\}$, $\bot = \emptyset$, $\Theta = \{\cup, \dot{\times}\}$, where $\dot{\times}$ (a restricted Cartesian Product) is defined as: $A \dot{\times} B = \{(x, y) \mid x \in A \subseteq D_1 \wedge y \in B \subseteq D_2\}$, $\Delta = \{\pi_1, \pi_2, \setminus\}$, where π_i is a projection of sets on the ith coordinate, formally defined as $\pi_i : 2^{D_i} \cup 2^{D_1 \times D_2}$ with $\pi_i(A) = A$ if $A \subseteq D_i$ and $\pi_1(A) = \{x \mid (x, y) \in A\}$, $\pi_2(A) = \{y \mid (x, y) \in A\}$ if $A \subseteq D_1 \times D_2$. Let $X = 2^{D_1} \cup 2^{D_2} \cup 2^{D_1 \times D_2}$, and let η satisfies: $\eta(A) = \hat{A}$ if $A \subseteq D_1 \cup D_2$, and $\eta(A) = \{\{x\} \mid (x, y) \in A\} \cup \{\{y\} \mid (x, y) \in A\}$ if $A \subseteq D_1 \times D_2$.

Note that in this case we have $X \setminus \{\emptyset\} = E^{\{\cup, \dot{\times}\}}$. For instance: $\{(a, 1), (b, 2)\} = \{a\} \dot{\times} \{1\} \cup \{b\} \dot{\times} \{2\}$.

Define $\sqsubseteq$ as follows: $A \sqsubseteq B \iff A \subseteq B \vee A \subseteq \pi_i(B)$, $i = 1, 2$.

One can show by inspection that $(X, E, \bot, \Theta, \Delta, \eta)$ is a *constructive, atomistic, and complete mereological system*, and for every $A \subseteq X$, $(X_A, \sqsubseteq_A)$ is a complete grid of degree (2,2) with $\sup X = \top = D_1 \times D_2$. As Figure 3 shows, the poset $(X, \sqsubseteq)$ is not a lattice. This example is a special case of a more general result that will be discussed in the next section.

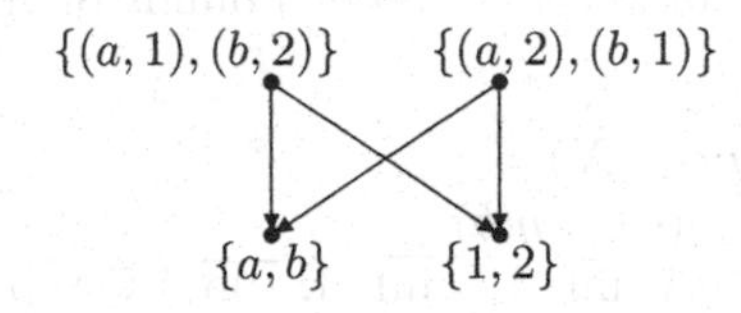

Fig. 3. Part of the poset from the Example 1 illustrating that this poset is not a lattice.

4 Mereological Systems for Direct Products

For simplicity we restrict our discussion to Direct Products, however the results could easily be extended to Heterogenious Relations.

Let T be a universal set of indexes and let $\{D_t \mid t \in T\}$ be an appropriate set of domains. We assume also that the set T is *finite*. From the viewpoint of applications in Software Engineering this is not a restriction at all ([1,13]).

For every $I \subseteq T$, let $D_I = \prod_{i \in I} D_i$, where $\prod_{i \in I}$ is a *Direct Product* over I. For example if $I = \{2, 5, 9\}$, then $D_I = \prod_{i \in I} D_i = D_2 \times D_5 \times D_9$. In other words, the set D_I is the set of all functions $f : I \to \bigcup_{i \in I} D_i$ such that $\forall i \in I.\ f(i) \in$

D_i. Usually, in particular for concrete cases, such functions are represented as vectors. For instance $f : \{2,5,9\} \to D_2 \cup D_5 \cup D_9$ with $f(2) = a_2$, $f(5) = a_5$, $f(9) = a_9$ is represented as a vector $(a_2, a_5, a_9) \in D_2 \times D_5 \times D_9$. However for theoretical reasoning the functional representation is more convenient. For every function $f : X \to Y$, and every $Z \subseteq X$, the symbol $f|_Z$ will denote the restriction of f to Z. For every function f, dom f will denote the domain of f. We will also assume that $D_i \cap D_j = \emptyset$ if $i \neq j$. This assumption allows us to identify every element of $a \in D_i$ with the function $f_a : \{i\} \to D_i$ where $f_a(i) = a$, which makes the notation more consistent and less ambiguous. We do not lose any generality here, moreover, in practical applications each D_i has a different interpretation anyway (for instance: input current, output current; Amperes in both cases but different meaning).

For every D_I, $I \subseteq T$, let $\mathcal{X}_I = \bigcup_{J \subseteq I} 2^{D_J}$. Also let $\mathcal{E} = \bigcup_{i \in T} \hat{D}_i$. Clearly $\mathcal{E} \subseteq \mathcal{X}_T$. Note also that for every $A \in \mathcal{X}_T$ and for every $f, g \in A$ we have $\mathrm{dom}\, f = \mathrm{dom}\, g$.

For every $A \in \mathcal{X}$, let $\tau(A) \subseteq T$, the *index set* of A, be defined as follows: $I = \tau(A) \iff A \subseteq D_I$. In other words, $f \in A \Rightarrow \mathrm{dom}\, f = \tau(A)$. We assume that $\tau(\emptyset) = I$ for any appropriate I. For instance if $A \subseteq D_2 \times D_5 \times D_7$ then $\tau(A) = \{2,5,7\}$. For every $A \in \mathcal{X}$ and every $K \subseteq T$, let $A|_K = \{f|_{K \cap \tau(A)} \mid f \in A\}$ if $K \cap \tau(A) \neq \emptyset$, and $A|_K = \emptyset$ if $K \cap \tau(A) = \emptyset$. Clearly $A|_K \subseteq D_{K \cap \tau(A)}$. We will write $A|_i$ instead of $A|_{\{i\}}$ for all $i \in T$.

By a *mereological system for direct products* we mean any mereological system with the domain $(\mathcal{X}_T, \mathcal{E})$. There can be many of them, dependending on the sets Θ and Δ. In this section we show the property of a *potential candidate for the universal mereological system for direct products.*

For every $I \subseteq T$, let $\pi_I : \mathcal{X}_T \to \mathcal{X}_I$, be mapping defined by: $\pi_I(A) = A|_I$. Every such mapping is called a *projection.*

Define $\Theta_T = \{\cup\} \cup \{\prod_{i \in I} \mid I \subseteq T\}$, $\Delta_T = \{\setminus\} \cup \{\pi_I \mid I \subseteq T\}$, and let $\eta_T : \mathcal{X}_T \to 2^{\mathcal{E}}$ be defined as follows: for every $A \in \mathcal{X}_T$ $\eta_T(A) = \bigcup_{i \in \tau(A)} \widehat{A|_i}$. For example: $\eta_T(\{(a,1),(b,2)\}) = \{\{a\},\{b\}\} \cup \{\{1\},\{2\}\} = \{\{a\},\{b\},\{1\},\{2\}\}$.

Consider the tuple $(\mathcal{X}_T, \mathcal{E}, \Theta_T, \Delta_T, \eta_T)$. Let $\sqsubseteq_T \subseteq \mathcal{X}_T \times \mathcal{X}_T$ be a relation defined by the equation (3) from the previous section.

Lemma 4. *1.* $(\mathcal{X}_T, \sqsubseteq_T)$ *is a bounded poset with* $\top = D_T$, $\bot = \emptyset$.
2. $\forall \mathcal{A} \subseteq \mathcal{X}_T$. $\overline{\mathrm{top}}^{\,2} \mathcal{A} = \bigcup_{A \in \overline{\mathrm{cmin}}\ \overline{\mathrm{ub}}\ \mathcal{A}} A$ *and* $\overline{\mathrm{bot}}^{\,2} \mathcal{A} = \bigcap_{A \in \overline{\mathrm{cmax}}\ \overline{\mathrm{lb}}\ \mathcal{A}} A|_J$, *where* $J = \bigcap_{A \in \overline{\mathrm{cmax}}\ \mathrm{lb}\ \mathcal{A}} \tau(A)$.
3. For every $\mathcal{A} \subseteq_T$, *the tuple* $((\mathcal{X}_T)\downarrow_{\mathcal{A}}, (\sqsubseteq_T)\downarrow_{\mathcal{A}})$ *is a grid of degree* $(2,2)$.

■

From Lemma 4 we can derive the following important result.

Theorem 1. *The 5-tuple* $(\mathcal{X}_T, \mathcal{E}, \Theta_T, \Delta_T, \eta_T)$ *is a complete, atomistic, and constructive mereological system.* ■

Proposition 1. *1. For every* $I \subseteq T$, $\mathcal{X}_I = \{A \mid A \sqsubseteq_I D_I\}$.

2. *For every* $A, B \in \mathcal{X}_T$, $A \sqsubseteq_T B \iff \tau(A) \subseteq \tau(B) \wedge A \subseteq B|_{\tau(A)}$.
3. $atoms_{\sqsubseteq_T}(\mathcal{X}_T) = \mathcal{E}$. ■

The formula from Proposition 1(2) *was used as a definition* in [11,12,13].

The fundamental principle behind a successful Tabular Expressions ([10,14]) specification technique is that most of relations may be described as $R = \bigcirc_{i\in I} R_i$, where $\bigcirc$ is an operation, or composition of operations, each R_i is easy to specify. A variety of operations was introduced and discussed (see [1,11,12,13,6]). We will not discuss most of them here due to lack of space. We will discuss only two operations denoted by $\uplus$ and $\otimes$. We shall show that the operator $\uplus$ corresponds to $\sqcup$, and that $\uplus$ could be defined in terms of more intuitive $\otimes$, where $\otimes$ corresponds to the well known join operator of Codd's relational data-base model [4].

Let $A, B \in \mathcal{X}_T$ and let $K = \tau(A) \cup \tau(B)$, $J = \tau(A) \cap \tau(B)$. We define the operations "$\uplus$", and "$\otimes$" as follows.

$$A \uplus B = \{f \mid \mathrm{dom}\, f = K \wedge ((f|_{\tau(A)} \in A \wedge f|_{\tau(B)\setminus\tau(A)} \in B|_{\tau(B)\setminus\tau(A)}) \vee ((f|_{\tau(B)} \in B \wedge f|_{\tau(A)\setminus\tau(B)} \in A|_{\tau(A)\setminus\tau(B)}))\},$$

$$A \otimes B = \{f \mid \mathrm{dom}\, f = K \wedge (f|_{\tau(A)} \in A \wedge f|_{\tau(B)} \in B)\}.$$

Let $A \subseteq D|_{\{1,3,5\}}$, $B \subseteq D|_{\{1,2,4\}}$. Then

$$A \uplus B = \{(x_1, x_2, x_3, x_4, x_5) \mid ((x_1, x_3, x_5) \in A \wedge (x_2, x_4) \in B|_{\{2,4\}}) \vee ((x_1, x_2, x_4) \in B \wedge (x_3, x_5) \in A|_{\{3,5\}})\},$$

$$A \otimes B = \{(x_1, x_2, x_3, x_4, x_5) \mid (x_1, x_3, x_5) \in A \wedge (x_1, x_2, x_4) \in B\}.$$

Let I be some index set, and let $\mathcal{A} = \{A_i \mid A_i \in \mathcal{P}_{\sqsubseteq}(D_T) \wedge i \in I\}$, $K = \bigcup_{i\in I} \tau(A_i)$, $J = \bigcap_{i\in I} \tau(A_i)$.

Let $\mathrm{Comp}_i\mathcal{A}$ be the set of all the *components*[3] of K that are NOT contained in $\tau(A_i)$.

For example if $I = \{1, 2\}$, $\tau(A_1)\setminus\tau(A_2) \neq \emptyset$ and $\tau(A_2)\setminus\tau(A_1) \neq \emptyset$ then there are three components of K generated by $\tau(A_1)$ and $\tau(A_2)$, namely $\tau(A_1)\cap\tau(A_2)$, $\tau(A_1) \setminus \tau(A_2)$, and $\tau(A_2) \setminus \tau(A_1)$, so $\mathrm{Comp}_1\{A_1, A_2\} = \{\tau(A_2) \setminus \tau(A_1)\}$, and $\mathrm{Comp}_2\{A_1, A_2\} = \{\tau(A_1) \setminus \tau(A_2)\}$.

We define the operations "$\biguplus_{i\in I}$" as:

$$\biguplus_{i\in I} A_i = \{f \mid \mathrm{dom}\, f = K \;\wedge\; \exists i \in I.\; (f|_{\tau(A)} \in A \;\wedge\; \forall C \in \mathrm{Comp}_i\mathcal{A}.\; f|_C \in \textstyle\bigcup_{j\neq i} A_j|_C)\}.$$

One can easily prove the following properties of the above operations.

Lemma 5 ([12]). $A \uplus B = (A \otimes B|_{\tau(B)\setminus\tau(A)}) \cup (B \otimes A|_{\tau(A)\setminus\tau(B)})$. ■

We may now formulate the second main result of this section.

Theorem 2. $\biguplus_{i\in I} A_i = \bigsqcup_{i\in I} A_i = \overline{\mathrm{top}}^{\,2}\{A_i \mid i \in I\}$ ■

---

[3] Let X be a set, $X_i \subseteq X$ for all $i \in I$. Define $X_i^0 = X_i$ and $X_i^1 = X \setminus X_i$. A nonempty set $A = \bigcap_{i\in I} X_i^{k_i}$, where $k_i = 0, 1$, is called a *component* of X generated by the sets X_i, $i \in I$. The components are disjoint and cover the entire set X (see [16]).

Unfortunately, it may happen that $A \uplus (B \uplus C) \neq (A \uplus B) \uplus C$. Consider $T = \{1,2\}$ and $D_T = \{a,b\} \times \{1,2\}$ and $A = \{a\}$, $B = \{1\}$ and $C = \{(b,2)\}$. We have $A \uplus (B \uplus C) = \{a\} \uplus (\{1\} \uplus \{(b,2)\}) = \{a\} \uplus \{(b,1),(b,2)\} = \{(a,1),(a,2),(b,1),(b,2)\}$, while $(A \uplus B) \uplus C = (\{a\} \uplus (\{1\}) \uplus \{(b,2)\} = \{(a,1)\} \uplus \{(b,2)\} = \{(a,1),(b,2)\}$.

We do not have any operational definition of "$\overline{\sqcap}$", in the style of "$\uplus$", and the properties of "$\overline{\sqcap}$" seem to occasionally be counter-intuitive. Consider again $D_T = \{a,b\} \times \{1,2\}$. One may prove that $\{(a,1),(b,1),(b,2)\} \sqcap \{(a,2),(b,1),(b,2)\} = \{b\}$, NOT equal to $\{(b,1),(b,2)\}$, as one might expect.

Some properties of "$\sqsubseteq$", and "$\otimes$" in the framework of tabular expressions but from the viewpoint of relational and cylindric algebras were analysed in [6,13].

5 Final Comment

The approach presented above could be seen as a refinement (for a specific application) of the Leonard-Goodman model [7,17]. We can define an overlaping operator "$\circ$" as $a \circ b \iff \eta(a) \cap \eta(b) \neq \emptyset$, and we can easily prove that the formulae $a \sqsubseteq b \;\Rightarrow\; \forall c.\ (c \circ a \Rightarrow c \circ b)$ and $a \circ b \iff \exists e.\forall d.\ (d \circ c \Rightarrow d \circ a \wedge d \circ b)$ do hold. We also can define a product and a sum (in the sense of [7,17]) as $a \cdot b = \overline{\text{bot}}\,\{a,b\}$ and $a + b = \overline{\text{top}}\,\{a,b\}$. A translation of the above results into Leśniewski's formalism (more or less in the style of [20]) is more complex but possible.

The approach presented above could also be applied to the relational database theory as there is a relationship between the operations considered in this paper and considered in [11,12,13] and the cylindric operations for relational databases as defined in [5,9].

In this paper we discussed only one mereological system for direct products (and consequently for heterogenous relations), a potential candidate for being universal. The approach allows us to create a variety of mereological systems, dependently on our needs. A family of possible operations is discussed in [12,13]. Recently the relation $\sqsubseteq$ from Section 4 was applied to define and to detect formality discrepancy between two Requirements Scenarios [15].

Acknowledgement

Hartmut Wedekind is thanked for pointing out the connection to Leonard-Goodman Calculus of Individuals.

References

1. R. Abraham, Evaluating Generalized Tabular Expressions in Software Documentation, CRL Report 346, McMaster University, Hamilton, Ontario, Canada, 1997. Available at http://www.crl.mcmaster.ca/SERG/serg.publications.html
2. T. A. Alspaugh, S. R. Faulk, K. Heninger-Britton, R. A. Parker, D. L. Parnas, J. E. Shore, Software Requirements for the A-7E Aircraft, NRL Memoramdum Report 3876, Naval Research Lab., Washington, DC, 1992.

3. C. Brink, W. Kahl, G. Schmidt (eds.), *Relational Methods in Computer Science*, Springer 1997.
4. E. F. Codd, A relational model of data for large shared data banks, *Comm. of the ACM*, 13 (1970) 377-388.
5. I. Düntsh, S. Mikulás, Cylindric structures and dependencies in data bases, *Theoretical Computer Science*, 269,1,2 (2001), 451-468.
6. J. Desharnais, R. Khédri, A. Mili, Towards a Uniform Relational Semantics for Tabular Expressions, Proc. RELMICS 98, Warsaw 1998.
7. N. Goodman, *The Structure of Appearance*, 3rd edition, D. Reidel, 1977 (first edition in 1951).
8. L. Henkin, J. D. Monk, A. Tarski, *Cylindric Algebras, Part I*, North Holland 1971.
9. T. Imielinski, W. Lipski, The relational model of data and cylindric algebras, *Journal of Computer and System Sciences*, 28 (1984), 80-102.
10. R. Janicki, Towards a Formal Semantics of Parnas Tables, *17th International Conference on Software Engineering* (*ICSE'95*), IEEE Computer Society, Seattle, WA, 1995, 231-240.
11. R. Janicki, On Formal Semantics of Tabular Expressions, CRL Report 355, McMaster University, Hamilton, Ontario, Canada 1997. Available at http://www.crl.mcmaster.ca/SERG/serg.publications.html
12. R. Janicki, Remarks on Mereology of Direct Products and Relations, in J. Desharnais (ed.), *Relational Methods in Computer Science*, Methodos Publ. 2002, to appear, early version avalable at http://www.crl.mcmaster.ca/SERG/serg.publications.html
13. R. Janicki, R. Khédri, On a Formal Semantics of Tabular Expressions, *Science of Computer Programming*, 39 (2001) 189-214.
14. R. Janicki, D. L. Parnas, J. Zucker, Tabular Representations in Relational Documents, in [3].
15. R. Khédri, Requirements Scenarios Formalization Technique: n Versions Towards One Good Version, Proc. of RELMICS'01, Genova 2001.
16. K. Kuratowski, A. Mostowski, *Set Theory*, North Holland, 1976.
17. H. Leonard, N. Goodman, The calculus of individuals and its uses, *Journal of Symbolic Logic*, 5 (1940), 45-55.
18. D. L. Parnas, Tabular representations of relations, CRL Report 260, McMaster University, Hamilton, Ontario, Canada 1992.
19. D. L. Parnas, J. Madey, Functional Documentation for Computer System Egineering, *Science of Computer Programming*, 24, 1 (1995), 41-61.
20. L. Polkowski, On Connection Synthesis via Rough Mereology, *Fundamenta Informaticae*, 46 (2001) 83-96.
21. L. Polkowski, A. Skowron, Rough Mereology; A New Paradigm for Approximate Reasoning, *Journal of Approximate Reasoning*, 15, 4 (1997), 316-333.
22. J. T. J. Strzednicki, V. F. Rickey (eds.), *Leśniewski's Systems*, Kluwer Academic, 1984.

An Optimal Lower Bound for Resolution with 2-Conjunctions*

Jan Johannsen and N.S. Narayanaswamy

Institut für Informatik, Ludwig-Maximilians-Universität München

Abstract. A lower bound is proved for refutations of certain clause sets in a generalization of Resolution that allows cuts on conjunctions of width 2. The hard clauses are the Tseitin graph formulas for a class of logarithmic degree expander graphs. The bound is optimal in the sense that it is truly exponential in the number of variables.

1 Introduction

The complexity of proof systems for propositional logic is a subject to which a lot of research effort has been successfully devoted recently. The interest in this topic comes from its relationship to the central questions of complexity theory (most notably, the $NP = co\text{-}NP$? problem) on the one hand, as well as to questions of efficiency of theorem proving and satisfiability testing algorithms on the other.

Much of this research has been concerned with the well-known Resolution proof system. The proof systems $Res(k)$, for $k \in \mathbb{N}$, were recently introduced by Krajíček [7] as a generalization of Resolution. Where Resolution only allows cuts on literals, i.e., variables and negated variables, $Res(k)$ allows cuts on conjunctions and disjunctions of k literals.

Besides the complexity-theoretical motivation, studying the complexity of $Res(k)$-proofs is also of interest from an algorithmic point of view. There is a well-known correspondence between Resolution proofs and the runs of search algorithms that branch on the value of a single variable, like Davis-Putnam or DLL procedures. In the same way, $Res(k)$-proofs corresponds to algorithms that branch on conditions that depend on the values of up to k variables. For $k = 2$, one such algorithm is Beigel and Eppstein's [3] algorithm for 3-coloring graphs. Also, the systems $Res(k)$ were recently used as a tool in a proof search algorithm for Resolution [1].

The first lower bounds for $Res(2)$ were given by Atserias et al. [2], who showed that the Weak Pigeonhole Principle $WPHP_n^{kn}$ requires $Res(2)$-proofs of size $\exp(n/(\log n)^{14})$, and random k-CNFs (of sufficiently large clause density) require $Res(2)$-proofs of size $\exp(n^{1/3}/(\log n)^2)$ with high probability. This paper also showed that $Res(2)$ is properly stronger than Resolution, by giving an example of a set of clauses having short $Res(2)$-refutations, but requiring superpolynomial Resolution refutations. The separation was improved to exponential

* Research supported by DFG grant no. Jo 291/2-2

K. Diks et al. (Eds): MFSC 2002, LNCS 2420, pp. 387–398, 2002.

by Atserias and Bonet [1] and Segerlind et al. [8], the latter more generally proved an exponential separation between $Res(k)$ and $Res(k+1)$ for every k.

All the known lower bounds for $Res(2)$ mentioned above are of the form $\exp(\Omega(N^{1-\epsilon}))$ for some ϵ, where N is the number of variables of the set of clauses in question. We prove the first lower bound for $Res(2)$ that is optimal in the sense that it is truly exponential in the number of variables, i.e., $\exp(\Omega(N))$.

Our hard examples are the Tseitin graph clauses [9] for a suitable expander graph of logarithmic degree. A lower bound for $Res(2)$-refutations of these clauses already follows from Ben-Sasson's [4] recent lower bound for proofs of the Tseitin clauses in much stronger bounded-depth Frege systems. Nevertheless, the lower bound that follows from his result for $Res(2)$-refutations of our clauses is of the form $\exp(\Omega(N^{1-\epsilon}))$, and thus does not achieve the same quality.

Our result is obtained by studying the effect of random restrictions on $Res(2)$-refutations, and their corresponding effect on random graphs. We show (in Section 3) that with high probability, a random restriction significantly reduces the width of small $Res(2)$-refutations. On the other hand, we identify (in Section 4) graphs for which the expansion remains large with high probability under a random restriction. Therefore there exists a graph with a $Res(2)$-refutation of the associated Tseitin tautologies whose width is smaller than the expansion, contradicting the known relationship between expansion and width [5].

2 Definitions and Results

A k-term is a conjunction of at most k literals, and a k-clause is a disjunction of k-terms. The proof system $Res(k)$ is defined as follows: The lines in a proof are k-clauses, and there are three inference rules:

$$\frac{A}{A \vee B} \qquad \text{(Weakening)}$$

$$\frac{A \vee T_1 \qquad B \vee T_2}{A \vee B \vee (T_1 \wedge T_2)} \qquad (\wedge\text{-Introduction})$$

$$\frac{A \vee (\ell_1 \wedge \ldots \ell_s) \qquad B \vee \bar{\ell}_1 \vee \ldots \vee \bar{\ell}_s}{A \vee B} \qquad \text{(Cut)}$$

where A and B are k-clauses, T_1 and T_2 are k-terms such that $(T_1 \wedge T_2)$ is again a k-term, $s \leq k$ and the ℓ_i are literals.

For a k-clause C, the width $w(C)$ is the number of variables occurring in C. The width $w(P)$ of a $Res(k)$-refutation P is the maximum of $w(C)$ over all k-clauses C in P.

A literal is called *free* in C, if it does not occur in a conjunction, i.e., it forms a 1-term of C. The *free width* of C, denoted by $fw(C)$ is the number of free literals in C.

Random Restrictions. A restriction ρ is a mapping of the set of variables to $\{0, 1, *\}$. For any set of variables, the probability distribution $\mathcal{R}_p$ of restrictions is defined as follows: A restriction $\rho \sim \mathcal{R}_p$ is picked by assigning independently to each variable the value $*$ with probability $1-p$, and each of the values $0, 1$ with probability $\frac{p}{2}$.

For a k-clause C and a restriction ρ, the k-clause $C\lceil_\rho$ is obtained by substituting $\rho(x)$ for x in C, for those variables x that occur in C with $\rho(x) \neq *$, and then simplifying the result.

For a *Res*(k)-refutation P, we obtain $P\lceil_\rho$ by replacing every k-clause C in P by $C\lceil_\rho$. If P is a *Res*(k)-refutation of some set of k-clauses F, $P\lceil_\rho$ is a *Res*(k)-refutation of $F\lceil_\rho$. We shall prove the following result below:

Theorem 1. *Let P be a Res(2)-refutation of size less than $\frac{1}{4}e^{L/512}$, and let $\rho \sim \mathcal{R}_{7/16}$. Then $w(P\lceil_\rho) < L$ with probability larger than $\frac{1}{2}$.*

Tseitin Clauses. The Tseitin formula for a graph G encodes the contradictory statement that there is an edge subgraph of G, i.e., a subgraph of G whose vertex set is the full vertex set of G, in which an odd number of vertices have odd degree, contradicting Eulers Theorem. To express this, a $0, 1$-valued vertex labeling function f is fixed that assigns 1 to an odd number of vertices, and it is claimed that for each vertex v, the parity of its degree is equal to $f(v)$.

Formally, let $G = (V, E)$ be a graph with $|V| = n$, and $f : V \to \{0, 1\}$ a function. The Tseitin clauses $\tau(G, f)$ is the set of clauses in the variables x_e for $e \in E$, consisting of the CNF of

$$\bigwedge_{v \in V} \bigoplus_{e \ni v} x_e \equiv f(v) .$$

The set $\tau(G, f)$ has $O(n \cdot 2^{\Delta(G)})$ clauses in at most $n \cdot \Delta(G)$ variables, of width at most $\Delta(G)$. Here $\Delta(G)$ denotes the maximum vertex degree in G. The function f is called *odd*, if $\sum_{v \in V} f(v)$ is odd. The following fact is well known [10].

Proposition 2. *$\tau(G, f)$ is unsatisfiable if G is connected and f is odd.*

The *expansion* $e(G)$ of a graph G is defined as

$$e(G) := \min\left\{ |E(U, V \setminus U)| ; \frac{n}{3} \leq |U| \leq \frac{2n}{3} \right\}.$$

The following result relating the expansion to the width of refutations of the Tseitin clauses was proved by Ben-Sasson and Wigderson [5]. Whereas they state it only for resolution refutations, it is easily verified that the proof works for any proof system in which every inference rule has at most two premises, as was already observed by Atserias et al. [2].

Theorem 3. *Let $G = (V, E)$ be a connected graph and $f : V \to \{0, 1\}$ odd. Then any Res(k)-refutation of $\tau(G, f)$ requires width at least $e(G)$.*

For a graph G and a restriction ρ, we define the graph $G\lceil_\rho := (V, E')$, where $E' := \{ e \in E ; \rho(x_e) = * \}$. In words, for a restriction ρ on the variables corresponding to the edges of G, $G\lceil_\rho$ is the graph obtained by discarding those edges that are assigned 0 or 1. The function $f\lceil_\rho$ is defined as

$$f\lceil_\rho(v) = f(v) \oplus \bigoplus_{\substack{e \ni v \\ \rho(e) \neq *}} \rho(x_e) \ .$$

i.e., $f\lceil_\rho$ is given by $f\lceil_\rho(v) = 1 - f(v)$ if an odd number of variables corresponding to edges incident on v are set to 1, and $f\lceil_\rho(v) = f(v)$ otherwise. Note that $f\lceil_\rho$ is odd iff f is, since every edge variable set to 1 by ρ flips the value of $f\lceil_\rho$ at 2 nodes. Clearly, it holds that $\tau(G, f)\lceil_\rho = \tau(G\lceil_\rho, f\lceil_\rho)$.

Definition 4. *We call a graph* $G = (V, E)$ useful, *if it satisfies the following three properties: for a restriction* $\rho \sim \mathcal{R}_{7/16}$,

1. $\Delta(G) \leq 4 \log n$
2. $G\lceil_\rho$ *is connected with probability at least* $\frac{3}{4}$.
3. $e(G\lceil_\rho) \geq \frac{1}{8} n \log n$ *with probability at least* $\frac{3}{4}$.

We shall prove the existence of a useful graph below. Our main result is the following:

Theorem 5. *Let* $G = (V, E)$ *be a useful graph, and* $f : V \to \{0, 1\}$ *an odd function. Any Res(2)-refutation of* $\tau(G, f)$ *requires size* $\exp(\Omega(n \log n))$.

Proof. Let P be a *Res*(2)-refutation of $\tau(G, f)$ of size $|P| < \frac{1}{4} e^{\frac{n \log n}{4096}}$. For a random restriction $\rho \sim \mathcal{R}_{7/16}$, we have

$$\Pr\big(w(P\lceil_\rho) < \frac{1}{8} n \log n\big) > \frac{1}{2},$$

by Theorem 1, and since G is useful, we have that

$$\Pr\big(e(G\lceil_\rho) \geq \frac{1}{8} n \log n \wedge G\lceil_\rho \text{ is connected}\big) \geq \frac{1}{2}$$

so there exists a restriction ρ such that $P\lceil_\rho$ is a refutation of $\tau(G\lceil_\rho, f\lceil_\rho)$ of width $w(P\lceil_\rho) < e(G\lceil_\rho)$, in contradiction to Theorem 3.

3 Width of Restricted Res(2)-Proofs

In this section we prove Theorem 1. To this end, we will prove the following.

Lemma 6. *For a 2-clause* C *with* $w(C) \geq L$, *and* $\rho \sim \mathcal{R}_{7/16}$,

$$\Pr_\rho\big(w(C\lceil_\rho) \geq L\big) \leq 2e^{-\frac{1}{512} L} \ .$$

The lemma implies Theorem 1 as follows: If a *Res*(2)-refutation P has less than S lines, then by the lemma and a union bound

$$\Pr_\rho\big(w(P\lceil_\rho) \geq L\big) \ \leq \sum_{\substack{C \in P \\ w(C) \geq L}} \Pr_\rho\big(w(C\lceil_\rho) \geq L\big) \ \leq \ S \cdot 2e^{-\frac{1}{512} L} \ ,$$

which is less than $\frac{1}{2}$ if $S < \frac{1}{4} e^{L/512}$.

To prove Lemma 6, we use the bound $\Pr_\rho(w(C\lceil_\rho) \geq L) \leq \Pr_\rho(C\lceil_\rho \neq 1)$. To upper bound the latter, we will apply a restriction $\rho_1 \sim \mathcal{R}_p$ to C, and then restrict the result $C\lceil_{\rho_1}$ by a further restriction $\rho_2 \sim \mathcal{R}_p$. The result $(C\lceil_{\rho_1})\lceil_{\rho_2}$, which we abbreviate by $C\lceil_{\rho_1,\rho_2}$, is the same as $C\lceil_\rho$ for a restriction $\rho \sim \mathcal{R}_{2p-p^2}$. For $p = \frac{1}{4}$, we obtain the required value $2p - p^2 = \frac{7}{16}$.

Without loss of generality, we can assume that no variable occurs both in a 2-term and in a free literal in C, due to the equivalences $x \vee (x \wedge y) = x$ and $x \vee (\bar{x} \wedge y) = x \vee y$.

We now define a graph Γ associated with C. The vertices of Γ are the variables of C, and there is an edge between two variables if they occur together in a term of C. To bound $\Pr_{\rho_1,\rho_2}(C\lceil_{\rho_1,\rho_2} \neq 1)$ we distinguish three cases according to the structure of Γ. Let m denote the maximal size of a matching in Γ.

Case 1: $fw(C) \geq \frac{L}{3}$, i.e., Γ has at least $\frac{L}{3}$ isolated vertices. We have

$$\Pr_{\rho_1,\rho_2}(C\lceil_{\rho_1,\rho_2} \neq 1) \leq \Pr_{\rho_1}(C\lceil_{\rho_1} \neq 1)$$

which is at most the probability that none of the free literals is set to 1 by ρ_1. Since these events are independent, in this case $\Pr_{\rho_1}(C\lceil_{\rho_1} \neq 1)$ is at most $(1 - \frac{p}{2})^{\frac{L}{3}} \leq e^{-\frac{p}{2}\frac{L}{3}} = e^{-\frac{1}{24}L}$.

Case 2: $m \geq \frac{L}{3}$, i.e., there are m pairwise variable-disjoint 2-terms in C. Again we bound

$$\Pr_{\rho_1,\rho_2}(C\lceil_{\rho_1,\rho_2} \neq 1) \leq \Pr_{\rho_1}(C\lceil_{\rho_1} \neq 1),$$

and since the events of these m terms being set to 1 by ρ_1 are independent, the latter is at most $(1 - \frac{p^2}{4})^{\frac{L}{3}} \leq e^{-\frac{p^2}{4}\frac{L}{3}} = e^{-\frac{1}{192}L}$.

Case 3: $fw(C) < \frac{L}{3}$ and $m < \frac{L}{3}$. This is the main case, where we actually use the fact that C is restricted twice. We obtain an upper bound on $\Pr_{\rho_1,\rho_2}(C\lceil_{\rho_1,\rho_2} \neq 1)$ by conditioning on the event that the first restriction creates many free literals. Let $\alpha = \frac{p(1-p)}{4}$, then we have

$$\begin{aligned}\Pr_{\rho_1,\rho_2}(C\lceil_{\rho_1,\rho_2} \neq 1) \leq\ & \Pr_{\rho_1,\rho_2}(C\lceil_{\rho_1,\rho_2} \neq 1 \wedge fw(C\lceil_{\rho_1}) \geq \alpha\frac{L}{3}) \\ & + \Pr_{\rho_1,\rho_2}(C\lceil_{\rho_1,\rho_2} \neq 1 \wedge fw(C\lceil_{\rho_1}) < \alpha\frac{L}{3}) \qquad (1)\end{aligned}$$

The first of these terms is easy to bound, since we can use the fact that $C\lceil_{\rho_1}$ has many free literals. We upper bound the second term by the probability that $C\lceil_{\rho_1}$ does not have many free literals.

The first term in equation (1) can be bounded as follows,

$$\begin{aligned}&\Pr_{\rho_1,\rho_2}(C\lceil_{\rho_1,\rho_2} \neq 1 \wedge fw(C\lceil_{\rho_1}) \geq \alpha\frac{L}{3}) \\ &\quad \leq \Pr_{\rho_1,\rho_2}(C\lceil_{\rho_1,\rho_2} \neq 1 \mid fw(C\lceil_{\rho_1}) \geq \alpha\frac{L}{3}) \leq (1 - \frac{p}{2})^{\alpha\frac{L}{3}} \leq e^{-\frac{p}{2}\alpha\frac{L}{3}} = e^{-\frac{1}{512}L}\end{aligned}$$

using the same analysis as in Case 1 for $C\lceil_{\rho_1}$.

The second term in (1) is bounded by

$$\Pr_{\rho_1}(C\lceil_{\rho_1}\neq 1 \wedge fw(C\lceil_{\rho_1}) < \alpha\frac{L}{3})$$

To upper bound this term, we consider a subclause C' of C such that for all

$$\Pr_{\rho_1}(C\lceil_{\rho_1}\neq 1 \wedge fw(C\lceil_{\rho_1}) < \alpha\frac{L}{3}) \leq \Pr_{\rho_1}(C'\lceil_{\rho_1}\neq 1 \wedge fw(C'\lceil_{\rho_1}) < \alpha\frac{L}{3}) \qquad (2)$$

C' is simpler and more structured than C, and we use its structure to upper bound $\Pr_{\rho_1}(C'\lceil_{\rho_1}\neq 1 \wedge fw(C'\lceil_{\rho_1}) < \alpha\frac{L}{3})$.

To define the subclause C', we now construct an edge subgraph Γ' of Γ, by choosing certain edges of Γ into Γ' as follows:

- Choose a maximum cardinality matching M in Γ, which is of size $m < \frac{L}{3}$.
- For each vertex u not in M, choose one edge $\{u, v\}$ from Γ, where v is a vertex in M.

The 2-clause C' is obtained by assigning to each edge of Γ' a unique 2-term of C. Note that an edge of Γ' can correspond to two distinct terms of C, e.g., the terms $x \wedge y$ and $\neg x \wedge \neg y$ correspond to the same edge.

We now claim a property of C' that establishes (2).

Claim. For any restriction ρ_1, $C\lceil_{\rho_1}\neq 1$ and $fw(C\lceil_{\rho_1}) < \alpha\frac{L}{3}$ implies that $C'\lceil_{\rho_1}\neq 1$ and $fw(C'\lceil_{\rho_1}) < \alpha\frac{L}{3}$.

Proof. Since C' is a subclause of C, $C\lceil_{\rho_1}\neq 1$ implies that $C'\lceil_{\rho_1}\neq 1$. We now show that $fw(C'\lceil_{\rho_1}) \leq fw(C\lceil_{\rho_1})$ for a restriction ρ_1 such that $C\lceil_{\rho_1}\neq 1$, which gives the statement of the claim.

For this purpose, let us consider a literal x that becomes free in $C'\lceil_{\rho_1}$. Then there must have been a 2-term $x \wedge y$ in C', such that $\rho_1(y) = 1$. Since $x \wedge y$ occurs in C also, x becomes also free in C. The only way for x not to be free in $C\lceil_{\rho_1}$ would be if $\bar{x}$ also became free in $C\lceil_{\rho_1}$, whence $C\lceil_{\rho_1}= 1$, which we assumed not to be the case. Thus if $C\lceil_{\rho_1}\neq 1$, then every free literal in $C'\lceil_{\rho_1}$ is also free in $C\lceil_{\rho_1}$.

To continue, we now use the bound

$$\Pr_{\rho_1}(C'\lceil_{\rho_1}\neq 1 \wedge fw(C'\lceil_{\rho_1}) < \alpha\frac{L}{3}) \leq \Pr_{\rho_1}(fw(C'\lceil_{\rho_1}) < \alpha\frac{L}{3} \mid C'\lceil_{\rho_1}\neq 1) .$$

To upper bound $\Pr_{\rho_1}(fw(C'\lceil_{\rho_1}) < \alpha\frac{L}{3} \mid C'\lceil_{\rho_1}\neq 1)$, note that all the vertices outside of M have degree 1 in Γ', and at least one vertex in each edge of M has degree 1 in Γ'. We define the set O as the set of vertices in Γ' of degree 1, except for edges in M where both ends have degree 1, in which case we put one of them into O arbitrarily. Thus every edge in M will have exactly one vertex in O.

Let $fw_O(C'\lceil_{\rho_1})$ denote the number of literals in C' corresponding to vertices of O that become free in $C'\lceil_{\rho_1}$. Clearly, $fw_O(C'\lceil_{\rho_1}) \leq fw(C'\lceil_{\rho_1})$. Moreover, since we are in the case where $m < \frac{L}{3}$ and $fw(C) < \frac{L}{3}$, we have $|O| > \frac{L}{3}$. Therefore it follows that

$$\Pr_{\rho_1}\left(fw_O(C'\lceil_{\rho_1}) < \alpha\frac{L}{3} \mid C'\lceil_{\rho_1} \neq 1\right) \leq \Pr_{\rho_1}\left(fw_O(C'\lceil_{\rho_1}) < \alpha|O| \mid C'\lceil_{\rho_1} \neq 1\right).$$

We will now prove an upper bound on this term. Let $O = \{x_1, \ldots x_{|O|}\}$. To each vertex x_i in O, let Y_i be the indicator variable of the event that x_i becomes free in $C'\lceil_{\rho_1}$, thus $Y := Y_1 + \ldots + Y_{|O|} = fw_O(C'\lceil_{\rho_1})$. Clearly, $\Pr(Y_i = 1) = \frac{p}{2}(1-p)$. By the linearity of expectation, $E(Y) = \frac{p}{2}(1-p)|O|$, and thus $fw_O(C'\lceil_{\rho_1}) < \alpha|O|$ is the event that $Y < \frac{1}{2}E(Y)$.

Nevertheless, we cannot use a Chernoff-Hoeffding bound to upper bound the probability of this event, as the random variables $Y_1, \ldots, Y_{|O|}$ are not independent. We obtain this upper bound below by using properties of Γ'.

Note that Γ' is a disconnected graph with m connected components, and each component is a star centered at a vertex in M. Let O_i be the set of vertices in O in the ith star, $d_i := |O_i|$, and let Z_i denote the number of variables in O_i that become free under ρ_1. Clearly, for a ρ_1 with $C'\lceil_{\rho_1} \neq 1$ we have $Y = Z_1 + \ldots + Z_m$, and $Z_1, \ldots, Z_m$ are independent random variables. This allows us to prove a Chernoff-like bound on the probability that Y is much smaller than expected.

Lemma 7. *Let Y be the random variable defined as above and $\mu = E(Y)$. Then for $0 < \delta < 1$,*

$$\Pr_{\rho_1}\left(Y < (1-\delta)\mu \mid C'\lceil_{\rho_1} \neq 1\right) \leq \left(\frac{p}{2}\right)^m e^{-\frac{\delta^2\mu}{2}}$$

Proof. The proof is by an analysis similar to the Chernoff bounds.

$$\begin{aligned} &\Pr_{\rho_1}\left(Y < (1-\delta)\mu \mid C'\lceil_{\rho_1} \neq 1\right) \\ &= \Pr_{\rho_1}(e^{-tY} > e^{-t(1-\delta)\mu} \mid C'\lceil_{\rho_1} \neq 1) \qquad \text{for any } t > 0 \\ &\leq \frac{E(e^{-tY} \mid C'\lceil_{\rho_1} \neq 1)}{e^{-t(1-\delta)\mu}} \qquad \text{by Markov's inequality} \qquad (3) \end{aligned}$$

To upper bound (3) we now evaluate $E(e^{-tY}C'\lceil_{\rho_1} \neq 1|)$. We know that $Y = \sum_{i=1}^m Z_i$ where $Z_1, \ldots, Z_m$ are independent random variables. Consequently,

$$\begin{aligned} E(e^{-tY} \mid C'\lceil_{\rho_1} \neq 1) &= E\left(e^{-t\sum_{i=1}^m Z_i} \mid C'\lceil_{\rho_1} \neq 1\right) \\ &= E\left(\prod_{i=1}^m e^{-tZ_i} \mid C'\lceil_{\rho_1} \neq 1\right) \\ &= \prod_{i=1}^m E(e^{-tZ_i} \mid C'\lceil_{\rho_1} \neq 1) \qquad (4) \end{aligned}$$

To upper bound (4) we evaluate $E(e^{-tZ_i} \mid C'\lceil_{\rho_1}\neq 1|)$. Recall that Z_i takes values in the range $\{0,\ldots,d_i\}$. Denote by F_i the set of variables in O_i that become free literals under ρ_1.

$$\begin{aligned} E(e^{-tZ_i} \mid C'\lceil_{\rho_1}\neq 1) &= \sum_{k=0}^{d_i} e^{-tk} \Pr_{\rho_1}(Z_i = k \mid |C'\lceil_{\rho_1}\neq 1) \\ &= \sum_{k=0}^{d_i} e^{-tk} \sum_{S\in\binom{O_i}{k}} \Pr_{\rho_1}(F_i = S \mid C'\lceil_{\rho_1}\neq 1) \end{aligned} \quad (5)$$

We now have to upper bound $\Pr_{\rho_1}(F_i = S \mid C'\lceil_{\rho_1}\neq 1)$ for a k element set S. The variables in S become free if $\rho_1(u) = *$ for every $u \in S$, they all form a conjunction with the same literal whose underlying variable is the center of the ith star, and this literal is set to 1 by ρ_1. This happens with probability at most $\frac{p}{2}(1-p)^k$. Then the literals corresponding to elements of $O_i \setminus S$ are not free if they are set to 0 by ρ_1. This happens with probability p^{d_i-k}, since they cannot be set to 1 due to the condition that $C'\lceil_{\rho_1}\neq 1)$. Consequently, $F_i = S$ with probability at most $\frac{p}{2}(1-p)^k p^{d_i-k}$. Substituting this into equation 5 we get

$$\begin{aligned} E(e^{-tZ_i} \mid C'\lceil_{\rho_1}\neq 1) &\leq \sum_{k=0}^{d_i} e^{-tk}\binom{d_i}{k}\frac{p}{2}(1-p)^k p^{d_i-k} \\ &= \frac{p}{2}((1-p)e^{-t}+p)^{d_i} \\ &= \frac{p}{2}(1+(1-p)(e^{-t}-1))^{d_i} \end{aligned}$$

Substituting this into equation 4 we get

$$\begin{aligned} &E(e^{-tY} \mid C'\lceil_{\rho_1}\neq 1) \\ &\quad = \prod_{i=1}^{m}\frac{p}{2}(1+(1-p)(e^{-t}-1))^{d_i} \\ &\quad = (\frac{p}{2})^m(1+(1-p)(e^{-t}-1))^{|O|} \\ &\quad \leq (\frac{p}{2})^m e^{(1-p)(e^{-t}-1)|O|} && \text{using } 1+x \leq e^x \\ &\quad = (\frac{p}{2})^m e^{\frac{2}{p}(e^{-t}-1)\mu} && \text{since } \mu = \tfrac{p(1-p)|O|}{2} \end{aligned}$$

Substituting this into (3), setting $t = \ln(\frac{1}{1-\delta})$, and using the fact that $(1-\delta)^{1-\delta} \geq e^{-\delta+\frac{\delta^2}{2}}$ for $0 < \delta < 1$, we get

$$\begin{aligned} \Pr_{\rho_1}(Y < (1-\delta)\mu \mid C'\lceil_{\rho_1}\neq 1) &\leq (\frac{p}{2})^m\frac{e^{\frac{2}{p}(e^{-t}-1)\mu}}{e^{-t(1-\delta)\mu}} \leq (\frac{p}{2})^m(\frac{e^{\frac{-2\delta}{p}}}{(1-\delta)^{1-\delta}})^{\mu} \\ &\leq (\frac{p}{2})^m e^{(\delta(1-\frac{2}{p})-\frac{\delta^2}{2})\mu} \leq (\frac{p}{2})^m e^{-\frac{\delta^2\mu}{2}} \end{aligned}$$

since $\delta(1-\frac{2}{p}) < 0$, hence the lemma is proved.

Thus, for $\delta = \frac{1}{2}$, we obtain an upper bound

$$\Pr_{\rho_1,\rho_2}\left(fw_O(C'\lceil_{\rho_1}) < \alpha|O| \mid C'\lceil_{\rho_1} \neq 1\right) < \left(\frac{p}{2}\right)^m e^{-\frac{\mu}{8}} \leq e^{-\frac{p(1-p)}{16}|O|} \leq e^{-\frac{1}{256}L},$$

using $\mu = \frac{p(1-p)}{2}|O|$ and $|O| > \frac{L}{3}$. Thus, the second term in (1) is at most $e^{-\frac{1}{256}L}$, and therefore, in case 3, we obtain

$$\Pr_{\rho_1,\rho_2}\left(C\lceil_{\rho_1,\rho_2} \neq 1\right) \leq e^{-\frac{1}{256}L} + e^{-\frac{1}{512}L} \leq 2e^{-\frac{1}{512}L}$$

which exceeds the bounds obtained in cases 1 and 2, thus the lemma is proven. □

4 Existence of *Useful* Graphs

Let $\mathcal{G}(n, q)$ be the distribution of random graphs on n vertices with edge probability q. We first study the connectivity and expansion of a graph from this distribution. We use the following well-known result concerning the connectivity of random graphs (see e.g. Bollobás' book [6] for a proof.)

Lemma 8. *If $qn - \log n \to \infty$ as $n \to \infty$, then for $G \sim \mathcal{G}(n,q)$, the probability of being connected tends to 1 as $n \to \infty$.*

Regarding the expansion of a random graph, we have the following result.

Lemma 9. *For $G \sim \mathcal{G}(n,q)$,*

$$\Pr\left(e(G) < \frac{qn^2}{9}\right) \leq 2^n e^{\frac{-qn^2}{36}}$$

Proof. For each set $U \subseteq V$, the expectation of $|E(U, V \setminus U)|$ is $q(n - |U|)|U|$. For a graph G, if $e(G) < (1-\delta)q\frac{2n^2}{9}$, then there is a $U \subset V$ with $\frac{n}{3} \leq |U| \leq \frac{2n}{3}$ such that $E(U, V \setminus U) \leq (1-\delta)q|U|(n - |U|)$. Consequently,

$$\begin{aligned}
\Pr\left(e(G) \leq (1-\delta)q\frac{2n^2}{9}\right) \\
&\leq \sum_{\frac{n}{3} \leq |U| \leq \frac{2n}{3}} \Pr\left(|E(U, V \setminus U)| \leq (1-\delta)q|U|(n-|U|)\right) \\
&\leq 2^n \max_{\frac{n}{3} \leq |U| \leq \frac{2n}{3}} \Pr\left(|E(U, V \setminus U)| \leq (1-\delta)q|U|(n-|U|)\right) \\
&\leq 2^n \max_{\frac{n}{3} \leq |U| \leq \frac{2n}{3}} e^{\frac{-\delta^2}{2}q(n-|U|)|U|}
\end{aligned}$$

where the last inequality follows by a Chernoff bound. For $\delta = \frac{1}{2}$, this yields the lemma, using the fact that $(n - |U|)|U|$ is lower bounded by $\frac{2n^2}{9}$.

We will now show that a useful graph G exists. To this end, we let $q := \frac{2\log n}{n}$, and we show that for sufficiently large n, a random graph $G \sim \mathcal{G}(n,q)$ is useful with positive probability. First, we study the degree of G.

Lemma 10. *For a random graph $G \sim \mathcal{G}(n,q)$, where $q = \dfrac{2\log n}{n}$,*

$$\Pr(\Delta(G) > 4\log n) \;\le\; n^{-\frac{1}{10}}$$

Proof. The expected degree of a vertex is $q(n-1) < 2\log n$, therefore by a Chernoff bound

$$\begin{aligned}\Pr(\Delta(G) > 4\log n) &\le \Pr(\Delta(G) > 2q(n-1)) \\ &\le n(\frac{e}{4})^{q(n-1)} = n2^{(\log e - 2)q(n-1)} \\ &\le n2^{-\frac{11}{10}\log n \frac{n-1}{n}} \le n^{-\frac{1}{10}}\end{aligned}$$

using the fact that $\log e - 2 < -\frac{11}{20}$.

Next, we need to study properties of G under a random restriction. Note that if G is randomly chosen from $\mathcal{G}(n,q)$ and $\rho \sim \mathcal{R}_p$ for any p, then $G\lceil_\rho$ is a random sample from $\mathcal{G}(n, q(1-p))$.

Lemma 11. *For a random graph $G \sim \mathcal{G}(n,q)$, where $q = \dfrac{2\log n}{n}$,*

$$\Pr_{\rho\sim\mathcal{R}_{7/16}} (G\lceil_\rho \textit{ is disconnected}) \ge \frac{1}{4}$$

with probability at most $\frac{1}{5}$ for sufficiently large n.

Proof. We define the random variable $D_G = \Pr_\rho(G\lceil_\rho \text{ is disconnected})$. Thus the probability we need to bound is $\Pr_G(D_G \ge \frac{1}{4})$. The expectation of D_G is

$$E(D_G) = \Pr_{G,\rho}(G\lceil_\rho \text{ is disconnected}) = \Pr_{G\sim\mathcal{G}(n,q')}(G \text{ is disconnected})$$

where $q' = \frac{9\log n}{8n}$. Since $q'n - \log n \to \infty$ as $n \to \infty$, this probability is less than $\frac{1}{20}$ for sufficiently large n, by Lemma 8. Therefore, for these n,

$$\Pr_G(D_G \ge \frac{1}{4}) \le \Pr_G(D_G \ge 5E(D_G)) \le \frac{1}{5}$$

by Markov's inequality.

Lemma 12. *For a random graph $G \sim \mathcal{G}(n,q)$, where $q = \dfrac{2\log n}{n}$,*

$$\Pr_{\rho\sim\mathcal{R}_{7/16}} (e(G\lceil_\rho) < \frac{1}{8}n\log n) \ge \frac{1}{4}$$

with probability at most 2^{-n+2} for sufficiently large n.

Proof. We define the random variable $X_G = \Pr_\rho(e(G\lceil_\rho) < \frac{1}{8}n\log n)$. Thus the probability we need to bound is $\Pr_G(X_G \geq \frac{1}{4})$. The expectation of X_G is

$$E(X_G) = \Pr_{G,\rho}\left(e(G\lceil_\rho) < \frac{1}{8}n\log n\right) = \Pr_{G\sim\mathcal{G}(n,q')}\left(e(G) < \frac{1}{8}n\log n\right)$$

where $q' = \frac{9\log n}{8n}$. Since $\frac{q'n^2}{9} = \frac{1}{8}n\log n$, Lemma 9 gives us the following bound on this probability

$$E(X_G) \leq 2^n e^{\frac{-q'n^2}{36}} \leq 2^n c^{\frac{-n\log n}{32}} \leq 2^{-n}$$

for sufficiently large n. Therefore we obtain

$$\Pr_G\left(X_G \geq \frac{1}{4}\right) \leq \Pr\left(X_G \geq 2^{-n+2}E(X_G)\right) \leq 2^{-n+2}$$

by Markov's inequality.

We now put the above calculations together to conclude that a useful graph exists.

Corollary 13. *For n sufficiently large, $G \sim \mathcal{G}(n,q)$, where $q = \dfrac{2\log n}{n}$, is useful with nonzero probability.*

Proof. By Lemmas 10, 11 and 12 and a union bound, the probability that G is not useful is bounded by $\frac{1}{5} + \frac{1}{n^{1/10}} + \frac{1}{2^{n-2}}$, which is strictly less than one for large n.

References

1. A. Atserias and M. L. Bonet. On the automatizability of resolution and related propositional proof systems. Technical Report ECCC TR02-010, Electronic Colloquium on Computational Complexity, 2002.
2. A. Atserias, M. L. Bonet, and J. L. Esteban. Lower bounds for the weak pigeonhole principle beyond resolution, 2002. To appear in *Information and Computation.* Preliminary Version in Proc. 28th International Colloquium on Automata Languages and Programming, 2001.
3. R. Beigel and D. Eppstein. 3-coloring in time $O(1.3446^n)$: a no-MIS algorithm. In *Proc. 36th IEEE Symposiom on Foundations of Computer Science*, pages 444–452, 1995.
4. E. Ben-Sasson. Hard examples for bounded-depth Frege. To appear in *Proc. 34th ACM Symposium on Theory of Computing*, 2002.
5. E. Ben-Sasson and A. Wigderson. Short proofs are narrow — resolution made simple. *Journal of the ACM*, 48:149–169, 2001. Preliminary Version in Proc. 31st Symposium on Theory of Computing, 1999.
6. B. Bollobás. *Random Graphs.* Academic Press, 1985.
7. J. Krajíček. On the weak pigeonhole principle. *Fundamenta Mathematicae*, 170:123–140, 2001.

8. N. Segerlind, S. R. Buss, and R. Impagliazzo. A switching lemma for small restrictions and lower bounds for k-DNF resolution. Submitted for publication, 2002.
9. G. Tseitin. On the complexity of derivation in propositional calculus. In A. O. Slisenko, editor, *Studies in Constructive Mathematics and Mathematical Logic, Part 2*, pages 115–125. Consultants Bureau, 1970.
10. A. Urquhart. Hard examples for resolution. *Journal of the ACM*, 34:209–219, 1987.

Improved Parameterized Algorithms for Planar Dominating Set*

Iyad A. Kanj** and Ljubomir Perković

DePaul University, School of CTI, 243 S. Wabash Avenue, Chicago, IL 60604,
ikanj@cs.depaul.edu, lperkovic@cs.depaul.edu.

Abstract. Recently, there has been a lot of interest and progress in lowering the worst-case time complexity for the PLANAR DOMINATING SET problem. In this paper, we present improved parameterized algorithms for the PLANAR DOMINATING SET problem. In particular, given a planar graph G and a positive integer k, we can compute a dominating set of size bounded by k or report that no such set exists in time $O(2^{27\sqrt{k}}n)$, where n is the number of vertices in G. Our algorithms induce a significant improvement over the previous best algorithm for the problem.

Keywords. planar dominating set, parameterized algorithms, NP-complete problems.

1 Introduction

In the PLANAR DOMINATING SET problem we are given a planar graph G and a positive integer k, and we are asked to compute a set of vertices D in G of cardinality bounded by k such that every vertex in G is either in D or adjacent to some vertex in D, or to report that no such set exists. It is well-known that the PLANAR DOMINATING SET problem is NP-hard [10], and hence cannot be solved in polynomial time unless P = NP.

Knowing the above fact does not obviate the need for solving the problem due to its theoretical and practical importance. To cope with the intractability of PLANAR DOMINATING SET several approaches have been proposed. The most famous among which are approximation algorithms [4] and parameterized algorithms [1,2,3,9].

A parameterized problem is said to be *fixed-parameter tractable* if given an instance I of the problem of length n, and a positive integer k (the parameter), the problem can be solved in time $O(f(k)n^c)$ for some function f independent of n and some positive constant c independent of k. The significance of the fixed-parameter tractability of a problem is that it indicates that the exponential growth in the running time of the algorithm for the problem is only limited to the parameter. Hence, if the parameter is relatively small (as it is the case in many applications), the running time of the algorithm can be moderate. Initiated by Downey and Fellows, the area of parameterized complexity has been receiving

* This work was supported in part by DePaul University Competitive Research Grant.
** The corresponding author.

K. Diks et al. (Eds): MFSC 2002, LNCS 2420, pp. 399–410, 2002.

a lot of attention recently, and many important NP-hard optimization problems have been shown to be fixed-parameter tractable [9]. To list few of these fixed-parameter tractable problems that have received a lot of attention, we mention the VERTEX COVER [8,11], PLANAR DOMINATING SET [1,2,3,9], and MAX-SAT [5,7] problems.

Whereas the DOMINATING SET problem on general graphs is W[2]-hard, and hence, not believed to be fixed-parameter tractable [9], the PLANAR DOMINATING SET is. Baker presented an algorithm that computes a maximum independent set in an l-outerplanar graph in time $O(8^l n)$ [4], and claimed that her algorithm for maximum independent set can be adapted to minimum dominating set without any increase in the running time. Baker's claim, together with the observation that a planar graph having a dominating set of size bounded by k can be at most $3k$-outerplanar, would imply an algorithm of running time $O(8^{3k}n)$ for the PLANAR DOMINATING SET problem [1]. As we shall explain in Section 3, Baker's algorithm as it is does not work for the minimum dominating set problem. However, it can be modified at the expense of increasing the running time to yield an algorithm of running time $O(27^{3k}n)$ for the PLANAR DOMINATING SET problem. Downey and Fellows presented an algorithm of running time $O(11^k n)$ for the problem [9]. However, it turned out that the algorithm was flawed. Their algorithm was recently modified and corrected to give an algorithm of running time $O(8^k n)$ [2]. Alber et al. were able to give an $O(4^{6\sqrt{34}\sqrt{k}}n)$ (approximately $O(2^{70\sqrt{k}}n)$) algorithm for the problem, thus significantly lowering down the exponent in the running time to yield the first algorithm for the problem with a sublinear exponent [1]. According to Cai and Juedes, PLANAR DOMINATING SET cannot be solved in time $O(2^{o(\sqrt{k})}n^c)$ unless 3-SAT $\in$ DTIME$(2^{o(n)})$ [6]. Hence, reducing the exponent in Alber et al's algorithm by more than a fractional constant seems unlikely. Although Alber et al.'s algorithm seems to have lowered the exponent significantly, the exponent is still large which makes the algorithm impractical, as was also observed by the authors in [1]. Even when k is moderately small ($k = 4$), the running time of the algorithm is computationally infeasible ($O(2^{140}n)$), and in such a case the $O(8^k n)$ algorithm is preferable.

In this paper we further pursue the efforts to lower down the running time of the algorithm for the PLANAR DOMINATING SET problem. Our first observation is on Baker's algorithm. We show how Baker's algorithm can be modified to deal with the PLANAR DOMINATING SET problem and a variation of the PLANAR DOMINATING SET problem that we call PLANAR RED-BLACK-WHITE DOMINATING SET. The second observation is on the size of the separator for the planar graph. By focusing on separating only large components, we are able to reduce the size of the separator established in [1] from $51k$ to $15k$. This directly induces a significant improvement on Alber et al's algorithm, giving an algorithm for PLANAR DOMINATING SET that runs in time $O(2^{33\sqrt{k}}n)$. We then show how the above observations, combined with a divide and conquer approach, give an algorithm of running time $O(2^{27\sqrt{k}}n)$ for the problem. This is again a significant improvement on Alber et al's algorithm. The exponent in our algorithm, though still not very practical, is much reduced.

2 Preliminaries and Background

In this section we present some basic definitions and results that are essential for the later discussion. Most of these definitions and results were given in [1], and the reader is referred to [1] for a more elaborate presentation.

Let $G = (V, E)$ be a planar graph given with an embedding in the plane. The *layer decomposition* of G with respect to the embedding, is a partitioning of V into disjoint layers $(L_1, \ldots, L_r)$ defined inductively as follows. Layer L_1 is the set of vertices that lie on the outer face of G, and layer L_i is the set of vertices that lie on the outer face of $G - \bigcup_{j=1}^{i-1} L_j$ for $1 < i \leq r$ (note that a layer may be disconnected). The graph G is called *r-outerplanar* if it has a layer decomposition consisting of r layers. If $r = 1$, G is simply said to be *outerplanar*. It is well-known that a layer decomposition of a planar graph G can be computed in linear time [1].

The statement of the following proposition was proved in [1] and can be easily verified by the reader.

Proposition 1. *([1]) If a planar graph G has a dominating set of size bounded by k, then any layer decomposition of G consists of at most $3k$ layers.*

A *separator* in a graph G is a set of vertices S whose removal disconnects G. Given a dominating set D of G of size bounded by k, it was shown in [1] how, for every layer L_i, one can construct a separator consisting of vertices in L_{i-1}, L_i, and L_{i+1}, that separates layer L_{i-1} from layer L_{i+2}. We briefly mention how this can be done and the reader is referred to [1] for further details.

For a layer L_i of G we associate three sets of vertices. The set of *upper triples*, *middle triples*, and *lower triples*. Let C be a non-empty component of layer L_{i+1}. Denote by $B(C)$ the unique shortest cycle in layer L_i that encloses C, and by $B(B(C))$ the unique shortest cycle in layer L_{i-1} that encloses $B(C)$.

Let x be a vertex in $D \cap L_{i-1}$ that has neighbors x_1 and x_2 on $B(C)$. Let y and z be the outermost neighbors of x on $B(C)$ in the walk that starts at x_1 and ends at x_2 going around x and visiting all its neighbors in $B(C)$. Then:

Definition 1. *The triple of vertices $\{x, y, z\}$ is an upper triple for layer L_i associated with x and layer component C.*

Let x be a vertex in $D \cap L_i$ that has neighbors y and z on $B(C)$ then:

Definition 2. *The triple of vertices $\{x, y, z\}$ is called a middle triple for layer L_i associated with x and layer component C.*

Let x be a vertex in $D \cap L_{i+1}$ whose boundary cycle $B(\{x\})$ encloses C. Let y and z be the two neighbors of x on $B(\{x\})$ such that the path from y to z is shortest, and such that C is enclosed by the region formed by the edges (x, y), (x, z), and the path between y and z. Then:

Definition 3. *The triple of vertices $\{x, y, z\}$ is called a lower triple for layer L_i associated with x and layer component C.*

It was shown in [1] that the set of upper, middle, and lower triples associated with a non-empty layer component C of layer L_{i+1} separates the layer components inside C from layer L_{i-1}. In particular, if we let S_i be the set of upper triples, middle triples, and lower triples of layer L_i, then it was shown in [1] that S_i is a separator that separates layer L_{i-1} from layer L_{i+2}. Let $d_j = |D \cap L_j|$ for $j = 1, \ldots, r$, and c_{i+1} be the number of non-empty components of layer L_{i+1}. It was also shown in [1] that the number of vertices in the set of upper triples, middle triples, and lower triples for layer L_i, is bounded by $5d_{i-1}+4c_{i+1}$, $5d_i+4c_{i+1}$, and $5d_{i+1}+4c_{i+1}$, respectively. Let $S = (S_1, \ldots, S_r)$, then S is called a *layerwise separator* of G. Using the above bounds, it was shown in [1] that $|S| \le |S_1| + \ldots + |S_r| \le 51k$. Assume, in general, that we are able to derive an upper bound of $h(k)$ on the size of S for some function h. From the separator S, we can construct three families each consisting of disjoint separators. The families are $F_1 = S_1 \cup S_4 \cup S_7 \cup \ldots$, $F_2 = S_2 \cup S_5 \cup S_8 \cup \ldots$, and $F_3 = S_3 \cup S_6 \cup S_9 \cup \ldots$. One of the these families, assume F_1 without loss of generality, must have size bounded by $h(k)/3$.

Based on the family of separators F_1, one can find a moderately small subset of separators of F_1, $(S_{i_1}, S_{i_2}, \ldots, S_{i_q})$, that separates the graph into *chunks* each containing a small number of consecutive layers. We follow the same notation of [3] and we call such a separator a *partial layerwise separator*. We do this in a similar fashion to that of [1,3]. However, there is a little, yet important, difference. Instead of bounding the size of $max\{|S_{i_j}| \mid 1 \le j \le q\}$, we are going to bound $\sum_{j=1}^{q} |S_{i_j}|$. Let $c' > 0$, and consider the subfamilies F_1^i, $i = 1, \ldots, \lceil c'\sqrt{k} \rceil$ of separators of F_1, where $F_1^i = S_{3i-2} \cup S_{3i-2+3\lceil c'\sqrt{k}\rceil} \cup S_{3i-2+6\lceil c'\sqrt{k}\rceil} \cup \ldots$. Note that the subfamilies F_1^i, $i = 1, \ldots, \lceil c'\sqrt{k} \rceil$ are mutually disjoint, and each consists of mutually disjoint separators. Since the size of F_1 is bounded by $h(k)/3$, and since F_1 was partitioned into $\lceil c'\sqrt{k} \rceil$ subfamilies, at least one of these subfamilies F_1^j must have size bounded by $h(k)/(3c'\sqrt{k})$. Moreover, any two consecutive separators in F_1^j are separated by at most $3c'\sqrt{k} + 9$ layers. Thus, we have constructed a partial layerwise separator $F_1^j = (S_{i_1}, S_{i_2}, \ldots, S_{i_q})$ of total size bounded by $h(k)/(3c'\sqrt{k})$ that separates the graph into chunks each containing at most $3c'\sqrt{k} + 9$ layers. Note that so far, we have showed the existence of a partial layerwise separator satisfying the above conditions, given that G has a dominating set of size bounded by k. However, we have not specified how a separator satisfying these properties can be constructed. Basically, once the separator S has been constructed, constructing the partial layerwise separator becomes a straightforward task. A layerwise separator S satisfying the above bounds can be constructed in time $O(\sqrt{k}n)$ by finding minimum-size separators for consecutive layers using maximum flow algorithms, as was observed and explained in [1,3]. Thus we have the following proposition.

Proposition 2. *Let G be a planar graph having a dominating set of size bounded by k, and having a layerwise separator of size bounded by $h(k)$ for some function h. Then for any constant $c' > 0$, we can construct a partial layerwise separator $S = (S_{i_1}, \ldots, S_{i_q})$ such that $\sum_{i=1}^{q} |S_i| \le h(k)/(3c'\sqrt{k})$, and that separates* *the graph into chunks* $(G_{i_0}, G_{i_1}, \ldots, G_{i_q})$ *each having at most* $3c'\sqrt{k}+9$ *layers. Moreover, this construction can be done in time* $O(\sqrt{k}n)$.

Remark. Note that if G has a dominating set of size bounded by k, then as shown in [1], the graph must have a layerwise separator of size bounded by $51k$ (we shall improve on this bound later). So when we construct a layerwise separator S, if $|S| > 51k$, then necessarily the graph does not have a dominating set of size bounded by k, and hence, the answer to the problem instance is negative.

3 A Note on Baker's Algorithm

Baker, in her seminal work [4], gave a general technique to devise approximation schemes for the maximum independent set problem on planar graphs. Her technique is based on a dynamic programming algorithm that for a planar graph of outerplanarity l, computes a maximum independent set in time $O(8^l n)$, where n is the number of vertices in the graph [4]. Baker also mentions in her paper that the $O(8^l n)$ algorithm, with a slight modification and without any increase in its running time, can be applied to solve the minimum dominating set problem on planar graphs ([4], pages 175 and 176-177). If Baker's algorithm works, then her algorithm would imply that the Planar Dominating Set problem is solvable in time $O(8^{3k} n)$, as observed in [1]. However, we do believe that Baker's algorithm in its current form cannot be applied to solve the minimum dominating set problem as we shall explain below. For the sake of simplicity, let us consider only the case when the graph is outerplanar. The discussion carries on to the case of an l-outerplanar graph for any l.

The main idea behind Baker's algorithm is to represent the structure of the graph by a tree with labeled nodes, where each node in the tree labeled (u, v) corresponds to a subgraph called a "slice" with boundary nodes u and v. The maximum independent set for the graph is computed in a bottom-up fashion using a dynamic programming approach that starts by computing the maximum independent sets for the leaves of the tree based on the status of the boundary nodes (i.e, whether each boundary node is in the maximum independent set of the slice or not), and then computes the maximum independent set for each node in the tree by merging the compatible maximum independent sets for its children based on the status of the boundary nodes. This computation is carried out until the maximum independent set for the root of the tree, which represents the whole graph, has been computed.

Whereas this approach works for maximum independent set, it does not work for minimum dominating set in the sense that it does not always give a minimum dominating set[1]. The reason is that the minimum dominating set problem, in its standard definition, is not amenable to the dynamic programming approach.

[1] This different behavior of the dominating set problem was also observed in [1,2,3] which lead to considering constrained versions of the Dominating Set problem as we do here.

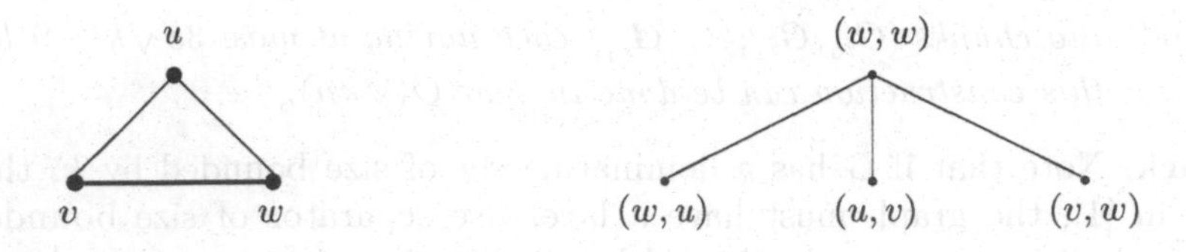

Fig. 1. A graph on which Baker's algorithm does not work

Figure 1 shows an outerplanar graph and its structure tree constructed by Baker's algorithm. The graph has a minimum dominating set of size 1, however, Baker's algorithm would give a dominating set of size 2 for this graph.

Baker's algorithm for maximum independent set can be modified to deal with the minimum dominating set problem. Instead of letting every vertex have two status possibilities: 0 meaning that the vertex is not in the minimum dominating set and 1 meaning that the vertex is in the minimum dominating set, we allow every vertex to have three status possibilities. The status 0 means that a vertex is not in the minimum dominating set and it may or may not be dominated, status 1 means that a vertex is in the dominating set, and status 2 means that the vertex is not in the dominating set but is dominated. Now we allow a set of vertices for a given slice not to be a dominating set for the slice, since its boundary nodes may need to be covered by the vertices in a different slice. However, when merging two slices we need to be careful. The reason is that we want to make sure that the set of vertices computed covers all the nodes in the slice, except possibly the boundary nodes (i.e., all other nodes except possibly the boundary nodes should have status 1 or 2). Thus, the set that we now compute for a certain slice under certain constraints is no longer a dominating set, but is basically a minimum set of nodes under the given constraints that covers all the nodes of the slice except possibly the boundary nodes.

The new rules for the computation can be easily written. The only major difference between the new computation and the old one is that status 0 for a vertex in a given slice is also compatible with status 2 for the same vertex in another slice. The details of the computation are skipped due to shortage of space.

If we modify Baker's algorithm as mentioned above, the algorithm works. Since now each node has three status bits, i.e., $r = 3$, the algorithm runs in time $O(r^{3l}n) = O(27^l n)$ ([4], page 175).

Theorem 1. *For an l-outerplanar graph, a minimum dominating set can be computed in time $O(27^l n)$, where n is the number of vertices in the graph.*

Define a variant of the minimum dominating set problem as follows.

> Planar Red-Black-White Dominating Set (Planar RBW-DS) problem. Given an undirected planar graph $G = (V, E)$ where $V = R \cup B \cup W$, compute a minimum dominating set consisting of vertices in $B \cup W$ that dominates all the vertices in $R \cup B$.

The above algorithm for minimum dominating set can be used to solve the PLANAR RBW-DS problem as follows. If a vertex is BLACK, then its status bit can take the values 0, 1, or 2. If the vertex is RED, then its status bit is restricted to be either 0 or 2 throughout the computation. If the vertex is WHITE, then its status bit is restricted to be 1 or 2 throughout the computation. The remaining computation is carried out in the same way as above. The details are omitted for lack of space. We have the following theorem.

Theorem 2. *For an l-outerplanar graph, the* PLANAR RBW-DS *problem can be solved in time $O(27^l n)$, where n is the number of vertices in the graph.*

4 Separation of Non-shallow Components

Let G be an r-outerplanar graph, and let $\mathcal{L} = (L_1, \ldots, L_r)$ be a layer decomposition of G. Throughout this section, we shall assume that G has a dominating set D of size bounded by k. We can also assume that $k \geq 2$, since there is a trivial linear time algorithm (in n and k) for the problem when $k = 1$. We define the *depth* of a connected component of layer L_i to be the number of layers in the interior of it. We call a component of layer L_i *deep* if its depth is greater than $\lceil\sqrt{k}\rceil + 1$, *medium* if its depth is exactly $\lceil\sqrt{k}\rceil + 1$, and *shallow* otherwise.

With the layer decomposition $\mathcal{L}$ of a connected graph G, we associate a tree $T_{\mathcal{L}}$ whose nodes correspond to the deep, medium, and shallow components of layers $L_1, \ldots, L_r$, and whose edges are defined as follows: the node v in $T_{\mathcal{L}}$ corresponding to a component C of layer L_i will have a parent edge to the node u in $T_{\mathcal{L}}$ corresponding to the unique component of layer L_{i-1} enclosing C. To simplify our discussion, we extend the definition of deep, medium, and shallow from the connected components of $\mathcal{L}$ to the corresponding nodes in $T_{\mathcal{L}}$. For a non-root node v we denote by $p(v)$ the parent of v in $T_{\mathcal{L}}$, and by $\mathcal{L}(v)$ the layer component corresponding to v in $\mathcal{L}$. For a deep node v in $T_{\mathcal{L}}$, we denote by $c(v)$ the set of deep and medium children of v in $T_{\mathcal{L}}$. Let m be the number of medium components in $\mathcal{L}$ (or alternatively, the medium nodes in $T_{\mathcal{L}}$). We have the following proposition.

Proposition 3.
(a) For a medium node v, the number of vertices in D that are in the interior of the component
$\mathcal{L}(v)$ is at least $\sqrt{k}/3$.
(b) $m \leq 3\sqrt{k}$.

Proof. Part (a) follows from Proposition 1 and the fact that a medium layer component contains $\lceil\sqrt{k}\rceil+1$ layers in its interior. Since the medium components are mutually disjoint, and G has a dominating set of size bounded by k, from part (a) it follows that the number or medium components, and hence medium nodes, m is bounded by $3\sqrt{k}$.

Definition 4. *A deep node v of $T_{\mathcal{L}}$ is said to be a complex node if it has at least three deep or medium children, otherwise v is said to be a simple node. A* *deep node v is said to be bad if the parent of v has at least three deep or medium grandchildren, otherwise v is said to be good.*

Let *Complex* be the set of complex nodes, *Simple* the set of simple nodes, *Bad* the set of bad nodes, and *Good* the set of good nodes in $T_{\mathcal{L}}$. We have the following proposition whose proof is purely combinatorial. We skip the proof due to lack of space.

Proposition 4.
(a) $|Complex| \leq \lfloor (m-1)/2 \rfloor$,
(b) $\sum_{v \in Complex} |c(v)| \leq \lfloor 3/2(m-1) \rfloor$,
(c) $\sum_{v \in Bad} |c(v)| \leq 3(m-1)$.

For every $v \in T_{\mathcal{L}}$, let d_v be the number of vertices of D in layer component $\mathcal{L}(v)$. We will construct separators that separate the non-shallow (i.e., deep or medium) components of $\mathcal{L}$. We will speak of lower, middle, and upper triples for the deep nodes of $T_{\mathcal{L}}$. These triples are constructed in the same way mentioned in Section 2. For instance, when speaking of a middle triple for a deep node v in $T_{\mathcal{L}}$, we mean basically a triple that is associated with a vertex a in $D \cap \mathcal{L}(v)$ and a non-shallow component corresponding to a child w of v. Similarly, an upper triple for a deep node v in $T_{\mathcal{L}}$ is associated with a vertex a in $D \cap \mathcal{L}(u)$, where u is the parent of v, and a component corresponding to a non-shallow child of v in $T_{\mathcal{L}}$. A lower triple for a deep node v associates a vertex a in $D \cap \mathcal{L}(w)$, where w is a child of v, and a component corresponding to a non-shallow sibling of w in $T_{\mathcal{L}}$. Note that a lower triple will associate a vertex a in D in a shallow component corresponding to node w in $T_{\mathcal{L}}$ with a non-shallow component that is a sibling of w in $T_{\mathcal{L}}$. This needs to be done in order to ensure that the set of lower, middle, and upper triples separates the non-shallow components of $\mathcal{L}$.

The upper, middle, and lower triples for the deep nodes of $T_{\mathcal{L}}$ are defined and constructed in the same way mentioned in Section 2, and hence, they are the upper, middle, and lower triples associated with the vertices of D and the non-shallow components of G. Let S_i be the set of vertices in the upper, middle, and lower triples in G associated with every deep node v in $T_{\mathcal{L}}$ such that $\mathcal{L}(v)$ is a component of layer L_i, then one can prove the following proposition by a similar argument to that in [1].

Proposition 5. *Let S_i be as defined above. Then S_i separates the non-shallow components of layer L_{i-1} from those of layer L_{i+2}.*

If we let $S = (S_1, \ldots, S_r)$, then S is a layerwise separator for the non-shallow components of G. We will derive an upper bound on the number of vertices in S by deriving upper bounds on the number of vertices in the upper, middle, and lower triples for the deep nodes in $T_{\mathcal{L}}$.

4.1 Middle Triples

If v is a simple node, i.e. v has at most two deep or medium children, then one can easily see that M_v, the number of vertices involved in the middle triples at node v, is not larger than $5d_v$. If v is a complex node, then we use the inequality mentioned in Section 2 to bound the number of vertices included in the middle triples namely: $M_v \le 5d_v + 4|c(v)|$ (note that by the definition of $c(v)$, the shallow components are excluded in this counting). Therefore, the total number of vertices included in all middle triples, over all deep (non-root) nodes of $T_{\mathcal{L}}$ is bounded by

$$\begin{aligned}\sum_{v \text{ deep}} M_v &= \sum_{v \in Simple} M_v + \sum_{v \in Complex} M_v \\ &\le \sum_{v \in Simple} 5d_v + \sum_{v \in Complex} (5d_v + 4|c(v)|) \\ &\le (\sum_{v \in Simple \cup Complex} 5d_v) + (4 \sum_{v \in Complex} |c(v)|) \\ &\le 5(k - m\sqrt{k}/3) + 4\lfloor 3/2(m-1) \rfloor \\ &\le 5k - 5m\sqrt{k}/3 + 6m - 6\end{aligned}$$

The second-to-last inequality follows from part (b) of Proposition 4. Also note that by part (a) of Proposition 3, each medium component contains at least $\sqrt{k}/3$ vertices in D that were not included in the counting. Thus, at least $m\sqrt{k}/3$ vertices in D can be subtracted from k in the counting.

4.2 Lower Triples

Let $child(v)$ be the set of all children of node v (including the shallow nodes). If v is a simple node, then L_v, the number of vertices involved in the lower triples, is bounded by $5\sum_{w \in child(v)} d_w$. If v is a complex node, then $L_v \le 5\sum_{w \in child(v)} d_w + 4|c(v)|$. Therefore, the total number of vertices included in all lower triples, over all deep (non-root) nodes of $T_{\mathcal{L}}$ is bounded by

$$\begin{aligned}\sum_{v \text{ deep}} L_v &= \sum_{v \in Simple} L_v + \sum_{v \in Complex} L_v \\ &\le 5(k - m\sqrt{k}/3) + 4 \sum_{v \in Complex} |c(v)| \\ &\le 5k - 5m\sqrt{k}/3 + 6m - 6\end{aligned}$$

4.3 Upper Triples

In order to bound U_v, the number of vertices involved in the upper triples at v, we need to be a bit more careful. Let u be the parent of v.

If v is a good node, then all deep children of u are good. In this case we have $\sum_{v \in c(u)} U_v \le 5d_u$. Suppose now v is bad, then again, all deep children of u are bad. We can bound $\sum_{v \in c(u)} U_v$ by $5d_u + 4\sum_{v \in c(u)} |c(v)|$. Therefore, the total number of vertices included in all upper triples, over all deep (non-root) nodes of $T_{\mathcal{L}}$ is bounded by

$$\begin{aligned}\sum_{v \text{ deep}} U_v &= \sum_{v \in Good} U_v + \sum_{v \in Bad} U_v \\ &\leq 5 \sum_{u \text{ deep}} d_u + 4 \sum_{v \in Bad} |c(v)| \\ &\leq 5k - 5m\sqrt{k}/3 + 12m - 12\end{aligned}$$

The last inequality follows from part (c) in Proposition 4. It follows from the above that

$$\begin{aligned}|S| \leq |S_1| + \ldots + |S_r| &\leq \sum_{v \text{ deep}} (U_v + M_v + L_v) \\ &\leq 15k - 5m\sqrt{k} + 24m - 24 \leq 15k + 34\sqrt{k}\end{aligned}$$

for $k \geq 2$. Based on this layerwise separator that separates the non-shallow components of G, we can construct a partial layerwise separator $(S_{i_1}, \ldots, S_{i_q})$ that separates the graph into chunks $(G_{i_0}, G_{i_1}, \ldots, G_{i_q})$. However, since the shallow components were not necessarily separated from the non-shallow components, a shallow component in chunk G_{i_j} may be connected to a non-shallow component in chunk G_{i_j}, or to a non-shallow component in chunk $G_{i_{j-1}}$, but not both (since the non-shallow components have been separated). Now if a shallow component is connected to a non-shallow component in the same chunk, the depth of the chunk does not increase. If it is connected to a non-shallow component in the above chunk, the depth of the above chunk might increase by at most $\lceil\sqrt{k}\rceil + 1$ layers. Thus, we have a dual proposition to Proposition 2.

Proposition 6. *Let G be a planar graph having a dominating set of size k. Suppose G has a layerwise separator of size bounded by $15k + 34\sqrt{k}$ that separates the non-shallow components of G. Then for any constant $c' > 0$, we can construct a partial layerwise separator for G of size bounded by $5\sqrt{k}/3c' + 12/c'$ that separates the graph into chunks each having at most $(3c'+1)\sqrt{k}+11$ layers. Moreover, this construction can be done in time $O(\sqrt{k}n)$.*

Proposition 6 implies a direct improvement on Alber et al.'s $O(2^{70\sqrt{k}}n)$ algorithm for Planar Dominating Set ([1]) as follows. Since each chunk has depth bounded by $(3c' + 1)\sqrt{k} + 11$, the treewidth of each chunk is bounded by $(9c' + 3)\sqrt{k} + 33$. Since the separator of the graph has size $5\sqrt{k}/c' + 12/c'$, we conclude that the treewidth of G, $tw(G)$, is bounded by $5\sqrt{k}/c' + (9c' + 3)\sqrt{k} + 12/c' + 33$. Choosing $c' = \sqrt{5}/3$, we get $tw(G) \leq 33\sqrt{k}/2 + 50$. This implies an algorithm of running time $O(4^{33\sqrt{k}/2}n) = O(2^{33\sqrt{k}}n)$ for Planar Dominating Set.

Theorem 3. *The* Planar Dominating Set *problem can be solved in time $O(2^{33\sqrt{k}}n)$.*

In the next section, we will show how a better algorithm can be obtained for the problem.

5 The Algorithm

In this section we show how to design a more efficient algorithm for PLANAR DOMINATING SET. From the previous section, we can assume that we have constructed a partial layerwise separator $S = (S_{i_1}, \ldots, S_{i_q})$ for G of size bounded by $5\sqrt{k}/3c' + 12/c'$ that separates G into chunks $(G_{i_0}, G_{i_1}, \ldots, G_{i_q})$ each having at most $(3c'+1)\sqrt{k}+11$ layers, where c' is a constant to be specified later. If we fail to construct such a separator, we know that G does not have a dominating set of size bounded by k, and hence, the answer to the problem instance is negative.

The basic idea behind the algorithm is to apply a simple divide and conquer strategy by removing the vertices in the separator and splitting the graph into chunks, then computing a minimum dominating set for each chunk using the algorithm for PLANAR RBW-DS. To do this, for each vertex v in the separator we will "guess" whether v is in the minimum dominating set for G or not (basically, what we mean by guessing is enumerating all sequences corresponding to the different possibilities). For each guess of all the vertices in the separator, we will solve the corresponding instance with respect to that guess. However, when we guess a status for a certain vertex, the vertex cannot be simply removed from the graph. The effect of the status of the guessed vertex needs to be reflected in the resulting graph. If we guess v to be in the minimum dominating set, then we are going to include v in the minimum dominating set, remove v from G, and color all its neighbors WHITE to reflect that these neighbors are now dominated. If v is guessed to be outside the minimum dominating set, then v has to be dominated. If v belongs to the separator that separates chunks $G_{i_{j-1}}$ and G_{i_j}, then v can be either dominated by its neighbors in chunk $G_{i_{j-1}}$ or those in chunk G_{i_j}. We will color v RED to indicate that v has to be dominated but cannot be included in the dominating set, and in one case append it to chunk $G_{i_{j-1}}$, and in the other to chunk G_{i_j}, each time removing the edges that connect v to its neighbors in the other chunk. Initially all vertices in G are colored BLACK. After guessing each vertex in the separator and updating the graph accordingly, the instance becomes an instance of the the PLANAR RBW-DS problem. We run the PLANAR RBW-DS algorithm to compute a minimum dominating set for each chunk, and hence for G under the restriction imposed by the guessed vertices in the separator. A minimum dominating set for the graph is then computed by taking the dominating set of minimum size over all possible guesses. Since we are enumerating all possible guesses, we are certain that the algorithm will determine a minimum dominating set for G. If this minimum dominating set of G has size bounded by k the answer to the problem instance is positive, otherwise the answer is negative.

To analyze the running time of the algorithm, let s be the size of the partial layerwise separator S. We need to compute the number of sequences that the algorithm enumerates. For each vertex $v \in S_{i_j}$, there are three possibilities: v is in the minimum dominating set, v is out and dominated by vertices in $G_{i_{j-1}}$, or v is out and dominated by vertices in G_{i_j}. It follows that the total number of sequences that need to be enumerated is bounded by 3^s. For each such sequence, we need to run the algorithm for PLANAR RBW-DS on at most $O(\sqrt{k})$ chunks each of depth bounded by $(3c'+1)\sqrt{k}+11$. Noting that $s \leq 5\sqrt{k}/c' + 12/c'$, we conclude that the running time of the algorithm is $O(\sqrt{k}3^{5\sqrt{k}/c'+12/c'}27^{(3c'+1)\sqrt{k}+11}n)$. Choosing $c' = \sqrt{5}/3$, we get that the running time of the algorithm is bounded by $O(3^{17\sqrt{k}}n) = O(2^{27\sqrt{k}}n)$. We have the following theorem.

Theorem 4. *The* PLANAR DOMINATING SET *problem can be solved in time* $O(2^{27\sqrt{k}}n)$.

This is clearly a significant improvement over Alber et al's $O(2^{70\sqrt{k}}n)$ algorithm for the problem.

References

1. J. ALBER, H. L. BODLAENDER, H. FERNEAU, AND R. NIEDERMEIER, Fixed parameter algorithms for Dominating Set and related problems on planar graphs, To appear in *Algorithmica* (2002). Earlier version appeared in *LNCS* **1851**, (2000), pp. 97-110.
2. J. ALBER, H. FAN, M. R. FELLOWS, H. FERNAU, R. NIEDERMEIER, F. ROSAMOND, AND U. STEGE, Refined search tree techniques for Dominating Set on planar graphs, in *LNCS* **2136**, (2001), pp. 111-122.
3. J. ALBER, H. FERNAU, AND R. NIEDERMEIER, Parameterized complexity: Exponential speed-up for planar graph problems, in *LNCS* **2076**, (2001), pp. 261-272.
4. B. S. BAKER, Approximation algorithms for NP-complete problems on planar graphs, *Journal of the ACM* **41**, (1994), pp. 153-180.
5. N. BANSAL AND V. RAMAN, Upper bounds for MAX-SAT further improved, *LNCS* **1741**, (1999), pp. 247-258.
6. L. CAI AND D. JUEDES, On the existence of subexponential-time parameterized algorithms, available at `http://www.cs.uga.edu/~cai/`.
7. J. CHEN, AND I. A. KANJ, Improved exact algorithms for Max-Sat, in *LNCS* **2286**, (2002).
8. J. CHEN, I. A. KANJ, AND W. JIA, Vertex cover: further observations and further improvement, *Journal of Algorithms* **41**, (2001), pp. 280-301.
9. R. G. DOWNEY AND M. R. FELLOWS, *Parameterized Complexity*, New York, New York: Springer, (1999).
10. M. GAREY AND D. JOHNSON, *Computers and Intractability: A Guide to the Theory of NP-completeness*, Freeman, San Francisco, 1979.
11. R. NIEDERMEIER AND P. ROSSMANITH, Upper bounds for vertex cover further improved, in *LNCS* **1563**, (1999), pp. 561-570.

Optimal Free Binary Decision Diagrams for Computation of EAR_n

Jan Kára and Daniel Král'

Department of Applied Mathematics and
Institute for Theoretical Computer Science*, Charles University,
Malostranské náměstí 25, 118 00 Prague 1, Czech Republic,
{kara,kral}@kam.mff.cuni.cz

Abstract. Free binary decision diagrams (FBDDs) are graph-based data structures representing Boolean functions with a constraint (additional to binary decision diagrams) that each variable is tested during the computation at most once. The function EAR_n is a Boolean function on $n \times n$ Boolean matrices; $\mathrm{EAR}_n(M) = 1$ iff the matrix M contains two equal adjacent rows. We prove that the size of optimal FBDDs computing EAR_n is $2^{\Theta(\log^2 n)}$.

1 Introduction

Graph-based data structures representing Boolean functions are important both from the practical (verification of circuits) and from the theoretical (combinatorial properties of Boolean functions) point of view. The sizes of minimal representations of different Boolean functions in a certain class of data structures and the relation between the sizes in different classes are intensively studied. We refer the reader to a recent monograph [6] on the topic by Wegener.

A binary decision diagram (BDD) is a directed graph where vertices are labelled with input variables and the two outgoing arcs with values 0 and 1 (we understand 0 to be false and 1 to be true). The computation is started in a special vertex called a source and guided in a natural way by the input to one of the two special vertices, called sinks – one of them is an accepting sink (1-sink) and the other one is a rejecting sink (0-sink). We refer the reader for a formal definition to Section 2. We study free binary decision diagrams (FBDDs) in this paper; they are BDDs with an additional constraint that each variable is tested during the computation at most once. FBDDs were introduced by Masek in [3] (he called them read-once branching programs) already in 1976. Lots of upper and lower bounds on the sizes of FBDDs have been proved since: The first lower bound was proved in [7,8] and further ones were proved later, e.g. [1,2,4,5].

The function EAR_n is defined on $n \times n$ Boolean matrices as follows: The value of $\mathrm{EAR}_n(M)$ is 1 iff M contains two adjacent equal rows, i.e. if there exists $1 \le i_0 < n$ such that for all $1 \le j \le n$ $M[i_0, j] = M[i_0 + 1, j]$ (throughout

* Institute for Theoretical Computer Science (ITI) is supported by Ministry of Education of Czech Republic as project LN00A056.

K. Diks et al. (Eds): MFSC 2002, LNCS 2420, pp. 411–422, 2002.

the paper the first coordinate always corresponds to the rows of the matrix). The problem to decide whether the function EAR_n has FBDDs of a polynomial size was mentioned as an open problem in [6] (Problem 6.17). We prove that the size of optimal FBDDs for the function $\text{EAR}_n(M)$ is $2^{\Theta(\log^2 n)}$. This settles the original problem. The interest in the size of FBDDs for EAR_n is amplified by the fact that the size of its optimal FBDDs is neither polynomial nor exponential.

The paper is structured as follows: We recall basic definitions related to (free) binary decision diagrams in Section 2. Next, we prove the upper bound $2^{O(\log^2 n)}$ on the size of FBDDs computing the function EAR_n in Section 3. The matching lower bound $2^{\Omega(\log^2 n)}$ is proved in Section 4.

2 Definitions and Notation

A *binary decision diagram (BDD, branching program)* $\mathcal{B}$ is an acyclic directed graph with three special vertices: a *source*, a 0*-sink* and a 1*-sink*. We call the vertices of $\mathcal{B}$ *nodes*. Each node except for the sinks has out-degree two and it is assigned one of the input variables; one of the two arcs leading from it is labelled with 0 and the other with 1. The out-degrees of the sinks are zero. The *size* of a BDD is the number of its nodes. The *computation path* for $x_1, \ldots, x_n$ in $\mathcal{B}$ is the (unique) path $v_0, \ldots, v_k$ with the following properties:

- The node v_0 is the source of $\mathcal{B}$.
- If the node $v_i, 0 \leq i \leq k-1$ has been assigned a variable x_j, then v_{i+1} is the unique node to which an arc labelled with the value of x_j leads from v_i to.
- The node v_k is either the 0-sink or the 1-sink.

The value of the function $f_{\mathcal{B}}(x_1, \ldots, x_n)$ is equal to 1 if the last node of the computation path for $x_1, \ldots, x_n$ is the 1-sink. We say that the function $f_{\mathcal{B}}$ is *computed* by $\mathcal{B}$ and the diagram $\mathcal{B}$ *represents* the functions $f_{\mathcal{B}}$. We say that $\mathcal{B}$ is *reduced* if each its node is on a computation path for some choice of the input variables and there are no parallel arcs in $\mathcal{B}$. It is straightforward (cf. [6]) to prove that for each binary decision diagram there exists one which is reduced and which computes the same function. We say that a binary decision diagram $\mathcal{B}$ is a *free binary decision diagram (FBDD, read-once branching program)* if for any choice of the input variables, the computation path for them does not contain two vertices with the same variable assigned, i.e. during the computation each variable is tested at most once.

Let $\mathcal{B}$ be a fixed (free) binary decision diagram in this paragraph. If the values of the variables $x_1, \ldots, x_n$ are fixed, then we define for a node v on the computation path for $x_1, \ldots, x_n$ the *test set* of v as the set of all the variables assigned to the nodes preceding v, i.e. the set of the variables which have been already tested during the computation. In case of (F)BDDs computing the function EAR_n, we understand the test sets as the set of the coordinates of the tested entries of the matrix.

3 Upper Bound

Let n be the size of the input matrix for the function EAR_n unless otherwise stated throughout this section. Consider the following algorithm (Algorithm 1): The algorithm is based on a function `test(row1,row2,column,bit1,bit2)` which tests whether the submatrix formed by the rows from `row1` to `row2` and the columns from `column` to n contain two equal adjacent rows; it assumes that `matrix[row1,column]=bit1` and `matrix[row2,column]=bit2`. The function starts sweeping the entries of the column `column` from the row `row1` to the row `row2`. If the function discovers two different entries, let say that `matrix[i,column]<>matrix[i+1,column]`, then two equal adjacent rows can be only among the rows from the row `row1` to the row i or among the rows from the row $i+1$ to the row `row2`. It is possible to call the function recursively at the moment to handle each of these two cases separately. However, we make a recursive call only for the smaller case and for the larger one, we "restart" the function with new parameters (this saves space used by the algorithm and the proof of Proposition 2 suggests why this is a good idea). The only role of parameters `bit1` and `bit2` is to prevent the function to access a single bit twice; the values of the two entries provided in these two variables were already accessed and the called functions need them in order to work properly (see the proof of Proposition 1 and the proof of Proposition 3 for details).

Algorithm 1.

```
Initial call: test(1, n, 1, matrix[1,1], matrix[n,1])

procedure test(row1, row2, column, bit1, bit2: integer);
var i: integer;
begin
restart:
  if row1+1=row2 then                          {Condition 1}
    begin
      if bit1<>bit2 then exit;
      for i:=column+1 to n do
        if matrix[row1,i]<>matrix[row2,i] then
          exit;
      accept
    end;
  i:=row1+1;
  while i<=row2-1 do
    if bit1=matrix[i,column] then              {Condition 2}
      i:=i+1
    else
      begin
        if row1=i-1 then                       {Condition 3}
          begin
            row1:=row1+1;
```

```
          bit1:=not(bit1);
          goto restart;
        end;
      if (i-1)-row1 < row2-i then    {Condition 4}
        begin
          if column=n then accept;
          test(row1,i-1,column+1,
               matrix[row1,column+1],matrix[i-1,column+1]);
          row1:=i;
          bit1:=not(bit1);
          goto restart
        end
      else
        begin
          test(i,row2,column,not(bit1),bit2);
          bit2:=bit1;
          row2:=i-1;
          goto next_column
        end;
    end;
next_column:
  if column=n then accept;
  if bit1<>bit2 then row2:=row2-1;
  column:=column+1;
  bit1:=matrix[row1,column];
  bit2:=matrix[row2,column];
  goto restart
end;
```

The following three propositions are proved by induction for `column` $= n, \ldots, 1$ and `row2` $-$ `row1` $= 1, \ldots, n-1$. We include their rather technical proofs for the completeness:

Proposition 1. *The function* `test` *from Algorithm 1 accesses only the entries of the matrix with the coordinates* $[x, y]$ *which satisfy one of the following:*

$$\texttt{row1} < x < \texttt{row2} \textit{ and } y = \texttt{column}$$

$$\texttt{row1} \leq x \leq \texttt{row2} \textit{ and } \texttt{column} < y \leq n$$

Moreover, the function `test` *accesses each such entry at most once.*

Proof. If Condition 1 applies, the statement of the proposition is obvious. Otherwise, the while-cycle is started; note that in this case, only the entries in the column `column` are accessed before the function is "restarted", i.e. a goto-statement to the label `restart` is executed. Once the function is restarted, the induction can be used to get the statement of the proposition. If Condition 2 is always true, the execution reaches the label `next_column`. Since in this case i is increased by one in each loop of the while-cycle, the statement of the lemma for the entries in the column `column` holds and the induction gives the statement for the entries in the columns $\texttt{column}+1,\ldots,n$.

Let us assume that Condition 2 is false for some i. If Condition 3 applies, the induction is used. Otherwise, a recursive call is made. In this case, the induction together with a careful comparison of the parameters of the recursive call to the function `test` and the parameters when the function is "restarted" gives the statement.

Proposition 2. *The depth of the recursive calls in the function* `test` *from Algorithm 1 is at most* $\log_2(\texttt{row2}-\texttt{row1}+1)$.

Proof. The difference $\texttt{row2}-\texttt{row1}$ is never increased during the execution of the function `test`. The recursive call is made only when the algorithm reaches Condition 4. If $(i-1)-\texttt{row1} < \texttt{row2}-i$, then $2i-1 < \texttt{row1}+\texttt{row2}$. Hence the expression $\texttt{row2}-\texttt{row1}+1$ in the recursive call, i.e., $(i-1)-\texttt{row1}$, is smaller or equal to the half of the expression $\texttt{row2}-\texttt{row1}+1$. In the case that $(i-1)-\texttt{row1} \geq \texttt{row2}-i$, a symmetric argument yields the analogous conclusion. We may conclude that at each call the expression $\texttt{row2}-\texttt{row1}+1$ is at least halved, i.e., it is at most half of this expression when the function `test` has been called and thus the recursion depth is at most $\log_2(\texttt{row2}-\texttt{row1}+1)$.

Proposition 3. *The function* `test` *accepts iff there are two equal adjacent rows in the submatrix of the input matrix formed by the rows from the row* `row1` *to the row* `row2` *and by the columns from the column* `column` *to the column* n.

Proof. If Condition 1 applies, the statement of the proposition is obvious. Otherwise, the while-cycle is started. If all the entries at positions $[i,\texttt{column}]$ for $\texttt{row1} < i < \texttt{row2}$ are equal to `bit1`, the execution of the function reaches the label `next_column`. If $\texttt{column} = n$, then there are two equal adjacent rows in the submatrix (recall that $\texttt{row1}+1 \leq \texttt{row2}$) and the function accepts. Otherwise, it is tested whether $\texttt{bit1} = \texttt{bit2}$. If so, then the entries of the rows from the row `row1` to the row `row2` in the column `column` are the same. The function is "restarted" with the value of `column` increased by one and the induction is used to get the claim. Otherwise, $\texttt{bit1} \neq \texttt{bit2}$. If there are two equal adjacent rows in the submatrix, then they must be among the rows from the row `row1` to the row $\texttt{row2}-1$; hence the value of `row2` is decreased by one, the value of `column` is increased by one, the function is "restarted" and the induction gives the statement.

Assume now that there exists i_0, $\texttt{row1} < i_0 < \texttt{row2}$, such that `bit1` differs from the entry at the position $[i_0,\texttt{column}]$. If $i_0 = \texttt{row1}+1$ (tested by Condition 3), we just increase `row1` by one and "restart" the function. The induction again gives the statement in this case. Otherwise, the adjacent rows can only be either among the rows from the row `row1` to the row i_0-1 or among the rows from the row i_0 to the row `row2` (note that it holds $i_0 < \texttt{row2}$ due to the condition of the while-cycle). The function is recursively called for the part consisting of the smaller number of rows. If the smaller part is the part containing the rows with already tested entries, the column `column` is cut from it. In this case, two equal adjacent rows (if exist) in this smaller submatrix (from which the column `column` was cut) together with two entries from the column `column` form two equal adjacent rows. We may conclude: If there are two adjacent rows in the smaller part of the submatrix, the function accepts due to the induction and there are two equal adjacent rows in the original submatrix (the case that the smaller part does not contain rows with already tested entries is trivial). If the smaller part does not contain two adjacent rows, then the function is "restarted" for the larger part of the submatrix and again the induction is used to get the statement in a similar fashion when the recursive call for the smaller part was made.

Theorem 1. *There is an FBDD $\mathcal{B}$ computing* EAR_n *of size* $2^{O(\log^2 n)}$.

Proof. Algorithm 1 stores during its computation at most $O(\log n)$ numbers from the range from 1 to n due to Proposition 2 (each call of the function `test` increases the number of the stored variables by a constant). Hence the algorithm uses at most $O(\log^2 n)$ bits and the number of different states which can be reached during the computation by the algorithm is bounded by $2^{O(\log^2 n)}$; the state includes the pointer to the instruction to be executed and the content of the memory. We can create an FBDD $\mathcal{B}$ with $2^{O(\log^2 n)}$ nodes which simulates the computation of Algorithm 1: Its nodes will correspond to the states of the algorithm just before accessing an entry of the input matrix and depending on its value the computation (in the diagram) continues to one of the consequent nodes. The computation reaches the 1-sink if Algorithm 1 accepts. Proposition 1 implies that $\mathcal{B}$ is actually a free binary decision diagram and Proposition 3 gives that $\mathcal{B}$ computes the function EAR_n.

4 Lower Bound

4.1 Notation Used in the Lower Bound Proof

Let n be fixed in this subsection and determine the size of the input matrix. The adversary generates an input matrix depending on the previous computation done by the diagram. The adversary strategy is described by a pair consisting of a binary tree T of depth $d := \lfloor \log_3 n \rfloor - 1$ (a tree consisting only of a root has depth 1 and an empty tree has depth 0) and a sequence of bits of length $(\lfloor \log_3 n \rfloor - 1)\lfloor \log n \rfloor^2$ (the bases of the logarithms are two unless written differently). If the adversary uses a labelled tree T and a bit vector b, we say that the computation is *guided* by (T, b). There is a natural one-to-one correspondence between the vertices of T and sequences of 0 and 1 of length at most d which can be defined inductively as follows: Let $a_1, \ldots, a_k$ be a sequence of zeroes and ones. If $k = 1$, then the corresponding vertex is the root. If $a_1 = 0$, resp. $a_1 = 1$, then the corresponding vertex is the vertex of the left, resp. right, subtree of the root corresponding to $a_2, \ldots, a_k$. Each vertex v of T is assigned an integer from a

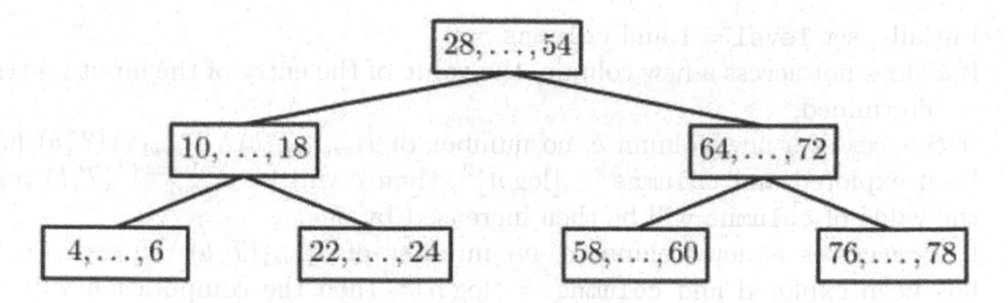

Fig. 1. Intervals for a labelled tree of order $n = 81$

certain interval: If the sequence $a_1, \ldots, a_d$ corresponds to v, then v is assigned an integer from the following interval (cf. Figure 1):

$$\left[\left(\frac{2a_1}{3^1} + \ldots + \frac{2a_d}{3^d} + \frac{1}{3^{d+1}}\right)n + 1, \left(\frac{2a_1}{3^1} + \ldots + \frac{2a_d}{3^d} + \frac{2}{3^{d+1}}\right)n\right]$$

We call a binary tree with integers assigned to its vertices from the intervals above a *labelled tree of order* n or a *labelled tree* if the value of n is clear from the context. We write $I_i(T)$ for the set of the integers assigned to the vertices of T in depth at most i, $0 \leq i \leq d$; $I_0(T)$ is an empty set. If T is a labelled tree of order n and of depth d, we define a column $c_k(T)$, $1 \leq k \leq d$ of n bits as follows: The i-th entry of $c_k(T)$ is 0 if the number of integers of $I_k(T)$ less than i is even and it is 1 otherwise. Let $b := b_k^l, 1 \leq k \leq d = \lfloor \log_3 n \rfloor - 1, 1 \leq l \leq \lfloor \log n \rfloor^2$, be the sequence of $(\lfloor \log_3 n \rfloor - 1)\lfloor \log n \rfloor^2$ bits which is a part of the description of the adversary strategy. We define $c_k^l(T, b)$ to be $c_k(T)$ if $b_k^l = 0$ and to be the negation of $c_k(T)$ otherwise ($1 \leq k \leq d$ and $1 \leq l \leq \lfloor \log n \rfloor^2$).

We say that the k-th pair of rows, $1 \leq k \leq n-1$, was *explored* during the computation if there is c, $1 \leq c \leq n$ such that the entries $M[k, c]$ and $M[k+1, c]$ of the input matrix M were already tested and $M[k, c] \neq M[k+1, c]$, i.e. the k-th pair of rows of the input matrix have been discovered to be different. If $\mathcal{B}$ is a FBDD computing the EAR$_n$ function and v is the node on the computation path in B, then the *exploration set* $E(v)$ of v is the set of all the integers k for which the k-th pair of rows was explored before the test at v. The numbers of $E(v)$ are called *explored*. We say that the exploration set $E(v)$ is *separated* if $|\{k, k+1\} \cap E(v)| \leq 1$ for any $1 \leq k \leq n-2$ and neither $1 \in E(v)$ nor $n - 1 \in E(v)$.

4.2 Adversary Strategy

Let n be a fixed (sufficiently large) integer, $\mathcal{B}$ a fixed FBDD computing EAR$_n$, T a fixed labelled tree of order n and b a fixed vector describing the adversary strategy in this subsection. The adversary creates the input matrix M depending on the computation performed by $\mathcal{B}$. The adversary selects a whole column each time when $\mathcal{B}$ accesses a variable from a column whose any entry has not been tested so far (we call such a column *new*). The adversary stops the computation at a certain node before reaching any of the sinks. The strategy is as follows:

- Initially, set $\texttt{level} = 1$ and $\texttt{columns} = 0$.
- If $\mathcal{B}$ does not access a new column, the value of the entry of the input matrix is determined.
- If $\mathcal{B}$ accesses a new column c, no number of $I_{\texttt{level}}(T,b) \setminus I_{\texttt{level}-1}(T,b)$ has been explored and $\texttt{columns} < \lfloor \log n \rfloor^2$, then c will be $c_{\texttt{level}}^{\texttt{columns}+1}(T,b)$ and the value of $\texttt{columns}$ will be then increased by one.
- If $\mathcal{B}$ accesses a new column c, no number of $I_{\texttt{level}}(T,b) \setminus I_{\texttt{level}-1}(T,b)$ has been explored and $\texttt{columns} = \lfloor \log n \rfloor^2$, then the computation will be stopped.
- If $\mathcal{B}$ accesses a new column c, there is a number of $I_{\texttt{level}}(T,b) \setminus I_{\texttt{level}-1}(T,b)$ which has been explored and the inequality $\texttt{level} < \lfloor \log_3 n \rfloor - 1$ holds, then c will be $c_{\texttt{level}+1}^{1}(T,b)$, the value of $\texttt{level}$ will be then increased by one and the value of $\texttt{columns}$ will be set to one.
- If $\mathcal{B}$ accesses a new column c, there is a number of $I_{\texttt{level}}(T,b) \setminus I_{\texttt{level}-1}(T,b)$ which has been explored and $\texttt{level} = \lfloor \log_3 n \rfloor - 1$, then the value of $\texttt{level}$ will be increased by one and the computation will be stopped.

The adversary wants to force $\mathcal{B}$ to either explore lots of (not too close) pairs of rows or to access lots of columns without exploring any pairs of rows. In the former case, $\mathcal{B}$ has to "remember" which pairs of rows have been explored; in the latter, $\mathcal{B}$ has to "remember" at least some values from lots of columns. In order to reach the former goal, the adversary provides to $\mathcal{B}$ columns c_k^l for $k = \texttt{level}$ until a number from $I_{\texttt{level}}(T,b) \setminus I_{\texttt{level}-1}(T,b)$ is explored. In order to reach the latter goal, the adversary provides $\mathcal{B}$ columns c_k^l increasing l. When one of the goals is reached, the adversary stops the computation.

If the computation was stopped when $\texttt{level} \le \lfloor \log_3 n \rfloor - 1$, we say that the computation was stopped *prematurely*. Note that at most $O(log^3 n)$ columns are accessed before the computation is stopped by the adversary (this is the point where we need that n is sufficiently large). Since the matrix consisting only of columns $c_k^l, 1 \le k \le d, 1 \le l \le \lfloor \log n \rfloor^2$, definitely contains two equal adjacent rows, the computation is stopped before reaching any of the sinks (if n is large).

We write $v(T,b)$ for the node of $\mathcal{B}$ where the adversary stopped the computation when guided by (T,b), $E(T,b)$ for the exploration set of $v(T,b)$ and $M(T,b)$ be the test set of $v(T,b)$.

4.3 Analysis of the Strategy

Let n be a fixed (sufficiently large) integer and $\mathcal{B}$ a fixed FBDD computing EAR_n in this subsection.

Lemma 1. *Let T be a labelled tree of order n and of depth d. For each $x_1, x_2 \in I_k(T), 1 \le k < d$, there exists a number x for which $x_1 < x < x_2$ and $x \in I_{k+1}(T) \setminus I_k(T)$. In addition, $\min I_{k+1}(T) \in I_{k+1}(T) \setminus I_k(T)$ and $\max I_{k+1}(T) \in I_{k+1}(T) \setminus I_k(T)$.*

Proof. This immediately follows from the choice of the intervals in the definition of a labelled tree.

Lemma 2. *Let the pair (T, b) be a description of the adversary strategy. Then the set $E(T, b)$ is separated.*

Proof. It is enough to realize that $E(T, b) \subseteq I_d(T)$ where d is the depth of T. The set $I_d(T)$ satisfies the condition $|\{k, k+1\} \cap I_d(T)| \leq 1$ for any $1 \leq k \leq n-2$ because the intervals from which the numbers are assigned to the vertices of T are disjoint and any two of them are separated by at least one integer.

Lemma 3. *Let the pairs (T, b) and (T', b') be descriptions of the adversary strategy. If $E(T, b) \neq E(T', b')$, then $v(T, b) \neq v(T', b')$.*

Proof. Suppose the opposite, i.e. $v(T, b) = v(T', b')$. We prove that $\mathcal{B}$ does not compute EAR_n. We may assume that there exists $i_0 \in E(T, b) \setminus E(T', b')$ because $E(T, b) \neq E(T', b')$ and the roles of (T, b) and (T', b') are symmetric. If n is large, then $M(T, b) \cup M(T', b')$ omits at least one column. We select this column, call it c_0, in such manner that the only possible pair of rows which can be equal is the i_0-th pair (i.e., its entries alternate except for the i_0-th and the (i_0+1)-th entry). We choose the untested entries of the matrix in such way that the i_0-th and (i_0+1)-th rows are equal; note that the entries of the columns previously defined by the adversary might be changed in this step. $\mathcal{B}$ accepts the created matrix. On the other hand, if we fill with the same entries the matrix partially discovered during the computation guided by (T, b) (note that $\mathcal{B}$ – when continuing the computation – can only access the entries of the matrix which were not tested during the computation guided by either (T, b) or (T', b') and hence this defines all the entries of the matrix which can be accessed when continuing the started computation), $\mathcal{B}$ accepts. But since $i_0 \in E(T, b)$ and only the i_0-th and (i_0+1)-th entries of c_0 are equal, $\mathcal{B}$ had to reject.

Lemma 4. *Let the pairs (T, b) and (T', b') determine the adversary strategy. If $M(T, b) \neq M(T', b')$, then $v(T, b) \neq v(T', b')$.*

Proof. If $E(T, b) \neq E(T', b')$, then $v(T, b) \neq v(T', b')$ due to Lemma 3. Assume $E(T, b) = E(T', b')$ and let $E_0 := E(T, b) = E(T', b')$. We may further assume that there exists $[x, y] \in M(T, b) \setminus M(T', b')$ because $M(T, b) \neq M(T', b')$ and the roles of (T, b) and (T', b') are symmetric. Since E_0 is separated (due to Lemma 2), $x-1 \notin E_0 \wedge x-1 \geq 1$ or $x \notin E_0 \wedge x \leq n-1$. It is enough to consider the case that $x \notin E_0$ and $x \leq n-1$ due to the symmetry. If n is large, then $M(T, b) \cup M(T', b')$ omits at least one column, say c_0. We complete the matrix from the computation guided by (T, b) in such manner that the x-th and the $(x+1)$-th row are equal (this is possible because $x \notin E_0$) and we choose c_0 to be a column whose entries alternates except for the pair formed by the x-th and the $(x+1)$-th one. The entries already defined by the adversary might be changed in this step. The rest of the matrix is completed arbitrarily. The diagram $\mathcal{B}$ accepts this matrix. On the other hand, if we fill with the same entries the matrix partially discovered during the computation guided by (T', b') (recall that $\mathcal{B}$ can only access the entries of the matrix which were not tested during the computation guided by either (T, b) or (T', b') when continuing the stopped computation and hence this defines all the entries of the matrix which $\mathcal{B}$ may access), $\mathcal{B}$ accepts. But $\mathcal{B}$ did not test the entry of the matrix with the coordinates $[x, y]$ and the x-th pair of rows is the only one which can form two equal adjacent rows. If we choose this entry to be different from the entry with the coordinates $[x+1, y]$, then $\mathcal{B}$ had to reject.

Lemma 5. *If there exists a labelled tree T such that for any bit vector b the computation guided by (T, b) stops prematurely, then the size of $\mathcal{B}$ is at least $2^{\lfloor \log n \rfloor^2}$.*

Proof. Let T be a labelled tree with the properties from the statement of the lemma. Let k_0 be the largest value of `level` obtained for some vector b_0 when the computation is stopped. Let B be the set of $2^{\lfloor \log n \rfloor^2}$ bit vectors which agree with b_0 for all the entries except for $b^l_{k_0}, 1 \leq l \leq \lfloor \log n \rfloor^2$. The value of `level` when the computation guided by (T, b) is stopped for $b \in B$ is k_0: It cannot be more due to the choice of b_0 and it cannot be less because reaching the level k_0 can be influenced only by the entries b^l_k for $k < k_0$. We prove that all the nodes $v(T, b)$ for $b \in B$ are mutually different.

Let b and b' be two vectors in B with $v(T, b) = v(T, b')$. It holds that $E(T, b) = E(T, b')$ due to Lemma 3 and $M(T, b) = M(T, b')$ due to Lemma 4; let E_0 and M_0 be their common values. Let l_0 be the smallest l for which b and b' differ; we may assume that $b^{l_0}_{k_0} = 0$ and $b'^{l_0}_{k_0} = 1$. Let $[x_0, y_0] \in M_0$ be the first entry of the matrix tested in the column defined by $c^{l_0}_{k_0}$; due to the choice of l_0, the entry $[x_0, y_0]$ is the same for the computation guided by (T, b) and (T, b'). On the other hand, since $b^{l_0}_{k_0} \neq b'^{l_0}_{k_0}$, the value of the entry $[x_0, y_0]$ is different when the computation is guided by (T, b) and when it is guided by (T', b'). Let x_1 and x_2 be (the uniquely determined) integers such that $x_1 \leq x_0 \leq x_2$, $[x, y_0] \in M_0$ for all $x_1 \leq x \leq x_2$, all the entries $[x, y_0]$ have the same value in either of the two matrices and x_1 is the smallest and x_2 is the largest integer with these properties. It cannot be that both $x_1 - 1 \in E_0 \vee x_1 = 1$ and $x_2 \in E_0 \vee x_2 = n$; otherwise, either $x_1 - 1 \in I_{k_0}(T) \setminus I_{k_0 - 1}(T)$ or $x_2 \in I_{k_0}(T) \setminus I_{k_0-1}(T)$ (Lemma 1) and the computation cannot be stopped prematurely with `level` $= k_0$. We assume that $x_2 \in E_0$ (the other case is symmetric).

If n is large, then M_0 omits at least one column. We complete the matrix from the computation guided by (T, b) in such manner that the x_2-th and the $(x_2 + 1)$-th row are equal (this is possible because $x_2 \notin E_0$) and we choose c_0 to be a column whose entries alternates except for the pair formed by the x_2-th and the $(x_2 + 1)$-th entry. The entries previously defined by the adversary might be changed in this step. The rest of the matrix is completed arbitrarily. The diagram $\mathcal{B}$ accepts this matrix. On the other hand, if we fill with the same entries the matrix partially discovered during the computation guided by (T, b') (note that this defines all the entries of the matrix which can be accessed by $\mathcal{B}$ when continuing the started computation), $\mathcal{B}$ accepts. But the entries $[x_2, y]$ and $[x_2 + 1, y]$ are different and only the x_2-th pair of rows might form two equal adjacent row due to c_0. Thus $\mathcal{B}$ had to reject.

4.4 The Bound

Theorem 2. *Let $\mathcal{B}$ be a FBDD computing* EAR$_n$*. The the size of $\mathcal{B}$ is at least* $2^{\Omega(\log^2 n)}$.

Proof. We assume that n is large enough to hold Lemma 3 and Lemma 5. We may assume: For each labelled binary tree T, there exists a bit vector b_T for which the computation guided by (T, b_T) does not stop prematurely. Otherwise, Lemma 5 would imply the lower bound. We prove that $|\bigcup_T \{v(T, b_T)\}| \geq 2^{\Omega(\log^2 n)}$. Let $E_0 = \bigcup_T \{E(T, b_T)\}$. Due to Lemma 3, it is enough to prove that $|E_0| \geq 2^{\Omega(\log^2 n)}$.

We will not use integer parts in the rest of the proof in order to improve clarity of the arguments. It is straightforward to check that this does not change the asymptotic of the obtained result. We create a bipartite graph G with one of its parts formed by vertices corresponding to the elements of E_0 and the other one formed by vertices corresponding to all the possible labelled trees of order n (we further actually identify the vertices with the elements to which they correspond). The number N of labelled trees of order n is:

$$N = \prod_{k=1}^{d} \left(\frac{n}{3^k}\right)^{2^{k-1}}$$

We join each labelled tree T to $E \in E_0$ such that $E \subseteq I(T)$; the degree of any vertex corresponding to a labelled tree is at least one, since it is joined at least to $E(T, b_T)$. Thus G contains at least N edges. On the other hand, it straightforward to verify that the degree of any vertex corresponding to $E \in E_0$ is at most the following number N' (each E contains at least one element from the k-th level of the tree, $1 \leq k \leq d$):

$$N' = \prod_{k=1}^{d} \left(\frac{n}{3^k}\right)^{2^{k-1}-1}$$

We may conclude that the size of E_0 is at least:

$$|E_0| \geq \frac{N}{N'} = \prod_{k=1}^{d} \frac{n}{3^k}$$

A straightforward computation gives the desired bound:

$$|E_0| \geq \prod_{k=1}^{d} \frac{n}{3^k}$$

$$\log |E_0| \geq \sum_{k=1}^{d} \log \frac{n}{3^k} \geq \sum_{k=1}^{d} \log \frac{3^d}{3^k} = \sum_{k=1}^{d} (d-k) = \Omega(d^2) = \Omega(\log^2 n)$$

$$|E_0| \geq 2^{\Omega(\log^2 n)}$$

Acknowledgement

The second author would like to thank Petr Savický for introducing binary decision diagrams to him during his outstanding lectures on the topic. Heplful comments and suggestions of both the referees are appreciated.

References

1. Babai, L., Hajnal, P., Szemerédi, E., Turán, G.: A lower bound for read-once only branching programs. Journal of Computer and System Sciences **35** (1987) 153–162
2. Kriegel, K., Waack, S.: Lower bounds on the complexity of real-time branching programs. RAIRO – Theoretical Informatics and Applications **22** (1988) 447–459.
3. Masek, W.: A fast algorithm for the string editing problem and decision graph complexity. M. Sc. Thesis, MIT (1976)
4. Savický, P., Žák, S.: A large lower bound for 1-branching programs. ECCC report 96-036 (1996)
5. Savický, P., Žák, S.: A read-once lower bound and $(1,+k)$-hierarchy for branching programs. Theoretical Computer Science **238(1-2)** (2000) 347–362
6. Wegener, I.: Branching Programs and Binary Decision Diagrams – Theory and Applications. SIAM Monographs on Discrete Mathematics and Applications 4 (2000)
7. Wegener, I.: On the complexity of branching programs and decision trees for clique functions, Journal of the ACM **35** (1988) 461–471
8. Žák, S.: An exponential lower bound for one-time-only branching programs. Proc. 11th International Symposium on Mathematical Foundations of Computer Science 1984, LNCS vol. **176** (1984) 562-566

Unification Modulo Associativity and Idempotency Is NP-complete*

Ondřej Klíma

Dept. of Mathematics, Masaryk University,
Janáčkovo nám. 2a, 662 95 Brno, Czech Republic,
klima@math.muni.cz

Abstract. We show that the unification problem for the theory of one associative and idempotent function symbol (AI-unification), i.e. solving word equations in free idempotent semigroups, is NP-complete.

Keywords: unification, idempotent semigroups, complexity

1 Introduction

Equational unification is a generalization of syntactic (Robinson) unification in which semantic properties of function symbols are taken into account. Applications of equational unification can be found in areas as term rewriting, logic programming and automated theorem proving. For an overview of unification theory consult e.g. [3,4].

From algebraic point of view, an equational theory corresponds to a variety of algebras of a given type and equational unification is solving of equations in the free algebra of this variety.

The AI-UNIFICATION problem (i.e. unification problem for the equational theory of one associative and idempotent binary function symbol) is of interest for unification theory also because it was the first "natural" example of a problem of unification type zero [1,11]. The AI-UNIFICATION problem is decidable since finitely generated idempotent semigroups are finite, but the complexity of the problem was unknown (see [7] for an overview about complexity aspects of unification problems). Later on Klíma and Srba [10] proved the NP-completeness of the AI-MATCHING problem (the matching problem is a special case of the unification problem where all right-hand sides of equations are variable free). Before this result there was only the result of Kapur and Narendran [6] that the GENERAL AI-MATCHING problem (i.e. problem where additional uninterpreted function symbols are allowed) is NP-hard. The decidability of the GENERAL AI-UNIFICATION problem has been proved by Baader and Schulz [2].

In this paper we prove that the AI-UNIFICATION problem is NP-complete. For the NP-hardness of the AI-UNIFICATION problem we refer to [10]. And to show that the AI-UNIFICATION problem is in NP we use a method different

* Supported by the Ministry of Education of the Czech Republic under the project MSM 143100009.

K. Diks et al. (Eds): MFSC 2002, LNCS 2420, pp. 423–432, 2002.

from the one employed in [10], where a polynomial upper bound for the length of the shortest solutions of a given solvable matching system was found. As a side result of this paper we obtain a reduction of the degree of such a polynomial for the AI-MATCHING problem. Note that there is no polynomial upper bound for the shortest solutions of a solvable unification system.

We also use a result from [2] to obtain the NP-completeness of the GENERAL AI-UNIFICATION problem as a consequence of our result.

2 Preliminaries

We denote the set of all positive integers by $\mathbb{N}$, the set of all subsets of a given set X by $\mathcal{P}(X)$ and the cardinality of this set X by $\|X\|$.

Let Σ be an arbitrary finite alphabet. Then Σ^+ is (as usually) the free semigroup over the set Σ and $\Sigma^* = \Sigma^+ \cup \{\epsilon\}$, where ϵ is the empty word, is the free monoid over the set Σ.

An idempotent semigroup is a semigroup where the identity $x^2 = x$ is satisfied. For a given set Σ we denote by $\sim_\Sigma$ the congruence on Σ^* such that for $p, q \in \Sigma^*$ we have $p \sim_\Sigma q$ if and only if $p = q$ is a consequence of the identity $x^2 = x$. In other words, $\sim_\Sigma$ is the symmetric and transitive closure of the relation $\{(uvw, uvvw) \mid u, v, w \in \Sigma^*\}$ on the set Σ^*. This means that $\Sigma^*/\!\sim_\Sigma$ is the free idempotent monoid over the set Σ. If we remove the class $\epsilon \sim_\Sigma = \{\epsilon\}$ from $\Sigma^*/\!\sim_\Sigma$, then we obtain the free idempotent semigroup over the set Σ. In what follows we write simply $\sim$ instead of $\sim_\Sigma$ if the set Σ is obvious from context.

In this paper we will work with a well-known description of the congruence $\sim$. We need the following notation for a formulation of the basic statement. For a word $u \in \Sigma^*$ we call the set of all letters occurring in u the *content* of u and denote it $\mathsf{c}(u)$. Further, $\sharp_u$ is the number of distinct letters occurring in the word u, i.e. $\sharp_u = \|\mathsf{c}(u)\|$. Now, for $u \in \Sigma^+$ we define

$0(u)$ – the longest prefix of u which contains only $\sharp_u - 1$ different letters,
$1(u)$ – the longest suffix of u which contains only $\sharp_u - 1$ different letters.

Proposition 1 ([5]). *Let $u, v \in \Sigma^+$. Then $u \sim v$ if and only if $\mathsf{c}(u) = \mathsf{c}(v)$, $0(u) \sim 0(v)$ and $1(u) \sim 1(v)$.*

By an invariant we mean a mapping from the set Σ^+ (or Σ^*) to some set Θ; in this paper the set Θ is always either Σ^* or $\mathcal{P}(\Sigma)$. Examples of invariants are mappings c, 0 and 1. Definitions of other invariants follow. For $i \in \mathbb{N}$ and a word $u \in \Sigma^+$ we define

$\mathrm{pref}_i(u)$ – the longest prefix of u which contains at most i different letters,
$\mathrm{suff}_i(u)$ – the longest suffix of u which contains at most i different letters.

We have $\mathrm{pref}_i(u) = u$ and $\mathrm{suff}_i(u) = u$ for all $i \geq \sharp_u$. For $i < \sharp_u$, $\mathrm{pref}_i(u)$ is the $(\sharp_u - i)$-th iteration of the invariant 0 on u and dually $\mathrm{suff}_i(u)$ is the $(\sharp_u - i)$-th iteration of the invariant 1 on u. For $u \in \Sigma^+$ we denote by $\overrightarrow{u}$ the sequence of the first occurrences of all letters from $\mathsf{c}(u)$ in u. So, for $u, v \in \Sigma^+$ we have $\overrightarrow{u} = \overrightarrow{v}$ if and only if $\sharp_u = \sharp_v$ and $\mathsf{c}(\mathrm{pref}_i(u)) = \mathsf{c}(\mathrm{pref}_i(v))$ for all $i = 1, \ldots, \sharp_u$. The sequence of the last occurrences of all letters from $\mathsf{c}(u)$ in u is denoted $\overleftarrow{u}$.

It follows from Proposition 1 that for $u, v \in \Sigma^+$ satisfying $u \sim v$ we have also $\mathrm{pref}_i(u) \sim \mathrm{pref}_i(v)$, $\mathrm{suff}_i(u) \sim \mathrm{suff}_i(v)$ for all $i \in \mathbb{N}$, $\overrightarrow{u} = \overrightarrow{v}$ and $\overleftarrow{u} = \overleftarrow{v}$.

We extend the definitions of the previous invariants also to the empty word in the following way: $0(\epsilon) = 1(\epsilon) = \overrightarrow{\epsilon} = \overleftarrow{\epsilon} = \epsilon$.

We use a standard notation from combinatorics on words. The length of a word $u \in \Sigma^*$ is denoted $|u|$. We say that $u \in \Sigma^*$ is a factor of $v \in \Sigma^*$ if and only if there are a prefix $v_1 \in \Sigma^*$ of v and a suffix $v_2 \in \Sigma^*$ of v such that $v = v_1 u v_2$.

Let $\mathcal{C}$ be a finite set of *constants* and X be a finite set of *variables* such that $\mathcal{C} \cap \mathrm{X} = \emptyset$. A *word equation* $L =^? R$ is a pair $(L, R) \in (\mathcal{C} \cup \mathrm{X})^* \times (\mathcal{C} \cup \mathrm{X})^*$. We say that an equation $L =^? R$ is trivial if $L = R$. A *system of word equations* S is a finite set of equations, i.e. $S = \{L_1 =^? R_1, \ldots, L_n =^? R_n\}$ for $n \geq 0$. A *solution* of the system S in the free idempotent monoid over $\mathcal{C}$ is a homomorphism $\alpha : (\mathcal{C} \cup \mathrm{X})^* \to \mathcal{C}^*$ which behaves as an identity on the set $\mathcal{C}$ and equates all the equations of S, i.e. $\alpha(L_i) \sim_{\mathcal{C}} \alpha(R_i)$ for all $i = 1, \ldots, n$. Such a homomorphism is fully established by a mapping $\alpha|_{\mathrm{X}} : \mathrm{X} \to \mathcal{C}^*$. We use the same symbol α for this mapping and its unique extension to a homomorphism $\alpha : (\mathcal{C} \cup \mathrm{X})^* \to \mathcal{C}^*$. A solution of the system S in the free idempotent semigroup over $\mathcal{C}$ is a solution in the free idempotent monoid such that $\alpha(x) \in \mathcal{C}^+$ for all $x \in \mathrm{X}$.

Definition 1. *Given a system of word equations $S = \{L_1 =^? R_1, \ldots, L_n =^? R_n\}$ with a set of variables* X *and a set of constants $\mathcal{C}$ as an instance of the* AIU-UNIFICATION *problem, the task is to decide whether this system has a solution in the free idempotent monoid over $\mathcal{C}$. The* AIU-MATCHING *problem is the modification of the* AIU-UNIFICATION *problem where only (so called matching) systems with $R_i \in \mathcal{C}^*$ are considered. We speak about the* AI-UNIFICATION *and* AI-MATCHING *problems if only solutions in the free idempotent semigroup are considered.*

What we call the AIU-UNIFICATION problem is usually called the AIU-UNIFICATION problem with constants in literature, because constants are allowed to occur in the system. The case when no constants occur in the system is called the ELEMENTARY AIU-UNIFICATION problem (in this case solutions are mappings $\alpha : \mathrm{X} \to \mathrm{X}^*$). The ELEMENTARY AIU-UNIFICATION problem is not interesting for us by reason that any system without constants has a trivial solution in the free idempotent monoid. The same holds for the ELEMENTARY AI-UNIFICATION problem.

Note that the AIU-UNIFICATION and AI-UNIFICATION problems are decidable since the free idempotent monoid over a finite set is finite by Proposition 1.

The last problem often dealt with in the unification theory is the general unification problem. Basically, in the GENERAL AI-UNIFICATION problem we consider some associative and idempotent binary function symbols and some additional uninterpreted function symbols. We refer the reader to the survey papers [3,4] for a precise definition. The NP-hardness of the GENERAL AI-MATCHING problem has been proved by Kapur and Narendran [6] and the decidability of the GENERAL AI-UNIFICATION problem follows from the decidability of the AI-UNIFICATION problem and a nice result of Baader and Schulz [2].

We turn our attention to the complexity of the matching problems.

Proposition 2 ([10]). *The* AIU-MATCHING *and* AI-MATCHING *problems are NP-complete.*

The non-deterministic polynomial algorithm constructed in [10] is based on the existence of a polynomial upper bound for the length of the shortest solutions of a given solvable matching system. But there is no such polynomial upper bound in general as Example 1 shows.

Example 1 (due to J. Srba). Let S be the system over the set of constants $\mathcal{C} = \{a_0, a_1, \ldots, a_n\}$ and the set of variables $\mathrm{X} = \{x_1, \ldots, x_n\}$ which contains for each $i \in \mathbb{N}$, $1 < i \leq n$, the equation $x_i =^? x_{i-1} a_i x_{i-1}$ and in addition $x_1 =^? a_0 a_1 a_0$. It is easy to see that the system S has a unique solution in the free monoid $\mathcal{C}^*$, which is the shortest solution in the free idempotent monoid over $\mathcal{C}$. For this solution α we have $|\alpha(x_i)| = 2^{i+1} - 1$. (These words $\alpha(x_i)$ are usually called the Zimin words in literature.)

Another special case of a unification problem is often discussed — the case of one equation. For some unification problems, for example the syntactic unification or A-unification problem, solving individual equations is as hard as solving systems of equations because for any system one can construct in polynomial time an equation which has the same set of all solutions. The situation is different for the AIU-UNIFICATION problem. The solvability of an equation is characterized by certain conditions which can be checked in polynomial time (see Lemma 1 in [8]) although AIU-UNIFICATION problem is NP-hard by Proposition 2.

3 Simple AIU-Unification Problem

Let $L =^? R$ be a word equation and $\alpha : \mathrm{X} \to \mathcal{C}^*$ be a mapping. Which information about α do we need to know if we want to decide whether α is a solution of the equation $L =^? R$? We would like to apply Proposition 1. We have to compute $\mathsf{c}(\alpha(L))$ and $\mathsf{c}(\alpha(R))$, hence we need to know $\mathsf{c}(\alpha(x))$ for all variables $x \in \mathrm{X}$. If we want to compute $\mathsf{c}(\mathrm{pref}_i(\alpha(L)))$ then we need to know $\overrightarrow{\alpha(x)}$ for some variables. Therefore, we will assume that $\overrightarrow{\alpha(x)}$ and $\overleftarrow{\alpha(x)}$ are given for all variables $x \in \mathrm{X}$. This assumption is natural; the AIU-UNIFICATION problem with this assumption is almost the same as without it as we show below.

For two mappings $\alpha, \beta : \mathrm{X} \to \mathcal{C}^*$ we write $\alpha \approx_{\mathrm{X}} \beta$ if and only if $\overrightarrow{\alpha(x)} = \overrightarrow{\beta(x)}$ and $\overleftarrow{\alpha(x)} = \overleftarrow{\beta(x)}$ for all $x \in \mathrm{X}$. For mappings $\alpha, \beta : \mathrm{X} \to \mathcal{C}^*$ and $I : \mathrm{X} \to \mathcal{P}(\{\to, \leftarrow\})$ we write $\alpha \approx^I_{\mathrm{X}} \beta$ if and only if for all $x \in \mathrm{X}$ we have $\overrightarrow{\alpha(x)} = \overrightarrow{\beta(x)}$ when $\to \in I(x)$ and $\overleftarrow{\alpha(x)} = \overleftarrow{\beta(x)}$ when $\leftarrow \in I(x)$.

A triple (S, ν, I) is called a *system with partial information about invariants* if $S = \{L_1 =^? R_1, \ldots, L_n =^? R_n\}$ is a system of word equations and $\nu : \mathrm{X} \to \mathcal{C}^*$ and $I : \mathrm{X} \to \mathcal{P}(\{\to, \leftarrow\})$ are mappings. A solution of the system with partial information about invariants (S, ν, I) is a solution $\alpha : \mathrm{X} \to \mathcal{C}^*$ of the system S in the free idempotent monoid over $\mathcal{C}$ such that $\alpha \approx^I_{\mathrm{X}} \nu$. (We can assume that $|\nu(x)| \leq 2 \cdot \|\mathcal{C}\|$ because we can always exchange $\nu(x)$ for the word $\overrightarrow{\nu(x)}\,\overleftarrow{\nu(x)}$.)

We speak about a *system with complete information about invariants* if $I(x) = \{\rightarrow, \leftarrow\}$ for all $x \in \mathrm{X}$. Such a system is denoted simply (S, ν) and a solution α of this system is a solution of the system S satisfying $\alpha \approx_{\mathrm{X}} \nu$.

Definition 2. *The* SIMPLE AIU-UNIFICATION *problem asks to decide whether a given system with complete information about invariants has a solution.*

Lemma 1. *Let (S, ν, I) be a system with partial information about invariants. Then one can construct (in polynomial time) a system S' such that the set of all solutions of (S, ν, I) and the set of all solutions of $S \cup S'$ coincide.*

Proof. The proof is not included due to space limitation — see the full version [9].

As a consequence of Lemma 1 we obtain a polynomial reduction from the SIMPLE AIU-UNIFICATION problem to the AIU-UNIFICATION problem. On the other hand, it is easy to see that there exist non-deterministic polynomial reductions from the AIU-UNIFICATION and AI-UNIFICATION problems to the SIMPLE AIU-UNIFICATION problem (they guess a mapping $\nu : \mathrm{X} \to \mathcal{C}^*$ satisfying $|\nu(x)| \leq 2 \cdot \|\mathcal{C}\|$). So, if we want to prove that the AIU-UNIFICATION and AI-UNIFICATION problems are in NP, it is enough to show that the SIMPLE AIU-UNIFICATION problem is in NP. We show even more:

Proposition 3. *The* SIMPLE AIU-UNIFICATION *problem is decidable in polynomial time.*

Proof. Let (S, ν) be a system with complete information about invariants over a finite set of constants $\mathcal{C}$ and a finite set of variables X.

We may assume that $\sharp_{\nu(x)} \geq 2$ for any $x \in \mathrm{X}$ because if $\sharp_{\nu(x)} \leq 1$ for some $x \in \mathrm{X}$ then $\alpha(x) \sim \nu(x)$ for any solution α of the system (S, ν) and we can replace all occurrences of the variable x in S with $\nu(x)$. We also assume that all constants from $\mathcal{C}$ and all variables from X occur in S. We denote m the cardinality of $\mathcal{C}$, i.e. $\|\mathcal{C}\| = m$. So, we have $2 \leq \sharp_{\nu(x)} \leq m$ for any $x \in \mathrm{X}$.

We extend the set of variables X. For each $x \in \mathrm{X}$ and $i = 1, \ldots, \sharp_{\nu(x)} - 1$ we introduce new variables x_i^{pref} and x_i^{suff}. Moreover, for each such new variable y we also introduce the variable $\overline{y}$. We denote the sets of these variables in the following way:

$$\mathrm{X}^{\mathrm{pref}} = \{x_i^{\mathrm{pref}} \mid x \in \mathrm{X},\ i = 1, \ldots, \sharp_{\nu(x)} - 1\},$$

$$\mathrm{X}^{\mathrm{suff}} = \{x_i^{\mathrm{suff}} \mid x \in \mathrm{X},\ i = 1, \ldots, \sharp_{\nu(x)} - 1\},$$

$$\overline{\mathrm{X}}^{\mathrm{pref}} = \{\overline{y} \mid y \in \mathrm{X}^{\mathrm{pref}}\}, \quad \overline{\mathrm{X}}^{\mathrm{suff}} = \{\overline{y} \mid y \in \mathrm{X}^{\mathrm{suff}}\}.$$

Note that these sets are pairwise disjoint and each of them has $\sum_{x \in \mathrm{X}} (\sharp_{\nu(x)} - 1)$ elements. We put $\mathrm{X}^{\mathrm{ext}} = \mathrm{X} \cup \mathrm{X}^{\mathrm{pref}} \cup \mathrm{X}^{\mathrm{suff}}$, $\overline{\mathrm{X}} = \overline{\mathrm{X}}^{\mathrm{pref}} \cup \overline{\mathrm{X}}^{\mathrm{suff}}$ and $\mathcal{X} = \mathrm{X}^{\mathrm{ext}} \cup \overline{\mathrm{X}}$. (By the way, we have $\|\mathcal{X}\| < 4 \cdot \|\mathrm{X}\| \cdot m$.) Moreover, for $k \in \mathbb{N}$ we define $\overline{\mathrm{X}}_k^{\mathrm{pref}}$ as the set of all elements from $\overline{\mathrm{X}}^{\mathrm{pref}}$ with index k, i.e. $\overline{\mathrm{X}}_k^{\mathrm{pref}} = \{\overline{x}_k^{\mathrm{pref}} \mid x \in \mathrm{X},\ \sharp_{\nu(x)} > k\}$. Analogicaly $\overline{\mathrm{X}}_k^{\mathrm{suff}} = \{\overline{x}_k^{\mathrm{suff}} \mid x \in \mathrm{X},\ \sharp_{\nu(x)} > k\}$ and $\overline{\mathrm{X}}_k = \overline{\mathrm{X}}_k^{\mathrm{pref}} \cup \overline{\mathrm{X}}_k^{\mathrm{suff}}$.

We say that a mapping $\alpha : \mathcal{X} \to \mathcal{C}^*$ is compatible on the set $\mathcal{X}$ if and only if the following conditions hold for all $x \in \mathrm{X}$ and $i = 1, \dots, \sharp_{\nu(x)} - 1$:

$$\alpha(x_i^{\mathrm{pref}}) = \mathrm{pref}_i(\alpha(x)), \qquad \alpha(\overline{x}_i^{\mathrm{pref}}) = 1(\alpha(x_i^{\mathrm{pref}})), \tag{1}$$
$$\alpha(x_i^{\mathrm{suff}}) = \mathrm{suff}_i(\alpha(x)), \qquad \alpha(\overline{x}_i^{\mathrm{suff}}) = 0(\alpha(x_i^{\mathrm{suff}})). \tag{2}$$

This definition explains the role of variables from the set $\mathcal{X}$. If α is compatible on $\mathcal{X}$ and $\alpha \approx_{\mathrm{X}} \nu$, then we have $\alpha(x_1^{\mathrm{pref}}) = \mathrm{pref}_1(\alpha(x)) \sim \mathrm{pref}_1(\overrightarrow{\alpha(x)}) = \mathrm{pref}_1(\overrightarrow{\nu(x)})$, so $\alpha(x_1^{\mathrm{pref}})$ is determined up to $\sim$. Moreover, it means $\alpha(\overline{x}_1^{\mathrm{pref}}) = \epsilon$.

We will define a certain system with partial information about invariants over the set of variables $\mathcal{X}$ such that the set of all compatible solutions on $\mathcal{X}$ of this system corresponds to the set of all solutions of the original system with complete information about invariants (S, ν) over the set of variables X. We define $I(x) = \{\to, \leftarrow\}$, $I(x_i^{\mathrm{pref}}) = \{\to\}$ and $I(x_i^{\mathrm{suff}}) = \{\leftarrow\}$ for all $x \in \mathrm{X}$ and $i = 1, \dots, \sharp_{\nu(x)} - 1$. We put $I(\overline{y}) = \emptyset$ for all $\overline{y} \in \overline{\mathrm{X}}$. Now we extend the domain of the mapping ν to the set $\mathcal{X}$. We put $\nu(x_i^{\mathrm{pref}}) = \mathrm{pref}_i(\nu(x))$ and $\nu(x_i^{\mathrm{suff}}) = \mathrm{suff}_i(\nu(x))$ for all $x \in \mathrm{X}$ and $i = 1, \dots, \sharp_{\nu(x)} - 1$. Finally, we define $\nu(\overline{x}_i^{\mathrm{pref}}) = 1(\nu(x_i^{\mathrm{pref}}))$ and $\nu(\overline{x}_i^{\mathrm{suff}}) = 0(\nu(x_i^{\mathrm{suff}}))$. So, ν is compatible on $\mathcal{X}$.

Now we can see that any solution $\alpha : \mathrm{X} \to \mathcal{C}^*$ of the system (S, ν) with complete information about invariants can be uniquely (by the rules (1–2)) extended to a compatible solution of the system (S, ν, I) with partial information about invariants over the set of variables $\mathcal{X}$. (Let us point out that $\mathsf{c}(\alpha(\overline{x}_i^{\mathrm{pref}}))$ and $\mathsf{c}(\nu(\overline{x}_i^{\mathrm{pref}}))$ are not necessarily the same while $\mathsf{c}(\alpha(x_i^{\mathrm{pref}})) = \mathsf{c}(\nu(x_i^{\mathrm{pref}}))$.) Conversely, the restriction of any (not necessarily compatible) solution $\alpha : \mathcal{X} \to \mathcal{C}^*$ of the system (S, ν, I) with partial information about invariants to the set of variables X is a solution of the original system (S, ν). From this moment the mappings ν and I are fixed and we concentrate only on compatible solutions of systems (T, ν, I) with partial information about invariants, where T is some system of equations over the set of constants $\mathcal{C}$ and the set of variables $\mathcal{X}$. We say that two such systems T_1 and T_2 are equivalent if and only if the systems with partial information about invariants (T_1, ν, I) and (T_2, ν, I) have the same compatible solutions.

We describe an algorithm which simplifies the system S to some equivalent system step by step. The basic idea of the simplification is that the system $T \cup \{L =^? R\}$ is exchanged for the equivalent system $T \cup \{\mathsf{o}(L) =^? \mathsf{o}(R), \mathsf{j}(L) =^? \mathsf{j}(R)\}$, where o, j are certain mappings with the properties $\alpha(\mathsf{o}(U)) = 0(\alpha(U))$, $\alpha(\mathsf{j}(U)) = 1(\alpha(U))$ for $U \in \{L, R\}$ and for any compatible mapping $\alpha : \mathcal{X} \to \mathcal{C}^*$.

We define the set $\mathcal{EF} \subseteq (\mathcal{C} \cup \mathrm{X}^{\mathrm{ext}})^*$, which has the property that both sides of any equation appearing in systems within the algorithm belong to the set $\mathcal{EF} \cup \overline{\mathrm{X}}$. At first, for an arbitrary word $U \in (\mathcal{C} \cup \mathrm{X})^*$ we define the set $\mathcal{EF}(U) \subseteq (\mathcal{C} \cup \mathrm{X}^{\mathrm{ext}})^*$ of *extended factors* of U in the following way: for $V \in (\mathcal{C} \cup \mathrm{X})^*$, $x, y \in \mathrm{X}$, $i \in \{1, \dots, \sharp_{\nu(x)} - 1\}$ and $k \in \{1, \dots, \sharp_{\nu(y)} - 1\}$, the set $\mathcal{EF}(U)$ contains the words $x_i^{\mathrm{suff}} V y_k^{\mathrm{pref}}$, $x_i^{\mathrm{suff}} V$, $V y_k^{\mathrm{pref}}$ and V if and only if xVy, xV, Vy and V respectively are factors of U. Finally, we denote by $\mathcal{EF}$ the union of all sets $\mathcal{EF}(U)$ such that U is a left-hand or right-hand side of some equation of S.

We define the mappings $\mathfrak{c}$, $\mathfrak{o}$, $\mathfrak{j}$ on the set $\mathcal{EF}$ which correspond to the invariants c, 0 and 1. We put $\mathfrak{c}(A) = \mathrm{c}(\nu(A))$ for a word $A \in \mathcal{EF}$. It is easy to see that for every mapping $\alpha : \mathrm{X} \to \mathcal{C}^*$ we have:

$$\forall A \in \mathcal{EF} : \alpha \approx^I_{\mathcal{X}} \nu \implies \mathfrak{c}(\alpha(A)) = \mathfrak{c}(A). \tag{3}$$

The following definition of the mapping $\mathfrak{o}$ from the set $\mathcal{EF} \setminus \{\epsilon\}$ to the set $\mathcal{EF} \cup \overline{\mathrm{X}}^{\mathrm{suff}}$ is correct because ν is compatible on the set $\mathcal{X}$, which implies $\mathfrak{c}(x_i^{\mathrm{pref}}) \subseteq \mathfrak{c}(x_j^{\mathrm{pref}}) \subset \mathfrak{c}(x)$ for all $x \in \mathrm{X}$ and $1 \le i < j < \sharp_{\nu(x)} = \|\mathfrak{c}(x)\|$.

Let $A \in \mathcal{EF}$ be a non-empty word and let $B \in \mathcal{EF}$ be the longest prefix of A such that $\mathfrak{c}(B) \neq \mathfrak{c}(A)$ and $b \in \mathcal{C} \cup \mathrm{X}^{\mathrm{ext}}$ be the next letter in A, i.e. Bb is also a prefix of A — the shortest prefix satisfying $\mathfrak{c}(Bb) = \mathfrak{c}(A)$.

- If $b \in \mathcal{C}$ then we put $\mathfrak{o}(A) = B$.
- If $b = x$ or $b = x_j^{\mathrm{pref}}$ for some $x \in \mathrm{X}$ and $j \in \mathbb{N}$, then we put $\mathfrak{o}(A) = Bx_i^{\mathrm{pref}}$, where $i \in \mathbb{N}$ is the largest number such that $\mathfrak{c}(Bx_i^{\mathrm{pref}}) \subset \mathfrak{c}(A)$. If such i does not exist, i.e. $\mathfrak{c}(Bx_1^{\mathrm{pref}}) = \mathfrak{c}(A)$, we put again $\mathfrak{o}(A) = B$.
- If $b = x_i^{\mathrm{suff}} \in \mathrm{X}^{\mathrm{suff}}$ (which means $B = \epsilon$, $\mathfrak{c}(A) = \mathfrak{c}(x_i^{\mathrm{suff}})$), we put $\mathfrak{o}(A) = \overline{x}_i^{\mathrm{suff}}$.

The mapping $\mathfrak{j} : \mathcal{EF} \setminus \{\epsilon\} \to \mathcal{EF} \cup \overline{\mathrm{X}}^{\mathrm{pref}}$ is defined dually. The following lemma is a trivial consequence of the definitions of $\mathfrak{o}$ and $\mathfrak{j}$.

Lemma 2. *Let $A \in \mathcal{EF} \setminus \{\epsilon\}$ and let $\alpha : \mathcal{X} \to \mathcal{C}^*$ be a compatible mapping such that $\alpha \approx^I_{\mathcal{X}} \nu$. Then $0(\alpha(A)) = \alpha(\mathfrak{o}(A))$ and $1(\alpha(A)) = \alpha(\mathfrak{j}(A))$.* □

As a consequence of Proposition 1, the property (3) and Lemma 2 we obtain:

Lemma 3. *Let $L, R \in \mathcal{EF} \setminus \{\epsilon\}$ and let α be a compatible mapping on $\mathcal{X}$ such that $\alpha \approx^I_{\mathcal{X}} \nu$. Then α is a solution of the equation $L =^? R$ if and only if $\mathfrak{c}(L) = \mathfrak{c}(R)$ and α is a solution of the system $\{\mathfrak{o}(L) =^? \mathfrak{o}(R), \mathfrak{j}(L) =^? \mathfrak{j}(R)\}$.* □

We have introduced all required technical notation, so we are ready to describe the algorithm. The algorithm exchanges certain equations $L =^? R$ from the current system satisfying $\mathfrak{c}(L) = \mathfrak{c}(R)$ for $\mathfrak{o}(L) =^? \mathfrak{o}(R)$ and $\mathfrak{j}(L) =^? \mathfrak{j}(R)$ in each step. It does this in such a way that a variable from the set $\overline{\mathrm{X}}$ appears at most one time in the process. It is possible because $\overline{x}_i^{\mathrm{pref}} \in \overline{\mathrm{X}}$ and $\overline{x}_i^{\mathrm{suff}} \in \overline{\mathrm{X}}$ can appear in the equations $\mathfrak{o}(L) =^? \mathfrak{o}(R)$ or $\mathfrak{j}(L) =^? \mathfrak{j}(R)$ only if $\|\mathfrak{c}(L)\| = \|\mathfrak{c}(R)\| = i$. That is why we exchange in every step exactly all equations $L =^? R$ with maximal sets $\mathfrak{c}(L) = \mathfrak{c}(R)$. If such an equation containing a variable from the set $\overline{\mathrm{X}}$ occurs after the exchange, then we put it aside to a new system $S^E \cup S^X$; more precisely, we put the equations of the form $\overline{x}_i^{\mathrm{pref}} =^? \overline{y}_i^{\mathrm{pref}}$ to the system S^E and the equations of the form $\overline{x}_i^{\mathrm{pref}} =^? A$, where $A \in \mathcal{EF}$, to the system S^X. In this way we guarantee that the current system S is always a subset of $\mathcal{EF} \times \mathcal{EF}$.

The Algorithm.

The algorithm uses systems of equations S, S^E, S^X, S', T and S_k^E, S_k^X for $k \in \{1, \ldots, m\}$ which are specified as follows. The system S is modified within the algorithm, but all the time $S \subseteq \mathcal{EF} \times \mathcal{EF}$. The systems S^E and S^X are constructed in the algorithm in such a way that $S^E = \bigcup_{k=1}^m S_k^E$ and $S^X = \bigcup_{k=1}^m S_k^X$, where S_k^E and S_k^X are constructed when the main cycle of the algorithm is performed for the parameter k and $S_k^E \subseteq \overline{\mathrm{X}}_k^{\mathrm{pref}} \times \overline{\mathrm{X}}_k^{\mathrm{pref}} \cup \overline{\mathrm{X}}_k^{\mathrm{suff}} \times \overline{\mathrm{X}}_k^{\mathrm{suff}}$, $S_k^X \subseteq \overline{\mathrm{X}}_k \times \mathcal{EF}$. We denote $S' = S \cup S^E \cup S^X$. Finally, the system T is auxiliary and is used locally during each iteration of the main cycle.

Initialization.
Put $T = \emptyset$ and $S_k^E = S_k^X = \emptyset$ for any $k \in \{1, \ldots, m\}$. Delete trivial equations and check whether $\mathfrak{c}(L) = \mathfrak{c}(R)$ holds for all equations $L =^? R$ in S. If $\mathfrak{c}(L) \neq \mathfrak{c}(R)$ for some equation, then stop with the answer "unsolvable".

Main Cycle.
For any k from m down to 1 perform the following three steps.

1. **Replacing of Equations.**
   Put to the system T all non-trivial equations $\mathfrak{o}(L) =^? \mathfrak{o}(R)$ and $\mathfrak{j}(L) =^? \mathfrak{j}(R)$ where $L =^? R$ is an equation in S satisfying $\|\mathfrak{c}(L)\| = k$. Then remove all equations $L =^? R$ with $\|\mathfrak{c}(L)\| = k$ from S.

   Remark 1. It follows directly from the definition of $\mathfrak{o}$ that the variable $\overline{x}_i^{\mathrm{pref}}$ can not occur in any equation $\mathfrak{o}(L) =^? \mathfrak{o}(R)$. If $\overline{x}_i^{\mathrm{pref}}$ occurs in some non-trivial equation $\mathfrak{j}(L) =^? \mathfrak{j}(R) \in T$ in this step, then $i = k$ and this equation is of the form $\mathfrak{j}(L) =^? \overline{x}_i^{\mathrm{pref}}$ or $\overline{x}_i^{\mathrm{pref}} =^? \mathfrak{j}(R)$ where $\overline{x}_i^{\mathrm{pref}} \notin \mathsf{c}(\mathfrak{j}(L))$, $\overline{x}_i^{\mathrm{pref}} \notin \mathsf{c}(\mathfrak{j}(R))$ respectively. Analogically $\overline{x}_i^{\mathrm{suff}}$ can appear in T only as one side of an equation $\mathfrak{o}(L) =^? \mathfrak{o}(R)$ and only if $i = k$.

2. **Construction of Systems S_k^E and S_k^X.**
   Define $S_k^E = T \cap (\overline{\mathrm{X}}_k \times \overline{\mathrm{X}}_k)$ and remove this subsystem from T. Let E_k be the equivalence relation on $\overline{\mathrm{X}}_k$ generated by S_k^E and let Q_k be the partition on $\overline{\mathrm{X}}_k$ corresponding to E_k. For every $q \in Q_k$ choose some $A \in \mathcal{EF}$ such that there is an equation $A =^? \overline{y}$ or $\overline{y} =^? A$ in T for some $\overline{y} \in q$; if such A does not exist, do nothing for q. Then put the equation $\overline{y} =^? A$ to S_k^X and replace all equations $B =^? \overline{z}$ and $\overline{z} =^? B$ in T, where $\overline{z} \in q$, $B \in \mathcal{EF}$, with $A =^? B$.

   Remark 2. Remark 1 gives $S_k^E \subseteq \overline{\mathrm{X}}_k^{\mathrm{pref}} \times \overline{\mathrm{X}}_k^{\mathrm{pref}} \cup \overline{\mathrm{X}}_k^{\mathrm{suff}} \times \overline{\mathrm{X}}_k^{\mathrm{suff}}, T \subseteq \mathcal{EF} \times \mathcal{EF}$.

3. **Negative Answer.**
   If there is an equation $L =^? R$ in T such that $\mathfrak{c}(L) \neq \mathfrak{c}(R)$, then stop with the answer "unsolvable", else transfer all non-trivial equations from T to S.

   Remark 3. If the algorithm transfers some equation $L =^? R$ from T into the system S, then $\mathfrak{c}(L) = \mathfrak{c}(R)$, $L, R \in \mathcal{EF}$ and $\|\mathfrak{c}(L)\| = k - 1$. This means that at the end of this step S contains only non-trivial equations $L =^? R$ such that $\mathfrak{c}(L) = \mathfrak{c}(R)$, $L, R \in \mathcal{EF}$ and $\|\mathfrak{c}(L)\| < k$. In particular, for $k = 1$ the resulting system S is empty.

Positive Answer.
Stop with the answer "solvable".

Lemma 4. *The algorithm stops with the correct answer in polynomial time. Moreover, if the original system S is solvable then it is equivalent to the resulting system S'.*

Proof. The detailed proof can be found in the full version of the paper [9]. Basically, it is not hard to see that any step of the algorithm reduces the current system $S' \cup T$ to an equivalent one and that the answer "unsolvable" is always correct. Here we only suggest how we can define a solution of the resulting system $S' = S^E \cup S^X$ when the algorithm answers "solvable". We do this inductively with respect to $\|\mathfrak{c}(\omega)\|$ for $\omega \in \mathcal{X}$.

At first, we put $\alpha(x_1^{\mathrm{pref}}) = \nu(x_1^{\mathrm{pref}})$ and $\alpha(x_1^{\mathrm{suff}}) = \nu(x_1^{\mathrm{suff}})$ for all $x \in \mathrm{X}$. For $x \in \mathrm{X}$ with $\|\mathfrak{c}(x)\| = i > 1$ we put $\alpha(x) = \alpha(x_{i-1}^{\mathrm{pref}})\, ab\, \alpha(x_{i-1}^{\mathrm{suff}})$ where a and b are uniquely determined by the conditions $a \in \mathfrak{c}(x) \setminus \mathfrak{c}(x_{i-1}^{\mathrm{pref}})$ and $b \in \mathfrak{c}(x) \setminus \mathfrak{c}(x_{i-1}^{\mathrm{suff}})$. For every $x \in \mathrm{X}$ we put $\alpha(x_i^{\mathrm{pref}}) = \alpha(x_{i-1}^{\mathrm{pref}})\, ab\, \alpha(\overline{x}_i^{\mathrm{pref}})$ where a and b are uniquely determined by the conditions $a \in \mathfrak{c}(x_i^{\mathrm{pref}}) \setminus \mathfrak{c}(x_{i-1}^{\mathrm{pref}})$ and $b \in \mathfrak{c}(x_i^{\mathrm{pref}}) \setminus \mathfrak{c}(\overline{x}_i^{\mathrm{pref}})$. We define $\alpha(x_i^{\mathrm{suff}})$ analogically.

Assume that we have determined $\alpha(y)$ for all $y \in \mathrm{X}^{\mathrm{ext}}$, $\|\mathfrak{c}(y)\| \leq i$. If for $\overline{x}_{i+1}^{\mathrm{pref}}$ there is an equation $\overline{z} =^? A$ in S^X, where $\overline{z} \in \overline{\mathrm{X}}_{i+1}^{\mathrm{pref}}$ belongs to the same class of the partition Q_{i+1} as $\overline{x}_{i+1}^{\mathrm{pref}}$, then such an equation is only one and $\|\mathfrak{c}(A)\| = i$, so $\alpha(y)$ is already given for all $y \in \mathfrak{c}(A) \subseteq \mathcal{C} \cup \mathrm{X}^{\mathrm{ext}}$. Hence we can correctly define $\alpha(\overline{x}_{i+1}^{\mathrm{pref}}) = \alpha(A)$. If such an equation does not exist, then we choose any word $u \in \mathcal{C}^*$ such that $\sharp_u = i$ and $\mathfrak{c}(u) \subset \mathfrak{c}(x_{i+1}^{\mathrm{pref}})$ and define $\alpha(\overline{z}) = u$ for every $\overline{z} \in \overline{\mathrm{X}}_{i+1}^{\mathrm{pref}}$ which belongs to the same class of the partition Q_{i+1} as $\overline{x}_{i+1}^{\mathrm{pref}}$. We define $\alpha(\overline{z})$ for $\overline{z} \in \overline{\mathrm{X}}_{i+1}^{\mathrm{suff}}$ analogically. It is not hard to see that α is really a compatible solution on $\mathcal{X}$ and that the algorithm stops in polynomial time. □

The previous lemma concludes the proof of the proposition. □

As an application of the method used to prove Lemma 4 we can improve the upper bound for the shortest solution of a matching system given in [10].

Proposition 4. *Let $S = \{L_1 =^? R_1, \ldots, L_n =^? R_n\}$ be a matching system over the set of constants $\mathcal{C}$, i.e. $R_i \in \mathcal{C}^*$. If there exists a solution of the system S, then there exists a solution α satisfying $|\alpha(x)| < 2 \cdot \|\mathcal{C}\| \cdot \max_i\{|R_i|\}$ for all $x \in \mathrm{X}$.*

Proof. Let ν be a solution of S and $r = \max_i\{|R_i|\} \geq 2$. Then the system (S, ν) with complete information about invariants is solvable. Let $S' = S^E \cup S^X$ be the system computed by the algorithm. It is easy to see that $S^E = \emptyset$. Any equation in S^X is of the form $\overline{y} = A$ where $\overline{y} \in \overline{\mathrm{X}}$ and A is a factor of some R_i such that $|A| \leq r - 2$. Hence, we can assume $|\alpha(\overline{y})| \leq r - 2$ for all $\overline{y} \in \overline{\mathrm{X}}$. Then we can inductively prove $|\alpha(x_i^{\mathrm{pref}})| \leq i \cdot r$ for a solution defined by the formulas from the proof of Lemma 4 (the same holds for x_i^{suff}). Altogether we have $|\alpha(x)| \leq 2 \cdot (\sharp_{\nu(x)} - 1) \cdot r + 2 < 2 \cdot \sharp_{\nu(x)} \cdot r \leq 2 \cdot \|\mathcal{C}\| \cdot r$. □

4 Conclusion

Theorem 1. *The* AIU-UNIFICATION *problem and the* AI-UNIFICATION *problem are NP-complete.*

Proof. The NP-hardness is a consequence of Proposition 2.

Let S be a system. We can guess a mapping $\nu : \mathrm{X} \to \mathcal{C}^*$ such that $|\nu(x)| \leq 2 \cdot \|\mathcal{C}\|$ for all $x \in \mathrm{X}$ and then check in polynomial time whether the system (S, ν) with complete information about invariants has a solution by Proposition 3. The difference between the AIU-UNIFICATION and AI-UNIFICATION problems is only in the additional assumption $\nu(x) \neq \epsilon$ for all $x \in \mathrm{X}$. □

Applying the result from [2] we obtain the following consequence.

Corollary 1. *The* GENERAL AIU-UNIFICATION *problem and the* GENERAL AI-UNIFICATION *problem are NP-complete.* □

Our method of solving of a system with complete information about invariants can be also used for counting the number of solutions (up to $\sim$) of such a system. We expect that this method should help to determine the complexity of the counting problem of AI-MATCHING.

Acknowledgments

I would like to thank Michal Kunc for his help with preparation of the paper.

References

[1] Baader F.: The Theory of Idempotent Semigroups is of Unification Type Zero, J. of Automated Reasoning **2** (1986) 283–286.

[2] Baader F., Schulz K. U.: Unification in the Union of Disjoint Equational Theories: Combining Decision Procedures, J. of Symbolic Computation **21** (1996) 211–243.

[3] Baader F., Siekmann J. H.: Unification Theory, Handbook of Logic in Artificial Intelligence and Logic Programming (1993), Oxford University Press.

[4] Baader F., Snyder W.: Unification Theory, Handbook of Automated Reasoning (2001), Elsevier Science Publishers.

[5] Green J. A., Rees D.: On Semigroups in which $x^r = x$, Proc. Camb. Phil. Soc. **48** (1952) 35–40.

[6] Kapur D., Narendran P.: NP–completeness of the Set Unification and Matching Problems, In Proceedings of CADE'86, Vol. 230 of LNCS (1986) 489–495.

[7] Kapur D., Narendran P.: Complexity of Unification Problems with Associative-commutative Operators, J. of Automated Reasoning **9**, No.2, (1992) 261–288.

[8] Klíma O.: On the solvability of equations in semigroups with $x^r = x$, Contributions to General Algebra **12**, Proc. of 58th Workshop on General Algebra, Verlag Johannes Heyn, Klagenfurt 2000 (237–246).

[9] Klíma O.: Unification Modulo Associativity and Idempotency is NP-complete, Technical report available at http://math.muni.cz/~klima.

[10] Klíma O., Srba J.: Matching Modulo Associativity and Idempotency is NP-complete, In Proceedings of MFCS'00, Vol. 1893 of LNCS (2000) 456–466.

[11] Schmidt-Schauss M.: Unification under Associativity and Idempotence is of Type Nullary, J. of Automated Reasoning **2** (1986) 277–281.

On the Complexity of Semantic Equivalences for Pushdown Automata and BPA

Antonín Kučera*[1] and Richard Mayr[2]

[1] Faculty of Informatics, Masaryk University,
Botanická 68a, 60200 Brno, Czech Republic,
tony@fi.muni.cz

[2] Department of Computer Science, Albert-Ludwigs-University Freiburg,
Georges-Koehler-Allee 51, D-79110 Freiburg, Germany,
mayrri@informatik.uni-freiburg.de

Abstract. We study the complexity of comparing pushdown automata (PDA) and context-free processes (BPA) to finite-state systems, w.r.t. strong and weak simulation preorder/equivalence and strong and weak bisimulation equivalence. We present a complete picture of the complexity of all these problems. In particular, we show that strong and weak simulation preorder (and hence simulation equivalence) is *EXPTIME*-complete between PDA/BPA and finite-state systems in both directions. For PDA the lower bound even holds if the finite-state system is fixed, while simulation-checking between BPA and any fixed finite-state system is already polynomial. Furthermore, we show that weak (and strong) bisimilarity between PDA and finite-state systems is *PSPACE*-complete, while strong (and weak) bisimilarity between two PDAs is *EXPTIME*-hard.

1 Introduction

Transition systems are a fundamental and widely accepted model of processes with discrete states and dynamics (such as computer programs). Formally, a transition system is a triple $\mathcal{T} = (S, Act, \rightarrow)$ where S is a set of *states* (or *processes*), Act is a finite set of *actions*, and $\rightarrow \subseteq S \times Act \times S$ is a *transition relation*. We write $s \stackrel{a}{\rightarrow} t$ instead of $(s, a, t) \in \rightarrow$ and we extend this notation to elements of Act^* in the natural way. A state t is *reachable* from a state s, written $s \rightarrow^* t$, iff $s \stackrel{w}{\rightarrow} t$ for some $w \in Act^*$.

In the *equivalence-checking* approach to formal verification, one describes the *specification* (the intended behavior) and the actual *implementation* of a given process as states in transition systems, and then it is shown that they are *equivalent*. Here the notion of equivalence can be formalized in various ways according to specific needs of a given practical problem (see, e.g., [15] for an overview). It seems, however, that *simulation* and *bisimulation* equivalence are of special importance as their accompanying theory has been developed very

* Supported by the Grant Agency of the Czech Republic, grant No. 201/00/0400.

K. Diks et al. (Eds): MFSC 2002, LNCS 2420, pp. 433–445, 2002.

intensively and found its way to many practical applications. Let $\mathcal{T} = (S, Act, \rightarrow)$ be a transition system. A binary relation $R \subseteq S \times S$ is a *simulation* iff whenever $(s,t) \in R$, then for each $s \stackrel{a}{\rightarrow} s'$ there is some $t \stackrel{a}{\rightarrow} t'$ such that $(s',t') \in R$. A process s is *simulated* by t, written $s \sqsubseteq t$, iff there is a simulation R such that $(s,t) \in R$. Processes s,t are *simulation equivalent*, written $s \simeq t$, iff they can simulate each other. A *bisimulation* is a symmetric simulation relation, and two processes s and t are *bisimilar* iff they are related by some bisimulation. In order to abstract from internal ('invisible') transitions of a given system, simulations and bisimulations are sometimes considered in their *weak* forms. Here, the silent steps are usually modeled by a distinguished action τ, and the *extended* transition relation $\Rightarrow \subseteq S \times Act \times S$ is defined by $s \stackrel{a}{\Rightarrow} t$ iff either $s = t$ and $a = \tau$, or $s \stackrel{\tau^i}{\rightarrow} s' \stackrel{a}{\rightarrow} t' \stackrel{\tau^j}{\rightarrow} t$ for some $i,j \in \mathbb{N}_0$ and $s',t' \in S$.

Simulations (and bisimulations) can also be viewed as *games* [12,14] between two players, the attacker and the defender. In a simulation game the attacker wants to show that $s \not\sqsubseteq t$, while the defender attempts to frustrate this. Imagine that there are two tokens put on states s and t. Now the two players, attacker and defender, start to play a *simulation game* which consists of a (possibly infinite) number of *rounds* where each round is performed as follows: The attacker takes the token which was put on s originally and moves it along a transition labeled by (some) a; the task of the defender is to move the other token along a transition with the same label. If one player cannot move then the other player wins. The defender wins every infinite game. It can be easily shown that $s \sqsubseteq t$ iff the defender has a universal winning strategy. The only difference between a simulation game and a *bisimulation game* is that the attacker can *choose* his token at the beginning of every round (the defender has to respond by moving the other token). Again we get that $s \sim t$ iff the defender has a winning strategy. Corresponding 'weak forms' of the two games are defined in the obvious way. We use the introduced games at some points to give a more intuitive justification for our claims. Simulations and bisimulations can also be used to relate states of *different* transition systems; formally, two systems are considered to be a single one by taking the disjoint union.

In this paper we mainly consider processes of *pushdown automata*, which are interpreted as a (natural) model of sequential systems with mutually recursive procedures. A pushdown automaton is a tuple $\Delta = (Q, \Gamma, Act, \delta)$ where Q is a finite set of *control states*, Γ is a finite *stack alphabet*, Act is a finite *input alphabet*, and $\delta : (Q \times \Gamma) \rightarrow \mathcal{P}(Act \times (Q \times \Gamma^*))$ is a *transition function* with finite image (here $\mathcal{P}(M)$ denotes the power set of M). We can assume (w.l.o.g.) that each transition increases the height (or length) of the stack by at most one (each PDA can be efficiently transformed to this kind of normal form). In the rest of this paper we adopt a more intuitive notation, writing $pA \stackrel{a}{\rightarrow} q\beta \in \delta$ instead of $(a,(q,\beta)) \in \delta(p,A)$. To Δ we associate the transition system $\mathcal{T}_\Delta$ where $Q \times \Gamma^*$ is the set of states (we write $p\alpha$ instead of (p,α)), Act is the set of actions, and the transition relation is determined by $pA\alpha \stackrel{a}{\rightarrow} q\beta\alpha$ iff $pA \stackrel{a}{\rightarrow} q\beta \in \delta$.

Let A, B be classes of processes. The problem whether a given process s of A is simulated (or weakly simulated) by a given process t of B is denoted by $A \sqsubseteq B$ (or

$A \sqsubseteq_w B$, respectively). Similarly, the problem if s and t are simulation equivalent, weakly simulation equivalent, bisimilar, or weakly bisimilar, is denoted by $A \simeq B$, $A \simeq_w B$, $A \sim B$, or $A \approx B$, respectively. The classes of all pushdown processes and finite-state processes (i.e., processes of finite-state transition systems) are denoted **PDA** and **FS**, respectively. **BPA** (basic process algebra), also called context-free processes, is the subclass of **PDA** where $|Q| = 1$, i.e., without a finite-control.

The State of the Art for Simulation

It has been known for some time that strong simulation preorder between **PDA** and **FS** is decidable in exponential time. This is because one can reduce the simulation problem to the model-checking problem with **PDA** and a fixed formula of the modal μ-calculus (see, e.g., [6,4]). As model checking **PDA** with the modal μ-calculus is *EXPTIME*-complete [17] the result follows. A *PSPACE* lower bound for the **FS** $\sqsubseteq$ **BPA** problem and a co-$\mathcal{NP}$ lower bound for the **BPA** $\sqsubseteq$ **FS** and **BPA** $\simeq$ **FS** problems have been shown in [6]. Furthermore, an *EXPTIME* lower bound for the **FS** $\sqsubseteq$ **PDA** and **FS** $\simeq$ **PDA** problems have been shown in [4], but in these constructions the finite-state systems were not fixed. The problems of comparing two different BPA/PDA processes w.r.t. simulation preorder/equivalence are all undecidable.

Our Contribution

We show that the problems **BPA** $\sqsubseteq$ **FS**, **FS** $\sqsubseteq$ **BPA** and **BPA** $\simeq$ **FS** are *EXPTIME*-complete, but polynomial for every fixed finite-state system. On the other hand, the problems **PDA** $\sqsubseteq$ **FS**, **FS** $\sqsubseteq$ **PDA** and **PDA** $\simeq$ **FS** are *EXPTIME*-complete, even for a fixed finite-state system. Here, the main point are the lower bounds, which require some new insights into the power of the defender in simulation games. The matching upper bounds are obtained by a straightforward extension of the above mentioned reduction to the model-checking problem with the modal μ-calculus.

The State of the Art for Bisimulation

It was known that strong and weak bisimulation equivalence between **PDA** and **FS** is decidable in exponential time, because one can construct (in polynomial time) characteristic modal μ-calculus formulae for the finite-state system and thus reduce the problem to model checking the **PDA** with a modal μ-calculus formula [11], which is decidable in exponential time [17]. The best known lower bound for the **PDA** $\approx$ **FS** problem was *PSPACE*-hardness, which even holds for a fixed finite state system [8]. The problem **PDA** $\sim$ **FS** is also *PSPACE*-hard, but polynomial in the size of the **PDA** for every fixed finite-state system [8]. Interestingly, the problem **BPA** $\approx$ **FS** (and **BPA** $\sim$ **FS**) is polynomial [7]. The symmetric problem of **PDA** $\sim$ **PDA** is decidable [9,13], but the complexity is not known. So far, the best known lower bound for it was *PSPACE*-hardness [8]. The decidability of the **PDA** $\approx$ **PDA** problem is still open.

Our Contribution

We show that the problems **PDA** $\sim$ **FS** and **PDA** $\approx$ **FS** are *PSPACE*-complete by improving the known *EXPTIME* upper bound to *PSPACE*. Furthermore, we show that the symmetric problem **PDA** $\sim$ **PDA** is *EXPTIME*-hard, by improving the known *PSPACE* lower bound to *EXPTIME*. This new *EXPTIME* lower bound even holds for the subclass of normed **PDA**.

Due to space constraints, several proofs are omitted. They can be found in the full version of the paper [5].

2 Lower Bounds

In this section we prove that all of the problems **BPA** $\sqsubseteq$ **FS**, **FS** $\sqsubseteq$ **BPA** and **BPA** $\simeq$ **FS** are *EXPTIME*-hard. The problems **PDA** $\sqsubseteq$ **FS**, **FS** $\sqsubseteq$ **PDA**, **PDA** $\simeq$ **FS** are *EXPTIME*-hard even for a *fixed* finite-state system. Moreover, we show *EXPTIME*-hardness of the **PDA** $\sim$ **PDA** problem.

An *alternating LBA* is a tuple $\mathcal{M} = (S, \Sigma, \gamma, s_0, \vdash, \dashv, \pi)$ where $S, \Sigma, \gamma, s_0, \vdash$, and $\dashv$ are defined as for ordinary non-deterministic LBA. In particular, S is a finite set of control states (we reserve 'Q' to denote a set of control states of pushdown automata), $\vdash, \dashv \in \Sigma$ are the left-end and right-end markers, respectively, and $\pi : S \to \{\forall, \exists, acc, rej\}$ is a function which partitions the control states of S into *universal*, *existential*, *accepting*, and *rejecting*, respectively. We assume (w.l.o.g.) that γ is defined so that

- for all $s \in S$ and $A \in \Sigma$ such that $\pi(s) = \forall$ or $\pi(s) = \exists$ we have that $|\gamma(s, A)| = 2$ (i.e., $\gamma(s, A) = \{s_1, s_2\}$ for some $s_1, s_2 \in S$). The first element of $\gamma(s, A)$ is denoted by $first(s, A)$, and the second one by $second(s, A)$. It means that each configuration of $\mathcal{M}$ where the control state is universal or existential has exactly two immediate successors (configurations reachable in one computational step).
- for all $s \in S$ and $A \in \Sigma$ such that $\pi(s) = acc$ or $\pi(s) = rej$ we have that $\gamma(s, A) = \emptyset$, i.e., each configuration of $\mathcal{M}$ where the control state is accepting or rejecting is 'terminated' (without any successors).

A *computational tree* for $\mathcal{M}$ on a word $w \in \Sigma^*$ is a finite tree T satisfying the following: the root of T is (labeled by) the initial configuration $s_0 \vdash w \dashv$ of $\mathcal{M}$, and if N is a node of $\mathcal{M}$ labeled by a configuration usv where $u, v \in \Sigma^*$ and $s \in S$, then the following holds:

- if s is accepting or rejecting, then T is a leaf;
- if s is existential, then T has one successor whose label is one of the two configurations reachable from usv in one step (here, the notion of a computational step is defined in the same way as for 'ordinary' Turing machines);
- if s is universal, then T has two successors labeled by the two configurations reachable from usv in one step.

$\mathcal{M}$ *accepts* w iff there is a computational tree T such that all leafs of T are accepting configurations. The acceptance problem for alternating LBA is known to be *EXPTIME*-complete.

In subsequent proofs we often use $M_\star$ to denote the set $M \cup \{\star\}$ where M is a set and $\star \notin M$ is a fresh symbol.

Theorem 1. *The problem* $\boldsymbol{BPA} \sqsubseteq \boldsymbol{FS}$ *is EXPTIME-hard.*

Proof. Let $\mathcal{M} = (S, \Sigma, \gamma, s_0, \vdash, \dashv, \pi)$ be an alternating LBA and $w \in \Sigma^*$ an input word. We construct (in polynomial time) a BPA system $\Delta = (\Gamma, Act, \delta)$, a finite-state system $\mathcal{F} = (S, Act, \rightarrow)$, and processes α and X of Δ and $\mathcal{F}$, resp., such that $\mathcal{M}$ accepts w iff $\alpha \not\sqsubseteq X$. Let n be the length of w. We put $\Gamma = S_\star \times \Sigma \cup S \times \Sigma_\star \times \{0, \cdots, n{+}2\} \cup S \times \Sigma \times \{W\} \cup \{T, Z\}$. Configurations of $\mathcal{M}$ are encoded by strings over $S_\star \times \Sigma$ of length $n+2$. A configuration usv, where $u, v \in \Sigma^*$ and $s \in S$, is written as $\langle\star, v(k)\rangle\langle\star, v(k-1)\rangle \cdots \langle\star, v(2)\rangle\,\langle s, v(1)\rangle\langle\star, u(m)\rangle \cdots \langle\star, u(1)\rangle$ where k and m are the lengths of v and u, resp., and $v(i)$ denotes the ith symbol of v (configurations are represented in a 'reversed order' since we want to write the top stack symbol on the left-hand side). Elements of $S \times \Sigma_\star \times \{0, \cdots, n+2\}$ are used as top stack symbols when pushing a new configuration to the stack (see below); they should be seen as a finite memory where we keep (and update) the information about the position of the symbol which will be guessed by the next transition (as we count symbols from zero, the bounded counter reaches the value $n+2$ after guessing the last symbol), about the control state which is to be pushed, and about the (only) symbol of the form $\langle s, a\rangle$ which was actually pushed. The Z is a special 'bottom' symbol which can emit all actions and cannot be popped. The role of symbols of $S \times \Sigma \times \{W\} \cup \{T\}$ will be clarified later. The set of actions is $Act = \{a, c, f, s, d, t\} \cup (S_\star \times \Sigma)$, and δ consists of the following transitions:

1. $(\langle s, \star\rangle, i) \xrightarrow{a} (\langle s, \star\rangle, i+1)\,\langle\star, A\rangle$	for all $A \in \Sigma$, $s \in S$, $0 \le i \le n+1$;
2. $(\langle s, \star\rangle, i) \xrightarrow{a} (\langle s, A\rangle, i+1)\,\langle s, A\rangle$	for all $A \in \Sigma$, $s \in S$, $0 \le i \le n+1$;
3. $(\langle s, A\rangle, i) \xrightarrow{a} (\langle s, A\rangle, i+1)\langle\star, B\rangle$	for all $A, B \in \Sigma$, $s \in S$, $0 \le i \le n+1$;
4. $(\langle s, A\rangle, n+2) \xrightarrow{c} (\langle s, A\rangle, W)$	for all $A \in \Sigma$, $s \in S$;
5. $(\langle s, A\rangle, W) \xrightarrow{d} \varepsilon$	for all $s \in S$, $A \in \Sigma$ such that s is not rejecting;
6. $(\langle s, A\rangle, W) \xrightarrow{f} (\langle s', \star\rangle, 0)$	for all $s, s' \in S$, $A \in \Sigma$ such that $\pi(s) \in \{\forall, \exists\}$ and $s' = \mathit{first}(s, A)$;
7. $(\langle s, A\rangle, W) \xrightarrow{s} (\langle s', \star\rangle, 0)$	for all $s, s' \in S$, $A \in \Sigma$ such that $\pi(s) \in \{\forall, \exists\}$ and $s' = \mathit{second}(s, A)$;
8. $(\langle s, A\rangle, W) \xrightarrow{f} (\langle s', \star\rangle, 0)$	for all $s, s' \in S$, $A \in \Sigma$ such that $\pi(s) = \exists$ and $s' = \mathit{second}(s, A)$;
9. $(\langle s, A\rangle, W) \xrightarrow{s} (\langle s', \star\rangle, 0)$	for all $s, s' \in S$, $A \in \Sigma$ such that $\pi(s) = \exists$ and $s' = \mathit{first}(s, A)$;
10. $(\langle s, A\rangle, W) \xrightarrow{y} T$	for all $s \in S$, $y \in \{f, s\}$ such that $\pi(s) = \mathit{acc}$;
11. $T \xrightarrow{t} T$	
12. $Z \xrightarrow{y} Z$	for all $y \in Act$;
13. $\langle x, A\rangle \xrightarrow{\langle x, A\rangle} \varepsilon$	for all $x \in S_\star$, $A \in \Sigma$.

The process α corresponds to the initial configuration of $\mathcal{M}$, i.e.,

$$\alpha \quad = \quad (\langle s_0, \vdash\rangle, n{+}2)\, \langle \star, \dashv\rangle\, \langle \star, w(n)\rangle \cdots \langle \star, w(2)\rangle\, \langle \star, w(1)\rangle\, \langle s_0, \vdash\rangle\, Z$$

The behavior of α can be described as follows: whenever the top stack symbol is of the form $(\langle s, A\rangle, W)$, we know that the previously pushed configuration contains the symbol $\langle s, A\rangle$. If s is *rejecting*, no further transitions are possible. Otherwise, $(\langle s, A\rangle, W)$ can either disappear (emitting the action d—see rule 5), or it can perform one of the f and s actions as follows:

- If s is *universal* or *existential*, $(\langle s, A\rangle, W)$ can emit either f or s, storing *first*(s, A) or *second*(s, A) in the top stack symbol, respectively (rules 6, 7).
- If s is *existential*, $(\langle s, A\rangle, W)$ can also emit f and s while storing *second*(s, A) and *first*(s, A), respectively (rules 8, 9).
- If s is *accepting*, $(\langle s, A\rangle, W)$ emits f or s and pushes the symbol T which can do the action t forever (rules 10, 11).

If $(\langle s, A\rangle, W)$ disappears, the other symbols stored in the stack subsequently perform their symbol-specific actions and disappear (rule 13). If s is not accepting and $(\langle s, A\rangle, W)$ emits f or s, a new configuration is guessed and pushed to the stack; the construction of δ ensures that

- exactly $n + 2$ symbols are pushed (rules 1–4);
- at most one symbol of the form $\langle s', B\rangle$ is pushed; moreover, the s' must be the control state stored in the top stack symbol. After pushing $\langle s', B\rangle$, the B is also remembered in the top stack symbol (rule 2);
- if no symbol of the form $\langle s', B\rangle$ is pushed, no further transitions are possible after guessing the last symbol of the configuration (there are no transitions for symbols of the form $(\langle s', *\rangle, n + 2)$);
- after pushing the last symbol, the action c is emitted and a 'waiting' symbol $(\langle s', B\rangle, W)$ is pushed.

Now we define the finite-state system $\mathcal{F}$. The set of states of $\mathcal{F}$ is given by

$$S = \{X, F, S, U, C_0, \cdots, C_n\} \cup \{C_0, \cdots, C_n\} \times \{0, \cdots, n+1\} \times (S_\star \times \Sigma)^4_\star.$$

Transitions of $\mathcal{F}$ are

1. $X \stackrel{a}{\to} X,\ X \stackrel{c}{\to} F,\ X \stackrel{c}{\to} S,\ X \stackrel{c}{\to} C_i$ for every $0 \le i \le n$;
2. $F \stackrel{f}{\to} X,\ F \stackrel{y}{\to} U$ for every $y \in Act - \{f\}$;
3. $S \stackrel{s}{\to} X,\ S \stackrel{y}{\to} U$ for every $y \in Act - \{s\}$;
4. $C_i \stackrel{d}{\to} (C_i, 0, \star, \star, \star, \star),\ C_i \stackrel{y}{\to} U$ for every $0 \le i \le n$, $y \in Act - \{d\}$;
5. $U \stackrel{y}{\to} U$ for every $y \in Act$;
6. $(C_i, j, \star, \star, \star, \star) \stackrel{y}{\to} (C_i, j{+}1, \star, \star, \star, \star)$ for all $0 \le i \le n$, $0 \le j < i$, and $y \in S_\star \times \Sigma$;
7. $(C_i, i, \star, \star, \star, \star) \stackrel{y}{\to} (C_i, i{+}1, y, \star, \star, \star)$ for all $0 \le i \le n$ and $y \in S_\star \times \Sigma$;
8. $(C_i, i{+}1, y, \star, \star, \star) \stackrel{z}{\to} (C_i, (i{+}2) \bmod (n{+}2), y, z, \star, \star)$ for all $0 \le i \le n$ and $y, z \in S_\star \times \Sigma$;

9. $(C_i, j, y, z, \star, \star) \xrightarrow{u} (C_i, (j{+}1) \bmod (n{+}2), y, z, \star, \star)$
   for all $0 \le i \le n$, $i{+}2 \le j \le n{+}1$, and $y, z \in S_\star \times \Sigma$;
10. $(C_i, j, y, z, \star, \star) \xrightarrow{u} (C_i, j{+}1, y, z, \star, \star)$
   for all $0 \le i \le n$, $0 \le j < i$, and $y, z, u \in S_\star \times \Sigma$;
11. $(C_i, i, y, z, \star, \star) \xrightarrow{u} (C_i, i{+}1, y, z, u, \star)$
   for all $0 \le i \le n$ and $y, z, u \in S_\star \times \Sigma$;
12. $(C_i, i{+}1, y, z, u, \star) \xrightarrow{v} (C_i, (i{+}2) \bmod (n{+}2), y, z, u, v)$
   for all $0 \le i \le n$ and $y, z, u, v \in S_\star \times \Sigma$;
13. $(C_i, (i{+}2) \bmod (n{+}2), y, z, u, v) \xrightarrow{x} U$
   for all $0 \le i \le n$, $x \in Act$, and $y, z, u, v \in S_\star \times \Sigma$ such that (y, z) and (u, v) are *not* compatible pairs (see below).

A fragment of $\mathcal{F}$ is shown in Fig. 1. The role of states of the form $(C_i, 0, \star, \star, \star, \star)$ and their successors (which are not drawn in Fig. 1) is clarified below.

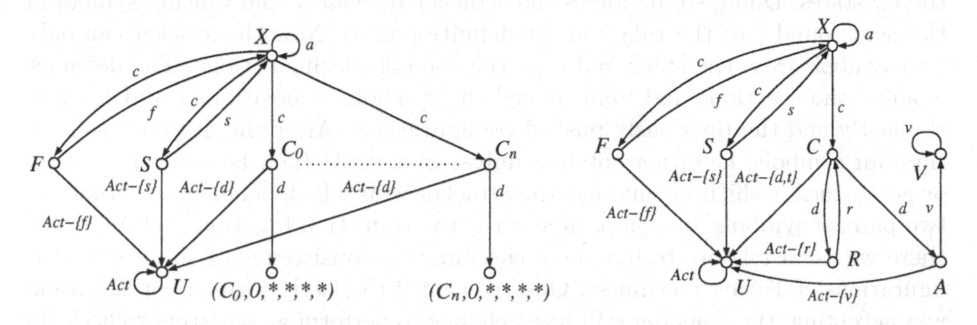

Fig. 1. The systems $\mathcal{F}$ and $\mathcal{F}'$ (successors of $(C_i, 0, \star, \star, \star, \star)$ in $\mathcal{F}$ are omitted).

Now we prove that $\mathcal{M}$ accepts w iff $\alpha \not\sqsubseteq X$. Intuitively, the simulation game between α and X corresponds to constructing a branch in a computational tree for $\mathcal{M}$ on w. The attacker (who plays with α) wants to show that there is an accepting computational tree, while the defender aims to demonstrate the converse. The attacker is therefore 'responsible' for choosing the appropriate successors of all existential configurations (selecting those for which an accepting subtree exists), while the defender chooses successors of universal configurations (selecting those for which no accepting subtree exists). The attacker wins iff the constructed branch reaches an accepting configuration. The choice is implemented as follows: after pushing the last symbol of a configuration, the attacker has to emit the c action and push a 'waiting' symbol (see above). The defender can reply by entering the state F, S, or one of the C_i states. Intuitively, he chooses among the possibilities of selecting the first or the second successor, or checking that the ith symbol of the lastly pushed configuration was guessed correctly (w.r.t. the previous configuration). Technically, the choice is done by forcing the attacker to emit a specific action in the *next* round—observe that if the defender performs, e.g., the $X \xrightarrow{c} F$, transition, then the attacker *must* use one of his

f transitions in the next round, because otherwise the defender would go immediately to the state U where he can simulate 'everything', i.e., the attacker loses the game. As the defender is responsible only for selecting the successors of *universal* configurations, the attacker has to follow his 'dictate' only if the lastly pushed configuration was universal; if it was existential, he can choose the successor according to *his* own will (see the rules 6–9 in the definition of δ). If the lastly pushed configuration was rejecting, the attacker cannot perform any further transitions from the waiting symbol, which means that the defender wins. If the configuration was accepting and the defender enters F of S via the action c, then the attacker wins; first he replaces the waiting symbol with T, emitting f or s, resp. (so that the defender has to go back to X) and then he does the action t. The purpose of the states C_i (and their successors) is to ensure that the attacker cannot gain anything by 'cheating', i.e., by guessing configurations incorrectly. If the defender is suspicious that the attacker has cheated when pushing the last configuration, he can 'punish' the attacker by going (via the action c) to one of the C_i states. Doing so, he forces the attacker to *remove* the waiting symbol in the next round (see the rule 5 in the definition of δ). Now the atacker can only pop symbols from the stack and emit the symbol-specific actions. The defender 'counts' those actions and 'remembers' the symbols at positions i and $i+1$ in the lastly and the previously pushed configurations. After the defender collects the four symbols, he either enters a universal state U (i.e., he wins the game), or gets 'stuck' (which means that the attacker wins). It depends on whether the two pairs of symbols are compatible w.r.t. the transition function γ of $\mathcal{M}$ or not (here we use a folklore technique of checking the consistency of successive configurations of Turing machines). Observe that if the lastly pushed configuration was accepting, the defender still has a chance to perform a consistency check (in fact, it is his 'last chance' to win the game). On the other hand, if the defender decides to check the consistency right at the beginning of the game (when the attacker plays the c transition from α), he inevitably loses because the attacker reaches the bottom symbol Z in $n+2$ transitions and then he can emit the action t. It follows that the attacker has a winning strategy iff $\mathcal{M}$ accepts w. □

Theorem 2. *The problem* ***PDA*** $\sqsubseteq$ ***FS*** *is EXPTIME-hard even for a fixed finite-state process.*

Proof. We modify the construction of Theorem 1. Intuitively, we just re-implement the cheating detection so that the compatibility of selected pairs of symbols is checked by the pushdown system and not by $\mathcal{F}$ (now we can store the four symbols in the finite control). However, it must still be the defender who selects the (position of the) pair. This can be achieved with a fixed number of states (see [5]).

Theorem 3. *The problem* ***FS*** $\sqsubseteq$ ***BPA*** *is EXPTIME-hard.*

Proof. The technique is similar to the one of Theorem 1 (see [5]).

Theorem 4. *The problem* ***FS*** $\sqsubseteq$ ***PDA*** *is EXPTIME-hard even for a fixed finite-state process.*

An immediate consequence of Theorem 1 and Theorem 2 is the following:

Corollary 5. *The problem* **BPA** $\simeq$ **FS** *is EXPTIME-hard. Moreover, the problem* **PDA** $\simeq$ **FS** *is EXPTIME-hard even for a fixed finite-state process.*

Proof. There is a simple (general) reduction from the $A \sqsubseteq B$ problem to the $A \simeq B$ problem (where A, B are classes of processes) which applies also in this case—given processes $p \in A$ and $q \in B$, we construct processes p', q' such that p' has only the transitions $p' \xrightarrow{a} p$, $p' \xrightarrow{a} q$, and q' has only the transition $q' \xrightarrow{a} q$. It follows immediately that $p' \simeq q'$ iff $p \sqsubseteq q$. □

The problem of **PDA** $\sim$ **PDA** is decidable, but the exact complexity is not known. The decision procedures described in [9,13] do not give any upper complexity bound. So far, the best known lower bound for this problem was *PSPACE*-hardness [8]. However, while the problem **PDA** $\sim$ **FS** is *PSPACE*-complete (see Section 3) the problem **PDA** $\sim$ **PDA** is at least *EXPTIME*-hard. This *EXPTIME* lower bound even holds for the subclass of normed **PDA** (a **PDA** is *normed* iff from every reachable configuration it is possible to empty the stack).

The proof of the following theorem uses a technique which can be traced back to Jančar [1]; a more explicit formulation is due to Srba [10] who used the technique in the different context of Basic Parallel Processes. The main idea is that in a bisimulation game the defender can force the attacker to do certain things according to the defender's choices. The full proof can be found in [5].

Theorem 6. *The problem* **PDA** $\sim$ **PDA** *is EXPTIME-hard, even for normed* **PDA***.*

Proof. (sketch) The proof is done by a polynomial-time reduction of the (*EXPTIME*-complete) acceptance problem of alternating LBA to the **PDA** $\sim$ **PDA** problem. The bisimulation game proceeds as follows. The attacker guesses LBA configurations and pushes them onto the stack. The defender is forced to copy these moves. At existential control states (of the LBA) the attacker chooses the successor control state, and at the universal control states (of the LBA) the defender gets to choose the successor control state (this requires the technique mentioned above where the defender forces the attacker to do certain things). At any time, the defender can force the attacker to enter a so-called check-phase. In this check-phase it is verified if the LBA configuration at the top of the stack is really a successor configuration (according to the transition rules of the LBA) of the LBA configuration that was pushed onto the stack before. If not, then the defender wins the bisimulation game. This construction forces the attacker to play 'honestly', i.e., to correctly simulate the behavior of the LBA. If an accepting configuration (of the LBA) is reached in this way then the attacker wins the bisimulation game (having proved, despite the interference of the defender's choices at the universal control states, that the alternating LBA accepts). Otherwise, the bisimulation game goes on forever and the defender wins. This construction ensures that the attacker has a winning strategy if and only if the alternating LBA accepts. Thus, the alternating LBA accepts iff the two PDAs are not bisimilar. □

3 Upper Bounds

The next theorem extends the result for strong simulation which appeared in [6]; the proof is based on the same idea, but the constructed formula φ is now completely fixed.

Theorem 7. *The problems* **PDA** $\sqsubseteq_w$ **FS**, **FS** $\sqsubseteq_w$ **PDA**, *and* **PDA** $\simeq_w$ **FS** *are in EXPTIME.*

Proof. All of the above mentioned problems are polynomially reducible to the model-checking problem with pushdown automata and a fixed formula φ of the modal μ-calculus (which is decidable in deterministic exponential time [17]).

Let $\varphi \equiv \nu X.\Box_a \Diamond_b \langle c \rangle X$, where $\Box_a \psi \equiv \nu Y.(\psi \wedge [a]Y)$ and $\Diamond_b \psi \equiv \nu Z.(\psi \vee \langle b \rangle Z)$. Intuitively, $\Box_a \psi$ says that each state which is reachable from a given process via a finite sequence of a-transitions satisfies ψ, and $\Diamond_b \psi$ says that a given process can reach a state satisfying ψ via a finite sequence of b-transitions. Hence, the meaning of φ can be explained as follows: a process satisfies φ iff after each finite sequence of a-transitions it can perform a finite sequence of b-transitions ended with one c-transition so that the state which is entered again satisfies φ (we refer to [3] for a precise definition of the syntax and semantics of the modal μ-calculus). Now let $\Delta = (Q, \Gamma, Act, \delta)$ be a pushdown system, $\mathcal{F} = (F, Act, \rightarrow)$ a finite-state system, $p\alpha$ a process of Δ, and f a process of $\mathcal{F}$. We construct a pushdown system $\Delta = (Q \times F \times Act \times \{0,1\}, \Gamma \cup \{Z\}, \{a,b,c\}, \delta')$ (where $Z \notin \Gamma$ is a new bottom symbol) which 'alternates' the $\stackrel{x}{\Rightarrow}$ transitions of Δ and $\mathcal{F}$, remembering the 'x' in its finite control. Formally, δ' is constructed as follows:

- for all $qA \stackrel{x}{\rightarrow} r\beta \in \delta$ and $g \in F$ we add $(q,g,\tau,0)A \stackrel{a}{\rightarrow} (r,g,x,0)\beta$ to δ';
- for all $qA \stackrel{\tau}{\rightarrow} r\beta \in \delta$, $x \in Act$, and $g \in F$ we add $(q,g,x,0)A \stackrel{a}{\rightarrow} (r,g,x,0)\beta$ to δ';
- for all $q \in Q$, $g \in F$, $x \in Act$, and $Y \in \Gamma \cup \{Z\}$ we add $(q,g,x,0)Y \stackrel{b}{\rightarrow} (q,g,x,1)Y$ to δ';
- for each transition $g \stackrel{x}{\rightarrow} g'$ of $\mathcal{F}$ and all $q \in Q$, $Y \in \Gamma \cup \{Z\}$ we add $(q,g,x,1)Y \stackrel{b}{\rightarrow} (q,g',\tau,1)Y$ to δ';
- for all $g \stackrel{\tau}{\rightarrow} g'$ of $\mathcal{F}$, $x \in Act$, $q \in Q$, and $Y \in \Gamma \cup \{Z\}$ we add $(q,g,x,1)Y \stackrel{b}{\rightarrow} (q,g',x,1)Y$ to δ';
- for all $q \in Q$, $g \in F$, and $Y \in \Gamma \cup \{Z\}$ we add $(q,g,\tau,1)Y \stackrel{c}{\rightarrow} (q,g,\tau,0)Y$ to δ';

We claim that $p\alpha \sqsubseteq_w f$ iff $(p,f,\tau,0)\alpha Z \models \varphi$. Indeed, each sequence of a-transitions of $(p,f,\tau,0)\alpha Z$ corresponds to some $\stackrel{x}{\Rightarrow}$ move of $p\alpha$ and vice versa; and after each such sequence, the 'token' can be switched from 0 to 1 (performing b), and now each sequence of b's ended with one c corresponds to a $\stackrel{x}{\Rightarrow}$ move of f. Then, the token is switched back to 0 and the computation proceeds in the same way. φ says that this can be repeated forever, unless we reach a state which cannot do any a when the token is set to 0. The new bottom symbol Z has been added to ensure that $(p,f,\tau,0)\alpha Z$ cannot get stuck just due to the emptiness of the stack. The **FS** $\sqsubseteq_w$ **PDA** direction is handled in a very similar way (the roles of $p\alpha$ and f are just interchanged). □

Corollary 8. *The problems* **BPA** $\sqsubseteq_w$ **FS**, **FS** $\sqsubseteq_w$ **BPA**, *and* **BPA** $\simeq_w$ **FS** *are decidable in polynomial time for (any)* fixed *finite-state process.*

Proof. The complexity result of [17] says that model-checking with any fixed formula of the modal μ-calculus and pushdown processes with a *fixed* number of control states is decidable in polynomial time. By synchronizing a given BPA process with a given (fixed) finite-state process as in Theorem 7 we obtain a pushdown system with a fixed number of control states, and the result follows. □

Now we show that the problem **PDA** $\approx$ **FS** is in *PSPACE*. First, we recall some results from [2]. A characteristic formula of a finite-state system F w.r.t. $\approx$ is a formula Θ_F s.t. for every general system G which uses the same set of actions as F we have that $G \models \Theta_F \iff G \approx F$. It has been shown in [2] that characteristic formulae for finite-state systems w.r.t. $\approx$ can be effectively constructed in the temporal logic EF (a simple fragment of CTL), by using the following theorem (here, $\approx_k$ denotes 'weak bisimilarity up-to k', which means that the defender has a strategy to defend for at least k rounds in the weak bisimulation game).

Theorem 9. *(taken from [2])*
Let F be a finite-state system with n states and G a general system. States $g \in G$ and $f \in F$ are weakly bisimilar iff the following conditions hold: (1) $g \approx_n f$ and (2) For each state g' which is reachable from g there is a state $f' \in F$ such that $g' \approx_n f'$.

One constructs characteristic formulae $\Phi_{k,f}$ for states f in F w.r.t. $\approx_k$ that satisfy $g \models \Phi_{k,f} \iff g \approx_k f$. The family of $\Phi_{k,f}$ formulae is defined inductively on k as follows:

$$\Phi_{0,f} := \mathit{true}$$

$$\Phi_{k+1,f} := \left(\bigwedge_{a \in Act} \bigwedge_{f' \in S(f,a)} \Diamond_a \Phi_{k,f'} \right) \wedge \left(\bigwedge_{a \in Act} (\neg \Diamond_a (\bigwedge_{f' \in S(f,a)} \neg \Phi_{k,f'})) \right)$$

where $S(f,a) = \{f' \mid f \stackrel{a}{\to} f'\}$ and $\Diamond_\tau$ means "reachable via a finite number of τ-transitions" and $\Diamond_a := \Diamond_\tau \langle a \rangle \Diamond_\tau$ for $a \neq \tau$.

Empty conjunctions are equivalent to *true*. Thus, by Theorem 9, the characteristic formula Θ_f for a process f of a finite-state system $\mathcal{F} = (F, Act, \to)$ with n states is

$$\Theta_f \equiv \Phi_{n,f} \wedge \neg \Diamond \left(\bigwedge_{f' \in F} \neg \Phi_{n,f'} \right)$$

So one can reduce the problem **PDA** $\approx$ **FS** to a model checking problem for pushdown automata and (a slight extension of) the logic EF. The following proof-sketch uses many results by Walukiewicz [16]. For a complete proof, it would be necessary to repeat many of these, so we just sketch the main ideas and the crucial modification of the algorithm from [16]. It has been shown by Walukiewicz in [16] that model checking pushdown automata with the logic EF is *PSPACE*-complete. But our result does not follow directly from that. First, our characteristic formulae use a slight extension of EF, because of the $\Diamond_\tau$ operator (normal EF has only the $\Diamond$ operator). However, the model checking algorithm of [16] can trivially be generalized to this extension of EF, without increasing its complexity. The second, and more important problem is that the size of the characteristic formula Θ_F is exponential in n (where n is the number of states of F). However, a closer analysis of the model checking algorithm presented in [16] reveals that its complexity does not depend directly on the size of the formula, but rather on the number of its distinct subformulae. More precisely, this algorithm uses a so-called assumption function that assigns sets of subformulae to every control-state of the PDA. Of course, each EF formula has only a polynomial number of subformulae and hence the assumption function can be represented in polynomial space. However, it is also true for our characteristic formula Θ_F — although its size is exponential in n, the number of its distinct subformulae $\Phi_{k,f}$ is bounded by $\mathcal{O}(n^2)$, because $0 \le k \le n$ and F has only n states. Hence, we can run the mentioned model-checking algorithm for EF. Instead of 'unwinding' the $\Phi_{k,f}$ subformulae, we keep the abbreviations $\Phi_{k,f}$ as long as possible and expand them only (on-the-fly) when necessary (using the inductive definitions above). Thus, the whole algorithm works in polynomial space and we obtain the following theorem.

Theorem 10. *The problem* **PDA** $\approx$ **FS** *is in PSPACE.*

4 Conclusions

The following table summarizes the complexity of all problems of comparing PDA and BPA to finite-state systems w.r.t. strong and weak simulation preorder/equivalence and strong and weak bisimilarity. **FS** means a finite-state system that is part of the input of the problem, while $\mathcal{F}$ means "any fixed finite-state system" for the upper complexity bounds and "some fixed finite-state system" for the lower complexity bounds.

	$\sqsubseteq_w$ **FS** $\sqsubseteq$ **FS**	$\sqsubseteq_w \mathcal{F}$ $\sqsubseteq \mathcal{F}$	**FS** $\sqsubseteq_w$ **FS** $\sqsubseteq$	$\mathcal{F} \sqsubseteq_w$ $\mathcal{F} \sqsubseteq$	$\simeq_w$ **FS** $\simeq$ **FS**	$\simeq_w \mathcal{F}$ $\simeq \mathcal{F}$	$\approx$ **FS** $\sim$ **FS**	$\sim \mathcal{F}$	$\approx \mathcal{F}$
BPA	EXPTIME complete	in P	EXPTIME complete	in P	EXPTIME complete	in P	in P	in P	in P
PDA	EXPTIME complete	EXPTIME complete	EXPTIME complete	EXPTIME complete	EXPTIME complete	EXPTIME complete	PSPACE complete	in P	PSPACE complete

Finally, we have also shown (in Theorem 6) that the problem **PDA** $\sim$ **PDA** of checking bisimilarity of two pushdown systems is *EXPTIME*-hard. Thus, it is harder than the problem **PDA** $\sim$ **FS** of checking bisimilarity of a pushdown system and a finite-state system, which is only *PSPACE*-complete.

References

[1] P. Jančar. High undecidability of weak bisimilarity for Petri nets. In *Proceedings of CAAP'95*, volume 915 of *LNCS*, pages 349–363. Springer, 1995.

[2] P. Jančar, A. Kučera, and R. Mayr. Deciding bisimulation-like equivalences with finite-state processes. *Theoretical Computer Science*, 258(1–2):409–433, 2001.

[3] D. Kozen. Results on the propositional μ-calculus. *Theoretical Computer Science*, 27:333–354, 1983.

[4] A. Kučera. On simulation-checking with sequential systems. In *Proceedings of ASIAN 2000*, volume 1961 of *LNCS*, pages 133–148. Springer, 2000.

[5] A. Kučera and R. Mayr. On the complexity of semantic equivalences for pushdown automata and BPA. Technical report FIMU-RS-2002-01, Faculty of Informatics, Masaryk University, 2002.

[6] A. Kučera and R. Mayr. Simulation preorder over simple process algebras. *Information and Computation*, 173(2):184–198, 2002.

[7] A. Kučera and R. Mayr. Weak bisimilarity between finite-state systems and BPA or normed BPP is decidable in polynomial time. *Theoretical Computer Science*, 270(1–2):677–700, 2002.

[8] R. Mayr. On the complexity of bisimulation problems for pushdown automata. In *Proceedings of IFIP TCS'2000*, volume 1872 of *LNCS*, pages 474–488. Springer, 2000.

[9] G. Sénizergues. Decidability of bisimulation equivalence for equational graphs of finite out-degree. In *Proceedings of 39th Annual Symposium on Foundations of Computer Science*, pages 120–129. IEEE Computer Society Press, 1998.

[10] J. Srba. Strong bisimilarity and regularity of basic parallel processes is PSPACE-hard. In *Proceedings of STACS 2002*, volume 2285 of *LNCS*, pages 535–546. Springer, 2002.

[11] B. Steffen and A. Ingólfsdóttir. Characteristic formulae for processes with divergence. *Information and Computation*, 110(1):149–163, 1994.

[12] C. Stirling. The joys of bisimulation. In *Proceedings of MFCS'98*, volume 1450 of *LNCS*, pages 142–151. Springer, 1998.

[13] C. Stirling. Decidability of DPDA equivalence. *Theoretical Computer Science*, 255:1–31, 2001.

[14] W. Thomas. On the ehrenfeucht-fraïssé game in theoretical computer science. In *Proceedings of TAPSOFT'93*, volume 668 of *LNCS*, pages 559–568. Springer, 1993.

[15] R. van Glabbeek. The linear time—branching time spectrum. *Handbook of Process Algebra*, pages 3–99, 1999.

[16] I. Walukiewicz. Model checking CTL properties of pushdown systems. In *Proceedings of FST&TCS 2000*, volume 1974 of *LNCS*, pages 127–138. Springer, 2000.

[17] I. Walukiewicz. Pushdown processes: Games and model-checking. *Information and Computation*, 164(2):234–263, 2001.

An Improved Algorithm for the Membership Problem for Extended Regular Expressions

Orna Kupferman* and Sharon Zuhovitzky

Hebrew University, School of Engineering and Computer Science,
Jerusalem 91904, Israel,
{orna,sharonzu}@cs.huji.ac.il

Abstract. Extended regular expressions (ERE) define regular languages using union, concatenation, repetition, intersection, and complementation operators. The fact ERE allow intersection and complementation makes them exponentially more succinct than regular expressions. The membership problem for extended regular expressions is to decide, given an expression r and a word w, whether w belongs to the language defined by r. Since regular expressions are useful for describing patterns in strings, the membership problem has numerous applications. In many such applications, the words w are very long and patterns are conveniently described using ERE, making efficient solutions to the membership problem of great practical interest.
In this paper we introduce alternating automata with synchronized universality and negation, and use them in order to obtain a simple and efficient algorithm for solving the membership problem for ERE. Our algorithm runs in time $O(m \cdot n^2)$ and space $O(m \cdot n + k \cdot n^2)$, where m is the length of r, n is the length of w, and k is the number of intersection and complementation operators in r. This improves the best known algorithms for the problem.

1 Introduction

Regular languages of finite words are naturally defined by repeated applications of closure operators. *Regular expressions* (*RE*, for short) contain *union* ($\vee$), *concatenation* ($\cdot$), and *repetition* ($*$) operators and can define all the regular languages. It turned out that enriching RE with *intersection* ($\wedge$) and *complementation* ($\neg$) makes the description of regular languages more convenient. Indeed, there are evidences for the succinctness of *semi-extended regular expressions* (*SERE*, for short, which extend RE with intersection) with respect to RE, and for the succinctness of *extended regular expressions* (*ERE*, for short, which extend RE with both intersection and complementation) with respect to SERE. For example, specifying the regular language of all words that contain a non-overlapping repetition of a subword of length n requires an RE of length $\Omega(2^n)$

* Supported in part by BSF grant 9800096.

K. Diks et al. (Eds): MFSC 2002, LNCS 2420, pp. 446–458, 2002.

and can be done with an SERE of length $O(n^2)$ [Pet02]. Also, specifying the singleton language $\{1^n\}$ requires an SERE of length $\Theta(n)$ and can be done with an ERE of length *polylog*(n) [KTV01]. In general, ERE are nonelementary more succinct than RE [MS73].

The membership problem for RE and their extensions is to decide, given an RE r and a word w, whether w belongs to the language defined by r. Since RE are useful for describing patterns in strings, the membership problem has many applications (see [KM95] for recent applications in molecular biology). In many of these applications, the words w are very long and patterns are described using SERE or ERE, making efficient solutions for the membership problem for them of great practical interest [Aho90].

Studies of the membership problem for RE have shown that an *automata-theoretic approach* is useful: a straightforward algorithm for membership in RE translates the RE to a nondeterministic automaton. For an RE of size m, the automaton is of size $O(m)$, thus for a word of length n, the algorithm runs in time $O(mn)$ and space $O(m)$ [HU79]. For SERE, a translation to nondeterministic automata involves an exponential blow-up, and until recently, the best known algorithms for membership in SERE were based on dynamic programing and ran in time $O(mn^3)$ and space $O(mn^2)$ [HU79,Hir89]. Myers describes a technique for speeding up the membership algorithms by $\log n$ [Mye92], but still one cannot expect an $O(mn)$ algorithm for membership in SERE, as the problem is LOGCFL-complete [Pet02], whereas the one for RE is NL-complete [JR91].

The translation of SERE to automata is not easy even when alternating automata are considered. The difficulty lies in expressions like $(r_1 \wedge r_2) \cdot r_3$, where the two copies of the alternating automaton that check membership of a guessed prefix in r_1 and r_2 should be synchronized in order to agree on the suffix that is checked for membership in r_3. Indeed, $(r_1 \wedge r_2) \cdot r_3$ is not equal to $(r_1 \cdot r_3) \wedge (r_2 \cdot r_3)$. In contrast, $(r_1 \vee r_2) \cdot r_3$ is equal to $(r_1 \cdot r_3) \vee (r_2 \cdot r_3)$, which is why RE can be efficiently translated to alternating automata, while SERE cannot [KTV01].

Several models of automata with *synchronization* are studied in the literature (c.f., [Hro86,Slo88,DHK+89,Yam00b]). Essentially, in synchronous alternating automata, some of the states are designated as synchronization states, and the spawned copies of the automaton have to agree on positions in the input word in which they visit synchronization states. In [Yam00a], Yamamoto introduces *partially input-synchronized alternating automata*, and shows an algorithm for membership in SERE that is based on a translation of SERE into the new model. While the new algorithm improves the known algorithms – it runs in time $O(m \cdot n^2)$ and space $O(m \cdot n + k \cdot n^2)$, where k is the number of $\wedge$ operators, the model of partially input-synchronized alternating automata is needlessly complicated, and we found the algorithm that follows very cryptic and hard to implement. On the other hand, Yamamoto's idea of using alternating automata with some type of synchronization as an automata-theoretic framework for solving the membership problem for SERE seems like a very good direction.

In this paper, we introduce *alternating automata with synchronized universality and negation* (ASUN, for short). ASUN are simpler than the model of Yamamoto, but have a richer structure. In addition to the existential and universal states of usual alternating automata, ASUN have *complementation states*, which enable a dualization of the automaton's behavior. This enables us to translate an ERE of length m into an ASUN of size $O(m)$, where synchronized universality is used to handle conjunctions and synchronized negation is used to handle complementation. We show that the membership problem for ASUN can be solved in time $O(m \cdot n^2)$ and space $O(m \cdot n + k \cdot n^2)$, where k is the number of $\wedge$ and $\neg$ operators. Thus, our bound coincides with that of Yamamoto's, but we handle a richer class of regular expressions, and the algorithm that follows is much simpler and easy to implement (in fact, we describe a detailed pseudo-code in one page).

2 Definitions

2.1 Extended Regular Expressions

Let Σ be a finite *alphabet*. A *finite word* over Σ is a (possibly empty) finite sequence $w = \sigma_1 \cdot \sigma_2 \cdots \sigma_n$ of concatenated letters in Σ. The length of a word w is denoted by $|w|$. The symbol ϵ denotes the empty word. We use $w[i,j]$ to denote the subword $\sigma_i \cdots \sigma_j$ of w. If $i > j$, then $w[i,j] = \epsilon$. *Extended Regular Expressions* (ERE) define languages by inductively applying union, intersection, complementation, concatenation, and repetition operators. Thus, ERE extend regular expressions with intersection and complementation operators. Formally, for an alphabet Σ, an ERE over Σ is one of the following.

- $\emptyset$, ϵ, or σ, for $\sigma \in \Sigma$.
- $r_1 \vee r_2$, $r_1 \wedge r_2$, $\neg r_1$, $r_1 \cdot r_2$, or r_1^*, for ERE r_1 and r_2.

We use $\mathcal{L}(r)$ to denote the language that r defines. For the base cases, we have $\mathcal{L}(\emptyset) = \emptyset$, $\mathcal{L}(\epsilon) = \{\epsilon\}$, and $\mathcal{L}(\sigma) = \{\sigma\}$. The operators $\vee$, $\wedge$, $\neg$, $\cdot$, and $*$ stand for union, intersection, complementation, concatenation, and repetition (also referred to as Kleene star), respectively. Formally,

- $\mathcal{L}(r_1 \vee r_2) = \mathcal{L}(r_1) \cup \mathcal{L}(r_2)$.
- $\mathcal{L}(r_1 \wedge r_2) = \mathcal{L}(r_1) \cap \mathcal{L}(r_2)$.
- $\mathcal{L}(\neg r_1) = \Sigma^* \setminus \mathcal{L}(r_1)$.
- $\mathcal{L}(r_1 \cdot r_2) = \{w_1 \cdot w_2 : w_1 \in \mathcal{L}(r_1) \text{ and } w_2 \in \mathcal{L}(r_2)\}$.
- Let $r_1^0 = \{\epsilon\}$ and let $r_1^i = r_1^{i-1} \cdot r_1$, for $i \geq 1$. Thus, $\mathcal{L}(r_1^i)$ contains words that are the concatenation of i words in $\mathcal{L}(r_1)$. Then, $\mathcal{L}(r_1^*) = \bigcup_{i \geq 0} r_1^i$.

2.2 Alternating Automata with Synchronized Universality

An *alternating automaton with synchronized universality* (*ASU*, in short) is a seven-tuple $\mathcal{A} = \langle \Sigma, Q, \mu, q_0, \delta, \psi, F \rangle$, where,

- Σ is a finite input alphabet.
- Q is a finite set of states.
- $\mu : Q \rightarrow \{\vee, \wedge\}$ maps each state to a branching mode. The function μ induces a partition of Q to the sets $Q_e = \mu^{-1}(\vee)$ and $Q_u = \mu^{-1}(\wedge)$ of *existential* and *universal* states, respectively.
- $q_0 \in Q$ is an initial state.
- $\delta : Q \times (\Sigma \cup \{\epsilon\}) \rightarrow 2^Q$ is the transition function.
- $\psi : Q_u \rightarrow Q$ is a *synchronization function.*
- $F \subseteq Q_e$ is a set of final states.

ASU run on finite words over Σ. Consider a word $w = \sigma_1, \ldots, \sigma_n$. For technical convenience, we also refer to $\sigma_{n+1} = \epsilon$. During the run of $\mathcal{A}$ on w, it splits into several *copies*. A *position* of a copy of $\mathcal{A}$ is a pair $\langle q, i \rangle \in Q \times \{0, \ldots, n\}$, indicating that the copy is in state q, reading the letter σ_{i+1}. If $q \in Q_e$, we say that $\langle q, i \rangle$ is an *existential position.* Otherwise, it is a *universal position.* The run of $\mathcal{A}$ starts with a single copy in the *initial position* $\langle q_0, 0 \rangle$. For $0 \leq i \leq n$, a position $\langle q', i' \rangle$ is a σ_{i+1}-*successor* of a position $\langle q, i \rangle$ if $q' \in \delta(q, \sigma_{i+1})$ and $i' = i + 1$. A position $\langle q', i' \rangle$ is an ϵ-*successor* of $\langle q, i \rangle$ if $q' \in \delta(q, \epsilon)$ and $i' = i$. Finally, $\langle q', i' \rangle$ is a *successor* of $\langle q, i \rangle$ if $\langle q', i' \rangle$ is a σ_{i+1}-successor or an ϵ-successor of $\langle q, i \rangle$. Note that nondeterministic automata with ϵ-moves are a special case of ASU where all states are existential, in which case the function ψ is empty.

Consider a copy of $\mathcal{A}$ in position $\langle q, i \rangle$. If q is an existential state, the copy can move to one of the states in $\delta(q, \sigma_{i+1}) \cup \delta(q, \epsilon)$, thus the new position of the copy is some σ_{i+1} successor or ϵ-successor of $\langle q, i \rangle$. If q is a universal state, the copy should move to all the states in $\delta(q, \sigma_{i+1}) \cup \delta(q, \epsilon)$. Thus, the copy splits into copies that together cover all σ_{i+1}-successors and ϵ-successors of $\langle q, i \rangle$. The computation graph of $\mathcal{A}$ on w embodies all the possible runs of $\mathcal{A}$ on w and is defined as follows.

Definition 1. *The computation graph of $\mathcal{A}$ on an input word $w = \sigma_1 \ldots \sigma_n$ is the directed graph $G = \langle V, E \rangle$, where $V = Q \times \{0, \ldots, n\}$, and $E(\langle q, i \rangle, \langle q', i' \rangle)$ iff $\langle q', i' \rangle$ is a successor of $\langle q, i \rangle$.*

Note that $|V| = m \cdot (n + 1)$, for $m = |Q|$. A *leaf* of G is a position $\langle q, i \rangle \in V$ such that for no $\langle q', i' \rangle \in V$ we have $E(\langle q, i \rangle, \langle q', i' \rangle)$. A *path* from $\langle q, i \rangle$ to $\langle q', i' \rangle$ in G is a sequence of positions $p_1, p_2, \ldots, p_k$, such that $p_1 = \langle q, i \rangle$, $p_k = \langle q', i' \rangle$ and $E(p_i, p_{i+1})$ for $1 \leq i < k$. A *run* of $\mathcal{A}$ on w is obtained from the computation graph by resolving the nondeterministic choices in existential positions. Thus, a run is obtained from G by removing all but one of the edges that leave each existential position. Formally, we have the following.

Definition 2. *Let $G = \langle V, E\rangle$ be the computation graph of $\mathcal{A}$ on w. A run of $\mathcal{A}$ on w is a graph $G_r = \langle V_r, E_r\rangle$ such that $V_r \subseteq V$, $E_r \subseteq E$ and the following hold.*

- $\langle q_0, 0\rangle \in V_r$.
- *Let $\langle q, i\rangle \in V_r$ be a universal position. Then for every position $\langle q', i'\rangle \in V$ such that $E(\langle q, i\rangle, \langle q', i'\rangle)$, we have $\langle q', i'\rangle \in V_r$ and $E_r(\langle q, i\rangle, \langle q', i'\rangle)$.*
- *Let $\langle q, i\rangle \in V_r$ be an existential position. Then either $i = n$ or there is a single position $\langle q', i'\rangle \in V_r$ such that $E_r(\langle q, i\rangle, \langle q', i'\rangle)$.*

Note that when a copy of $\mathcal{A}$ is in an existential position $\langle q, n\rangle$ and $\delta(q, \epsilon) \neq \emptyset$, the copy can choose between moving to an ϵ-successor of q or having q as its final state. In contrast, if $\langle q, n\rangle$ is universal, the copy must continue to all ϵ-successors of q. Note also that the computation graph G and a run G_r may have cycles. However, these cycles may contain only ϵ-transitions.

Recall that the synchronization function ψ maps each universal state q to a state in Q. Intuitively, whenever a copy of $\mathcal{A}$ in state q is split into several copies, the synchronization function forces all these copies to visit the state $\psi(q)$ and to do so simultaneously. We now formalize this intuition. Let G_r be a run of an ASU $\mathcal{A}$ on an input word w. Let $q \in Q$ be a universal state and $s = \psi(q)$. Consider a position $\langle q, i\rangle$. We say that a position $\langle s, j\rangle$, for $j \geq i$, *covers* $\langle q, i\rangle$, if there is a path from $\langle q, i\rangle$ to $\langle s, j\rangle$ and for every position $\langle s', j'\rangle$ on this path, we have $s' \neq s$. In other words, $\langle s, j\rangle$ is the first instance of s on a path leaving $\langle q, i\rangle$. We say that $\langle q, i\rangle$ is *good* in G_r if there is exactly one position $\langle s, j\rangle$ that covers $\langle q, i\rangle$ and all the paths in G_r that leave $\langle q, i\rangle$ eventually reach $\langle s, j\rangle$.

A run $G_r = \langle V, E_r\rangle$ of an ASU $\mathcal{A}$ on an input word $w = \sigma_1 \ldots \sigma_n$ is *accepting* if for all leaves $\langle q, i\rangle \in V$, we have $q \in F$ and $i = n$, and all the universal positions are good in G_r. A word w is *accepted* by $\mathcal{A}$ if there is an accepting run of $\mathcal{A}$ on w. The *language* of $\mathcal{A}$, denoted $L(\mathcal{A})$, is the set of all words accepted by the ASU $\mathcal{A}$.

Example 1. Let $\Sigma = \{0, 1\}$. Consider the language $L \subseteq \Sigma^*$, where a word $w \in \Sigma^*$ is in L iff the $(i+1)$-th letter in w is 0, for some $i = 7 \pmod{12}$. Thus, if we take $\tau = 0 + 1$, then $L = ((\tau^{12})^* \cdot \tau^7) \cdot 0 \cdot \tau^*$. A nondeterministic automaton that recognizes L has at least 13 states. For an integer n, we have $n = 7 \pmod{12}$ if $n = 1 \pmod 3$ and $n = 3 \pmod 4$. Thus, L can be expressed by the ERE $r = (((\tau\tau\tau)^* \cdot \tau) \wedge ((\tau\tau\tau\tau)^* \tau\tau\tau)) \cdot 0 \cdot \tau^*$. For example, the word $w_1 = 1^7 01$ is in L while the word $w_2 = 10101111$ is not. Note that w_2 satisfies both conjuncts of r, but not in the same prefix. We show now an ASU $\mathcal{A}$ with 10 states that recognizes L. $\mathcal{A} = \langle \{0, 1\}, \{q_0, \ldots, q_9\}, \mu, q_0, \delta, \psi, \{q_9\}\rangle$, where q_0 is the only universal state, with $\psi(q_0) = q_8$, and the function δ is described in Figure 1.

A run of $\mathcal{A}$ on a word w splits in q_0 into two copies. One copy makes a nondeterministic move to q_8 after reading a prefix of length $1 \pmod 3$ and the other copy does the same after reading a prefix of length $3 \pmod 4$. If the two copies reach q_8 at the same place in the input word, then $\langle q_0, 0\rangle$ is good and if

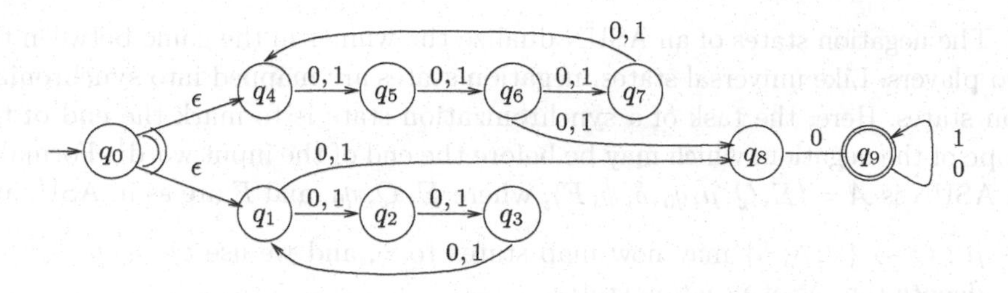

Fig. 1. An ASU for $(((0+1)^{12})^* \cdot (0+1)^7) \cdot 0 \cdot (0+1)^*$

the next letter in the input is 0, the automaton moves to an accepting sink. If the two copies reach q_8 eventually but not at the same time, it means that both conjuncts were satisfied but not necessarily in the same prefix, and the run is not accepting.

2.3 Alternating Automata with Synchronized Universality and Negation

Alternating automata with synchronized universality and negation (*ASUN*, in short) extend ASU by having, in addition to existential and universal states, also *negation states.* It is easy to understand the task of negation states and how an ASUN runs on an input word by taking the *game-theoretic approach* to alternating automata. Let us first explain this approach for ASU. Consider an ASU $\mathcal{A} = \langle \Sigma, Q, \mu, q_0, \delta, \psi, F\rangle$ and an input word $w = \sigma_1 \cdots \sigma_n$. We can view the behavior of $\mathcal{A}$ on w as a game between two players: player 1, who wants to prove that $\mathcal{A}$ accepts w, and player 2, who wants to prove that $\mathcal{A}$ rejects w. For two positions $\langle q, i\rangle$ and $\langle s, j\rangle$, with $j \geq i$, we say that player 1 *wins the game from* $\langle q, i\rangle$ *to* $\langle s, j\rangle$, denoted $\langle q, i\rangle \to \langle s, j\rangle$, if $\mathcal{A}$ with initial state q and final state s accepts the word $w[i+1, j]$. Otherwise, player 1 *loses* the game from $\langle q, i\rangle$ to $\langle s, j\rangle$, denoted $\langle q, i\rangle \not\to \langle s, j\rangle$. We also use $\langle q, i\rangle \mapsto \langle s, j\rangle$ to indicate that $\mathcal{A}$ with initial state q and final state s accepts the word $w[i+1, j]$ with only a single visit to s. Note that $\mathcal{A}$ accepts w if there is $s \in F$ such that $\langle q_0, 0\rangle \to \langle s, n\rangle$.

The relation $\to$ can be defined inductively as follows.

- For $q \in Q_e$, we have that $\langle q, i\rangle \to \langle s, j\rangle$ iff $j = i$ and $s = q$, or $j = i$ and $s \in \delta(q, \epsilon)$, or $j = i+1$ and $s \in \delta(q, \sigma_{i+1})$, or there is $\langle q', i'\rangle$ such that $\langle q, i\rangle \to \langle q', i'\rangle$ and $\langle q', i'\rangle \to \langle s, j\rangle$.
- For $q \in Q_u$, we have that $\langle q, i\rangle \to \langle s, j\rangle$ iff there is $i \leq j' \leq j$ such that $\langle \psi(q), j'\rangle \to \langle s, j\rangle$ and for all q' and i', if $i' = i$ and $q' \in \delta(q, \epsilon)$, or $i' = i+1$ and $q' \in \delta(q, \sigma_{i+1})$, then $\langle q', i'\rangle \mapsto \langle \psi(q), j'\rangle$.

Intuitively, when $q \in Q_e$, we only have to find a successor of $\langle q, i\rangle$ from which an accepting run continues. On the other hand, when $q \in Q_u$, we have to find a position $\langle \psi(q), j'\rangle$ that covers $\langle q, i\rangle$, witnesses that $\langle q, i\rangle$ is good, and from which an accepting run continues.

The negation states of an ASUN dualize the winner in the game between the two players. Like universal states, negation states are mapped into synchronization states. Here, the task of a synchronization state is to mark the end of the scope of the negation, which may be before the end of the input word[1]. Formally, an ASUN is $\mathcal{A} = \langle \Sigma, Q, \mu, q_0, \delta, \psi, F \rangle$, where Σ, Q, q_0, and F are as in ASU, and

- $\mu : Q \to \{\vee, \wedge, \neg\}$ may now map states to $\neg$, and we use $Q_n = \mu^{-1}(\neg)$ to denote the set of *negation* states.
- The transition function $\delta : Q \times (\Sigma \cup \{\epsilon\}) \to 2^Q$ is such that for all $q \in Q_n$, we have $\delta(q, \epsilon) = \{q'\}$, for some $q' \neq q$, and $\delta(q, \sigma) = \emptyset$ for all $\sigma \in \Sigma$. Thus, each negation state has a single ϵ-successor, and no other successors. For states in $Q_e \cup Q_u$, the transition function is as in ASU.
- The synchronization function $\psi : Q_u \cup Q_n \to Q$ now applies to both universal and negation states.

The computation graph G of an ASUN $\mathcal{A}$ on an input word w is defined exactly as the computation graph for ASU. We define acceptance by an ASUN in terms of the game between players 1 and 2. The relation $\to$ is defined as above for states in Q_e and Q_u. In addition, for every $q \in Q_n$ with $\delta(q, \epsilon) = \{q'\}$, we have that $\langle q, i \rangle \to \langle s, j \rangle$ iff there is $i \leq j' \leq j$ such that $\langle q', i \rangle \not\to \langle \psi(q), j' \rangle$ and $\langle \psi(q), j' \rangle \to \langle s, j \rangle$. Thus, $\mathcal{A}$ with initial state q and final state s accepts $w[i+1, j]$ if there is $i \leq j' \leq j$ such that $\mathcal{A}$ with initial state q' and final state $\psi(q)$ rejects $w[i+1, j']$ and $\mathcal{A}$ with initial state $\psi(q)$ and final state s accepts $w[j'+1, j]$. We then say that the position $\langle \psi(q), j' \rangle$ *covers* $\langle q, i \rangle$.

3 ASUN for ERE

Let Σ be a finite alphabet. Given an ERE r of length m over Σ, we build an ASUN $\mathcal{A}_r$ over Σ with $O(m)$ states such that $L(r) = L(\mathcal{A}_r)$. The construction is similar to the one used for translating regular expression into nondeterministic automata with ϵ-transitions [HMU00]. The treatment of conjunctions is similar to the one described by Yamamoto in [Yam00a], adjusted to our simpler type of automata. The treatment of negations is by negation states.

Theorem 1. *Given an ERE r of length m, we can construct an ASUN $\mathcal{A}_r$ of size $O(m)$ such that $L(r) = L(\mathcal{A}_r)$. Furthermore, $\mathcal{A}_r$ has the following properties:*

1. *$\mathcal{A}_r$ has exactly one accepting state, and there are no transitions out of the accepting state.*

[1] Readers familiar with alternating automata know that it is easy to complement an alternating automaton by dualizing the function μ and the acceptance condition. A similar complementation can be defined for ASU, circumventing the need for negation states. We found it simpler to add negation states with synchronization, as they enable us to keep the structure of the ASUN we are going to associate with ERE very restricted (e.g., a single accepting state), which leads to a simple membership algorithm.

2. *There are no transitions into the initial state.*
3. *The function ψ is one-to-one.*
4. *Every universal state has exactly two ϵ-successors, and no other successors.*
5. *For every existential state q, exactly one of the following holds: q has no successors, q has one or two ϵ-successors, or q has one σ-successor for exactly one $\sigma \in \Sigma$.*
6. *For every state q, exactly one of the following holds: q is the initial state, q is a σ-successor of a single other state, or q is an ϵ-successor of one or two other states.*

Proof: The construction of $\mathcal{A}_r$ is inductive, and we describe it in Figure 2. For the basic three cases of $r = \emptyset, \epsilon$, or σ, the only states of $\mathcal{A}_r$ are q_{in} and q_{fin}. For the cases $r = r_1 \vee r_2, r_1 \wedge r_2, r_1 \cdot r_2, r_1^*$, or $\neg r_1$, we refer to the ASUN $\mathcal{A}_1$ and $\mathcal{A}_2$ of the SERE r_1 and r_2. In particular, the state space of $\mathcal{A}_1$ is Q_1, it has an initial state q_1^{in} and final state q_1^{fin}, and similarly for $\mathcal{A}_2$.

The initial state of the ASUN associated with $r_1 \wedge r_2$ is universal, and the two copies has to synchronize in its final state; thus $\psi(q_{in}) = q_{fin}$. This guarantees that the two copies proceed on words of the same length. Similarly, the initial state of the ASUN associated with $\neg r_1$ is a negation state, and the scope of the negation is bounded to the ASUN $\mathcal{A}_1$; thus $\psi(q_{in}) = q_{fin}$. All the other states are existential. □

Each state of $\mathcal{A}_r$ is associated with a subexpression of r. In particular, the universal and negation states of $\mathcal{A}_r$ are associated with conjunctions and negations, respectively. We refer to $\wedge$ and $\neg$ as *special operators*, refer to states $q \in Q_u \cup Q_n$

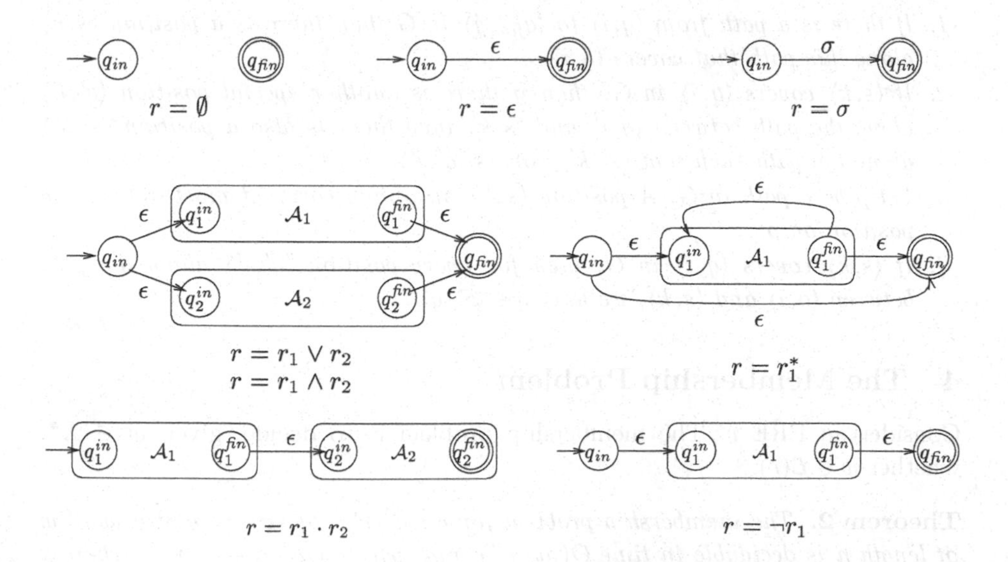

Fig. 2. The ASU $\mathcal{A}_r$ for the ERE r.

as *special states*, and refer to positions $\langle q, i \rangle \in (Q_u \cup Q_n) \times \{0, \ldots, n\}$ as *special positions*. In order to analyze the structure of $\mathcal{A}_r$, we introduce the function $\varphi : Q \to Q_u \cup Q_n \cup \{\bot\}$. Intuitively, $\varphi(q)$, for a state $q \in Q$, is the special state associated with innermost special operator in r in which the subexpression associated with q is strictly nested. If no such special operator exists, then $\varphi(q) = \bot$. Note that we talk about strict nesting, thus $\varphi(q) \neq q$. The formal definition of φ is inductive, and we use the notations in Figure 2. Thus, if $r = \emptyset, \epsilon$, or σ, we refer to q_{in} and q_{fin}, and if r that has r_1 or r_2 as immediate subexpressions, we also refer to Q_1 and Q_2 and the functions φ_1 and φ_2 defined for the ASUN $\mathcal{A}_1$ and $\mathcal{A}_2$. Now, for $r = \epsilon$, $\emptyset$, σ, $r_1 \vee r_2$, $r_1 \cdot r_2$, or r_1^*, and a state $q \in \mathcal{A}_r$, we have

$$\varphi(q) = \begin{bmatrix} \bot & \text{If } q = q_{in} \text{ or } q = q_{fin}. \\ \varphi_i(q) & \text{If } q \in Q_i, \text{ for } i \in \{1,2\}. \end{bmatrix}$$

Then, for $r = r_1 \wedge r_2$ or $r = \neg r_1$, and a state $q \in \mathcal{A}_r$, we have

$$\varphi(q) = \begin{bmatrix} \bot & \text{If } q = q_{in} \text{ or } q = q_{fin}. \\ \varphi_i(q) & \text{If } q \in Q_i, \text{ for } i \in \{1,2\}, \text{ and } \varphi_i(q) \neq \bot. \\ q_{in} & \text{If } q \in Q_i, \text{ for } i \in \{1,2\}, \text{ and } \varphi_i(q) = \bot. \end{bmatrix}$$

Let $\varphi^1(q) = \varphi(q)$, and let $\varphi^{i+1}(q) = \varphi(\varphi^i(q))$, for $i \geq 1$. Then, $\varphi^*(q) = \{\varphi^i(q) : i \geq 1\}$.

Lemma 1. *Consider the computation graph G of an ASUN $\mathcal{A}_r$ constructed in Theorem 1 on an input word w. Let q be a special state and let $\langle q, i \rangle$ be a position in G. The following hold.*

1. *If there is a path from $\langle q, i \rangle$ to $\langle q_{fin}, j \rangle$ in G then there is a position $\langle s, k \rangle$ along this path that covers $\langle q, i \rangle$.*
2. *If $\langle s, k \rangle$ covers $\langle q, i \rangle$ in G, then if there is another special position $\langle q', i' \rangle$ along the path between $\langle q, i \rangle$ and $\langle s, k \rangle$ then there is also a position $\langle s', k' \rangle$ along the path such that $\langle s', k' \rangle$ covers $\langle q', i' \rangle$.*
3. *Let p be a path in G. A position $\langle s, k \rangle$ on p may cover at most one special position on p.*
4. *If $\langle s, k \rangle$ covers $\langle q, i \rangle$ in G, then for every position $\langle q', i' \rangle$ along the path between $\langle q, i \rangle$ and $\langle s, k \rangle$, we have $q \in \varphi^*(q')$.*

4 The Membership Problem

Consider an ERE r. The membership problem is to decide, given $w \in \Sigma^*$, whether $w \in \mathcal{L}(r)$.

Theorem 2. *The membership problem for an ERE r of size m and a word w of length n is decidable in time $O(m \cdot n^2)$ and space $O(m \cdot n + k \cdot n^2)$, where k is the number of special operators in r.*

Proof: We describe an algorithm that runs in time $O(m \cdot n^2)$ and space $O(m \cdot n + k \cdot n^2)$, and determines the membership of w in $\mathcal{L}(r)$. A pseudo-code for the algorithm appears in Figure 3. Given an ERE r, we build the ASUN $\mathcal{A} = \langle \Sigma, Q, \mu, q_{in}, \delta, \psi, \{q_{fin}\} \rangle$ according to Theorem 1. The ASUN $\mathcal{A}$ has $O(m)$ states. Next, we build the computation graph G of $\mathcal{A}$ on w. As mentioned before, G has $O(m \cdot n)$ nodes. According to Theorem 1, every state $q \in Q$ has at most two ϵ-successors or exactly one σ-successor, for exactly one $\sigma \in \Sigma$. Therefore, G has $O(m \cdot n)$ edges.

The algorithm operates on G. It begins by determining a full order $\leq$ on the special positions of G. The full order is an extension of the partial order $\leq'$, where $\langle q', i' \rangle \leq' \langle q, i \rangle$ iff $\varphi(q') = q$. Note that according to Lemma 1(4), in every path in G from a special position $\langle q, i \rangle$ to a covering position $\langle s, j \rangle$ of it, if there is another special position $\langle q', i' \rangle$ on this path, then $\langle q', i' \rangle \leq \langle q, i \rangle$. In addition, according to Lemma 1(2), the path also includes some position $\langle s', j' \rangle$ that covers $\langle q', i' \rangle$.

The algorithm proceeds by calling the function *Find_Sync* for each of the special positions, by the order of $\leq$ (starting with the least element according to $\leq$; that is, innermost states are processed first). The function *Find_Sync*($\langle q, i \rangle$) constructs the set $S_{\langle q,i \rangle}$, which consists of indices of covering positions of $\langle q, i \rangle$. Recall that universal positions have two ϵ-successors, denoted *left-child* and *right-child*, while negation positions have exactly one ϵ-successor, denoted *left-child*. The construction of $S_{\langle q,i \rangle}$ begins by initializing two empty sets, $S_{\langle q,i,left \rangle}$ and $S_{\langle q,i,right \rangle}$ (in the case where q is a negation state, the second set is redundant). Next, the procedure *Update*($\langle s, j \rangle, \langle q, i \rangle, d$) is called, with d being either *left* or *right*. This procedure searches G in DFS manner for covering positions of $\langle q, i \rangle$ starting from the position $\langle s, j \rangle$. The indices of the covering positions are accumulated in the set $S_{\langle q,i,d \rangle}$. Thus, if q is a universal state, *Update* is called for the left and right children of $\langle q, i \rangle$, and the results of the two calls are intersected, forming the set $S_{\langle q,i \rangle}$, which consists of covering positions reachable on both sides. If q is a negation state, then only one call for *Update* is necessary. In this case, the negation is achieved by complementing the set retrieved by *Update* with respect to the set $\{i, \ldots, n\}$ of all potential indices of covering positions for $\langle q, i \rangle$. Note that during the search held by *Update*, if a special position $\langle q', i' \rangle$ is found then we already have $S_{\langle q',i' \rangle}$. Therefore the search may continue from positions in $S_{\langle q',i' \rangle}$, if there are any. This point is crucial for the efficiency of the algorithm.

After the calls for *Find_Sync* for all special positions are completed, the function *Find_Path* is called. This function starts at the initial position $\langle q_{in}, 0 \rangle$ searching the graph for the accepting position $\langle q_{fin}, n \rangle$ in a DFS manner. Like the *Update* procedure, whenever a special position $\langle q, i \rangle$ is found, the search continues from positions in $S_{\langle q,i \rangle}$, if there are any. The algorithm returns the result of the function *Find_Path*, which is *true* iff the accepting position was found.

```
function Membership_Check(G)
   let ⟨q_1, i_1⟩ ≤ ⟨q_2, i_2⟩ ≤ ··· ≤ ⟨q_l, i_l⟩ be a full order on the special positions of G,
      such that ≤ extends the partial order ≤' given by ⟨q, i⟩ ≤' ⟨q', i'⟩ iff φ(q) = q'.
   for all q ∈ Q, 0 ≤ i ≤ n do
         ⟨q, i⟩.index = −1;
         ⟨q, i⟩.visited = false;
   for j = 1 to l do S_⟨q_j,i_j⟩ = Find_Sync(⟨q_j, i_j⟩);
   return(Find_Path(⟨q_in, 0⟩));

function Find_Sync(⟨q_j, i_j⟩)
   if q_j is a universal state then
      S_⟨q_j,i_j,left⟩ := ∅; S_⟨q_j,i_j,right⟩ := ∅;
      Update(left_child(⟨q_j, i_j⟩), ⟨q_j, i_j⟩, left);
      Update(right_child(⟨q_j, i_j⟩), ⟨q_j, i_j⟩, right);
      return(S_⟨q_j,i_j,left⟩ ∩ S_⟨q_j,i_j,right⟩);
   else // q_j is a negation state
      S_⟨q_j,i_j,left⟩ := ∅; Update(child(⟨q_j, i_j⟩), ⟨q_j, i_j⟩, left);
      return({i_j, ..., n} \ S_⟨q_j,i_j,left⟩);

procedure Update(⟨s, j⟩, ⟨q, i⟩, d)
   if ⟨s, j⟩.index = i or ⟨s, j⟩ is a leaf then return;
   ⟨s, j⟩.index := i;
   if ⟨s, j⟩ is existential then
      if |δ(s, ε)| = 2 then
         Update(left_child(⟨s, j⟩), ⟨q, i⟩, d);
         Update(right_child(⟨s, j⟩), ⟨q, i⟩, d);
      else if child(⟨s, j⟩) = ⟨ψ(q), j⟩ then S_⟨q,i,d⟩ := S_⟨q,i,d⟩ ∪ {j}
            else Update(left_child(⟨s, j⟩), ⟨q, i⟩, d);
   else // ⟨s, j⟩ is special
      for every j' in S_⟨s,j⟩ do Update(⟨ψ(q), j'⟩, ⟨q, i⟩, d);
   return;
function Find_Path(⟨q, i⟩)
   if ⟨q, i⟩ is accepting then return(true);
   if ⟨q, i⟩.visited or ⟨q, i⟩ is a leaf then return(false);
   ⟨q, i⟩.visited := true;
   if ⟨q, i⟩ is existential then
      if |δ(q, ε)| = 2 then
         return(Find_Path(right_child(⟨q, i⟩)) or Find_Path(left_child(⟨q, i⟩)));
      return(Find_Path(child(⟨q, i⟩)));
   else: // ⟨q, i⟩ is special
      for every j in S_⟨q,i⟩ do
         if Find_Path(⟨ψ(q), j⟩) then return(true);
      return(false);
```

Fig. 3. The membership-checking algorithm.

For every position $\langle q, i\rangle$ of G, we keep a boolean flag $\langle q, i\rangle.visited$. *Find_Path* sets the flag for every position it visits, thus avoiding any type of repetitions. In addition, for every position $\langle q, i\rangle$ of G, we keep an integer variable $\langle q, i\rangle.index$. This variable is used in the *Update* procedure and it maintains the index of the special position that we are currently trying to cover. This allows the algorithm to avoid cycling and repetition of bad paths while trying to cover a certain special position. However, multiple checks of a position are allowed when done in different contexts, that is, while trying to cover two different special positions. The reason for allowing this kind of repetition is the possibility of having a position $\langle s, j\rangle$ that might be visited in paths from two special positions $\langle q, i\rangle$ and $\langle q, i'\rangle$. In this case we do not have previous knowledge about covering positions, and we need to go further with the check again. This makes our algorithm quadratic in n rather than linear in n.

The correctness of the algorithm follows from the following claim.

Claim. Let $S_{\langle q,i\rangle}$ be the set constructed for $\langle q, i\rangle$ in *Find_Sync*($\langle q, i\rangle$). For every $j \in \{0, \ldots, n\}$, we have that $\langle q, i\rangle \mapsto \langle \psi(q), j\rangle$ iff $j \in S_{\langle q,i\rangle}$.

It is left to show that the algorithm runs in time $O(m \cdot n^2)$ and space $O(m \cdot n + k \cdot n^2)$. For keeping the computation graph, the algorithm requires $O(m \cdot n)$ space. If there are k special operators in r, then there are k special states is $\mathcal{A}$. Since every state in $\mathcal{A}$ corresponds to at most $n+1$ positions in G, we have $O(k \cdot n)$ special positions in G. For each special position we keep at most two sets of at most $n+1$ indices. Therefor, the total space required for storing these sets is $O(k \cdot n^2)$, resulting in overall of $O(m \cdot n + k \cdot n^2)$ space.

Let q be a special state in $\mathcal{A}$. In each call to *Update*($\langle s, j\rangle, \langle q, i\rangle, d$), we have $\varphi(s) = q$. Therefore, there are $O(n^2)$ calls for *Update* involving q and s. Note that the first call to *Update*($\langle s, j\rangle, \langle q, i\rangle, d$) changes $\langle s, j\rangle.index$ to i. Hence, the next calls would return immediately as $\langle s, j\rangle.index = i$. We have $O(m)$ states s, thus the *Update* procedure is called $O(m \cdot n^2)$ times. The function *Find_Path* is called at most twice for each position, as the second call returns immediately. Therefore, there are only $O(m \cdot n)$ calls to *Find_Path*. As mentioned before, the construction of $\mathcal{A}$ and G can be done in time $O(m \cdot n)$. Hence, the overall running time of the algorithm is $O(m \cdot n^2)$. □

References

[Aho90] A.V. Aho. Algorithms for finding patterns in strings. *Handbook of Theoretical Computer Science*, pages 255–300, 1990.

[DHK+89] J. Dassow, J. Hromkovic, J. Karhumaki, B. Rovan, and A. Slobodova. On the power of synchronization in parallel computing. In *Proc. 14th International Symp. on Mathematical Foundations of Computer Science*, volume 379 of *Lecture Notes in Computer Science*, pages 196–206. Springer-Verlag, 1989.

[Hir89] S. Hirst. A new algorithm solving membership of extended regular expressions. Technical report, Basser Department of Computer Science, The University of Sydney, 1989.

[HMU00] J.E. Hopcroft, R. Motwani, and J.D. Ullman. *Introduction to Automata Theory, Languages, and Computation (2nd Edition).* Addison-Wesley, 2000.

[Hro86] J. Hrokovic. Tradeoffs for language recognition on parallel computing models. In *Proc. 13th Colloq. on Automata, Programming, and Languages*, volume 226 of *Lecture Notes in Computer Science*, pages 156–166. Springer-Verlag, 1986.

[HU79] J.E. Hopcroft and J.D. Ullman. *Introduction to Automata Theory, Languages, and Computation.* Addison-Wesley, 1979.

[JR91] T. Jiang and B. Ravikumar. A note on the space complexity of some decision problems for finite automata. *Information Processing Letters*, 40:25–31, 1991.

[KM95] James R. Knight and Eugene W. Myers. Super-pattern matching. *Algorithmica*, 13(1/2):211–243, 1995.

[KTV01] O. Kupferman, A. Ta-Shma, and M.Y. Vardi. Counting with automata. Submitted, 2001.

[MS73] A.R. Meyer and L.J. Stockmeyer. Word problems requiring exponential time: Preliminary report. In *Proc. 5th ACM Symp. on Theory of Computing*, pages 1–9, 1973.

[Mye92] G. Myers. A four russians algorithm for regular expression pattern matching. *Journal of the Association for Computing Machinery*, 39(4):430–448, 1992.

[Pet02] H. Petersen. The membership problem for regular expressions with intersection is complete in LOGCFL. In *Proc. 18th Symp. on Theoretical Aspects of Computer Science*, Lecture Notes in Computer Science. Springer-Verlag, 2002.

[Slo88] A. Slobodova. On the power of communication in alternating machines. In *Proc. 13th International Symp. on Mathematical Foundations of Computer Science*, volume 324 of *Lecture Notes in Computer Science*, pages 518–528, 1988.

[Yam00a] H. Yamamoto. An automata-based recognition algorithm for semi-extended regular expressions. In *Proc. 25th International Symp. on Mathematical Foundations of Computer Science*, volume 1893 of *Lecture Notes in Computer Science*, pages 699–708. Springer-Verlag, 2000.

[Yam00b] H. Yamamoto. On the power of input-synchronized alternating finite automata. In *Proc. 6th International Computing and Combinatorics Conference*, volume 1858 of *Lecture Notes in Computer Science*, pages 457–466, 2000.

Efficient Algorithms for Locating the Length-Constrained Heaviest Segments, with Applications to Biomolecular Sequence Analysis (Extended Abstract)

Yaw-Ling Lin, Tao Jiang, and Kun-Mao Chao

[1] Department of Computer Science and Information Management,
Providence University,
200 Chung Chi Road, Shalu, Taichung County, Taiwan 433,
yllin@pu.edu.tw

[2] Department of Computer Science and Engineering,
University of California Riverside,
Riverside, CA 92521-0144, USA,
jiang@cs.ucr.edu

[3] Department of Life Science,
National Yang-Ming University,
Taipei, Taiwan 112,
kmchao@ym.edu.tw

Abstract. We study two fundamental problems concerning the search for interesting regions in sequences: (i) given a sequence of real numbers of length n and an upper bound U, find a consecutive subsequence of length at most U with the maximum sum and (ii) given a sequence of real numbers of length n and a lower bound L, find a consecutive subsequence of length at least L with the maximum average. We present an $O(n)$-time algorithm for the first problem and an $O(n \log L)$-time algorithm for the second. The algorithms have potential applications in several areas of biomolecular sequence analysis including locating GC-rich regions in a genomic DNA sequence, post-processing sequence alignments, annotating multiple sequence alignments, and computing length-constrained ungapped local alignment. Our preliminary tests on both simulated and real data demonstrate that the algorithms are very efficient and able to locate useful (such as GC-rich) regions.

Keywords: Algorithm, efficiency, maximum consecutive subsequence, length constraint, biomolecular sequence analysis, ungapped local alignment.

1 Introduction

With the rapid expansion in genomic data, the age of large-scale biomolecular sequence analysis has arrived. An important line of research in sequence analysis is to locate biologically meaningful segments, *e.g.* *conserved* segments and

K. Diks et al. (Eds): MFSC 2002, LNCS 2420, pp. 459–470, 2002.

GC-rich regions, in DNA sequences. Conserved segments of a DNA sequence are slow changing sequences that form strong candidates for functional elements both in protein coding and regulatory regions of genes [7,10,15]. Regions of a DNA sequence that are rich in nucleotides C and G are usually significant in gene recognition. In order to locate these interesting segments, many combinatorial and probabilistic techniques have been proposed. Perhaps the most popular ones are window-based. That is, a window of a fixed length is moved down the sequence/alignment and the content statistics are calculated at each position that the window is moved to [12,14]. Since an optimal region could span several windows, the window-based approach might fail in finding the exact locations of some interesting regions.

In this paper, we study two fundamental problems concerning the search for the "heaviest" segment of a numerical sequence that naturally arises in the above applications. Our main results, as described below, are efficient algorithms for locating the length-constrained heaviest segments in a given sequence or alignment. The algorithms have potential applications in locating GC-rich regions in a genomic DNA sequence, post-processing sequence alignments, annotating multiple sequence alignments, and computing length-constrained ungapped local alignment.

Let $A = \langle a_1, a_2, \ldots, a_n \rangle$ be a sequence of real numbers and $U \leq n$ a positive integer, the objective of our first problem is to find a consecutive subsequence of A of length *at most* U such that the sum of the numbers in the subsequence is maximized. By using a technique of partitioning each suffix of A into minimal *left-negative* (consecutive) subsequences, we propose an $O(n)$-time algorithm for finding the length-constrained maximum sum consecutive subsequence of A. The algorithm can be used to find GC-rich regions and efficiently construct ungapped local alignments with length constraints in $O(mn)$ time, where m, n are the lengths of the two input sequences being aligned, as explained in the next section. We note in passing that a linear-time algorithm for finding the maximum sum consecutive subsequence with length at least L can be easily obtained [11] by extending the dynamically algorithm for the standard maximum sum consecutive subsequence problem in [6].

An alternative measure of the weight of the target segment that we consider is as follows. Given a sequence of real numbers, $A = \langle a_1, a_2, \ldots, a_n \rangle$, and a positive integer $L \leq n$, the goal is to find a consecutive subsequence of A of length *at least* L such that the average of the numbers in the subsequence is maximized. We propose a novel technique to partition each suffix of A into *right-skew* segments of strictly decreasing averages, and based on this partition, we devise an $O(n \log L)$-time algorithm for locating the maximum average consecutive subsequence of length at least L. [1] The algorithm is expected to have applications in finding GC-rich regions in a genomic DNA sequence, postprocessing sequence alignments, and annotating multiple sequence alignments.

[1] Note that, when there is no length constraint, finding the maximum average consecutive subsequence is equivalent to finding the maximum element.

Observe that both problems studied in this paper have straightforward dynamic programming algorithms with running time proportional to the product of the input sequence length n and the length constraint (*i.e.* U or L). Such algorithms are perhaps fast enough for sequences of small lengths, but can be too slow for instances in some biomolecular sequence analysis applications, such as finding GC-rich regions and post-processing sequence alignments, where long genomic sequences are involved. Our above algorithms would be able to handle genomic sequences of length up to millions of bases with satisfactory speeds, as demonstrated in the preliminary experiments. The heaviest segment problems that we study here are mostly motivated by their applications in several areas of biomolecular sequence analysis, such as locating GC-rich regions in gene recognition and comparative genomics [12,14,11,8], post-processing sequence alignments [4,5,13,3,17,18], annotating multiple sequence alignments [15,3,1], as well as computing ungapped local alignments with length constraints [1,3,2].

The rest of the paper is organized as follows. We present the algorithm for the length-constrained maximum sum consecutive subsequence problem in Section 2 and the algorithm for the length-constrained maximum average consecutive subsequence problem in Section 3. Some preliminary experiments on the speed and performance of the algorithms are given in Section 4. Section 5 concludes the paper with a few remarks.

Due to the page limit, many applications of biomolecular sequence analysis and proofs of all lemmas and corollaries are omitted in the extended abstract and provided in the appendix.

2 Maximum Sum Consecutive Subsequence with Length Constraints

Given a sequence of real numbers, $A = \langle a_1, a_2, \ldots, a_n \rangle$, and a positive integer $U \leq n$, the goal is to find a consecutive subsequence of A of length at most U such that the sum of the numbers in the subsequence is maximized. It is straightforward to design a dynamic programming algorithm for the problem with running time $O(nU)$. We also note in passing that since there is an $O(n \log^2 n)$-time algorithm for finding the maximum sum path on a tree with length at most U [16], the above problem can also be solved in $O(n \log^2 n)$ time. Here, we present an algorithm running in $O(n)$.

Let $A_1, A_2, \ldots, A_k$ be disjoint (consecutive) subsequences of A forming a *partition* of A, *i.e.* $A = A_1 A_2 \cdots A_k$. A_i is called the ith segment of the partition. Denote $w(A) = \sum_{a_i \in A} a_i$ as the sum of the sequence. The following definition is a key of our linear-time construction.

Definition 1. *A real sequence $A = \langle a_1, a_2, \ldots, a_n \rangle$ is* left-negative *if and only if the sum of each proper prefix $\langle a_1, a_2, \ldots, a_i \rangle$ is negative or zero for all $1 \leq i \leq n-1$; that is, $w(\langle a_1, a_2, \ldots, a_i \rangle) \leq 0$ for all $1 \leq i \leq n-1$. A partition of the sequence $A = A_1 A_2 \cdots A_k$ is* minimal left-negative *if each $A_i, 1 \leq i \leq k$, is left-negative, and, for each $1 \leq i \leq k-1$, the sum of A_i is positive, i.e. $w(A_i) > 0$.*

For example, the sequence $\langle -4, 1, -2, 3\rangle$ is left-negative while the sequence $\langle 5, -3, 4, -1, 2, -6\rangle$ is not. On the other hand, the partition $\langle 5\rangle\langle -3, 4\rangle\langle -1, 2\rangle\langle -6\rangle$ of the latter sequence is minimal left-negative. Note that any singleton sequence is trivially left-negative by definition. Furthermore, it can be shown that any sequence can be uniquely partitioned into minimal left-negative segments.

Lemma 1. *Every sequence of real numbers can be uniquely partitioned into minimal left-negative segments.*

For any sequence $A = \langle a_1, a_2, \ldots, a_n\rangle$, each suffix sequence of A, $\langle a_i, \ldots, a_n\rangle$, defines a minimal left-negative partition, denoted as $A_1^{(i)} A_2^{(i)} \cdots A_k^{(i)}$, for some $k \geq 1$. Suppose that $A_1^{(i)} = \langle a_i, \ldots, a_{p[i]}\rangle$. Then, $p[i]$ is called the *left-negative pointer* of index i. Note that the left-negative pointers of A implicitly encode the minimal left-negative partition of each suffix $\langle a_i, \ldots, a_n\rangle$ of A. An efficient algorithm for computing the left-negative pointers as well as the minimal left-negative partition of each suffix of A is illustrated in Figures 1 and 2.

Lemma 2. *The algorithm* MLN-POINT *given in Figure 1 finds all left-negative pointers for a length n sequence in $O(n)$ time.*

We are ready to show that the length-constrained maximum sum consecutive subsequence problem can be solved in linear time.

Theorem 1. *Given a length n real sequence, finding the consecutive subsequence of length at most U with the maximum sum can be done in $O(n)$ time.*

Proof. We propose an $O(n)$ time algorithm, MSLC(A, U), as shown in Figure 3. In the algorithm, the variable i is the current working pointer scanning elements

---

MLN-POINT(A)
Input: A real sequence $A = \langle a_1, a_2, \ldots, a_n\rangle$.
Output: n left-negative pointers of A, encoded by array $p[\cdot]$.
1 **for** $i \leftarrow n$ **downto** 1 **do**
2 $\quad p[i] \leftarrow i; w[i] \leftarrow a_i;$ $\quad \triangleright$ Each $\langle a_i\rangle$ alone is left-negative.
3 $\quad$ **while** $(p[i] < n)$ and $w[i] \leq 0$ **do**
4 $\quad\quad w[i] \leftarrow w[i] + w[p[i] + 1]$
5 $\quad\quad p[i] \leftarrow p[p[i] + 1]$

Fig. 1. Set up the left-negative pointers.

REPORT-MLN-PART(i)
Input: i denoting the suffix sequence $\langle a_i, a_{i+1}, \ldots, a_n\rangle$.
Output: the minimal left-negative partition of the suffix.
1 **while** $i \leq n$ **do** $\quad \triangleright$ Reports (i, j) as a left-negative segment $\langle a_i, \ldots, a_j\rangle$.
2 $\quad$ OUTPUT $(i, p[i]); i \leftarrow p[i] + 1$

Fig. 2. Compute the minimal left-negative partition of a suffix sequence.

MSLC(A, U)
Input: A real sequence $A = \langle a_1, a_2, \ldots, a_n \rangle$, and an upper bound U.
Output: The maximum consecutive subsequence of A with length at most U.
1 $i \leftarrow 1$
2 **while** $a_i \leq 0$ and $i \leq n$ **do** $i \leftarrow i + 1$
3 **if** $i = n$ **then** ▷ Elements $a_1, a_2, \ldots, a_{n-1}$ are all negative.
4 Find the maximum element in A and return.
5 MLN-POINT(A) ▷ Compute left-negative pointers. See Fig 1.
6 $j \leftarrow i$; $ms \leftarrow 0$ ▷ Initialization.
7 **while** $i \leq n$ **do**
8 **while** $a_i \leq 0$ and $i \leq n$ **do** $i \leftarrow i + 1$
9 $j \leftarrow \max(i, j)$
10 **while** $j < n$ and $p[j+1] < i + U$ and $w[j+1] > 0$ **do** $j \leftarrow p[j+1]$
11 **if** SUM(i, j) $> ms$ **then** $mi \leftarrow i; mj \leftarrow j; ms \leftarrow$ SUM(i, j) ▷ Update max.
12 $i \leftarrow i + 1$
13 **return** (mi, mj, ms)

SUM(i, j)
Output: The sum of the subsequence $\langle a_i, a_{i+1}, \ldots, a_j \rangle$, $\sum_{x=i}^{j} a_x$, is just $s_j - s_{i-1}$.
The prefix sums, $s_k = \sum_{i=1}^{k} a_i$, $s_0 = 0$, can be pre-computed in $O(n)$ time.

Fig. 3. Finding the maximum sum consecutive subsequence with length constraint.

of A from left to right. The pair (i, j) represents a consecutive subsequence of A, $\langle a_i, \ldots, a_j \rangle$, currently being considered as a candidate maximum sum consecutive subsequence satisfying the length constraint. The algorithm essentially looks at every positive a_i and identifies its corresponding *good partner*, a_j, such that (i, j) constitutes a candidate solution.

Note that the sum of any proper prefix of a left-negative segment is negative by definition. The correctness of the algorithm then follows from the fact that a left-negative segment is *atomic* in the sense that when it is combined with preceding left-negative segments, it is always combined *as a whole*; for otherwise the addition of any proper prefix of the segment would only decrease the sum of the combined segment. This observation justifies the condition checking and grouping in Step 10 of the algorithm.

The time complexity of the algorithm is $O(n)$ because the good-partner pointer j only advances forward as the scanning pointer i advances. It follows that the total work spent on Step 10 is bounded by $O(n)$. It is not hard to verify that the remaining part of the algorithm spends at most $O(n)$ time. □

The algorithm MSLC can be combined with Huang's technique [11] to yield a linear-time algorithm that could handle both a length upper bound and a length lower bound simultaneously.

Corollary 1. *Given a length n real sequence and positive integers $L \leq U$, finding the consecutive subsequence of length between L and U with the maximum sum can be done in $O(n)$ time.*

3 Maximum Average Consecutive Subsequence with Length Constraints

Given a sequence of real numbers, $A = \langle a_1, a_2, \ldots, a_n \rangle$, and a positive integer L, $1 \le L \le n$, our goal is now to find a consecutive subsequence of A with length at least L such that the average value of the numbers in the subsequence is maximized.

Recall that $w(A) = \sum_{i=1}^{n} a_i$ is the sum of elements of A. Furthermore, let $d(A) = |A| = n$, be the length of the sequence A. The *average* of A is defined as $\mu(A) = w(A)/d(A)$. The definition below is the key to our construction.

Definition 2. *A sequence $A = \langle a_1, a_2, \ldots, a_n \rangle$ is* right-skew *if and only if the average of any prefix $\langle a_1, a_2, \ldots, a_i \rangle$ is always less than or equal to the average of the remaining suffix $\langle a_{i+1}, a_{i+2}, \ldots, a_n \rangle$. A partition $A = A_1 A_2 \cdots A_k$ is* decreasingly right-skew *if each segment A_i of the partition is right-skew and $\mu(A_i) > \mu(A_j)$ for any $i < j$.*

The following are some useful properties of right-skew segments and their averages.

Lemma 3 (Combination). *Let A, B be two sequences with $\mu(A) < \mu(B)$. Then $\mu(A) < \mu(AB) < \mu(B)$.*

Lemma 4. *Let A, B be two right-skew sequences with $\mu(A) \le \mu(B)$. Then the sequence AB is also right-skew.*

Lemma 5. *Every real sequence $A = \langle a_1, a_2, \ldots, a_n \rangle$ has a unique decreasingly right-skew partition.*

For a sequence $A = \langle a_1, a_2, \ldots, a_n \rangle$, each suffix of A, $\langle a_i, \ldots, a_n \rangle$, defines a decreasingly right-skew partition, denoted as $A_1^{(i)} A_2^{(i)} \cdots A_k^{(i)}$, for some $k \ge 1$. Suppose that $A_1^{(i)} = \langle a_i, \ldots, a_{p[i]} \rangle$, where $p[i]$ is called the *right-skew pointer* of index i. Note that the right-skew pointers of A implicitly encode the decreasingly right-skew partitions for each suffix $\langle a_i, \ldots, a_n \rangle$ of A. Given the right-skew pointers, one can easily report the decreasingly right-skew partitions of a suffix as illustrated in Figure 4. Interestingly, we can compute all right-skew pointers in linear time.

REPORT-DRS-PART$(i, p[\cdot])$
Input: i denoting the suffix sequence $\langle a_i, a_{i+1}, \ldots, a_n \rangle$; $p[\cdot]$: right-skew pointers of A.
Output: The decreasingly right-skew partition of the suffix.
1 **while** $i \le n$ **do** ▷ Reports $\langle a_i, \ldots, a_j \rangle$ as a right-skew segment.
2 OUTPUT $(i, p[i]); i \leftarrow p[i] + 1$

Fig. 4. Report the decreasingly right-skew partition of a suffix sequence.

DRS-POINT(A)
Input: A sequence $A = \langle a_1, a_2, \ldots, a_n \rangle$.
Output: n right-skew pointers of A, encoded by array $p[\cdot]$.
1 **for** $i \leftarrow n$ **downto** 1 **do**
2 $\quad p[i] \leftarrow i; w[i] \leftarrow w(a_i); d[i] \leftarrow d(a_i);$ ▷ Each $\langle a_i \rangle$ alone is right-skew.
3 $\quad$ **while** $(p[i] < n)$ and $(w[i]/d[i] \leq w[p[i]+1]/d[p[i]+1])$ **do**
4 $\quad\quad w[i] \leftarrow w[i] + w[p[i]+1]$
5 $\quad\quad d[i] \leftarrow d[i] + d[p[i]+1]$
6 $\quad\quad p[i] \leftarrow p[p[i]+1]$

Fig. 5. Set up the right-skew pointers in $O(n)$ time.

Lemma 6. *The algorithm* DRS-POINT *given in Figure 5 computes all right-skew pointers for a length n sequence in $O(n)$ time.*

The next lemma is first presented in [11]. We include it here for completeness.

Lemma 7. *Given a real sequence A, let B denote the shortest consecutive subsequence of A with length at least L such that the average is maximized. Then the length of B is at most $2L - 1$.*

In searching for the maximum average consecutive subsequence, our construction will need to locate, for each element a_i, its corresponding partner, a_j, such that the segment $\langle a_i, \ldots, a_j \rangle$ constitutes a candidate solution. Suppose that segment $A = \langle a_i \ldots a_j \rangle$ is being currently considered a candidate solution, where $j - i + 1 \geq L$, and $B = \langle a_{j+1}, \cdots, a_{p[j+1]} \rangle$ is the first right-skew segment to the right of A. We consider if the segment A should be extended to include some prefix (or the whole) of the segment B. The following lemma shows that A should be combined with the segment B *as a whole* if and only if $\mu(A) < \mu(B)$. In other words, the segment $B = \langle a_{j+1}, \cdots, a_{p[j+1]} \rangle$ is *atomic* (for A).

Lemma 8 (Atomic). *Let A, B, C be three real sequences with $\mu(A) < \mu(B) < \mu(C)$. Then $\mu(AB) < \mu(ABC)$.*

The next lemma allows us to perform binary search in the decreasingly right-skew partition of a suffix sequence when trying to find the "optimal" extension from a candidate solution segment.

Lemma 9 (Bitonic). *Let P be a (prefix) real sequence, and $A_1 A_2 \cdots A_m$ the decreasingly right-skew partition of a sequence A. Suppose that $\mu(PA_1 \cdots A_k) = \max\{\mu(PA_1 \cdots A_i) | \ 0 \leq i \leq m\}$. Then $\mu(PA_1 \cdots A_i) > \mu(A_{i+1})$ if and only if $i \geq k$.*

Now we are ready to state the main result of this section.

Theorem 2. *Given a length n real sequence, finding the consecutive subsequence of length at least L with the maximum average can be done in $O(n \log L)$ time.*

Proof. We propose an $O(n \log L)$ time algorithm, MAXAVGSEQ(A, L), as shown in Figure 6. The pointer i scans elements of A from left to right. The pair (i, j) represents a segment of A, $\langle a_i, \ldots, a_j \rangle$, currently being considered as the candidate solution. For each element a_i, the algorithm finds its corresponding *good partner*, a_j, such that (i, j) constitutes a candidate solution.

MAXAVGSEQ(A, L)
Input: A real sequence $A = \langle a_1, a_2, \ldots, a_n \rangle$ and a lower bound L.
Output: The maximum average consecutive subsequence of A of length at least L.

1 DRS-POINT(A) $\triangleright$ Compute the right-skew pointers, see Fig 5.
2 **for** $i \leftarrow 1$ **to** $n - L + 1$ **do**
3 $\quad j \leftarrow i + L - 1$
4 $\quad$ **if** $\mu(i, j) < \mu(j + 1, p[j + 1])$ **then** $j \leftarrow$ LOCATE(i, j) $\triangleright$ Move j.
5 $\quad g[i] \leftarrow j$
6 **return** The maximum $\mu(i, g[i])$ pair.

$\mu(i, j) = \text{SUM}(i, j)/(j - i + 1)$ is the average of segment $\langle a_i, \ldots, a_j \rangle$.
LOCATE(i, j): Binary search in the list: $\langle \mu(i, j^{(0)}), \ldots, \mu(i, j^{(L)}) \rangle$, where $j^{(k)}$ is defined recursively: $j^{(0)} = j$, $j^{(k)} = \min\{p[j^{(k-1)} + 1], n\}$.

Fig. 6. Finding the maximum average consecutive subsequence with length constraint.

Observe that right-skew segments are *atomic* in the sense that it is always better to add a whole right-skew segment in an extension process than to add a proper prefix, as shown in Lemma 8. Thus the possible good partners will be the right endpoints of the right-skew segments in the decreasingly right-skew partition of the suffix sequence $\langle a_{j+1}, \ldots, a_n \rangle$.

Let $j^{(k)}$ denote the right endpoint of the kth right-skew segment in the suffix sequence $\langle a_{j+1}, \ldots, a_n \rangle$. Note that $j^{(k)}$ can be defined recursively using the formula: $j^{(0)} = j$ and $j^{(k)} = \min\{p[j^{(k-1)} + 1], n\}$. By Lemma 7, there exists a maximum average segment whose length is at most $2L - 1$. Thus, the correctness of algorithm MAXAVGSEQ(A, L) follows if LOCATE(i, j) correctly computes the optimal j^* such that $\mu(i, j^*) = \max\{\mu(i, j^{(k)}) | 0 \leq k \leq L\}$, where $\mu(i, j)$ denotes the average of segment $\langle a_i, \ldots, a_j \rangle$. This is explained along with the following time complexity analysis of algorithm LOCATE.

To prove that the algorithm MAXAVGSEQ runs in $O(n \log L)$ time, it suffices to prove that algorithm LOCATE finds the (restricted) good partner j^* of i in $O(\log L)$ time. The key idea used in the algorithm is as follows. Although exploring the entire list $\langle j^{(1)}, \ldots, j^{(L)} \rangle$ to find the (restricted) good partner requires $O(L)$ time, Lemma 9 suggests that we may be able to find j^* by a binary search *without* having to generate the entire list $\langle j^{(1)}, \ldots, j^{(L)} \rangle$. To do so, we need maintain $\lceil \log L \rceil$ *pointer-jumping tables* $p^{(k)}[1..n], 1 \leq k \leq \lceil \log L \rceil$. Let $p^{(0)}[i] = p[i]$ and $p^{(k+1)}[i] = \min\{p^{(k)}[p^{(k)}[i] + 1], n\}$ be defined recursively. Intuitively, one pointer jump from j to $p^{(k)}[j+1]$ is equivalent to 2^k original pointer jumps from

LOCATE(i, j)
Input: A prefix subsequence $\langle a_i, \ldots, a_j \rangle$ of A.
Output: The maximum average subsequence with prefix $\langle a_i, \ldots, a_j \rangle$ and length at most $2L - 1$.
1 **for** $k \leftarrow \lceil \log L \rceil$ **downto** 0 **do**
2 **if** $j \geq n$ or $\mu(i, j) \geq \mu(j + 1, p[j + 1])$ **then return** j
3 **if** $p^{(k)}[j + 1]) < n$ and $\mu(i, p^{(k)}[j + 1]) < \mu(p^{(k)}[j + 1] + 1, p[p^{(k)}[j + 1] + 1])$ **then** $j \leftarrow p^{(k)}[j + 1]$
4 **if** $j < n$ and $\mu(i, j) < \mu(j + 1, p[j + 1])$ **then** $j \leftarrow p[j + 1]$ ▷ Final step.
5 **return** $j^* = j$

Fig. 7. Finding the maximum average consecutive subsequence with prefix $\langle a_i, \ldots, a_j \rangle$ and length at most $2L - 1$.

j to $j^{(2^k)}$. Note that, these $p^{(k)}[1..n]$ tables can be pre-computed with an overall time complexity of $O(n \log L)$.

Now we explain how the binary search performed in Steps 1 through 3 of LOCATE(i, j) for finding j^* works. Let $j^* = j^{(\ell)}$ for some $0 \leq \ell \leq L$. Then the problem of finding j^* can be thought of identifying an unknown binary string (the binary encoding of ℓ) of at most $\lceil \log L \rceil$ bits. In the algorithm, we identify the bits one by one from the ($\lceil \log \rceil - 1$)th (the most significant bit) down to the 0th (the lowest) bit, and for each kth bit, we check if $\mu(i, p^{(k)}[j + 1]) < \mu(p^{(k)}[j + 1] + 1, p[p^{(k)}[j + 1] + 1])$ using the pointer-jumping tables. The bitonicity property in Lemma 9 can be used to determine whether the current index $j^{(\ell)}$ under consideration has surpassed the desired j^*. Note that, Step 4 of LOCATE(i, j) makes a final check on the result since the value of index j at the moment can be one step short of the optimal index value $j^* = j^{(\ell)}$ for some even number ℓ.

Therefore, LOCATE(i, j) finds a (restricted) good partner of i in $O(\log L)$ time. It follows that the algorithm MAXAVGSEQ(A, L) runs in at most $O(n \log L)$ time since Step 4 of the algorithm takes $O(\log L)$ time, and the precomputation of the jumping tables also takes at most $O(n \log L)$ time. □

4 Implementation and Preliminary Experiments

We have implemented a family of programs for locating the length-constrained heaviest segments, based on the algorithms described in this paper. Specifically, five programs are discussed below:

- *mslc*: Given a real sequence of length n and an upper bound U, this program locates the maximum-sum subsequence of length at most U in $O(n)$-time.
- *mslc_slow*: A brute-force $O(nU)$-time version of *mslc*.
- *mavs*: Given a real sequence of length n and a lower bound L, this program locates the maximum-average subsequence of length at least L in $O(n \log L)$.
- *mavs_slow*: A brute-force $O(nL)$-time version of *mavs*.

– *mavs_linear*: Instead of finding a good partner by binary search, as done in *mavs*, this program linearly scan right-skew segments for the good partnership. In the worst case, the time complexity is $O(nL)$. However, our empirical tests showed that it ran faster than *mavs* in most cases.

Table 1 summarizes the comparative evaluation of the five programs on a random integer sequence ranged from -50 to 50 of length 1,000,000. These experiments were carried out on a Sun Enterprise 3000 UltraSPARC based system. Several length lower and upper bounds were used to illustrate their performance. For example, with $L=U=$5,000, *mslc* ran in 1.08 seconds, while *mslc_slow* took 578.45 seconds. It is not surprising to see that the running time of *mslc* was independent of U, and the running time of *mavs* increased slightly for larger L, whereas *mslc_slow* and *mavs_slow* grew proportionally to U and L, respectively. It is worth mentioning that *mavs_linear*, which scans right-skew segments linearly, ran even faster than *mavs*, which performs binary search among right-skew segments. The main reason was that the length of the maximum average consecutive subsequence seems usually quite close to L. Thus, *mavs_linear* could quickly locate the good partners by a linear scan.

Table 1. Comparative evaluation of the five methods on a random integer sequence ranged from -50 to 50 of length 1,000,000. The time unit is second.

		Maximum Sum		Maximum Average		
n	L, U	*mslc*	*mslc_slow*	*mavs*	*mavs_slow*	*mavs_linear*
1,000,000	100	1.14	12.67	8.55	46.72	3.15
1,000,000	500	1.12	57.36	9.63	232.17	3.29
1,000,000	1,000	1.15	122.97	9.11	471.64	3.06
1,000,000	5,000	1.08	578.45	10.92	2331.52	3.36
1,000,000	10,000	1.12	1270.11	11.92	4822.25	3.13

We have also used the programs to analyze the homo sapiens 4q sequence contig of size 459kb from position 114130kb to 114589kb (sequenced by YMGC and WUGSC, GenBank accession number NT_003253). For instance, we found that the regions with the highest C+G ratio of length at least 200, 5000, and 10000 are 390396–390604 (C+G ratio 0.866), 389382–394381 (C+G ratio 0.513), and 153519–163520 (C+G ratio 0.475), respectively. This might suggest further biological experiments to better understand these GC-rich regions.

Huang's *LCP* program [11] is very efficient in finding in a sequence all GC-rich regions of length at least L. These GC-rich regions can be refined by locating their subregions with the highest C+G ratio by using our programs *mavs* or *mavs_linear*. To illustrate this approach, we studied the rabbit α-like globin gene cluster sequence of 10621bp, which is available from GenBank by accession number M35026 [9]. The length cutoff L considered was 50, and the minimum ratio p was chosen at 0.7. Table 2 summarizes the empirical results. *LCP* found

Table 2. Refining the regions found by program *LCP*.

LCP				*mavs*			
Start	End	Length	C+G Ratio	Start	End	Length	C+G Ratio
3372	3444	73	0.740	3395	3444	50	0.740
6355	6713	359	0.805	6619	6671	53	0.943
7830	7933	104	0.779	7861	7912	52	0.808
8029	8080	52	0.769	8029	8081	52	0.769
8483	8578	96	0.760	8483	8532	50	0.800
9557	10167	611	0.782	9644	9695	52	0.981

six interesting GC- rich regions. Take the region starting from position 6355 and ending at position 6713 for example. The length of this region is 359bp, and its C+G ratio is 0.805. Using the program *mavs*, we were able to find a subregion (of length 53bp) with the highest C+G ratio, which starts from position 6619 and ends at position 6671 with C+G ratio 0.943. Table 2 presents more examples of such refinements.

5 Concluding Remarks

In this paper, two fundamental problems concerning the search for the heaviest segment of a sequence with length constraints are considered. The first problem is to find a consecutive subsequence of length at most U with the maximum sum and the second is to find a consecutive subsequence of length at least L with the maximum average. We have presented a linear-time algorithm for the first and an $O(n \log L)$-time algorithm for the second. Our results also imply efficient solutions for finding a maximum sum consecutive subsequence of length within a certain range and length-constrained ungapped local alignment. The algorithms have applications to several important problems in biomolecular sequence analysis.

It would be interesting to know if there is a linear-time algorithm to find a maximum average consecutive subsequence of length at least L. It also remains open to develop an efficient algorithm for locating the maximum average consecutive subsequence of length between bounds L and U.

Acknowledgements

We thank Xiaoqiu Huang for his freely available program *LCP*. We also thank Wen-Lian Hsu, Ming-Yang Kao, Ming-Tat Ko, and Hsueh-I Lu for helpful conversations. Y.-L. Lin was supported in part by grant NSC 89-2218-E-126-006 from the National Science Council, Taiwan. T. Jiang was supported in part by a UCR startup grant and NSF Grants CCR-9988353 and ITR-0085910. K.-M. Chao was supported in part by grant NSC 90-2213-E-010-003 from the National Science Council, Taiwan, and by the Medical Research and Advancement Foundation in Memory of Dr. Chi-Shuen Tsou.

References

1. N.N. Alexandrov and V.V. Solovyev. Statistical significance of ungapped alignments. *Pacific Symposium on Biocomputing (PSB-98)*, pages 463–472, 1998.
2. A. Arslan and Ö Eğecioğlu. Algorithms for local alignments with constraints. *Manuscript*, 2001.
3. A. Arslan, Ö Eğecioğlu, and P. Pevzner. A new approach to sequence comparison: Normalized sequence alignment. *Bioinformatics*, 17:327–337, 2001.
4. V. Bafna and D.H. Huson. The conserved exon mothod for gene finding. *Proc. Int. Conf. Intell. Syst. Mol. Biol. (ISMB)*, 8:3–12, 2000.
5. S. Batzoglou, L. Pachter, J. Mesirov, B. Berger, and E. Lander. Comparative analysis of mouse and human DNA and applications to exon prediction. *Proc. Int. Conf. Comp. Mol. Biol. (RECOMB)*, 4, 2000.
6. J. Bentley. *Programming Pearls.* Addison-Wesley, Reading, Massachusetts, 1986.
7. M.S. Boguski, R.C. Hardison, S. Schwartz, and W. Miller. Analysis of conserved domains and sequence motifs in cellular regulatory proteins and locus control regions using new software tools for multiple alignment and visualization. *New Biol.*, 4:247–260, 1992.
8. S. Hannenhalli and S. Levy. Promoter prediction in the human genome. *Bioinformatics*, 17:S90–S96, 2001.
9. R.C. Hardison, D. Krane, C. Vandenbergh, J.-F.F. Cheng, J. Mansberger, J. Taddie, S. Schwartz, X. Huang, and W. Miller. Sequence and comparative analysis of the rabbit alpha-like globin gene cluster reveals a rapid mode of evolution in a G+C rich region of mammalian genomes. *J. Mol. Biol.*, 222:233–249, 1991.
10. R.C. Hardison, J.L. Slighton, D.L. Gumucio, M. Goodman, N. Stojanovic, and W. Miller. Locus control regions of mammalian beta-globin gene clusters: combining phylogenetic analyses and experimental results to gain functional insights. *Gene*, 205:73–94, 1997.
11. X. Huang. An algorithm for identifying regions of a DNA sequence that satisfy a content requirement. *CABIOS*, 10:219–225, 1994.
12. A. Nekrutenko and W.-H. Li. Assessment of compositional heterogeneity within and between eukaryotic genomes. *Genome Research*, 10:1986–1995, 2000.
13. P.S. Novichkov, M.S. Gelfand, and A.A. Mironov. Prediction of the exon-intron structure by comparison sequences. *Mol. Biol.*, 34:200–206, 2000.
14. P. Rice, I. Longden, and A. Bleasby. EMBOSS: the European molecular biology open software suite. *Trends Genet.*, 16:276–277, 2000.
15. N. Stojanovic, L. Florea, C. Riemer, D. Gumucio, J. Slightom, M. Goodman, W. Miller, and R. Hardison. Comparison of five method for finding conserved sequences in multiple alignments of gene regulatory regions. *Nucleic Acids Research*, 27:3899–3910, 1999.
16. B.Y. Wu, K.-M. Chao, and C. Y. Tang. An efficient algorithm for the length-constrained heaviest path problem on a tree. *Infomation Processing Letters*, 69:63–67, 1999.
17. Z. Zhang, P. Berman, and W. Miller. Alignments without low-scoring regions. *J. Comput. Biol.*, 5:197–200, 1998.
18. Z. Zhang, P. Berman, T. Wiehe, and W. Miller. Post-processing long pairwise alignments. *Bioinformatics*, 15:1012–1019, 1999.

Derivation of Rational Expressions with Multiplicity

Sylvain Lombardy and Jacques Sakarovitch

Laboratoire Traitement et Communication de l'Information (CNRS URA 820),
Ecole Nationale Supérieure des Télécommunications,
46, rue Barrault, 75634 Paris Cedex 13, France,
{lombardy,sakarovitch}@enst.fr

Abstract. This paper introduces a generalization of the *partial derivatives* of rational expressions, due to Antimirov, to rational expressions with multiplicity. We define the derivation of a rational expression with multiplicity in such a way that the result is a *polynomial of expressions*. This amounts to interpreting the addition symbol at the upper level in the semiring of coefficients.
Former results of Brzozowski and of Antimirov are then expressed in that framework that allows to deal with rational power series, and automata and expressions with multiplicity as well.

1 Introduction

The purpose of this paper is to generalize the definition and constructions, due to V. Antimirov ([1]), of the so-called *partial derivatives* – and that we shall call here *derived terms* – from rational expressions and languages to rational expressions and languages *with multiplicity*. As it turns out, this is another example where taking multiplicities into account and considering series rather than languages yield simplification and better understanding of constructions and results on languages (as advocated by S. Eilenberg in the preface of [6]).

In 1964, J. Brzozowski defined the *derivatives* of a rational expression ([3]). He showed that, *modulo* the axioms of associativity, commutativity, and idempotency of the addition (on the set of languages) – the ACI properties – the set of derivatives of a given expression is *finite*, yielding both a new proof for (one direction of) Kleene's Theorem and an algorithm turning an expression into a deterministic finite automaton.

This problem (of turning an expression into a finite automaton) has attracted much attention since the beginning of the theory ([7, 10]) and is an area of active research since then. In 1996, V. Antimirov made a fundamental contribution by defining the "*partial derivatives*" of an expression ([1]). Using his own words, "the idea behind [his] construction is that it allows to take into account the ACI-properties of only those occurrences of "+" in [rational] terms which appear at the very upper level" . Roughly speaking, the derivation proposed by Antimirov has two effects. First, it performs the "normal" derivation and, second, it breaks the result into "parts" , hence the name *partial derivatives*, such that this result is

K. Diks et al. (Eds): MFSC 2002, LNCS 2420, pp. 471–482, 2002.

the sum of the parts. This construction has a number of outcomes: the number of partial derivatives is not only finite but " small": smaller than or equal to the number of occurences of letters in the expression; they are easier to compute than the classical derivatives and they yield a *non-deterministic* finite automaton with (almost) the same number of states; finally, the subset construction applied to that automaton gives back the deterministic one computed by Brzozowski's algorithm. The computation of Antimirov's automaton has been made really efficient by Champarnaud and Ziadi [5].

What is presented here is at the same time a *formalization* and a *generalization* of Antimirov's construction. We first define rational expressions with multiplicity in a semiring $\mathbb{K}$, or $\mathbb{K}$-expressions. For sake of simplicity in dealing with those expressions, we suppose that $\mathbb{K}$ is *commutative*, although most of our constructions and formulae are independent of this assumption.

We then define the derivation of a $\mathbb{K}$-expression with respect to a letter and then to a word. The main feature of our definition, that indeed realizes Antimirov's main idea, is that the result of the derivation of a $\mathbb{K}$-expression is not a $\mathbb{K}$-expression anymore but a linear combination of $\mathbb{K}$-expressions with coefficients in $\mathbb{K}$.

The generalization of Antimirov's results is then straightforward and gives our main result (Theorem 1 and Theorem 3). For any $\mathbb{K}$-expression E, there exists a finite set of $\mathbb{K}$-expressions, called the *derived terms* of E, such that the derivation of E with respect to any word is a linear combination with coefficients in $\mathbb{K}$ of derived terms. The number of derived terms is smaller than or equal to the *litteral length* of E. When $\mathbb{K} = \mathbb{B}$, the Boolean semiring, the derived terms of E are exactly the partial derivatives of E.

An automaton with multiplicity, and which recognizes the series denoted by E, is then built, the states of which are the derived terms. In particular, let us note that this technique overcomes the problem that the addition is not idempotent anymore, and that Brzozowski's theorem does not hold in that setting.

In the last section, we study a phenomenon that can arise if the coefficients belong to a semiring which is not positive; in this case, the computation may generate some useless derived terms – called *shadow terms*.

A direct consequence, or corollary, of these results is that a rational series belongs to a finitely generated $\mathbb{K}$-module which is *stable* by derivatives (or residuals); but it should be stressed that this new proof of a well-known (and fundamental) result is *not* the main account, or purpose of this paper. Rather, what we show here is that, when a rational series s is defined by an expression E, a *symbolic computation* performed on E *effectively* gives – through their expressions – a finite set of generators of the stable module that contains this series s.[1]

This work has been done shortly after the participation of the authors to the CIAA and DCGARS workshops held in London (Ontario) in July 2000. There,

[1] Because of the limited size of this paper, some proofs have been skipped or shortened. A full version of the paper is available at the following URL: `http://www.enst.fr/~jsaka/PUB/DREM.ps`. A large acount of it will also appear in [12]

Caron and Flouret presented a generalization of the Glushkov construction to expressions with multiplicity, yielding a $\mathbb{K}$-automaton that plays the same role with respect to ours as the so-called *position automaton* do with respect to Antimirov's automaton in the Boolean case ([4]). At the WATA workshop held in Dresden (RFA) in March 2002, Jan Rutten draw our attention on the fact that our main theorem (Theorem 2) was indeed already stated and proved – in slightly different setting – in a previous work of him ([11]).

2 Rational Expressions with Multiplicity

Let A be a finite alphabet and let $\mathbb{K}$ be a semiring. The *associative* and *commutative* addition of $\mathbb{K}$ is denoted by $\oplus$, its multiplication simply by concatenation.

The semiring of formal power series over A^* with multiplicity in $\mathbb{K}$ is denoted by $\mathbb{K}\langle\langle A^*\rangle\rangle$. The inherited *commutative* addition in $\mathbb{K}\langle\langle A^*\rangle\rangle$ is denoted by $\oplus$, the (Cauchy) product by concatenation. For (rational) power series, their definitions, their notations and the related results, we refer to Berstel and Reutenauer's book [2] which we basically follow. But for one point: in [2], the semiring of coefficients is always equipped with the *discrete* topology; as it can easily be seen, this is an unnecessary assumption and $\mathbb{K}$ may well be equipped with *any* (metric or order) topology. For instance, in $\mathbb{Q}$, with the usual topology, $(1/2)^* = 1 + 1/2 + (1/2)^2 + \cdots = 2$ (*cf.* [8]). The semiring $\mathbb{K}\langle\langle A^*\rangle\rangle$ is then supposed to be equipped with the product topology derived from the topology on $\mathbb{K}$.

The coefficient of a word f of A^* in a series s of $\mathbb{K}\langle\langle A^*\rangle\rangle$ is denoted by $\langle s, f\rangle$. The *constant term* of s is the coefficient of the empty word 1_{A^*} in s and is denoted by $\mathsf{c}(s)$. A series is *proper* if $\mathsf{c}(s) = 0_{\mathbb{K}}$. The *proper part* of s, denoted by s_p, is the proper series which coincide with s on all words f different from 1_{A^*} and one can write $s = \mathsf{c}(s)\, 1_{A^*} \oplus s_p$. The star of a proper series is always well defined. The star of a series which is not proper may be defined or not, and the decision between the both cases is given by the following proposition:

Lemma 1. *[2, Exer I.3.4] The star of a series s is defined iff the star of $\mathsf{c}(s)$ is defined (in $\mathbb{K}$) and it holds: $s^* = (\mathsf{c}(s)^*\, s_p)^* \mathsf{c}(s)^*$.*

Rational Expressions. From now on, and for reasons we shall explain later, *we suppose that the semiring $\mathbb{K}$ is commutative.*

The definition of *rational expressions over* $\mathbb{K}$ goes as the one of classical rational expressions: it amounts to the construction of a set of well-formed formulae.

Let $\{0, 1, +, \cdot, *\}$ be a set of operations. Naturally, $+$ and $\cdot$ are binary, $*$ is unary, and $\mathsf{0}$ and $\mathsf{1}$ are "0-ary" operations, that is, they are constants. Moreover, for every k in $\mathbb{K}$, there is a unary operation, again denoted by k.

i) $\mathsf{0}$, $\mathsf{1}$, and a, for every a in A, are rational expressions (the *atomic* expressions or formulae).

ii) If E is a rational expression and k is in $\mathbb{K}$, then $(k\,\mathsf{E})$ is a rational expression.

iii) If E and F then $(\mathsf{E} + \mathsf{F})$, $(\mathsf{E} \cdot \mathsf{F})$, and (E^*) are rational expressions.

We denote by $\mathbb{K}\,\mathsf{RatE}\,A^*$ the set of rational expressions over A with multiplicity in $\mathbb{K}$.

The complexity of a rational expression can be described by different parameters. The *litteral length*, denoted by $\ell(\mathsf{E})$, is the number of atomic formulae in E that are letters. Remark that $\ell((\mathsf{E}+\mathsf{F})) = \ell((\mathsf{E}\cdot\mathsf{F})) = \ell(\mathsf{E})+\ell(\mathsf{F})$ and that $\ell((\mathsf{E}^*)) = \ell((k\,\mathsf{E})) = \ell(\mathsf{E})$. This is the parameter on which the automaton we build in section 5 depends. Whereas, many of our proofs are actually by induction on the depth of an rational expression. The *depth*[2] of an expression is inductively defined by:

$$\mathsf{d}(0) = \mathsf{d}(1) = 0, \qquad \forall a \in A \quad \mathsf{d}(a) = 0,$$
$$\mathsf{d}((k\,\mathsf{E})) = \mathsf{d}((\mathsf{E}^*)) = 1+\mathsf{d}(\mathsf{E}), \quad \mathsf{d}((\mathsf{E}\cdot\mathsf{F})) = \mathsf{d}((\mathsf{E}+\mathsf{F})) = 1+\max(\mathsf{d}(\mathsf{E}),\mathsf{d}(\mathsf{F})).$$

The constant term of an expression E is defined by induction on the depth of the expression E:

$$\mathsf{c}(0) = 0_{\mathbb{K}}\,, \quad \mathsf{c}(1) = 1_{\mathbb{K}}\,, \quad \forall a \in A \qquad \mathsf{c}(a) = 0_{\mathbb{K}}\,,$$
$$\mathsf{c}((k\,\mathsf{E})) = k\,\mathsf{c}(\mathsf{E})\,, \quad \mathsf{c}((\mathsf{E}+\mathsf{F})) = \mathsf{c}(\mathsf{E}) \oplus \mathsf{c}(\mathsf{F})\,, \quad \mathsf{c}((\mathsf{E}\cdot\mathsf{F})) = \mathsf{c}(\mathsf{E})\,\mathsf{c}(\mathsf{F})$$
$$\text{and} \quad \mathsf{c}((\mathsf{E}^*)) = \mathsf{c}(\mathsf{E})^* \quad \text{iff the latter is defined in } \mathbb{K}.$$

A rational expression E is a formula. It can either be *valid* and denotes a rational series, or not. We say that an expression is valid if $\mathsf{c}(\mathsf{E})$ is defined.

The series *denoted by a valid expression* E, which we note as $|\mathsf{E}|$, is defined by induction on the depth of the expression E as well:

$$|0| = 0_{\mathbb{K}}\,, \quad |1| = 1_{A^*}\,, \quad |a| = a\,, \quad \text{for every } a \text{ in } A\,, \quad |(k\,\mathsf{E})| = k\,|\mathsf{E}|\,,$$
$$|(\mathsf{E}+\mathsf{F})| = |\mathsf{E}| \oplus |\mathsf{F}|\,, \quad |(\mathsf{E}\cdot\mathsf{F})| = |\mathsf{E}|\,|\mathsf{F}|\,, \quad \text{and} \quad |(\mathsf{E}^*)| = |\mathsf{E}|^*.$$

Two valid expressions are *equivalent* if they denote the same series.

The definition of the constant term of an expression is consistent with the one of series denoted by an expression. Actually, $\mathsf{c}(\mathsf{E})$ and $\mathsf{c}(|\mathsf{E}|)$ are equal if E is an atomic expression, and the constant term and the interpretation of an expression are defined by the same induction. Hence $\mathsf{c}(\mathsf{E})$ is equal to $\mathsf{c}(|\mathsf{E}|)$. The last equation is besides made consistent by Lemma 1: $|(\mathsf{E}^*)|$ is defined iff $\mathsf{c}(\mathsf{E})^*$ is defined thus iff (E^*) is valid.

Remark that Lemma 1 allows to define the star of a series with a constant term denoted by an expression E without requiring any information on the expression that denotes $|\mathsf{E}|_p$.

Example 1. In this example, $A = \{a,b\}$ and $\mathbb{K} = \mathbb{Q}$. Let E_1 be the expression: $\mathsf{E}_1 = (((\frac{1}{6}\,a^*) + (\frac{1}{3}\,b^*))^*)$. Let $\mathsf{F}_1 = ((\frac{1}{6}\,a^*) + (\frac{1}{3}\,b^*))$. It comes $\mathsf{c}(\mathsf{F}_1) = \frac{1}{2}$, hence $\mathsf{c}(\mathsf{F}_1)^* = 2$. Thus, although $|\mathsf{F}_1|$ is not proper, the series denoted by E_1 is well-defined and E_1 is valid.

The first reason why the semiring $\mathbb{K}$ has been supposed to be *commutative* is to keep the definition of rational expressions with multiplicity simple enough, while the basic property still holds:

Proposition 1. *A series of* $\mathbb{K}\langle\langle A^*\rangle\rangle$ *is* $\mathbb{K}$*-rational iff it is denoted by a rational expression with multiplicity in* $\mathbb{K}$.

[2] We choose "depth" rather than "height" in order to avoid any confusion with the "star height" of a rational expression (that we are dealing with other papers).

Trivial Identities. The following identities trivially hold on rational expressions with multiplicity:

$$(k\,0) \equiv 0\,, \quad (0_{\mathbb{K}}\,\mathsf{E}) \equiv 0\,, \quad (0 \cdot \mathsf{E}) \equiv (0 \cdot \mathsf{E}) \equiv 0\,, \tag{1}$$

$$0 + \mathsf{E} \equiv \mathsf{E} + 0 \equiv \mathsf{E}\,, \quad (1_{\mathbb{K}}\,\mathsf{E}) \equiv \mathsf{E}\,, \quad (k\,1) \cdot \mathsf{E} \equiv \mathsf{E} \cdot (k\,1) \equiv (k\,\mathsf{E})\,. \tag{2}$$

One can see the trivial identities as rewriting rules (that consist in replacing every sub-expression that have the same form as a left term of these equalities with the corresponding right term), and it should be clear that this leads to a "reduced form" for every rational expression which is unique and which can be computed in a time proportional to the length of the expression, provided the multiplication in $\mathbb{K}$ is seen as an operation with fixed cost.

Remark 1. The trivial identities have nothing to do with the associativity and commutativity axioms for the $+$ operation: $(a+(b+c))$, $((a+b)+c)$ and $(a+(c+b))$ are three different expressions, nor with the associativity of "$\cdot$" , in spite of the simplifications we have used in the examples.

3 Derivatives

We now introduce polynomials of expressions and their derivatives. The set $\mathbb{K}\langle \mathbb{K}\,\mathsf{RatE}\,A^* \rangle$ of linear combinations of rational expressions, or *polynomials of expressions*, is a $\mathbb{K}$-semimodule; the addition is commutative and the multiplication by an element of $\mathbb{K}$ is distributive:

$$k\,\mathsf{E} \oplus k'\,\mathsf{F} = k'\,\mathsf{F} \oplus k\,\mathsf{E}\,, \qquad k\,\mathsf{E} \oplus k'\,\mathsf{E} = [k \oplus k']\,\mathsf{E}\,. \tag{3}$$

We define a multiplication law on the monomials (*i.e.* on the elements of the base of the semimodule), which is generalized to polynomials by a distributivity axiom.

$$[k\,\mathsf{E}][k'\,\mathsf{F}] \equiv [k\,k']\,(\mathsf{E} \cdot \mathsf{F})\,, \tag{4}$$

$$([\mathsf{E} \oplus \mathsf{E}'] \cdot \mathsf{F}) \equiv (\mathsf{E} \cdot \mathsf{F}) \oplus (\mathsf{E}' \cdot \mathsf{F})\,, \quad (\mathsf{E} \cdot [\mathsf{F} \oplus \mathsf{F}']) \equiv (\mathsf{E} \cdot \mathsf{F}') \oplus (\mathsf{E} \cdot \mathsf{F}')\,. \tag{5}$$

In the following, $[k\,\mathsf{E}]$ or $k\,\mathsf{E}$ is a monomial whereas $(k\,\mathsf{E})$ is an expression. The series denoted by a polynomial of rational expressions is obtained by extending by linearity the interpretation defined on rational expressions.

Remark 2. If $\mathbb{K}$ is not commutative, the interpretation of the left handside and the right handside of the identity (4) may differ. This is the main reason for our assumption of commutativity. However this difficulty can be overcome, and it will be done in a forthcoming work.

Remark 3. The set of polynomials of rational expressions is not a semialgebra. The multiplication that is defined by (4) is not associative because we want to keep the symbolic aspect of computation: *i.e.* $((\mathsf{E} \cdot \mathsf{F}) \cdot \mathsf{G}) \neq (\mathsf{E} \cdot (\mathsf{F} \cdot \mathsf{G}))$. This, however, does not cause any problem in the sequel.

Definition 1. *Let* E *be in* $\mathbb{K}\,\mathsf{RatE}\,A^*$ *and let* a *be in* A. *The* derivative *of* E *with respect to* a, *denoted by* $\frac{\partial}{\partial a}\mathsf{E}$, *is* a polynomial of rational expressions *with coefficients in* $\mathbb{K}$, *defined inductively by the following formulae.*

$$\frac{\partial}{\partial a}0 = \frac{\partial}{\partial a}1 = 0\,, \quad \frac{\partial}{\partial a}b = \begin{cases} 1 & \textit{if} \ \ b = a \\ 0 & \textit{otherwise} \end{cases} \tag{6}$$

$$\frac{\partial}{\partial a}(k\,\mathsf{E}) = k\,\frac{\partial}{\partial a}\mathsf{E} \tag{7}$$

$$\frac{\partial}{\partial a}(\mathsf{E}{+}\mathsf{F}) = \frac{\partial}{\partial a}\mathsf{E} \oplus \frac{\partial}{\partial a}\mathsf{F} \tag{8}$$

$$\frac{\partial}{\partial a}(\mathsf{E}\cdot\mathsf{F}) = \left(\left[\frac{\partial}{\partial a}\mathsf{E}\right]\cdot\mathsf{F}\right) \oplus \mathsf{c}(\mathsf{E})\,\frac{\partial}{\partial a}\mathsf{F} \tag{9}$$

$$\frac{\partial}{\partial a}(\mathsf{E}^*) = \mathsf{c}(\mathsf{E})^*\left(\left[\frac{\partial}{\partial a}\mathsf{E}\right]\cdot(\mathsf{E}^*)\right) \tag{10}$$

The derivative of a polynomial of expressions is defined by linearity:

$$\frac{\partial}{\partial a}\left(\bigoplus_{i\in I} k_i\,\mathsf{E}_i\right) = \bigoplus_{i\in I} k_i\,\frac{\partial}{\partial a}\mathsf{E}_i \tag{11}$$

Implicitely, the (polynomials of) expressions are reduced by trivial identities (*e.g.* if $\mathsf{c}(\mathsf{E}) = 0_{\mathbb{K}}$, then $\frac{\partial}{\partial a}(\mathsf{E}\cdot\mathsf{F})$ is equal to $\left(\left[\frac{\partial}{\partial a}\mathsf{E}\right]\cdot\mathsf{F}\right)$ and not to $\left(\left[\frac{\partial}{\partial a}\mathsf{E}\right]\cdot\mathsf{F}\right)\oplus 0_{\mathbb{K}}\,\frac{\partial}{\partial a}\mathsf{F}$). If we compare the equations (7) to (10) to the classical ones, it is the replacement of a "+" by a "$\oplus$" in the right handside of (8) and (9) that realizes the generalization of the idea of Antimirov. Equations (7), (10) and (11) are the natural ones that are necessary for the generalization to expressions with multiplicity. Notice that equation (10) is defined only if (E^*) is a valid expression.

In contrast to the Boolean case, the number of polynomials obtained by iterating the derivative process can be infinite. Theorem 3 will state that all these different polynomials are linear combinations of a fixed finite number of expressions.

Example 2. (Ex. 1 continued)

$$\begin{aligned}\frac{\partial}{\partial a}\mathsf{E}_1 &= \frac{\partial}{\partial a}(\mathsf{F}_1{}^*) = 2\,\frac{\partial}{\partial a}\left(\tfrac{1}{6}\,a^*\right)\cdot\mathsf{F}_1{}^* \oplus 2\,\frac{\partial}{\partial a}\left(\tfrac{1}{3}\,b^*\right)\cdot\mathsf{F}_1{}^* \\ &= \tfrac{1}{3}\,(a^*\cdot\mathsf{F}_1{}^*)\,, \\ \frac{\partial}{\partial b}\mathsf{E}_1 &= 2\,\frac{\partial}{\partial b}\left(\tfrac{1}{3}\,b^*\right)\cdot\mathsf{F}_1{}^* = \tfrac{2}{3}\,(b^*\cdot\mathsf{F}_1{}^*)\end{aligned}$$

$$\begin{aligned}\frac{\partial}{\partial a}(a^*\cdot\mathsf{F}_1{}^*) &= \left(\frac{\partial}{\partial a}a^*\right)\cdot\mathsf{F}_1{}^* \oplus \mathsf{c}(a^*)\,\frac{\partial}{\partial a}(\mathsf{F}_1{}^*) \\ &= (a^*\cdot\mathsf{F}_1{}^*) \oplus \tfrac{1}{3}\,(a^*\cdot\mathsf{F}_1{}^*) = \tfrac{4}{3}\,(a^*\cdot\mathsf{F}_1{}^*)\,, \\ \frac{\partial}{\partial b}(a^*\cdot\mathsf{F}_1{}^*) &= \left(\frac{\partial}{\partial b}a^*\right)\cdot\mathsf{F}_1{}^* \oplus \mathsf{c}(a^*)\,\frac{\partial}{\partial b}(\mathsf{F}_1{}^*) = \tfrac{2}{3}\,(b^*\cdot\mathsf{F}_1{}^*)\,, \\ \frac{\partial}{\partial a}(b^*\cdot\mathsf{F}_1{}^*) &= \left(\frac{\partial}{\partial a}b^*\right)\cdot\mathsf{F}_1{}^* \oplus \mathsf{c}(b^*)\,\frac{\partial}{\partial a}(\mathsf{F}_1{}^*) = \tfrac{1}{3}\,(a^*\cdot\mathsf{F}_1{}^*)\,,\end{aligned}$$

$$\frac{\partial}{\partial b}(b^* \cdot \mathsf{F_1}^*) = \left(\frac{\partial}{\partial b} b^*\right) \cdot \mathsf{F_1}^* \oplus \mathsf{c}(b^*) \frac{\partial}{\partial b}(\mathsf{F_1}^*)$$
$$= (b^* \cdot \mathsf{F_1}^*) \oplus \tfrac{2}{3}(b^* \cdot \mathsf{F_1}^*) = \tfrac{5}{3}(b^* \cdot \mathsf{F_1}^*) \ .$$

The derivative of an expression with respect to a *word* f is defined by induction on the length of f (by convention, the derivation with respect to the empty word is the identity):

$$\forall f \in A^* \,,\ \forall a \in A \qquad \frac{\partial}{\partial fa}\mathsf{E} = \frac{\partial}{\partial a}\left(\frac{\partial}{\partial f}\mathsf{E}\right) . \tag{12}$$

The derivative of an expression with respect to a word corresponds to the *(left) quotient* of a series by this word. Recall that if s is a series in $\mathbb{K}\langle\langle A^*\rangle\rangle$, the left quotient of s by a word f in A^* is the series $f^{-1}s$ defined by $\langle f^{-1}s, g\rangle = \langle s, fg\rangle$, for every g in A^*. The link between derivative and quotient is expressed in the following theorem, that can be proved directly, but that will appear as a corollary of more precise properties of the derivation of a rational expression.

Theorem 1. $\forall f \in A^* \qquad |\frac{\partial}{\partial f}(\mathsf{E})| = f^{-1}|\mathsf{E}|$.

Lemma 2. $\forall f, g \in A^* \qquad \frac{\partial}{\partial fg}\mathsf{E} = \frac{\partial}{\partial g}\left(\frac{\partial}{\partial f}\mathsf{E}\right)$. □

The following identities hold on derivatives with respect to a word :

Proposition 2. *For every word f in A^+,*

i) $\frac{\partial}{\partial f}(k\,\mathsf{E}) = k\,\frac{\partial}{\partial f}\mathsf{E}$; ii) $\frac{\partial}{\partial f}(\mathsf{E}+\mathsf{F}) = \frac{\partial}{\partial f}\mathsf{E} \oplus \frac{\partial}{\partial f}\mathsf{F}$;

iii) $$\frac{\partial}{\partial f}(\mathsf{E}\cdot\mathsf{F}) = \left[\frac{\partial}{\partial f}\mathsf{E}\right]\cdot\mathsf{F} \oplus \left[\bigoplus_{\substack{f=g\,h \\ g\in A^*,\, h\in A^+}} \mathsf{c}(\frac{\partial}{\partial g}\mathsf{E})\,\frac{\partial}{\partial h}\mathsf{F}\right] ;$$

iv) $$\frac{\partial}{\partial f}(\mathsf{E}^*) = \bigoplus_{\substack{f=g_1g_2\cdots g_n \\ g_1,g_2,\ldots,g_n\in A^+}} \mathsf{c}(\mathsf{E})^* \prod_{i=1}^{n-1}\left(\mathsf{c}(\frac{\partial}{\partial g_i}\mathsf{E})\,\mathsf{c}(\mathsf{E})^*\right)\left[\frac{\partial}{\partial g_n}\mathsf{E}\right]\cdot(\mathsf{E}^*) \ .$$ □

Proposition 3. *Let* E *be in* $\mathbb{K}\,\mathsf{RatE}\,A^*$. *It then holds:* $\langle|\mathsf{E}|, f\rangle = \mathsf{c}(\frac{\partial}{\partial f}\mathsf{E})$.

Proof. The proof goes by induction on the depth of the expression and makes use of Proposition 2.

The result is true for 0 and 1: the derivation with respect to any word f in A^+ is null and the coefficient of f in the series 0 and 1_{A^*} is actually null.

The proof for $(k\,\mathsf{E})$, $(\mathsf{E}+\mathsf{F})$ and $(\mathsf{E}\cdot\mathsf{F})$ directly follows from the linearity of $\mathsf{c}(\mathsf{E})$. To prove the result on E^*, and in order to avoid an infinite sum, we use

Lemma 1 before applying the same arguments:

$$\begin{aligned}
\langle\!\langle (\mathsf{E}^*)|, f\rangle\!\rangle &= \langle\!\langle (\mathsf{c}(\mathsf{E})^* \, |\mathsf{E}|_p)^* \, \mathsf{c}(\mathsf{E})^*, f\rangle\!\rangle = \langle\!\langle (\mathsf{c}(\mathsf{E})^* \, |\mathsf{E}|_p)^*, f\rangle\!\rangle \, \mathsf{c}(\mathsf{E})^* \\
&= \bigoplus_{\substack{f=g_1g_2\cdots g_n \\ g_1,g_2,\ldots,g_n\in A^+}} \langle \mathsf{c}(\mathsf{E})^* \, |\mathsf{E}|_p, g_1\rangle \ldots \langle \mathsf{c}(\mathsf{E})^* \, |\mathsf{E}|_p, g_n\rangle \, \mathsf{c}(\mathsf{E})^* \\
&= \bigoplus_{\substack{f=g_1g_2\cdots g_n \\ g_1,g_2,\ldots,g_n\in A^+}} \mathsf{c}(\mathsf{E})^* \, \langle\!\langle \mathsf{E}|_p, g_1\rangle \ldots \mathsf{c}(\mathsf{E})^* \, \langle\!\langle \mathsf{E}|_p, g_n\rangle \, \mathsf{c}(\mathsf{E})^* \, .
\end{aligned}$$

As the g_i are all *different* from 1_{A^*}, $\langle\!\langle \mathsf{E}|_p, g_i\rangle = \langle\!\langle \mathsf{E}|, g_i\rangle$.

$$\begin{aligned}
\langle\!\langle (\mathsf{E}^*)|, f\rangle\!\rangle &= \bigoplus_{\substack{f=g_1g_2\cdots g_n \\ g_1,g_2,\ldots,g_n\in A^+}} \mathsf{c}(\mathsf{E})^* \, \mathsf{c}(\frac{\partial}{\partial g_1}\mathsf{E}) \ldots \mathsf{c}(\mathsf{E})^* \, \mathsf{c}(\frac{\partial}{\partial g_n}\mathsf{E}) \, \mathsf{c}(\mathsf{E})^* \\
&= \mathsf{c}\left(\bigoplus_{\substack{f=g_1g_2\cdots g_n \\ g_1,g_2,\ldots,g_n\in A^+}} \mathsf{c}(\mathsf{E})^* \, \mathsf{c}(\frac{\partial}{\partial g_1}\mathsf{E}) \ldots \mathsf{c}(\mathsf{E})^* \left[\frac{\partial}{\partial g_n}\mathsf{E}\right] \cdot (\mathsf{E}^*) \right) \\
&= \mathsf{c}(\frac{\partial}{\partial f}\mathsf{E}^*) \, . \qquad \square
\end{aligned}$$

Proof. (of Theorem 1) For every pair of words f and g in A^*,

$$\langle f^{-1}|\mathsf{E}|, g\rangle = \langle\!\langle \mathsf{E}|, fg\rangle = \mathsf{c}(\frac{\partial}{\partial fg}\mathsf{E}) = \mathsf{c}\left(\frac{\partial}{\partial g}\left[\frac{\partial}{\partial f}\mathsf{E}\right]\right) = \langle\!\langle \frac{\partial}{\partial f}\mathsf{E}|, g\rangle \, . \qquad \square$$

4 Derived Terms

We state now our main theorem, from which the generalization of Antimirov's result ensues.

Theorem 2. *Let* E *be in* $\mathbb{K}\,\mathsf{RatE}\,A^*$. *There exist an integer* n, $n \leqslant \ell(\mathsf{E})$, *and* n *rational expressions* K_1, K_2, …, K_n *such that for every letter* a *in* A, *there exist* n *coefficients,* $k_1^{(a)}$, $k_2^{(a)}$, …, $k_n^{(a)}$, *and* n^2 *coefficients* $\{z_{i,j}^{(a)}\}_{i,j\in[n]}$ *in* $\mathbb{K}$ *such that:*

$$\text{i)}\ \frac{\partial}{\partial a}\mathsf{E} = \bigoplus_{i\in[n]} k_i^{(a)}\,\mathsf{K}_i\,; \qquad \text{ii)}\ \forall i\in[n] \quad \frac{\partial}{\partial a}\mathsf{K}_i = \bigoplus_{j\in[n]} z_{i,j}^{(a)}\,\mathsf{K}_j\,. \tag{13}$$

Definition 2. *We call* derived term *of* E *any of the expressions* K_i, *the existence of which is asserted in Theorem 2.*

In the case where $\mathbb{K} = \mathbb{B}$, these K_i are exactly what Antimirov called "*partial derivatives*" of E, with the explanation that they are "*parts*" of the *derivatives* of E ([1]).

Example 3. (Ex. 1 continued) The derived terms of E_1 are $(a^*{\cdot}\mathsf{F}_1{}^*)$ and $(b^*{\cdot}\mathsf{F}_1{}^*)$.

Proof. By induction on the depth of the expression E (not on its *litteral* length). The statement obviously holds for 0 and 1 and for $\mathsf{E} = a$, $a \in A$. We then successively show:

a) If it is true for E, it is true for $(k\,\mathsf{E})$, $k \in \mathbb{K}$. Obvious from (7). The derived terms of $(k\,\mathsf{E})$ are the same as those of E.

b) If it is true for E and F, it is true for $(\mathsf{E}+\mathsf{F})$. It holds:

$$\frac{\partial}{\partial a}(\mathsf{E}+\mathsf{F}) = \frac{\partial}{\partial a}\,\mathsf{E} \oplus \frac{\partial}{\partial a}\,\mathsf{F} = \bigoplus_{i\in[n]} k_i^{(a)}\,\mathsf{K}_i \oplus \bigoplus_{p\in[s]} l_p^{(a)}\,\mathsf{L}_p$$

with obvious notation. The set of derived terms of $(\mathsf{E}+\mathsf{F})$ is the union of those of E and F and this set clearly satisfies the proposition.

c) If it is true for E and F, it is true for $(\mathsf{E}\cdot\mathsf{F})$. It holds:

$$\frac{\partial}{\partial a}(\mathsf{E}\cdot\mathsf{F}) = \left(\left[\frac{\partial}{\partial a}\,\mathsf{E}\right]\cdot\mathsf{F}\right) \oplus \mathsf{c}(\mathsf{E})\,\frac{\partial}{\partial a}\,\mathsf{F} = \bigoplus_{i\in[n]} k_i^{(a)}(\mathsf{K}_i\cdot\mathsf{F}) \oplus \bigoplus_{p\in[s]} \left(\mathsf{c}(\mathsf{E})\,l_p^{(a)}\right)\mathsf{L}_p$$

The set of derived terms of $(\mathsf{E}\cdot\mathsf{F})$ is the union of the set $\{(\mathsf{K}_i\cdot\mathsf{F})\}_{i\in[n]}$ and of $\{\mathsf{L}_p\}_{p\in[s]}$ and one verifies that, for every i in $[n]$ and every a in A, it holds:

$$\frac{\partial}{\partial a}(\mathsf{K}_i\cdot\mathsf{F}) = \bigoplus_{j\in[n]} z_{i,j}^{(a)}\,(\mathsf{K}_j\cdot\mathsf{F}) \oplus \bigoplus_{p\in[s]} \left(\mathsf{c}(\mathsf{K}_i)\,l_p^{(a)}\right)\,\mathsf{L}_p\,.$$

d) If it is true for E, it is true for (E^*) . It holds/

$$\frac{\partial}{\partial a}(\mathsf{E}^*) = \mathsf{c}(\mathsf{E})^*\left(\left[\frac{\partial}{\partial a}\,\mathsf{E}\right]\cdot(\mathsf{E}^*)\right) = \bigoplus_{i\in[n]} \left(\mathsf{c}(\mathsf{E})^*\,k_i^{(a)}\right)\,(\mathsf{K}_i\cdot\mathsf{E}^*)$$

The set of derived terms of (E^*) is $\{(\mathsf{K}_i\cdot\mathsf{E}^*)\}_{i\in[n]}$. And one verifies that, for every i in $[n]$ and every a in A, it holds:

$$\frac{\partial}{\partial a}(\mathsf{K}_i\cdot\mathsf{E}^*) = \bigoplus_{j\in[n]} z_{i,j}^{(a)}\,(\mathsf{K}_j\cdot\mathsf{E}^*) \oplus \bigoplus_{j\in[n]} \left(\mathsf{c}(\mathsf{K}_i)\,\mathsf{c}(\mathsf{E})^*\,k_j^{(a)}\right)\,(\mathsf{K}_j\cdot\mathsf{E}^*)\,. \qquad \square$$

From Theorem 2 directly follows Theorem 3 theorem which is the generalization of Antimirov's result.

Theorem 3. *Let* E *be in* $\mathbb{K}\,\mathsf{RatE}\,A^*$. *There exist an integer* m, $m \leqslant \ell(\mathsf{E})$, *and* m *rational expressions* $\mathsf{K}_1, \mathsf{K}_2, \ldots, \mathsf{K}_m$ *such that for every word* f *in* A^+, *there exist* m *coefficients in* $\mathbb{K}$, $k_1^{(f)}, k_2^{(f)}, \ldots, k_m^{(f)}$, *such that*

$$\frac{\partial}{\partial f}\,\mathsf{E} = \bigoplus_{i=1}^{i=m} k_i^{(f)}\,\mathsf{K}_i\,. \qquad \square$$

Remark 4. The derivative and the left quotient are right actions of A^* on the set of polynomials of rational expressions and the set of rational series respectively. Theorem 3 says that the orbit of a rational expression with multiplicity under the action of A^* belongs to a finitely generated $\mathbb{K}$-module. The function which maps a polynomial of expressions P onto the rational power series $|P|$ is a morphism of actions. Therefore, Theorem 3 implies that the orbit of a rational series under the action of A^* belongs to a finitely generated $\mathbb{K}$-module as well, and provides a new proof for this classical result [2]. Derived terms are an explicit representation of a set of generators of this $\mathbb{K}$-module.

5 The Automaton of Derived Terms

To any rational expression with multiplicity E in $\mathbb{K}\,\mathsf{RatE}\,A^*$, we associate a $\mathbb{K}$-automaton (*i.e.* an automaton over the alphabet A with multiplicity in $\mathbb{K}$) in the following way. Let $P = \{\mathsf{K}_1, \mathsf{K}_2, \ldots, \mathsf{K}_n\}$ be the set of the derived terms of E. Let $\mathsf{K}_0 = \mathsf{E}$ and let P_E be the union of P and K_0 (K_0 may already belong to P). The *automaton of derived terms of* E is the $\mathbb{K}$-automaton $\mathcal{A}_\mathsf{E} = \langle P_\mathsf{E}, A, Z, I, T\rangle$ defined by:

$$I_{\mathsf{K}_i} = \begin{cases} 1_\mathbb{K} & \text{if } \mathsf{K}_i = \mathsf{K}_0 \\ 0_\mathbb{K} & \text{otherwise} \end{cases}, \qquad Z_{\mathsf{K}_i,\mathsf{K}_j} = \bigoplus_{a\in A} z_{i,j}^{(a)}\, a\,, \qquad T_{\mathsf{K}_j} = \mathsf{c}(\mathsf{K}_j),$$

where the $z_{i,j}^{(a)}$ have been defined at Proposition 3 (we put $z_{0,j}^{(a)} = k_j^{(a)}$ if $\mathsf{K}_0 \notin P$).

Theorem 4. *Let* E *be in* $\mathbb{K}\,\mathsf{RatE}\,A^*$. *The series realized by the automaton of derived terms of* E *is equal to the series denoted by* E: $|\mathcal{A}_\mathsf{E}| = |\mathsf{E}|$.

Proof. The definition of the $\mathbb{K}$-automaton $\mathcal{A}_\mathsf{E}$ is indeed equivalent to the definition of a "$\mathbb{K}$-*representation*" (I, ζ, T). This is a part of the proof of the so-called Kleene-Schützenberger Theorem (*cf.* [2]). In the representation (I, ζ, T), I and T are two vectors of dimension n (where $n = \mathsf{Card}(P_\mathsf{E})$) with entries in $\mathbb{K}$ defined above and $\zeta \colon A^* \to \mathbb{K}^{n\times n}$ is the morphism from A^* into the monoid of $n \times n$-matrices with entries in $\mathbb{K}$ defined by $(a)\zeta_{i,j} = z_{i,j}^{(a)}$, for every a in A and every pair of integers i and j smaller than n.

The *series realized* by the representation (I, ζ, T) (and thus by the automaton $\mathcal{A}_\mathsf{E}$) is, by definition:

$$|\mathcal{A}_\mathsf{E}| = \bigoplus_{f\in A^*} (I \cdot (f)\zeta \cdot T)\, f.$$

It is easy to verify by induction on the length of f that the folowing holds:

$$\forall f \in A^*,\ \forall a \in A,\ \forall i \in [n] \qquad k_i^{(fa)} = \sum_{j\in[n]} k_i^{(f)}\, z_{i,j}^{(a)}$$

that is:

$$\forall f \in A^+,\ \forall i \in [n] \qquad (I \cdot (f)\zeta)_i = k_i^{(f)}, \tag{20}$$

and then, by Proposition 3,
$\forall f \in A^+ \quad \langle\!|\mathcal{A}_\mathsf{E}|, f\rangle\!\rangle = \bigoplus_{i[n]} k_i^{(f)}\, \mathsf{c}(\mathsf{K}_i) = \langle\!|\mathsf{E}|, f\rangle\!\rangle$.

The theorem is then trivially verified, since the coefficient of the empty word is

$$\langle\!|\mathcal{A}_\mathsf{E}|, 1_{A^*}\rangle\!\rangle = \mathsf{c}(\mathsf{K}_0) = \mathsf{c}(\mathsf{E}). \qquad \square$$

Example 4. The opposite figure shows the automaton of derived terms of E_1.

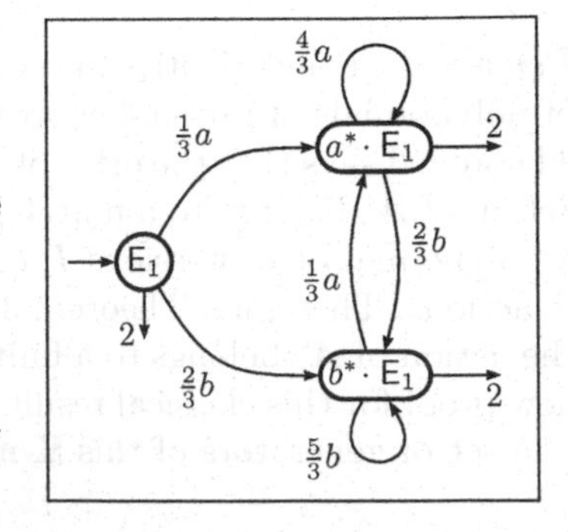

6 Shadow Derived Terms

In the statement of Theorem 2 and 3 the number of expressions K_i has been given two different "names": n and m respectively. It may indeed happen that these two numbers are different, m being smaller than or equal to n. This is the case when there exists a derived term K_i such that, for every word f, $(I \cdot (f)\zeta)_i = 0_{\mathbb{K}}$. And then, by (20), the term K_i does not appear in the sum of Theorem 3.

Example 5. $\mathsf{E}_2 = (a\,b\,a = (a(a - b\,a)))$. For simplicity, we write $a\,b$ for $(a \cdot b)$, $a\,b\,a$ for $((a \cdot b) \cdot a)$ and $(a - b \cdot a)$ for $(a + (-1_{\mathbb{K}}(b \cdot a)))$. It comes:

$$\frac{\partial}{\partial a} \mathsf{E}_2 = b\,a \oplus (a - b \cdot a)\,, \qquad \frac{\partial}{\partial aa} \mathsf{E}_2 = \frac{\partial}{\partial a} b\,a \oplus \frac{\partial}{\partial a}(a - b \cdot a) = 1\,,$$

$$\frac{\partial}{\partial b} \mathsf{E}_2 = 0\,, \qquad \frac{\partial}{\partial ab} \mathsf{E}_2 = \frac{\partial}{\partial b} b\,a \oplus \frac{\partial}{\partial b}(a - b \cdot a) = 1 = a \oplus (-1_{\mathbb{K}})a = 0\,.$$

The derived of E_2 are $b\,a$, $(a - b \cdot a)$, a and 1: $a = \frac{\partial}{\partial b}(a - b \cdot a)$ is a shadow derived term.

Proposition 4. *Let* E *be in* $\mathbb{K}\,\mathsf{RatE}\,A^*$ *and let* $\mathcal{A}_{\mathsf{E}}$ *be its automaton of derived terms. The* $\mathbb{K}$*-automaton obtained from* $\mathcal{A}_{\mathsf{E}}$ *by erasing the states that correspond to shadow derived terms and by trimming the result realizes the same series* $|\mathcal{A}_{\mathsf{E}}| = |\mathsf{E}|$.

Proof. If K_i is a shadow derived term, then, by (20), $k_i^{(f)} = (I \cdot (f)\zeta)_i = 0_{\mathbb{K}}$, for every f in A^*. This state K_i can be erased in the representation (I, ζ, T) without changing the realized series. The further trimming of the corresponding automaton does not change the realized series either. □

Example 6. The figure (a) below shows the automaton of derived terms of E_2; the figure (b) shows the effect of, first, the suppression of the state a and then the trimming that erases the state $b\,a$.

(a) The automaton of derived terms of E_2 (b) The automaton of non-shadow derived terms of E_2

The effective computation of the shadow derived terms depends on the semiring of coefficients. In many cases – and in particular in all classical cases – this computation does not bring any problem. On the one hand, if the semiring is positive, there is no shadow term. For instance, Boolean semiring, sub-semirings of $\mathbb{R}_+$, (max, +)-semirings, $(\mathcal{P}(A^*), \cup, .)$,*etc.* are positive. On the other hand, K_i is a shadow term if and only if the series realized by the automaton $\mathcal{A}_i = \langle P_{\mathsf{E}}, A, Z, I, \{\mathsf{K}_i\}\rangle$ is equal to zero. This can be easily decided if the semiring of coefficients is a sub-semiring of a field (*cf.* [6], Equality theorem). Examples of such semirings are $\mathbb{N}$, $\mathbb{Z}$, $\mathbb{Q}$, $\mathbb{Z}[X]$, $\mathbb{N}\langle\langle A^*\rangle\rangle$ *etc.* In any case, this computation is not really necessary as the automaton of derived terms is already a *finite* automaton that realizes the series denoted by an expression E and that can be effectively computed from E.

7 Further Developments

As we mentionned earlier, the techniques developped in this paper can easily be extended to the case of non commutative semiring $\mathbb{K}$ of coefficients.

It should be noted that our definition of rational expressions with multiplicity is slightly different from the one taken by Rutten [11]. We have identified the elements of $\mathbb{K}$ with *operators* whereas Rutten consider them as *atomic formulae* with the same status as the letters of A. This has the advantage it takes directly into account the non commutative case. On the other hand, our point of view relates our construction with Anisimov's one and yields the bound on the states of the automaton. Moreover, by playing with the definition of "the derivation with respect to the empty word", it is possible to associate different automata to an expression with various properties ([9]). This will be developped in a forthcoming work.

References

1. V. Antimirov, Partial derivatives of regular expressions and finite automaton constructions. *Theoret. Comput. Sci. 155* (1996), 291–319.
2. J. Berstel and Ch. Reutenauer, *Les séries rationnelles et leurs langages.* Masson, 1984. Translation: *Rational Series and their Languages.* Springer, 1986.
3. J. A. Brzozowski, Derivatives of regular expressions. *J. Assoc. Comput. Mach. 11* (1964), 481–494.
4. P. Caron and M. Flouret, Glushkov construction for multiplicities. *Pre-Proceedings of CIAA'00*, M. Daley, M. Eramian and S. Yu, eds, Univ. of Western Ontario, (2000), 52–61.
5. J.-M. Champarnaud and D. Ziadi, New finite automaton constructions based on canonical derivatives. *Pre-Proceedings of CIAA'00*, M. Daley, M. Eramian and S. Yu, eds, Univ. of Western Ontario, (2000), 36–43.
6. S. Eilenberg, *Automata, Languages and Machines, volume A.* Academic Press, 1974.
7. V. Glushkov, The abstract theory of automata. *Russian Mathematical Surveys 16* (1961), 1–53.
8. W. Kuich, and A. Salomaa, *Semirings, Automata, Languages.* Springer, 1986.
9. S. Lombardy, Approche structurelle de quelques problèmes de la théorie des automates. Thèse Doc. ENST, 2001.
10. R. McNaughton, and H. Yamada, Regular expressions and state graphs for automata. *IRE Trans. on electronic computers 9* (1960), 39–47.
11. J. Rutten, Automata, power series, and coinduction: taking input derivatives seriously. *Proceedings ICALP'99* (J. Wiedermann, P. van Emde Boas and M. Nielsen, Eds) Lecture Notes in Comput. Sci. **1644** (1999), 645–654.
12. J. Sakarovitch, *Eléments de théorie des automates.* Vuibert, 2002, to appear. Eng. trans. Cambridge University Press, to appear.

Hypothesis-Founded Semantics for Datalog Programs with Negation

Yann Loyer* and Nicolas Spyratos

[1] Istituto di Elaborazione della Informazione, Consiglio Nazionale delle Ricerche, Area della Ricerca CNR di Pisa, Via Moruzzi 1, I-56124 Pisa, Phone (39) 050 315 2901, Fax (39) 050 315 2810, loyer@lri.fr,

[2] Laboratoire de Recherche en Informatique, UMR 8623, Université de Paris Sud, Bat. 490, 91405 Orsay, Phone (33) 1 69 15 66 24, spyratos@lri.fr

Abstract. A precise meaning or semantics must be associated with any logic program or deductive database, and that even in presence of incomplete information. The different semantics that can be assigned to a logic program correspond to different assumptions made concerning the atoms whose logical values cannot be inferred from the rules. Thus, the well-founded semantics corresponds to the assumption that every such atom is false, while the Kripke-Kleene semantics corresponds to the assumption that every such atom is unknown. However, these assumptions are uniform in the sense that they always assign the same default value to all atoms: either everything is supposed to be false by default (closed world assumption) or everything is supposed to be unknown by default (open world assumption). In several application environments, however, including information retrieval and information integration, such uniformity is not realistic. In this paper, we propose to unify and extend the assumption-based approaches by allowing assumptions to be non-uniform. To deal with such assumptions, we extend the concept of unfounded set of Van Gelder to the notion of support of a hypothesis. Based on the support of a hypothesis, we define our hypothesis-founded semantics and show that this semantics generalizes both the Kripke-Kleene semantics and the well-founded semantics of Datalog programs with negation.

Keywords: semantics of logic programs, non-monotonic reasoning, hypothesis.

1 Introduction

In the middle of the 1970's, Minsky [20] and McCarthy [16] pointed out that pure classical logic is inadequate to represent the commonsense nature of human reasoning, mainly because of the non-monotonic character of human reasoning. Using Przymusinski's words [23]: "Non-monotonicity of human reasoning is caused

* Corresponding author.

K. Diks et al. (Eds): MFSC 2002, LNCS 2420, pp. 483–494, 2002.

by the fact that our knowledge about the world is almost always incomplete and therefore we are forced to reason in the absence of complete information and as a result we often have to revise our conclusions, when new information becomes available".

Logic programming, deductive databases and non-monotonic reasoning are closely related since the first two implement negation using different non-monotonic negation operators, but can also be used as inference engines to implement other non-monotonic formalisms [19,23,29].

Despite the absence of complete information, a precise meaning or semantics must be associated with any logic program or deductive database. Two main approaches have been developed. The first is based on the Clark's predicate completion of the program [7,8,13,14] and has lead to the Kripke-Kleene semantics of Fitting [8]. In the second, different semantics have been proposed: the stable model semantics [12] based on autoepistemic logic, the default model semantics [2] based on default logic, the weakly perfect model semantics [26] based on circumscription and, finally, the well-founded semantics that extends the previous ones to all logic programs and is equivalent to suitable forms of all four major formalizations of non-monotonic reasoning (McCarthy's circumscription [17,18], Reiter's closed world assumption [27], Moore's autoepistemic logic [21] and Reiter's default theory [28]). The well-founded semantics is one of the most widely studied and most commonly accepted approaches to negation in logic programming [1,4,5,6,9,34]. It approximates the stable-models semantics [12], another major approach to logic program with negation [10,32], and is useful for an efficient computation of the stable models [22,30].

In [15], we showed that the different semantics that can be assigned to a logic program or deductive database correspond to different assumptions made concerning the atoms whose logical values cannot be inferred from the rules. For example, the well-founded semantics corresponds to the assumption that every such atom is false (Closed World Assumption), while the Kripke-Kleene semantics corresponds to the assumption that every such atom is unknown (Open World Assumption). We think that we should not be limited to these two approaches, and that we should be able to associate to a Datalog program a semantics based on any given hypothesis representing our default or assumed knowledge. While modeling human commonsense reasoning, it does not seem realistic to be limited to suppose by default that everything is false, or that everything is unknown. The interest of using non-uniform hypotheses for assigning a semantics to a logic program has already been shown in the domain of information retrieval in [11]. In that paper is shown the necessity of using a CWA for some predicates, e.g. for defining a relation *author*, and an OWA for some predicates, e.g. for defining a relation *friends*, so the program is modified in order to simulate a combination of the open world and closed world assumptions. Another field in which the possible use of non-uniform hypotheses should be important and natural is that of integration of information coming from different sources. The following "toy example" illustrates a problem of information integration, where the information consists of facts that a central server collects from a number of autonomous sources and then tries to combine them using (a) a set of logical rules, i.e. a logic program, and (b) a hypothesis, representing the server's own estimate.

Example 1. Consider a legal case where a judge (the "central server") has to decide whether to charge a person named John accused of murder. To do so, the judge first collects facts from two different sources: the public prosecutor and the person's lawyer. The judge then combines the collected facts using a set of rules in order to reach a decision. For the sake of our example, let us suppose that the judge has collected a set of facts F that he combines using a set of rules R as follows: $F = \{\neg \text{witness}(John), \text{friends}(John, Ted)\}$, and

$$R \begin{cases} \text{suspect(X)} & \leftarrow \text{motive}(X) \\ \text{suspect(X)} & \leftarrow \text{witness}(X) \\ \text{suspect(X)} & \leftarrow \text{suspect'(X)} \\ \text{innocent}(X) & \leftarrow \text{alibi}(X,Y) \wedge \neg\text{friends}(X,Y) \\ \text{innocent}(X) & \leftarrow \text{innocent'}(X) \\ \text{friends}(X,Y) & \leftarrow \text{friends}(Y,X) \\ \text{friends}(X,Y) & \leftarrow \text{friends}(X,Z) \wedge \text{friends}(Z,Y) \\ \text{charge}(X) & \leftarrow \text{suspect}(X) \\ \text{charge}(X) & \leftarrow \neg\text{innocent}(X) \end{cases}$$

The first fact of F says that there is no witness against John, i.e. the fact witness($John$) is false. The second fact of F says that Ted is a friend of John, i.e. the fact friends($John, Ted$) is true.

Turning now to the set of rules, the three first rules of R describe how the prosecutor works: in order to support the claim that a person X is a suspect, the prosecutor tries to provide a motive (first rule) or a witness against X (second rule) or is convinced by other evidence that the person is suspect (third rule).

The fourth and fifth rules of R describe how the lawyer works: in order to support the claim that X is innocent, the lawyer tries to provide an alibi for X by a person who is not a friend of X (fourth rule), or is convinced by other evidence that the person is innocent (fifth rule). The fourth rule depends on the sixth and seventh rules which define the relation *friends*.

Finally, the last two rules of R are the "decision making rules" and describe how the judge works: in order to reach a decision as to whether to charge X, the judge examines the premisses *suspect*(X) and $\neg$*innocent*(X). As explained earlier, the values of these premisses come from two different sources: the prosecutor and the lawyer. Each of these premisses can have the value true or false. However, it is also possible that the value of a premise is unknown. For example, if no motive is known, no witness has been found and there is no other evidence that X is suspect, then the value of suspect(X) will be unknown.

In view of these observations, the question is what value is appropriate to associate with charge(X). X should be charged if it has been *explicitly* proved that he is not innocent or that he is suspect. It follows that there are three possible values for charge(X) : *unknown*, *true*, *false*.

The value *unknown* for a premise means that the premise is true or false but its actual value is currently unknown. For the purposes of this paper we shall assume that any premise whose value is not known is associated with the value *unknown*. We note that the value *unknown* is related to the so-called "null values" of attributes in database theory. In database theory, however, a distinction is made between two types of null values [33]: (i) the attribute value exists but is currently unknown; (ii) the attribute value does not exist. An example of the first type is the Department-value for an employee that has just been hired but has not yet been assigned to a specific department, and an example of the second type is the maiden name of a male employee. The value *unknown* corresponds to the first type of null value.

Returning now to our example, the decision whether to charge John depends on the value that charge(John) will receive when collecting the values of the premisses together. Looking at the facts of F and the rules of R (and using intuition) we can see that suspect(John) and innocent(John) both receive the value *unknown* and so then does charge(John). This is clearly a case where the judge cannot decide whether to actually charge John!

In the context of decision making, however, one has to reach a decision (based on the available facts and rules) even if some values are not known. This can be accomplished by *assuming* values for some or all unknown premisses. Such an assignment of values to unknown premisses is what we call a *hypothesis*.

Thus in our example, the judge should assume the innocence of John, and then John should not be charged. We note that this is precisely what happens in real life under similar circumstances, i.e. the defendant is *assumed* innocent until proved guilty.

Now, we can easily remark that uniform hypotheses, as the closed word assumption and the open world assumption, do not fit with the reality in this example and that a non-uniform hypothesis is needed. If we follow the closed world assumption and assign to all atoms the default value false, then the judge will infer that John is not innocent and must be charged. If we follow the open world assumption and assign to all atoms the default value unknown, then the atoms suspect($John$), innocent($John$) and charged($John$) are assigned the value unknown and the judge cannot take a decision. An intuitively appealing hypothesis in this situation is to assume by default that the atoms motive($John$), witness($John$), suspect'($John$) are false, that the atom innocent'($John$) is true and that the others are unknown. With such a hypothesis, the judge could infer that John is innocent, not suspect and should not be charged.

Clearly, when hypothesizing on unknown premisses we would like our hypothesis to be "reasonable" in some sense, with respect to the available information, i.e., with respect to the given facts and rules. Roughly speaking, we define a hypothesis H to be "reasonable" or *sound* using the following test : (1) if there is no contradiction between H and F, then add H to F to produce a new set of facts $F' = F \cup H$; (2) apply the rules of R to F' to produce a new assignment of values H'; (3) if the literals present in H that correspond to atoms which appear as head of some rules in R are also present in H' then H is sound, otherwise H is not sound. That is, if there is no fact of H that has changed value as a result of rule application then H is a sound hypothesis; otherwise H is unsound.

In our example, for instance, consider the following hypothesis:

$$H_1 = \{\neg\text{motive}(John), \neg\text{witness}(John), \neg\text{suspect'}(John), \neg\text{suspect}(John), \neg\text{alibi}(John), \neg\text{innocent'}(John), \neg\text{innocent}(John), \neg\text{charge}(John)\}$$

Applying the above test we find the following values for the facts of H_1:

$$H_1' = \{\neg\text{motive}(John), \text{friends}(John, Ted), \text{friends}(Ted, John), \neg\text{suspect}(John), \neg\text{innocent}(John), \text{charge}(John)\}$$

As we can see, the fact charge(John) that appears as head of some rules in R has changed value, i.e. this fact was true in H_1 and now is false in H_1'. Therefore, H_1 is not a sound hypothesis.

Next, consider the following hypothesis:

$$H_2 = \{\neg\text{motive}(John), \neg\text{witness}(John), \neg\text{suspect'}(John), \neg\text{suspect}(John), \text{innocent'}(John), \text{innocent}(John), \neg\text{charge}(John)\}$$

Applying again our test we find :

$$H_2' = \{\neg\text{motive}(John), \text{friends}(John, Ted), \text{friends}(Ted, John), \neg\text{suspect}(John), \text{innocent}(John), \neg\text{charge}(John)\}$$

That is, the values of the facts of H_2 that appear as head of some rules in R remain unchanged in H_2', thus H_2 is a sound hypothesis. Intuitively, a sound hypothesis means that what we have assumed is compatible with the given facts and rules. From now on let us use the notation $\mathcal{P} = \langle F, R\rangle$, where F is the set of facts and R is the set of rules, and let us call $\mathcal{P}$ a program.

In principle, we may assume or hypothesize any value for every possible ground atom. However, given a program $\mathcal{P}$ and a hypothesis H, we cannot expect H to be sound with respect to $\mathcal{P}$, in general. What we can expect is that some subset of H is sound with respect to $\mathcal{P}$. It is then natural to ask, given program $\mathcal{P}$ and hypothesis H, what is the maximal subset of H that is sound with respect to $\mathcal{P}$. We show that this maximal subset is unique and propose a method to compute it. We call it the *support* of H by $\mathcal{P}$ and we denote it by $s^H_{\mathcal{P}}$. Intuitively, the support of H indicates how much of H can be assumed safely, i.e., remaining compatible with the facts and rules of $\mathcal{P}$.

In what follows, we show that the support $s^H_{\mathcal{P}}$ can be used to define a hypothesis-founded semantics of $\mathcal{P} = \langle F, R\rangle$, denoted by $HFS^H_{\mathcal{P}}$. This is done by a fixpoint computation using an immediate consequence operator $T_{\mathcal{P}}$ and the notion of support of a given hypothesis with respect to a sequence of programs as follows: $F_0 = F$, $F_{i+1} = T_{\langle F_i, R\rangle}(F_i) \cup s^H_{\langle F_i, R\rangle}$.

We also show that there is an interesting connection between hypothesis based semantics and the usual semantics of Datalog programs with negation. More precisely, we show that if $\mathcal{P}$ is a Datalog program with negation then (i) if H is the closed world assumption then $HFS^H_{\mathcal{P}}$ coincides with the well-founded semantics of $\mathcal{P}$ [32], and (ii) if H is the open world assumption then $HFS^H_{\mathcal{P}}$ coincides with the Kripke-Kleene semantics of $\mathcal{P}$ [8].

The remaining of the paper is organized as follows. In Section 2 we recall very briefly some definitions and notations from well-founded semantics and Kripke-Kleene semantics. We then proceed, in Section 3, to define sound hypotheses and their support by a logic program $\mathcal{P}$; we also discuss computational issues and we present algorithms for computing the support of a hypothesis by a program $\mathcal{P}$. In Section 4 we define the notion of hypothesis-founded semantics of a program $\mathcal{P}$ and show that the notion of support actually unifies and extends the notions of well-founded semantics and Kripke-Kleene semantics. Section 5 contains concluding remarks and suggestions for further research.

2 Preliminaries

A Datalog program with negation is a finite set of formulas, called rules, of the form $A \leftarrow L_1, ..., L_n$ where A is an atom and the L_i's are positive or negative literals.[1] A is called the head of the rule and $L_1, ..., L_n$ the body.

An (partial) interpretation of a Datalog program with negation $\mathcal{P}$ is a set I of ground literals such that there is no atom A such that $\{A, \neg A\} \subset I$.

Two interpretations I and J are *compatible* if $I \cup J$ is an interpretation. Given an interpretation I, we denote by $def(I)$ the set of all ground atoms A such that $A \in I$ or $\neg A \in I$, i.e. the set of ground atoms that are not unknown in I. Moreover, if S is any set of ground atoms, we define the *restriction* of I to S, denoted by $I_{/S}$ as follows: $I_{/S} = I \cap (S \cup \neg.S)$, where $\neg.S = \{\neg A \mid A \in S\}$.

A Datalog program with negation can be seen as a pair $\langle F, R\rangle$ where F is a set of facts, that can be seen more generally as an interpretation that is equivalent to the set of rules of the form $A \leftarrow true$ or $A \leftarrow false$, and R a set of rules. In the remaining of this paper, in order to simplify the presentation, we use the term "program" to mean "Datalog program with negation", and we assume that all programs are instantiated programs.

2.1 Well-Founded Semantics

Well-founded semantics of logic programs were first proposed in [32]. The well-founded semantics of a Program $\mathcal{P}$ is based on the closed world assumption, i.e. every atom is supposed to be $false$ by default. If we consider an instantiated program P defined as in [32], its well-founded semantics is defined using the following two operators on partial interpretations I : `(i)` the immediate consequence operator T_P, defined by $T_P(I) = \{head(r) \mid r \in P \ \ and \ \ \forall B \in body(r), B \in I\}$, and `(ii)` the unfounded operator U_P, where $U_P(I)$ is defined to be the greatest unfounded set with respect to the partial interpretation I. We recall that a set of instantiated atoms U is said to be unfounded with respect to I if for all instantiated atoms $A \in U$ and for all rules $r \in P$ the following holds: $head(r) = A \Rightarrow \exists B \in body(r) \ (\neg B \in I \ \ or \ \ B \in U)$.

[1] if B is an atom, then B and $\neg B$ are literals. A literal is ground if all its variables are instantiated.

The operator W_P, called the well-founded operator, is then defined by $W_P(I) = T_P(I) \cup \neg U_P(I)$ and is shown to be monotone with respect to set inclusion. The well-founded semantics of P is defined to be the least fixpoint of W_P [32].

2.2 Kripke-Kleene Semantics

The Kripke-Kleene semantics was introduced in [8]. In the approach of [8], a valuation is a function from the Herbrand base to the set of logical values $\{true,\ false,\ unknown\}$. Now, given an instantiated program $\mathcal{P}$ defined as in [8], its Kripke-Kleene semantics is defined using an operator $\Phi_{\mathcal{P}}$ on valuations, defined as follows : given a ground atom A, if there is a rule in $\mathcal{P}$ with head A such that the truth value of its body under v is $true$, then $\Phi_{\mathcal{P}}(v)(A) = true$, else if there is a rule in $\mathcal{P}$ with head A and for every rule in $\mathcal{P}$ with head A the truth value of the body under v is false, then $\Phi_{\mathcal{P}}(v)(A) = false$, else $\Phi_{\mathcal{P}}(v)(A) = unknown$.

The Kripke-Kleene semantics of a Program $\mathcal{P}$ is based on the open world assumption (i.e. every atom is supposed to be *unknown* by default), and is defined to be the iterated fixpoint of $\Phi_{\mathcal{P}}$ obtained by beginning the iteration with the everywhere unknown valuation.

3 Hypothesis Testing

3.1 The Support of a Hypothesis

Given a program $\mathcal{P} = \langle F, R\rangle$, we consider two ways of inferring information from $\mathcal{P}$. First by activating the rules of R in order to derive new facts from those of F, through an immediate consequence operator $T_{\mathcal{P}}$. Second, by a kind of default reasoning based on a given hypothesis.

Definition 1 (Immediate Consequence Operator $T_{\mathcal{P}}$). *The immediate consequence operator $T_{\mathcal{P}}$ takes as input an interpretation I and returns an interpretation $T_{\mathcal{P}}(I)$, defined as follows:*

$$\begin{aligned} T_{\mathcal{P}}(I) = \{A \mid \exists A \leftarrow L_1, ..., L_n \in \mathcal{P}\ (\forall L_i\ (L_i \in I))\} \\ \cup\{\neg A \mid \exists A \leftarrow L_1, ..., L_n \in \mathcal{P} \\ \textit{and}\ \forall A \leftarrow L'_1, ..., L'_n \in \mathcal{P}\ (\exists L'_i\ (\neg L'_i \in I))\} \end{aligned}$$

What we call a *hypothesis* is actually just an interpretation H. However, we use the term "hypothesis" to stress the fact that the values assigned by H to the atoms of the Herbrand base are *assumed* values — and *not* values that have been computed using the facts and rules of the program. As such, a hypothesis H must be tested against the "sure" knowledge provided by $\mathcal{P}$. The test consists in "adding" H to F, then activating the rules of $\mathcal{P}$ (using $T_{\mathcal{P}}$) to derive an interpretation H'. More formally, let $H_{/Heads(\mathcal{P})}$ denote the restriction of H to the set $Head(\mathcal{P})$, where $Head(\mathcal{P}) = \{A \mid \exists A \leftarrow L_1, ..., L_n \in \mathcal{P}\}$. If $H_{/Heads(\mathcal{P})} \subset H'$ then the hypothesis H is sound, i.e. the values defined by H are not in contradiction with those defined by $\mathcal{P}$. Hence the following definition:

Definition 2 (Sound Hypothesis). *Let $\mathcal{P} = \langle F, R\rangle$ be a program and H a hypothesis. H is sound w.r.t. $\mathcal{P}$ if F and H are compatible and $H_{/Head(\mathcal{P})} \subset T_{\mathcal{P}}(F \cup H)$.*

We use the restriction of H to $Head(\mathcal{P})$ before making the comparison with $T_{\mathcal{P}}(F \cup H)$ because all atoms which are not head of any rule of $\mathcal{P}$ will be unknown in $T_{\mathcal{P}}(F \cup H)$. Then H and $T_{\mathcal{P}}(F \cup H)$ are compatible on these atoms.

The following example illustrates the definition of sound hypothesis with respect to a logic program.

Example 2. We consider the program $\mathcal{P} = \langle \mathcal{F}, \mathcal{R}\rangle$ such that R is the set of rules given in example 1 and $F = \{\text{witness(John)}\}$.

Let H be the hypothesis defined by $H = \{\neg\text{motive}(John), \neg\text{witness}(John), \neg\text{suspect}(John)\}$. We can easily note that H is not sound with respect to $\mathcal{P}$. The atom witness(John) is defined in H and in F, but with different values, so H and F are not compatible.

Let H' be the hypothesis defined by $H' = \{\neg\text{motive(John)}, \neg\text{suspect(John)}\}$. F and H' are compatible, so it is possible to collect the knowledge defined by these two interpretations in a new one without creating inconsistencies. $F \cup H' = \{\text{witness(John)}, \neg\text{motive(John)}, \neg\text{suspect(John)}\}$. Then we activate the rules of R on the interpretation $F \cup H'$: $T_{\mathcal{P}}(F \cup H') = \{\text{witness(John), suspect(John)}\}$. We observe that H' is not sound with respect to $\mathcal{P}$ because $H'_{/Heads(\mathcal{P})}$ is not a subset of $T_{\mathcal{P}}(F \cup H)$ and is in contradiction with the derived knowledge.

Even if a hypothesis H is not sound w.r.t. $\mathcal{P}$, it may be that some subset of H is sound w.r.t. $\mathcal{P}$. Of course, we are interested to know what is the maximal subset of H (i.e. the maximal subset of H with respect to set inclusion) that is sound w.r.t. $\mathcal{P}$. We shall call this maximal subset the "support" of H. To see that the maximal subset of H is unique (and thus that the support is a well-defined concept), we give the following lemma:

Lemma 1. *If H_1 and H_2 are two sound subsets of H w.r.t. $\mathcal{P}$, then $H_1 \cup H_2$ is sound w.r.t. $\mathcal{P}$.*

Thus the maximal sound subset of H is defined by $\bigcup\{H' \mid H' \leq H$ and H' is sound w.r.t. $\mathcal{P}\}$.

Definition 3 (Support). *Let $\mathcal{P}$ be a program and H a hypothesis. The support of H w.r.t. $\mathcal{P}$, denoted $s^H_{\mathcal{P}}$, is the maximal sound subset of H w.r.t. $\mathcal{P}$ (where maximality is understood w.r.t. set inclusion).*

Example 3. Let $\mathcal{P}$ be the program and H the hypothesis defined in the example 2, then the support of H with respect to $\mathcal{P}$ is $s^H_{\mathcal{P}} = \{\neg\text{motive(John)}\}$.

We can remark that the support of a hypothesis with respect to a program $\mathcal{P} = \langle F, R\rangle$ is compatible with the interpretation obtained by activating the rules of R on the facts of F, i.e. $T_{\mathcal{P}}(F)$ and $s^H_{\mathcal{P}}$ are compatible.

We now give an algorithm for computing the support $s^H_{\mathcal{P}}$ of a hypothesis H w.r.t. a program $\mathcal{P}$. Consider the following sequence $\langle PF_i\rangle$, $i \geq 0$:

- $PF_0 = \emptyset$;
- $PF_i = def(T_{\mathcal{P}}(F \cup H_{i-1}) \setminus H)$

where $H_{i-1} = H \setminus \{A, \neg A \mid A \in PF_{i-1} \text{ or } \{A, \neg A\} \subset F \cup H\}$, i.e. the maximal subset of H that is compatible with F and contains no facts corresponding to atoms in PF_{i-1}.

The intuition here is that we want to evaluate step by step the atoms that could potentially have a logical value different than their values in H. We have the following results:

Proposition 1. *The sequence $\langle PF_i \rangle$, $i \geq 0$ is increasing with respect to set inclusion and it has a limit reached in a finite number of steps. This limit is denoted PF.*

If an atom of the Herbrand base is not in PF, then it means that, with respect to $\mathcal{P}$, there is no way of inferring for that atom a logical value different than its value in H.

Theorem 1. *Let $\mathcal{P}$ be a logic program and H a hypothesis, we have*

$$s_{\mathcal{P}}^{H} = H \setminus \{A, \neg A \mid A \in PF \text{ or } \{A, \neg A\} \subset F \cup H\}$$

4 Hypothesis Founded Semantics

As we explained earlier, given a program $P = \langle F, R \rangle$, we derive information in two ways: by activating the rules (i.e. by applying the immediate consequence operator $T_{\mathcal{P}}$) and by making a hypothesis H and computing its support $s_{\mathcal{P}}^{H}$ w.r.t. $\mathcal{P}$. In the whole, the information that we derive comes from $T_{\mathcal{P}}(F) \cup s_{\mathcal{P}}^{H}$ which is an interpretation. Roughly speaking, the semantics that we would like to associate with a program $\mathcal{P}$ is the maximum of information that we can derive from $\mathcal{P}$ under a hypothesis H but *without* any other information. To implement this idea we proceed as follows:

1. As we do not want any extra information (other than $\mathcal{P}$ and $s_{\mathcal{P}}^{H}$), we begin our computation with the facts F.
2. In order to actually derive the maximum of information from $\mathcal{P}$, we collect together the knowledge inferred by activating the rules of R, i.e. by applying the operator $T_{\mathcal{P}}$, and as much of assumed knowledge as possible, i.e. the support of H w.r.t. $\mathcal{P}$.
3. We add the new facts that we derive to the program and define a new program $\mathcal{P}_i$ on which we apply the same operations as at step 2 (application of $T_{\mathcal{P}_i}$ and computation of $s_{\mathcal{P}_i}^{H}$), and continue by iteration until we reach a limit.

Proposition 2. *The sequence $\langle F_n \rangle$, $n \geq 0$ defined by :*

- $F_0 = F$, *and*
- $F_{n+1} = T_{\langle F_n, R \rangle}(F_n) \cup s_{\langle F_n, R \rangle}^{H}$,

is increasing with respect to set inclusion and has a limit denoted by $HFS^H_{\mathcal{P}}$.

Proposition 3. *The interpretation* $HFS^H_{\mathcal{P}}$ *is a model of* $\mathcal{P}$.

This justifies the following definition of semantics for $\mathcal{P}$.

Definition 4 (Hypothesis Founded Semantics of $\mathcal{P}$). *The interpretation* $HFS^H_{\mathcal{P}}$ *is defined to be the semantics of* $\mathcal{P}$ *w.r.t.* H *or the* H*-founded semantics of* $\mathcal{P}$.

The following example illustrates the computation of that semantics.

Example 4. Let $H = \{A, \neg B, D, E\}$ and $\mathcal{P}$ be the program defined by $F = \emptyset$ and $R = \{A \leftarrow \neg B; B \leftarrow \neg A; C \leftarrow \neg A; D \leftarrow E; E \leftarrow D, \neg C\}$. Then we have

- $F_1 = T_{\langle F,R\rangle}(F) \cup s^H_{\langle F,R\rangle} = \emptyset \cup \{A, \neg B\}$;
- $F_2 = T_{\langle F_1,R\rangle}(F_1) \cup s^H_{\langle F_1,R\rangle} = \{A, \neg B, \neg C\} \cup \{A, \neg B\}$;
- $F_3 = T_{\langle F_2,R\rangle}(F_2) \cup s^H_{\langle F_2,R\rangle} = \{A, \neg B, \neg C\} \cup \{A, \neg B, D, E\}$;
- $F_4 = T_{\langle F_3,R\rangle}(F_3) \cup s^H_{\langle F_3,R\rangle} = \{A, \neg B, \neg C, D, E\} \cup \{A, \neg B, D, E\} = HFS^H_{\mathcal{P}}$;

Following this definition, any given program $\mathcal{P}$ can be associated with different semantics, one for each possible hypothesis H. Theorem 2 below asserts that the usual semantics of Datalog programs with negation correspond to particular hypothesis-founded semantics. Thus our approach is conservative, generalizes and extends the unfounded sets of Van Gelder [32] to any given hypothesis H (not just the closed world assumption). Moreover, it unifies the computation of those semantics and allows us to verify easily that the Kripke-Kleene semantics is included in the well-founded semantics.

Theorem 2. *Let* $\mathcal{P}$ *be a Datalog program with negation.*

1. *If* $H_{\mathcal{F}} = \neg.\mathcal{HB}_{\mathcal{P}}$, *i.e. if* $H_{\mathcal{F}}$ *is the closed world assumption, then* $HFS^{H_{\mathcal{F}}}_{\mathcal{P}}$ *coincides with the well-founded semantics of* $\mathcal{P}$;
2. *If* $H_{\mathcal{U}} = \emptyset$, *i.e. if* $H_{\mathcal{U}}$ *is the open world assumption, then* $HFS^{H_{\mathcal{U}}}_{\mathcal{P}}$ *coincides with the Kripke-Kleene semantics of* $\mathcal{P}$.

5 Concluding Remarks

We have defined a formal framework based on hypothesis testing for reasoning about non-monotonicity and incompleteness in the context of Datalog programs with negation, that allows us to consider any given knowledge as assumption for the missing information. A basic concept of this framework is the support provided by a program $\mathcal{P} = \langle F, R\rangle$ to a hypothesis H. The support of the hypothesis H is the maximal subset of H that does not contradict the facts of F or the facts derived from F using the rules of R. We have then used the concept of support to define hypothesis-founded semantics for Datalog programs with negation. We have given algorithms for computing these supports and semantics. Finally, we have shown that our semantics generalizes the well-founded semantics and the Kripke-Kleene semantics that can be seen now as particular hypothesis-founded semantics.

We believe that hypothesis-founded semantics can be useful not only in the contexts of information integration and information retrieval as proposed in introduction, but also in the context of explanation-based systems. Indeed, assume that a given hypothesis H turns out to be a part of the H-semantics of a program $\mathcal{P}$. Then $\mathcal{P}$ can be seen as an "explanation" of the hypothesis H. We are currently investigating several aspects of this explanation oriented viewpoint, and working on the application of that approach to logic-based information retrieval.

References

1. Alferes, J.J., Damásio, C.V., Pereira, L.M., *A logic programming system for non-monotonic reasoning*, Journal of Automated Reasoning, 14: 97–147, 1995.
2. Bidoit N., Froideveaux C., *General logical databases and programs: Default logic semantics and stratification*, Journal of Information and Computation, 1988.
3. Bidoit N., Froideveaux C., *Negation by default and unstratifiable logic programs*, TCS, 78, 1991.
4. Brass, S., Dix, J., *Characterizations of the disjunctive well-founded semantics: confluent calculi and iterated GCWA*, Journal of Automated Reasoning, 20(1): 143–165, 1998.
5. Chen, W., Swift, T., Warren, D.S., *Efficient top-down computation of queries under the well-founded semantics*, Journal of Logic Programming, 24(3): 161–199, 1995.
6. Chen, W., Warren, D.S., *Tabled evaluation with delaying for general logic programs*, Journal of the ACM, 43(1): 20–74, 1996.
7. Clark, K. L., *Negation as failure*, in H. Gallaire and J. Minker, editors, Logic and Databases, Plenum Press, New York, 293-322, 1978.
8. Fitting, M. C., *A Kripke/Kleene Semantics for Logic Programs*, J. Logic Programming, 2:295-312, 1985.
9. Fitting, M. C., *The family of stable models*, J. Logic Programming, 17:197–225, 1993.
10. Fitting, M. C., *Fixpoint semantics for logic programming—a survey*, Theoretical Computer Science, To appear.
11. Fuhr, N. and Rölleke, T., *HySpirit – a Probabilistic Inference Engine for Hypermedia Retrieval in Large Databases*, in: Schek, H.-J.; Saltor, F.; Ramos, I.; Alonso, G. (eds.). Proceedings of the 6th International Conference on Extending Database Technology (EDBT), 24-38, 1997.
12. Gelfond, M. and Lifschitz, V., *The Stable Model Semantics for Logic Programming*, in: R. Kowalski and K. Bowen (eds.), Proceedings of the Fifth Logic Programming Symposium MIT Press, Cambridge, MA, 978-992, 1988.
13. Kunen, K., *Negation in Logic Programming*, J. Logic Programming, 4(4):289-308, 1987.
14. Lloyd, J. W., *Foundations of Logic Programming*, Springer Verlag, New York, first edition, 1984.
15. Loyer, Y. Spyratos, N. and Stamate, D., *Computing and Comparing Semantics of Programs in Four-Valued Logics*, in: M. Kutylowski, L. Pacholski and T. Wierzbicki (eds.), Proceedings of the 24th International Symposium on Mathematical Foundations of Computer Science (*MFCS'*99), LNCS 1672: 59-69, 1999.

16. McCarthy, J., *Epistemological problems in artificial intelligence*, in Proceedings of IJCAI'77, American Association for Artificial Intelligence, Morgan Kaufmann, Los Altos, CA, 1038-1044, 1977.
17. McCarthy, J., *Circumscription - a form of non-monotonic reasoning*, Journal of Artificial Intelligence, 13:27-39, 1980.
18. McCarthy, J., *Applications of circumscription to formalizing common sense knowledge*, Journal of Artificial Intelligence, 28:89-116, 1986.
19. Minker, J., *An overview of non-monotonic reasoning and logic programming*, Journal of Logic Programming, Special Issue, 17, 1993.
20. Minsky, M., *A framework for representing knowledge*, in P. Winston, editor, The Psychology of Computer Vision, MIT Press, New York, 1975.
21. Moore, R. C., *Semantics considerations on non-monotonic logic*, Journal of Artificial Intelligence, 25:75-94, 1985.
22. Niemela, I. and Simons, P., *Efficient implementation of the well-founded and stable model semantics*, Proceedings of JICSLP'96, MIT Press, 1996.
23. Przymusinski, T. C., *Non-monotonic formalisms and logic programming*, Proceedings of the Sixth International Conference on Logic Programming, 1989.
24. Przymusinski, T. C., *Extended Stable Semantics for Normal and Disjunctive Programs*, in D. H. D. Warren and P. Szeredi (eds.), Proceedings of the Seventh International Conference on Logic Programming, MIT Press, Cambridge, MA, 459-477, 1990.
25. Przymusinski, T. C., *Well-Founded Semantics Coincides with Three-Valued Stable Semantics*, Fund. Inform., 13:445-463, 1990.
26. Przymusinska, H. and Przymusinski, T. C., *Weakly perfect model semantics for logic programs*, in R. Kowalski and K. Bowen, editors, Proceedings of the Fifth Logic Programming Symposium, MIT Press, 1106-1122, 1988.
27. Reiter, R., *On closed-world databases* , in H. Gallaire and J. Minker, editors, Logic and Databases, Plenum Press, New York, 55-76, 1978.
28. Reiter, R., *A logic for default theory*, Journal of Artificial Intelligence, 13:81-132, 1978.
29. Reiter, R., *Non-monotonic reasoning*, Annual Reviews of Computer Science, 1986.
30. Subrahmanian, V.S., Nau, D., Vago, C., *WFS + branch bound = stable models*, IEEE Transactions on Knowledge and Data Engineering, 7:362-377, 1991.
31. Van Gelder, *The Alternating Fixpoint of Logic Programs with Negation*, in: Proceedings of the Eighth Symposium on Principles of Database Systems, ACM, Philadelphia, 1-10, 1989.
32. Van Gelder, A., Ross, K. A., Schlipf, J. S., *The Well-Founded Semantics for General Logic Programs*, J. ACM, 38:620-650, 1991.
33. Zaniolo, C., *Database Relations with Null Values*, Journal of Computer and System Sciences, 28: 142-166, 1984.
34. Zukowski, U., Brass, S., Freitag, B., *Improving the alternated fixpoint: the transformation approach*, in: Proceedings of LPNMR'97, LNCS 1265, pages 40–59, 1997.

On the Problem of Scheduling Flows on Distributed Networks*

Thomas Lücking, Burkhard Monien, and Manuel Rode**

Department of Mathematics and Computer Science, University of Paderborn,
Fürstenallee 11, 33102 Paderborn, Germany,
{luck,bm,rode}@uni-paderborn.de

Abstract. We consider the problem of scheduling a given flow on a synchronous distributed network. This problem arises using diffusion-based approaches to compute a balancing flow, which has to be scheduled afterwards. We show that every distributed scheduling strategy requires at least $\frac{3}{2}$ times the minimum number of rounds. Furthermore we give a distributed algorithm for flows on tree networks, which requires at most two times the optimal number of rounds.

1 Introduction

Load balancing is an essential task for the efficient use of parallel computer systems. It can be observed that the work loads of many parallel applications have dynamic behavior and may change dramatically during runtime. To achieve an efficient use of the processor network, the load has to be balanced between the processors during runtime. Obviously, the balancing scheme is supposed to be highly efficient itself in order to ensure an overall benefit.

One major field of application are parallel adaptive finite element simulations where a geometric space, discretized using a mesh, is partitioned into sub-regions. The computation proceeds on the mesh elements in each sub-region independently [5]. As the computation is carried out, the mesh refines and coarsens, depending on the problem characteristics such as turbulence or shocks (in the case of fluid dynamics simulations, for example) and the size of the sub-regions (in terms of the numbers of elements) has to be balanced. The problem of parallel finite element simulation has been extensively studied – see the book [5] for an excellent selection of applications, case studies and references.

Much work has been done on the topic of *load balancing*. The approaches depend on the model used to describe the interprocessor communication. In this paper we consider synchronous distributed processor networks. In each round, a processor of the network can send and receive messages to/from all its neighbors

* Partly supported by the DFG-Sonderforschungsbereich 376 Massive Parallelität: Algorithmen, Entwurfsmethoden, Anwendungen, by the IST Programme of the EU under contract number IST-1999-14186 (ALCOM-FT), and by the IST Programme of the EU under contract number IST-2001-33116 (FLAGS).

** Graduate School of Dynamic Intelligent Systems

K. Diks et al. (Eds): MFSC 2002, LNCS 2420, pp. 495–505, 2002.

simultaneously. Furthermore, we assume that the situation is fixed, i.e., no load is generated or consumed during the balancing process, and the network does not change.

In order to balance the network, it is necessary to migrate parts of the processors' loads during runtime. We assume that the load consists of independent load units, called *tokens*. One possible approach to balance the network is based on a 2-step model.

First, a balancing flow, that is, a flow of tokens putting the network into a balanced state where all processors keep the same load up to one token, is calculated. This can be done efficiently with help of *diffusion* (for more details see [1]), yielding a balancing flow optimal with respect to the l_2-norm [2,4]. The flow can be represented by a directed acyclic graph $G = (V, E)$, having $|V| = n$ vertices and $|E| = m$ edges, with flow function f.

Second, the load items are migrated according to that flow [2,3,8]. The goal is to use the minimum number κ of rounds to reach the balanced state. This *flow scheduling problem* trivially leads to a formulation as a linear program with $\kappa(m+n)$ unknowns and $m + 2\kappa n$ equations. Such a system is solvable in $\mathcal{O}(\kappa^5(m+n)^5)$ rounds using e.g. Karmarkar's algorithm [7]. Diekmann et. al. [2] show that the $(\kappa-1)$-commodity flow problem on bipartite graphs can be reduced to the problem of deciding whether the flow can be scheduled in κ rounds.

Moreover they propose a *distributed* scheduling strategy which needs at most $\mathcal{O}(\sqrt{n})$ times the minimum number of rounds, and they give a worst-case example on a tree network yielding $\Omega(\log(n))$ times the minimum number of rounds. We are not aware of any other results on this problem.

The upper bound of $\mathcal{O}(\sqrt{n})$ is due to missing information on the structure of the flow graph. In this paper we introduce a distributed algorithm for flow graphs on tree networks. It investigates the structure of the flow graph before sending tokens. We show that the algorithm requires at most 2κ rounds.

In Section 2 we introduce the notation used in this paper. We then show in Section 3 that every distributed scheduling strategy requires at least $\frac{3}{2}$ times the minimum number of rounds. In Section 4 we propose a distributed algorithm which schedules flows on tree networks. We prove that it requires at most 2 times the minimum number of rounds. Furthermore, we show that this upper bound is tight.

2 Definitions

We consider a *synchronous processor network*. In order to describe this network formally, we use the model defined in [6]. The network $N = (V, C)$ consists of $|V|$ **processors** and $|C|$ undirected **channels**. Execution of the entire system begins with all processors in arbitrary start states, and all channels empty. Then the processors, in lock-step, repeatedly perform the following two steps:

1. Generate the messages to be sent to the neighbors. Put these messages in the appropriate channels.

2. Compute the new state from the current state and the incoming messages. Remove all messages from the channels.

The combination of the two steps is called **round.**

The load situation on N is given by the **load function** $l : V \to \mathbb{N}_0$, representing the number of unit sized tokens on the processors. A **flow** on this processor network N is a directed acyclic graph $G = (V, E)$ with $|E|$ **flow edges**, and a **flow function** $f : E \to \mathbb{N}_0$. We denote by $s : E \to V$ and $t : E \to V$ functions, defining source and target of each edge. For each $v \in V$ we call $\text{in}(v) = \{e \in E \mid t(e) = v\}$ the **set of incoming edges** of v, and $\text{out}(v) = \{e \in E \mid s(e) = v\}$ the **set of outgoing edges** of v. For every $v \in V$ the flow property holds, that is, we have

$$l(v) + \sum_{e \in \text{in}(v)} f(e) \geq \sum_{e \in \text{out}(v)} f(e). \tag{1}$$

We call (G, l, f) a **flow graph.**

Definition 1 (Schedule). *A* **schedule** *S for a flow graph (G, l, f) using k rounds is a decomposition of the flow function f into k flow functions $f^1, \ldots, f^k$, determining the flow in each round, i.e., $S = (f^1, \ldots, f^k)$ with*

$$\forall e \in E : f(e) = \sum_{i=1}^{k} f^i(e) \quad \textit{and} \tag{2}$$

$$\forall v \in V, j \in \{1, \ldots, k\} : l(v) + \sum_{i=1}^{j-1} \sum_{e \in \text{in}(v)} f^i(e) \geq \sum_{i=1}^{j} \sum_{e \in \text{out}(v)} f^i(e). \tag{3}$$

Inequality (3) ensures that in every round each processor has sufficient load.

Definition 2 (Flow Scheduling Problem). *We call the problem of finding a schedule S for a flow graph (G, l, f), which requires the minimum number of rounds $\kappa((G, l, f))$, the* **flow scheduling problem.**

In the following we restrict our discussions to tree networks N. Furthermore, we assume that the network is rooted, that is, exactly one processor is assigned to be the root of the tree, and each other processor v knows its **parent** with respect to this root, denoted by $\text{parent}(v)$. We call the graph G of a flow on N a **directed tree** if each edge of G is directed away from the root of N, i.e., $s(e) = \text{parent}(t(e))$ for each $e \in E$.

3 Worst-Case Analysis for Distributed Scheduling Strategies

In this section we show that for every distributed scheduling strategy there exists a flow graph on which it requires at least $\frac{3}{2}$ times the minimum number of rounds. The flow graph that will be used to prove this result is illustrated in Figure 1.

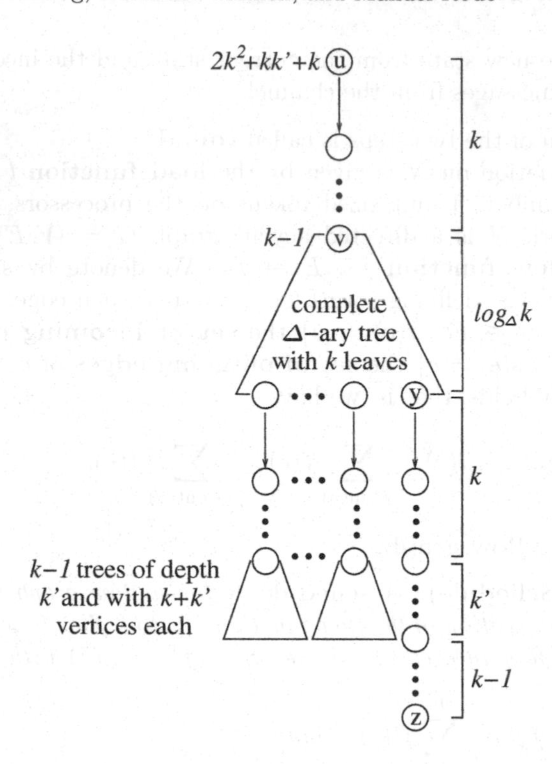

Fig. 1. Flow graph used in the proof of Theorem 1. The terms on the left of vertex u and vertex v give their initial loads. The expressions on the right hand side of the graph denote the lengths of the paths or depths of the trees.

Theorem 1. *Let $\Delta \in \mathbb{N}$ with $\Delta \geq 2$. Then for every distributed scheduling strategy there exists a flow graph (G, l, f) with $\deg(G) = \Delta + 1$ on which it requires at least $\frac{3}{2}$ times the minimum number of rounds.*

Proof. Let k be a power of Δ, and choose $k' \in \mathbb{N}_0$ minimal with $\frac{\Delta^{k'+1}-1}{\Delta-1} \geq k+k'$. Consider the processor network of degree $\Delta+1$ in Figure 1. It consists of a path of length k from vertex u to vertex v, followed by a complete Δ-ary tree of depth $\log_\Delta k$ with root v and k leaves. To each of the first $k-1$ leaves, a path of length k is attached whose last vertex is the root of a Δ-ary tree with $k + k'$ vertices and depth k'. To the remaining leaf y, a path of length $2k + k' - 1$ is attached.

Initially, only two vertices have some load. Vertex v has an amount of $k - 1$ tokens, and u keeps as much load as necessary to provide exactly one token to each vertex of the network in the balanced state, that is, an amount of $2k^2 + kk' + k$. Since the network is a tree, the balancing flow is uniquely determined and therefore also l_2-optimal.

If the initial load of vertex v is sent towards the path below y, then the minimum number of migration rounds is attained. At the same time u's load is sent downwards. After $2k + k' + \log_\Delta k$ rounds each processor holds one token, and the migration phase is finished.

When using a distributed scheduling strategy, each vertex initially only knows its own load and the flow on its incident edges. Among all vertices with distance at most $k + \log_\Delta k$ to v, the incoming edges of those being at the same distance to v have equal flow. Hence, all subtrees below each inner vertex of the complete tree are equal up to a depth of k. Thus during the first k rounds these vertices cannot gather information that helps to decide in which direction to send the load they might get.

As no helpful information is available during the first k rounds, we can assume that y is the leaf to which the least amount of load has been sent. The only source of load in a distance of at most k from the k leaves is v with load $k-1$. Hence, the least amount of load cannot exceed $\frac{k-1}{k} < 1$. Thus in the worst-case none of the tokens reaches the path at y during the first k rounds of any distributed scheduling algorithm, and the path has to be filled up by tokens stemming from u. One of these tokens has to travel a distance of $3k + k' + \log_\Delta k - 1$ (from u to z, the last vertex of the path). So the ratio between the worst case number of scheduling rounds of any distributed algorithm and the maximum distance a token would have to travel in an optimal schedule is

$$\frac{3k + k' + \log_\Delta k - 1}{2k + k' + \log_\Delta k}.$$

As $\frac{k'}{k}$ becomes 0 for $k \to \infty$, the ratio approaches $\frac{3}{2}$ as k increases. □

Note that the result in Theorem 1 is a lower bound on the worst-case performance of all distributed scheduling strategies. For a certain strategy there might exist a flow graph on which it performs much worse.

4 The Distributed Algorithm

In this section we give a distributed scheduling algorithm for flow graphs (G, l, f) on tree networks. We start with a distributed algorithm for flow graphs with directed tree G. Using this result, we give a distributed algorithm for any flow graph on tree networks.

4.1 Flow Graphs (G, l, f) on Tree Networks with Directed Tree G

Let (G, l, f) be a flow graph with directed tree G. For each $e \in E$ we recursively define for all $i \in \mathbb{N}_0$

$$d_0(e) = f(e),$$

$$d_{i+1}(e) = \max\left\{ \sum_{\tilde{e} \in \mathrm{out}(t(e))} d_i(\tilde{e}) - l(t(e)), 0 \right\}.$$

Note that $d_i(e)$ is the number of tokens that have to be sent along edge e to vertices at distance of at least i from vertex $t(e)$.

Theorem 2. *Let (G, l, f) be a flow graph with directed tree G, and let $k \in \mathbb{N}_0$ be minimal such that $d_k(e) = 0$ for all $e \in E$. Then an optimal schedule requires exactly k rounds, that is, $\kappa((G, l, f)) = k$.*

Proof. We prove the theorem in two steps:

1. There exists a valid schedule using k rounds.
   We construct a schedule $S = (f^1, \ldots, f^k)$ as follows. In round i, every vertex v with incoming edge e sends $f^i(\tilde{e}) = d_{k-i}(\tilde{e}) - d_{k-i+1}(\tilde{e})$ tokens along each edge $\tilde{e} \in \text{out}(v)$. After i rounds

$$\sum_{j=1}^{i} f^j(\tilde{e}) = \sum_{j=1}^{i} d_{k-j}(\tilde{e}) - d_{k-j+1}(\tilde{e}) = d_{k-i}(\tilde{e})$$

   tokens have been sent along edge $\tilde{e}$ for all $\tilde{e} \in E$. Denote by $l_i(v)$, $0 \le i < k$, the load on vertex v after i rounds. Then we have

$$\begin{aligned} l_i(v) &= l(v) + d_{k-i}(e) - \sum_{\tilde{e} \in \text{out}(v)} d_{k-i}(\tilde{e}) \\ &= l(v) + \max\left\{ \sum_{\tilde{e} \in \text{out}(v)} d_{k-i-1}(\tilde{e}) - l(v), 0 \right\} - \sum_{\tilde{e} \in \text{out}(v)} d_{k-i}(\tilde{e}) \\ &\ge \sum_{\tilde{e} \in \text{out}(v)} (d_{k-i-1}(\tilde{e}) - d_{k-i}(\tilde{e})) = \sum_{\tilde{e} \in \text{out}(v)} f^{i+1}(\tilde{e}). \end{aligned}$$

   Hence the described schedule is feasible, and it requires k rounds.

2. Any valid schedule takes at least k rounds.
   We now show by induction that the schedule described above takes the minimum number of rounds, i.e., that for any flow graph for which an optimal schedule requires k rounds, $d_k(e) = 0$ for all $e \in E$. If the optimal schedule takes $k = 1$ round, then each vertex has sufficient load to fulfill all its outgoing edges at once, and $d_1(e) = 0$ for all $e \in E$. So assume that the described schedule is optimal for all flow graphs (G, l, f) with minimum number of rounds $\kappa((G, l, f)) \le k \in \mathbb{N}_0$. Now let (G, l, f) be a flow graph with minimum number of rounds $\kappa((G, l, f)) = k + 1$.
   Denote by f' the flow computed in the first round of the optimal schedule, and let f'' be the remaining flow. Then we have $f(e) = f'(e) + f''(e)$ for each edge $e \in E$. After the first round the load for each $v \in V$ is

$$l'(v) = l(v) + \sum_{\tilde{e} \in \text{in}(v)} f'(\tilde{e}) - \sum_{\tilde{e} \in \text{out}(v)} f'(\tilde{e}).$$

   Applying the schedule construction to the load function l and flow function f', yielding $d'_0(e), d'_1(e), \ldots$, and to the load function l' and flow function

f'', yielding $d''_0(e), d''_1(e), \ldots$ for all $e \in E$, we get two schedules using 1 and k rounds, respectively, due to the induction hypothesis. So, $d'_1(e) = 0$ and $d''_k(e) = 0$ for all $e \in E$. By induction we show that applying the construction to the original problem with load l and flow f yields

$$d_i(e) = \begin{cases} d''_i(e) + f'(e) & \text{if } 0 \le i < k, \\ \max\{f'(e) - \xi(e), 0\} & \text{if } i = k, \\ 0 & \text{else,} \end{cases}$$

with $\xi(e) = l'(t(e)) - \sum_{\tilde{e} \in \mathrm{out}(t(e))} d''_{k-1}(\tilde{e}) \ge 0$ for all $e \in E$.
We have $d_0(e) = f(e) = f'(e) + f''(e) = f'(e) + d''_0(e)$ which proves the claim for $i = 0$. So let $i \ge 1$. Using the induction hypothesis we get

$$\begin{aligned} d_i(e) &= \max\left\{ \sum_{\tilde{e} \in \mathrm{out}(t(e))} d_{i-1}(\tilde{e}) - l(t(e)), 0 \right\} \\ &= \max\left\{ \sum_{\tilde{e} \in \mathrm{out}(t(e))} (d''_{i-1}(\tilde{e}) + f'(\tilde{e})) \right. \\ &\qquad \left. -(l'(t(e)) - f'(e) + \sum_{\tilde{e} \in \mathrm{out}(v)} f'(\tilde{e})), 0 \right\} \\ &= \begin{cases} d''_i(e) + f'(e) & \text{if } 0 < i < k, \\ \max\{f'(e) - \xi(e), 0\} & \text{if } i = k, \end{cases} \end{aligned}$$

$$\begin{aligned} d_{k+1}(e) &= \max\left\{ \sum_{\tilde{e} \in \mathrm{out}(t(e))} (f'(\tilde{e}) - \xi(\tilde{e})) - l(t(e)), 0 \right\} \\ &= \max\left\{ \sum_{\tilde{e} \in \mathrm{out}(t(e))} d'_0(\tilde{e}) - l(t(e)) - \sum_{\tilde{e} \in \mathrm{out}(t(e))} \xi(\tilde{e}), 0 \right\} = 0. \end{aligned}$$

Hence, as $d_{k+1}(e) = 0$ for all $e \in E$, the schedule takes $k + 1$ rounds. □

From Theorem 2 we get a distributed Algorithm A which schedules flow graphs (G, l, f) with directed tree G. The algorithm works as follows. Each vertex v keeps a variable i which is initialized with 0.

Then vertex v successively sends $d_1(e), \ldots, d_{k_e}(e)$ along its incoming edge $e = (\mathrm{parent}(v), v)$ to its parent, where k_e is minimal with $d_{k_e}(e) = 0$. Note that $d_1(e)$ can be computed from the flows on the outgoing edges, and that in the ith round each vertex v can compute $d_i(e)$, $1 < i \le k_e$, from the values received in the previous round. The value $d_i(\tilde{e})$ of an outgoing edge $\tilde{e}$ is assumed to be 0 if no information has been sent along $\tilde{e}$. Due to Theorem 2 we have $k_e \le \kappa((G, l, f))$ for all $e \in E$.

Algorithm A then proceeds as follows. If the number of tokens on v is at least $\sum_{\tilde{e} \in out(v)} (d_{k_e-i-1}(\tilde{e}) - d_{k_e-i}(\tilde{e}))$, then v sends $(d_{k_e-i-1}(\tilde{e}) - d_{k_e-i}(\tilde{e}))$ tokens via each of its outgoing edges $\tilde{e}$ and increments i. Otherwise no tokens are sent. This step is repeated until $i = k_e$. We now prove that the schedule implicitly defined by Algorithm A requires at most $2 \cdot \kappa((G,l,f))$ rounds.

Corollary 1. *Let (G,l,f) be a flow graph with directed tree G. Then Algorithm A schedules the flow graph in $2 \cdot \kappa((G,l,f))$ rounds.*

Proof. Let $\kappa = \kappa((G,l,f))$, let $v \in V$ arbitrary, and denote by i_r the value of i after r rounds of Algorithm A on v. Obviously, after r rounds v has sent an overall amount of $d_{k_e - i_r}(\tilde{e})$ tokens along each outgoing edge $\tilde{e}$, whereas in the schedule constructed in Theorem 2, $d_{k-r}(\tilde{e})$ tokens with $\kappa = \max\{k_e | e \in E\}$ would have been sent. Note that after at most κ rounds every vertex v has received the values $d_1(\tilde{e}), \ldots, d_\kappa(\tilde{e})$ for all $\tilde{e} \in \text{out}(v)$. We show that $k_e - i_{\kappa+r} \leq \kappa - r$ for all $r \geq 0$. Since the sequence $d_0(e), \ldots, d_\kappa(e)$ decreases monotonically, this implies $d_{k_e - i_{\kappa+r}}(\tilde{e}) \geq d_{\kappa-r}(\tilde{e})$.

If $r = 0$, then no load has been sent, and the claim holds. Now assume the claim holds for $r \geq 0$. If $k_e - i_{\kappa+r} < \kappa - r$, then we immediately get $k_e - i_{\kappa+r+1} \leq \kappa - (r+1)$. If $k_e - i_{\kappa+r} = \kappa - r$, then

$$\begin{aligned} d_{\kappa-r}(e) + l(v) &= \max\left\{ \sum_{\tilde{e} \in out(v)} d_{\kappa-r-1}(\tilde{e}), l(v) \right\} \\ &\geq \sum_{\tilde{e} \in out(v)} d_{k_e - i_{\kappa+r} - 1}(\tilde{e}) \\ &= \sum_{\tilde{e} \in out(v)} d_{k_e - i_{\kappa+r}}(\tilde{e}) + \sum_{\tilde{e} \in out(v)} \left(d_{k_e - i_{\kappa+r} - 1}(\tilde{e}) - d_{k_e - i_{\kappa+r}}(\tilde{e}) \right) . \end{aligned}$$

Hence the load $l'(v)$ kept by vertex v after r rounds is

$$\begin{aligned} l'(v) &= l(v) + d_{\kappa-r}(e) - \sum_{\tilde{e} \in out(v)} d_{k_e - i_{\kappa+r}}(\tilde{e}) \\ &\geq \sum_{\tilde{e} \in out(v)} \left(d_{k_e - i_{\kappa+r} - 1}(\tilde{e}) - d_{k_e - i_{\kappa+r}}(\tilde{e}) \right) . \end{aligned}$$

Thus v has sufficient load to send $\left(d_{k_e - i_{\kappa+r} - 1}(\tilde{e}) - d_{k_e - i_{\kappa+r}}(\tilde{e}) \right)$ tokens via each of its outgoing edges $\tilde{e}$. Therefore $i_{\kappa+r}$ is incremented and the claim remains true. □

4.2 Arbitrary Flow Graphs (G, l, f) on Tree Networks

We now consider an arbitrary flow graph (G,l,f) on a tree network.

Theorem 3. *Let (G,l,f) be a flow graph on a tree network. Then there exists a distributed algorithm which schedules the flow graph in $2 \cdot \kappa((G,l,f))$ rounds.*

Proof. We construct an algorithm which we call Algorithm B. We distinguish between **upward edges** and **downward edges**. An edge $e \in E$ is called upward if parent$(s(e)) = t(e)$ holds, otherwise it is called downward. Since the underlying network is a tree, every vertex has at most one outgoing edge which is upward. Algorithm B works as follows.

In each round every vertex v sends all available load via its outgoing upward edge until it is saturated. Note that a vertex with an outgoing upward edge receives load via all its incoming edges while sending load only via its sole outgoing upward edge. As a consequence all upward edges are saturated after at most $\kappa((G,l,f))$ rounds.

Simultaneously, as in Algorithm A, each vertex v connected with its parent by a downward edge e sends the values $d_1(e), \ldots, d_{k_e}(e)$ along e, where k_e is minimal with $d_{k_e} = 0$. Here, $d_0(e), \ldots, d_{k_e}(e)$ are calculated according to the remaining flow with saturated upward edges. Therefore v has to know the load $l'(v)$ it will have when all upward edges are saturated. This load can easily be computed:

$$l'(v) = l(v) + \sum_{\substack{e \in \text{in}(v) \\ v = \text{parent}(s(e))}} f(e) - \sum_{\substack{e \in \text{out}(v) \\ t(e) = \text{parent}(v)}} f(e)$$

After v has received $d_1, \ldots, d_{k_{\tilde{e}}}$ for all outgoing downward edges $\tilde{e}$, and all incoming upward edges are saturated, which is the case after at most $\kappa((G,l,f))$ rounds, vertex v sends tokens along its outgoing edges in the same way as in Algorithm A.

Similar to the proof of Corollary 1, it can be proven that this requires $2 \cdot \kappa((G,l,f))$ rounds. □

Note that the proof of Theorem 3 does not depend on the choice of the root of the tree network. If the processor network is not rooted, then parents can be determined successively, starting at the leaves. The number of required rounds is bounded by half the diameter of N. The following example shows that the bound proved in Theorem 3 is tight.

Example 1. Consider the flow graph (G,l,f) with root r in Figure 2. Using Algorithm B described in the proof of Theorem 3, vertex v sends its k tokens to its parent r. As a result, one of the tokens from u has to be sent along the path from u to w of length $2k$. However, sending the k tokens of v toward w leads to $k+2$ rounds. So the ratio between the number of rounds used by Algorithm A and the minimum number of rounds is

$$\frac{2k}{k+2} \to 2 \text{ for } k \to \infty.$$

5 Conclusion

In this paper we showed that no distributed scheduling strategy can have a worst-case performance lower than $\frac{3}{2}$, even on tree networks. Furthermore, we

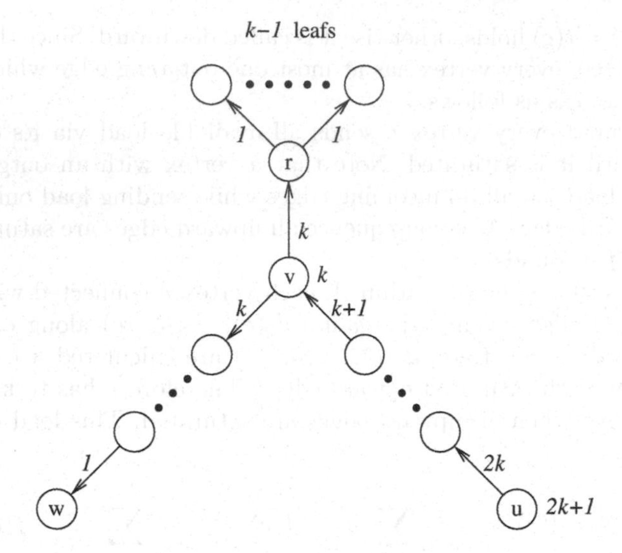

Fig. 2. Flow graph for which Algorithm B needs twice the minimum number of rounds.

presented a distributed algorithm for flows on tree networks yielding a worst-case performance of 2, whereas the local scheduling strategy in [2] takes at least $\Omega(\log(n))$ times the minimum number of rounds at the worst. We showed that the factor of 2 is tight. A natural question is whether the gap between the lower and the upper bound can be closed. Moreover, an open question is what factor can be achieved for arbitrary flow graphs.

Acknowledgment

The authors would like to thank Robert Elsässer, Torsten Fahle, Ulf-Peter Schroeder and Meinolf Sellmann for fruitful discussions.

References

1. G. Cybenko. Load balancing for distributed memory multiprocessors. *Journal of Parallel and Distributed Computing*, 7:279–301, 1989.
2. R. Diekmann, A. Frommer, and B. Monien. Efficient schemes for nearest neighbor load balancing. *Parallel Computing*, pages 789–812, 1999.
3. R. Diekmann, F. Schlimbach, and C. Walshaw. Quality balancing for parallel adaptive fem. In *Solving Irregulary Structured Problems in Parallel (IRREGULAR'98)*, pages 170–181, 1998.
4. R. Elsässer, B. Monien, and R. Preis. Diffusive load balancing schemes on heterogeneous networks. In *Proceedings of the 12th ACM Symposium on Parallel Algorithms and Architectures (SPAA'00)*, pages 30–38, 2000.
5. G. Fox, R. Williams, and P. Messina. *Parallel Computing Works!* Morgan Kaufmann Publishers, Inc., 1994.

6. N.A. Lynch. *Distributed Algorithms.* Morgan Kaufmann Publishers, Inc., 1996.
7. A. Schrijver. *Theory of Linear and Integer Programming.* Wiley, 1994.
8. C. Walshaw, M. Cross, and M. Everett. Dynamic load-balancing for parallel adaptive unstructured meshes. In *Proceedings of the 8th SIAM Conference on Parallel Processing for Scientific Computing*, 1997.

Unit Testing for CASL Architectural Specifications

Patricia D.L. Machado[1] and Donald Sannella[2]

[1] Systems and Computing Department, Federal University of Paraiba
[2] Laboratory for Foundations of Computer Science, University of Edinburgh

Abstract. The problem of testing modular systems against algebraic specifications is discussed. We focus on systems where the decomposition into parts is specified by a CASL-style architectural specification and the parts (*units*) are developed separately, perhaps by an independent supplier. We consider how to test such units without reference to their context of use. This problem is most acute for generic units where the particular instantiation cannot be predicted.

1 Introduction

Formal testing is concerned with deriving test cases from formal specifications and checking whether programs satisfy these test cases for a selected finite set of data, as a practical alternative to formal proof. Much work has focused on test case selection and (automatic) interpretation of test results. Testing from algebraic specifications has been investigated, focusing on both "flat" specifications [Ber89,Gau95,LA96,Mac99] and structured specifications [LA96,Mac00b,DW00]. It is often assumed that the structure of the specification matches the structure of the program [LeG99,DW00] although it is possible to take specification structure into account without making this strong assumption [Mac00b].

Tests are not interesting unless we can interpret their results in terms of correctness. *Oracles* are decision procedures for interpreting the results of tests. The *oracle problem* arises whenever a finite executable oracle cannot be defined; this may stem e.g. from the semantic gap between specification and program values. Guaranteeing correctness by testing requires tests covering all possible interactions with the system, usually an infinite and impractical activity. When only finite sets of interactions are considered, successful testing can accept incorrect programs and unsuccessful testing can reject correct programs. The latter must be avoided since the costs of finding and fixing errors are too high to waste effort on non-errors. Accepting incorrect programs is not a problem as long as test coverage is adequate, since testing is not expected to guarantee correctness.

This paper addresses testing of modular systems where parts are developed independently from CASL-style architectural specifications [ABK+03,BST02]. In this context, the problem reduces to that of testing units independently in such a way that their integration can be checked in a cost-effective way. An architectural specification consists of a list of *unit declarations*, naming the units (components)

K. Diks et al. (Eds): MFSC 2002, LNCS 2420, pp. 506–518, 2002.

that are required with specifications for each of them, together with a *unit term* that describes the way in which these units are to be combined.

When testing modular systems and their parts, different perspectives need to be considered. From a *supplier's* point of view, units have to be checked independently of their contexts of use. Checking generic units poses special problems since the particular instantiation is unknown in general, and the set of all possible instantiations is almost always infinite. In this paper, we present styles of testing non-generic and generic units that address these problems. The main novelty is in the case of generic units. *Integration testing* is left to future work. A extended version of this paper with examples demonstrating most of the main definitions and results is available as an Edinburgh technical report.

2 Preliminary Definitions

Algebraic Specifications. We assume familiarity with basic concepts and notation of algebraic specification, where programs are modelled as algebras. Each signature $\Sigma = (S, F, Obs)$ is equipped with a distinguished set $Obs \subseteq S$ of *observable sorts*. *Sign* is the category of signatures. We consider specifications with axioms of first-order equational logic, and write $Sen(\Sigma)$ for the set of all Σ-sentences. We restrict to algebras with non-empty carriers, and write $Alg(\Sigma)$ for the class of all such Σ-algebras. $[A]$ denotes the reachable subalgebra of $A \in Alg(\Sigma)$.

Behavioural Equality. For a non-observable sort, it is not appropriate to assume that equality on values is the usual set-theoretical one. Equality on values of a Σ-algebra A can be interpreted by an appropriate *behavioural equality*, a partial Σ-congruence $\approx_A = (\approx_{A,s})_{s \in S}$ with $Dom(\approx_A) = \{a \mid a \approx_A a\}$. The partial *observational equality* $\approx_{Obs,A} = (\approx_{Obs,A,s})_{s \in S}$ is one example of a behavioural equality, where related elements are those that cannot be distinguished by observable computations.[1] A *Σ-behavioural equality* is defined as $\approx_\Sigma = (\approx_{\Sigma,A})_{A \in Alg(\Sigma)}$, one behavioural equality for each Σ-algebra A.

Approximate Equality. Behavioural equality can be difficult to test. Consider e.g. observational equality on non-observable sorts, defined in terms of a set of contexts that is usually infinite. One approach involves the use of *approximate equalities* [Mac99,Mac00c] which are binary relations on values of the algebra. An approximate equality is *sound* with respect to a behavioural equality if all values (in $Dom(\approx_A)$) that it identifies are indeed equal, or *complete* if all equal values are identified. Any *contextual equality* $\sim_{\mathcal{C},A}$ defined from a subset $\mathcal{C}$ of the observable computations[2] is complete with respect to observational equality, al-

---

[1] Let C_{Obs} be the set of all Σ-contexts $T_\Sigma(X \cup \{z_s\})$ of observable sorts with context variable z_s of sort s. Then values a and b of sort s are *observationally equal*, $a \approx_{Obs,A,s} b$, iff $a, b \in [A]$ and $\forall C \in C_{Obs} \cdot \forall \alpha : X \to [A] \cdot \alpha_a^\#(C) = \alpha_b^\#(C)$, where $\alpha_a, \alpha_b : X \cup \{z_s\} \to [A]$ are the extensions of α defined by $\alpha_a(z_s) = a$ and $\alpha_b(z_s) = b$.

[2] Let $\mathcal{C} \subseteq C_{Obs}$. Values a and b are *contextually equal w.r.t* $\mathcal{C}$ iff $a, b \in [A]$ and $\forall C \in \mathcal{C} \cdot \forall \alpha : X \to [A] \cdot \alpha_a^\#(C) = \alpha_b^\#(C)$. Obviously, if $\mathcal{C} = C_{Obs}$, then $\sim_{\mathcal{C},A} = \approx_{Obs,A}$.

though it need not be a partial congruence. The set-theoretical equality is sound – in programming terms, this is equality on the underlying data representation.

Families of Equalities. The signature of different parts of a structured specification may differ, and the interpretation of equality (e.g. observational equality) may depend on the signature. So, to deal with structured specifications, we use *families* of equalities indexed by signatures. Let $\approx = (\approx_\Sigma)_{\Sigma \in Sign}$ be a family of behavioural equalities and let $\sim = (\sim_\Sigma)_{\Sigma \in Sign}$ and $\triangleq = (\triangleq_\Sigma)_{\Sigma \in Sign}$ denote arbitrary families of approximate equalities. The family $\sim$ is complete (sound) w.r.t. $\approx$ if $\sim_\Sigma$ is complete (sound) w.r.t. $\approx_\Sigma$ for all Σ. The *reduct* of $\sim_\Sigma$ by $\sigma : \Sigma' \to \Sigma$ is $(\sim_\Sigma)|_\sigma = ((\sim_{\Sigma,A})|_\sigma)_{A \in Alg(\Sigma)}$ where $(\sim_{\Sigma,A})|_\sigma = ((\sim_{\Sigma,A})_{\sigma(s)})_{s \in S'}$; then the family $\sim$ is *compatible* with signature morphisms if for all $\sigma : \Sigma' \to \Sigma$ and all $A \in Alg(\Sigma)$, $\sim_{\Sigma',A|_\sigma} = (\sim_{\Sigma,A})|_\sigma$. For the results below having compatibility as a condition, it is enough to consider the signatures arising in the structure of a given specification (for a normalized specification, this includes intermediate signatures arising during normalization). Compatibility may fail if these signatures have different sets of observers for the same sort. The family of set-theoretical equalities is always compatible, but it is easy to check that the family of observational equalities is not compatible in general. One may restrict the introduction of new observers to avoid this problem, see e.g. [BH99].

3 Formal Testing from Algebraic Specifications

Testing from algebraic specifications boils down to checking if axioms are satisfied by programs [Gau95]. Oracles are usually active procedures which drive the necessary tests and interpret the results. *Test cases* are extracted from specifications together with *test sets* defined at specification level and associated with axioms. Then, oracles are defined for each test case or group of test cases. A *test obligation* combines a test case, test data, a test oracle and a program to be tested. When there is no chance of confusion, this is referred to simply as a *test*.

Test Cases and Test Data. For simplicity, each axiom is regarded as a separate test case. Techniques exist to simplify test cases or make them more practical, but this is a separate topic. Test data sets are usually defined from specifications rather than from programs. Here, test sets will be taken to be sets of ground terms [Gau95,Mac99] corresponding to sets of values in the program under test.

Test Oracle Design. The oracle problem often reduces to the problem of comparing values of a non-observable sort for equality; when equality is interpreted as behavioural equality, e.g. observational equality, it may be difficult or impossible to decide. Quantifiers ranging over infinite domains make the oracle problem more difficult. An approach to defining oracles that addresses these problems is presented in [Mac99], where equality is computed using two approximate equalities, one sound and one complete. These equalities are applied according to the context in which equations occur, positive or negative[3]. To handle the quantifier problem, restrictions are placed on the contexts in which they can occur. An *approximate oracle* is then a procedure that decides whether a given axiom is satisfied by a program or not. Such an oracle computes a "testing satisfaction" relation (given below) which differs from the standard one in the way equality is computed and also because quantifiers range only over given test sets.

Definition 3.1 (Testing Satisfaction). *Let Σ be a signature, $T \subseteq T_\Sigma$ be a Σ-test set, $\sim, \simeq$ be two Σ-approximate equalities, A be a Σ-algebra and $\alpha : X \to Dom(\approx_A)$ be a valuation.* Testing satisfaction $\models^T_{\sim,\simeq}$ *is defined as follows.*

1. $A, \alpha \models^T_{\sim,\simeq} t = t'$ *iff* $\alpha^\#(t) \sim_A \alpha^\#(t')$;
2. $A, \alpha \models^T_{\sim,\simeq} \neg\psi$ *iff* $A, \alpha \models^T_{\simeq,\sim} \psi$ *does not hold;*
3. $A, \alpha \models^T_{\sim,\simeq} \psi_1 \wedge \psi_2$ *iff both* $A, \alpha \models^T_{\sim,\simeq} \psi_1$ *and* $A, \alpha \models^T_{\sim,\simeq} \psi_2$ *hold;*
4. $A, \alpha \models^T_{\sim,\simeq} \forall x{:}s \cdot \psi$ *iff* $A, \alpha[x \mapsto {}^\#(t)] \models^T_{\sim,\simeq} \psi$ *holds for all* $t \in T_s$

where $\alpha[x \mapsto v]$ is the valuation α superseded at x by v. Satisfaction of formulae involving $\vee$, $\Rightarrow$, $\Leftrightarrow$, $\exists$ is given using their definitions in terms of $\neg$, $\wedge$, $\forall$. Note that $\sim$ is always applied in positive contexts and $\simeq$ is always applied in negative contexts: the approximate equalities are reversed when negation is encountered.

The following theorem relates testing satisfaction to usual behavioural satisfaction ($\models_\approx$), where equality is interpreted as behavioural equality ($\approx$) and quantification is over all of $Dom(\approx)$.

Theorem 3.2 ([Mac99]). *If $\sim$ is complete, $\simeq$ is sound, and ψ has only positive occurrences of $\forall$ and negative occurrences of $\exists$, then $A, \alpha \models_\approx \psi$ implies $A, \alpha \models^T_{\sim,\simeq} \psi$.* □

The restriction to positive $\forall$ and negative $\exists$ here and in later results is not a problem in practice, since it is satisfied by most common specification idioms.

4 Testing from Non-generic Unit Specifications

This section discusses testing non-generic program units against specifications without considering the internal modular structure of the units. The styles of testing presented can be used to test the individual units of a modular system.

Good practice requires units to be checked independently of their contexts of use. For formal (functional) testing, this corresponds to checking whether the unit satisfies its specification. Testing from flat specifications can follow directly the approach presented in Sect. 3 for each test case. However, once structured

[3] A context is *positive* if it is formed by an even number of applications of negation (e.g. ϕ is in a positive context in both $\phi \wedge \psi$ and $\neg\neg\phi$). Otherwise, the context is *negative*. Note that ϕ is in a negative context and ψ is in a positive context in $\phi \Rightarrow \psi$ since it is equivalent to $\neg\phi \vee \psi$. A formula or symbol *occurs positively* (resp. *negatively*) in ϕ if it occurs in a positive (resp. negative) context within ϕ.

specifications are considered, there are additional complications. First, the structure has to be taken into account when interpreting test results w.r.t. specification axioms. Also, in order to check axioms that involve "hidden symbols" it is necessary to provide an additional implementation for these symbols as the program under test is not required to implement them.

Structured specifications are built using structuring primitives like renaming, union, exporting and extension. These provide a powerful mechanism for reusing and adapting specification as requirements evolve. In *structured specifications with testing interface* [Mac00b], test sets are incorporated into specifications.

Definition 4.1 (Structured Specifications with Testing Interface). *The syntax and semantics of structured specifications are inductively defined as follows. Each specification SP is assigned a signature $Sig(SP)$ and two classes of $Sig(SP)$-algebras. $Mod_{\approx}(SP)$ is the class of* "real" models *of SP w.r.t. the family of Σ-behavioural equalities $\approx = (\approx_\Sigma)_{\Sigma \in Sign}$, and $ChMod_{\sim,\hat{=}}(SP)$ is the class of* "checkable" models *of SP determined by testing w.r.t. the families of approximate equalities $\sim = (\sim_\Sigma)_{\Sigma \in Sign}$ and $\hat{=} = (\hat{=}_\Sigma)_{\Sigma \in Sign}$ and the test sets associated with each axiom.*

1. (**Basic**) $SP = \langle \Sigma, \Psi \rangle$ *with* $\Psi \subseteq \{(\psi, T) \mid \psi \in Sen(\Sigma)$ and $T \subseteq T_\Sigma\}$.
   - $Sig(SP) = \Sigma$
   - $Mod_{\approx}(SP) = \{A \in Alg(\Sigma) \mid \bigwedge_{(\psi,T)\in\Psi} A \models_{\approx} \psi\}$
   - $ChMod_{\sim,\hat{=}}(SP) = \{A \in Alg(\Sigma) \mid \bigwedge_{(\psi,T)\in\Psi} A \models^T_{\sim,\hat{=}} \psi\}$
2. (**Union**) $SP = SP_1 \cup SP_2$, *where SP_1 and SP_2 are structured specifications, with $Sig(SP_1) = Sig(SP_2)$.*
   - $Sig(SP) = Sig(SP_1) \quad [= Sig(SP_2)]$
   - $Mod_{\approx}(SP) = Mod_{\approx}(SP_1) \cap Mod_{\approx}(SP_2)$
   - $ChMod_{\sim,\hat{=}}(SP) = ChMod_{\sim,\hat{=}}(SP_1) \cap ChMod_{\sim,\hat{=}}(SP_2)$
3. (**Renaming**) $SP = \mathsf{translate}\ SP'\ \mathsf{with}\ \sigma$, *where* $\sigma : Sig(SP') \to \Sigma$.
   - $Sig(SP) = \Sigma$
   - $Mod_{\approx}(SP) = \{A \in Alg(\Sigma) \mid A|_\sigma \in Mod_{\approx}(SP')\}$
   - $ChMod_{\sim,\hat{=}}(SP) = \{A \in Alg(\Sigma) \mid A|_\sigma \in ChMod_{\sim,\hat{=}}(SP')\}$
4. (**Exporting**) $SP = SP'|_\Sigma$, *where Σ is a subsignature of $Sig(SP')$.*
   - $Sig(SP) = \Sigma$
   - $Mod_{\approx}(SP) = \{A'|_\Sigma \mid A' \in Mod_{\approx}(SP')\}$
   - $ChMod_{\sim,\hat{=}}(SP) = \{A'|_\Sigma \mid A' \in ChMod_{\sim,\hat{=}}(SP')\}$

Operations presented in Definition 4.1 are primitive ones and, in practice, more complex operations, defined from their combination, are found in CASL and other languages. Extension – "**then**" in CASL – is defined in terms of renaming and union: $SP'\ \mathsf{then\ sorts}\ S\ \mathsf{opns}\ F\ \mathsf{axioms}\ \Psi \stackrel{\text{def}}{=} \langle \Sigma, \Psi \rangle \cup \mathsf{translate}\ SP'\ \mathsf{with}\ \sigma$, where SP' is a structured specification, S is a set of sorts, F is a set of function declarations, $\Sigma = Sig(SP') \cup (S, F)$, $\sigma : Sig(SP') \hookrightarrow \Sigma$ is the inclusion, and Ψ is a set of axioms over Σ with their associated test sets. The union of specifications over possibly different signatures – "**and**" in CASL – can be expressed as: $SP_1\ \mathsf{and}\ SP_2 \stackrel{\text{def}}{=} \mathsf{translate}\ SP_1\ \mathsf{with}\ \sigma_1 \cup \mathsf{translate}\ SP_2\ \mathsf{with}\ \sigma_2$, where $\Sigma = Sig(SP_1) \cup Sig(SP_2)$ and $\sigma_1 : Sig(SP_1) \hookrightarrow \Sigma$, $\sigma_2 : Sig(SP_2) \hookrightarrow \Sigma$.

To handle the oracle problem for structured specifications, two styles of testing are suggested in [Mac00b]: structured and flat testing. *Structured testing* of a Σ-algebra A against a structured specification SP corresponds to membership in the class of checkable models of SP, i.e., $A \in ChMod_{\sim,\simeq}(SP)$. Structured testing is based on the structure of SP; it may involve more than one set of test obligations (see Definition 4.1) and may demand additional implementation of symbols not in A (not exported by SP).

On the other hand, flat testing takes a monolithic approach based on an unstructured view of the specification without considering non-exported symbols and using a single pair of approximate equalities on the overall signature of SP. More specifically, *flat testing* corresponds to testing satisfaction of axioms extracted from SP, i.e., $\bigwedge_{(\psi,T)\in TAx(SP)} A \models^{T}_{\sim_\Sigma,\simeq_\Sigma} \psi$ where $\Sigma = Sig(SP)$ and $TAx(SP)$ are the *visible axioms of* SP, defined as follows.

Definition 4.2 (Visible Axioms). *The set of* visible axioms *together with corresponding test sets of a specification* SP *is defined as follows.*

1. $TAx(\langle\Sigma,\Psi\rangle) = \Psi$
2. $TAx(SP_1 \cup SP_2) = TAx(SP_1) \;\cup\; TAx(SP_2)$
3. $TAx(\textsf{translate } SP' \textsf{ with } \sigma) = \sigma(TAx(SP'))$
4. $TAx(SP'|_\Sigma) = \{(\phi, T \cap T_\Sigma) \mid (\phi,T) \in TAx(SP') \text{ and } \phi \in Sen(\Sigma)\}$

The visible axioms of a specification SP exclude those that refer to non-exported symbols, translating the rest to the signature of SP.

Whether structured or flat testing is performed, we must ask: under which conditions are correct models not rejected by testing? Results in [Mac00b,Mac00c] show that under certain assumptions, structured testing and flat testing do not reject correct models, even though incorrect ones can be accepted.

Theorem 4.3 ([Mac00b]). *If* $\sim$ *is complete,* $\simeq$ *is sound, and the axioms of* SP *have only positive occurrences of* $\forall$ *and negative occurrences of* $\exists$*, then* $A \in Mod_{\approx}(SP)$ *implies* $A \in ChMod_{\sim,\simeq}(SP)$. □

Theorem 4.4 ([Mac00b]). *If* $\sim$ *is complete and compatible and* $\simeq$ *is sound and compatible and the axioms of* SP *have only positive occurrences of* $\forall$ *and negative occurrences of* $\exists$*, then* $A \in Mod_{\approx}(SP)$ *implies* $\bigwedge_{(\psi,T)\in TAx(SP)} A \models^{T}_{\sim,\simeq} \psi$. □

Structured testing is more flexible than flat testing in the sense that fewer assumptions are made. The family of observational equalities is not compatible as mentioned in Sect. 2, so the additional assumption is a strong one. On the other hand, flat testing is simpler. Both theorems cover a prevalent use of quantifiers. Their duals also hold, but are less interesting. There are also variants of these theorems that substitute assumptions on test sets for assumptions on quantifiers.

Structured testing and flat testing are two extremes. In practice, we may combine them. This can be done via normalization, where a structured specification SP is transformed into a specification $\boldsymbol{nf}(SP)$ of the form $\langle\Sigma',\Psi'\rangle|_\Sigma$ [BCH99] having the same signature and class of models. This procedure groups axioms, taking hidden symbols into account, so that the result is a flat specification which exports visible symbols. The usual normalization procedure is easily extended to structured specifications with testing interface (see [Mac00a] for details) and then we obtain the following result:

Theorem 4.5 ([Mac00a]). *If $\sim$ and $\doteq$ are compatible, then $ChMod_{\sim,\doteq}(SP) = ChMod_{\sim,\doteq}(\mathbf{nf}(SP))$.* □

The main advantage of normal form is to allow a combination of compositional and non-compositional testing, namely *semi-structured* testing, where normal forms are used to replace *parts* of a specification, especially when these parts are combined by union. The result is to reduce the number of different experiments that are performed. Then, the resulting specification can be checked by structured testing and a result analogous to Theorems 4.3 and 4.4 is obtained:

Theorem 4.6 ([Mac00a]). *If $\sim$ is complete, $\doteq$ is sound, $\approx$ is compatible and the axioms of SP have only positive occurrences of $\forall$ and negative occurrences of $\exists$, then $A \in Mod_{\approx}(SP)$ implies $A \in ChMod_{\sim,\doteq}(\mathbf{nf}(SP))$.* □

Even though we have shown theoretical results regarding structured, flat and semi-structured testing, these styles of testing can be infeasible in practice when the structure of the program is not taken into account, since it may be necessary to decompose the program to reflect certain signatures in the structure of the specification and/or re-test the whole program every time it is modified.

5 Testing from Generic Unit Specifications

This section is concerned with testing generic units independent of particular instantiations. This is a difficult task for testing since we have to anticipate the behaviour of a generic unit when instantiated by specific units, but the set of all possible instantiations is almost always infinite. Not all units having the right signature are correct implementations of the argument specification, and correctness cannot generally be determined by testing. However, testing a generic unit using an incorrect implementation of the argument specification may lead to rejection of correct generic units.

The syntax and semantics of generic unit specifications are as follows. First, let $Alg(\Sigma' \to \Sigma) = \{F : Alg(\Sigma') \rightharpoonup Alg(\Sigma) \mid \forall A \in Dom(F), F[A]|_{\Sigma'} = A\}$ be the class of persistent partial functions taking Σ'-algebras to Σ-algebras, where $\Sigma' \subseteq \Sigma$. These functions on algebras model generic units; they are partial since a generic unit is only required to produce a result when instantiated by an algebra that satisfies the argument specification.

Definition 5.1 (Generic Unit Specifications). *Let $Sig(SP) = \Sigma$ and $Sig(SP') = \Sigma'$, such that SP extends SP', i.e. $\Sigma' \subseteq \Sigma$ and for all $A \in Mod_{\approx}(SP)$, $A|_{\Sigma'} \in Mod_{\approx}(SP')$.*

- *$Sig(SP' \to SP) = \Sigma' \to \Sigma$*
- *$Mod_{\approx}(SP' \to SP) = \{F \in Alg(\Sigma' \to \Sigma) \mid \forall A \in Mod_{\approx}(SP'), A \in Dom(F)$ and $F[A] \in Mod_{\approx}(SP)\}$*

– $ChMod_{\sim,\doteq}(SP' \rightarrow SP) = \{F \in Alg(\Sigma' \rightarrow \Sigma) \mid \forall A \in ChMod_{\sim,\doteq}(SP'),$ $A \in Dom(F)$ and $F[A] \in ChMod_{\sim,\doteq}(SP)\}$

Let $SP' \rightarrow SP$ be a generic unit specification and let F be a generic unit that is claimed to correctly implement this specification. In contrast to testing from non-generic unit specifications, membership in the class of checkable models does not give rise to a feasible style of testing from generic unit specifications. First, the class of models of SP' may be infinite. Moreover, the class of checkable models of $SP' \rightarrow SP$ cannot be directly compared to its class of "real" models. Suppose $F \in Mod_{\approx}(SP' \rightarrow SP)$ and $A \in ChMod_{\sim,\doteq}(SP')$, but $A \notin Mod_{\approx}(SP')$ and $F[A] \notin ChMod_{\sim,\doteq}(SP)$. Then, $F \notin ChMod_{\sim,\doteq}(SP' \rightarrow SP)$ – a correct F is rejected due to bugs in A.

As explained earlier, a testing method should ensure that correct models are not rejected. This requires that incorrect models of the parameter specification are not used in testing. Suppose another class of models, named *strong models*, is defined as an alternative to the class of checkable models.

Definition 5.2 (Strong Models). *The class of* strong models *of $SP' \rightarrow SP$ is defined as $SMod_{\sim,\doteq}(SP' \rightarrow SP) = \{F \in Alg(\Sigma' \rightarrow \Sigma) \mid \forall A \in Mod_{\approx}(SP'),$ $A \in Dom(F)$ and $F[A] \in ChMod_{\sim,\doteq}(SP)\}$.*

This represents the class of models which are successfully tested when only correct implementations of SP' are considered. We then have:

Theorem 5.3. *If $\sim$ is complete, $\doteq$ is sound, and the axioms of SP have only positive occurrences of $\forall$ and negative occurrences of $\exists$, then $F \in Mod_{\approx}(SP' \rightarrow SP)$ implies $F \in SMod_{\sim,\doteq}(SP' \rightarrow SP)$.*

Proof. Suppose $F \in Mod_{\approx}(SP' \rightarrow SP)$. Then, $\forall A \in Mod_{\approx}(SP')$, $A \in Dom(F)$ and $F[A] \in Mod_{\approx}(SP)$. By Theorem 4.3, $F[A] \in ChMod_{\sim,\doteq}(SP)$. Hence, $F \in SMod_{\sim,\doteq}(SP' \rightarrow SP)$. □

This means that if we can test for membership in the class of strong models, then correct models of generic unit specifications are not rejected. Obviously, incorrect models can be accepted. As with Theorem 4.3, the dual also holds.

In practice, testing membership in $SMod_{\sim,\doteq}(SP' \rightarrow SP)$ (this also applies to $ChMod_{\sim,\doteq}(SP' \rightarrow SP)$) is not possible, since as already noted the class $Mod_{\approx}(SP')$ is almost always infinite and in any case membership in this class is not testable in general. A feasible approach to test whether generic units are models of $SP' \rightarrow SP$ should only rely on a finite subset of $Mod_{\approx}(SP')$. So let $\mathcal{C}$ be a set of units ("stubs") chosen according to some coverage criteria. Then we define a "weak" class of models of $SP' \rightarrow SP$ as follows.

Definition 5.4 (Weak Models). *Let $\mathcal{C} \subseteq Mod_{\approx}(SP')$. The class of* weak models *of $SP' \rightarrow SP$ is defined as $WMod_{\sim,\doteq,\mathcal{C}}(SP' \rightarrow SP) = \{F \in Alg(\Sigma' \rightarrow \Sigma) \mid \forall A \in \mathcal{C}$, $A \in Dom(F)$ and $F[A] \in ChMod_{\sim,\doteq}(SP)\}$.*

The intention here is to select a class $\mathcal{C}$ which is finite and has a reasonable size such that F can be tested with this class and useful information gained. The following theorem shows that under certain assumptions, correct generic units are not rejected by testing w.r.t. the class of weak models.

Theorem 5.5. *If $\sim$ is complete, $\doteq$ is sound, and the axioms of SP have only positive occurrences of $\forall$ and negative occurrences of $\exists$, then $F \in Mod_{\approx}(SP' \to SP)$ implies $F \in WMod_{\sim,\doteq,\mathcal{C}}(SP' \to SP)$ for any $\mathcal{C} \subseteq Mod_{\approx}(SP')$.*

Proof. Suppose $F \in Mod_{\approx}(SP' \to SP)$. Then, $\forall A \in \mathcal{C} \subseteq Mod_{\approx}(SP')$, $A \in Dom(F)$ and $F[A] \in Mod_{\approx}(SP)$. By Theorem 4.3, $F[A] \in ChMod_{\sim,\doteq}(SP)$. Hence, $F \in WMod_{\sim,\doteq,\mathcal{C}}(SP' \to SP)$. □

The class of weak models is comparable to the classes of checkable and strong models, i.e., for any $\mathcal{C} \subseteq Mod_{\approx}(SP')$, $F \in ChMod_{\sim,\doteq}(SP' \to SP)$ implies $F \in WMod_{\sim,\doteq,\mathcal{C}}(SP' \to SP)$, provided the assumptions of Theorem 4.3 hold for SP', and $F \in SMod_{\sim,\doteq}(SP' \to SP)$ implies $F \in WMod_{\sim,\doteq,\mathcal{C}}(SP' \to SP)$. Moreover, for any algebra A used to test membership of F in the class of weak models of $SP' \to SP$, we can conclude that $F[A]$ is indeed a checkable model of SP, i.e., if $A \in \mathcal{C}$ and $F \in WMod_{\sim,\doteq,\mathcal{C}}(SP' \to SP)$ then $F[A] \in ChMod_{\sim,\doteq}(SP)$, by Definition 5.4. However, what if $A \in Mod_{\approx}(SP')$, but $A \notin \mathcal{C}$? Is $F[A] \in ChMod_{\sim,\doteq}(SP)$? How do we select an appropriate *finite* set of $Sig(SP')$-algebras so that an answer to the above question can be given?

Even though, under the assumptions of Theorem 5.5, testing membership in the class of weak models does not reject correct programs, not all sets $\mathcal{C}$ of stubs are equally interesting. It is desirable that a generic unit F be tested without regard to the units that are going to be used to instantiate it subsequently. Then, if we can conclude that $F[A]$ is a checkable model for some A, it may be possible to avoid re-testing F when A is replaced by a different unit. In other words, we need to select $\mathcal{C}$ to be a representative subset of $Mod_{\approx}(SP')$ so that given a correct realisation A of SP' ($A \in Mod_{\approx}(SP')$) and a generic unit F in the class of weak models of $SP' \to SP$ ($F \in WMod_{\sim,\doteq,\mathcal{C}}(SP' \to SP)$), we can conclude that $F[A]$ is a realisation of SP ($F[A] \in ChMod_{\sim,\doteq}(SP)$).

One possible answer to the above questions might be to pick one representative of every equivalence class w.r.t. an equivalence relation $\equiv$ on algebras when defining $\mathcal{C}$. This might be the observational equivalence on algebras [BHW95] or an approximation to it. The idea is similar to equivalence partitioning of test sets and the uniformity hypothesis in black-box testing [Ber91].

Let $\equiv$ be an equivalence relation and define the *closure of $\mathcal{C}$ under* $\equiv$ as $Cl_{\equiv}(\mathcal{C}) = \{A \in Alg(\Sigma) \mid \exists B \in \mathcal{C} \cdot A \equiv B\}$. In particular, we might pick $\mathcal{C}$ so that $Cl_{\equiv}(\mathcal{C})$ coincides with $Mod_{\approx}(SP)$. Following [BHW95], we focus on equivalence relations that are "factorizable" by partial congruences of interest. The idea comes from automata theory: two finite state machines are equivalent (i.e. accept the same language) iff quotienting each one by the Nerode equivalence on states yields isomorphic machines. (In fact, we will require only right factorizability.)

Definition 5.6 (Factorizability). *An equivalence $\equiv \subseteq Alg(\Sigma) \times Alg(\Sigma)$ is* factorizable *by a family of partial Σ-congruences $\approx = (\approx_A)_{A \in Alg(\Sigma)}$ if $A \equiv B$ iff $A/{\approx_A} \cong B/{\approx_B}$; $\equiv$ is* right factorizable *by $\approx$ if $A \equiv B$ implies $A/{\approx_A} \cong B/{\approx_B}$.*

It is shown in [BHW95] that various definitions of observational equivalence are factorizable by corresponding observational equalities. We will need an equivalence that is right factorizable by an approximate equality that is complete with respect to our chosen behavioural equality. Complete equalities are coarser than $\approx$, and if $\approx$ is observational equality then equivalences that are factorizable by such equalities are coarser than observational equivalence. Requiring *right* factorizability permits the equivalence to be *finer* than the factorizable one, including observational equivalence and equivalences finer than that.

The following theorem relates behavioural satisfaction of a sentence and ordinary satisfaction of the same sentence in a quotient algebra.

Theorem 5.7 ([BHW95]). *Let* $\approx = (\approx_A)_{A \in Alg(\Sigma)}$ *be a family of partial* Σ-*congruences. Then* $A/{\approx_A} \models \psi$ *iff* $A \models_{\approx} \psi$. □

Putting these together:

Corollary 5.8. *Let* $\equiv$ *be right factorizable by* $\approx$. *Then* $A \equiv B$ *implies* $A \models_{\approx} \psi$ *iff* $B \models_{\approx} \psi$.

Proof. $A \models_{\approx} \psi$ iff $A/{\approx_A} \models \psi$ (Theorem 5.7) iff $B/{\approx_B} \models \psi$ ($A \equiv B$, right factorizability, preservation of satisfaction by $\cong$) iff $B \models_{\approx} \psi$ (Theorem 5.7). □

Theorem 5.9. *Let* $\equiv$ *be right factorizable by* $\sim$. *Then* $A \equiv B$ *implies* $A \in ChMod_{\sim,\sim}(SP)$ *iff* $B \in ChMod_{\sim,\sim}(SP)$.

Proof. By induction on the structure of SP, using Corollary 5.8 for specifications of the form $\langle \Sigma, \Psi \rangle$. Since test sets are finite sets of ground terms, $A \models^{T}_{\sim,\sim} \psi$ is equivalent to $A \models_{\sim} \psi'$ where ψ' is obtained from ψ by replacing each subformula of the form $\forall x : s \cdot \varphi$ by $\bigwedge_{t \in T_s} \varphi[t/x]$ and each subformula of the form $\exists x : s \cdot \varphi$ by $\bigvee_{t \in T_s} \varphi[t/x]$. □

A further assumption will be that generic units preserve $\equiv$, i.e. are "stable":

Definition 5.10 (Stability). *A generic unit* $F \in Alg(\Sigma' \to \Sigma)$ *is* stable *with respect to equivalences* $\equiv_{\Sigma'} \subseteq Alg(\Sigma') \times Alg(\Sigma')$ *and* $\equiv_{\Sigma} \subseteq Alg(\Sigma) \times Alg(\Sigma)$ *if for any* $A \in Dom(F)$, $A \equiv_{\Sigma'} B$ *implies* $B \in Dom(F)$ *and* $F[A] \equiv_{\Sigma} F[B]$.

Stability with respect to observational equivalence is a reasonable assumption for generic units expressed in a programming language, since stability is closely related to the security of the data encapsulation mechanisms in that language, see [Sch87] and [ST97]. For an equivalence that is only an approximation to observational equivalence, stability seems reasonable as a hypothesis in the context of testing. We then have the main result of this section:

Theorem 5.11. *Let* $\mathcal{C} \subseteq Mod_{\approx}(SP')$. *If* $F \in WMod_{\sim,\sim,\mathcal{C}}(SP' \to SP)$, $A \in Cl_{\equiv_{\Sigma'}}(\mathcal{C})$, $\equiv$ *is right factorizable by* $\sim$ *and* F *is stable with respect to* $\equiv_{\Sigma'}$ *and* $\equiv_{\Sigma}$, *then* $F[A] \in ChMod_{\sim,\sim}(SP)$.

Proof. $A \in Cl_{\equiv_{\Sigma'}}(\mathcal{C})$ means that $A \equiv_{\Sigma'} B$ for some $B \in \mathcal{C}$, and then $F[B] \in ChMod_{\sim,\sim}(SP)$. By stability, $F[A] \equiv_{\Sigma} F[B]$. Then, by Theorem 5.9, $F[A] \in ChMod_{\sim,\sim}(SP)$. □

Corollary 5.12. *Let* $\mathcal{C} \subseteq Mod_{\approx}(SP')$. *If* $F \in WMod_{\sim,\sim,\mathcal{C}}(SP' \to SP)$, $\equiv$ *is right factorizable by* $\sim$ *and* F *is stable with respect to* $\equiv_{\Sigma'}$ *and* $\equiv_{\Sigma}$, *then* $F \in WMod_{\sim,\sim,Cl_{\equiv}(\mathcal{C})}(SP' \to SP)$.

Theorem 5.11 and Corollary 5.12 are useful "amplification" results. They allow information gained from testing particular instantiations (the set of stubs $\mathcal{C}$) to be extrapolated to give information about instantiations that have not actually been tested. Theorem 5.11 relates to the practice of replacing a module in a working system (in this case, the parameter of a generic unit) with another version. If the two versions can be shown to be equivalent ($\equiv$), then the overall system will continue to work provided the assumptions in Theorem 5.11 are met. In Corollary 5.12, if we choose $\mathcal{C}$ so that $Cl_{\equiv}(\mathcal{C}) = Mod_{\approx}(SP)$, then the conclusion is equivalent to membership in the class of strong models of $SP' \rightarrow SP$. In most cases, this ideal will not be achievable; nevertheless, we can aim to include in $\mathcal{C}$ representatives of equivalence classes related to the particular class of applications for which F is intended to be used.

In order for testing to avoid rejecting correct units, according to Theorem 5.5 we need to restrict to specifications with only positive occurrences of $\forall$ and negative occurrences of $\exists$. A much more serious restriction when combining Theorems 5.5 and 5.11 is that $\sim$ needs to be both sound *and* complete. However, an analysis of the proof shows that it is sufficient if $\sim$ is complete for equations in positive positions and sound for equations in negative positions. If $\approx$ is observational equality then this can be achieved by taking $\sim$ to be a contextual equality (and hence complete) and restricting equations in negative positions to be of observable sorts only, for which $\sim$ is sound and complete.

6 Concluding Remarks

We have presented ideas relating to testing modular systems against CASL-style architectural specifications. Our overall objective is to support independent development and verification of program components.

The problem of testing against architectural specifications reduces to:

1. testing non-generic units against structured specifications;
2. testing generic units against specifications of the form $SP' \rightarrow SP$; and
3. "integration testing" for unit terms that avoids re-testing.

Solutions to (1) are presented in Sect. 4, where previous results for testing against structured specifications are reviewed and discussed in the context of architectural specifications. Then, based on previous research on behavioural implementations, ideas concerning (2) are presented in Sect. 5. Since the class of possible parameter units ("stubs") is almost always infinite, we suggest that a representative finite class be selected, allowing testing results to be extrapolated to equivalent units. For this, stability of generic units is assumed. Problem (3) is future work, although it is already clear that the results in Sect. 5 and [Mac00b,Mac00a,Mac00c] are relevant.

Other questions for the future concern the circumstances under which our approximation to stability holds, as well as the connection between equivalence on algebras and testing satisfaction, including the question of how this equivalence can be effectively checked. We also aim to extend the results to specifications of higher-order generic units. Finally, a general method of applying the ideas presented along with practical case studies are needed.

Acknowledgments

Thanks to Marie-Claude Gaudel for encouragement and to members of IFIP WG 1.3 for interesting questions. This research was partly funded by PROTEM-CC/CNPq.

References

[ABK+03] E. Astesiano, M. Bidoit, H. Kirchner, B. Krieg-Brückner, P. Mosses, D. Sannella and A. Tarlecki. CASL: The common algebraic specification language. *Theoretical Computer Science.* To appear (2003).

[BB01] H. Baumeister and D. Bert. Algebraic specification in CASL. In *Software Specification Methods – An Overview Using a Case Study.* Springer (2001).

[Ber89] G. Bernot. A formalism for test with oracle based on algebraic specifications. Report 89-4, LIENS/DMI, Ecole Normale Supérieure, Paris (1989).

[Ber91] G. Bernot. Testing against formal specifications: a theoretical view. *Proc. TAPSOFT'91*, Brighton. Springer LNCS 494, 99–119 (1991).

[BCH99] M. Bidoit, M.V. Cengarle and R. Hennicker. Proof systems for structured specifications and their refinements. *Algebraic Foundations of Systems Specifications*, chapter 11. Springer (1999).

[BH99] M. Bidoit and R. Hennicker. Observational logic. *Proc. AMAST'98*, Manaus. Springer LNCS 1548, 263–277 (1999).

[BHW95] M. Bidoit, R. Hennicker and M. Wirsing. Behavioural and abstractor specifications. *Science of Computer and Programming*, 25:149–186 (1995).

[BST02] M. Bidoit, D. Sannella and A. Tarlecki. Architectural specifications in CASL. *Formal Aspects of Computing.* To appear (2002).

[DW00] M. Doche and V. Wiels. Extended institutions for testing. *Proc. AMAST 2000.* Springer LNCS 1816, 514–528 (2000).

[Gau95] M.-C. Gaudel. Testing can be formal, too. *Proc. TAPSOFT'95*, Aarhus. Springer LNCS 915 (1995).

[LeG99] P. LeGall. *Vers une spécialisation des logiques pour spécifier formellement et pour tester des logiciels.* Habilitation thesis, Université d'Evry (1999).

[LA96] P. LeGall and A. Arnould. Formal specification and test: correctness and oracle. *Proc. WADT'95*, Oslo. Springer LNCS 1130 (1996).

[Mac99] P.D.L. Machado. On oracles for interpreting test results against algebraic specifications. *Proc. AMAST'98*, Manaus. Springer LNCS 1548, 502–518 (1999).

[Mac00a] P.D.L. Machado. The rôle of normalisation in testing from structured algebraic specifications. *Proc. WADT'99*, Bonas. Springer LNCS 1827, 459–476 (2000).

[Mac00b] P.D.L. Machado. Testing from structured algebraic specifications. *Proc. AMAST 2000.* Springer LNCS 1816, 529–544 (2000).

[Mac00c] P.D.L. Machado. *Testing from Structured Algebraic Specifications: The Oracle Problem.* PhD thesis, LFCS, University of Edinburgh (2000).

[ST97] D. Sannella and A. Tarlecki. Essential concepts of algebraic specification and program development. *Formal Aspects of Computing*, 9:229–269 (1997).

[Sch87] O. Schoett. *Data Abstraction and the Correctness of Modular Programming.* PhD thesis, LFCS, University of Edinburgh (1987).

Symbolic Semantics and Analysis for Crypto-CCS with (Almost) Generic Inference Systems*

Fabio Martinelli

Istituto di Informatica e Telematica – C.N.R., Pisa, Italy.

Abstract. Crypto-CCS is a formal description language for distributed protocols which is suitable to abstractly model the cryptographic ones. Indeed, this language adopts a message-manipulating rule which may be used to mimic some features of cryptographic functions. We equip the Crypto-CCS calculus with a symbolic operational semantics. Moreover, we provide a mechanized method to analyze the security properties of cryptographic protocols (with finite behaviour), symbolically. Our work extends the previous one on symbolic verification techniques for cryptographic protocols modeled with process algebras since it deals with (almost) generic inference systems instead of fixed ones.

1 Introduction

In the last years, several formal languages for describing distributed and communicating protocols have been refined to model at a high level of abstraction some specific features of cryptographic protocols.

Cryptography is usually modeled by representing encryptions as terms of an algebra, e.g., $E(m, k)$ may represent the encryption of a message m with a key k. Usually, the so-called perfect encryption abstraction is adopted: encryptions are considered as injective functions which can be inverted only by knowing the correct information, i.e. the decryption key. For instance, common inference rules for modeling the behavior of the encryption and decryption (in a shared-key schema) are the followings:

$$\frac{m \quad k}{E(m,k)} \qquad \frac{E(m,k) \quad k}{m} \tag{1}$$

which should be read as: from a message m and a key k we can build the encryption $E(m, k)$; from an encryption $E(m, k)$ and a decryption key k we can obtain the encrypted message m. The long standing tradition of modeling the specific

* Work partially supported by Microsoft Research Europe (Cambridge); by MIUR project "MEFISTO"; by MIUR project "Tecniche e strumenti software per l'analisi della sicurezza delle comunicazioni in applicazioni telematiche di interesse economico e sociale"; by CNR project "Strumenti, ambienti ed applicazioni innovative per la società dell'informazione" and finally by CSP with the project "SeTAPS".

K. Diks et al. (Eds): MFSC 2002, LNCS 2420, pp. 519–531, 2002.

features of cryptographic functions as term-rewriting rules met the powerful verification techniques developed for process algebras. As a matter of fact, several formal languages for describing communication protocols, for instance LOTOS [9] and CSP [10], have been exploited for representing cryptographic protocols without changes in syntax or semantics: the inference rules have been given at the meta-level of the verification. Instead others, like the π–calculus [1] and the CCS [13,14,11], have been effectively refined: the π–calculus have been equipped with two pattern matching constructs for modeling message splitting and shared-key decryption, respectively; the CCS has been equipped with a term-rewriting construct that permits to infer new messages from others.

A main source of difficulty in the analysis cryptographic protocols is the uncertainty about the possible enemies or malicious users that may try to interfere with the normal execution of a protocol. Typically, the verification problem for a security property may be expressed as a verification problem for open systems [6,13], i.e., for every possible enemy X we have that

$$S \,|\, X \quad \mathbf{sat} \quad S_{spec} \tag{2}$$

Usually, S_{spec} is a process and **sat** a relation among processes. Roughly, if we imagine that S_{spec} denotes a secure system, then the previous statement requires that the system S in composition with whatever intruder X is still secure. The universal quantification makes it difficult to deal with such verification problems. A common solution is to consider a particular intruder –the "most powerful" one– which is able to impersonate each possible attacker on a protocol for a security property. Thus, after fixing a specific enemy, the resulting system may be analyzed by using standard tools as state-exploration or model checking.

However, these analysis methods mainly suffer of two problems:

– several computation states need to be explicitly represented and analyzed;
– some restrictions are usually necessary on the size of messages, i.e., the number of encryption/decryption operations that the intruder can perform, to make the analysis feasible. Indeed, in principle, the intruder could send to the protocol partecipants whatever message it can infer from the ones it knows or it has intercepted during the attack. This means that, with the usual semantics with explicit substitutions, we should consider infinitely many computation states, one for each message that the intruder can build and send and a protocol participant is willing to receive.

Many authors noted that symbolic description methods [7] may be useful to face both problems, e.g. see [2,4,5]. In a symbolic semantics, usually the set of possible messages that an agent may receive in input is implicitly denoted by some sort of symbolic representation. On this representation several operations are possible that mimic the concrete semantics (usually based on explicit replacing of variables with messages). Clearly, there must be an agreement between the concrete semantics and the symbolic one. A symbolic semantics is usually more technically challenging than the corresponding concrete one. However, analyzing with the symbolic representation is usually more efficient than with the explicit one; moreover, as in the case of cryptographic protocols, the symbolic semantics often allows to finitely describe infinite sets of messages that otherwise could not be explicitly treated.

Our work improves the previous one on symbolic techniques for process algebras for cryptographic protocols. Indeed, our symbolic analysis is able to manage (almost) generic inference systems. There is an assumption on the format of such inference systems; however, commonly used inference systems enjoy this assumption. Usually, symbolic analysis approaches in the literature are given for fixed inference systems. Moreover, we enhance also the results in the specific cryptosystem modeling shared-key used by Amadio and Lugiez [2] and Boreale [4], since ours allows generic messages as keys. Fiore and Abadi also [5] proposed a symbolic analysis framework for shared-key cryptography but they only conjecture the completeness of such system w.r.t. generic keys. Our approach may be regarded as an extension and generalization to the framework of process algebras of the one of Millen and Shmatikov [15]. They proposed a constraint solving system which works for shared-key (public-key) crypto-systems with generic keys; however, their computation model, i.e. the strand spaces [17], is specific for security analysis. An interesting work, even though not directly related to process algebras, is the one of Blanchet [3] which uses generic prolog rules both for describing the protocol behavior and the inference system. This makes his theory rather general; however, his approach does not always guarantee termination of the analysis. Another difference with the previous work is that we do not have a built-in notion of the intruder. In our framework, an intruder could be any term of the algebra, even though a specific one may be useful for the security analysis but does not significantly influence the symbolic semantics of the language. This makes our approach, in principle, not uniquely tailored for security analysis.

As a future work we plan to implement (an optimized version of) our analysis in order to check its feasibility in practice.

The rest of the paper is organized as follows. Section 2 defines the symbolic operational semantics for Crypto-CCS based on lists of constraints. Section 3 gives the necessary technical tools to manage such a symbolic semantics. Finally, Section 4 presents a symbolic analysis method for checking security protocols together with an example.

2 Crypto-CCS: An Operational Language for the Description of Cryptographic Protocols

This section presents the language we use for the description of security protocols, which is a slight modification of *CCS* process algebra [16] using cryptography-modeling constructs and dealing with secret (confidential) values (hence the name Crypto-CCS). The model consists of a set of sequential agents able to communicate among each other by exchanging messages.

2.1 Messages and Inference Systems

Here, we define the data handling part of language. Messages are the data manipulated by agents. We introduce also the notion of inference system which models the possible operations on messages.

Messages. Consider a collection of symbols $\mathcal{F} = \{F^1, \ldots, F^l, \ldots\}$ that represent the constructors for messages. Each constructor has an arity k_i, for $1 \leq i$ which returns the number of arguments of the constructor F^i. When k_i is 0 we say that F^i is a basic message (or a constant). We assume that the set of symbols which are not constants is finite. Moreover, assume to have a countable set V of message variables, ranged over by $x, y, z \ldots$. Then, the set *Msgs* of messages is the least set s.t. $V \subseteq Msgs$ and if $m_1, \ldots, m_{k_i}$ are in $Msgs$ then also $F^i(m_1, \ldots, m_{k_i}) \in Msgs$. A message $m \in Msgs$ without variables is said to be *closed*.

Inference System. Agents are able to obtain new messages from the set of messages produced or received through an inference system. This system consists in a set of inference schemata. An inference schema can be written as $r = \frac{m_1 \quad \ldots \quad m_n}{m_0}$ where $m_1, \ldots, m_n$ is a set of premises (possibly empty) and m_0 is the conclusion. Consider a substitution ρ, with $\rho : V \mapsto Msgs$, then let $m\rho$ be the message m where each variable x is replaced with $\rho(x)$. Given a sequence of closed messages $m'_1, \ldots, m'_n$, we say that a closed message m can be inferred from $m'_1, \ldots, m'_n$ through the application of the schema r (written as $m'_1, \ldots, m'_n \vdash_r m$) if there exists a substitution ρ s.t. $m_0\rho = m$ and $m_i\rho = m'_i$, for $i \in \{1, \ldots, n\}$. Given an inference system, we can define an inference function $\mathcal{D}$ s.t. if ϕ is a finite set of closed messages, then $\mathcal{D}^R(\phi)$ is the set of closed messages that can be deduced starting from ϕ by applying only rules in R.

In Table 1, we present the usual inference system for modeling shared-key encryption within process algebras (e.g., see [10]). Rule (*pair*) is used to construct *pairs* while Rules (*fst*) and (*snd*) are used to split pairs; Rule (*enc*) is used to construct encryptions and Rule (*dec*) is used to decrypt messages.

Table 1. A simple inference system.

$$\frac{x \quad y}{Pair(x,y)}(pair) \quad \frac{Pair(x,y)}{x}(fst) \quad \frac{Pair(x,y)}{y}(snd)$$

$$\frac{x \quad y}{E(x,y)}(end) \qquad \frac{E(x,y) \quad y}{x}(dec)$$

2.2 Agents and Systems

We define the control part of our language for the description of cryptographic protocols. Basically, we consider (compound) systems which consists of sequential agents running in parallel. The terms of our language are generated by the following grammar:

(COMPOUND SYSTEMS:) $S ::= S \setminus L \mid S_1 \mid S_2 \mid A_\phi$
(SEQUENTIAL AGENTS:) $A ::= \mathbf{0} \mid p.A \mid [m_1 \dots m_n \vdash_r x]A_1$
(PREFIX CONSTRUCTS:) $p ::= c!m \mid c?x$

where $m, m_1, \dots, m_n$ are *closed* messages or variables, x is a variable, C is a finite set of channels with $c \in C$, ϕ is a finite set of messages, L is a subset of C.

We briefly give the informal semantics of sequential agents and compound systems as well as some static constraints on the terms of the language.

SEQUENTIAL AGENTS:

- $\mathbf{0}$ is the process that does nothing.
- $p.A$ is the process that can perform an action according to the particular prefix construct p and then behaves as A:
  - $c!m$ allows the message m to be sent on channel c.
  - $c?x$ allows messages m to be received on channel c. The message received substitutes the variable x.
- $[m_1 \dots m_n \vdash_r x]A_1$ is the inference construct. If, applying a case of inference schema r with the premises $m_1 \dots m_n$, a message m can be inferred, then the process behaves as A_1 (where m substitutes x). This is the message-manipulating construct of the language for modeling cryptographic operations.

COMPOUND SYSTEMS:

- A compound system $S \setminus L$ allows only external actions whose channel is not in L. The internal action τ, which denotes an internal communication among agents, is never prevented.
- A compound system $S \mid S_1$ performs an action a if one of its sub-components performs a. A synchronization or internal action, denoted by the symbol τ, may take place whenever S and S_1 are able to perform two complementary, i.e. send-receive, actions.
- Finally, the term A_ϕ represents a system which consists of a single sequential agent whose knowledge, i.e. the set of messages which occur in its term A, is described by ϕ. The agent's knowledge increases as it receives messages (see rule (?)), infers new messages from the messages it knows (see rule $\mathcal{D}_1$). Sometimes we omit to represent agent's knowledge when this can be easily inferred from the context.

Note 2.1. We do not explictly define equality check among messages in the syntax. However, this can be implemented through the usage of the inference construct. For instance consider the rule $equal \doteq \dfrac{x \quad x}{Equal(x, x)}$. Then $[m = m']A$ (with the expected semantics) may be equivalently expressed as $[m \quad m' \vdash_{equal} y]A$ where y does not occur in A. ■

STATIC CONSTRAINTS:

We assume on our terms some well-formedness conditions that can be statically checked. In particular:

- *Bindings.* The inference construct $[m_1 \dots m_n \vdash_r x]A_1$ binds the variable x in A_1. The prefix construct $c?x.A$ binds the variable x in A. We assume that each variable may be bound at most once.
- *Agent's knowledge.* For every sequential agent A_ϕ, we require that all the closed messages and free variables (i.e., not bound) that appear in the term A belong to its knowledge ϕ. A sequential agent is said to be closed when ϕ only consists of closed messages.

2.3 Symbolic Operational Semantics

In the usual semantics, variables store values. However, we can imagine that variables stores sets of possible values. We need only to find a suitable representation for such a set of values. We consider a language for describing set of substitutions, where a substitution records for each variable the message it stores. Thus, a set of substitutions denote for each variable a set of possible values.

For instance, we can denote that a variable x may be the application of the message constructor F to every message in $Msgs$ by simply requiring $x \in F(y)$. Indeed, variable y is not constrained and so its value can range over all possible messages. We can also join two constraints: e.g., $y \in m, x \in F(y)$ (where m is closed) simply represents that x must be equal to $F(m)$ and y to m.

During the analysis of a security protocol is also useful to be able to express that a variable may store whatever message one can deduce from a certain set of messages. This enable us to model what an intruder may send to the system under attack by using the set of messages it has already acquired during the computation, i.e. its knowledge ϕ. Indeed, we use $x \in \mathsf{D}^R(\{t_1, \dots, t_n\})$ to express that x can be deduced by applying the rules in R from the messages $\{t_1, \dots, t_n\}$. Such contraints on variables may be extended to constraints on terms, i.e. we may require that $F(x) \in \mathsf{D}^R(\{t_1, \dots, t_n\}), G(z) \in y$ and so on.

The lists of constraints on the set of possible substitutions, ranged over by $M, N, \dots$, may be defined as follow:

$$M ::= C, M \mid \epsilon \;\text{ with }\; C ::= \mathbf{false} \mid \mathbf{true} \mid t \in t' \mid t \in \mathsf{D}^R(\{t_1, \dots, t_n\})$$

where $t, t', t_1, \dots, t_n$ are messages (possibly open). Lists of constraints are used to represents (possibly infinite) sets of substitutions from variables to closed messages. Let *Substs* be such set of substitutions. The semantics $[\![M]\!]$ of list of constraints M is defined as follows:

$$[\![\epsilon]\!] \doteq \mathit{Substs}, \quad [\![C, M]\!] \doteq [\![C]\!] \cap [\![M]\!], \quad [\![\mathbf{true}]\!] \doteq \mathit{Substs}, \quad [\![\mathbf{false}]\!] \doteq \emptyset$$
$$[\![t \in t']\!] \doteq \{\sigma \mid t\sigma \in \{t'\sigma\}\}$$
$$[\![t \in (\mathsf{D}^R(\{t_1, \dots, t_n\}))]\!] \doteq \{\sigma \mid t\sigma \in \mathcal{D}^R(\{t_1\sigma, \dots, t_n\sigma\})\}$$

The symbolic semantics for Crypto-CCS is given in Table 2, where $S \xrightarrow{M:\alpha}_{\mathsf{s}} S'$ means that S may evolve through an action α in S' when the constrains in M are satisfied. The function $constr(r, (m_1, \dots, m_n), x)$, with $r \doteq m'_1 \dots m'_n \vdash m'_0$, defines the list of constraints imposed by the rule r and is defined as $m''_1 \in$

Table 2. Symbolic operational semantics, where the symmetric rules for $|_{1s}, |_{2s}, \backslash_{1s}$ are omitted.

$$(!_s)\,\frac{}{(c!m.A)_\phi \xrightarrow{\mathbf{true}:c!m}_s (A)_\phi} \qquad (?_s)\,\frac{}{(c?x.A)_\phi \xrightarrow{\mathbf{true}:c?x}_s (A)_{\phi\cup\{x\}}}$$

$$(\mathcal{D}_s)\,\frac{(A_1)_{\phi\cup\{x\}} \xrightarrow{M:\alpha}_s (A_1')_{\phi'}}{([m_1 \dots m_n \vdash_r x]A_1)_\phi \xrightarrow{N,M:\alpha}_s (A_1')_{\phi'}} \quad N = \mathrm{constr}(r,(m_1,\dots,m_n),x)$$

$$(|_{1s})\,\frac{S \xrightarrow{M:\alpha}_s S'}{S \,|\, S_1 \xrightarrow{M:\alpha}_s S' \,|\, S_1} \quad (|_{2s})\,\frac{S \xrightarrow{M:c!m}_s S' \quad S_1 \xrightarrow{N:c?x}_s S_1'}{S \,|\, S_1 \xrightarrow{M,N,x\in m:\tau}_s S' \,|\, S_1'} \quad (\backslash_{1s})\,\frac{S \xrightarrow{M:c!m}_s S' \quad c \notin L}{S \backslash L \xrightarrow{M:c!m}_s S' \backslash L}$$

$m_1, \dots, m_n'' \in m_n, x \in m_0''$ where $m_i'' = m_i'\sigma'$ and σ' is a renaming of the variables in the terms m_i with totally fresh ones (we omit here the technical details for the sake of clarity). Given a ground substitution σ, we write $S\sigma$ for the system where each free occurrence of a variable x is replaced with $\sigma(x)$. There is a correspondence between the standard transitions semantics, i.e. $\longrightarrow$ (see [11]) and the symbolic one, i.e. $\longrightarrow_s$. In particular, if there is a substitution σ which enjoys the constraint M and $S \xrightarrow{M:\alpha}_s S'$ then $S\sigma \xrightarrow{\alpha\sigma} S'\sigma$. Similary, if $S\sigma \xrightarrow{a} S'$ then there is $S \xrightarrow{M:\alpha}_s S''$ s.t. $\sigma \in [\![M]\!], a = \alpha\sigma$ and $S'' = S'\sigma$. As a notation we also use $S \overset{M,\gamma}{\mapsto}_s S'$ if γ is a finite sequence of actions $\alpha_i, 1 \le i \le n$ and $M = M_1, \dots, M_n$ is a list of constraints s.t. $S = S_0 \xrightarrow{M_1:\alpha_1}_s \dots \xrightarrow{M_n:\alpha_n}_s S_n = S'$. $Sort(S)$ indicates all the channels syntactically occurring in the term of S.

3 Symbolic Analysis

In order to develop a decidable verification theory, we need to compute when a list of constraints is consistent or not, i.e. there exists a substitution σ which enjoys it. To obtain such result we will assume some restrictions on the class of inference systems. Here we use a terminology similar to [8] about inference systems. Given a well-founded measure on messages, we say that a rule $r \doteq \frac{m_1 \;\dots\; m_n}{m_0}$ is a S-rule (*shrinking* rule), whenever the conclusion has a smaller size than one of the premises (call such premises *principal*); moreover all the variables occurring in the conclusion must be in all the principal premises. The rule r is a G-rule (*growing* rule) whenever the conclusion is larger than each of the premises, and each variable occurring in the premises must also be in the conclusion.

Definition 3.1. *We say that an inference system enjoys a G/S property if it consists only of G-rules and S-rules, moreover whenever a message can be deduced through a S-rule, where one of the principal premises is derived by means of a G-rule, then the same message may be deduced from the premises of the G-rule, by using only G-rules.*

Several of the inference systems used in the literature for describing cryptographic systems enjoy this restriction[1].

The main advantage of considering systems enjoying this restriction is that proofs for messages may have a normal form, either:

- these consist of a sequence of applications of G-rules, or
- these consist of a sequence of applications of G-rules and S-rules, where each S-rules is applied with principal premises either in ϕ or obtained through S-rules.

Indeed, using G-rules for inferring the principal premises of an S-rules, is not useful. Thus, shrinking rules may be significantly applied only to messages in ϕ and to messages obtained by S-rules. However, since the measure for classifying the S-rules is well founded then such a *shrinking* phase eventually terminates.

Proposition 3.2. *If the inference system enjoys the G/S restriction then* $\mathcal{D}^R(\phi)$ *is decidable when* ϕ *is finite.* ■

We start by defining a simple kind of lists of constraints whose consistency may be easily decided.

Normal Forms. We say that a list M of constraints is in *normal form* (*nf* for short) whenever the followings hold:

- the left hand side of each constraint is a variable,
- each variable occurs in the left hand side at most once,
- each deduction which occurs in the right hand side only consists of G-rules.

We say that a list M of constraints is in *simple* normal form (*snf* for short) if it is in normal form and moreover all the constraints are of the form $x \in t$.

Consistency Check. We will be interested in establishing whether or not a list of constraints in normal form is satisfiable. We first study the consistency problem lists in simple normal form. We can define a dependency relation between the variables that occur in lists of constraints in simple normal form: we say $x \leq_M^1 y$ ($x \leq_M^0 y$) whenever there exists in M a constraints like $x \in F(t_1, \dots t_n)$ ($x \in y$) and y occurs in t_i, for $i \in \{1, \dots n\}$. Consider the relation $\leq_M$ which is obtained by the transitive closure of the relation $\leq_M^1 \cup \leq_M^0$ where at least one step of closure is obtained through a pair of the relation $\leq_M^1$. Roughly, $x \leq_M y$ means that $\sigma(y)$ is a proper subterm of $\sigma(x)$, for any $\sigma \in [\![M]\!]$.

We say that a list of constraints is *not contradictory* (*nc* for short) whenever $\leq_M$ is an acyclic relation, i.e. there exists no variable x s.t. $x \leq_M x$. It is not difficult to check that a list of constraints in simple normal form is consistent, i.e. $[\![M]\!] \neq \emptyset$, if and only if M is *nc* and no **false** occurs in M.

[1] It is worthy noticing that in [8] a restriction, called S/G, has been defined. However, this is rather different from ours and it is mainly suitable for deductions of logical formulas which has several specific features w.r.t. message deductions we deal with here.

The unique difference between the normal forms and the simple ones is in the presence of deduction constraints. In the normal form list of constraints we allow only G-rules. Thus, the size of the term denoted by the variable on the left hand side will be larger than the ones of the terms denoted by variables occurring in the deduction constraint. Moreover, each application of a G-rule would grow the set of possible dependencies of the variable in the left hand side. Thus, if it is not possible to construct a substitution σ by non-deterministically choosing a term in the eduction constrain to match with the variable then it definitely not possible to assign $\sigma(x)$ in any other way. Thus, we may define $ct(M)$ that given a list on normal form returns true if and only if M is consistent. If **false** belongs to M then $ct(M)$ is false, otherwise we have:

$$ct(M, x \in \mathsf{D}^G(t'_1, \ldots, t'_n), M_1) \doteq \bigvee_{i \in \{1,\ldots,n\}} ct(M, x \in t'_i, M_1)$$
$$ct(M) \doteq \begin{cases} true & \text{if } M \text{ is in } snf \text{ and } nc \\ false & \text{if } M \text{ is in } snf \text{ and not } nc \end{cases}$$

Covering. We say that a set of lists of constraints $\{M_1, \ldots, M_l\}$ is a covering for a list M whenever $[\![M]\!] = \bigcup_{i \in \{1,\ldots,l\}} [\![M_i]\!]$. The notion of covering naturally extends to sets of lists of constraints.

We can define a procedure nfc (see Table 3) that given a list of constraints returns a normal form covering of such list, under certain assumptions. Together with the function that checks the consistency of lists of constraints in normal form, nfc gives a procedure for checking the consistency of the lists of constraints that will be generated during our security analysis. The nfc procedure performs basic transformations yet preserving the covering. It assumes that the list of constraints is monotonic, i.e. if $M = M_1, t \in \mathsf{D}^R(T), M_2, t' \in \mathsf{D}^R(T'), M_3$ then $T \subseteq T'$; moreover, it assumes that M is also well defined, i.e. if $M = M_1, x \in t(x \in D^R(T)), M_2$ then the variables in t (T) occurs in the left hand side of a constraint in M_1. (The class of constraints generated during security analysis is of this kind.) Note that in a call $nfc(C; C')$ C is already in nf. The most interesting cases are the ones which deal with deduction constraints.

- The case $nfc(C; F(t_1, \ldots, t_n) \in \mathsf{D}^G(T'), C_1)$ is justified from the fact that a term like $F(t_1, \ldots, t_n)$ may be only generated by applying one of the G-rules or it is directly in the set of terms of the deduction constraint.
- A constraint like form $x \in \mathsf{D}^G(T), x \in \mathsf{D}^G(T')$ may be expressed as $x \in \mathsf{D}^G(T \cap T')$ when $T \subseteq T'$ or $T' \subseteq T$. Indeed, the previous constraint requires that x belongs to the intersection of $\mathsf{D}^G(T)$ and $\mathsf{D}^G(T')$. (Recall that we consider monotonic[2] lists of constraints.)
- The case about general deductions is justified by the Prop. 3.2. Indeed, if $t \in \mathsf{D}^R(\{t_1, \ldots, t_n\})$ then we may have either

[2] Monotonicity assumption may be dropped by imposing another restriction on the inference systems, i.e. that each message may unify with the conclusion of at most one G-rule. In this case the first rule for $x \in \mathsf{D}^G(T)$ becomes: $nfc(C, x \in \mathsf{D}^G(T), C'; x \in \mathsf{D}^G(T'), C_1) \doteq \bigcup_{\{t_1..t_l\} \subseteq T, \{t'_1..t'_k\} \subseteq T'} nfc(C, x \in \mathsf{D}^G(\{t_1, .., t_l, .., t'_1, .., t'_k\}), C'; t_1 \in \mathsf{D}^G(T'), .., t_l \in \mathsf{D}^G(T'), t'_1 \in \mathsf{D}^G(T), .., t'_k \in \mathsf{D}^G(T), C_1)$.

Table 3. Normal form covering procedure (the order of applications matters).

$$
\begin{array}{ll}
nfc(C;) & \doteq \{C\} \\
nfc(C;\mathbf{true},C_1) & \doteq nfc(C;C_1) \\
nfc(C;\mathbf{false},C_1) & \doteq \{\mathbf{false}\} \\
nfc(C;t \in t,C_1) & \doteq nfc(C;C_1) \\
nfc(C,x \in \mathsf{D}^G(T),C';x \in t,C_1) & \doteq nfc(C,C';x \in t,x \in \mathsf{D}^G(T),C_1) \\
nfc(C;x \in t,C_1) & \doteq \begin{cases} \{\mathbf{false}\} & x \text{ occurs in } t \\ nfc(C[t/x],x \in t;C_1[t/x]) & \text{otherwise} \end{cases} \\
nfc(C;t \in x,C_1) & \doteq nfc(C;x \in t,C_1) \\
nfc(C;F(t_1,\ldots,t_m) \in G(t'_1,\ldots,t'_n),C_1) & \doteq \{\mathbf{false}\} \\
nfc(C;F(t_1,\ldots,t_n) \in F(t'_1,\ldots,t'_n),C_1) & \doteq nfc(C;t_1 \in t'_1,\ldots,t_n \in t'_n,C_1) \\
nfc(C;F(t_1,\ldots,t_n) \in \mathsf{D}^G(T'),C_1) & \doteq \bigcup_{t' \in T'} nfc(C;F(t_1,\ldots,t_n) \in t'_i,C_1) \cup \\
 & \quad \cup_{C' \in G\text{-steps }(T,F(t_1,\ldots,t_n))} nfc(C;C',C_1) \\
nfc(C,x \in \mathsf{D}^G(T'),C';x \in \mathsf{D}^G(T''),C_1) & \doteq nfc(C,x \in \mathsf{D}^G(T' \cap T''),C';C_1) \\
nfc(C;x \in \mathsf{D}^G(T'),C_1) & \doteq nfc(C,x \in \mathsf{D}^G(T');C_1) \\
nfc(C;t \in \mathsf{D}^{R,app}(T),C_1) & \doteq \bigcup_{(C_2) \in S\text{-steps}(T,t,app,C_1)} nfc(C;C_2) \\
 & \quad \cup nfc((C;t \in \mathsf{D}^G(T)))
\end{array}
$$

where:

G-steps$(T,t) = \{m_0 \in t, m_1 \in \mathsf{D}^G(T), .., m_k \in \mathsf{D}^G(T) \mid r \in G\text{-rules}, r = m_1..m_k \vdash m_0\}$

S-steps$(T,t,app) = \{(C', t \in \mathsf{D}^{R,app \cup \{(r,t')\}}(T \cup \{m_0\}), C'_1) \mid r \in S\text{-rules}, t' \in T,$
$(r,t') \notin app, (t', D^G(T'))$ not in $C, r = m_1..m_p..m_k \vdash m_0, C' = m_p \in t', m_i \in \mathsf{D}^G(T),$
C'_1 is C_1 with each $t_1 \in \mathsf{D}^{R,app_1}(T'')$ replaced with $t_1 \in \mathsf{D}^{R,app_1 \cup \{(r,t')\}}(T'' \cup \{m_0\})\}$

- $t \in \mathsf{D}^G(\{t_1,\ldots,t_n\})$, or
- there is a shrinking rule $r = m_1 \ldots m_n \vdash_r m_0$ s.t. $m_p \in t_l$ (with $p \in \{1,\ldots,n\}$) m_p principal premise of r and $m_i \in \mathsf{D}^G(\{t_1,\ldots,t_n\})$, where $i \in \{1,\ldots n\} \setminus \{p\}$. We decorate deductions with a set of pairs (rule, term) which basically record which rule have been already applied with a given term as principal premise. Note that applying twice the same rule with the same principal premise is not useful (by assuming that each S-rule has exactly one principal premise).

Lemma 3.3. *We have that $nfc(;M)$ is a covering in normal form of M, i.e., $[\![M]\!] = \cup_{M' \in nfc(;M)} [\![M']\!]$ provided that M is monotonic and well defined.* ■

4 Security Analysis: The Most General Intruder Approach

In this section, we extend our language to include a special agent, the so-called most general intruder. This agent is used to model every possible attack that can be performed by an agent that can be expressed as a Crypto-CCS term. This allow us to perform the so-called secrecy analysis of Dolev-Yao[3]. We simply consider a family of constants $Top^{L,n}$, where n is an integer and L is a set of channels where Top may communicate. The symbolic semantics of such terms is the following:

$$\frac{c \in L}{(Top^{L,n})_\phi \stackrel{x_n \in \mathsf{D}^R(\phi):c!x_n}{\longrightarrow_{\mathsf{s}}} (Top^{L,n+1})_\phi} \quad \frac{c \in L}{(Top^{L,n})_\phi \stackrel{\mathbf{true}:c?x_n}{\longrightarrow_{\mathsf{s}}} (Top^{L,n+1})_{\phi \cup \{x_n\}}}$$

The Top process may receive every message and store it in its knowledge or send a message which may denote every possible message deducible from its knowledge. (The standard operational semantics is similar.) In the Dolev-Yao model, a message is said to remain a secret among the agent of a system S, whenever this cannot be acquired by an agent that is supposed to have a certain initial knowledge, say ϕ. The usual verification scenario is the following: we check whether for each possible agent X_ϕ and for each possible γ s.t. $(S \mid X_\phi) \setminus L \stackrel{M,\gamma}{\mapsto_{\mathsf{s}}} (S' \mid X'_{\phi'}) \setminus L$ we have that $[\![M, m \in \mathsf{D}^R(\phi')]\!] = \emptyset$, i.e. m is never discovered by the intruder. The set of actions L represents the set of all channels where S can communicate. This means that the system S can only interact with the intruder.

Using the Top process we may avoid the necessity of considering the quantification over all possible terms.

Proposition 4.1. *Given a closed system S, consider a finite set of closed messages $\phi \cup \{m\}$. There exists a term X_ϕ (with $Sort(X) \subseteq L$) s.t. $(S \mid X_\phi) \setminus L \stackrel{M:\gamma}{\mapsto_{\mathsf{s}}} (S' \mid X'_{\phi'}) \setminus L$ and $[\![M, m \in \mathsf{D}^R(\phi')]\!] \neq \emptyset$ if and only if $(S \mid Top_\phi^{L,0}) \setminus L \stackrel{M':\gamma'}{\mapsto_{\mathsf{s}}} (S' \mid Top_{\phi'}^{L,n'}) \setminus L$ and $[\![M', m \in \mathsf{D}^R(\phi')]\!] \neq \emptyset$.* ■

Moreover, the possible symbolic computations $(S \mid Top^{L,0}) \setminus L \stackrel{M:\gamma}{\mapsto_{\mathsf{s}}}$ are in a bounded number. This is due to the restriction on the channels in L which forces Top to interact with S and the fact S may perform only finite computations. The list of constraints obtained during the computation is monotonic and well defined. This makes decidable the secrecy analysis for finite Crypto-CCS terms, since the consistency of such lists of constraints is decidable. As a matter of fact, for each symbolic computation $(S \mid Top_\phi^0) \setminus L \stackrel{M:\gamma}{\mapsto_{\mathsf{s}}} (S' \mid Top_{\phi'}^{n'}) \setminus L$ we only need to check the consistency of $M, m \in \mathsf{D}^R(\phi')$.

Example 4.2. Consider a protocol where an user A sends a message m encrypted with a key k known only by A and B. The protocol could be described by the Crypto-CCS term $A \mid B$, where $A \doteq [m \quad k \vdash_{enc} z]c!z.\mathbf{0}$ and $B \doteq c?y[y \quad k \vdash_{dec} z].\mathbf{0}$. Suppose an intruder that initially knows nothing, i.e. its initial knowledge ϕ is the empty-set $\emptyset$. For the sake of brevity we inspect only one of the symbolic computations of $((A \mid B) \mid Top^0) \setminus \{c\}$, in particular we analyze: $((A \mid B) \mid Top^0) \setminus \{c\} \stackrel{z \in E(m,k):\tau}{\longrightarrow_{\mathsf{s}}} ((\mathbf{0} \mid B) \mid Top^1_{\{z\}}) \setminus L$ which derives from a communication of A with the intruder (actually, this may be seen as the intruder which tries to

[3] The interested reader is referred to [12] for other security properties that can be reduced to the inspection of the intruder knowledge.

impersonate the user B). Now, we need to check the consistency of the list of constraints $C = z \in E(m,k), m \in \mathsf{D}^R(\{z\})$. We can apply the nfc procedure to obtain a normal form covering, we give here a piece of the computation:

$$\begin{array}{c} nfc(; z \in E(m,k), m \in \mathsf{D}^R(\{z\})) = \\ nfc(z \in E(m,k); m \in \mathsf{D}^R(\{E(m,k)\})) = \ldots \\ nfc(z \in E(m,k); m \in \mathsf{D}^G(\{E(m,k)\})) = \ldots \\ nfc(z \in E(m,k); m \in E(m,k)) = \{\mathbf{false}\} \end{array}$$

After checking the whole computation, we find that $nfc(; C) = \{\mathbf{false}\}$. Thus, as expected, the intruder is not able to discover the exchanged message m because it does not know the decryption key. ∎

Acknowledgments

We would like to thank the anonymous referees for their helpful comments.

References

[1] M. Abadi and A. D. Gordon. A calculus for cryptographic protocols: The spi calculus. *Information and Computation*, 148(1):1–70, 1999.

[2] R. M. Amadio and D. Lugiez. On the reachability problem in cryptographic protocols. In Proc. of CONCUR, volume 1877 of LNCS, pages 380–394, 2000.

[3] B. Blanchet. An efficient cryptographic protocol verifier based on prolog rules. In Proc. of CSFW, IEEE Press, pages 82–96, 2001.

[4] M. Boreale. Symbolic trace analysis of cryptographic protocols. In Proc. of ICALP, LNCS, pages 667–681, 2001.

[5] M. Fiore and M. Abadi. Computing symbolic models for verifying cryptographic protocols. In Proc. of CSFW, IEEE Press, pages 160–173, 2001.

[6] R. Focardi and F. Martinelli. A uniform approach for the definition of security properties. In Proc. of FM'99, volume 1708 of *LNCS*, pages 794–813, 1999.

[7] A. Ingólfsdóttir and H. Lin. A symbolic approach to value-passing processes. In J. Bergstra, A. Ponse, and S. Smolka, editors, *Handbook of Process Algebra*, pages 427–478. North-Holland, 2001.

[8] D. Kindred and J. M. Wing. Fast, automatic checking of security protocols. In Proc. of *Second USENIX Workshop on Electronic Commerce*, pages 41–52, Oakland, California, 1996.

[9] G. Leduc and F. Germeau. Verification of security protocols using LOTOS - method and application. *Computer Communications*, 23(12):1089–1103, 2000.

[10] G. Lowe. Breaking and fixing the Needham Schroeder public-key protocol using FDR. In Proc. of TACAS, volume 1055 of LNCS, pages 147–166, 1996.

[11] F. Martinelli. Analysis of security protocols as open systems. Tech. Rep. IAT-B4-2001-06. To appear on TCS.

[12] F. Martinelli. Encoding several authentication properties as properties of the intruder's knowledge. Tech. Rep. IAT-B4-2001-20. Submitted for publication.

[13] F. Martinelli. *Formal Methods for the Analysis of Open Systems with Applications to Security Properties.* PhD thesis, University of Siena, Dec. 1998.

[14] F. Martinelli. Languages for description and analysis of authentication protocols. In Proc. of ICTCS, pages 304–315. World Scientific, 1998.

[15] J. Millen and V. Shmatikov. Constraint solving for bounded-process cryptographic protocol analysis. In Proc. of CCS, ACM Press, 2001.

[16] R. Milner. *Communication and Concurrency.* International Series in Computer Science. Prentice Hall, 1989.

[17] Thayer, Herzog, and Guttman. Honest ideals on strand spaces. In Proc. of CSFW. IEEE Computer Society Press, 1998.

The Complexity of Tree Multicolorings

Dániel Marx*

Dept. of Computer Science and Information Theory,
Budapest University of Technology and Economics,
dmarx@cs.bme.hu

Abstract. The multicoloring problem is that given a graph G and integer demands $x(v)$ for every vertex v, assign a set of $x(v)$ colors to vertex v, such that neighboring vertices have disjoint sets of colors. In the *preemptive sum multicoloring problem* the *finish time* of a vertex is defined to be the highest color assigned to it. The goal is to minimize the sum of the finish times. The study of this problem is motivated by applications in scheduling. Answering a question of Halldórsson et al. [4], we show that the problem is strongly **NP**-hard in binary trees. As a first step toward this result we prove that list multicoloring of binary trees is **NP**-complete.

1 Introduction

Graph multicoloring problems are often used to model scheduling of dependent jobs. Given a set of jobs, one has to assign a set of time slots to every job. The constraints are the following: every job has a length, which is the number of time slots it requires, and there are interfering pairs of jobs which cannot be active in the same time slot. In the *preemptive* scheduling model it is assumed that the jobs can be interrupted arbitrarily, the time slots assigned to a job do not have to be consecutive. This scheduling problem can be translated into a multicoloring problem on graphs as follows. The vertices of a graph correspond to the jobs and two jobs are connected if they cannot be executed at the same time. The colors correspond to the time slots and every vertex has a color requirement $x(v)$, which is the length of the job. In a multicoloring $x(v)$ colors have to be assigned to every vertex v such that neighboring vertices have disjoint sets of colors. Clearly, there is one to one correspondence between the feasible preemptive schedulings of the jobs and the feasible multicolorings of the graph.

One traditional optimization goal is to minimize the total completion time (makespan) of the scheduling, that is, the highest color assigned to the vertices (or, equivalently, the total number of different colors assigned). This problem is called *multicoloring* or *weighted coloring*. Another well-studied optimization goal is to minimize the average completion time of the jobs, which is the same as to minimize the sum of the completion times. This problem, *preemptive minimum sum multicoloring*, will be studied in this paper. It can be stated formally as follows:

* Research supported by grant OTKA 30122 of the Hungarian National Science Fund.

K. Diks et al. (Eds): MFSC 2002, LNCS 2420, pp. 532–542, 2002.

Preemptive Sum Multicoloring (pSMC)
Input: A graph $G(V, E)$ and a *demand function* x: $V \to \mathbb{N}$
Output: A *multicoloring* Ψ: $V \to 2^{\mathbb{N}}$ such that $|\Psi(v)| = x(v)$ for every $v \in V$, and $\Psi(u) \cap \Psi(v) = \emptyset$ if u and v are neighbors in G. *Goal:* Let the *finish time* of vertex v in coloring Ψ be the highest color assigned to it, $f_\Psi(v) = \max\{i \in \Psi(v)\}$. The goal is to minimize $\sum_{v \in V} f_\Psi(v)$, the *sum* of the coloring Ψ.

If every demand is 1, i.e., $x(v) \equiv 1$, then we obtain the *chromatic sum* problem as a special case. The study of chromatic sums were started in [9,11,10]. The complexity and approximability of the chromatic sum in certain restricted classes of graphs were investigated in several papers [2,6,12,13].

Approximation results for arbitrary demand function $x(v)$ on general and k-colorable graphs were given by Bar-Noy et al. [1]. A polynomial time approximation scheme for preemptive minimum sum multicoloring is known for trees [4], for partial k-trees and planar graphs [3]. In [4] it is shown that the problem can be solved optimally in polynomial time in trees if every demand is bounded by a fixed constant. However, in general, the complexity of the problem in trees (and in paths) remained an open question. The main result of the paper is to show that the problem is **NP**-hard on binary trees, even if every demand is polynomially bounded. As a first step, we also prove the **NP**-completeness of another variant of multicoloring, the so-called list multicoloring.

In Section 2, we introduce some notations and present the result on list multicoloring. Section 3 defines penalty gadgets, which are the most important tools of the reduction in Section 4.

2 Preliminaries

We slightly extend the problem by allowing $x(v) = 0$. Clearly this does not make the problem more difficult, but it will be needed for technical reasons. If $x(v) = 0$, then define $f_\Psi(v) = 0$ in every coloring Ψ. Notice that by using this definition the trivial inequality $f_\Psi(v) \geq x(v)$ holds even if $x(v) = 0$.

Let us introduce some notations. If $V' \subseteq V$ and Ψ is a coloring then let $f_\Psi(V') = \sum_{v \in V'} f_\Psi(v)$. Similarly, $x(V') = \sum_{v \in V'} x(v)$. The sum of the optimum coloring of (G, x) is denoted by $\mathrm{OPT}(G, x)$, or by $\mathrm{OPT}(G)$ if the function $x(v)$ is clear from the context. The notation $[a, b]$ stands for the set $\{a, a+1, \dots, b\}$ if $a \leq b$, otherwise it is the empty set.

The size of the input to the multicoloring problem is the size of the graph, and it does not include the size of the demand function.

Instead of the preemptive sum multicoloring problem, we start with the **NP**-completeness of another multicoloring problem. The following is the obvious common generalization of list coloring and multicoloring (for a thorough overview on list coloring and related problems, see [14]):

List Multicoloring
Input: A graph $G(V, E)$, a *demand function* x: $V \to \mathbb{N}$, a set of colors C and a *color list* L: $V \to 2^C$ for each vertex
Question: Is there a *multicoloring* Ψ: $V \to 2^C$ such that $|\Psi(v)| = x(v)$, $\Psi(v) \subseteq L(v)$ for every $v \in V$, and $\Psi(u) \cap \Psi(v) = \emptyset$ if u and v are neighbors in G?

Clearly, this problem is **NP**-complete in every class of graphs where either multicoloring or list coloring is **NP**-complete. List coloring is **NP**-complete in bipartite graphs [5,8], but both problems can be solved in polynomial time in trees (see [7] for a linear time list coloring algorithm in trees). On the other hand, list multicoloring of trees is **NP**-complete:

Theorem 2.1. *The list multicoloring problem remains* **NP***-complete restricted to trees.*

Proof. The reduction is from the maximum independent set problem. For every graph $G(V, E)$ and integer k, we will construct a tree T (in fact, a star), a demand function, and a color list for each node, such that the tree can be colored with the lists if and only if G has an independent set of size k. The colors correspond to the vertices of G, the leaves of the star correspond to the edges of G. The construction will ensure that the colors given to the central node correspond to an independent set in G.

Let $e_1, e_2, \ldots, e_m$ be the edges of G and denote by $u_{i,1}$ and $u_{i,2}$ the two end vertices of edge e_i. The tree T is a star with a central node v and m leaves $v_1, \ldots, v_m$. The demand of v is k and the demand of every leaf is 1. The set of colors C corresponds to the vertex set V. The color list of the central node v is the set C, the list of node v_i is the set $\{u_{i,1}, u_{i,2}\}$.

Assume that there is a proper list coloring of T. It assigns k colors to v. The corresponding set of k vertices will be independent in G: at least one end vertex of each edge e_i is not contained in this set since node v_i must be colored with either $u_{i,1}$ or $u_{i,2}$. On the other hand, if there is an independent set of size k in G, then we can assign this k colors to v and extend the coloring to the nodes v_i: either $u_{i,1}$ or $u_{i,2}$ is not contained in the independent set, thus it can be assigned to v_i. □

There are two main difficulties in adapting these ideas for the minimum sum coloring problem.

- We want to prove **NP**-completeness in binary trees. The central node of the star has high degree.
- There are no lists in the minimum sum coloring problem. How can we forbid a node from using certain colors?

The first problem can be solved quite easily with a 'color copying' trick. To demonstrate this, we present a stronger form of Theorem 2.1:

Theorem 2.2. *The list multicoloring problem remains* **NP***-complete restricted to binary trees.*

Proof. The proof is essentially the same as in Theorem 2.1, but the degree m central node of the star is replaced by a path $v'_1, v'_2, \ldots, v'_{2m-1}$ of $2m-1$ nodes. The m neighbors of v are connected to the m nodes $v'_1, v'_3, \ldots, v'_{2m-1}$ one by one. The list of every node v'_i is C, the demands are $x(v'_{2i+1}) = k$ and $x(v'_{2i}) = |C| - k$. It is easy to see that in every proper multicoloring of the tree, the nodes $v'_1, v'_3, \ldots, v'_{2m-1}$ receive the same set of k colors. Furthermore, as in the previous proof, this set corresponds to an independent set in G.

□

To solve the second problem, certain 'penalty gadgets' will be constructed, Section 3 is devoted to this task.

3 The Penalty Gadgets

The goal of the penalty gadgets is that by connecting such a gadget to a node v, we can force v not to use certain colors: if node v uses a forbidden color, then the gadget can be colored only with a 'very large' penalty.

For offset t, demand size d and penalty C we define a tree $T_{t,d,C}$. The root r of this tree will be connected to some node v. When the root r of this tree uses the set $[t+1, t+d]$, then the tree can be colored optimally. On the other hand, if v uses even one color from $[t+1, t+d]$, then r cannot have the set $[t+1, t+d]$ and so $f_\Psi(T_{t,d,C}) \geq OPT(T_{t,d,C}, x) + C$. When C is sufficiently large, then this will force v to use colors not in $[t+1, t+d]$.

Proposition 3.1. *For integers $d, C > t \geq 0$ there is a binary tree $T_{t,d,C}$ and a demand function $x(v)$ such that*

1. *The root r has demand $x(r) = d$.*
2. *$\Psi(r) = [t+1, t+d]$ in every optimum coloring Ψ.*
3. *If $\Psi(r) \neq [t+1, t+d]$ for a coloring Ψ, then $f_\Psi(T_{t,d,C}) \geq OPT(T_{t,d,C}, x) + C$.*
4. *The demand x of every vertex is polynomially bounded by d and C.*

Furthermore, there is an algorithm which, given t, d and C, outputs the tree $T_{t,d,C}$, the demand function x and the value $OPT(T_{t,d,C}, x)$ in time polynomial in d and C.

Proof. Let $k = \lceil \log_2(C+t) \rceil$ and $\widehat{C} = 2^k$. Obviously, $C + t \leq \widehat{C} < 2(C+t)$. The tree $T_{t,d,C}$ consists of a complete binary tree and some attached paths. The complete binary tree T_0 has $k+1$ levels, the root r is on level 1 and the leaves, $\ell_1, \ell_2, \ldots, \ell_{\widehat{C}}$, are on level $k+1$. Attach a path of $k+3$ nodes to every leaf: node ℓ_i $(1 \leq i \leq \widehat{C})$ is connected to path P_i: $a_{i,k+2}, a_{i,k+1}, \ldots, a_{i,2}, a_{i,1}, a_{i,0}$ (nodes ℓ_i and $a_{i,k+2}$ are neighbors). Figure 1 shows the construction for $t=2$, $d=4$, $C=6$. Clearly, $T_{t,d,C}$ has $2\widehat{C} - 1 + (k+3)\widehat{C}$ nodes, which is polynomially bounded in C.

We say that a node is of type j if it is either on the jth level of T_0 or it is an $a_{i,j}$ for some $1 \leq i \leq \widehat{C}$.

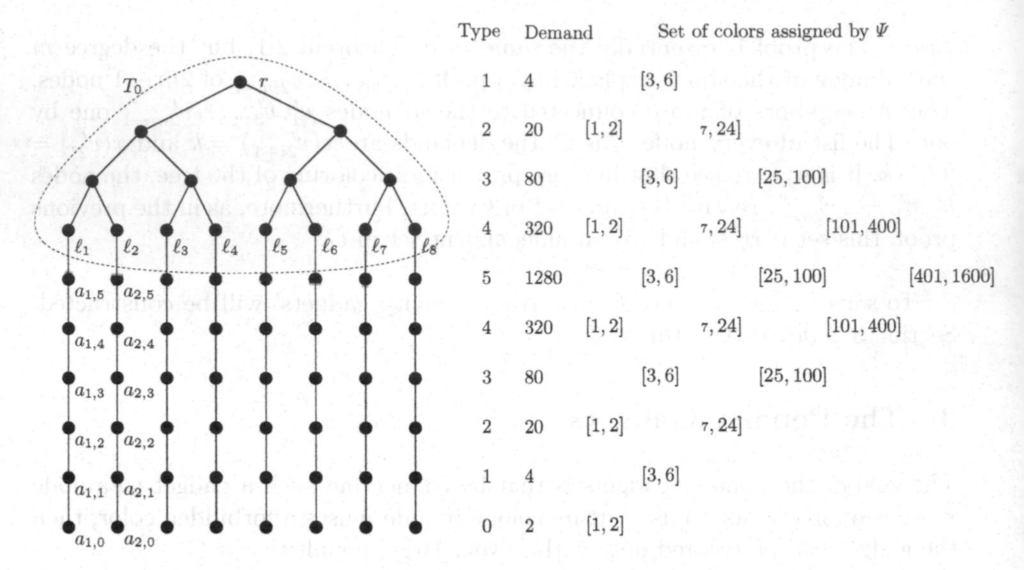

Fig. 1. The tree $T_{t,d,C}$ for $t = 2$, $d = 4$ and $C = 6$. The nodes on the same level have the same type. On the right are the demands and also the colors assigned by the optimum coloring.

The demand $x(v)$ will depend only on the type of node v. Let

$$g(0) = t,$$
$$g(1) = d,$$
$$g(n) = (3d + t + C) \cdot 4^{n-2} \text{ for } n \geq 2.$$

Obviously, $g(n)$ is monotone and it is easy to see that

$$g(i+1) \geq 3g(i) + C + t \geq g(i-1) + C + t$$

for all $i \geq 1$ (these inequalities will be used later).

For a node v of type i let $x(v) = g(i)$. This implies that $x(r) = g(1) = d$ for the root r. The maximum value of $x(v)$ is $g(k+2) = (3d + t + C) \cdot 4^k$, which is bounded by a polynomial of d and C.

We describe a proper multicoloring Ψ, which will turn out to be the unique optimum solution. The same color set is assigned to the nodes of the same type. Start with $\Psi(v) = [1, t]$ for every node v of type 0. Then color the different types in increasing order: assign to the nodes of type i the first $g(i)$ colors not used by the type $i-1$ nodes. This gives a proper coloring since the already colored neighbors of type i nodes are type $i-1$ nodes. Notice that the root r receives the set $[t+1, t+d]$, as required. It is easy to prove that the finish time of a node v of type i is $f_\Psi(v) = g(i) + g(i-1) = x(i) + x(i-1)$ since there will be exactly $g(i-1)$ 'skipped' colors and the finish time of nodes of type i is greater then the finish time of the nodes of type $i-1$ because $g(i) > g(i-2)$. The following simple observation will be used later: if u is a type i node and v is its type $i+1$ neighbor, then in every coloring Φ, the equalities $\Phi(u) = \Psi(u)$ and $f_\Phi(v) = f_\Psi(v) = g(i+1) + g(i)$ imply $\Phi(v) = \Psi(v)$. This follows directly from the definition of Ψ: there is just one way of choosing the first $x(v) = g(i+1)$ colors not used by u.

The following three lemmas show that Ψ is an optimum coloring, and if a coloring Φ assigns to r a set different from $\Psi(r) = [t+1, t+d]$, then $f_\Phi(T_{t,d,C}) \geq f_\Psi(T_{t,d,C}) + C$.

Lemma 3.2. *(a) $f_\Phi(T_0) \geq f_\Psi(T_0) - t$ holds for every coloring Φ of $(T_{t,d,C}, x)$.*
(b) If $\Phi(r) = \Psi(r)$, then $f_\Phi(T_0) \geq f_\Psi(T_0)$.
(c) If there is a $v \in T_0 \setminus \{r\}$ such that $f_\Phi(v) < f_\Psi(v)$, then $f_\Phi(T_0) \geq f_\Psi(T_0) + C$.

Proof. Let $L = \{v \in T_0 : f_\Phi(v) < f_\Psi(v)\}$ and let $H = T_0 \setminus L$ be its complement in T_0. We note that L is an independent set. To see this, let v and u be neighbors of type i and $i+1$, respectively. The sum of their demand is $g(i) + g(i+1)$, thus at least one of them must have finish time not smaller than $g(i) + g(i+1)$. Clearly this makes it impossible to have $f_\Phi(v) < f_\Psi(v) = g(i) + g(i-1)$ and $f_\Phi(u) < f_\Psi(u) = g(i) + g(i+1)$ simultaneously.

Partition the vertices of T_0 as follows. Define a subset S_v for every node $v \in H$. Let $v \in S_v$ for every $v \in H$, and $u \in L$ is in S_v iff v is the parent of u. When the root r is in L then r forms a set itself, $S^* = \{r\}$. It is clear that this defines a partition, every vertex is in exactly one subset. Apart from S^*, every subset contains a node from H and zero, one or two nodes from L.

Assume that the set S_v contains no node from L. Then $f_\Phi(S_v) \geq f_\Psi(S_v)$ follows from the definition of H and L. Now consider a set S_v which has at least one node from L. It contains a type i node v from H and one or two type $i+1$ nodes (u_1, u_2) from L. Since v and u_z $(z = 1, 2)$ are neighbors and the sum of their demand is $g(i) + g(i+1)$, at least one of them must have finish time at least $g(i) + g(i+1)$. Since u_z is in L, we have $f_\Phi(u_z) < f_\Psi(u_z) = g(i) + g(i+1)$, thus $f_\Phi(v) \geq g(i) + g(i+1)$. Therefore, $f_\Phi(v) - f_\Psi(v) \geq (g(i) + g(i+1)) - (g(i-1) + g(i)) = g(i+1) - g(i-1)$. Since $f_\Psi(u_z) = g(i+1) + g(i)$ and $x(u_z) = g(i+1)$, clearly $f_\Phi(u_z) - f_\Psi(u_z) \geq -g(i)$. Now

$$f_\Phi(S_v) - f_\Psi(S_v) \geq (g(i+1) - g(i-1)) - 2g(i) \geq g(i+1) - 3g(i) \geq C + t,$$

where the last inequality follows from $g(i+1) \geq 3g(i) + C + t$.

If r is in S^*, then $f_\Phi(S^*) = f_\Psi(S^*) + (f_\Phi(r) - f_\Psi(r))$ holds. Therefore $f_\Phi(T_0) \geq f_\Psi(T_0) + (f_\Phi(r) - f_\Psi(r)) \geq f_\Psi(T_0) - t$, since $f_\Phi(r) \geq d$. This proves statement (a), and (b) also follows because $\Phi(r) = \Psi(r)$ implies $f_\Phi(r) - f_\Psi(r) = 0$. Furthermore, if $f_\Phi(u) < f_\Psi(u)$ for some $u \in T_0 \setminus \{r\}$, then $f_\Phi(S_v) \geq f_\Psi(S_v) + C + t$ for the set S_v of the partition that contains u. This proves statement (c). □

Lemma 3.3. *$f_\Phi(P_i) > f_\Psi(P_i)$ holds for every coloring $\Phi \neq \Psi$ of $T_{t,d,C}$ and for every $1 \leq i \leq \widehat{C}$.*

Proof. Assume that $f_\Phi(P_i) \leq f_\Psi(P_i)$, define $L = \{v \in P_i : f_\Phi(v) < f_\Psi(v)\}$ and $H = P_i \setminus L$. If $f_\Phi(P_i) \leq f_\Psi(P_i)$ and Φ is different from Ψ, then there is a $v \in P_i$

such that $f_\Phi(v) < f_\Psi(v)$, thus L is not empty. As in Lemma 3.2, it is easy to see that L is an independent set. The nodes of P_i are partitioned into $|H|$ classes: if $v \in H$ then v is in S_v, if $u \in L$ then u is in S_v, where v is the child of u. Notice that $a_{i,0} \in H$ since $f_\Psi(a_{i,0}) = x(a_{i,0}) = g(0) \leq f_\Phi(a_{i,0})$.

We prove that $f_\Phi(S_v) \geq f_\Psi(S_v)$ for every S_v. If $S_v = \{v\}$, then it is clear that $f_\Phi(S_v) \geq f_\Psi(S_v)$ holds. Assume that $S_v = \{u, v\}$, node $u \in L$ is type $j+1$, and $v \in H$ (its child) is type $j \geq 0$. The finish time of node v is at least $x(u) + x(v) = g(j+1) + g(j)$, therefore

$$f_\Phi(S_v) \geq x(u) + (x(u) + x(v)) = g(j+1) + (g(j+1) + g(j))$$

holds. On the other hand, if $j \geq 1$, then $f_\Psi(S_v) = (g(j+1) + g(j)) + (g(j) + g(j-1))$, thus $f_\Phi(S_v) > f_\Psi(S_v)$ follows from $g(j+1) > g(j) + g(j-1)$. In the case $j = 0$, we have $f_\Psi(S_v) = t + (t+d) = g(j) + (g(j) + g(j+1)) < f_\Phi(S_v)$, since $f_\Phi(S_v) \geq g(j+1) + (g(j) + g(j+1)) = d + (t+d)$ (recall that $t < d$). Since H is not empty, there is at least one subset S_v in the partition with $f_\Phi(S_v) > f_\Psi(S_v)$, contradicting $f_\Phi(P_i) \leq f_\Psi(P_i)$. □

Lemma 3.4. *If $\Phi(r) \neq \Psi(r) = [t+1, t+d]$, then $f_\Phi(T_{t,d,C}) \geq f_\Psi(T_{t,d,C}) + C$.*

Proof. Denote by $P^* = \bigcup_{i=1}^{\widehat{C}} P_i = T_{t,d,C} \setminus T_0$ the union of the paths. If there is a node $v \in T_0 \setminus \{r\}$ with $f_\Phi(v) < f_\Psi(v)$, then by part (c) of Lemma 3.2 $f_\Phi(T_0) \geq f_\Psi(T_0) + C$, and by Lemma 3.3 $f_\Phi(P^*) \geq f_\Psi(P^*)$ follows, which implies $f_\Phi(T_{t,d,C}) \geq f_\Psi(T_{t,d,C}) + C$, and we are ready. Therefore it can be assumed that $f_\Phi(v) \geq f_\Psi(v)$ for every node $v \in T_0 \setminus \{r\}$. Furthermore, if there is a $v \in T_0$ with $f_\Phi(v) \geq f_\Psi(v) + C + t$, then $f_\Phi(T_0) \geq f_\Psi(T_0) + C$, thus $f_\Phi(P^*) \geq f_\Psi(P^*)$ implies $f_\Phi(T_{t,d,C}) \geq f_\Psi(T_{t,d,C}) + C$. In the following, it will be assumed that $f_\Psi(v) \leq f_\Phi(v) \leq f_\Psi(v) + C + t$ holds for every $v \in T_0 \setminus \{r\}$.

Call a vertex v 'changed' in Φ if $\Phi(v) \neq \Psi(v)$. The goal is to show that if the root r is changed, then all the nodes $a_{1,k+2}, a_{2,k+2}, \ldots, a_{\widehat{C},k+2}$ are changed. Let v be a node of type i in T_0 and let u be one of its children, a node of type $i+1$. If v is changed, then there is a color $j \in \Phi(v)$ and $j \notin \Psi(v)$. We consider two cases. If $j \leq f_\Psi(u)$, then by the fact that $j \notin \Psi(v)$ and the way Ψ was defined $j \in \Psi(u)$ follows. Therefore u is also changed since $j \in \Phi(v)$ implies $j \notin \Phi(u)$. In the second case, where $j > f_\Psi(u) = g(i+1) + g(i)$ we have

$$\begin{aligned} f_\Phi(v) \geq j > g(i+1) + g(i) &= (g(i+1) - g(i-1)) + (g(i) + g(i-1)) \\ &= g(i+1) - g(i-1) + f_\Psi(v) \geq f_\Psi(v) + C + t, \end{aligned}$$

contradicting the assumption $f_\Phi(v) \leq f_\Psi(v) + C + t$.

Assume that $f_\Phi(T_{t,d,C}) < f_\Psi(T_{t,d,C}) + C$. By applying the previous result inductively, one finds that all the leaves ℓ_i and their children $a_{i,k+2}$ $(1 \leq i \leq \widehat{C})$ are changed. Lemma 3.3 ensures that Φ is not an optimum coloring of P_i, thus $f_\Phi(P_i) \geq f_\Psi(P_i) + 1$ and $f_\Phi(P^*) \geq f_\Psi(P^*) + \widehat{C} \geq f_\Psi(P^*) + C + t$. By Lemma 3.2, $f_\Phi(T_0) \geq f_\Psi(T_0) - t$, hence $f_\Phi(T_{t,d,C}) \geq f_\Psi(T_{t,d,C}) + C$. □

To prove Prop. 3.1, we have to show that requirements 2 and 3 hold. If $\Phi(r) = \Psi(r)$, then by part (b) of Lemma 3.2 and by Lemma 3.3, $f_\Phi(T_{t,d,C}) \geq f_\Psi(T_{t,d,C})$. If $\Phi(r) \neq \Psi(r)$, then by Lemma 3.4, $f_\Phi(T_{t,d,C}) \geq f_\Psi(T_{t,d,C}) + C$. Therefore the coloring Ψ is an optimum coloring and the tree satisfies the requirements of the proposition.

Clearly, the described tree $T_{t,d,C}$ and the demand function x can be constructed in polynomial time. The sum of the optimum solution can be also calculated, by adding the appropriate finish time of every node. □

4 The Reduction

We will reduce the maximum independent set problem to the minimum sum coloring problem in binary trees. In the decision version of the minimum sum coloring problem, the input is a graph G, a demand function $x(v)$, and an integer K, the question is whether there exists a multicoloring Ψ with sum less than K. The reduction is based on the proof of Theorem 2.2. The penalty gadgets $T_{t,d,C}$ of Section 3 are used to imitate the effect of the color lists.

More precisely, the penalty gadget is used in two different ways: as a lower penalty gadget and as an upper penalty gadget. The *lower penalty gadget* $T^L_{d,C}$ is a tree $T_{0,d,C}$. By connecting the root of such a tree to a node v, the node v is forced to use only colors greater than d: otherwise the gadget can be colored only with a penalty C. A tree will be called a tree of type T^L if it is the tree $T^L_{d,C}$ for some d and C.

The *upper penalty gadget* $T^U_{d,C}$ is a tree $T_{d,C,C}$. If this gadget is connected to a node v, then this forces v to use only colors not greater than d. If v uses only colors not greater than d, then its finish time is at most d, and the gadget can be colored optimally. If v uses a color greater than d but not greater than $d+C$, then the gadget can be colored only with a penalty of C. If v uses colors greater than $d+C$, then it has finish time at least $d+C$, which is a penalty of at least C compared to the case when v uses only colors at most d.

Theorem 4.1. *The minimum sum preemptive multicoloring problem is* **NP**-*complete on binary trees when the value of the demand function is polynomially bounded.*

Proof. Let a graph $G(V, E)$ and an integer k be given. Denote $n = |V|$, $m = |E|$ and let $C = 8mn$. Let integers $u_{i,1} < u_{i,2}$ denote the two end vertices of the ith edge in G.

We define a binary tree T, which consists of a core $\widehat{T}$ and some attached subtrees of type T^L and T^U. We start with a path of $2m-1$ nodes, $a_1, b_1, a_2, b_2, \ldots,$ a_{m-1}, b_{m-1}, a_m. Define $x(a_i) = k$ $(1 \leq i \leq m)$ and $x(b_i) = C + n - k$ $(1 \leq i \leq m-1)$. For every $1 \leq i \leq m$ attach a path of 6 nodes to a_i. Let these nodes be $c_{i,1}, d_{i,1}, c_{i,2}, d_{i,2}, c_{i,3}, d_{i,3}$. Let $x(c_{i,j}) = 1$, $x(d_{i,j}) = C+n-1$ $(j = 1, 2)$ and $x(c_{i,3}) = 1$, $x(d_{i,3}) = u_{i,2} - u_{i,1} - 1$. Clearly, $x(v) \geq 0$ for every node v. This completes the definition of $\widehat{T}$. Now attach trees of type T^L and T^U to $\widehat{T}$ as follows (see Figure 2):

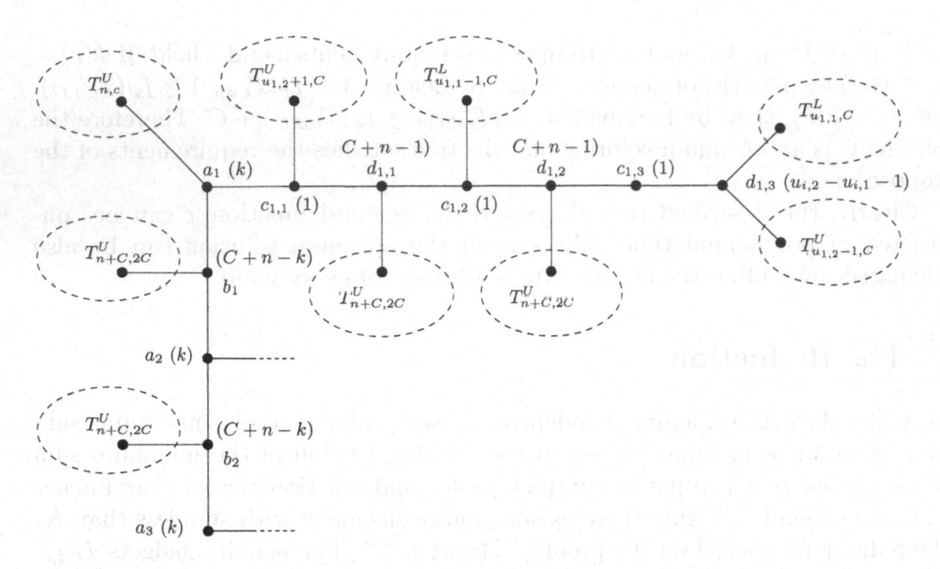

Fig. 2. The tree T for $m = 3$. For the sake of clarity, the nodes $c_{i,j}, d_{i,j}$ for $i \geq 2$ and the subtrees connected to these nodes are omitted. The numbers in parentheses are the demand of the vertices.

– a $T^U_{C+n,2C}$ to every node b_i $(1 \leq i \leq m-1)$,
– a $T^U_{n,C}$ to the node a_1,
– a $T^U_{C+n,2C}$ to every node $d_{i,j}$ $(1 \leq i \leq m,\ j = 1,2)$,
– a $T^U_{u_{i,2}+1,C}$ to every node $c_{i,1}$ $(1 \leq i \leq m)$,
– a $T^L_{u_{i,1}-1,C}$ to every node $c_{i,2}$ $(1 \leq i \leq m)$,
– a $T^L_{u_{i,1},C}$ and a $T^U_{u_{i,2}-1,C}$ to every node $d_{i,3}$ $(1 \leq i \leq m)$.

It is clear that the size of the resulting tree T is polynomial in n, the number of vertices of G, because $\widehat{T}$ has $8m-1$ nodes and we attach $7m$ trees to it, each of size bounded by a polynomial in $C+n$.

As required by Prop. 3.1, the algorithm that constructs the trees of type T^U and T^L also outputs the minimum sum of these $7m$ trees, that is, the value of $\mathrm{OPT}(T \setminus \widehat{T})$. Let $K = \mathrm{OPT}(T \setminus \widehat{T}) + x(\widehat{T}) + C$.

The intuition behind the construction is that in a 'well-behaved' solution, when the coloring of the T^L and T^U trees are optimal, for every i, the three nodes $c_{i,1}, c_{i,2}, c_{i,3}$ have the same color. The trees attached to these nodes ensure that this color must be either $u_{i,1}$ or $u_{i,2}$, one of the end nodes of the ith edge in G. This color cannot appear in a_i, this is the reason why the k colors assigned to the nodes a_i form an independent set, at least one end node of each edge is not in the set.

First we prove that if there is an independent set S of size k, then T can be colored with sum smaller than K. Let $\widehat{u}_i \in \{u_{i,1}, u_{i,2}\}$, $\widehat{u}_i \notin S$ be an end node of the ith edge. Assume that Ψ colors all the trees of type T^U and T^L optimally, i.e., $f_\Psi(T \setminus \widehat{T}) = \mathrm{OPT}(T \setminus \widehat{T})$ and let

- $\Psi(a_i) = S$ $(1 \leq i \leq m)$,
- $\Psi(b_i) = [1, C+n] \setminus S$ $(1 \leq i \leq m-1)$,
- $\Psi(c_{i,j}) = \{\widehat{u}_i\}$ $(1 \leq i \leq m, j = 1, 2, 3)$,
- $\Psi(d_{i,j}) = [1, C+n] \setminus \{\widehat{u}_i\}$ $(1 \leq i \leq m,\ j = 1, 2)$,
- $\Psi(d_{i,3}) = [u_{i,1}+1, u_{i,2}-1]$ $(1 \leq i \leq m)$.

It is straightforward to verify that Ψ is a proper coloring of T. Notice that $f_\Psi(v) \leq x(v) + n$ holds for every node v of $\widehat{T}$, thus $f_\Psi(\widehat{T})$ can be bounded by $x(\widehat{T}) + |\widehat{T}|n$. Therefore $f_\Psi(T) = f_\Psi(T \setminus \widehat{T}) + f_\Psi(\widehat{T}) \leq \mathrm{OPT}(T \setminus \widehat{T}) + x(\widehat{T}) + |\widehat{T}|n = \mathrm{OPT}(T \setminus \widehat{T}) + x(\widehat{T}) + (8m-1)n < \mathrm{OPT}(T \setminus \widehat{T}) + x(\widehat{T}) + C = K$, what we had to show.

To prove the other direction, we will show that when there is a coloring Ψ with sum $f_\Psi(T) < K$, then there is a set of k independent vertices in G. Obviously $f_\Psi(T) = f_\Psi(\widehat{T}) + f_\Psi(T - \widehat{T}) \geq x(\widehat{T}) + \mathrm{OPT}(T \setminus \widehat{T})$. If there is even one node $v \in \widehat{T}$ such that $f_\Psi(v) \geq x(v) + C$, then $f_\Psi(\widehat{T}) \geq x(\widehat{T}) + C$ and $f_\Psi(T) \geq \mathrm{OPT}(T \setminus \widehat{T}) + x(\widehat{T}) + C = K$. Thus it can be assumed that $f_\Psi(v) < x(v) + C$ for every $v \in \widehat{T}$. Now consider a tree T_v of type T^L or T^U attached to some node $v \in \widehat{T}$. If $f_\Psi(T_v) \geq \mathrm{OPT}(T_v) + C$, then $f_\Psi(T) \geq x(\widehat{T}) + \mathrm{OPT}(T \setminus \widehat{T}) + C = K$. Thus it can be assumed that $f_\Psi(T_v) < \mathrm{OPT}(T_v) + C$. Therefore, by the definition of T_v, if it is a $T^L_{d,C}$ (resp. $T^U_{d,C}$) tree, then Ψ assigns to its root the set $[1, d]$ (resp. $[d+1, d+C]$). Obviously, it follows that the node v cannot use the colors in this set.

By the argument in the previous paragraph, $f_\Psi(a_1) < x(a_1) + C \leq n + C$ and $\Psi(a_1) \cap [n+1, n+C] = \emptyset$, which implies that $\Psi(a_1)$ contains only colors not greater than n. Similarly, $f_\Psi(b_1) < x(b_1) + C \leq n + 2C$ and $\Psi(b_1) \cap [n+C+1, n+3C] = \emptyset$, which implies that the $n - k + C$ colors in $\Psi(b_1)$ are not greater than $n + C$. This set of colors must be disjoint from the k colors in $\Psi(a_1)$, therefore we have $\Psi(b_1) = [1, n+C] \setminus \Psi(a_1)$. Furthermore, $f_\Psi(a_2) < x(a_2) + C \leq n + C$, hence it must use the k colors not used by b_1, therefore $\Psi(a_2) = \Psi(a_1)$. Continuing on this way, we get $\Psi(a_i) = \Psi(a_1) = S$ for all $2 \leq i \leq m$ and S contains k colors not greater than n.

Assume that the set S is not independent, that is, both end vertices of some edge of G is in this set, $u_{i,1}, u_{i,2} \in S$. From the assumption $f_\Psi(T) < K$ follows that $c_{i,1}$ cannot use either of these colors.

We have seen that $f_\Psi(c_{i,1}) < 1 + C$ and $\Psi(c_{i,1}) \cap [u_{i,2}+1, u_{i,2}+C] = \emptyset$ follow from the assumption $f_\Psi(T) < K$, which implies that the color of $c_{i,1}$ is at most $u_{i,2} \leq n$. Moreover, since $f_\Psi(d_{i,1}) < 2C + n - 1$ and $\Psi(d_{i,1}) \cap [n+C+1, n+3C] = \emptyset$, thus node $d_{i,1}$ must use the first $C + n - 1$ colors missing from $c_{i,1}$, therefore we have $\Psi(d_{i,1}) = [1, C+n] \setminus \Psi(c_{i,1})$. Similarly as in the case of the nodes a_j and b_j, it follows that $\Psi(c_{i,1}) = \Psi(c_{i,2}) = \Psi(c_{i,3}) = \{u\}$. Furthermore, notice that $u \geq u_{i,1}$, since $c_{i,2}$ cannot use the colors below $u_{i,1}$: these colors are assigned to the root of the attached tree $T^L_{u_{i,1}-1,C}$. Similarly, u cannot be in $[u_{i,2}+1, u_{i,2}+C]$ since $c_{i,1}$ cannot use these colors. Finally, observe that $d_{i,3}$ must have the colors $[u_{i,1}+1, u_{i,2}-1]$ which forbids $c_{i,3}$ from using a color between $u_{i,1}$ and $u_{i,2}$. Since u is a color not greater than C, thus it must be either $u_{i,1}$ or $u_{i,2}$.

If the demands are polynomially bounded, then the problem is obviously in **NP**: a proper coloring with the given sum is a polynomial size certificate, which finishes the proof of **NP**-completeness. □

Acknowledgments

I'm grateful to Katalin Friedl for useful discussions and for helpful comments, which considerably improved the presentation of the paper. The comments of Judit Csima were also very valuable.

References

1. Amotz Bar-Noy, Magnús M. Halldórsson, Guy Kortsarz, Ravit Salman, and Hadas Shachnai. Sum multicoloring of graphs. *J. Algorithms*, 37(2):422–450, 2000.
2. Amotz Bar-Noy and Guy Kortsarz. Minimum color sum of bipartite graphs. *J. Algorithms*, 28(2):339–365, 1998.
3. Magnús M. Halldórsson and Guy Kortsarz. Multicoloring planar graphs and partial k-trees. In *Randomization, approximation, and combinatorial optimization (Berkeley, CA, 1999)*, pages 73–84. Springer, Berlin, 1999.
4. Magnús M. Halldórsson, Guy Kortsarz, Andrzej Proskurowski, Ravit Salman, Hadas Shachnai, and Jan Arne Telle. Multi-coloring trees. In *Computing and combinatorics (Tokyo, 1999)*, pages 271–280. Springer, Berlin, 1999.
5. M. Hujter and Zs. Tuza. Precoloring extension. II. Graph classes related to bipartite graphs. *Acta Mathematica Universitatis Comenianae*, 62(1):1–11, 1993.
6. Klaus Jansen. The optimum cost chromatic partition problem. In *Algorithms and complexity (Rome, 1997)*, pages 25–36. Springer, Berlin, 1997.
7. Klaus Jansen and Petra Scheffler. Generalized coloring for tree-like graphs. *Discrete Appl. Math.*, 75(2):135–155, 1997.
8. J. Kratochvíl. Precoloring extension with fixed color bound. *Acta Mathematica Universitatis Comenianae*, 62(2):139–153, 1993.
9. Ewa Kubicka. *The Chromatic Sum of a Graph*. PhD thesis, Western Michigan University, 1989.
10. Ewa Kubicka, Grzegorz Kubicki, and Dionisios Kountanis. Approximation algorithms for the chromatic sum. In *Computing in the 90's (Kalamazoo, MI, 1989)*, pages 15–21. Springer, Berlin, 1991.
11. Ewa Kubicka and Allan J. Schwenk. An introduction to chromatic sums. In *Proceedings of the ACM Computer Science Conf.*, pages 15–21. Springer, Berlin, 1989.
12. S. Nicoloso, M. Sarrafzadeh, and X. Song. On the sum coloring problem on interval graphs. *Algorithmica*, 23(2):109–126, 1999.
13. Tibor Szkaliczki. Routing with minimum wire length in the dogleg-free Manhattan model is NP-complete. *SIAM J. Comput.*, 29(1):274–287, 1999.
14. Zsolt Tuza. Graph colorings with local constraints—a survey. *Discuss. Math. Graph Theory*, 17(2):161–228, 1997.

On Verifying Fair Lossy Channel Systems

Benoît Masson* and Ph. Schnoebelen

Lab. Spécification & Vérification, ENS de Cachan & CNRS UMR 8643,
61, av. Pdt. Wilson, 94235 Cachan Cedex France,
phs@lsv.ens-cachan.fr

Abstract. Lossy channel systems are systems of finite state automata that communicate via unreliable unbounded fifo channels. They are an important computational model because of the role they play in the algorithmic verification of communication protocols.
In this paper, we show that fair termination is decidable for a large class of these systems.

1 Introduction

Channel Systems are systems of finite state automata that communicate via asynchronous unbounded fifo channels (see example on Fig. 1). They are a natural model for asynchronous communication protocols and constitute the semantical basis for ISO protocol specification languages such as SDL and Estelle.

Automated Verification of Channel Systems. Formal verification of channel systems is important since even the simplest communication protocols can have tricky behaviors and hard-to-find bugs. But channel systems are Turing powerful [1], and no verification method for them can be general and fully algorithmic.

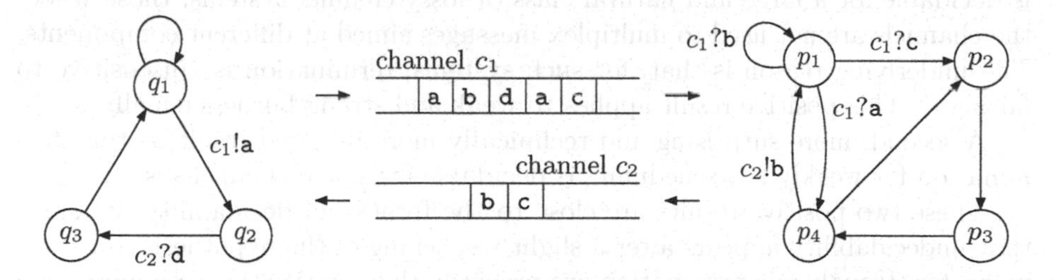

Fig. 1. A channel system with two component automata and two channels

---

* Now at Dept. Comp. Sci., ENS de Lyon. Email: bmasson@ens-lyon.fr. The research described in this paper was conducted while B. Masson was at LSV.

[1] A Turing machine is easily simulated (with polynomial-time overhead) by a single-channel system that stores in its channel the contents of the Turing machine work tape plus a marker for the current position of the reading head [BZ81].

K. Diks et al. (Eds): MFSC 2002, LNCS 2420, pp. 543–555, 2002.

Lossy Channels. A few years ago, Abdulla and Jonsson identified *lossy channel systems* as a very interesting model: in lossy channel systems messages can be lost while they are in transit, without any notification [2]. These lossy systems are the natural model for fault-tolerant protocols where the communication channels are not supposed to be reliable (see [ABJ98,AAB99] for applications). Surprisingly, some verification problems become decidable when one assumes channels are lossy: reachability, safety properties over traces, and inevitability properties over states are decidable for lossy channel systems [Fin94,AK95,CFP96,AJ96b].

One should not believe that lossy channel systems are trivial models where no interesting behavior can be enforced (since messages can always be lost), so that most verification problems would be vacuously decidable. Quite the opposite is true, and many problems are undecidable for these systems: recurrent reachability properties are undecidable, so that temporal logic model-checking is undecidable too [AJ96a]. Furthermore, boundedness is undecidable [May00], as well as all behavioral equivalences [Sch01]. Finally, all the known decidable problems have nonprimitive recursive complexity [Sch02] and are thus much harder than most decidable verification problems.

Fairness Properties. The most important undecidable problem for lossy channel system is *recurrent control state reachability* (RCS), shown undecidable by Abdulla and Johnson [AJ96a]. RCS asks whether there exists a run visiting a given control state infinitely often (i.e. an infinite run satisfying a Büchi acceptance condition). The undecidability of RCS is often summarized by the slogan "*fairness properties are undecidable for lossy channel systems*".

Our Contribution. In this paper we show that, in fact, there exist natural fairness properties that are decidable for lossy channel systems. Indeed, we show that termination under the assumption of fair scheduling ("fair termination") is decidable for a large and natural class of lossy channel systems: those where the channels are not used to multiplex messages aimed at different components. The underlying reason is that, for such systems, termination is "insensitive to fairness". This positive result applies to weak and strong fairness equally.

A second, more surprising and technically more involved result, is that termination for weakly fair scheduling is decidable for single-channel systems.

These two positive results are close to the frontier of decidability: we show that undecidability appears after a slight weakening of the hypothesis. Furthermore, for strongly fair termination, we precisely characterize the communication layouts that ensure decidability, showing that multiplexed channels really are the central issue.

Finally, beyond termination, there is only one other decidable problem that can meaningfully be investigated under the assumption of fair scheduling, namely *inevitability properties.* We show that these properties immediately become undecidable when fair scheduling is assumed.

---

[2] These systems are very close to the *completely specified protocols* independently introduced by Finkel [Fin94].

Plan of the Paper. We first recall the necessary notions in Section 2. Then we study fair termination for systems without multiplexed channels in Section 3, and for single-channel systems in Section 4. Characterization of the layouts ensuring decidability is done in Section 5. Finally, Section 6 discusses fair inevitability. Proofs omitted for lack of space are available in the longer version of this paper.

2 Channel Systems

Given a finite alphabet $\Sigma = \{a, b, \ldots\}$, we let $\Sigma^* = \{u, v, \ldots\}$ denote the set of all finite words over Σ. For $u, v \in \Sigma^*$, we write $u.v$ (also uv) for the *concatenation* of u and v. We write ε for the empty word and Σ^+ for $\Sigma^* \setminus \{\varepsilon\}$. The length of $u \in \Sigma^*$ is denoted $|u|$.

The *subword relation*, denoted $u \sqsubseteq v$, relates any two words u and v s.t. u can be obtained by erasing some (possibly zero) letters from v. For example abba $\sqsubseteq$ <u>ab</u>racada<u>b</u>r<u>a</u>, (the underlined letters are not erased). We write $u \sqsubset v$ when $u \sqsubseteq v$ and $v \not\sqsubseteq u$, that is when $u \sqsubseteq v$ and $|u| < |v|$.

When C is a finite index set, $\Sigma^{*C} = \{U, V, \ldots\}$ is the set of mappings from C to Σ^*, i.e. the set of C-indexed tuples of Σ-words. Concatenation and subword ordering extend to tuples from Σ^{*C} in the obvious way.

2.1 (Perfect) Channel Systems

In this paper we adopt the *extended* model of (lossy) channel systems *where emptiness of channels can be tested for* (and where several messages can be read and written on several channels in a single step). Testing channels for emptiness was allowed in [Sch01] (inspired by [May00]) and we observed that known decidability results do not depend on whether this extension is allowed or not. This remains the case in this paper and the reader will observe that our undecidability proofs do not rely on the extension.

Definition 2.1 (Channel System). *A* channel system *(with n components and m channels) is a tuple $S = \langle \Sigma, C, A_1, A_2, \ldots, A_n \rangle$ where*

- $\Sigma = \{a, b, \ldots\}$ *is a finite alphabet of* messages,
- $C = \{c_1, \ldots, c_m\}$ *is a finite set of m* channels,
- *for $1 \leq k \leq n$, $A_k = \langle Q_k, \Delta_k \rangle$ is the kth* component *of the system:*
  - $Q_k = \{r, s, \ldots\}$ *is a finite set of* control states,
  - $\Delta_k \subseteq Q_k \times \Sigma^{*C} \times Q_k \times \Sigma^{*C} \cup Q_k \times C \times Q_k$ *is a finite set of* rules.

A rule $\delta \in \Delta_k$ of the form (s, U, r, V) is written $s \xrightarrow{?U\,!V} r$ and means that A_k can move from s to r by consuming U (i.e. consuming $U(c)$ on each channel $c \in C$) and writing V ($V(c)$ on each c). This assumes that U is available in the channels. A rule of the form (s, c, r) is written $s \xrightarrow{c=\varepsilon?} r$ and means that A_k can move from s to r after checking that channel c is empty.

Formally, the behavior of S is given via a transition system: a *global state* of S is a tuple $\sigma \in Q_1 \times \cdots \times Q_n$ of control states, one for each component of S.

For $1 \leq k \leq n$, we let $\sigma(k)$ denote the kth component of σ. A *configuration* of S is a pair (σ, W) of a global state and a channel contents $W \in \Sigma^{*C}$ ($W(c) = u$ means that c contains u).

The possible moves between configurations are given by the rules of S. For two configurations (σ, W) and (σ', W') of S we write $\sigma, W \xrightarrow{k:\delta}_{\text{perf}} \sigma', W'$ when:

- δ is some $r \xrightarrow{?U\,!V} s$, $\sigma(k) = r$, there is a W'' s.t. $W = UW''$ and $W' = W''V$. Furthermore $\sigma' = \sigma[k \mapsto s]$, i.e. $\sigma'(k) = s$ and $\sigma'(i) = \sigma(i)$ for all $i \neq k$.
- δ is some $r \xrightarrow{c=\varepsilon?} s$, $\sigma(k) = r$, $W(c) = \varepsilon$ (and further $W' = W$ and $\sigma' = \sigma[k \mapsto s]$).

We write $\sigma, W \xrightarrow{k}_{\text{perf}}$ and say that A_k *is enabled in configuration* (σ, W) when there exists some $\sigma, W \xrightarrow{k:\delta}_{\text{perf}} \sigma', W'$. Otherwise we say A_k is not enabled and write $\sigma, W \not\xrightarrow{k}_{\text{perf}}$.

2.2 Lossy Channel Systems

The notation "$\rightarrow_{\text{perf}}$" stresses that we just defined **perf**ect steps, i.e. steps where no message is lost. Lossy channel systems are channel systems where steps need not be perfect. Instead, any number of messages can be lost from the channels, without any notification.

In Abdulla and Jonsson's model a lossy step is a perfect step possibly preceded and followed by arbitrary losses from the channels. Formally, we write $\sigma, W \xrightarrow{k:\delta}_{\text{loss}} \sigma', W'$ when there exist channel contents V and V' s.t. $W \sqsupseteq V$, $\sigma, V \xrightarrow{k:\delta}_{\text{perf}} \sigma, V'$ and $V' \sqsupseteq W'$. Perfect steps are lossy steps (with no losses). Below we omit writing explicitely the loss subscript for lossy steps, and are simply careful of writing $\rightarrow_{\text{perf}}$ for all perfect steps.

A *run* π of S (from some initial configuration (σ_0, W_0) often left implicit) is a maximal sequence of steps, of the form $\sigma_0, W_0 \xrightarrow{k_1:\delta_1} \sigma_1, W_1 \xrightarrow{k_2:\delta_2} \sigma_2, W_2 \xrightarrow{k_3:\delta_3} \sigma_3, W_3 \cdots$ Maximality implies that π is either infinite, or finite and ends with a blocked configuration, i.e. a configuration from which no more step is possible. A *perfect* run (also, a *faithful run*) is a run where all steps are perfect (no losses).

By "termination", we mean the absence of any infinite run starting from some given initial configuration. We recall that

Theorem 2.2 ([AJ96b,Fin94]). *Termination is decidable for lossy channel systems.*

2.3 Fair Scheduling

There exist many different notions of fairness [Fra86]. Here we consider *fair scheduling of the components*, which is the most natural fairness assumption for asynchronous protocols.

A run of some system S is obtained by interleaving steps from the different components $A_1, \ldots, A_n$. The intuition is that a fair run is a run where all components are fairly treated in their contribution to the run. Formally, given an infinite run $\pi = \sigma_0, W_0 \xrightarrow{k_1:\delta_1} \sigma_1, W_1 \xrightarrow{k_2:\delta_2} \cdots$, we say that:

- π is *weakly fair w.r.t. component* k iff either $k_i = k$ for infinitely many i, or $\sigma_i, W_i \not\xrightarrow{k}_{\text{perf}}$ for infinitely many i. That is, iff component A_k moves infinitely often in π, or is infinitely often not enabled.
- π is *strongly fair w.r.t. component* k iff either $k_i = k$ for infinitely many i, or $\sigma_i, W_i \not\xrightarrow{k}_{\text{perf}}$ for almost all i. That is, iff component A_k moves infinitely often in π, or is eventually never enabled.

Additionaly, all finite runs are (vacuously) fair. We say a run is *weakly fair* (resp. *strongly fair*) if it is weakly (resp. strongly) fair w.r.t. all components $A_1, \ldots, A_n$ of S. Clearly, a strongly fair run is also weakly fair.

Remark 2.3. Observe that we do not consider that a component is enabled when it can only perform lossy steps. This definition makes our decidability proofs a bit more involved, but we find it more consistent with the role losses may or may not play in the fairness of scheduling.

2.4 Communication Layouts

The *communication layout*, or more simply "the layout", of a channel system $S = \langle \Sigma, C, A_1, \ldots, A_n \rangle$ is a graph depicting which components read from, and write to, which channels. Formally $L(S)$ is the bipartite directed graph having the channels and the components of S as vertices, having an edge from A_k to c if there is a rule $r \xrightarrow{?U\,!V} s$ in Δ_k that writes to c (i.e. $V(c) \neq \varepsilon$), and an edge from c to A_k if there is a rule in Δ_k that reads from c (i.e. $U(c) \neq \varepsilon$). Additionaly, $L(S)$ has an edge from c to A_k if Δ_k has a rule $r \xrightarrow{c=\varepsilon?} s$ that checks c for emptiness.

For example, L_1 in Fig. 2 is the layout of the system from Fig. 1.

Note that such a layout only describes possible reads and writes (those present in the rules) that are not necessarily actual reads and writes from actual runs.

The layouts of channel systems provide an abstract view of their architecture and are helpful in classifying them. Below we say that a channel c is *multiplexed* if two (or more) components read from it. E.g. any system having L_2 or L_3 (from Fig. 2) as layout has a multiplexed channel since two components read from c_1. (Observe that situations where several components *write* to a same channel are not considered a case of multiplexing.) Many systems have a simple

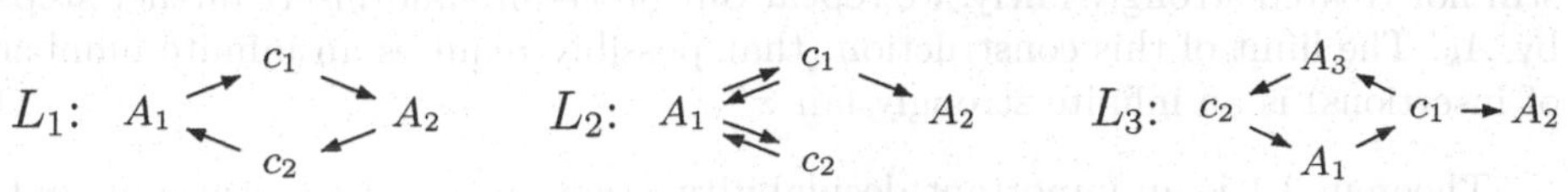

L_1: $A_1 \nearrow c_1 \searrow A_2$, $A_2 \swarrow c_2 \nwarrow A_1$

L_2: $A_1 \rightleftarrows c_1 \searrow A_2$, $A_1 \rightleftarrows c_2$

L_3: $c_2 \nearrow A_3 \nwarrow c_1$, $c_2 \searrow A_1 \nearrow c_1$, $c_1 \to A_2$

Fig. 2. Three communication layouts

layout like L_1 and have no multiplexed channel. Also many systems use different channels for connecting different sender-receiver pairs, leading to layouts without multiplexed channel.

3 Fair Termination without Multiplexed Channel

For a system S, *strongly fair termination* (resp. weakly) is the property that S has no strongly (resp. weakly) fair infinite run.

Theorem 3.1. *Strongly fair termination and weakly fair termination are decidable for lossy channel systems without multiplexed channel.*

This first positive result is a consequence of the fact that, for systems without multiplexed channel, termination is insensitive to whether the system is fairly scheduled or not (and is therefore decidable by Theorem 2.2). This is proved in Lemma 3.3 after we introduce the necessary definitions.

Definition 3.2. *A lossy channel system S is* insensitive to fairness *(for termination) if the equivalences "S has a strongly fair infinite run iff S has a weakly fair infinite run iff S has an infinite run" hold.*
A communication layout L is insensitive to fairness *if all systems having L as layout are insensitive to fairness.*

Lemma 3.3. *If $S = \langle \Sigma, C, A_1, \ldots, A_n \rangle$ is a system with no multiplexed channel, then S is insensitive to fairness.*

Proof. Assume $\pi = \sigma_0, W_0 \xrightarrow{k_1:\delta_1} \sigma_1, W_1 \xrightarrow{k_2:\delta_2} \sigma_2, W_2 \ldots$ is an infinite run of S. Let $I \subseteq \{1, \ldots, n\}$ be the set of (indexes of) components that are not treated strongly fairly in π, i.e. $k \in I$ iff $k = k_i$ for finitely many i and A_k is infinitely often enabled along π. We let $C_I \subseteq C$ be the set of channels that are read by components in I. Let $l \in \mathbb{N}$ be large enough so that $k_i \notin I$ for all $i \geq l$ and let π' be π where every W_i for $i \geq l$ has been replaced by $W'_i \stackrel{\text{def}}{=} W_i[C_I \mapsto \varepsilon]$, a variant of W_i where channels from C_I have been emptied. π' is a valid run of S since losses can explain the changes in the channel contents, and since only components from I (that never move after l) would have been affected by these changes. We write I' for the set of components that are not treated strongly fairly in π' (observe that $I' \subseteq I$). Now we can build a strongly fair run π'' by inserting, for every $k \in I'$, a step $\sigma_i, W_i \xrightarrow{k:\delta} \sigma'_i, W_i$ at a position i beyond l where A_k is enabled (one such position exists). Such a step does not change W_i (possibly by losing what δ would write) but it modifies $\sigma_i(k)$ and we propagate this change of A_k's control state on all further σ_j. If, when in state $\sigma_i(k)$, A_k is still not treated strongly fairly, we repeat our procedure and insert further steps by A_k. The limit of this construction (that possibly requires an infinite number of insertions) is an infinite strongly fair π''. □

Theorem 3.1 is an important decidability result since systems without multiplexed channel are natural and very common. In fact, these systems are so common (see the examples in [ABJ98,AAB99]) that we feel allowed to claim that, in most practical cases, termination of lossy channel systems does not depend on fair scheduling.

In Section 5 we show more precisely how the absence of multiplexed channels is a necessary condition for Theorem 3.1. Before that, we show the undecidability of strongly fair and weakly fair termination in the general case.

Theorem 3.4. *Strongly fair termination is undecidable for lossy channel systems whose communication layout contains* $A_1 \rightleftarrows c \rightarrow A_2$.

Proof. With a Turing machine M we associate the system S_M depicted in Fig. 3. The intended behavior of S_M is the following: first A_1 fills c with some number of blank symbols $\Box_1$ using one # to mark the intended beginning of the string: adding one $\Box_1$ requires two full rotations of the contents of c, replacing all $\Box_1$'s with $\Box_2$'s and then replacing the $\Box_2$'s by $\Box_1$'s. Then A_1 non-deterministically decides that c is full enough and proceeds to the cleaning state (where the $\Box_2$'s are replaced by plain $\Box$'s) that prepares for the *start* state where M is simulated using the contents of c as a bounded workspace, until the *accept* state is eventually reached (if M accepts). At this stage, S_M writes a parasitic character @ on its channel, replaces every other letter by a $\Box$ (i.e. cleans the contents of c) and starts the simulation anew. A_2 does nothing useful but it can consume from c (and will eventually under fair scheduling).

We claim that S_M has a strongly fair run iff M accepts, which proves undecidability. Clearly, if M accepts, S_M has fair infinite runs where it fills c with enough blanks before simulating M an infinite number of times. The parasitic @ that comes up between two successful simulations of M will be either removed by losses, or consumed by A_2 as a way to ensure strong fairness.

Now the more delicate part is to prove that if S_M has a fair infinite run then M accepts. So we assume there is a strongly fair run π (possibly lossy). This run has to eventually move to the *start* state: indeed, if π avoids the *start* state forever, then strong fairness implies that the # marker will eventually be read

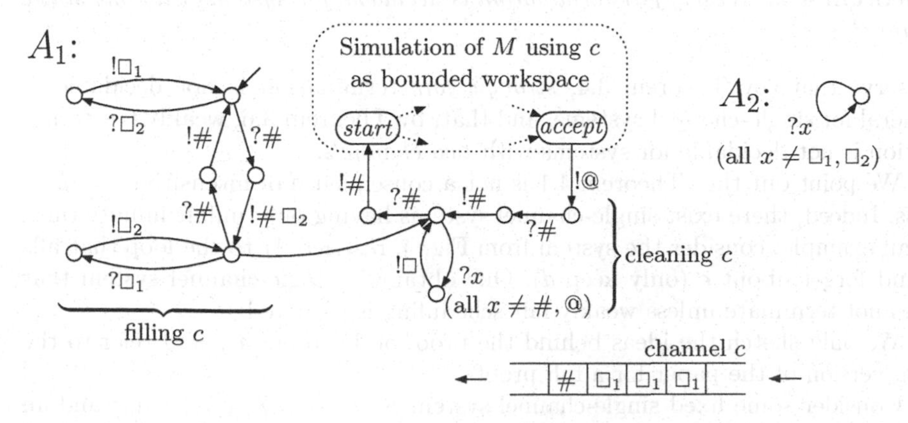

Fig. 3. Structure of S_M in Theorem 3.4

by A_2 and then the system will block (thanks to the $\Box_1 \leftrightarrow \Box_2$ swaps). Once π starts simulating M, the contents of c cannot increase in size: it will diminish through losses (or through reads from A_2). After some time, a lower bound is reached and no more loss will ever occur. From now on, strong fairness can only be ensured by having A_2 read the parasitic @, so that the simulation from *start* to *accept* must be performed an infinite number of times. Since no loss occurs, these simulations of M are faithful and prove that M accepts. □

Remark 3.5. The above proof does not describe in more details how the space-bounded Turing machine is simulated because this is standard (since [BZ81]), and because we do not really need to use Turing machines anyway: it is possible to reduce from perfect channel systems. Replace the simulation of M in Fig. 3 by a single-component single-channel system S_1 that works in bounded space (it does not modify the number of messages stored in c). Then the system we built has a strongly fair infinite run iff there exists a number m s.t. S_1, *running as a perfect channel system*, accepts when started with m messages in its channel.

Theorem 3.6. *Weakly fair termination is undecidable for lossy channel systems with two channels.*

Proof (sketch). We prove undecidability for systems whose layout contains the pattern L_2 (from Fig. 2). As in the proof of Theorem 3.4, we associate a system S_M with a Turing machine M in such a way that S_M has a weakly fair infinite run iff M accepts.

Here filling c can only proceed as long as d contains one $\#_1$ and one $\#_2$, but since A_2 can always read one of these characters, weak fairness requires that, eventually, filling c stops and S_M proceeds to the simulation of M. From this point, the reasoning goes on as in the earlier proof. □

4 Weakly Fair Termination for Single-Channel Systems

Theorem 4.1. *Weakly fair termination is decidable for systems with one single channel.*

Observe that, by Theorem 3.4, *strongly fair* termination is not decidable in general for single-channel systems, and that, by Theorem 3.6, weakly fair termination is not decidable for systems with *two channels*.

We point out that Theorem 4.1 is not a consequence of insensitivity to fairness. Indeed, there exist single-channel systems having only unfair infinite runs: as an example, consider the system from Fig. 4, restrict A_1 to the loop that fills c and forget about c (only keep d). One obtains a single-channel system that does not terminate unless weakly fair scheduling is assumed.

We only sketch the ideas behind the proof of Theorem 4.1 and refer to the long version of the paper for a full proof.

Consider some fixed single-channel system $S = \langle \Sigma, \{c\}, A_1, \ldots, A_n \rangle$ and an initial configuration σ_0, w_0. We say an infinite run $\pi = \sigma_0, w_0 \xrightarrow{k_1:\delta_1} \sigma_1, w_1 \xrightarrow{k_2:\delta_2}$

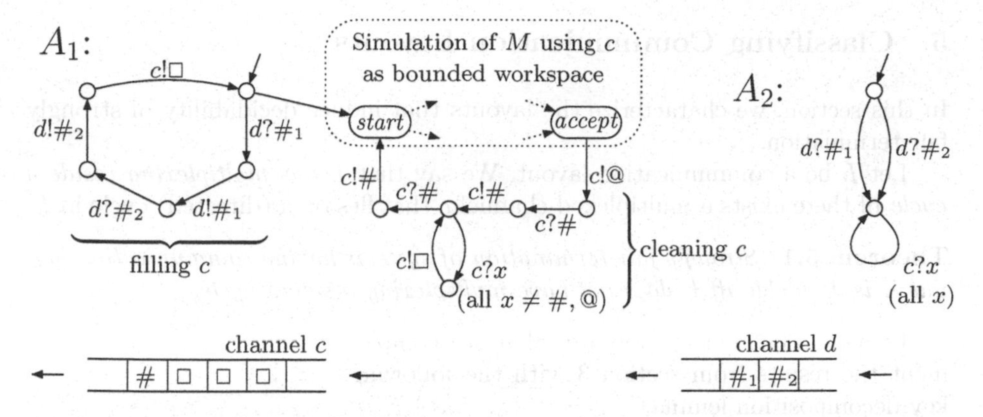

Fig. 4. Structure of S_M in Theorem 3.6

$\cdots$ is a *bounded run* if there exists some maximal size $K \in \mathbb{N}$ s.t. $|w_i| \leq K$ for all i. Otherwise π is *unbounded*. We say π is *ultimately periodic* if there are two numbers $l, p > 0$ s.t. $\sigma_i, w_i, k_i, \delta_i = \sigma_{i+p}, w_{i+p}, k_{i+p}, \delta_{i+p}$ for all $i \geq l$.

We say a step $\sigma, w \xrightarrow{k:\delta} \sigma', w'$ is *back-lossy* [3] if δ is some $r \xrightarrow{?u\,!v} s$, w is some uu' and w' is $u'v'$ for some $v' \sqsubseteq v$ (that is, losses may only occur during the writing of v at the back of c, and not inside c). Also, all steps with δ of the form $r \xrightarrow{c=\varepsilon?} s$ are (vacuously) back lossy. A run is *back-lossy* if all its steps are. It is *ultimately back-lossy* if after some point all its steps are back-lossy.

The next three lemmas exhibit a sequence of transformations that yield an ultimately periodic weakly fair run out of an unbounded run, entailing Corollary 4.5. The proofs of these lemmas rely on the same extraction and modification techniques on runs we used earlier.

Lemma 4.2. *If S has an unbounded run, then it has an unbounded run that is ultimately back-lossy.*

Lemma 4.3. *If S has an unbounded run that is ultimately back-lossy, then it has a weakly fair infinite run (perhaps not back-lossy).*

Lemma 4.4. *If S has a weakly fair infinite run, then it has a weakly fair infinite run that is ultimately periodic.*

Corollary 4.5. *Either S only has bounded runs, or it has an ultimately periodic weakly fair infinite run (or both).*

Now the proof of Theorem 4.1 is easy: Boundedness of single-channel systems is not decidable, but it is obviously semi-decidable, and for a bounded S weakly fair termination is easily checked after the finite graph of configurations has been constructed. Similarly, the existence of a weakly fair ultimately periodic run π is easily seen to be semi-decidable since it suffices to exhibit a finite prefix of π. Combining these two semi-decision methods, we obtain a decision algorithm.

[3] The terminology "back-lossy" appears in [Sch01] and was inspired by the front-lossy systems of [CFP96].

5 Classifying Communication Layouts

In this section, we characterize the layouts that induce decidability of strongly fair termination.

Let L be a communication layout. We say that L *has multiplexing inside a cycle* iff there exists a multiplexed channel c that lies on a (directed) cycle in L.

Theorem 5.1. *Strongly fair termination of systems having communication layout L is decidable iff L does not have multiplexing inside a cycle.*

Proving Theorem 5.1 requires that we complement the results from section 3 with the following key decomposition lemma.

Let L be a layout s.t. some channel c does not lie on a (directed) cycle in L. Then L can be seen as $L_1 \oplus_c L_2$, i.e. the gluing via c of two disjoint layouts L_1 and L_2 (both of them containing c), as illustrated in Fig. 5.

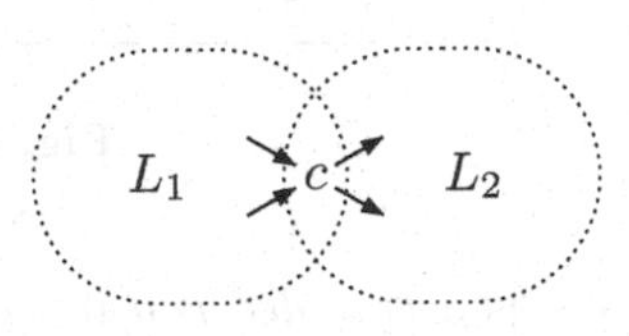

Fig. 5. L is $L_1 \oplus_c L_2$

Lemma 5.2. *L is insensitive to fairness iff L_1 and L_2 are.*

Proof. We only need to prove the ($\Leftarrow$) direction. For this we consider a system S with $L(S) = L$, and prove S is insensitive to fairness. Let S_1 and S_2 be the subsystems obtained from S by keeping only the components (and the channels) from L_1 (resp. L_2). Let π be an infinite run by S. There are two cases:

π **Contains Infinitely Many Steps from S_1:** then S_1 has an infinite run (since steps by S_2 cannot influence S_1) and it has an infinite fair run (since it is insensitive to fairness). Inserting as many steps by S_2 as necessary, one turns this run into a π' that is fair w.r.t. all components of S.

π **Contains Finitely Many Steps from S_1:** then π can be written as the concatenation $\pi_1.\pi_2$ of a finite prefix π_1 where all steps by S_1 can be found, followed by an infinite run π_2 of S_2 (from some starting configuration). By insensitivity, π_2 can be replaced by a fair π_2' (fair w.r.t. S_2). We obtain a run fair w.r.t. all of S by inserting in π_2' (that is, after π_1) as many steps by S_1 as necessary, using losses to make sure these extra steps do not add to c. □

Corollary 5.3. *Layouts without multiplexing inside a cycle are insensitive to fairness.*

Proof. By induction on the number of multiplexed channels in the layout. Lemma 5.2 lets us reduce to the base case where the multiplexed channels (if any) are *degenerate*, i.e. no component writes to them (e.g. in the above picture, c is degenerate in L_2). Insensitivity for systems with degenerate multiplexed channels is proved exactly like with Theorem 3.1. □

Proof (of Theorem 5.1). The ($\Leftarrow$) direction was proved as Corollary 5.3. The ($\Rightarrow$) direction is an easy extension of Theorem 3.4. Assume that L has a cycle $A_1 \to c_1 \to A_2 \to c_2 \cdots A_n \to c_n \to A_1$ s.t. one channel, say c_n, is multiplexed. Then c_n is read by some component A distinct from A_1. If A itself is not on the cycle, then it is easy to adapt the proof of Theorem 3.4 and prove undecidability. Otherwise A is some A_i for $i > 1$ and we can find a shorter cycle $A_i \to c_i \to A_{i+1} \cdots A_n \to c_n \to A_i$, where this time the outside component A is A_1, and we conclude as before. □

Remark 5.4. Lemma 5.2 and Corollary 5.3 apply to strong and weak fairness equally. The reason why the characterization provided by Theorem 5.1 does not hold for weakly fair termination is that Theorem 3.4 only deals with strong termination (which cannot be avoided, see Theorem 4.1).

6 Other Verification Problems with Fair Scheduling

Termination is not the only verification problem that is known to be decidable for lossy channel systems, but problems like reachability only consider *finite* runs. The other known decidable problem for which fairness assumptions are meaningful is *inevitability* (shown decidable in [AJ96b]). Here one asks whether all runs eventually visit a configuration belonging to a given set G.

Termination is a special case of inevitability (with G being the set of blocked configurations), but our positive results for fair termination do not generalize to fair inevitability:

Theorem 6.1. *Inevitability under strongly fair or weakly fair scheduling is undecidable for systems with (more than one component and) a communication layout containing* $A_1 \rightleftarrows c$.

Proof (sketch). By a reduction from RCS.

Let A_1 be any given component having some control state r. We build a system S by associating A_1 and the system A_2 from Fig. 6 (observe that A_2 does not use the channel and is always enabled). Now let G be the set of all configurations $(\langle s, s' \rangle, w)$ of S s.t. $s \neq r$ and $s' = q_2$. Then all fair runs of S inevitably visit G iff A_1 does not have a run visiting r infinitely often (unless there is a self-loop on r). □

This proof idea can be adapted to layouts that contains a cycle, so that inevitability under fair scheduling is undecidable for all "interesting" layouts.

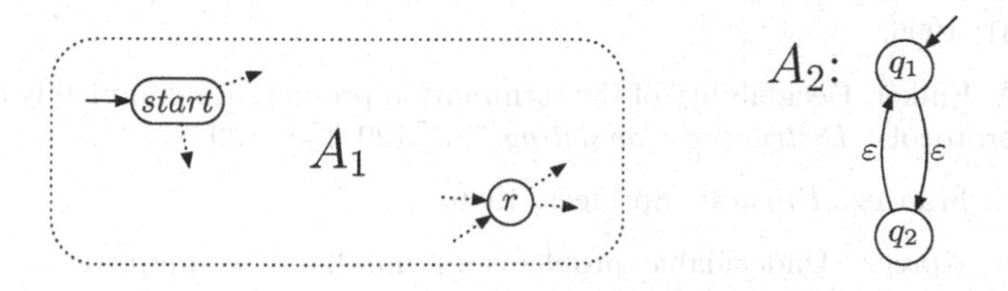

Fig. 6. Reducing RCS to fair inevitability

7 Conclusions

We studied the decidability of termination under strongly and weakly fair scheduling. We showed that, when systems have no multiplexed channels, termination does not depend on whether scheduling is fair or not. In practice, most systems do not have multiplexed channels since they use distinct channels for any pair of components that communicate.

We also showed that, for systems where an arbitrary number of components communicate through a single channel, weakly fair termination is decidable.

These results are technically involved, and are close to the border of decidability. Indeed, two channels make weakly fair termination undecidable, and strongly fair termination is decidable iff no multiplexed channel occurs inside a communication cycle.

References

[AAB99] P. A. Abdulla, A. Annichini, and A. Bouajjani. Symbolic verification of lossy channel systems: Application to the bounded retransmission protocol. In *Proc. 5th Int. Conf. Tools and Algorithms for the Construction and Analysis of Systems (TACAS'99)*, LNCS 1579, pages 208–222. Springer, 1999.

[ABJ98] P. A. Abdulla, A. Bouajjani, and B. Jonsson. On-the-fly analysis of systems with unbounded, lossy FIFO channels. In *Proc. 10th Int. Conf. Computer Aided Verification (CAV'98)*, LNCS 1427, pages 305–318. Springer, 1998.

[AJ96a] P. A. Abdulla and B. Jonsson. Undecidable verification problems for programs with unreliable channels. *Information and Computation*, 130(1):71–90, 1996.

[AJ96b] P. A. Abdulla and B. Jonsson. Verifying programs with unreliable channels. *Information and Computation*, 127(2):91–101, 1996.

[AK95] P. A. Abdulla and M. Kindahl. Decidability of simulation and bisimulation between lossy channel systems and finite state systems. In *Proc. 6th Int. Conf. Theory of Concurrency (CONCUR'95)*, LNCS 962, pages 333–347. Springer, 1995.

[BZ81] D. Brand and P. Zafiropulo. On communicating finite-state machines. Research Report RZ 1053, IBM Zurich Research Lab., June 1981. A short version appears in J.ACM 30(2):323–342, 1983.

[CFP96] G. Cécé, A. Finkel, and S. Purushothaman Iyer. Unreliable channels are easier to verify than perfect channels. *Information and Computation*, 124(1):20–31, 1996.

[Fin94] A. Finkel. Decidability of the termination problem for completely specificied protocols. *Distributed Computing*, 7(3):129–135, 1994.

[Fra86] N. Francez. *Fairness*. Springer, 1986.

[May00] R. Mayr. Undecidable problems in unreliable computations. In *Proc. 4th Latin American Symposium on Theoretical Informatics (LATIN'2000)*, LNCS 1776, pages 377–386. Springer, 2000.

[Sch01] Ph. Schnoebelen. Bisimulation and other undecidable equivalences for lossy channel systems. In *Proc. 4th Int. Symp. Theoretical Aspects of Computer Software (TACS'2001)*, LNCS 2215, pages 385–399. Springer, 2001.

[Sch02] Ph. Schnoebelen. Verifying lossy channel systems has nonprimitive recursive complexity. *Information Processing Letters*, 2002. To appear. Available at http://www.lsv.ens-cachan.fr/~phs.

Parameterized Counting Problems

Catherine McCartin

School of Mathematical and Computing Sciences, Victoria University,
Wellington, New Zealand,
mccartin@mcs.vuw.ac.nz

Abstract. Parameterized complexity has, so far, been largely confined to consideration of computational problems as decision or search problems. However, it is becoming evident that the parameterized point of view can lead to new insight into counting problems. The goal of this article is to introduce a formal framework in which one may consider parameterized counting problems.

1 Introduction

In practice, many situations arise where controlling one aspect, or parameter, of the input can significantly lower the computational complexity of a problem. For instance, in database theory, the database is typically huge, say of size n, whereas queries are typically small; the relevant parameter being the size of an input query $k = |\varphi|$. If n is the size of a relational database, and k is the size of the query, then determining whether there are objects described in the database that have the relationship described by the query can be solved trivially in time $O(n^k)$. On the other hand, for some tasks it may be possible to devise an algorithm with running time say $O(2^k n)$. This would be quite acceptable while k is small.

This was the basic insight of Downey and Fellows [8]. They considered, for instance, the following two well-known graph problems: VERTEX COVER and DOMINATING SET. Both these problems take as input a graph $G = (V, E)$ and a positive integer k that is considered to be the parameter. The question is whether there exists a set $V' \subseteq V$ that is, respectively, a *vertex cover* or a *dominating set*. A *vertex cover* is a set of vertices $V' \subseteq V$ such that, for every edge $uv \in E$, $u \in V'$ or $v \in V'$. A *dominating set* is a set of vertices $V' \subseteq V$ such that, for every vertex $u \in V$, there exists $v \in V'$ such that $uv \in E$.

They observed that, although both problems are NP-complete, the parameter k contributes to the complexity of these two problems in two qualitatively different ways.

They showed that VERTEX COVER is solvable in time $0(2^k n)$, where $n = |V|$, for a fixed k [8]. After many rounds of improvement, the current best known algorithm for VERTEX COVER runs in time $O(1.285^k + kn)$ [6]. In contrast, the best known algorithm for DOMINATING SET is still just the brute force algorithm of trying all k-subsets, with running time $O(n^{k+1})$.

K. Diks et al. (Eds): MFSC 2002, LNCS 2420, pp. 556–567, 2002.

These observations are formalized in the framework of *parameterized complexity theory* [9].

The notion of *fixed-parameter tractability* is the central concept of the theory. Intuitively, a problem is fixed-parameter tractable if we can somehow confine the any 'bad' complexity behaviour to some limited aspect of the problem, the parameter.

More formally, we consider a *parameterized language* to be a subset $L \subseteq \Sigma^* \times \Sigma^*$. If L is a parameterized language and $\langle \sigma, k \rangle \in L$ then we refer to σ as the *main part* and k as the *parameter*. A parameterized language, L, is said to be fixed-parameter tractable (FPT) if membership in L can be determined by an algorithm whose running time on instance $\langle \sigma, k \rangle$ is bounded by $f(k)|\sigma|^\alpha$, where f is an arbitrary function and α is a constant not depending on the parameter k.

Following naturally from the concept of fixed-parameter tractability is the appropriate notion of parameterized problem reduction. Apparent fixed-parameter *intractability* is established via a completeness program.

The main sequence of parameterized complexity classes is

$$FPT \subseteq W[1] \subseteq W[2] \subseteq \cdots \subseteq W[t] \cdots \subseteq W[P] \subseteq AW[P] \subseteq XP$$

This sequence is commonly termed the W-hierarchy. The complexity class $W[1]$ is the parametric analog of NP. The k-STEP NDTM HALTING PROBLEM is complete for $W[1]$. In the same sense that NP-completeness of the $q(n)$-STEP NDTM HALTING PROBLEM provides us with strong evidence that no NP-complete problem is likely to be solvable in polynomial time, $W[1]$-completeness of the k-STEP NDTM HALTING PROBLEM provides us with strong evidence that no $W[1]$-complete problem is likely to be fixed-parameter tractable. DOMINATING SET is, in fact, $W[2]$-complete. It is conjectured that all of the containments here are proper, but all that is currently known is that FPT is a proper subset of XP.

Parameterized complexity theory has been well-developed during the last 10 years. It is widely applicable, particularly because of hidden parameters such as treewidth, pathwidth, and other graph 'width metrics', that have been shown to significantly affect the computational complexity of many fundamental problems modelled on graphs. We refer the reader to the monograph of Downey and Fellows [9] for more details.

Of course, classical complexity is not only concerned with decision problems, but also with search, enumeration, and counting problems. The goal of the present paper is to introduce a formal framework in which to address issues of parameterized *counting complexity*.

Counting complexity is a very important branch of complexity, with many applications. It is also very hard. The consensus is that while decision problems can often have good approximation algorithms, it is generally thought that counting problems are very hard indeed (see, for example, [19]). Counting problems tend not to have approximation algorithms, randomized algorithms, PTAS's, or the like.

We think that parameterized complexity has a lot to say about the enumeration of small structures.

Raman and Arvind [2] have modified the method of bounded search trees (one of the most common methods used to demonstrate fixed-parameter tractability for decision problems) to show that the problem of counting all size k vertex covers of G is solvable in time $O(2^{k^2+k}k + 2^k n)$. The approach used in [2] appears promising as a method for demonstrating parametric tractability of many paameterized counting problems corresponding to well-studied decision problems.

Courcelle, Makowsky and Rotics [7] have considered counting and evaluation problems on graphs where the range of counting is definable in monadic second order logic. They show that these problems are fixed-parameter tractable, where the parameter is the treewidth of the graph. Andrzejak [1] , and Noble [15] have shown that for graphs $G = (V, E)$ of treewidth at most k, the Tutte polynomial can be computed in polynomial time and evaluated in time $O(|V|)$, despite the fact that the general problem is $\#P$-complete. Makowsky [14] has shown that the same bounds hold for the colored Tutte polynomial on coloured graphs $G = (V, E, c)$ of treewidth at most k. Finally, Grohe and Frick have introduced the notion of locally tree-decomposable classes of structures, and have shown that counting problems definable in first order logic can be solved in fixed-parameter linear time on such structures, see [10], [11].

Thus, it is clear that parameters can greatly alter the overall complexity of a problem, even in the realm of counting problems. All of the above have been ad hoc observations. As stated, the main goal of the present paper is to introduce a general framework for parameterized counting complexity, extending the framework introduced by Downey and Fellows for decision problems.

As well as introducing basic definitions for tractability and the notion of a parameterized counting reduction, we also look at a basic hardness class. We define $\#W[1]$, the parameterized analog of Valiant's class $\#P$. Our core problem here is #SHORT TURING MACHINE ACCEPTANCE, where the input is a non-deterministic Turing machine M and a string x, the parameter is a positive integer k, and the output is the number of k-step accepting computations of M on x. We show that #SHORT TURING MACHINE ACCEPTANCE is complete for $\#W[1]$. We also determine $\#W[1]$-completeness, or $\#W[1]$-hardness, for several other parameterized counting problems.

We note that we have obtained some similar results for the parameterized counting classes $\#W[2]$, $\#W[t]$ and $\#W[P]$. These are not included here.

We remark that parameterized complexity is intertwined with non-approximativity. For example, Bazgan [3] (and independently Cesati and Trevisan [5]) showed that associated with every optimization problem is a parametric decision problem. Moreover, if the optimization problem has an efficient PTAS then the parametric decision problem is fixed-parameter tractable. Thus, an optimization problem has no efficient PTAS if the associated parametric decision problem is $W[1]$-hard (unless the W-hierarchy collapses). Razborov and Alekhnovich [17] have shown that various axiom systems have no resolution proofs unless the W-hierarchy collapses. Thus, parametric hardness results are of more than just academic interest, since they have real ramifications in classical complexity too.

We would anticipate that our methods will have similar applications in counting complexity. For example, if one could show that the (parameterized) problem of counting Euler cycles in graphs of bounded degree was $\#W[1]$-hard then it would show that there is likely no polynomial time method for solving the classical problem; even without considering the open question of whether this problem is $\#P$ complete.

We finally remark that some of the results here were independently obtained by Grohe and Flum. They have recently shown that the problem of counting small cycles in a graph is $\#W[1]$-hard [12].

2 Classical Counting Problems and #P

The fundamental class of *counting problems* in classical complexity theory, $\#P$, was proposed by Valiant [18] in the late 1970's. His definition uses the notion of a witness function:

Definition 1 (Witness Function). *Let $w : \Sigma^* \to \mathcal{P}(\Gamma^*)$, and let $x \in \Sigma^*$. We refer to the elements of $w(x)$ as witnesses for x. We associate a decision problem $A_w \subseteq \Sigma^*$ with w:*

$$A_w = \{\, x \in \Sigma^* \mid w(x) \neq \emptyset \,\}.$$

In other words, A_w is the set of strings that have witnesses.

Definition 2 ($\#P$). *The class $\#P$ is the class of witness functions w such that:*

1. *there is a polynomial-time algorithm to determine, for given x and y, whether $y \in w(x)$;*
2. *there exists a constant $k \in \mathcal{N}$ such that for all $y \in w(x)$, $|y| \leq |x|^k$. (The constant k can depend on w).*

$\#P$ is the class of witness functions naturally associated with decision problems in the class NP.

The counting problem associated with a particular computational problem is to determine the *number* of solutions to the problem for any given instance. That is, given the witness function associated with the decision problem, what we are really interested in is $|w(x)|$ for any input x.

This leads us to other definitions for $\#P$, where $\#P$ is treated as a class of functions of the form $f : \Sigma^* \to \mathcal{N}$. For instance, see [16], [19].

We now need to establish how counting problems v and w are related under the process of reduction. For this purpose we introduce the notions of *counting reduction* and *parsimonious reduction.*

Definition 3 (Counting Reduction). *Let*

$$w : \Sigma^* \to \mathcal{P}(\Gamma^*)$$
$$v : \Pi^* \to \mathcal{P}(\Delta^*)$$

be counting problems, in the sense of [18]. A polynomial-time many-one counting reduction from w *to* v *consists of a pair of polynomial-time computable functions*

$$\sigma : \Sigma^* \to \Pi^*$$
$$\tau : \mathcal{N} \to \mathcal{N}$$

such that

$$|w(x)| = \tau(|v(\sigma(x))|).$$

When such a reduction exists we say that w *reduces to* v.

Intuitively, if one can easily count the number of witnesses of $v(y)$, then one can easily count the number of witnesses of $w(x)$.

There is a particularly convenient kind of reduction that preserves the number of solutions to a problem exactly. We call such a reduction *parsimonious.*

Definition 4 (Parsimonious Reduction). *A counting reduction* σ, τ *is parsimonious if* τ *is the identity function.*

Armed with the notion of a counting reduction, we can define the class of *#P-complete* problems that represent the 'hardest' problems in the class #P, that is, those problems to which all others in the class may be reduced.

One famous result is the following:

Theorem 1 (Valiant). *The problem of counting the number of perfect matchings in a bipartite graph is #P-complete.*

Despite the fact that a perfect matching can be found in polynomial time, counting the number of them is as hard as counting the number of satisfying assignments to a Boolean formula. Details of the proof of this theorem may be found in [13] or [16].

3 #W[1] – A Parameterized Counting Class

In order build a framework in which to consider parameterized counting problems, we first need to establish a some basic definitions. We recall:

Definition 5 (Parameterized Language). *A parameterized language* L *is a subset* $L \subseteq \Sigma^* \times \Sigma^*$. *If* L *is a parameterized language and* $\langle x, y\rangle \in L$, *then we will refer to* x *as the main part and* y *as the parameter. It is often convenient to consider that* y *is an integer, and to define a parameterized language to be a subset of* $\Sigma^* \times \mathcal{N}$.

As in the classical case, we use the notion of a witness function to formalize the association between parameterized counting problems and their corresponding decision problems.

Definition 6 (Parameterized Witness Function). *Let* $w : \Sigma^* \times \mathcal{N} \to \mathcal{P}(\Gamma^*)$, *and let* $\langle \sigma, k\rangle \in \Sigma^* \times \mathcal{N}$. *The elements of* $w(\langle \sigma, k\rangle)$ *are witnesses for* $\langle \sigma, k\rangle$. *We associate a parameterized language* $L_w \subseteq \Sigma^* \times \mathcal{N}$ *with* w:

$$L_w = \{ \langle \sigma, k\rangle \in \Sigma^* \times \mathcal{N} \mid w(\langle \sigma, k\rangle) \neq \emptyset \} .$$

L_w *is the set of problem instances that have witnesses.*

Definition 7 (Parameterized Counting Problem). *Let $w : \Sigma^* \times \mathcal{N} \to \mathcal{P}(\Gamma^*)$ be a parameterized witness function. The corresponding parameterized counting problem can be considered as a function $f_w : \Sigma^* \times \mathcal{N} \to \mathcal{N}$ that, on input $\langle \sigma, k \rangle$, outputs $|w(\langle \sigma, k \rangle)|$.*

We note here that 'easy' parameterized counting problems can be considered to be those in the class that we might call 'FFPT', the class of functions of the form $f : \Sigma^* \times \mathcal{N} \to \mathcal{N}$ where $f(\langle \sigma, k \rangle)$ is computable in time $g(k)|\sigma|^\alpha$, where g is an arbitrary function and α is a constant not depending on k.

To consider 'hard' parameterized counting problems, we need some more definitions:

Definition 8 (Parameterized Transformation). *A parameterized transformation from a parameterized language L to a parameterized language L' is an algorithm that computes, from input consisting of a pair $\langle \sigma, k \rangle$, a pair $\langle \sigma', k' \rangle$ such that:*

1. *$\langle \sigma, k \rangle \in L$ if and only if $\langle \sigma', k' \rangle \in L'$,*
2. *$k' = g(k)$ is a function only of k, and*
3. *the computation is accomplished in time $f(k)n^\alpha$, where $n = |\sigma|$, α is a constant independent of both f and k, and f is an arbitrary function.*

Definition 9 (Parameterized Counting Reduction). *Let*

$$w : \Sigma^* \times \mathcal{N} \to \mathcal{P}(\Gamma^*)$$
$$v : \Pi^* \times \mathcal{N} \to \mathcal{P}(\Delta^*)$$

be (witness functions for) parameterized counting problems. A parameterized counting reduction from w to v consists of a parameterized transformation

$$\rho : \Sigma^* \times \mathcal{N} \to \Pi^* \times \mathcal{N}$$

and a function

$$\tau : \mathcal{N} \to \mathcal{N}$$

running in time $f(k)n^\alpha$ (where $n = |\sigma|$, α is a constant independent of both f and k, and f is an arbitrary function) such that

$$|w(\langle \sigma, k \rangle)| = \tau(|v(\rho(\langle \sigma, k \rangle))|).$$

When such a reduction exists we say that w reduces to v.

As in the classical case, if one can easily count the number of witnesses of $v(\langle \sigma', k' \rangle)$, then one can easily count the number of witnesses of $w(\langle \sigma, k \rangle)$.

Definition 10 (Parsimonious Parameterized Counting Reduction). *A parameterized counting reduction ρ, τ is parsimonious if τ is the identity function.*

#SHORT TURING MACHINE ACCEPTANCE

Input: A nondeterministic Turing machine M, and a string x.
Parameter: A positive integer k.
Output: $acc_M(x,k)$, the number of $\leq k$-step accepting computations of machine M on input x.

Our main goal is to establish $\#W[1]$, the parameterized analogue of $\#P$. In the next section we define the parameterized counting class $\#W[1]$. In the following section we prove that the following fundamental parameterized counting problem is $\#W[1]$-complete:

The point here is that we are proving a parametric counting analog of Cook's theorem. For example, the $W[1]$-completeness of INDEPENDENT SET means that deciding whether a graph has an independent set of size k is as hard as deciding whether a non-deterministic Turing machine has an accepting path of length k on some input. We aim to show that for $\#W[1]$-complete problems *counting* the number of solutions is as hard as counting the number of 'short' accepting paths of a non-deterministic Turing machine on some input.

3.1 A Definition for $\#W[1]$

We first review the original definition for $W[1]$, presented in [9]. We will need some preliminary definitions.

A 3CNF formula can be considered as a circuit consisting of one input (of unbounded fanout) for each variable, possibly inverters below the variable, and structurally a large *and* of small *or*'s (of size 3) with a single output line. More generally, it is convenient to consider the model of a *decision circuit*. This is a circuit consisting of large and small gates with a single output line, and no restriction on the fanout of gates. For such a circuit, the *depth* is the maximum number of gates on any path from the input variables to the output line, and the $weft$ is the 'large gate depth.'

Formally, we define the weft of a circuit as follows:

Definition 11 (Weft). *Let C be a decision circuit. The $weft$ of C is defined to be the maximum number of large gates on any path from the input variables to the output line. (A gate is called large if it's fanin exceeds some pre-determined bound. None of the following results depend upon what the bound actually is).*

Let $\mathcal{F} = \{C_1, ..., C_n, ...\}$ be a family of decision circuits. Associated with $\mathcal{F}$ is a basic parameterized language

$$L_{\mathcal{F}} = \{\langle C_i, k\rangle : C_i \text{ has a weight } k \text{ satisfying assignment}\} .$$

Notation: We will denote by $L_{\mathcal{F}(t,h)}$ the parameterized language associated with the family of weft t depth h decision circuits.

Definition 12 ($W[1]$). *We define a language L to be in the class $W[1]$ iff there is a parametric transformation from L to $L_{\mathcal{F}(1,h)}$ for some h.*

#WEIGHTED WEFT t DEPTH h CIRCUIT SATISFIABILITY (WCS(t, h))

Input: A weft t depth h decision circuit C.
Parameter: A positive integer k.
Output: The number of weight k satisfying assignments for C.

Now consider the following generalized parametric counting problem:
Let $w_{\mathcal{F}(t,h)} : \Sigma^* \times \mathcal{N} \to \mathcal{P}(\Gamma^*)$ be the standard parameterized witness function associated with this counting problem:

$$w_{\mathcal{F}(t,h)}(\langle C, k\rangle) = \{ \text{ weight } k \text{ satisfying assignments for } C \ \} .$$

Definition 13 (#W[1]). *We define a parametric counting problem, f_v, to be in #W[1] iff there is a parameterized counting reduction from v, the parameterized witness function for f_v, to $w_{\mathcal{F}(1,h)}$.*

3.2 A Fundamental Complete Problem for #W[1]

In this section we outline the proof of the following theorem:

Theorem 2. *#SHORT TURING MACHINE ACCEPTANCE is complete for #W[1].*

The proof makes use of a series of parametric transformations described in [9], with alterations where required to ensure that each of these may be considered as a parsimonious parametric counting reduction. The main contribution of the current author is the reworking of the transformation used in the in the proof of lemma 1. Most of the transformations described in [9] are, in fact, parsimonious, we just need to argue that this is the case.

We begin with some preliminary *normalization* results about weft 1 circuits. We turn our attention to circuits having depth 2 and a particularly simple form, consisting of a single output *and* gate which receives arguments from *or* gates having fanin bounded by a constant s. Each such circuit is isomorphically represented by a boolean expression in conjunctive normal form having clauses with at most s literals. We will say that a circuit having this form is *s-normalized*, and let $F(s)$ denote the family of s-normalized circuits.

We show that there is a parsimonious parametric counting reduction from $w_{\mathcal{F}(1,h)}$, the standard parameterized witness function for $\mathcal{F}(1, h)$, to $w_{\mathcal{F}(s)}$, the standard parameterized witness function for $\mathcal{F}(s)$, where $s = 2^h + 1$. Thus, any parametric counting problem $f_v \in$ #W[1] can, in fact, be reduced to the following problem (where s is fixed in advance and depends on f_v):

Lemma 1. *$w_{\mathcal{F}(1,h)}$ reduces to $w_{\mathcal{F}(s)}$, where $s = 2^h + 1$, via a parsimonious parametric counting reduction.*

Proof Outline:

#WEIGHTED s-NORMALIZED CIRCUIT SATISFIABILITY

Input: An s-normalized decision circuit C.
Parameter: A positive integer k.
Output: The number of weight k satisfying assignments for C.

Let $C \in \mathcal{F}(1, h)$ and let k be a positive integer. We describe a parametric transformation that, on input $\langle C, k \rangle$, produces a circuit $C' \in \mathcal{F}(s)$ and an integer k' such that for every weight k input accepted by C there exists a unique weight k' input accepted by C'. The transformation proceeds in four stages, and follows that given in [9], with substantial alterations to the last step in order to ensure parsimony.

The first three steps culminate in the production a tree circuit, C', of depth 4, that corresponds to a Boolean expression, E, in the following form. (We use product notation to denote logical $\wedge$ and sum notation to denote logical $\vee$.)

$$E = \prod_{i=1}^{m} \sum_{j=1}^{m_i} E_{ij}$$

where:
(1) m is bounded by a function of h,
(2) for all i, m_i is bounded by a function of h,
(3) for all i, j, E_{ij} is either:

$$E_{ij} = \prod_{k=1}^{m_{ij}} \sum_{l=1}^{m_{ijk}} x[i,j,k,l]$$

or

$$E_{ij} = \sum_{k=1}^{m_{ij}} \prod_{l=1}^{m_{ijk}} x[i,j,k,l],$$

where the $x[i,j,k,l]$ are literals (i.e., input Boolean variables or their negations) and for all i, j, k, m_{ijk} is bounded by a function of h. The family of circuits corresponding to these expressions has weft 1, with the large gates corresponding to the E_{ij}. (In particular, the m_{ij} are *not* bounded by a function of h.)

Note that the witnesses for C' are exactly the witnesses for our original circuit C.

In the fourth step, we employ additional nondeterminism.

Let C denote the normalized depth 4 circuit received from the previous step, corresponding to the Boolean expression E described above.

We produce an expression E' in product-of-sums form, with the size of the sums bounded by $2^h + 1$, that has a unique satisfying truth assignment of weight

$$k' = k + (k+1)(1+2^h)2^{2^h} + m + \sum_{i=1}^{m} m_i$$

corresponding to each satisfying truth assignment of weight k for C.

The idea is to employ extra variables in E' so that τ', a weight k' satisfying truth assignment for E', encodes both τ, a weight k satisfying truth assignment for E and a 'proof' that τ satisfies E. Thus, τ' in effect guesses τ and also checks that τ satisfies E.

We build E' so that the *only* weight k' truth assignments that satisfy E' are the τ''s that correspond to τ's satisfying E. □

Lemma 1 allows us to now state the following theorem:

Theorem 3.

$$\#W[1] = \bigcup_{s=1}^{\infty} \#W[1,s]$$

where $\#W[1,s]$ is the class of parametric counting problems whose associated witness functions reduce to $w_{\mathcal{F}(s)}$, the standard parameterized witness function for $\mathcal{F}(s)$, the family of s-normalized decision circuits.

We now want to show that $\#W[1]$ collapses to $\#W[1,2]$. We will need some more definitions:

A circuit C is termed *monotone* if it does not have any *not* gates. Equivalently, C corresponds to a boolean expression having only positive literals.

We define a circuit C to be *antimonotone* if all the input variables are negated and the circuit has no other inverters. Thus, in an antimonotone circuit, each fanout line from an input node goes to a *not* gate and in the remainder of the circuit there are *no* other *not* gates. The restriction to families of antimonotone circuits yields the classes of parametric counting problems #ANTIMONOTONE $W[t]$ and #ANTIMONOTONE $W[1,s]$.

Theorem 4. $\#W[1,s] =$ #ANTIMONOTONE $W[1,s]$ *for all* $s \geq 2$.

Using theorem 4 we can prove the following:

Theorem 5. $\#W[1] = \#W[1,2]$.

Theorem 6. *The following are complete for* $\#W[1]$*:*
(i) *#INDEPENDENT SET*
(ii) *#CLIQUE*

Proofs of these theorems follow those given in [9] for the corresponding theorems in the decision context. We just need to argue that the transformations used are, in fact, parsimonious. We are now in a position to establish theorem 2, which we restate here.

Theorem 2. *#SHORT TURING MACHINE ACCEPTANCE is complete for* $\#W[1]$.

Proof Outline:

To show hardness for $\#W[1]$ we provide a parsimonious parametric counting reduction from #CLIQUE. Again, this follows [9] with a slight alteration. $\#W[1]$-hardness will then follow by Theorem 6 (ii).

To show membership in $\#W[1]$ we describe a parsimonious parametric counting reduction from #SHORT TURING MACHINE ACCEPTANCE to

#WEIGHTED WEFT 1 DEPTH h CIRCUIT SATISFIABILITY (#WCS($1, h$)). □

We can now state an alternative definition for $\#W[1]$.

Definition 14 ((#*W*[1], Turing Machine Characterization). *Let $acc_{[M,k]}(x)$ be the number of k-step accepting computations of machine M on input $x \in \Sigma^*$.*

$$\#W[1] = \{f : \Sigma^* \to \mathcal{N} \mid f = acc_{[M,k]} \text{ for some NDTM } M\} .$$

4 Populating #*W*[1]

In this section, we note the complexity of some other parametric counting problems relative to $\#W[1]$ (see [9] for definitions). The first of these is #PERFECT CODE, which we have shown to be $\#W[1]$-complete.

The decision version of this problem was shown to be $W[1]$-hard in [9], and it was conjectured that the problem could be of intermediate difficulty between $W[1]$ and $W[2]$. Until recently, there was no evidence that the problem belonged to $W[1]$, although it can easily be shown to belong to $W[2]$. In [4] PERFECT CODE has been shown to be, in fact, $W[1]$-complete. A consequence of our argument proving $\#W[1]$-completeness for #PERFECT CODE is an alternative proof of membership in $W[1]$ for PERFECT CODE.

We also note that the following problems can be shown to be $\#W[1]$-hard:

#SIZED SUBSET SUM
#EXACT CHEAP TOUR
#SHORT CHEAP TOUR
(if we relax our notion of reducibility from $\leq^s_m$ to $\leq^s_T$).

5 Conclusions

Our goal in this paper was to introduce a formal framework in which to address issues of parameterized counting complexity. We have defined what it means for a parameterized counting problem to be 'easy' or tractable, via the class 'FFPT' of FPT-time computable functions. We have made a start on establishing a completeness program to exhibit (likely) intractability of 'hard' parameterized counting problems, by introducing the notion of a parameterized counting reduction and the class $\#W[1]$, the parameterized analog of Valiant's class $\#P$. We have shown that our definition of $\#W[1]$ is a 'sensible' one by showing that the fundamental problem #SHORT TURING MACHINE ACCEPTANCE is complete for $\#W[1]$.

We remark here that we have obtained similar results for other parameterized counting classes. For example, #DOMINATING SET is $\#W[2]$-complete, #WEIGHTED t-NORMALIZED SATISFIABILITY is $\#W[t]$-complete.

References

1. A. Andrzejak: *An algorithm for the Tutte Polynomials of Graphs of Bounded Treewidth.* Discrete Math. 190, pp 39-54, (1998).
2. V. Arvind and V. Raman: *Approximate Counting Small Subgraphs of Bounded Treewidth and Related Problems.* manuscript.
3. C. Bazgan: *Schémas d'approximation et complexité paramétrée.* Rapport de stage de DEA d'Informatique à Orsay, (1995).
4. M. Cesati: *Perfect Code is $W[1]$-complete.* Information Processing Letters 81(3), pp 163-168, (2002).
5. M. Cesati and L. Trevisan: *On the Efficiency of Polynomial Time Approximation Schemes.* Information Processing Letters 64(4), pp 165-171, (1997).
6. J. Chen, I.A. Kanj and W. Jia: *Vertex Cover: Further Observations and Further Improvements.* Journal of Algorithms 41, pp 280-301, (2001).
7. B. Courcelle, J.A. Makowsky and U. Rotics: *On the Fixed Parameter Complexity of Graph Enumeration Problems Definable in Monadic Second Order Logic.* Discrete Applied mathematics, Vol. 108, No.1-2, pp. 23-52, (2001).
8. R. G. Downey and M. R. Fellows: *Fixed-parameter tractability and completeness.* Congressus Numeratium, Vol 87, pp 161-187, (1992).
9. R. G. Downey and M. R. Fellows: *Parameterized Complexity.* Springer-Verlag (1999).
10. Markus Frick *Easy Instances for Model-Checking.* Ph.D. Thesis, Laboratory for Foundations of Computer Science, The University of Edinburgh, (2001).
11. Markus Frick and Martin Grohe: *Deciding First-Order Properties of Locally Tree-Decomposable Graphs.* Proceedings of the 26th International Colloquium on Automata, Languages and Programming, Lecture Notes in Computer Science 1644, pp 331-340, Springer-Verlag, (1999).
12. M. Grohe: personal communication.
13. D. Kozen: *The Design and Analysis of Algorithms.* Springer-Verlag (1991).
14. J.A. Makowsky: *Colored Tutte Polynomials and Kauffman Brackets for Graphs of Bounded Tree Width.* Proceedings of the 12th Annual ACM-SIAM Symposium on Discrete Algorithms, Washington DC, pp. 487-495, (2001).
15. S.D. Noble: *Evaluating the Tutte Polynomial for graphs of bounded tree-width.* Combin. Probab. Comput. 7, pp 307-321, (1998).
16. C. H. Papdimitriou: *Computational Complexity.* Addison Wesley (1994).
17. A. Razborov and M. Alekhnovich: *Resolution is Not Automatizable Unless W[P] is Tractable.* Proc. of the 42nd IEEE FOCS, pp 210-219, (2001).
18. L. Valiant: *The complexity of computing the permanent.* Theoret. Comput. Sci., Vol 8, pp 189-201, (1979).
19. D. Welsh and A. Gale: *The Complexity of Counting Problems.* Aspects of Complexity, editors R. Downey and D.Hirschfeldt, de Gruyter Series in Logic and Its Applications, pp 115-154, (2001).

On the Construction of Effective Random Sets

Wolfgang Merkle and Nenad Mihailović

Ruprecht-Karls-Universität Heidelberg, Mathematisches Institut,
Im Neuenheimer Feld 294, D–69120 Heidelberg, Germany,
{merkle|mihailovic}@math.uni-heidelberg.de

Abstract. We give a direct and rather simple construction of Martin-Löf random and rec-random sets with certain additional properties. First, reviewing the result of Gács and Kučera, given any set X we construct a Martin-Löf random set R from which X can be decoded effectively. Second, by essentially the same construction we obtain a Martin-Löf random set R that is computably enumerable selfreducible. Alternatively, using the observation that a set is computably enumerable selfreducible if and only if its associated real is computably enumerable, the existence of such a set R follows from the known fact that every Chaitin Ω real is Martin-Löf random and computably enumerable. Third, by a variant of the basic construction we obtain a rec-random set that is weak truth-table autoreducible.

The mentioned results on self- and autoreducibility complement work of Ebert, Merkle, and Vollmer [7,8,9], from which it follows that no Martin-Löf random set is Turing-autoreducible and that no rec-random set is truth-table autoreducible.

1 Introduction

In what follows we give a direct and rather simple construction of Martin-Löf random and rec-random sets with certain additional properties. First, reviewing the result of Gács and Kučera, we demonstrate that for any given set X there is a Martin-Löf random set R from which X can be decoded effectively.

Second, by essentially the same construction we obtain a Martin-Löf random set R that is computably enumerable selfreducible; i.e. R is Martin-Löf random and there is an oracle Turing machine that on input x queries its oracle only at places $z < x$ and, in case the oracle is indeed R, eventually outputs 1 in case x is in R and does not terminate in case x is not in R. Alternatively, using the observation that a set is computably enumerable selfreducible if and only if its associated real is computably enumerable, the existence of such a set R follows from the known fact that every Chaitin Ω real is Martin-Löf random and computably enumerable.

Third, by a variant of the basic construction we obtain a rec-random set that is weak truth-table autoreducible, i.e., is reducible to itself by an oracle Turing machine with a computable upper bound on its queries that never queries the oracle at the current input.

K. Diks et al. (Eds): MFSC 2002, LNCS 2420, pp. 568–580, 2002.

The mentioned results on self- and autoreducibility do not extend to slightly less powerful reducibilities. More precisely, no Martin-Löf random set is Turing-autoreducible and no rec-random set is truth-table autoreducible. These assertions can be obtained as corollaries to work of Ebert, Merkle, and Vollmer [7,8,9], where it is actually shown that such autoreductions are not possible even if one just requires that in the limit the reducing machine computes the correct value for a constant nonzero fraction of all places, while signalling ignorance about the correct value for the other places.

1.1 Notation

The notation used in the following is mostly standard, for unexplained notation refer to the surveys and textbooks cited in the bibliography [2,4,15].

If not explicitly stated differently, the terms *set* and *class* refer to sets of natural numbers and to sets of sets of natural numbers, respectively. For any set A, we write $A(x)$ for the value of the characteristic function of A at x, i.e., $A(x) = 1$ if x is in A, and $A(x) = 0$ otherwise. We identify A with its characteristic sequence $A(0)A(1)\ldots$.

We consider words over the binary alphabet $\{0, 1\}$. Words are ordered by the usual length-lexicographical ordering and the $(i + 1)$st word in this ordering is denoted by s_i, hence for example s_0 is the empty word λ. Occasionally, we identify words with natural numbers via the mapping $i \mapsto \mathrm{s}_i$. An *assignment* is a (total) function from some subset of the natural numbers to $\{0, 1\}$. An assignment is *finite* iff its domain is finite. An assignment with domain $\{0, \ldots, n - 1\}$ is identified in the natural way with a word of length n. The restriction of an assignment β to a set I is denoted by $\beta|I$, thus, in particular, for any set X, the assignment $X|I$ has domain I and agrees there with X. We call a subset of the natural numbers an *interval* if it is equal to $\{n, n + 1, \ldots, n + k\}$ for some natural numbers n and k.

The class of all sets is referred to as *Cantor space* and is denoted by $\{0, 1\}^\infty$. The class of all sets that have a word x as common prefix is called the *cylinder generated by* x and is denoted by $x\{0, 1\}^\infty$. For a set W, let $W\{0, 1\}^\infty$ be the union of all the cylinders $x\{0, 1\}^\infty$ where the word x is in W.

Recall the definition of the *uniform measure* (or *Lebesgue measure*) on Cantor space, which describes the distribution obtained by choosing the individual bits of a set by independent tosses of a fair coin.

1.2 Reducibilities

We briefly review some reducibilities, for a more detailed account we refer to Odifreddi [18] and Soare [21].

Recall the concept of an oracle Turing machine, i.e., a Turing machine that receives natural numbers as input, outputs binary values, and may ask during its computations queries of the form "$z \in X$?", where the set X, the *oracle*, can be conceived as an additional input to the computation. We write $M(X, x)$ for the binary output of an oracle Turing machine M on input x and oracle X, and we say $M(X,x)$ is undefined in case M does not terminate on input x and oracle X. Furthermore, we let $\mathrm{Q}(M,X,x)$ be the set of query words occurring during the computation of M on input x and with oracle X.

A set A is *Turing-reducible* to a set B if there is an oracle Turing machine M such that $M(B,x) = A(x)$ for all x. The definition of *truth-table-reducibility* is basically the same, except that in addition we require that M is total, i.e., for all oracles X and for all inputs x, the computation of $M(X,x)$ eventually terminates. By a result due to Nerode and to Trakhtenbrot [18, Proposition III.3.2], for any $\{0,1\}$-valued total oracle Turing machine there is an equivalent one that is again total and queries its oracle nonadaptively (i.e., M computes a list of queries that are asked simultaneously and after receiving the answers, M is not allowed to access the oracle again). A set A is *weakly truth-table-reducible* to a set B if A is Turing-reducible to B by an oracle Turing machine such that there is an computable function g that bounds its use, i.e., such that for all sets X, the set $\mathrm{Q}(M,X,x)$ contains only numbers less than or equal to $g(x)$. A set A is *computably enumerable* in a set B if there is an oracle Turing machine M such that $M(B,x) = 1$ in case $x \in A$ and $M(B,x)$ is undefined otherwise. For r in {tt, wtt, T, c.e.}, we say A is r-reducible to B, or $A \leq_{\mathrm{r}} B$ for short, if A is reducible to B with respect to truth-table, weak truth-table, Turing, or computably enumerable reducibility, respectively. By the definitions of these reducibilities it is immediate that

$$A \leq_{\mathrm{tt}} B \Rightarrow A \leq_{\mathrm{wtt}} B \Rightarrow A \leq_{\mathrm{T}} B \Rightarrow A \leq_{\mathrm{c.e.}} B$$

and in fact it can be shown that all these implications are strict.

In what follows, we consider reductions of a set to itself. Of course, reducing a set to itself is easy if one does not further restrict the oracle Turing machine performing the reduction. This leads to the concepts of autoreducibility and selfreducibility. A set is *T-autoreducible* if it can be reduced to itself by an oracle Turing machines that is not allowed to query the oracle at the current input, and a set is *T-selfreducible* if it can be reduced to itself by an oracle Turing machines that may only query the oracle at places strictly less than the current input. For reducibilities other than Turing reducibility, the concepts of auto- and selfreducibility are defined in the same manner. E.g., a set is *wtt-autoreducible* if it is T-autoreducible by an oracle Turing machine with a computable bound on its use, and a set A is *c.e.-selfreducible* if there is an oracle Turing machine that on input x queries its oracle only at places $z < x$ and such that $M(B,x) = 1$ in case $x \in A$, and $M(B,x)$ is undefined otherwise.

2 Random Sets

In this section, we review effective random sets and related concepts that are used in the following. For more comprehensive accounts of effective random sets and effective measure theory, we refer to the surveys cited in the bibliography [1,2,15].

Imagine a player that successively places bets on the individual bits of the characteristic sequence of an unknown set A. The betting proceeds in rounds

$i = 1, 2, \ldots$. During round i, the player receives as input the length $i-1$ prefix of A and then, first, decides whether to bet on the ith bit being 0 or 1 and, second, determines the stake that shall be bet. The stake might be any fraction between 0 and 1 of the capital accumulated so far, i.e., in particular, the player is not allowed to incur debts. Formally, a player can be identified with a *betting strategy*

$$b : \{0,1\}^* \to [-1, 1]$$

where on input w the absolute value of $b(w)$ is the fraction of the current capital that shall be at stake and the bet is placed on the next bit being 0 or 1 depending on whether $b(w)$ is negative or nonnegative.

The player starts with strictly positive, finite capital $d_b(\lambda)$. At the end of each round, in case the current guess has been correct, the capital is increased by this round's stake and, otherwise, is decreased by the same amount. So given a betting strategy b, we can inductively compute the corresponding *payoff function*, or *martingale*, d_b by applying the equations

$$d_b(w0) = d_b(w) - b(w) \cdot d_b(w), \qquad\qquad d_b(w1) = d_b(w) + b(w) \cdot d_b(w) .$$

Intuitively speaking, the payoff $d_b(w)$ is the capital the player accumulates till the end of round $|w|$ by betting on a set that has the word w as a prefix.

Conversely, any function d from strings to nonnegative reals that for all strings w satisfies the fairness condition

$$d(w) = \frac{d(w0) + d(w1)}{2} \tag{1}$$

induces canonically a betting function b. By the preceding discussion it follows for gambles as described above that for any martingale there is an equivalent betting strategy and vice versa. We will frequently identify martingales and betting strategies via this correspondence and, if appropriate, notation introduced for martingales will be extended to the induced betting strategies.

Definition 1. *A betting strategy b* succeeds *on a set A if the corresponding payoff function d_b is unbounded on the prefixes of A, i.e., if*

$$\limsup_{m \to \infty} d_b(A|\{0, \ldots, m\}) = \infty .$$

A betting strategy is computable if it is confined to rational values and there is a Turing machine that on input w outputs an appropriate finite representation of $b(w)$. In the context of recursion theory, usually computable betting strategies are considered [1,19,20,22], while in connection with complexity classes one considers betting strategies that in addition are computable within appropriate resource-bounds [2,14,15,17]. Observe that for a computable betting strategy with rational-valued initial capital the corresponding martingale is computable and vice versa.

Definition 2. *A set is* rec-random *if no computable betting strategy succeeds on it.*

Besides rec-random sets, we consider Martin-Löf random sets [16]. Let $\mathrm{W}_0, \mathrm{W}_1, \ldots$ be the standard enumeration of the computably enumerable sets [21].

Definition 3. *A class* $\mathcal{N}$ *is called a* Martin-Löf null class *if there exists a computable function* $g\colon \mathbb{N} \to \mathbb{N}$ *such that for all* i

$$\mathcal{N} \subseteq \mathrm{W}_{g(i)}\{0,1\}^\infty \quad \textit{and} \quad \mathrm{Prob}[\mathrm{W}_{g(i)}\{0,1\}^\infty] < \frac{1}{2^i} .$$

A set is Martin-Löf random *if it is not contained in any Martin-Löf null class.*

By definition, a class $\mathcal{N}$ has uniform measure 0 if there is a, not necessarily computable, function g as in Definition 3. Thus the concept of a Martin-Löf null class is indeed an effective variant of the classical concept of a class that has uniform measure 0. In particular, any Martin-Löf null class has uniform measure 0. By σ-additivity and since there are only countably many computable functions, also the union of all Martin-Löf null classes has uniform measure 0, hence the class of Martin-Löf random sets has uniform measure 1.

Martin-Löf random sets have been characterized in terms of martingales by Schnorr [20]. A set R is Martin-Löf random if and only if there is no subcomputable martingale which succeeds on R. A martingale d is *subcomputable*, sometimes also called *lower semi-computable*, if and only if there is a computable function $\widetilde{d}$ in two arguments such that for all words w, the sequence $\widetilde{d}(w,0), \widetilde{d}(w,1), \ldots$ is nondecreasing and converges to $d(w)$.

Remark 4. The class of Martin-Löf random sets is properly contained in the class of rec-random sets.

From the definitions of Martin-Löf random and rec-random sets in terms of subcomputable and computable betting strategies, it is immediate that the Martin-Löf random sets are contained in the rec-random sets. The strictness of this inclusion was implicitly shown by Schnorr [20]. For a proof, it suffices to recall that the prefixes of a Martin-Löf random set cannot be compressed by more than a constant while a corresponding statement for rec-random sets is false [13, Theorem 3.6.1 and Exercise 2.5.13].

Remark 5. It can be shown that the union of all Martin-Löf null classes is again a Martin-Löf null class; accordingly, there is a subcomputable martingale that succeeds on every set that is not Martin-Löf random [5, Section 6.2].

Remark 6. Given any martingale d, word w, and natural number k, we have

$$d(w) \;=\; \frac{1}{2^k} \sum_{u \in \{0,1\}^k} d(wu) \;, \tag{2}$$

as follows by an easy inductive argument that uses the fairness condition (1). Observe that conversely (1) is a special case of (2) where $k = 1$.

3 Every Set Is Reducible to a Martin-Löf Random Set

On first sight, it might appear that it is impossible to decode effectively from a Martin-Löf random set any meaningful information; such decoding seems to presuppose certain regularities in the given set, which in turn might be exploited in order to come up with a computable or subcomputable martingale that succeeds on the set. So it comes as a slight surprise that in fact any set is wtt-reducible to a Martin-Löf random set. This celebrated result has been obtained independently by Gács [10] and Kučera [11,12]. They state the result for T-reducibility, however the reductions constructed in their proofs are already wtt-reductions [22, Section 6.1]. In what follows, we give an alternate proof for their result. Subsequently, we adjust the construction in the proof in order to obtain c.e.-selfreducible Martin-Löf random sets and wtt-autoreducible rec-random sets.

The proof given by Gács is considerably more involved than ours, however his coding is much less redundant, see Remark 9. Compared to the proof of Kučera, our proof is somewhat more intuitive because it works by direct diagonalization against a universal subcomputable martingale as in Remark 5.

In the proof of Theorem 8 we use the following technical remark.

Remark 7. Given a rational $\delta > 1$ and a natural number k, we can compute a length $l(\delta, k)$ such that for any martingale d and any word v we have

$$|\{w \in \{0,1\}^{l(\delta,k)} : d(vw) \leq \delta d(v)\}| \geq k.$$

That is, for any martingale d and for any interval I of length $l(\delta, k)$ there are at least k assignments w on I such that d increases its capital by at most a factor of δ while betting on I, no matter how the restriction v of the unknown set to the places to the left of I looks like.

For a proof observe that by the generalized fairness condition (2), for any given v and l the average of $d(vw)$ over all words w of length l is just $d(v)$; hence by the Markov inequality we have

$$\frac{|\{w \in \{0,1\}^l : d(vw) > \delta d(v)\}|}{2^l} < \frac{1}{\delta}.$$

By $\delta > 1$, we have $1 - 1/\delta > 0$, hence it suffices to let $l(\delta, k) \geq \log \frac{k}{1-\frac{1}{\delta}}$.

Theorem 8 (Gács, Kučera). *Every set is wtt-reducible to a Martin-Löf random set.*

Proof. Fix a decreasing sequence $\delta_0, \delta_1, \ldots$ of rationals with $\delta_i > 1$ for all i such that the sequence $\beta_0, \beta_1, \ldots$ converges where

$$\beta_s = \prod_{i \leq s} \delta_i \, .$$

In addition, assume that given i we can compute an appropriate representation of δ_i. For $s = 0, 1, \ldots$, let $l_s = l(\delta_s, 2)$, where $l(.,.)$ is the function from

Remark 7. Partition the natural numbers into consecutive intervals $I_0, I_1, \ldots$ of length $l_0, l_1, \ldots$, respectively. For further use note that by choice of the l_s, for any word v and any martingale d, there are at least two words w of length l_s where

$$d(vw) \leq \delta_s d(v) . \quad (3)$$

Let X be any set. We construct a set R to which X is wtt-reducible, where the construction is done in stages $s = 0, 1, \ldots$. During stage s we specify the restriction of R to I_s. We ensure that R is Martin-Löf random as follows. According to Remark 5, fix a subcomputable martingale d that succeeds on all sets that are not Martin-Löf random. At stage s, call a word w of length l_s an *admissible extension* in case $s = 0$ if $d(w) \leq \beta_0$ and in case $s > 0$ if

$$d(vw) \leq \beta_s \quad \text{where } v = R|(I_0 \cup \ldots \cup I_{s-1}) .$$

During each stage s, we let $R|I_s$ be equal to some admissible extension. Since the β_s are bounded, this implies that d does not succeed on R, hence R is Martin-Löf random.

We will argue in a minute that at each stage there are at least two admissible extensions. Assuming the latter, the set X can be coded into R as follows. During stage s let $R|I_s$ be equal to the greatest admissible extension in case s is in X, and let $R|I_s$ be equal to the least admissible extension otherwise. An oracle Turing machine M that wtt-reduces X to R works as follows. On input s, M queries its oracle in order to obtain the restrictions v_s and w_s of the oracle to the sets $I_0 \cup \ldots \cup I_{s-1}$ and I_s, respectively. Then M runs two subroutines in parallel. Subroutine 0 simulates in parallel enumerations of $d(v_s w)$ for all $w < w_s$ and terminates if the simulation shows that $d(v_s w) > \beta_s$ for all these w, i.e., Subroutine 0 terminates if the simulation shows that no such w is an admissible extension of v_s. Subroutine 1 does the same for all $w > w_s$.

In case Subroutine i terminates before Subroutine $1-i$, then M outputs i. By construction, with oracle R for every s exactly one of the subroutines terminates and M correctly computes $X(s)$.

It remains to show that at each stage there are at least two possible extensions to choose from. Fix any stage s and by induction assume that the restriction v_s of R to the intervals I_0 through I_{s-1} could be defined by choosing admissible extensions at the previous stages and that hence we have $d(v_s) \leq \beta_{s-1}$. Then by (3) there are at least two words w of length l_s where

$$d(v_s w) \leq \delta_s d(v_s) \leq \delta_s \beta_{s-1} = \beta_s ,$$

i.e., at stage s there at least two admissible extensions. □

Remark 9. It can be shown that an infinite product of the form $\prod(1 + \alpha_i)$ with $\alpha_i \geq 0$ converges if and only if the series $\sum \alpha_i$ converges [3, Theorem 8.52]. Hence in the proof of Theorem 8 the β_i are bounded if we let $\delta_i = 1 + (i+1)^{-2}$. For this choice of δ_i, the interval length $l_i = l(\delta_i, 2)$ and thus the number of bits used to encode the bit $X(i)$ are bounded above by $4 + 2\log i$.

The proof given by Gács is considerably more involved, however his coding is much less redundant and uses only i bits in order to code $i - O(i^{1/2} \log i)$ bits.

4 Selfreductions of Martin-Löf Random Sets and Computably Enumerable Reals

In the proof of Theorem 8, we have constructed a Martin-Löf random set where bit $X(i)$ of the given set X has been coded into interval I_i by choosing either the least or the greatest admissible extension. We argue next that if we adjust the construction such that in each interval simply the least admissible extension is chosen, we obtain a set that is Martin-Löf random and c.e.-selfreducible. The existence of such a set has been shown before in a setting of computably enumerable reals; the connection between our construction and the previous result becomes clear by Proposition 11, which asserts that a set is c.e.-selfreducible if and only if its associated real is computably enumerable. Furthermore, in Remark 13, we argue that the existence of a c.e.-selfreducible Martin-Löf random set complements the fact that Martin-Löf random sets cannot be T-autoreducible.

The real associated with a set B is $0.b_0b_1\dots$ where $b_i = B(i)$. A real is called *Martin-Löf random* if it is associated to a Martin-Löf random set.

Definition 10. *A computably enumerable real,* c.e. real *for short, is a real that is the limit of a nondecreasing computable sequence of rationals.*

Computably enumerable reals are also called *left computable reals.*

Proposition 11. *A set is c.e.-selfreducible if and only if its associated real is c.e.*

Proof. Fix any set A and let $a_0a_1\dots$ be its characteristic sequence. The equivalence asserted in the proposition is immediate in case the characteristic sequence is eventually constant, i.e., if $a_j = a_{j+1} = \dots$ for some j. So assume otherwise.

First let A be c.e.-selfreducible by an oracle Turing machine M. Let α denote the real that is associated with A. We define inductively a computable sequence $\alpha_0, \alpha_1, \dots$ of rational numbers that converges nondecreasingly to α and where α_s can be written in the form

$$\alpha_s = 0.a_0^s \dots a_s^s, \quad a_j^i \in \{0,1\}.$$

Let $M_s(X,x)$ be the approximation to $M(X,x)$ obtained by running M for s steps on input x and oracle X; i.e., $M_s(X,x) = M(X,x)$ if M terminates within s computation steps, and $M_s(X,x)$ is undefined otherwise. For a start, let $a_0^0 = 0$, i.e., $\alpha_0 = 0$. In order to define α_s for $s > 0$, we distinguish two cases. In case for some $j < s$, we have

$$a_j^{s-1} = 0 \quad \text{and} \quad M_s(a_0^{s-1} \dots a_{j-1}^{s-1} 0^\infty, j) = 1,$$

then let j_s be the least such j and let

$$\alpha_s = 0.a_0^{s-1} \dots a_{j_s-1}^{s-1} 1 0^{s-j_s}.$$

In case there is no such j, let

$$\alpha_s = 0.a_0^{s-1} \dots a_{s-1}^{s-1} 0.$$

By construction, the sequence $\alpha_0, \alpha_1, \ldots$ is nondecreasing. Furthermore, an easy induction argument shows that $a_0^s a_1^s \ldots$ converges pointwise to $a_0 a_1 \ldots$ as s goes to infinity, and consequently the α_s converge to α.

Next assume that the real $\alpha = 0.a_0 a_1 \ldots$ is c.e. Let $\alpha_0, \alpha_1, \ldots$ be a computable sequence of rationals that converges nondecreasingly to α. Then the set A is c.e.-selfreducible by an oracle Turing machine M that works as follows. On input s, M queries its oracle in order to obtain the length s prefix $z_0 \ldots z_{s-1}$ of the oracle. Then M checks successively for $i = 0, 1, \ldots$ whether

$$\alpha_i > 0.z_0 \ldots z_{s-1}1; \tag{4}$$

if eventually such an index i is found, M outputs 1 while otherwise, if there is no such i, M does not terminate.

Now suppose that M is applied to oracle A and any input s. If $a_s = 0$, then (4) is false for all i, hence M does not terminate. On the other hand, if $a_s = 1$ then α is strictly larger than the righthand side of (4) because by case assumption there is some $j > s$ such that $a_j = 1$. Hence (4) is true for almost all i and M eventually outputs 1. □

Corollary 12. *There is a real that is Martin-Löf random and c.e. (Equivalently, there is a set that is Martin-Löf random and c.e.-selfreducible.)*

Proof. The two formulations of the assertion of the corollary are equivalent by Proposition 11. The construction in the proof of Theorem 8 yields a Martin-Löf random set in case the chosen extensions are always admissible. Thus it suffices to show that the set R that is obtained by always choosing the least admissible extension is c.e.-selfreducible. A machine M witnessing that R is c.e.-selfreducible works as follows. On input x, first M queries its oracle at all places strictly less than x and receives as answer the length x prefix α_x of its oracle. Then M computes the index s such that x is in the interval I_s, and lets v_s be the prefix of α_x of length $l_0 + \ldots + l_{s-1}$. Note that $R(x) = 1$ if and only if during stage s of the construction there has been no admissible extension w such that $v_s w$ extends $\alpha_x 0$ and recall that an extension w is admissible if $d(v_s w) \leq \beta_s$. So M may simply try to prove $d(v_s w) > \beta_s$ for all w where $v_s w$ extends $\alpha_x 0$ by approximating d from below, then outputting a 1 in case of success. □

Corollary 12 has been demonstrated before by Chaitin. A Chaitin Ω real is the halting probability of some universal prefix-free Turing machine. Chaitin could show that every Chaitin Ω real is Martin-Löf random and c.e. Indeed, also the reverse implication holds, i.e., the Chaitin Ω reals are just the reals that are Martin-Löf random and c.e. For a proof of this characterization and for references see Calude [6], where the equivalence is attributed to work of Calude, Hertling, Khoussainov, and Wang, of Chaitin, of Kučera and Slaman, and of Solovay.

By Corollary 12, there are Martin-Löf random sets that are c.e.-selfreducible. By the following remark, this result does not extend to the less powerful T-reducibility, i.e., no Martin-Löf random set is T-autoreducible.

Remark 13. Consider the following, more liberal variant of T-autoreducibility. A set A is infinitely often (i.o.) T-autoreducible if there is an oracle Turing machine that on input x eventually outputs either the correct value $A(x)$ or a special symbol that signals ignorance about the correct value; in addition, the correct value is computed for infinitely many inputs. The concept of i.o. tt-autoreducibility is defined accordingly, i.e., we require in addition that the machine performing the reduction is total.

Ebert, Merkle, and Vollmer [7,8,9] showed that every Martin-Löf random set is i.o. tt-autoreducible; in fact, it can be arranged that the fraction of correctly computed places up to input x exceeds $r(x)$ where r is any given computable rational-valued function that goes nonascendingly to 0. On the other hand, no Martin-Löf random set R is i.o. T-autoreducible in such a way that in the limit the fraction of places where $R(x)$ is computed correctly is a nonzero constant. The latter assertion remains true with Martin-Löf random and i.o. T-autoreducible replaced by rec-random and i.o. tt-autoreducible. In particular, no Martin-Löf random set is T-autoreducible and no rec-random set is tt-autoreducible.

5 Autoreductions of Rec-random Sets

Theorem 14. *There is a set that is rec-random and wtt-autoreducible.*

Proof. A set R as required can be obtained by a construction similar to the one used in the proof of Theorem 8. Choose $\delta_0, \delta_1, \ldots$ and $\beta_0, \beta_1, \ldots$ as in that proof and let $\varepsilon_s = (\delta_s - 1)/2$. Again, partition the natural numbers into consecutive intervals $I_0, I_1, \ldots$, however now interval I_s has length

$$l_s = l\left(1 + \frac{\varepsilon_s}{2}, 3\right).$$

Let $d_0, d_1, \ldots$ be an appropriate effective enumeration of all partial computable functions from $\{0,1\}^*$ to the rational numbers and let

$$E = \{e \colon d_e \text{ is a (total) martingale with initial capital } d_e(\lambda) = 1\}\,.$$

For the sake of simplicity we assume that 0 is in E. Furthermore, let

$$\widetilde{d}_s = \sum_{\{e \in E \colon e \leq s\}} \tau_e d_e \quad \text{where} \quad \tau_s = \frac{\varepsilon_s}{2^{l_0 + \ldots + l_s + 1}}\,.$$

The set R is constructed in stages $s = 0, 1, \ldots$ where during stage s we specify the restriction of R to the interval I_s. At stage s call a word w of length l_s an *admissible extension* if $s = 0$ or if $s > 0$ and we have

$$\widetilde{d}_{s-1}(vw) \leq \left(1 + \frac{\varepsilon_s}{2}\right) \widetilde{d}_{s-1}(v) \quad \text{where} \quad v = R|I_0 \cup \ldots \cup I_{s-1}. \tag{5}$$

Again at every stage s we will let $R|I_s$ be equal to some admissible extension and we argue that this way the set R automatically becomes rec-random. For a proof of the latter it suffices to show for all s we have

$$\widetilde{d}_s(R|I_0 \cup \ldots \cup I_s) \leq \beta_s. \tag{6}$$

Assuming that some d_i succeeded on R contradicts (6) because the β_i are bounded and because $\widetilde{d}_s \geq \tau_i d_i$ for $s \geq i$.

Inequality (6) follows by an inductive argument. For $s = 0$ we have

$$\widetilde{d}_0(R|I_0) = \tau_0 d_0(R|I_0) \leq \tau_0 2^{l_0} = \frac{\varepsilon_0}{2} \leq \delta_0 = \beta_0.$$

In the induction step, let v and w be the restriction of R to $I_0 \cup \ldots \cup I_{s-1}$ and to I_s, respectively. By the definition of admissible extension and by the induction hypothesis, we have

$$\widetilde{d}_{s-1}(vw) \leq \left(1 + \frac{\varepsilon_s}{2}\right) \widetilde{d}_{s-1}(v) \leq \left(1 + \frac{\varepsilon_s}{2}\right) \beta_{s-1}\,.$$

By definition, the values of $\widetilde{d}_{s-1}(vw)$ and of $\widetilde{d}_s(vw)$ are the same in case s is not in E, while otherwise they differ by

$$\tau_s d_s(vw) \leq \tau_s 2^{|vw|} \leq \frac{\varepsilon_s}{2}\,,$$

where the inequalities follow because a martingale can at most double at each step and by the definition of τ_s. In summary we have

$$\widetilde{d}_s(vw) \leq \left(1 + \frac{\varepsilon_s}{2}\right) \beta_{s-1} + \frac{\varepsilon_s}{2} \leq \delta_s \beta_{s-1} = \beta_s\,.$$

It remains to show that we can arrange that R is wtt-autoreducible. At stage s, let (w_0, w_1) be the least pair of admissible extensions such that w_0 and w_1 differ at least at two places. Then let the restriction of R to I_s be equal to w_0 in case $s+1 \notin E$ and be equal to w_1 otherwise. Observe that there is always such a pair because by choice of l_s there are at least 3 admissible extensions, hence there are at least two admissible extensions that differ in at least two distinct places. (Indeed, given three mutually distinct words w, w', w'' of the same length, then if w and w' differ only at one place, w'' must differ from w' at some other place, hence w' and w'' or w and w'' differ in at least two places.)

The set R is wtt-autoreducible by an oracle Turing machine M that works as follows. For simplicity, we describe the behavior of M for the case where its oracle is indeed R and omit the straightforward considerations for other oracles; anyway it should be clear from the description that M is of wtt-type.

On input x, first M determines the index s such that x is in I_s. Then M queries the oracle at all places in $I_0 \cup \ldots \cup I_s$ except at x; this way M obtains in particular the restrictions $v_0, \ldots, v_{s-1}$ of R to $I_0, \ldots, I_{s-1}$, respectively. Next M computes successively $E(j)$ for $j = 1, \ldots, s-1$. Observe that given the v_i and $E|\{0, \ldots, j-1\}$, it is possible to compute the admissible words and the words w_0 and w_1 of stage j; by comparing the two latter words to v_j one can then compute $E(j)$. Finally, M determines $R(x)$ by computing the words w_0 and w_1 of stage s and by comparing them to the known part of v_s. The last step exploits that w_0 and w_1 differ at least at two places and thus differ on $I_s \setminus \{x\}$. □

Acknowledgments

We like to thank Klaus Ambos-Spies, Antonín Kučera, and Frank Stephan for helpful discussion.

References

1. K. Ambos-Spies and A. Kučera. Randomness in computability theory. In P. A. Cholak et al. (eds.), *Computability Theory and Its Applications. Current Trends and Open Problems.* Contemporary Mathematics 257:1–14, American Mathematical Society (AMS), 2000.
2. K. Ambos-Spies and E. Mayordomo. Resource-bounded measure and randomness. In A. Sorbi (ed.), *Complexity, Logic, and Recursion Theory*, p. 1-47. Dekker, New York, 1997.
3. T. M. Apostol. *Mathematical Analysis*, third edition. Addison Wesley, 1978.
4. J. L. Balcázar, J. Díaz, and J. Gabarró. *Structural Complexity I*, Springer, 1995.
5. C. Calude, *Information and Randomness*, Springer-Verlag, 1994.
6. C. S. Calude. *A characterization of c.e. random reals.* Theoretical Computer Science 271:3-14, 2002.
7. T. Ebert. *Applications of Recursive Operators to Randomness and Complexity.* Ph.D. Thesis, University of California at Santa Barbara, 1998.
8. T. Ebert and H. Vollmer. *On the autoreducibility of random sequences.* In: M. Nielsen and B. Rovan (eds.), Mathematical Foundations of Computer Science 2000, Lecture Notes in Computer Science 1893:333–342, Springer, 2000.
9. T. Ebert and W. Merkle. *Autoreducibility of random sets: a sharp bound on the density of guessed bits.* Mathematical Foundations of Computer Science 2002, this volume.
10. P. Gács. *Every sequence is reducible to a random one.* Information and Control 70:186–192, 1986.
11. A. Kučera. Measure, Π^0_1-classes and complete extensions of PA. In: H.-D. Ebbinghaus et al. (eds.), *Recursion Theory Week.* Lecture Notes in Mathematics 1141:245-259, Springer, 1985.
12. A. Kučera. On the use of diagonally nonrecursive functions. In: H.-D. Ebbinghaus et al. (eds.), *Logic Colloquium '87.* Studies in Logic and the Foundations of Mathematics 129:219-239, North-Holland, 1989.
13. M. Li and P. Vitányi. *An Introduction to Kolmogorov Complexity and Its Applications*, second edition. Springer, 1997.
14. J. H. Lutz. Almost everywhere high nonuniform complexity. *Journal of Computer and System Sciences*, 44:220–258, 1992.
15. J. H. Lutz. The quantitative structure of exponential time. In L. A. Hemaspaandra and A. L. Selman (eds.), *Complexity Theory Retrospective* II, p. 225–260, Springer, 1997.
16. P. Martin-Löf. The definition of random sequences. *Information and Control* 9(6):602–619, 1966.
17. E. Mayordomo. *Contributions to the Study of Resource-Bounded Measure.* Doctoral dissertation, Universitat Politècnica de Catalunya, Barcelona, Spain, 1994.
18. P. Odifreddi. *Classical Recursion Theory.* North-Holland, Amsterdam, 1989.
19. C.-P. Schnorr. A unified approach to the definition of random sequences. *Mathematical Systems Theory*, 5:246–258, 1971.

20. C.-P. Schnorr. *Zufälligkeit und Wahrscheinlichkeit.* Lecture Notes in Mathematics 218, Springer, 1971.
21. R. I. Soare. *Recursively Enumerable Sets and Degrees.* Springer, 1987.
22. S. A. Terwijn. *Computability and Measure.* Doctoral dissertation, Universiteit van Amsterdam, Amsterdam, Netherlands, 1998.
23. B. A. Trakhtenbrot, *On autoreducibility.* Soviet Math. Doklady, 11:814-817, 1970.

On the Structure of the Simulation Order of Proof Systems
(Extended Abstract)

Jochen Messner

Abteilung Theoretische Informatik, Universität Ulm,
89069 Ulm, Germany,
messner@informatik.uni-ulm.de

Abstract. We examine the degree structure of the simulation relation on the proof systems for a set L. As observed, this partial order forms a distributive lattice. A greatest element exists iff L has an optimal proof system. In case L is infinite there is no least element, and the class of proof systems for L is not presentable. As we further show the simulation order is dense. In fact any partial order can be embedded into the interval determined by two proof systems f and g such that f simulates g but g does not simulate f. Finally we obtain that for any non-optimal proof system h an infinite set of proof systems that are pairwise incomparable with respect simulation and that are also incomparable to h.

1 Introduction

In order to compare the relative efficiency of the various known propositional proof systems Cook and Reckhow introduced the notion of simulation and p-simulation in [4,5]. There, the intuitive notion of a proof system is formalized by considering any polynomial time computable function h as a *proof system* for its range L. In this setting an h-proof for $y \in L$ is provided by a word w with $h(w) = y$. Now simulation and p-simulation are defined as a kind of many-one reducibility between proof systems. Formally, for two (partial) polynomial time computable functions $h, g \in \mathsf{FP}$ let us denote by $g \leq_{\mathrm{m}}^{\mathrm{np}} h$ that there is an at most polynomially length increasing function $f : \Sigma^* \to \Sigma^*$ such that $h(f(w)) = g(w)$ for any word w in the domain of g; if additionally $f \in \mathsf{FP}$ we denote this by $g \leq_{\mathrm{m}}^{\mathrm{p}} h$. Now if h and g are proof systems for L, and $g \leq_{\mathrm{m}}^{\mathrm{np}} h$ (resp. $g \leq_{\mathrm{m}}^{\mathrm{p}} h$) then h is said to *(p-)simulate* g. In other words: for simulation it is required that h-proofs are at most polynomially longer than the corresponding g-proofs, whereas for p-simulation it is additionally required that there is an efficient procedure translating g-proofs to the corresponding h-proofs. It is straightforward to see that $\leq_{\mathrm{m}}^{\mathrm{np}}$ and $\leq_{\mathrm{m}}^{\mathrm{p}}$ are both preorders, i.e. reflexive and transitive relations.

Especially for the set of propositional tautologies TAUT many concrete proof systems have been studied in the literature. Starting with the work of Tseitin [16] where it is observed that regular resolution does not simulate extended resolution, the investigation of the simulation relations that exist between these numerous known propositional proof systems has developed into a vivid field of

K. Diks et al. (Eds): MFSC 2002, LNCS 2420, pp. 581–592, 2002.

research. This research has in the meanwhile revealed a rich structure of these preorders, see e.g. [14,4,5,17,6].

In spite of these interesting results about concrete proof systems, we don't know much about the general structure of these preorders. Only the question on the existence of greatest elements—so called (p-)optimal proof systems—received some attention. It is known that the existence of a (p-)optimal proof system for TAUT is related to difficult open complexity theoretical questions (see [9,13,8,12]). When considering proof systems for other sets L it is easy to see that any set in NP (resp. P) has a (p-)optimal proof system (cf. [8]). On the other hand, no set that is hard for coNEXP (resp. EXP) with respect to polynomial time many-one reducibility has a (p-)optimal proof system (see [12]).

The purpose of this paper is to study the further order theoretic properties of these preorders (apart from the existence of a greatest element). Moreover, we will consider proof systems for any set L. Hence, for a recursively enumerable set L we will investigate the preorders $(\mathsf{FP}_{=L}, \leq_{\mathrm{m}}^{\mathrm{np}})$, and $(\mathsf{FP}_{=L}, \leq_{\mathrm{m}}^{\mathrm{p}})$ where $\mathsf{FP}_{=L}$ denotes the class of proof systems for L (i.e. the class of functions from FP with range L). In the following, basically all our observations hold for $\leq_{\mathrm{m}}^{\mathrm{np}}$, and $\leq_{\mathrm{m}}^{\mathrm{p}}$ interchangeably, and therefore we will use $r \in \{\mathrm{p}, \mathrm{np}\}$, and $\leq_{\mathrm{m}}^{r}$ to denote that a statement holds for simulation as well as for p-simulation. To use standard order theoretic notions it is useful (and usual, see e.g. [10,1]) to switch to the corresponding partial orders. Here for some proof system $f \in \mathsf{FP}_{=L}$ let the $\leq_{\mathrm{m}}^{r}$-*degree* of f be defined by $\mathsf{deg}_r(f) = \{g \mid g =_{\mathrm{m}}^{r} f\}$ where $g =_{\mathrm{m}}^{r} f$ denotes that $g \leq_{\mathrm{m}}^{r} f$ and $f \leq_{\mathrm{m}}^{r} g$. Let $\mathbf{FP}_L^r$ denote the class of all such degrees and let $\leq$ denote the partial order on $\mathbf{FP}_L^r$ induced by $\leq_{\mathrm{m}}^{r}$ on the degrees.

It turns out that the partial order $(\mathbf{FP}_L^r, \leq)$ is a lattice that therefore may be equivalently described by the algebra $(\mathbf{FP}_L^r, \sqcap, \sqcup)$ (cf. [3,7]) where $\sqcap$ ($\sqcup$) denotes the greatest lower (least upper) bound of two degrees with respect to $\leq$. Moreover, this lattice is distributive. Then we show that $(\mathbf{FP}_L^r, \leq)$ is a dense order. In fact, we show that for any two degrees $\mathfrak{a} < \mathfrak{b}$ from $\mathbf{FP}_L^r$ any countable distributive lattice $\mathcal{L}$ can be embedded into the sublattice $[\mathfrak{a}, \mathfrak{b}] = \{\mathfrak{c} \mid \mathfrak{a} \leq \mathfrak{c} \leq \mathfrak{b}\}$ of $(\mathbf{FP}_L^r, \sqcap, \sqcup)$ such that the greatest element of $\mathcal{L}$ (if it exists) is mapped to $\mathfrak{b}$ and the minimal element of $\mathcal{L}$ (if it exists) is mapped to $\mathfrak{a}$. Further we show that any non-greatest degree of $\mathbf{FP}_L^r$ is a member of an infinite set of pairwise incomparable degrees. Hence, for any non-(p-)optimal proof system h for L, there is an infinite number of proof systems that are pairwise incomparable and that are also incomparable to h (with respect to (p-)simulation).

To obtain these result we apply techniques developed in the theory of the polynomial time degrees of sets (see [10,11,15,2,1]). It is observed in [11] (cf. also [15,1]) that for the investigation of polynomial time degrees of sets the notion of a recursively presentable class plays an important and useful role. The same statement holds when considering the degree structure of proof systems. We therefore also study presentability questions for classes of proof systems. As an interesting result we obtain here that in case L is infinite the class of the degrees $\mathbf{FP}_L^r$ is not presentable which means that there is no recursive enumeration of

Turing machines $M_1, M_2, \ldots$ computing the functions $f_1, f_2, \ldots$ such that $\mathsf{FP}^r_L = \{\deg_r(f_i) \mid i \geq 1\}$.

The rest of this paper is structured as follows: In the next section we provide some basic notions from complexity and lattice theory. Then in Section 3 we show that $(\mathsf{FP}^r_L, \leq)$ is a distributive lattice that also uniquely determines the preorder $(\mathsf{FP}_{=L}, \leq^r_{\mathrm{m}})$. In Section 4 we examine presentability questions. Finally Section 5 contains the main results on the structure of $(\mathsf{FP}^r_L, \leq)$. Due to the space restrictions several proofs had to be omitted in this extended abstract.

2 Preliminaries

As a basic alphabet we consider $\Sigma = \{0, 1\}$ (this means that problem instances like, e.g., propositional formulas have to be suitably encoded as binary words). By Σ^* we denote the set of all binary words. We identify a word $x \in \Sigma^*$ with the positive integer $n \in \mathbb{N} = \{1, 2, 3, \ldots\}$ that has $1x$ as binary representation. Moreover we consider the words totally ordered by the standard order $\leq$ on $\mathbb{N}$. The length of a word $w \in \Sigma^*$ is denoted by $|w|$. By $\Sigma^{\leq n}$ we denote the set of all words of length n or less.

A Turing machine M computes a partial function $f : \Sigma^* \to \Sigma^*$ defined by $f(x) = M(x)$. Here, $M(x) = y$ denotes that M on input x reaches a final state and outputs y (i.e. stops with y written on its output tape), and $M(x) = \perp$ denotes that M on input x does not halt or otherwise halts in a non-final state. The set $f(\Sigma^*) = \{f(x) \mid x \in \Sigma^* \wedge f(x) \neq \perp\}$ is called the *range* of f. A Turing machine M is called *polynomial time machine* (*PTM*, for short) if a polynomial shut of clock is attached to M that stops any computation of M on input x after a polynomial number of steps. By FP we denote the class of functions computed by some PTM. Further let $\mathsf{FP}_{=L}$, $\mathsf{FP}_{\subseteq L}$, resp. $\mathsf{FP}_{\supseteq L}$ denote the classes of functions $f \in \mathsf{FP}$ such that $f(\Sigma^*) = L$, $f(\Sigma^*) \subseteq L$, resp. $f(\Sigma^*) \supseteq L$. The composition of functions f, g is denoted by $f \circ g$ (here we assume the conventions that $f(\perp) = \perp$). A total function $t : \mathbb{N} \to \mathbb{N}$ with $t(n) > n$ is called *time-constructible* if there is a Turing machine M with $\mathrm{time}_M(0^n) = t(n)$ where $\mathrm{time}_M(x)$ denotes the number of steps of M on input x. It is known that for any total recursive function $r : \mathbb{N} \to \mathbb{N}$ there is a time-constructible function t with $r(n) \leq t(n)$ for any n. A function $f : \Sigma^* \to \Sigma^*$ is called *at most polynomially length increasing* if there is a polynomial p such that $|f(w)| \leq p(|w|)$ for any word w. By P we denote the class of the polynomial time computable sets.

We will use standard order and lattice theoretic notions from [3,7]. A pair $(V, \leq)$ is called a *preorder* if V is a nonempty set, and the relation $\leq \subseteq V \times V$ is reflexive and transitive. When it is clear from the context which order relation is used, we will shortly use the underlying set V to denote the preorder $(V, \leq)$. For $u, v \in V$ we denote by $u < v$ that $u \leq v$ but not $v \leq u$. Two elements $u, v \in V$ are called *incomparable* if neither $u \leq v$ nor $v \leq u$. A subset $A \subseteq V$ is called *antichain* if its elements are pairwise incomparable. In a preorder $(V, \leq)$ an element $u \in V$ is called an *upper (lower) bound* for $U \subseteq V$ if $v \leq u$ (resp. $u \leq v$) for any $v \in U$. If additionally for any upper (lower) bound u' for U it holds $u \le u'$ ($u' \le u$), u is called a *least upper* (*greatest lower*) bound for U. If $v \in V$ is an upper bound (lower bound) for V it is called a *greatest* (*least*) element of $(V, \le)$.

A preorder $(V, \le)$ is called a *partial order* if the relation $\le$ is antisymmetric. A partial order is called *dense* if for any $u, w \in V$ such that $u < v$ there is a $w \in V$ with $u < w$ and $w < v$. In a partial order the least upper (greatest lower) bound of any subset $U \subseteq V$ is unique, if it exists. If in a partial order $(V, \le)$, for any two elements $\{v, w\} \subseteq V$ a least upper bound (denoted $v \sqcup w$) as well as a greatest lower bound (denoted $v \sqcap w$) exists, then $(V, \le)$ is called a *lattice*. As observed in [3,7] a lattice $(V, \le)$ is equally characterized by the algebra $(V, \sqcap, \sqcup)$. A lattice is called *distributive* if $u \sqcap (v \sqcup w) = (u \sqcap v) \sqcup (u \sqcap w)$ for any $u, v, w \in V$ (this already implies $u \sqcup (v \sqcap w) = (u \sqcup v) \sqcap (u \sqcup w)$). For two elements $u, v \in V$ let $[u, v] = \{w \in V \mid u \le w \le v\}$. Notice that in case $(V, \le)$ is a lattice then $([u, v], \le)$ is a sublattice, i.e. for any $w_1, w_2 \in [u, v]$ it holds that $w_1 \sqcap w_2 \in [u, v]$, and $w_1 \sqcup w_2 \in [u, v]$.

Given two preorders, $(V_1, \le_1)$, $(V_2, \le_2)$, an injective function $f : V_1 \to V_2$ is called an *order embedding* if it holds that $u \le_1 v$ iff $f(u) \le_2 f(v)$. If $(V_1, \le_1)$, $(V_2, \le_2)$ are lattices then an injective function f is called a *lattice embedding* if $f(u \sqcup_1 v) = f(u) \sqcup_2 f(v)$, and $f(u \sqcap_1 v) = f(u) \sqcap_2 f(v)$ for all $u, v \in V_1$. Notice that any lattice embedding is also an order embedding. We say that an embedding f *preserves the greatest and the least element* if in case v is a greatest element in $(V_1, \le_1)$ then $f(v)$ is a greatest element in $(V_2, \le_2)$, and in case v is a least element in $(V_1, \le_1)$ then $f(v)$ is a least element in $(V_1, \le_1)$.

Observe that $(\mathsf{P}, \subseteq_{\text{ae}})$ is a preorder where $A \subseteq_{\text{ae}} B$ denotes that $A \setminus B$ is finite. Let $(\mathsf{P}^*, \subseteq^*)$ denote the induced partial order, where P^* consists of the equivalence classes $[A] = \{B \in \mathsf{P} \mid B \subseteq_{\text{ae}} A \wedge A \subseteq_{\text{ae}} B\}$ for $A \in \mathsf{P}$, and where $[A] \subseteq^* [B]$ is defined by $A \subseteq_{\text{ae}} B$. In fact, $(\mathsf{P}^*, \subseteq^*)$ is a distributive lattice where for $[A], [B] \in \mathsf{P}^*$ we have $[A] \sqcup [B] = [A \cup B]$ and $[A] \sqcap [B] = [A \cap B]$. The lattice $(\mathsf{P}^*, \subseteq^*)$ is useful for our purposes due to the following observation

Lemma 1 ([1]). *Any countable partial order can be order embedded into* $(\mathsf{P}^*, \subseteq^*)$. *And any countable distributive lattice can be lattice embedded into the lattice* $(\mathsf{P}^*, \subseteq^*)$ *such that the greatest and the least element are preserved.*

Hence, in order to show that any countable distributive lattice can be lattice embedded into a lattice $(V, \le)$ such that the greatest and the least element are preserved it suffices to show that there is an embedding of $(\mathsf{P}^*, \subseteq^*)$ into $(V, \le)$ with these properties. Notice that since any class $[A] \in \mathsf{P}^*$ contains infinitely many sets, Lemma 1 implies that any countable preorder can be order embedded into the preorder $(\mathsf{P}, \subseteq_{\text{ae}})$.

3 Basic Observations

Since our original intention was to study the preorder $(\mathsf{FP}_{=L}, \le_{\text{m}}^{r})$ whereas most of our results are stated for the partial order $(\mathbf{FP}_{L}^{r}, \le_{\text{m}}^{r})$, it is worth mentioning that the structure of $(\mathsf{FP}_{=L}, \le_{\text{m}}^{r})$ is uniquely determined by the structure of

$(\mathsf{FP}^r_L, \leq^r_{\mathrm{m}})$ with the neglectable exception $L = \emptyset$ where there is only one proof system (namely the totally undefined function). Due to the limited space we omitted the straightforward proofs in this section.

Proposition 1. *If $L \neq \emptyset$ then any degree $\mathfrak{a} \in \mathsf{FP}^r_L$ contains an infinite number of proof systems.*

It is obvious that a proof system f for a set L exists iff L is recursively enumerable (short: r.e.). Hence FP^r_L is nonempty iff L is r.e. We will now see that $(\mathsf{FP}^r_L, \leq)$ is a lattice for L r.e.

For two functions f and g let the function $f \oplus g : \Sigma^* \to \Sigma^*$, be defined as follows.

$$f \oplus g : w \mapsto \begin{cases} f(v) \text{ if } w = 0v, \\ g(v) \text{ if } w = 1v. \end{cases}$$

Further, let $f \otimes g : \Sigma^* \to \Sigma^*$ be given by

$$f \otimes g : w \mapsto \begin{cases} \varphi \text{ if } w = \langle x, y \rangle \text{ and } f(x) = g(y) = \varphi, \\ \perp \text{ else.} \end{cases}$$

Clearly, if f and g are proof systems for L then $f \oplus g$ as well as $f \otimes g$ are proof systems for L. In fact, $f \oplus g$ is a least upper, and $f \otimes g$ is a greatest lower bound for f and g with respect to $\leq^r_{\mathrm{m}}$.

Lemma 2. *Let f and g be proof systems for L. Then*

- *$f \oplus g$ is a least upper bound for f and g with respect to $\leq^r_{\mathrm{m}}$.*
- *$f \otimes g$ is a greatest lower bound for f and g with respect to $\leq^r_{\mathrm{m}}$.*

We obtain immediately

Theorem 1. *Let L be r.e. Then $(\mathsf{FP}^r_L, \leq)$ is a lattice, where for any $\mathfrak{a}, \mathfrak{b} \in \mathsf{FP}^r_L$, $f \in \mathfrak{a}$ and $g \in \mathfrak{b}$ the least upper bound $\mathfrak{a} \sqcup \mathfrak{b}$ and the least lower bound $\mathfrak{a} \sqcap \mathfrak{b}$ is given by*

$$\mathfrak{a} \sqcup \mathfrak{b} = \mathsf{deg}_r(f \oplus g),$$
$$\mathfrak{a} \sqcap \mathfrak{b} = \mathsf{deg}_r(f \otimes g)\,.$$

Moreover one verifies that this lattice is distributive in a straightforward way.

Theorem 2. *Let L be r.e. Then $(\mathsf{FP}^r_L, \leq)$ is a distributive lattice.*

4 Presentability

A class F of functions is called *presentable*, if there is a recursive enumeration $M_1, M_2, \ldots$ of (encodings of) Turing machines that compute the functions in F. If additionally any M_i halts on every input, we call F *recursively presentable.*

If further any M_i is a PTM then we call F FP-*presentable.* Let us call a class $\mathbf{F} \subseteq \mathbf{FP}^r_L$ of degrees *(recursively,* FP-*) presentable* if there is a (recursively, FP-) presentable class F such that $\mathbf{F} = \{\deg_r(f) \mid f \in \mathsf{F}\}$.

In the following lemma we provide some basic observations about the presentability of function classes that will be used in the next section. Similar observations for classes of sets can be found in [1]. The proof is straightforward.

Lemma 3.

1. *If $\mathsf{F} \subseteq \mathsf{FP}$ is finite and nonempty then F is* FP-*presentable.*
2. *If F is (recursively) presentable then $\mathsf{G} = \{h \in \mathsf{FP} \mid \exists g \in \mathsf{F}\ h \leq^r_{\mathrm{m}} g\}$ is (recursively) presentable.*
3. *If F and G are (recursively) presentable then $\mathsf{F} \cup \mathsf{G}$ is (recursively) presentable.*

Let us now consider the interesting question, whether the class of proof systems $\mathsf{FP}_{=L}$ is presentable. In Proposition 2 we summerize some simple observations about the presentability of $\mathsf{FP}_{\subseteq L}$, $\mathsf{FP}_{\supseteq L}$, and $\mathsf{FP}_{=L}$.

Proposition 2.

1. *$\mathsf{FP}_{\subseteq L}$ is* FP-*presentable iff L has a p-optimal proof system.*
2. *If L is recursive then $\mathsf{FP}_{\subseteq L}$ is recursively presentable.*
3. *If L is r.e. then $\mathsf{FP}_{\subseteq L}$ is presentable.*
4. *If L is finite then $\mathsf{FP}_{\supseteq L}$ and $\mathsf{FP}_{=L}$ are* FP-*presentable.*

Complementing the results in Proposition 2 we will show in the following that the classes $\mathsf{FP}_{=L}$, and $\mathsf{FP}_{\supseteq L}$ are not presentable if L is infinite.

When L is an infinite r.e. set, one easily constructs an FP-presentable infinitely descending chain $\mathfrak{a}_1 > \mathfrak{a}_2 > \mathfrak{a}_3 > \cdots$ that may start at any degree $\mathfrak{a} = \mathfrak{a}_1$.

Theorem 3. *If L is infinite then for any $\mathfrak{a} \in \mathbf{FP}^r_L$ there exist $\mathfrak{a}_i \in \mathbf{FP}^r_L$ for $i \in \mathbb{N}$ such that*

$$\mathfrak{a} = \mathfrak{a}_1 > \mathfrak{a}_2 > \mathfrak{a}_3 > \cdots$$

and $\{\mathfrak{a}_i \mid i \geq 1\}$ is FP-*presentable.*

The following notion will be useful to simplify some proofs below: For proof systems h, g for L let us say that h *simulates g almost everywhere* if there is an at most polynomially length increasing function f such that for almost every $y \in L$ it holds that $h(f(w)) = y$ for any w with $g(w) = y$. If additionally $f \in \mathsf{FP}$ we say that h *p-simulates g almost everywhere.*

In a straightforward way one obtains the following lemma.

Lemma 4. *Let h, g be proof systems for L. Then h (p-)simulates g almost everywhere iff h (p-)simulates g.*

If L is finite, then any $h \in \mathsf{FP}_{=L}$ p-simulates any $g \in \mathsf{FP}_{=L}$ almost everywhere trivially via the totally undefined function. Combining this observation with Theorem 3 we obtain

Corollary 1. *If L is finite then FP^r_L consists of one degree. If L is infinite r.e. then FP^r_L contains infinitely many degrees.*

Corollary 2. *There is a least degree in FP^r_L iff L is finite.*

The proof of the following Theorem shows how to construct a lower bound for any presentable class $\mathsf{F} \subseteq \mathsf{FP}_{=L}$.

Theorem 4. *Any presentable class $\mathsf{F} \subseteq \mathsf{FP}_{=L}$ has a lower bound in $\mathsf{FP}_{=L}$ with respect to $\leq^r_{\mathrm{m}}$.*

Theorem 5. *Let L be r.e. The following statements are equivalent.*

1. *L is infinite.*
2. *$\mathsf{FP}_{=L}$ is not presentable.*
3. *$\mathsf{FP}_{\supseteq L}$ is not presentable.*

Proof. By Proposition 2 $\mathsf{FP}_{=L}$, and $\mathsf{FP}_{\supseteq L}$ are presentable if L is finite. Now assume that L is infinite. If $\mathsf{FP}_{=L}$ were presentable then Theorem 4 would provide a lower bound for $\mathsf{FP}_{=L}$ which would contradict Corollary 2. Hence $\mathsf{FP}_{=L}$ is not presentable for infinite L. For the remaining implication observe that $\mathsf{FP}_{=L}$ is presentable if $\mathsf{FP}_{\supseteq L}$ is: from a presentation $M_1, M_2, \ldots$ of $\mathsf{FP}_{\supseteq L}$ and an acceptor M for L, a presentation $M'_1, M'_2, \ldots$ for $\mathsf{FP}_{=L}$ is obtained as follows. On input x, the machine M'_i outputs $M_i(x)$ if M accepts $M_i(x)$, and otherwise rejects.

Actually the proof of the equivalence *1* $\iff$ *2* of Theorem 5 already shows the following sharper result.

Theorem 6. *FP^r_L is not presentable iff L is infinite.*

5 The Embedding Theorem

In this section we show that any countable distributive lattice can be embedded into the degree structure between two proof systems g and f for L such that f (p-)simulates g but g does not (p-)simulate f. Moreover the embedding can be constructed such that certain intermediate degrees can be avoided. This allows us to show that any non-(p-)optimal proof system is a member of an infinite antichain. In order to obtain an intermediate proof system in between g and f, the informal idea is to construct a new proof system g' that allows all proofs from g, but also all the proofs from f for some of the $y \in L$ such that no efficient translation of those proofs to g-proofs is possible. To achieve this, we will use the gap-language technique that has been developed in order to study the polynomial time degrees of sets (cf. [11,15,2,1]).

We first generalize the notions from [11] (see also [15,2,1]). Given a set $S \subseteq \mathbb{N}$, and a time-constructible function $t : \mathbb{N} \to \mathbb{N}$ such that $t(n) > n$ we define

$$\mathrm{G}_t(S) \;=\; \{x \in \Sigma^* \mid t^{(n)}(0) \leq |x| < t^{(n+1)}(0) \text{ for some } n \in S\}$$

where

$$t^{(n)} \text{ denotes } \underbrace{t \circ t \circ \cdots \circ t}_{n \text{ times}}.$$

It is easy to see that $G_t(S) \in P$ if $S \in P$ (cf. [15,1]). Moreover $G_t(S) = \bigcup_{i \in S} G_t(\{i\})$, and $G_t(\{i\}) \cap G_t(\{j\}) = \emptyset$ if $i \neq j$.

In the following we will construct functions t such that each gap $G_t(\{l\})$ for some l will contain some $y \in L$ that diagonalizes against some proof translation (as mentioned in the informal introduction of this section). This way one obtains for any S such that neither S nor $\mathbb{N} \setminus S$ is finite, an intermediate proof system g' if in addition to the g-proofs one allows all f-proofs for those $y \in L$ that are in $G_t(S)$. More formally, g' will be defined as $g \oplus (\mathrm{id}_t^S \circ f)$, where id_t^S denotes the identity function restricted to $G_t(S)$:

$$\mathrm{id}_t^S(x) = \begin{cases} x & \text{if } x \in G_t(S) \\ \perp & \text{else.} \end{cases}$$

Clearly, $\mathrm{id}_t^S \in FP$ if $G_t(S) \in P$.

Lemma 5. *Let f be a proof systems for L and let F be a recursively presentable class of functions such that for any $h \in F$ there are infinitely many $y \in L$ for which a w exists with $y = f(w) \neq h(1w)$. Then there is a time constructible function t such that for any A that is infinite $g \oplus (\mathrm{id}_t^A \circ f) \notin F$ for any function g.*

Proof. Let f_1, f_2, ... denote a recursive enumeration of F, and let $b(i,n)$ be a total recursive function such that

$$b(i,n) = |f(w)| \text{ where } w = \min\{w \mid |f(w)| \geq n \wedge f(w) \neq f_i(1w)\}.$$

Where $\min S$ for some set S of words denotes the least word in S with respect to the standard order. Notice that the assumptions on f and F guarantee that b is total recursive. Now let

$$b'(n) = \max\{b(i,n) \mid i \leq n\},$$

and let t be a time constructible function with $t(n) > b'(n)$.

Let A be infinite. Fix $i > 0$ and let $n \in A$ with $n \geq i$. Let $s = t^{(n)}(0)$. By definition of b there is an y of length $b(i,s)$ such that for some w with $f(w) = y$ it holds $f_i(1w) \neq y$. Now $t^{(n)}(0) = s \leq b(i,s) = |y| \leq b(s) < t(s) = t^{(n+1)}(0)$, and therefore $y \in G_t(\{n\}) \subseteq G_t(A)$. This shows $g \oplus (\mathrm{id}_t^A \circ f) \neq f_i$.

We will apply Lemma 5 most often to recursively presentable classes F where F is the $\leq_m^r$-closure of a class G of proof systems for L (i.e. $F = \{h \in FP \mid \exists g \in G\ h \leq_m^r g\}$). As seen in the following lemma such classes F with $f \notin F$ fulfill the premise of Lemma 5 if $f \notin F$.

Lemma 6. *Let F be the $\leq_m^r$-closure of a recursively presentable class $G \subseteq FP_{=L}$, and assume $f \notin F$ for some $f \in FP_{=L}$. Then F is a recursively presentable class such that for any $h \in F$ there are infinitely many $y \in L$ for which a w exists with $y = f(w) \neq h(1w)$.*

Proof. By Lemma 3 F is recursively presentable since G is recursively presentable. For contradiction assume now that there is a $h \in \mathsf{F}$ such that there are only finitely many $y \in L$ for which a w exists with $y = f(w) \neq h(1w)$. As $h \in \mathsf{F}$ there is a proof system $g \in \mathsf{G}$ for L such that $h \leq^r_{\mathrm{m}} g$. Now the assumptions about f and h imply that g (p-)simulate f almost everywhere. By Lemma 4 $f \leq^r_{\mathrm{m}} g$, hence $f \in \mathsf{F}$.

We now come to the main embedding theorem.

Theorem 7. *Let $f, g \in \mathsf{FP}_{=L}$ such that $g <^r_{\mathrm{m}} f$, and let G be a recursively presentable class of functions such that for any $h \in \mathsf{G}$ there are infinitely many $y \in L$ for which a w exists with $y = f(w) \neq h(1w)$. Then there is an embedding*

$$\begin{aligned} \mathsf{P} &\to \{h \mid g \leq^r_{\mathrm{m}} h \leq^r_{\mathrm{m}} f\} \\ A &\mapsto h_A \end{aligned}$$

such that $h_{\Sigma^*} =^r_{\mathrm{m}} f$ and $h_\emptyset =^r_{\mathrm{m}} g$, and for any sets $A, B \in \mathsf{P}$,

$$A \subseteq_{\mathrm{ae}} B \text{ iff } h_A \leq^r_{\mathrm{m}} h_B\,, \tag{1}$$

$$h_{A\cup B} =^r_{\mathrm{m}} h_A \oplus h_B\,, \tag{2}$$

$$h_{A\cap B} =^r_{\mathrm{m}} h_A \otimes h_B\,. \tag{3}$$

Further, $h_A \notin \mathsf{G}$ if A is infinite.

Proof. Assume that f, g, and G fulfill the premises. Let F' be the $\leq^r_{\mathrm{m}}$-closure of $\{g\}$, and let $\mathsf{F} = \mathsf{F}' \cup \mathsf{G}$. By the Lemmata 3, 6, the premises of Lemma 5 are fulfilled by f and F.

Applying Lemma 5 to f and F we obtain a time constructible function t. For any $A \in \mathsf{P}$ let

$$h_A = g \oplus (\mathrm{id}^A_t \circ f)\,.$$

Lemma 5 provides that $h_A \notin \mathsf{F}$, hence $h_A \notin \mathsf{G}$, if A is infinite.

Clearly, if $A \subseteq_{\mathrm{ae}} B$ then the h_A-proofs of almost all $y \in L$ are also h_B-proofs of y. Therefore by Lemma 4

$$h_A \leq^{\mathrm{p}}_{\mathrm{m}} h_B \text{ if } A \subseteq_{\mathrm{ae}} B\,.$$

This also implies that

$$h_{A\cap B} \leq^{\mathrm{p}}_{\mathrm{m}} h_A, h_B \leq^{\mathrm{p}}_{\mathrm{m}} h_{A\cup B}\,. \tag{4}$$

So $h_A \oplus h_B \leq^{\mathrm{p}}_{\mathrm{m}} h_{A\cup B}$ by Lemma 2. Now observe that also $h_{A\cup B} \leq^{\mathrm{p}}_{\mathrm{m}} h_A \oplus h_B$: to translate a $h_{A\cup B}$-proof w for y to a $h_A \oplus h_B$-proof for y, first check whether $w = 0v$ or $w = 1v$. In the first case v is a g-proof for y and therefore $00v$ is an $h_A \oplus h_B$-proof for y. In the second case $f(v) = y$ and $y \in \mathrm{G}_t(A \cup B) = \mathrm{G}_t(A) \cup \mathrm{G}_t(B)$. If $y \in \mathrm{G}_t(A)$ then $11v$ is an $h_A \oplus h_B$-proof for y; if otherwise $y \in \mathrm{G}_t(B)$ then $01v$ is an $h_A \oplus h_B$-proof for y. Hence,

$$h_A \oplus h_B =^{\mathrm{p}}_{\mathrm{m}} h_{A\cup B}\,. \tag{5}$$

By (4) and Lemma 2 $h_{A\cap B} \leq_{\mathrm{m}}^{\mathrm{p}} h_A \otimes h_B$. For a p-simulation in the reverse direction notice that any $h_A \otimes h_B$-proof for y is of the form $\langle u, v\rangle$ where $h_A(u) = h_B(v) = y$. If $u = 0u'$ or $v = 0v'$ then u', resp. v', is a g-proof for y in which case u, resp. v, is a $h_{A\cap B}$-proof for y. Otherwise $u = 1u'$ and $v = 1v'$ for some u', v' with $f(u') = y \in \mathsf{G}_t(A)$ and $f(v') = y \in \mathsf{G}_t(B)$. Hence $y \in \mathsf{G}_t(A \cap B)$, which shows that u is an $h_{A\cap B}$-proof for y. Hence,

$$h_A \otimes h_B =_{\mathrm{m}}^{\mathrm{p}} h_{A\cap B}\,. \tag{6}$$

It remains to show that $h_A \leq_{\mathrm{m}}^{r} h_B$ implies $A \subseteq_{\mathrm{ae}} B$. Fix two sets $A, B \in \mathsf{P}$, and assume that $h_A \leq_{\mathrm{m}}^{r} h_B$. Let $S = A \setminus B$. Clearly, $S \in \mathsf{P}$. Since $S \subseteq A$, $h_S \leq_{\mathrm{m}}^{\mathrm{p}} h_A$ by (4). By transitivity $h_S \leq_{\mathrm{m}}^{r} h_B$. By Lemma 2 and reflexivity, $h_S \leq_{\mathrm{m}}^{r} h_S \otimes h_B$, hence by (6) and transitivity $h_S \leq_{\mathrm{m}}^{r} h_{S\cap B} = h_\emptyset$. Observe $h_\emptyset(w) = g(v)$ if $w = 1v$ and $h_\emptyset(w) = \perp$ otherwise. So $h_\emptyset =_{\mathrm{m}}^{\mathrm{p}} g$, and therefore $h_S \leq_{\mathrm{m}}^{\mathrm{p}} g$. This implies $h_S \in \mathsf{F}$. By Lemma 5 S is finite. So $A \subseteq_{\mathrm{ae}} B$.

Notice that (1) implies that the embedding $A \mapsto h_A$ induces an order embedding of the lattice $(\mathsf{P}^*, \subseteq^*)$ into $([\mathsf{deg}_r(g), \mathsf{deg}_r(f)], \leq)$. By (2), (3) this order embedding is a lattice embedding. So we obtain

Corollary 3. *Let $\mathfrak{a}, \mathfrak{b} \in \mathbf{FP}_L^r$, and let $\mathbf{C} \subseteq \mathbf{FP}_L^r$ be a recursively presentable class of degrees such that $\mathfrak{a} < \mathfrak{b}$ and $\mathfrak{c} < \mathfrak{b}$ for any $\mathfrak{c} \in \mathbf{C}$. Then $(\mathsf{P}^*, \subseteq^*)$ can be lattice embedded into the sublattice $[\mathfrak{a}, \mathfrak{b}]$ such that the greatest and the least element is preserved. Moreover no element different from the least element $[\emptyset] \in \mathsf{P}^*$ is mapped to an element $\leq \mathfrak{c}$ for any $\mathfrak{c} \in \mathbf{C}$.*

Proof. Assume that $\mathbf{C}$ is recursively presentable via $\mathsf{F} \subseteq \mathsf{FP}_{=L}$. Let $g \in \mathfrak{a}$, $f \in \mathfrak{b}$, and let G be the $\leq_{\mathrm{m}}^{r}$-closure of $\mathsf{F} \cup \{g\}$ and apply Theorem 7 (notice that by the Lemmata 3, 6 the premises of Theorem 7 are fulfilled).

Since by Lemma 1 any countable distributive lattice can be embedded into $(\mathsf{P}^*, \subseteq_{\mathrm{ae}})$ such that greatest and least elements are preserved we obtain

Corollary 4. *Let $\mathfrak{a}, \mathfrak{b} \in \mathbf{FP}_L^r$, and let $\mathbf{C} \subseteq \mathbf{FP}_L^r$ be a recursively presentable class of degrees such that $\mathfrak{a} < \mathfrak{b}$ and $\mathfrak{c} < \mathfrak{b}$ for any $\mathfrak{c} \in \mathbf{C}$. Then any countable distributive lattice $\mathcal{L}$ can be lattice embedded into the sublattice $[\mathfrak{a}, \mathfrak{b}]$ such that the maximal and the minimal element is preserved. Moreover no element different from a least element in $\mathcal{L}$ is mapped to an element $\leq \mathfrak{c}$ for any $\mathfrak{c} \in \mathbf{C}$.*

By setting $\mathbf{C} = \{\mathfrak{a}\}$ we obtain as a consequence the following embedding result that is mentioned in the introduction.

Corollary 5. *Let $\mathfrak{a}, \mathfrak{b} \in \mathbf{FP}_L^r$ such that $\mathfrak{a} < \mathfrak{b}$. Then any countable distributive lattice $\mathcal{L}$ can be lattice embedded into the sublattice $[\mathfrak{a}, \mathfrak{b}]$ such that the maximal and the minimal element is preserved.*

By an embedding of the total order consisting of three elements into $[\mathfrak{a}, \mathfrak{b}]$ we obtain a degree $\mathfrak{c}$ with $\mathfrak{a} < \mathfrak{c} < \mathfrak{b}$.

Corollary 6. $\mathbf{FP}_L^r$ *is a dense order.*

The following theorem shows that for any non-optimal degree there is an incomparable degree.

Theorem 8. *Let* $\mathsf{c} \in \mathbf{FP}^r_L$ *and assume that* c *is not a greatest element in* $\mathbf{FP}^r_L$*. Then there is a degree* $\mathsf{d} \in \mathbf{FP}^r_L$ *such that neither* $\mathsf{c} \leq \mathsf{d}$ *nor* $\mathsf{d} \leq \mathsf{c}$.

An antichain is a subset $\mathsf{C} \subseteq \mathbf{FP}^r_L$ such that any two elements in C are incomparable. Hence, two incomparable elements form an antichain of size two. We now obtain that any nontrivial antichain C can be extended to an infinite antichain $\mathsf{D} \supseteq \mathsf{C}$.

Theorem 9. *Let* $\mathbf{C} \subseteq \mathbf{FP}^r_L$ *be a finite antichain consisting of at least two elements. Then there is an infinite antichain* $\mathsf{D} \subseteq \mathbf{FP}^r_L$ *with* $\mathsf{C} \subseteq \mathsf{D}$.

Combining Theorems 8, 9 we obtain

Corollary 7. *Let* $\mathsf{c} \in \mathbf{FP}^r_L$ *and assume that* c *is not a greatest element in* $\mathbf{FP}^r_L$*. Then* c *is a member of an infinite antichain.*

Let us at the end restate some of the main results in this section in terms of the preorder $(\mathsf{FP}_{=L}, \leq^r_{\mathrm{m}})$.

Corollary 8.

1. *Let f and h be proof systems for L such that $f <^r_{\mathrm{m}} h$. Then there is a proof system g for L such that $f <^r_{\mathrm{m}} g <^r_{\mathrm{m}} h$.*
2. *Let f and h be proof systems for L such that $f <^r_{\mathrm{m}} h$. Then any countable preorder can be order embedded into the interval $\{g \in \mathsf{FP} \mid f \leq^r_{\mathrm{m}} g \leq^r_{\mathrm{m}} h\}$.*
3. *Let f be a proof system for L that is not (p-)optimal. Then there is an infinite set F of proof systems for L that are pairwise incomparable and that are also incomparable to f (with respect to (p-)simulation).*

6 Conclusion

We saw that the lattice structure of $\mathbf{FP}^r_L$ is quite rich for infinite r.e. sets L. Moreover the structure seems to be very regular, and appears to be basically similar for any set L. Actually we encountered only two isomorphism types for $\mathbf{FP}^r_L$ when L is an infinite r.e. set, depending on whether a greatest element exists. It would be interesting to know whether more isomorphism types exist.

References

1. K. Ambos-Spies. Sublattices of the polynomial time degrees. *Information and Control*, 65:63–84, 1985.
2. J. L. Balcázar, J. Díaz, and J. Gabarró. *Structural Complexity I*. Springer-Verlag, 1988.
3. G. Birkhof. *Lattice Theory*, volume XXV of *Colloquium Publications*. American Mathematical Society, 3. edition, 1967.

4. S. A. Cook and R. A. Reckhow. On the lengths of proofs in the propositional calculus (preliminary version). In *Proceedings of the ACM Symposium on the Theory of Computing' 74*, pp. 135–148, 1974.
5. S. A. Cook and R. A. Reckhow. The relative efficiency of propositional proof systems. *The Journal of Symbolic Logic*, 44(1):36–50, 1979.
6. J. L. Esteban, N. Galesi, J. Messner. On the complexity of resolution with bounded conjunctions. To appear in the *Proceedings of ICALP 2002, Lecture Notes in Computer Science*, Springer-Verlag, 2002.
7. G. Grätzer. *General lattice theory.* Birkhäuser, Basel, 2nd edition, 1998.
8. J. Köbler and J. Messner. Complete problems for promise classes by optimal proof systems for test sets. In *Proceedings of the 13th Conf. on Computational Complexity*, pp. 132–140. IEEE, 1998.
9. J. Krajíček and P. Pudlák. Propositional proof systems, the consistency of first order theories and the complexity of computations. *The Journal of Symbolic Logic*, 54(3):1063–1079, 1989.
10. R. E. Ladner. On the structure of polynomial time reducibility. *Journal of the ACM*, 22:155–171, 1975.
11. L. H. Landweber, R. J. Lipton, and E. L. Robertson. On the structure of sets in NP and other complexity classes. *Theoretical Computer Science*, 15:181–200, 1981.
12. J. Messner. On optimal algorithms and optimal proof system. In *Proceedings of the 16th Symp. on Theoretical Aspects of Computing (STACS'99)*, volume 1563 of *Lecture Notes in Computer Science*, pp. 541–550. Springer-Verlag, 1999.
13. J. Messner and J. Torán. Optimal proof systems for propositional logic and complete sets. In *Proceedings of the 15th Symp. on Theoretical Aspects of Computing (STACS'98)*, volume 1373 of *Lecture Notes in Computer Science*, pp. 477–487. Springer-Verlag, 1998.
14. R. A. Reckhow. *On the lengths of proofs in the propositional calculus.* PhD thesis, University of Toronto, Department of Computer Science, 1976.
15. U. Schöning. A uniform approach to obtain diagonal sets in complexity classes. *Theoretical Computer Science*, 18:95–103, 1982.
16. G. S. Tseitin. On the complexity of proofs in propositional logics. In J. Siekmann and G. Wrightson, editors, *Automation of Reasoning*, volume 2, pp. 466–483. Springer-Verlag, 1983. Reprint of Seminars in mathematics, pp. 115–125, V.A. Seklov Math. Institut, Leningrad, 1970.
17. A. Urquhart. The complexity of propositional proofs. *Bulletin of Symbolic Logic*, 1(4):425–467, 1995.

Comorphism-Based Grothendieck Logics

Till Mossakowski

BISS, Dept. of Computer Science, University of Bremen

Abstract. In order to obtain a semantic foundation for heterogeneous specification, we extend Diaconescu's morphism-based Grothendieck institutions to the case of comorphisms. This is not just a dualization, because we obtain more general results, especially concerning amalgamation properties. We also introduce a proof calculus for structured heterogeneous specifications and study its soundness and completeness (where amalgamation properties play a rôle for obtaining the latter).

1 Introduction and Motivation

For the specification of large software systems, heterogeneous multi-logic specifications are needed, since complex problems have different aspects that are best specified in different logics. A combination of all the used logics would become too complex in many cases. Moreover, specialized languages and tools often have their strengths in particular aspects. Using heterogeneous specification, these strengths can be combined with comparably small effort.

In the literature, several approaches to heterogeneous specification have been developed [7, 8, 17, 23]. The most prominent approach is CafeOBJ with its cube of eight logics and twelve projections (formalized as *institution morphisms*) among them [9], having a semantics based on the notion of Grothendieck institution [8]. However, not only projections between logics, but also logic encodings (formalized as so called *comorphisms*) are relevant to heterogeneous specification [23, 17]. Moreover, besides these model theoretic approaches, also the need of integrating different proof calculi via "bridges" has been stressed [7]. The goal of this paper is to extend the CafeOBJ resp. Grothendieck institution approach to cover these aspects.

2 Institutions, Logics, Morphisms and Comorphisms

We now recall several notions mentioned in the introduction. *Institutions* [14] capture the model theory of a logic, *entailment systems* [16] capture proof theory, while *logics* [16] combine both. *Institution morphisms* [14] capture the intuition that one logic is *built upon*, or *projected onto* another one, while *institution comorphisms* [13], also called *institution representations* [22] or *maps of institutions* [16] capture the intuition that one logic is *encoded into* another one. Both notions also can be extended to full logics. We begin with the components of institutions, which are called *rooms*, and typically provide a satisfaction system local to some signature.

K. Diks et al. (Eds): MFSC 2002, LNCS 2420, pp. 593–604, 2002.

An *institution room* $(S, M, \models)$ consists of

- a set of S of *sentences*,
- a category M of *models*, and
- a satisfaction relation $\models \subseteq |M| \times S$.

Rooms are connected via corridors (which model change of notation within one logic, as well as translations between logics).

An *institution corridor* $(\alpha, \beta): (S_1, M_1, \models_1) \longrightarrow (S_2, M_2, \models_2)$ consists of

- a sentence translation function $\alpha: S_1 \longrightarrow S_2$, and
- a model reduction functor $\beta: M_2 \longrightarrow M_1$, such that

$$m_2 \models_2 \alpha(\varphi_1) \Leftrightarrow \beta(m_2) \models_1 \varphi_1$$

holds for each $m_2 \in M_2$ and each $\varphi_1 \in S_1$ (*satisfaction condition*).

Semantic entailment in an institution room is defined as usual: for $\Gamma \subseteq S, \varphi \in S$, we write $\Gamma \models \varphi$, if all models satisfying Γ also satisfy φ.

A *logic room* $(S, M, \models, \vdash)$ is an institution room $(S, M, \models)$ equipped with an *entailment relation* $\vdash \subseteq \mathcal{P}(S) \times S$, such that the following conditions are satisfied:

1. *reflexivity:* for any $\varphi \in S$, $\{\varphi\} \vdash \varphi$,
2. *monotonicity:* if $\Gamma \vdash \varphi$ and $\Gamma' \supseteq \Gamma$ then $\Gamma' \vdash \varphi$,
3. *transitivity:* if $\Gamma \vdash \varphi_i$, for $i \in I$, and $\Gamma \cup \{\varphi_i \mid i \in I\} \vdash \psi$, then $\Gamma \vdash \psi$,
4. *soundness:* for any $\Gamma \subseteq S$ and $\varphi \in S$,

$$\Gamma \vdash \varphi \text{ implies } \Gamma \models \varphi.$$

A logic room will be called *complete* if, in addition, the converse of the above implication holds.

A *logic corridor* $(\alpha, \beta): (S_1, M_1, \models_1, \vdash_1) \longrightarrow (S_2, M_2, \models_2, \vdash_2)$ is an institution corridor $(\alpha, \beta): (S_1, M_1, \models_1) \longrightarrow (S_2, M_2, \models_2)$ such that if $\Gamma \vdash_1 \varphi$, then $\alpha(\Gamma) \vdash_2 \alpha(\varphi)$ ($\vdash$*-translation*). Together with obvious notions of composition and identity, this gives us categories **InsRoom** and **LogRoom**.

Generally, sentences and models depend on a given vocabulary of non-logical symbols provided by a *signature*. Therefore, an *institution* is a functor I: **Sign** $\longrightarrow$ **InsRoom** (where **Sign** is called the category of *signatures*), and a *logic* is a functor L: **Sign** $\longrightarrow$ **LogRoom**. We will also use the more standard notation $(\mathbf{Sen}^I(\Sigma), \mathbf{Mod}^I(\Sigma), \models_\Sigma)$ for the institution room $I(\Sigma)$. $\mathbf{Mod}^I$ and $\mathbf{Sen}^I$ can easily be seen to be functors, and we arrive at the standard definition of institution as a quadruple $(\mathbf{Sign}, \mathbf{Sen}, \mathbf{Mod}, \models)$. We will freely use this notation whenever needed, and also write $m|_\sigma$ for $\mathbf{Mod}(\sigma)(m)$.

For the morphisms between institutions and logics, there are two obvious choices: Given institutions $I_1: \mathbf{Sign}_1 \longrightarrow \mathbf{InsRoom}$ and $I_2: \mathbf{Sign}_2 \longrightarrow \mathbf{InsRoom}$, an *institution morphism* $(\Psi, \mu): I_1 \longrightarrow I_2$ consists of a functor $\Psi: \mathbf{Sign}_1 \longrightarrow \mathbf{Sign}_2$ and a natural transformation $\mu: I_2 \circ \Psi \longrightarrow I_1$. In contrast, an *institution comorphism* $(\Phi, \rho): I_1 \longrightarrow I_2$ consists of a functor $\Phi: \mathbf{Sign}_1 \longrightarrow \mathbf{Sign}_2$ and a natural transformation $\rho: I_1 \longrightarrow I_2 \circ \Psi$. Together with obvious compositions and identities, this gives us categories **Ins** and **coIns**. Logic morphisms and comorphisms are defined analogously, leading to categories **Log** and **coLog**.

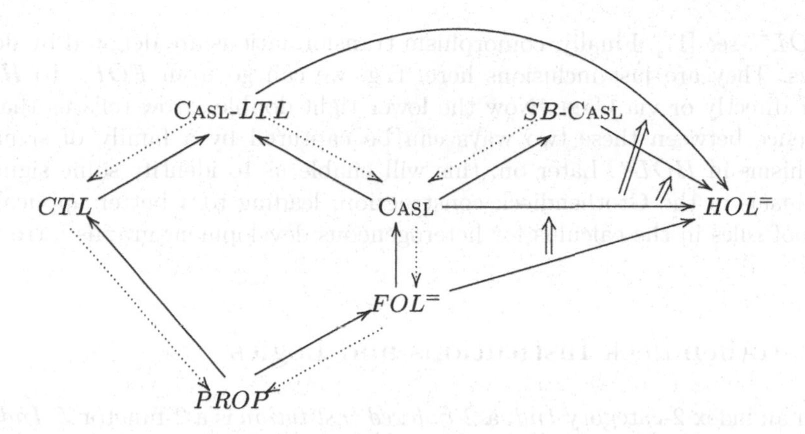

Fig. 1. Some institutions, morphisms, comorphisms and comorphism transformations

Given institution morphisms $(\Psi, \mu): I_1 \longrightarrow I_2$ and $(\Psi', \mu'): I_1 \longrightarrow I_2$, an *institution morphism transformation* $\theta: (\Psi, \mu) \longrightarrow (\Psi', \mu')$ is just a natural transformation $\theta: \Psi \longrightarrow \Psi'$ such that $\mu = \mu' \circ (I_2 \cdot \theta)$. [1] Similarly, given institution comorphisms $(\Phi, \rho): I_1 \longrightarrow I_2$ and $(\Phi', \rho'): I_1 \longrightarrow I_2$, an *institution comorphism transformation* $\theta: (\Phi, \rho) \longrightarrow (\Phi', \rho')$ is a natural transformation $\theta: \Phi \longrightarrow \Phi'$ such that $(I_2 \cdot \theta) \circ \rho = \rho'$.

The corresponding notions for logics are entirely analogous. This shows that **Ins**, **coIns**, **Log** and **coLog** are indeed 2-categories. In the sequel, most of the definitions and results hold for both institutions and logics, although we will not repeat this every time (indeed, most constructions work for an arbitrary category of rooms).

Consider, for example, the institutions in Fig. 1. *PROP* is propositional logic, $FOL^=$ is first-order logic, $HOL^=$ higher-order logic, both with equality. CASL [19] extends $FOL^=$ with partiality, subsorting and generation constraints (some form of induction). CASL-LTL [20] is an extension of CASL with a CTL-like labeled transition logic. We also have included *CTL* itself, a temporal logic featuring both temporal modalities and path quantifiers, with model-checkers available. SB-CASL [4] is an extension of CASL that follows the abstract state machine paradigm, where states correspond to algebras. There are obvious projection morphisms (denoted by dotted arrows), that always come in pair with inclusion comorphisms (denoted as solid arrows). Generally, in heterogeneous specification, these morphisms can be simulated by the corresponding comorphisms[2]. Some of these logics are represented via more complex comorphisms[3]

[1] In [8], a modification between μ and $\mu' \circ (I_2 \cdot \theta)$ is allowed, using a 2-category of rooms, but we do not think that this extra generality is of much practical use.

[2] More precisely, the corridors provided by these morphisms can be expressed as corridors induced by signature morphisms in the comorphism-based Grothendieck institution.

[3] Strictly speaking, these are not comorphisms, but simple theoroidal comorphisms in the sense of [13]. For simplicity, we will ignore this difference here.

in $HOL^=$, see [17]. Finally, comorphism transformations are denoted by double arrows. They are just inclusions here. E.g. we can go from $FOL^=$ to $HOL^=$ either directly or via CASL. Now the lower right double arrow tells us that the difference between these two ways can be captured by a family of signature morphisms in $HOL^=$. Later on, this will enable us to identify some signature morphisms in the Grothendieck construction, leading to a better applicability of proof rules in the calculus for heterogeneous development graphs introduced below.

3 Grothendieck Institutions and Logics

Given an index 2-category *Ind*, a *2-indexed institution* is a 2-functor $\mathcal{I}: Ind^* \longrightarrow \mathbf{Ins}$[4] into the 2-category of institutions, institution morphisms and institution morphism transformations. In cases where the 2-categorical structure is not needed, we omit the prefix "2-".

Similarly, a *2-indexed coinstitution* is a 2-functor $\mathcal{I}: Ind^* \longrightarrow \mathbf{coIns}$.

The Grothendieck construction for indexed institution has been described in [8]; we develop its dual here. In an indexed coinstitution $\mathcal{I}$, we use the notation $\mathcal{I}^i$ for $\mathcal{I}(i)$, (Φ^d, ρ^d) for the comorphism $\mathcal{I}(d)$ an so on.

Definition 3.1. *Given a* 2-indexed coinstitution $\mathcal{I}: Ind^* \longrightarrow \mathbf{coIns}$, *define the* Grothendieck institution $\mathcal{I}^\#$ *as follows:*

- *signatures in $\mathcal{I}^\#$ are pairs (Σ, i), where $i \in |Ind|$ and Σ a signature in $\mathcal{I}^i$,*
- *signature morphisms $(\sigma, d): (\Sigma_1, i) \longrightarrow (\Sigma_2, j)$ consist of a morphism $d: j \longrightarrow i \in Ind$ and a signature morphism $\sigma: \Phi^d(\Sigma_1) \longrightarrow \Sigma_2$,*
- *composition is given by $(\sigma_2, d_2) \circ (\sigma_1, d_1) = (\sigma_2 \circ \Phi^{d_2}(\sigma_1), d_1 \circ d_2)$,*
- *$\mathcal{I}^\#(\Sigma, i) = \mathcal{I}^i(\Sigma)$, and $\mathcal{I}^\#(\sigma, d) = \mathcal{I}^i(\Sigma_1) \xrightarrow{\rho^d} \mathcal{I}^j(\Phi^d(\Sigma_1)) \xrightarrow{\mathcal{I}^j(\sigma)} \mathcal{I}^j(\Sigma_2)$.*

Finally, institution comorphism transformations lead to a congruence on Grothendieck signature morphisms: the congruence is generated by

$$(\theta^u_\Sigma: \Phi^{d'}(\Sigma) \longrightarrow \Phi^d(\Sigma), d': j \longrightarrow i) \equiv (id: \Phi^d(\Sigma) \longrightarrow \Phi^d(\Sigma), d: j \longrightarrow i)$$

for $u : d \Rightarrow d' \in Ind$, $d, d': j \longrightarrow i \in Ind$, $\Sigma \in \mathbf{Sign}^i$.

Proposition 3.2. *$\equiv$ is contained in the kernel of $\mathcal{I}^\#$ (considered as a functor).*

Corollary 3.3. $\mathcal{I}^\#: \mathbf{Sign}^\# \longrightarrow \mathbf{InsRoom}$ *leads to a quotient Grothendieck institution* $\mathcal{I}^\# / \equiv: \mathbf{Sign}^\# / \equiv \longrightarrow \mathbf{InsRoom}$.

The theory of Grothendieck institutions for indexed institutions has been developed by Diaconescu [8]. Actually, the corresponding theory for indexed coinstitutions turns out to be much simpler. Here, we focus on cocompleteness and amalgamation results, since these are needed for doing structured proofs [18].

[4] *Ind** is the 2-categorical dual of *Ind*, where both 1-cells and 2-cells are reversed.

Proposition 3.4. *Let $\mathcal{I}: Ind^{op} \longrightarrow$ **coIns** be an indexed coinstitution such that Ind is J-complete for some small category J, Φ^d is cocontinuous for each $d: i \longrightarrow j \in Ind$, and the indexed category of signatures of $\mathcal{I}$ is locally J-cocomplete (the latter meaning that $\mathbf{Sign}^i$ is J-cocomplete for each $i \in |Ind|$). Then the signature category of the Grothendieck institution has J-colimits.*

Given an institution I and a diagram $D: J \longrightarrow \mathbf{Sign}^I$, a family of models $(m_j)_{j\in|J|}$ is called *D-consistent* if $m_k|_{D(\delta)} = m_j$ for each $\delta: j \longrightarrow k \in J$. A cocone $(\Sigma, (\mu_j)_{j\in|J|})$ over the diagram in $D: J \longrightarrow \mathbf{Sign}^I$ is called *weakly amalgamable* if for each D-consistent family of models $(m_j)_{j\in|J|}$, there is a Σ-model m with $m|_{\mu_j} = m_j$ ($j \in |J|$). If this model is unique, the cocone is called *amalgamable.*

These notions also extend to diagrams $D: J \longrightarrow Ind$ and cones over these: the rôle of the intra-institution model reductions above is now played by the inter-institution model translations.

An institution I is called *(weakly) semi-exact*, if any pushout of signatures is (weakly) amalgamable. An institution comorphism $(\Phi, \rho): I_1 \longrightarrow I_2$ is called (weakly) exact if, for each signature morphism $\sigma: \Sigma_1 \longrightarrow \Sigma_2$ in I_1, the naturality diagram

$$\begin{array}{ccc} \mathbf{Mod}^I(\Sigma_1) & \xleftarrow{\beta_{\Sigma_1}} & \mathbf{Mod}^J(\Phi(\Sigma_1)) \\ \uparrow{\scriptstyle \mathbf{Mod}^I(\sigma)} & & \uparrow{\scriptstyle \mathbf{Mod}^J(\Phi(\sigma))} \\ \mathbf{Mod}^I(\Sigma_2) & \xleftarrow{\beta_{\Sigma_2}} & \mathbf{Mod}^J(\Phi(\Sigma_2)) \end{array}$$

is (weakly) amalgamable, i.e. a (weak) pullback.

An indexed coinstitution $\mathcal{I}: Ind^{op} \longrightarrow$ **coIns** is called *(weakly) locally semi-exact*, if each institution I^i is (weakly) semi-exact ($i \in |Ind|$). It is called *(weakly) semi-exact* if for each pullback in Ind

$$\begin{array}{ccc} i & \xleftarrow{m1} & j1 \\ \uparrow{\scriptstyle m2} & & \uparrow{\scriptstyle n1} \\ j2 & \xleftarrow{n2} & k \end{array} \quad \text{the square} \quad \begin{array}{ccc} \mathbf{Mod}^i(\Sigma) & \xleftarrow{\beta_\Sigma^{m1}} & \mathbf{Mod}^{j1}(\Phi^{m_1}(\Sigma)) \\ \uparrow{\scriptstyle \beta_\Sigma^{m2}} & & \uparrow{\scriptstyle \beta_\Sigma^{n1}} \\ \mathbf{Mod}^{j2}(\Phi^{m_2}(\Sigma)) & \xleftarrow{\beta_\Sigma^{n2}} & \mathbf{Mod}^k(\Phi^{n_i}(\Phi^{m_i}(\Sigma))) \end{array}$$

is a (weak) pullback for each signature Σ in $\mathbf{Sign}^i$.

Proposition 3.5. *The Grothendieck institution $\mathcal{I}^\#$ of an indexed coinstitution $\mathcal{I}: Ind^{op} \longrightarrow$ **coIns** consisting of comorphisms with cocontinuous signature translation is (weakly) semi-exact if and only if*

- *$\mathcal{I}$ is (weakly) locally semi-exact,*
- *$\mathcal{I}$ is (weakly) semi-exact, and*
- *all institution comorphisms in $\mathcal{I}$ are (weakly) exact.*

Diaconescu proves the above results for so-called *embedding-indexed institutions*, which means that each signature translation Ψ^d has a left adjoint Φ^d. But these left adjoints lead to a corresponding indexed coinstitution, and in fact, strictly speaking Diaconescu uses this induced indexed coinstitution in his proofs. This shows that indexed coinstitutions are simpler and more general than embedding-indexed institutions (and only for these, Diaconescu has results about exactness and amalgamability). In particular, a simpler proof of Diaconescu's results can be obtained by reducing them to the above results via the following generalization of a result from [1]:

Proposition 3.6. *Given an embedding-indexed institution $\mathcal{I}: Ind^{op} \longrightarrow \mathbf{Ins}$, define the indexed coinstitution $\mathcal{I}^{co}: Ind^{op} \longrightarrow \mathbf{coIns}$ by*

$$\mathcal{I}^{co}(i) := \mathcal{I}(i) \text{ and } \mathcal{I}^{co}(d: i \longrightarrow j) := (\Phi^d, (\mu^d \cdot \Phi^d) \circ (\mathcal{I}^j \cdot \eta^d)),$$

where η^d is the unit of the adjunction between Φ^d and Ψ^d.
Then $\mathcal{I}^{\#}$ is isomorphic to $(\mathcal{I}^{co})^{\#}$.

Diaconescu already notes that the assumptions of (his more special version of) Prop. 3.5 are too strong to be met in practice. E.g. the CASL institution is not weakly locally semi-exact, and its encoding into $HOL^{=}$ is neither exact, nor does it have a cocontinuous signature translation. Below, we will try to weaken the assumptions by working with weakly amalgamable cocones rather than with amalgamable colimits: call an institution I *quasi-exact* if for each diagram $D: J \longrightarrow \mathbf{Sign}^I$, there is some weakly amalgamable cocone over D. An indexed coinstitution $\mathcal{I}: Ind^{op} \longrightarrow \mathbf{coIns}$ is called *locally quasi-exact*, if each institution $\mathcal{I}^i$ is quasi-exact ($i \in |Ind|$). It is called *quasi-exact*, if for each diagram $D: J \longrightarrow Ind$, there is some cone $(l, (d_j)_{j \in |J|})$ over D that is weakly amalgamable. *Quasi-semi-exactness* is the restriction of these notions to diagrams of shape $\bullet \longleftarrow \bullet \longrightarrow \bullet$.

4 Heterogeneous Development Graphs

We now come to proof theory:

Proposition 4.1. *The Grothendieck logic $\mathcal{L}^{\#}$ of an indexed cologic $\mathcal{L}: Ind^{op} \longrightarrow$ $\mathbf{coLog}$ is complete if and only if $\mathcal{L}$ is locally complete (i.e. each individual logic is complete).*

In many cases, not every institution will come with a complete entailment system; hence, it is difficult to apply the above proposition in practice. In [17], we therefore have represented $\mathcal{L}^{\#}$ in some expressive "universal" institution with good proof support. However, often it is crucial to use specific tools designed for specific logics to obtain good results. Hence, we need heterogeneous proving as well. A first attempt in this direction are *heterogeneous bridges* [7] which, however, have no clear semantical basis. We therefore aim at clear semantical basis for *heterogeneous specification and proofs*, where in the extreme for each proof goal the best-suited logic could be chosen individually. For this end, we need some heterogeneous structuring concept, namely *heterogeneous development graphs.*

In [17], we have defined heterogeneous development graphs to be homogeneous development graphs over a Grothendieck institution. We here directly define development graphs and their proof calculus over a Grothendieck institution, since this allows us to design a specialized rule called *Model expansion*, which together with a rule *Borrowing* allows us to deal with the re-use of entailment relations across comorphisms [6].

Fix an arbitrary 2-indexed coinstitution $\mathcal{I}: Ind^* \longrightarrow \mathbf{coIns}$. Let $(\mathbf{Sign}^\#/\equiv, \mathbf{Sen}^\#/\equiv, \mathbf{Mod}^\#/\equiv, \models^\#/\equiv)$ denote the components of the quotient Grothendieck institution $\mathcal{I}^\#/\equiv$. We need the congruence $\equiv$ induced by the 2-categorical structure (and the corresponding quotient), because with it we can prove more things in the calculus introduced below.

We further assume that *some* of the institutions come as logics. Moreover, we assume that some of the arrows $d \in Ind$ are marked with the fact that $\mathcal{I}^d$ has the model expansion property (i.e. the model translation components of the corridors are surjective on objects). Finally, we assume that there is a computable partial function yielding a weakly amalgamable (co)cone for some of the diagrams in $\mathbf{Sign}^\#/\equiv$. These assumptions are reasonable in practice.

Definition 4.2. *A* heterogeneous development graph *(over $\mathcal{I}$) is an acyclic directed graph $\mathcal{S} = \langle \mathcal{N}, \mathcal{L} \rangle$.*

$\mathcal{N}$ is a set of nodes. Each node $N \in \mathcal{N}$ is a tuple (Σ^N, Γ^N) such that $\Sigma^N \in \mathbf{Sign}^\#/\equiv$ is a signature and $\Gamma^N \subseteq \mathbf{Sen}^\#/\equiv(\Sigma^N)$ is the set of local axioms *of N.*

$\mathcal{L}$ is a set of directed links, so-called definition links, *between elements of $\mathcal{N}$. Each definition link from a node M to a node N is either*

- global *(denoted $M \xrightarrow{\sigma} N$), annotated with a signature morphism $\sigma : \Sigma^M \to \Sigma^N \in \mathbf{Sign}^\#/\equiv$, or*
- hiding *(denoted $M \xrightarrow[h]{\sigma} N$), annotated with a signature morphism $\sigma : \Sigma^N \to \Sigma^M \in \mathbf{Sign}^\#/\equiv$* going against the direction of the link. *Typically, σ will be an inclusion, and the symbols of Σ^M not in Σ^N will be hidden.*

Definition 4.3. *Given a node $N \in \mathcal{N}$, its associated class $\mathbf{Mod}_\mathcal{S}(N)$ of models (or N-models for short) consists of those Σ^N-models n for which*

- *n satisfies the local axioms Γ^N,*
- *for each $K \xrightarrow{\sigma} N \in \mathcal{S}$, $n|_\sigma$ is a K-model, and*
- *for each $K \xrightarrow[h]{\sigma} N \in \mathcal{S}$, n has a σ-expansion k (i.e. $k|_\sigma = n$) which is a K-model.*

The notion of *global reachability* is defined inductively: A node M is globally reachable from a node N via a signature morphism σ, $N \xrightarrow{\sigma}\!\!\!\!\!\rightarrow M$ for short, iff either $N = M$ and $\sigma = id$, or $N \xrightarrow{\sigma'} K \in \mathcal{S}$, and $K \xrightarrow{\sigma''}\!\!\!\!\!\rightarrow M$, with $\sigma = \sigma'' \circ \sigma'$.

Complementary to definition and hiding links, which *define* the theories of related nodes, we introduce the notion of a *theorem link* with the help of which we are able to *postulate* relations between different theories. Global and theorem links (denoted by $N \overset{\sigma}{- - \rightarrow} M$ and $N \overset{\sigma}{- - \rightarrow} M$, resp., where $\sigma: \Sigma^N \longrightarrow \Sigma^M$)

are the central data structure to represent proof obligations arising in formal developments. We also need theorem links $N \overset{\sigma}{\underset{h\,\theta}{-\,-\,\blacktriangleright}} M$ (where for some Σ, $\theta: \Sigma \longrightarrow \Sigma^N$ and $\sigma: \Sigma \longrightarrow \Sigma^M$) involving hiding.

Definition 4.4. *Let $\mathcal{S}$ be a development graph. $\mathcal{S}$ implies a global theorem link $N \overset{\sigma}{-\,-\blacktriangleright} M$ (denoted $\mathcal{S} \models N \overset{\sigma}{-\,-\blacktriangleright} M$), iff for all $m \in \mathbf{Mod}_{\mathcal{S}}(M)$, $m|_\sigma \in \mathbf{Mod}_{\mathcal{S}}(N)$. $\mathcal{S}$ implies a local theorem link $N \overset{\sigma}{-\,-\,\blacktriangleright} M$, if for all $m \in \mathbf{Mod}_{\mathcal{S}}(M)$, $m|_\sigma \models \Gamma^N$. Finally, $\mathcal{S}$ implies a hiding theorem link $N \overset{\sigma}{\underset{h\,\theta}{-\,-\blacktriangleright}} M$, iff for all $m \in \mathbf{Mod}_{\mathcal{S}}(M)$, $m|_\sigma$ has a θ-expansion to an N-model.*

A global definition link $M \xrightarrow{\sigma} N$ in a development graph is a *conservative extension* if every M-model can be expanded along σ to an N-model. We will allow annotating a global definition link as $M \xrightarrow[c]{\sigma} N$, which expresses that it is a conservative extension. These annotations can be seen as another kind of proof obligations.

There are quite a number of institution independent languages for structured specifications [21, 12, 10, 15, 11, 19], one of which also has been extended to the heterogeneous case [23]. Most of their constructs can be translated into the formalism of development graphs, which hence can be seen as a core formalism for structured and heterogeneous theorem proving. For the language CASL, such a translation has been laid out explicitly in [2]. An exception are freeness constraints, which are currently not present in development graphs, because a logic-independent proof theory for them is missing yet (if feasible at all).

We now extend our proof calculus for development graphs from [18] to the heterogeneous case. The rules are typically applied backwards, thereby possibly adding some new nodes and edges to the development graph.

The central rule of the proof system is the rule *Theorem-Hide-Shift* (cf. Fig. 2). It is used to get rid off hiding definition links going into the *target* of a global theorem link. Since it is quite powerful, we need some preliminary notions. Given a node N in a development graph $\mathcal{S} = \langle \mathcal{N}, \mathcal{L} \rangle$, the idea is that we unfold the subgraph below N into a tree and form a diagram with this tree. More formally, define the *diagram* $D: J \longrightarrow \mathbf{Sign}$ *associated with* N together with a map $G: |J| \longrightarrow \mathcal{N}$ inductively as follows:

- $\langle N \rangle$ is an object in J, with $D(\langle N \rangle) = \Sigma^N$. Let $G(\langle N \rangle)$ be just N.
- if $i = \langle\, M \xrightarrow{l_1} \cdots \xrightarrow{l_n} N \,\rangle$ is an object in J with $l_1, \ldots, l_n$ definition links in $\mathcal{L}$, and $l = K \xrightarrow{\sigma} M$ is a global definition link in $\mathcal{L}$, then
$$j = \langle\, K \xrightarrow{l} M \xrightarrow{l_1} \cdots \xrightarrow{l_n} N \,\rangle$$
is an object in J with $D(j) = \Sigma^K$, and l is a morphism from j to i in J with $D(l) = \sigma$. We set $G(j) = K$.
- if $i = \langle\, M \xrightarrow{l_1} \cdots \xrightarrow{l_n} N \,\rangle$ is an object in J with $l_1, \ldots, l_n$ definition links in $\mathcal{L}$, and $l = K \xrightarrow[h]{\sigma} M$ is a hiding definition link in $\mathcal{L}$, then

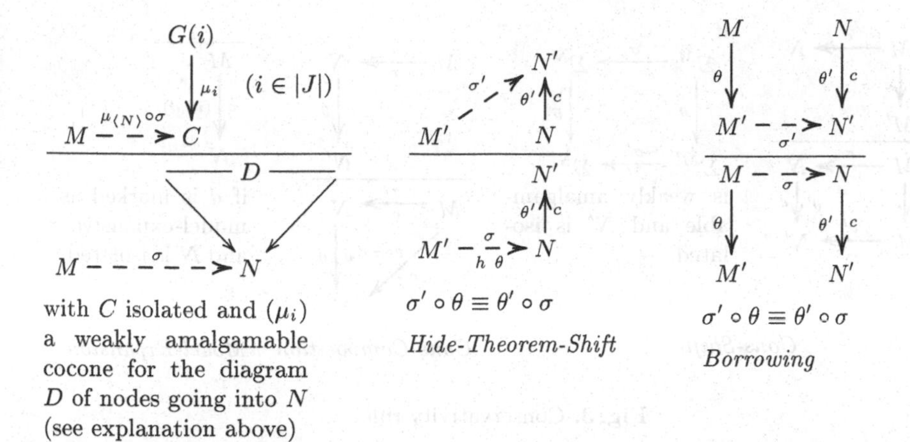

Theorem-Hide-Shift

$$\frac{\begin{array}{l} K \xrightarrow{\sigma\circ\sigma'} M \text{ for each } K \xrightarrow{\sigma'} N \\ L \xrightarrow[h\ \theta]{\sigma\circ\sigma'} M \text{ for each } L \xrightarrow[h]{\theta} K \text{ and } K \xrightarrow{\sigma'} N \end{array}}{N \xrightarrow{\sigma} M}$$

Glob-Decomposition

Fig. 2. Structural rules

$$j = \langle\, K \xrightarrow{l} M \xrightarrow{l_1} \cdots \xrightarrow{l_n} N \,\rangle$$

is an object in J with $D(j) = \Sigma^K$, and l is a morphism from i to j in J with $D(l) = \sigma$. We set $G(j) = K$.

Now in order to apply *Theorem-Hide-Shift*, $(\Sigma, (\mu_i : D(i) \longrightarrow \Sigma)_{i \in |J|})$ has to be a weakly amalgamable cocone for D, and C has to be a new isolated node with signature Σ and with ingoing global definition links $G(i) \xrightarrow{\mu_i} C$ for $i \in |J|$. Here, an isolated node is one with no local axioms and no ingoing definition links other than those shown in the rule.

In order to get rid off hiding links going into the *source* of a global theorem link, one first applies *Glob-Decomposition*, ending up with some local and hiding theorem links. The rule *Hide-Theorem-Shift* allows to prove the latter, using conservativeness of definition links. *Borrowing* is mainly used for shifting a proof goal into a different logic; it also exploits conservativity of definition links. We therefore also need rules dealing with conservativity:

Finally, we have a set of decomposition rules not interacting with hiding, and a rule *Basic Inference* allowing to reduce local theorem links to inference in the calculus of some of the logics in the indexed coinstitution:

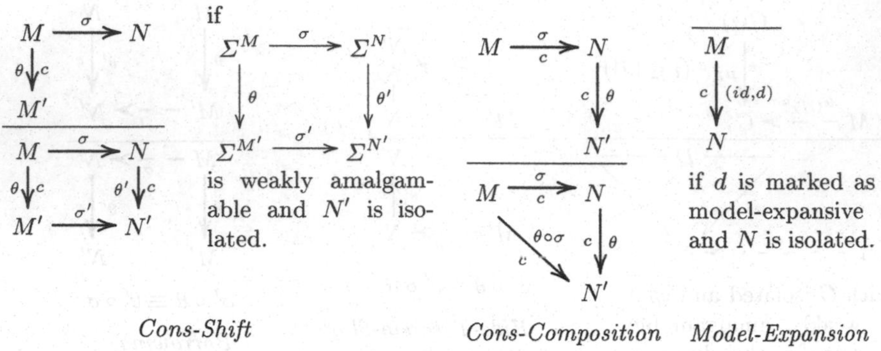

$$\frac{\begin{array}{c} M \xrightarrow{\sigma} N \\ \theta \downarrow c \\ M' \end{array}}{\begin{array}{ccc} M & \xrightarrow{\sigma} & N \\ \theta \downarrow c & & \theta' \downarrow c \\ M' & \xrightarrow{\sigma'} & N' \end{array}}$$

if $\begin{array}{ccc} \Sigma^M & \xrightarrow{\sigma} & \Sigma^N \\ \downarrow \theta & & \downarrow \theta' \\ \Sigma^{M'} & \xrightarrow{\sigma'} & \Sigma^{N'} \end{array}$ is weakly amalgamable and N' is isolated.

Cons-Shift

$$\frac{\begin{array}{ccc} M & \xrightarrow[c]{\sigma} & N \\ & & c \downarrow \theta \\ & & N' \end{array}}{\begin{array}{ccc} M & \xrightarrow[c]{\sigma} & N \\ & \searrow^{\theta\circ\sigma}_{c} & c \downarrow \theta \\ & & N' \end{array}}$$

Cons-Composition

$$\frac{}{\begin{array}{c} M \\ c \downarrow (id,d) \\ N \end{array}}$$

if d is marked as model-expansive and N is isolated.

Model-Expansion

Fig. 3. Conservativity rules

Subsumption:

$$\frac{N \xrightarrow{\sigma} M}{N \overset{\sigma}{- - \rightarrow} M}$$

Loc-Decomposition I:

$$\frac{K \overset{\sigma}{- - \rightarrow} L}{K \overset{\sigma''}{- - \rightarrow} M} \text{ if } L \xrightarrow{\sigma'} M \text{ and } \sigma''(\Gamma^K) = \sigma'(\sigma(\Gamma^K))$$

Loc-Decomposition II:

$$\frac{N \xrightarrow{\sigma'} M}{N \overset{\sigma}{- - \rightarrow} M} \text{ if } \sigma(\Gamma^N) = \sigma'(\Gamma^N)$$

Basic Inference:

$$\frac{Th_{\mathcal{S}}(M) \vdash_{\Sigma^M} \sigma(\varphi) \text{ for each } \varphi \in \Gamma^N}{N \overset{\sigma}{- - \rightarrow} M}$$

Here, $Th_{\mathcal{S}}(M)$ is inductively defined to be

$$\Gamma^M \cup \bigcup_{K \xrightarrow{\sigma} M \in \mathcal{S}} \sigma(Th_{\mathcal{S}}(K))$$

This is well-defined because development graphs have to be acyclic.

Theorem 4.5. *For a 2-indexed coinstitution $\mathcal{I}$: $Ind^* \longrightarrow$* **coIns** *(some of which come as logics), the proof calculus for heterogeneous development graphs is sound for $\mathcal{I}^\# / \equiv$. If, moreover,*

- *$\mathcal{I}$ is quasi-exact,*
- *all institutions comorphisms in $\mathcal{I}$ are weakly exact and model-expansive,*

- *there is a set* $\mathcal{L}$ *of institutions in* $\mathcal{I}$ *that come as* complete *logics,*
- *the rule system is extended with a (sound and complete) oracle for conservative extension for each logic in* $\mathcal{L}$*,*
- *all institutions in* $\mathcal{L}$ *are quasi-semi-exact,*
- *from each institution in* $\mathcal{I}$*, there is some comorphism in* $\mathcal{I}$ *going into some logic in* $\mathcal{L}$*, and*
- *hiding links are only used with signature morphisms whose comorphism component is model-bijective (i.e. the model translation is bijective on objects),*

then the proof calculus for heterogeneous development graphs is sound and complete for $\mathcal{I}^{\#}/\equiv$.

Note that due to the Gödel incompleteness theorem, one cannot expect to drop the oracle for conservative extensions, see [18]. The crucial achievement here is to restrict the oracle to *intra-logic* conservativity.

Further note that in contrast to Prop. 3.5, we need *neither* cocontinuity *nor* exactness of the comorphism signature translations here. Moreover, we need quasi-exactness only for some of the logics; this allows us to include logics which are not quasi-exact, such as CASL.

5 Conclusion and Related Work

We have dualized the Grothendieck institution approach to heterogeneous specification to the case of *institution comorphisms*, which leads to simpler and more general results concerning exactness and weak amalgamation.

We have extended the proof calculus for development graphs with hiding to this setting, and we have studied conditions for its soundness and completeness, which are related to various forms of exactness and weak amalgamation conditions. These conditions are so mild that they hold in typical practical examples; in particular, they are considerably weaker than both the exactness conditions for Grothendieck institutions in [8] and the Craig interpolation property needed for completeness of calculi for structured specification [5]. MAYA [3] implements development graphs and the calculus, but without heterogeneity yet.

Acknowledgments

Thanks to Andrzej Tarlecki, Serge Autexier and Dieter Hutter for useful cooperation and discussions, and to Lutz Schröder and the anonymous referees for valuable hints. This work has been supported by the project MULTIPLE of the *Deutsche Forschungsgemeinschaft* under Grant KR 1191/5-1.

References

1. M. Arrais and J. L. Fiadeiro. Unifying theories in different institutions. In M. Haveraaen, O. Owe, and O.-J. Dahl (eds.), *Proc. 11th WADT*, LNCS 1130, p. 81–101. Springer Verlag, 1996.

2. S. Autexier, D. Hutter, H. Mantel, and A. Schairer. Towards an evolutionary formal software-development using CASL. In C. Choppy and D. Bert (eds.), *Proc. 14th WADT*, LNCS 1827, p. 73–88. Springer-Verlag, 2000.
3. S. Autexier and T. Mossakowski. Integrating HOL-CASL into the development graph manager MAYA. In *FroCoS 2002*, LNCS 2309, p. 2–17. Springer, 2002.
4. H. Baumeister and A. Zamulin. State-based extension of CASL. In *Proceedings IFM 2000*, LNCS 1945. Springer-Verlag, 2000.
5. T. Borzyszkowski. Logical systems for structured specifications. *Theoretical Computer Science*, to appear.
6. M. Cerioli and J. Meseguer. May I borrow your logic? (transporting logical structures along maps). *Theoretical Computer Science*, 173:311–347, 1997.
7. S. Coudert, G. Bernot, and P. Le Gall. Hierarchical heterogeneous specifications. In J. L. Fiadeiro (ed.), *Proc. 13th WADT*, LNCS 1589, p. 106–120. Springer, 1999.
8. R. Diaconescu. Grothendieck institutions. *Applied categorical structures*. to appear.
9. R. Diaconescu and K. Futatsugi. Logical semantics of CafeOBJ. Technical report, JAIST, 1996. IS-RR-96-0024S.
10. R. Diaconescu, J. Goguen, and P. Stefaneas. Logical support for modularisation. In G. Huet and G. Plotkin (eds.), *Proc. of a Workshop on Logical Frameworks*, 1991.
11. F. Durán and J. Meseguer. Structured theories and institutions. In M. Hofmann, G. Rosolini, and D. Pavlovic (eds.), *CTCS '99*, ENTCS 29, 1999.
12. H. Ehrig and B. Mahr. *Fundamentals of Algebraic Specification 2*. Springer Verlag, Heidelberg, 1990.
13. J. Goguen and G. Rosu. Institution morphisms. Formal aspects of computing, to appear, 2001.
14. J. A. Goguen and R. M. Burstall. Institutions: Abstract model theory for specification and programming. *JACM*, 39:95–146, 1992.
15. J. A. Goguen and W. Tracz. An implementation-oriented semantics for module composition. In G. T. Leavens and M. Sitaraman (eds.), *Foundations of Component-Based Systems*, p. 231–263. Cambridge University Press, 2000.
16. J. Meseguer. General logics. In *Logic Colloquium 87*, p. 275–329. North Holland, 1989.
17. T. Mossakowski. Heterogeneous development graphs and heterogeneous borrowing. In M. Nielsen et al. (eds.), *Fossacs 2002*, LNCS 2303, p. 326–341. Springer, 2002.
18. T. Mossakowski, S. Autexier, and D. Hutter. Extending development graphs with hiding. In H. Hußmann (ed.), *FASE 2001*, LNCS 2029, p. 269–283. Springer, 2001.
19. P. D. Mosses. CoFI: The Common Framework Initiative for Algebraic Specification and Development. In *TAPSOFT '97*, LNCS 1214, p. 115–137. Springer, 1997.
20. G. Reggio, E. Astesiano, and C. Choppy. CASL-LTL - a CASL extension for dynamic reactive systems - summary. Technical Report of DISI - Università di Genova, DISI-TR-99-34, Italy, 2000.
21. D. Sannella and A. Tarlecki. Specifications in an arbitrary institution. *Information and Computation*, 76:165–210, 1988.
22. A. Tarlecki. Moving between logical systems. In M. Haveraaen, O. Owe, and O.-J. Dahl (eds.), Proc. 11th WADT, LNCS 1130, p. 478–502. Springer Verlag, 1996.
23. A. Tarlecki. Towards heterogeneous specifications. In D. Gabbay and M. d. Rijke (eds.), *FroCos 98*, Studies in Logic and Computation, p. 337–360. RSP, 2000.

Finite Test-Sets for Overlap-Free Morphisms
(Extended Abstract)

Gwenael Richomme and Francis Wlazinski

LaRIA, Université de Picardie Jules Verne,
5 rue du Moulin Neuf, 80000 Amiens, France,
{richomme,wlazinsk}@laria.u-picardie.fr

Abstract. We study finite test-sets for overlap-freeness of morphisms from A^* to B^*. When $\text{Card}(A) = \text{Card}(B) = 2$, Berstel and Séébold showed there exist such test-sets, and, Richomme and Séébold characterized all of them. Here, we complete these works and characterize test-sets for overlap-freeness of morphisms first for arbitrary alphabets A and B, and, second in the restricted case of uniform morphisms.

Keywords: Combinatorics on words, overlap-free words, morphisms, test-sets.

1 Introduction

At the beginning of the century, Thue [20,21] (see also [3]) worked on repetitions in words. Among other results, he showed the existence of an overlap-free infinite word over a binary alphabet. Since these works, many other results on repetitions in words have been achieved (see [6] for a survey, and [11] for related works).

Using a morphism, Thue [21] generated an infinite overlap-free word over a two-letter alphabet (called the Thue-Morse word since the works of Morse [13]). Morphisms are widely used to generate infinite words. To obtain an infinite word with some property P, one very often uses morphisms preserving the property P. A lot of studies concern such morphisms: Sturmian morphisms (see [12] for a recent survey), power-free morphisms [9], square-free morphisms [2,7], overlap-free binary morphisms [1,8,18,19,20,21]... Our paper deals with overlap-free morphisms.

For arbitrary alphabets A and B, the decidability of overlap-freeness of morphisms from A^* to B^* is not known. One way to decide it is to use finite test-sets for overlap-freeness of morphisms on A, that is, finite subsets T of A^* such that, given any morphism f defined on A, f is overlap-free if and only if $f(T)$ is overlap-free. This technique has already been used to obtain characterization for other property-free morphisms. It is for instance the case for the study of Sturmian morphisms [5,14], square-free morphisms [7], cube-free morphisms [10,16], k-power-free morphisms [16,17,22].

In [4], Berstel and Séébold have shown that an endomorphism f on $\{a,b\}$ is overlap-free if and only if the images of all overlap-free words of length at

K. Diks et al. (Eds): MFSC 2002, LNCS 2420, pp. 605–614, 2002.

most 3 are overlap-free or, equivalently, if $f(abbabaab)$ is overlap-free. In [15], Richomme and Séébold have improved this result showing that an endomorphism f on $\{a, b\}$ is overlap-free if and only if $f(bbabaa)$ is overlap-free. More precisely, they characterize all the finite test-sets for overlap-freeness of binary endomorphisms. All these results concern only endomorphisms and binary case. Here we characterize test-sets for overlap-freeness of morphisms in all the other cases.

In Section 2, we recall notions on words and morphisms. Let A and B be two alphabets. In Section 3, we recall that, when $\text{Card}(B) < \text{Card}(A)$, there do not exist non-empty overlap-free morphisms from A^* to B^*. In Section 4, we complete the works of Richomme and Séébold [15]. We characterize the finite test-sets for overlap-freeness of morphisms in case $\text{Card}(B) > \text{Card}(A) = 2$. In Section 5, we show that such a test-set does not exist when $\text{Card}(B) \geq \text{Card}(A) \geq 3$.

After that, we study test-sets in order to determine if a uniform morphism is overlap-free. Contrarily to the general case, in Section 6, we show that there exist test-sets for overlap-freeness of uniform morphisms from A^* to B^* when $\text{Card}(B) \geq \text{Card}(A) \geq 3$. We characterize all of them. Such test-sets exist also when $\text{Card}(B) \geq \text{Card}(A) = 2$. We characterize them in Section 7.

In Section 8, we end by taking into consideration the test-sets that are singleton when $\text{Card}(A) = 2$.

Note that by lack of place, we summarize the results and we only rarely give details on the combinatorial and technical proofs.

2 Words and Morphisms

We assume the reader is familiar with the notions of alphabet (here always non-empty and finite), letter and words (see, e.g., [11,12]). We precise our notation. Given a finite set X, we denote by $\text{Card}(X)$ its *cardinality*. The *empty word* is denoted ε, and A^* is the set of words over A. We denote by $|u|$ the length of a word u. A word u is a *factor* of a word v if there exist two (possibly empty) words p and s such that $v = pus$. We will also say that v *contains* the word u (as a factor). Given a set X of words, we denote by $\text{Fact}(X)$ the set of factors of words in X. An *overlap* is a word of the form $\alpha v \alpha v \alpha$ with α a letter, v a word. A word is said *overlap-free* if none of its factors is an overlap.

Let A, B be two alphabets. A *morphism* f from A^* to B^* is a mapping from A^* to B^* such that for all words u, v over A, $f(uv) = f(u)f(v)$. When B has no importance, we will say that f is a morphism on A or that f is defined on A. When $B = A$, f is called an *endomorphism* (on A). A morphism on A is entirely known by the images of the letters of A. Given a set X of words over A, we denote by $f(X)$ the set $\{f(w) \mid w \in X\}$. When $\text{Card}(A) = 2$, f is said to be a *binary* morphism (or endomorphism if $B = A$). The morphism f is *uniform* if there exist an integer $L \geq 0$ such that for each letter a in A we have $|f(a)| = L$.

A morphism f on A is *overlap-free*, if $f(w)$ is overlap-free for all overlap-free words w over A. A morphism f on A is *overlap-free* up to an integer n, if $f(w)$ is overlap-free for all overlap-free words w over A with $|w| \leq n$. The *empty morphism* ϵ ($\forall a \in A$, $\epsilon(a) = \varepsilon$) is overlap-free. As example of overlap-free morphisms, let recall [21] that the non-empty overlap-free endomorphisms on $\{a, b\}$ are exactly the morphisms obtained by compositions of the two overlap-free endomorphisms E and μ defined by $E(a) = b$, $E(b) = a$, $\mu(a) = ab$, and, $\mu(b) = ba$.

A *non-erasing* morphism is a morphism for which $f(a) \neq \varepsilon$ for all $a \in A$. The empty morphism ϵ is the only morphism which is both erasing and overlap-free. Indeed for any non-empty erasing morphism f, there exist two different letters a and b in A such that $f(abaa)$ contains an overlap.

A *strongly biprefix* morphism is a non-erasing morphism such that, for all different letters a and b in A, $f(a)$ and $f(b)$ start with different letters and end with different letters. Any non-erasing overlap-free morphism f is strongly biprefix (see [8]). Otherwise there exist two different letters a and b in A such that $f(aab)$ or $f(abb)$ contains an overlap.

3 Test-Sets for Overlap-Freeness

Let A, B be two alphabets. A set T ($\subseteq A^*$) is a *test-set for overlap-freeness* of morphisms from A^* to B^* if, for each morphism f from A^* to B^*, f is overlap-free if and only if $f(w)$ is overlap-free for all w in T. Let us first remark that all the words in a test-set for overlap-freeness of morphisms must be overlap-free.

If $\mathrm{Card}(A) = 1$, there are only three overlap-free words over A: ε, a and aa. Thus $\{aa\}$ is a test-set for overlap-freeness of morphisms on $\{a\}$.

Assume $\mathrm{Card}(B) < \mathrm{Card}(A)$. In this case, no morphism from A^* and B^* is strongly biprefix. So no non-empty morphism from A^* and B^* is overlap-free. More precisely, if f is a non-empty morphism from A^* to B^*, since we necessarily have $\mathrm{Card}(A) \geq 2$, there exist two letters x, y such that either $f(x)$ and $f(y)$ start with the same letter, or, $f(x) \neq \varepsilon$ and $f(y) = \varepsilon$. Thus $f(xxyx)$ contains an overlap. In such a case, $\{xxyx \mid x, y \in A, x \neq y\}$ is a finite test-set for overlap-freeness of morphisms from A^* to B^*.

From now on, A and B are two alphabets such that $\mathrm{Card}(B) \geq \mathrm{Card}(A) \geq 2$.

4 Binary Morphisms

When the starting alphabet A is of cardinality 2, there exist some finite test-sets for overlap-freeness of morphisms. This has been already proved in case $\mathrm{Card}(B) = 2$ in [4]. Moreover, when $\mathrm{Card}(B) = 2$, finite test-sets for morphisms from A^* to B^* have been characterized in [15] (see Theorem 7.3). Considering $A = \{a, b\}$ and the set $T_{\mathrm{B}}(\{a, b\}) = \{aba, bab, abba, baab\}$, we characterize test-sets for overlap-freeness of morphisms from A^* to B^* when $\mathrm{Card}(B) \geq 3$.

Theorem 4.1. *Given an alphabet B with $\mathrm{Card}(B) \geq 3$, a set T of overlap-free words over $\{a, b\}$ is a test-set for overlap-freeness of non-erasing morphisms from $\{a, b\}^*$ to B^* if and only if $T_{\mathrm{B}}(\{a, b\}) \subseteq \mathrm{Fact}(T)$.*

In particular, the set $T_{\rm B}(\{a,b\})$ is itself a test-set for overlap-freeness of non-erasing morphisms from $\{a,b\}^*$ to B^*.

The fact that any word of $T_{\rm B}(\{a,b\})$ belongs to any test-set for overlap-freeness of morphisms from $\{a,b\}^*$ to B^* is due to the following result.

Lemma 4.2. *Let A, B be two alphabets with* $\mathrm{Card}(B) > \mathrm{Card}(A)$. *Let $c \in A$. Let u be a non-empty overlap-free word over $A\backslash\{c\}$. The word cuc must be a factor of any test-set for overlap-freeness of morphisms from A^* to B^*.*

Proof. Without loss of generality, we can assume that $A \subset B$. Let $x \in B\backslash A$. Let f be the morphism from A^* to $(A \cup \{x\})^*$ defined by

$$\begin{cases} f(c) = cucxc, \\ f(b) = b, \text{ for all } b \in A\backslash\{c\}. \end{cases}$$

This morphism verifies for any overlap-free word w over A: $f(w)$ contains an overlap if and only if cuc is a factor of w. So, to show that f is not overlap-free, a test-set for overlap-freeness of morphisms must contain cuc as a factor of one of its word. Indeed if T is a set of overlap-free words with $cuc \notin \mathrm{Fact}(T)$, then $f(T)$ is overlap-free: Since f is not overlap-free, T cannot be a test-set for overlap-freeness. □

When $A = \{a,b\}$, in the previous lemma, the possible couples (c,u) are (a,ε), (a,b), (a,bb), (b,ε), (b,a), (b,aa). It follows that aba, $abba$, bab, $baab$ must be factors of any test-set for overlap-freeness of morphisms from $\{a,b\}^*$ to B^*.

Theorem 4.1 considers non-erasing morphisms. In the general case, we have:

Corollary 4.3. *Given two letters a and b, and an alphabet B with* $\mathrm{Card}(B) \geq 3$, *a set T of overlap-free words over $\{a,b\}$ is a test-set for overlap-freeness of morphisms from $\{a,b\}^*$ to B^* if and only if $T_{\rm B}(\{a,b\}) \subseteq$* $\mathrm{Fact}(T)$ *and there exist two words u and v in T such that $|u|_a \geq 3$, $|v|_b \geq 3$.*

Proof. Let T be a test-set for overlap-freeness of morphisms from $\{a,b\}^*$ to B^*. It is a test-set for overlap-freeness of non-erasing morphisms from $\{a,b\}^*$ to B^*. Thus $T_{\rm B}(\{a,b\}) \subseteq \mathrm{Fact}(T)$. Let $x \in B$. Let f be the morphism defined by $f(a) = x$ and $f(b) = \varepsilon$. For all overlap-free word w with $|w|_a \leq 2$, $f(w)$ is overlap-free. Thus T must contain a word u with $|u|_a \geq 3$. Considering $f \circ E$ instead of f, similarly, T must contain a word v with $|v|_b \geq 3$.

Conversely, let T be a set of overlap-free words over $\{a,b\}$ containing two words u and v with $|u|_a \geq 3$ and $|v|_b \geq 3$, and such that $T_{\rm B}(\{a,b\}) \subseteq \mathrm{Fact}(T)$. Let f be a non-overlap-free morphism on $\{a,b\}$. If f is erasing, $f(u)$ or $f(v)$ is not overlap-free since it contains $f(a)f(a)f(a)$ or $f(b)f(b)f(b)$. If f is non-erasing, by Theorem 4.1, there exists a word w in $T_{\rm B}(\{a,b\})$ (and thus in $\mathrm{Fact}(T)$) such that $f(w)$ contains an overlap. Thus, the set T is a test-set for overlap-freeness of morphisms from $\{a,b\}^*$ to B^*. □

5 Non-binary Morphisms

When A is an alphabet containing at least three letters, given a letter c in A, $A\backslash\{c\}$ contains at least two letters. From the works of Thue [21], we know that there exist an infinite number of overlap-free words u over $A\backslash\{c\}$. Thus Lemma 4.2 implies the next theorem in case $\text{Card}(B) > \text{Card}(A) \geq 3$.

Theorem 5.1. *Given two alphabets A and B with $\text{Card}(B) \geq \text{Card}(A) \geq 3$, there is no finite test-set for overlap-freeness of morphisms from A^* to B^*.*

When $\text{Card}(B) = \text{Card}(A) \geq 3$, Theorem 5.1 is a consequence of:

Lemma 5.2. *Let A be an alphabet containing at least three pairwise different letters a, b and c. Let u be a word over $\{a,b\}$ such that aua is overlap-free. The word $cauac$ must be a factor of any test-set for overlap-freeness of endomorphisms on A.*

Proof. The morphism π_3 from $\{a,b,c\}^*$ to $\{a,b,c\}^*$ defined by $\pi_3(a) = abc$, $\pi_3(b) = bca$, $\pi_3(c) = cab$ is overlap-free (see, e. g., [19]).

Let f be the endomorphism on A defined by

$$\begin{cases} f(a) = \pi_3(a) = abc, \\ f(b) = \pi_3(b) = bca, \\ f(c) = \pi_3(caua)caab = cababcf(u)abccaab, \\ f(d) = d \text{ for all } d \in A\backslash\{a,b,c\}. \end{cases}$$

This morphism verifies for any overlap-free word w over A: $f(w)$ contains an overlap if and only if $cauac$ is a factor of w. □

6 Uniform Non-binary Morphisms

In the previous section, we start studying the existence of test-sets for overlap-freeness of non-binary morphisms. We saw that there do not exist such finite test-sets in the general case. Here we prove that the situation is radically different in the case of uniform morphisms (Theorem 6.1): there always exist finite test-sets, and their characterization depends on $\text{Card}(A)$ and $\text{Card}(B)$. Let us first consider the case $\text{Card}(B) \geq \text{Card}(A) \geq 3$.

Let $T_{\text{U}}(A)$ be the set defined by $T_{\text{U}}(A) = T_{\text{U}1}(A) \cup T_{\text{U}2}(A)$ where $T_{\text{U}1}(A)$ is the set $\{xw_0x \mid x \in A, w_0 \in A^*, \forall a \in A, |xw_0|_a \leq 1\}$ (Note that "$\forall a \in A, |xw_0|_a \leq 1$" means that the letters occurring in xw_0 are pairwise different), and, where $T_{\text{U}2}(A)$ is the set of all words $xw_1\beta w_2y$ with $x, y, \beta \in A, w_1, w_2 \in A^*$, $\forall a \in A, |w_1\beta w_2|_a \leq 1$, and $|w_1| = |w_2| \geq 1, |\beta w_2|_x = 0 = |w_1\beta|_y = 0$.

Theorem 6.1. *Given two alphabets A and B with $\text{Card}(B) \geq \text{Card}(A) \geq 3$, a set T of overlap-free words over an alphabet A is a test-set for overlap-freeness of uniform morphisms from A^* to B^* if and only if $T_{\text{U}}(A) \subseteq \text{Fact}(T)$.*

Since $\forall w \in T_{\text{U}}(A), |w| \leq \text{Card}(A) + 2$, if we only want a bound on the length of words to check, we have:

Corollary 6.2. *Given two alphabets A and B with $\text{Card}(B) \geq \text{Card}(A) \geq 3$, a uniform morphism from A^* to B^* is overlap-free if and only if it is overlap-free up to $\text{Card}(A) + 2$.*

Theorem 3 states, in particular, that $T_U(A)$ is itself a test-set for overlap-freeness of uniform morphisms (when $\text{Card}(B) \geq \text{Card}(A) \geq 3$). There exist some smaller test-sets than $T_U(A)$. First we can delete from $T_U(A)$ the words that are factors of other words in $T_U(A)$. Any word xw_0x of $T_{U1}(A)$ ($x \in A$, $w_0 \in A^*$) such that $|w_0x| \leq \lfloor\frac{\text{Card}(A)-1}{2}\rfloor$, or such that $|w|$ is odd, is a factor of a word in $T_{U2}(A)$. Indeed, from $\text{Card}(A) \geq 2|w_0x| + 1$, taking $p = |w_0x|$, we know that there exist $p+1$ pairwise different letters $\{c_1, \dots, c_p, \beta\}$ that do not occur in w_0x. It follows that xw_0x is a factor of the word $c_1c_1 \dots c_p\beta xw_0x$ which belongs to $T_{U2}(A)$. Thus we can get a smaller test-set than $T_U(A)$ by only taking the words in $T_{U2}(A)$ and the words xw_0x of even length in $T_{U1}(A)$ with $|w_0x| > \lfloor\frac{\text{Card}(A)-1}{2}\rfloor$. We can observe that no other words in $T_{U1}(A)$ is a factor of a word in $T_{U2}(A)$ and that no word in $T_{U2}(A)$ is a factor of a word of odd length in $T_{U1}(A)$ except the words $xw_1\beta w_2y$ with $|w_1\beta w_2y|_x = 0$. Smaller test-sets for overlap-freeness of uniform morphisms (than the previous described) can be get taking overlap-free words containing several words of $T_U(A)$ as factors. Theorem 6.1 shows that there is no other way to obtain smaller test-sets.

We do not give details of the technical proof that $T_U(A)$ is a test-set. The other part of Theorem 6.1 means that any word in $T_U(A)$ must be a factor of any test-set for overlap-freeness. This is a consequence of the four following lemmas.

Lemma 6.3. *Let A and B be two alphabets such that* $\text{Card}(B) \geq \text{Card}(A) \geq 3$. *For any different letters a and b in A, the word aba must be a factor of any test-set for overlap-freeness of morphisms from A^* to B^*.*

Proof. Without loss of generality, we can assume $A \subseteq B$. Let a, b, c be three pairwise different letters in A. Let f be the morphism from A^* to B^* defined by:

$$\begin{cases} f(a) = acbabca, \\ f(b) = bacbcab, \\ f(d) = dbbabbd \text{ for all letters } d \in A\setminus\{a, b\}. \end{cases}$$

This morphism verifies for any overlap-free word w over A: $f(w)$ contains an overlap if and only if aba is a factor of w. □

Lemma 6.4. *Let $p \geq 1$ be an integer. Let A be an alphabet containing at least $2p+2$ letters, and let B be an alphabet with* $\text{Card}(B) \geq \text{Card}(A)$. *Given any pairwise different letters $c_1, \dots, c_p, d_1, \dots, d_p, \beta, \gamma$ in A, given any letter x in $\{c_1, \cdots, c_p, \gamma\}$, and any letter y in $\{d_1, \cdots, d_p, \gamma\}$, the word $xc_1 \cdots c_p\beta d_1 \cdots d_py$ must be a factor of any test-set for overlap-freeness of morphisms from A^* to B^*.*

Proof. Without loss of generality, we can assume $A \subseteq B$. Let p, A, c_1, ..., c_p, d_1, ..., d_p, β, γ, x, y as in the hypotheses of Lemma 6.4. In what follows, $c_{p+1} = d_0 = \beta$. We consider the morphism f from A^* to B^* defined by:

$$\begin{cases} f(c_i) = c_i\ d_{i-1}\ d_i\ \beta\ c_i\ \beta\ \beta\ c_i\ (\forall\ 1 \leq i \leq p), \\ f(d_i) = d_i\ \beta\ c_i\ \beta\ \beta\ c_i\ c_{i+1}\ d_i\ (\forall\ 1 \leq i \leq p), \\ f(\gamma) = \gamma\ \beta\ d_1\ \beta\ c_1\ \beta\ \beta\ \gamma, \\ f(\beta) = \beta\ d_p\ y\ \beta\ \beta\ x\ c_1\ \beta, \\ f(a) = a\ d_1\ d_1\ c_1\ d_1\ c_1\ c_1\ a\ (\forall a \in A\setminus\{c_1, \dots, c_p, \beta, d_1, \dots, d_p, \gamma\}). \end{cases}$$

This morphism verifies for any overlap-free word w over A: $f(w)$ contains an overlap if and only if w contains $xc_1 \cdots c_p\beta d_1 \cdots d_p y$ as a factor. □

Lemma 6.5. *Let $p \geq 1$ be an integer. Let A be an alphabet containing at least $2p+1$ letters, and let B be an alphabet with* $\mathrm{Card}(B) \geq \mathrm{Card}(A)$. *Given any pairwise different letters $c_1, \ldots, c_p, d_1, \ldots, d_p, \beta$ in A, given any letter x in $\{c_1, \cdots, c_p\}$, and any letter y in $\{d_1, \cdots, d_p\}$, $xc_1 \cdots c_p\beta d_1 \cdots d_p y$ must be a factor of any test-set for overlap-freeness of morphisms from A^* to B^*.*

The proof of Lemma 6.5 is similar to that of Lemma 6.4 considering the same morphism without the definition of $f(\gamma)$.

Lemma 6.6. *Let $p \geq 1$ be an integer. Let A be an alphabet containing at least $2p+1$ letters, and let B be an alphabet with* $\mathrm{Card}(B) \geq \mathrm{Card}(A)$. *Given any pairwise different letters $c_1, \ldots, c_p, d_1, \ldots, d_p, \gamma$ in A, the word $\gamma c_1 \cdots c_p d_1 \cdots d_p \gamma$ must be a factor of any test-set for overlap-freeness of morphisms from A^* to B^*.*

Proof. Without loss of generality, assume $A \subseteq B$. Let p, A, $c_1, \ldots, c_p, d_1, \ldots, d_p$, γ as in the hypotheses of Lemma 6.6. In what follows, $c_0 = d_{p+1} = \gamma$. We consider the morphism f from A^* to B^* defined by:

$$\begin{cases} f(c_i) = c_i\ d_i\ d_{i+1}\ d_{i+1}\ d_i\ d_{i+1}\ c_i\ (\forall\, 1 \leq i \leq p), \\ f(d_i) = d_i\ c_{i-1}\ c_i\ d_i\ d_{i+1}\ d_{i+1}\ d_i\ (\forall\, 1 \leq i \leq p), \\ f(\gamma) \ = \gamma\ c_p\ d_1\ c_p\ d_1\ d_1\ \gamma, \\ f(a) \ \ = a\ c_1\ c_1\ d_1\ c_1\ c_1\ a\ (\forall a \in A \backslash \{c_1, \ldots, c_p, d_1, \ldots, d_p, \gamma\}). \end{cases}$$

This morphism verifies for any overlap-free word w over A: $f(w)$ contains an overlap if and only if w contains $\gamma c_1 \cdots c_p d_1 \cdots d_p \gamma$ as a factor. □

7 Uniform Binary Morphisms

Now we come to the case of uniform morphisms with $\mathrm{Card}(A) = 2$. The characterization of test-sets for overlap-freeness depends on $\mathrm{Card}(B)$.

Theorem 7.1. *Let $A = \{a, b\}$ and let B be an alphabet with* $\mathrm{Card}(B) > 2$. *A set T of overlap-free words over A is a test-set for overlap-freeness of uniform morphisms from A^* to B^* if and only if the three following properties are verified:*

1. $\{aa, bb, aba, bab\} \subseteq \mathrm{Fact}(T)$,
2. $\{aab, bba, ababb, babaa\} \cap \mathrm{Fact}(T) \neq \emptyset$,
3. $\{baa, abb, bbaba, aabab\} \cap \mathrm{Fact}(T) \neq \emptyset$.

The necessity of the conditions comes from the following facts (we use the notation of the theorem):

- The morphism f defined by $f(a) = aa$, $f(b) = bc$ verifies: for all overlap-free words w, $f(w)$ is not overlap-free if and only if aa is a factor of w. Thus $aa \in \mathrm{Fact}(T)$. Considering $f \circ E$, we get $bb \in \mathrm{Fact}(T)$.

- The morphism f defined by $f(a) = aba$, $f(b) = bcb$ verifies: for all overlap-free words w, $f(w)$ is not overlap-free if and only if bab is a factor of w. Thus $bab \in \mathrm{Fact}(T)$. Considering $f \circ E$, we get $aba \in \mathrm{Fact}(T)$.
- The morphism f defined by $f(a) = ab$, $f(b) = ac$ verifies: for all overlap-free words w, $f(w)$ is not overlap-free if and only if $\{aab, bba, ababb, babaa\} \cap \mathrm{Fact}(w) \neq \emptyset$. Consequently $\{aab, bba, ababb, babaa\} \cap \mathrm{Fact}(T) \neq \emptyset$.
- Finally using the morphism f defined by $f(a) = ba$, $f(b) = ca$ we can state that $\{baa, abb, bbaba, aabab\} \cap \mathrm{Fact}(T) \neq \emptyset$.

Theorem 7.2. *A set T of overlap-free words over $\{a, b\}$ is a test-set for overlap-freeness of uniform endomorphisms on $\{a, b\}$ if and only if the four following properties are verified:*

1. $\{ab, ba\} \subseteq \mathrm{Fact}(T)$,
2. $\{aa, bb\} \cap \mathrm{Fact}(T) \neq \emptyset$,
3. $\{aab, bba, ababb, babaa\} \cap \mathrm{Fact}(T) \neq \emptyset$,
4. $\{baa, abb, bbaba, aabab\} \cap \mathrm{Fact}(T) \neq \emptyset$.

The necessity of the conditions comes from the following facts (we use the notation of the theorem):

- The morphism f defined by $f(a) = baa$, $f(b) = aab$ verifies: for all overlap-free words w, $f(w)$ is not overlap-free if and only if ab is a factor of w. Consequently $ab \in \mathrm{Fact}(T)$. Considering $f \circ E$, we get similarly $ba \in \mathrm{Fact}(T)$.
- The morphism f defined by $f(a) = aa$, $f(b) = bb$ verifies: for all overlap-free words w, $f(w)$ is not overlap-free if and only if aa or bb is a factor of w. Consequently $\{aa, bb\} \subseteq \mathrm{Fact}(T)$.
- The morphism f defined by $f(a) = aba$, $f(b) = abb$ verifies: for all overlap-free words w, $f(w)$ is not overlap-free if and only if aab, bba, $ababb$ or $babaa$ is a factor of w. Consequently $\{aab, bba, ababb, babaa\} \subseteq \mathrm{Fact}(T)$.
- The morphism f defined by $f(a) = aba$, $f(b) = bba$ verifies: for all overlap-free words w, $f(w)$ is not overlap-free if and only if baa, abb, $bbaba$ or $aabab$ is a factor of w. Consequently $\{baa, abb, bbaba, aabab\} \subseteq \mathrm{Fact}(T)$.

In order to show the effect of the "uniform" hypothesis on test-sets, we recall the following theorem to be compared with Theorem 7.2.

Theorem 7.3. [15, Theorem 5] *A set T of overlap-free words over $\{a, b\}$ is a test-set for overlap-freeness of endomorphisms on $\{a, b\}$ if and only if there exist two words u and v in T such that $|u|_a \geq 3$ and $|v|_b \geq 3$ and the six following properties are verified:*

1. $\{aa, bb\} \subseteq \mathrm{Fact}(T)$.
2. $\{aba, bab\} \cap \mathrm{Fact}(T) \neq \emptyset$,
3. $\{aab, bba, ababb, babaa\} \cap \mathrm{Fact}(T) \neq \emptyset$,
4. $\{baa, abb, bbaba, aabab\} \cap \mathrm{Fact}(T) \neq \emptyset$.
5. $\{aba, abba, bbab, babb\} \cap \mathrm{Fact}(T) \neq \emptyset$,
6. $\{bab, baab, aaba, abaa\} \cap \mathrm{Fact}(T) \neq \emptyset$,

If we want a result like Corollary 6.2, from Theorems 7.1 and 7.2, we can state

Corollary 7.4. *Given two alphabets A and B with* $\mathrm{Card}(B) \geq \mathrm{Card}(A) = 2$, *a uniform morphism from A^* to B^* is overlap-free if and only if it is overlap-free up to 3.*

This result was already known in the case $\mathrm{Card}(B) = \mathrm{Card}(A) = 2$ [4].

Proof. By definition, any overlap-free morphism is overlap-free up to 3.

Conversely if $A = \{a, b\}$, and f is an overlap-free morphism up to 3, then $f(\{aa, bb, aba, bab, aab, baa\})$ is overlap-free. The set $\{aa, bb, aba, bab, aab, baa\}$ verifies the three properties of Theorem 7.1 and the four properties of Theorem 7.2. It is a test-set for overlap-freeness of uniform morphisms defined on a two-letter alphabet. Thus, if f is uniform, then f is overlap-free. □

8 Test-Words

A *test-word for overlap-freeness* of morphisms is a word w such that $\{w\}$ is a test-set for overlap-freeness of morphisms.

In [4], Berstel and Séébold proved that the word *abbabaab* is a test-word for overlap-freeness of morphisms from $\{a, b\}^*$ to $\{a, b\}^*$. In [15], Richomme and Séébold proved that a smaller test-word for overlap-freeness of morphisms from $\{a, b\}^*$ to $\{a, b\}^*$ is *bbabaa*.

A natural question is: what about test-words when we consider morphisms from $\{a, b\}^*$ to B^* with $\mathrm{Card}(B) \geq 3$, or, when we consider uniform morphisms?

As a consequence of Corollary 4.3, we can see that *abbabaab* is a test-word for overlap-freeness of morphisms from $\{a, b\}^*$ to B^* where B is any alphabet with $\mathrm{Card}(B) \geq 3$. However, no word of length seven or less can be such a test-word. There are seven other test-words of length eight: *abaabbab*, *ababbaab*, *abbaabab*, *baababba*, *baabbaba*, *babaabba* and *babbaaba*.

When considering uniform morphisms from $\{a, b\}^*$ to B^* with $\mathrm{Card}(B) > 2$, from Theorem 7.1, there is no test-word for overlap-freeness of length 5 or less, and there exist two test-words of length 6: *aababb*, *bbabaa*. When considering uniform endomorphisms on $\{a, b\}$, from Theorem 7.2, there is no test-word for overlap-freeness of length 3 or less, and there exist two test-words of length 4: *abba*, *baab*. In this case, *aababb* and *bbabaa* are also test-words.

Acknowledgment

Thanks to P. Séébold for his remarks and his encouragments.

References

1. J.-P. Allouche and J. Shallit. Sums of digits, overlaps and palindromes. *Discret Math. and Theoret. Comput. Sci.*, 4:1–10, 2000.

2. J. Berstel. Mots sans carré et morphismes itérés. *Discrete Applied Mathematics*, 29:235–244, 1980.
3. J. Berstel. Axel thue's papers on repetition in words: a translation. Technical Report 20, Laboratoire de Combinatoire et d'Informatique Mathématique, 1995.
4. J. Berstel and P. Séébold. A characterization of overlap-free morphisms. *Discrete Applied Mathematics*, 46:275–281, 1993.
5. J. Berstel and P. Séébold. A characterization of Sturmian morphisms. In *MFCS'93*, volume 711 of *Lecture Notes in Computer Science*, pages 281–290, 1993.
6. C. Choffrut and J. Karhumäki. *Handbook of Formal Languages*, volume 1, chapter Combinatorics of Words. Springer, 1997.
7. M. Crochemore. Sharp characterizations of squarefree morphisms. *Theoretical Computer Science*, 18:221–226, 1982.
8. J. Karhumäki. On strongly cube-free ω-words generated by binary morphisms. In *FCT'81*, volume 117 of *Lecture Notes in Computer Science*, pages 182–191. Springer-Verlag, 1981.
9. M. Leconte. A characterization of power-free morphisms. *Theoretical Computer Science*, 38:117–122, 1985.
10. M. Leconte. *Codes sans répétition.* PhD thesis, LITP Université P. et M. Curie, october 85.
11. M. Lothaire. *Combinatorics on words*, volume 17 of *Encyclopedia of Mathematics.* Addison-Wesley, 1983. reprinted in 1997 by Cambridge University Press in the Cambridge Mathematical Library.
12. M. Lothaire. Algebraic combinatorics on words. *Cambridge University Press*, To appear. See http://www-igm.univ-mlv.fr/~berstel.
13. M. Morse. Recurrent geodesics on a surface of negative curvature. *Transactions Amer. Math. Soc.*, 22:84–100, 1921.
14. G. Richomme. Test-words for Sturmian morphisms. *Bulletin of the Belgian Mathematical Society*, 6:481–489, 1999.
15. G. Richomme and P. Séébold. Characterization of test-sets for overlap-free morphisms. *Discrete Applied Mathematics*, 98:151–157, 1999.
16. G. Richomme and F. Wlazinski. About cube-free morphisms. In H. Reichel and S. Tison, editors, *STACS'2000*, volume 1770 of *Lecture Notes in Computer Science*, pages 99–109. Springer-Verlag, 2000.
17. G. Richomme and F. Wlazinski. Some results on k-power-free morphisms. *Theoretical Computer Science*, 273:119–142, 2002.
18. P. Séébold. Sequences generated by infinitely iterated morphisms. *Discrete Applied Mathematics*, 11:255–264, 1985.
19. P. Séébold. On some generalizations of the Thue-Morse morphism. Technical Report 2000-14, LaRIA, november 2000. To appear in Theoretical Computer Science, special number on 60th birthday of J. Berstel.
20. A. Thue. Uber unendliche zeichenreihen. *Kristiania Videnskapsselskapets Skrifter Klasse I. Mat.-naturv*, 7:1–22, 1906.
21. A. Thue. Uber die gegenseitige Lage gleigher Teile gewisser Zeichenreihen. *Kristiania Videnskapsselskapets Skrifter Klasse I. Mat.-naturv*, 1:1–67, 1912.
22. F. Wlazinski. A test-set for k-power-free binary morphisms. Technical Report 2001-05, LaRIA, July 2001. To appear in Theoretical Informatics and Applications.

Characterizing Simpler Recognizable Sets of Integers

Michel Rigo

University of Liège, Institute of Mathematics,
Grande Traverse 12 (B 37), B-4000 Liège, Belgium,
M.Rigo@ulg.ac.be

Abstract. For the k-ary numeration system, we characterize the sets of integers such that the corresponding representations make up a star-free regular language. This result can be transposed to some linear numeration systems built upon a Pisot number like the Fibonacci system and also to k-adic numeration systems. Moreover we study the problem of the base dependence of this property and obtain results which are related to Cobham's Theorem.

1 Introduction

In formal language theory, the study of numeration systems gives rise to the following question: "for a given numeration system, is it possible to determine if a set of non-negative integers has a simple representation". Otherwise stated, is it possible for a given set $X \subseteq \mathbb{N}$, to find a "simple" algorithm – i.e., a finite automaton – testing membership in X? This relationship between arithmetic properties of sets of integers and syntaxical properties of the corresponding languages of representations has given rise to a lot of papers dealing with the so-called recognizable sets [1,2,3,5,6]. A subset X of $\mathbb{N}$ is said to be k*-recognizable* if the language made up of the k-ary expansions of all the elements in X is regular.

In [8,9], A. de Luca and A. Restivo investigate the same question but for a particular subset of regular languages. The "simplest" regular languages are certainly the star-free languages because the automata accepting those languages are counter-free. So instead of considering arbitrary regular languages, they considered k*-star-free* sets – i.e., sets of integers whose the k-ary representations made up a star-free language. One of the main results of [9] is that if a l-recognizable set X has its density function bounded by $c(\log n)^d$, for some constants c and d, then there exists a base k such that X is k-star-free.

In the present paper, we answer most of the open questions addressed in [9]. Especially, we give a complete characterization of the k-star-free sets using the first-order logical characterization of the star-free languages given by R. McNaughton and S. Papert [10]. Next we extend this result to more general numeration systems like the Fibonacci system.

The celebrated Theorem of Cobham [3] stipulates that the recognizability of a set depends on the base of the numeration system. If k and l are two multiplicatively independent integers then the only subsets of $\mathbb{N}$ which are simultaneously

K. Diks et al. (Eds): MFSC 2002, LNCS 2420, pp. 615–624, 2002.

k-recognizable and l-recognizable are exactly the ultimately periodic sets. When studying k-star-free sets, the problem of the base dependence is still present. We show that if X is an ultimately periodic set of period $s > 1$ then this set is k-star-free only for some k depending on the prime factors of s. We also show that a set is k-star-free iff it is k^m-star-free.

Finally, we study the same question about star-free representations for the unambiguous k-adic numeration system. It is worth noticing that the unique k-adic representation of an integer is not computed through the greedy algorithm and therefore this system differs from the other systems encountered in this paper. From Frougny's results on the normalization function, it appears that the star-free sets with respect to this latter system are exactly the k-star-free sets.

We assume the reader has some knowledge in formal language theory nad in automata theory, see for instance [4,11]. To fix notation, a deterministic finite automaton (DFA) over the alphabet Σ is a 5-tuple $\mathcal{M} = (Q, q_0, \Sigma, \delta, F)$.

2 Characterization of Star-Free Languages

In this section, we recall some well-known results about star-free languages. First, we state without proof a theorem about equivalent definitions of star-free languages. Next, we consider a running example to explain the terminology introduced in the statement of this theorem.

Theorem 1. *Let $L \subseteq \Sigma^*$ be a language. The following assertions are equivalent:*

i. The language L is star-free, i.e., it can be generated by an extended regular expression using only finite sets with a finite number of boolean operations $\cup$, $\cap$, $\setminus$ and concatenation products.
ii. There exists N such that for all $u, v, w \in \Sigma^*$, $uv^N w \in L \Leftrightarrow uv^{N+i} w \in L$ for all integers $i \geq 0$.
iii. The language L is aperiodic, i.e., the syntactic monoid of L has no non-trivial subgroups.
iv. The minimal automaton of L is permutation free.

A proof of this result can be found in [10]. Concerning aperiodic languages, the interested reader can also see [12].

Remark 1. In Theorem 1, it is easy to check that the second assertion is equivalent to the following one. There exists N such that for all $u, v, w \in \Sigma^*$,

$$uv^N w \in L \Leftrightarrow uv^{N+1} w \in L.$$

Roughly speaking, this means that a DFA accepting L is counter free. It can count the number of factors up to a threshold N but cannot count modulo an integer.

Example 1. Let us consider the language L of words over $\Sigma = \{0,1\}$ which do not contain two consecutive 1's. This language is star-free, we have

$$L = \Sigma^* \setminus \Sigma^*\{11\}\Sigma^* = \overline{\emptyset} \setminus \overline{\emptyset}\{11\}\overline{\emptyset} \quad \text{with } \emptyset = \{0\} \cap \{00\}$$

where we denote by $\overline{X}$ the complement $\Sigma^* \setminus X$ of X. To avoid any factor of the form 11, in Theorem 1 we simply take $N = 2$. So for all $u, v, w \in \Sigma^*$, $uv^2w \in L \Leftrightarrow uv^{2+i}w \in L$ for all integers $i \geq 0$. Recall that the *syntactic monoid* of L is the quotient of Σ^* by the congruence $\equiv_L$ on Σ^* defined by $u \equiv_L v$ iff for all $x, y \in \Sigma^*$, $xuy \in L \Leftrightarrow xvy \in L$. If $\mathcal{M} = (Q, q_0, \Sigma, \delta, F)$ is the minimal automaton of L, then the syntactic monoid of L can be obtained as the set of mappings $\{f_w : Q \to Q : q \mapsto \delta(q, w) \mid w \in \Sigma^*\}$ equipped with the composition product. The minimal automaton of L and the corresponding mappings $f_w : Q \to Q$ are given in Fig. 1. The multiplication table of the syntactic monoid of L is given in Fig. 2 (where f_w is simply denoted by w). One can check that it has only trivial subgroups. To conclude this example, we recall that a DFA is *permutation free* if there is no word w that makes a non-trivial permutation of any subset of the set of states. Clearly, the automaton given in Fig. 1 is permutation free.

There is one more characterization of the star-free languages. For the sake of simplicity, let us consider the alphabet $\Sigma_2 = \{0,1\}$. A word w in Σ_2^+ can be identified as a finite model $\mathfrak{M}_w = (M, <, P_1)$ where $M = \{1, \ldots, |w|\}$, $<$ is the usual binary relation on M and P_1 is a unary predicate for the set of positions in w carrying the letter 1. For our convenience, positions are counted from right to left. As an example, the word $w = 1101001$ corresponds to the model $\mathfrak{M}_w = (M, <, P_1)$ where $M = \{1, \ldots, 7\}$ and $P_1 = \{1, 4, 6, 7\}$. (Notice that as in [13] such a model could be expanded with its maximal element *max* or also other definable constants.) If $\varphi(x_0, \ldots, x_n)$ is a formula having at most

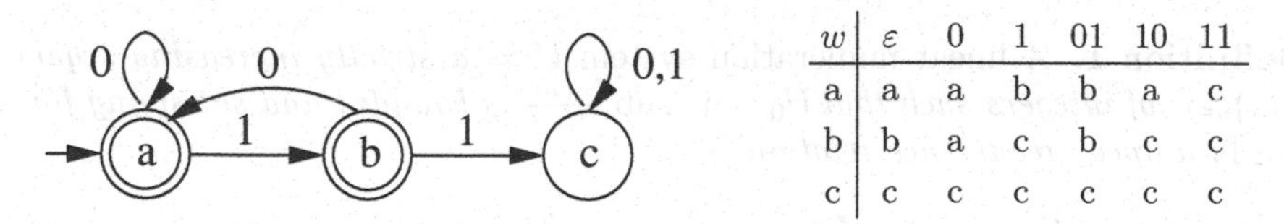

w	ε	0	1	01	10	11
a	a	a	b	b	a	c
b	b	a	c	b	c	c
c	c	c	c	c	c	c

Fig. 1. The minimal automaton of L and the corresponding functions f_w.

	ε	0	1	01	10	11
ε	ε	0	1	01	10	11
0	0	0	01	01	0	11
1	1	10	11	1	11	11
01	01	0	11	01	11	11
10	10	10	1	1	10	11
11	11	11	11	11	11	11

Fig. 2. The syntactic monoid of L.

$x_0, \ldots, x_n$ as free variables, the interpretation of φ in a word-model $\mathfrak{M}_w$ having M as domain and $r_0, \ldots, r_n$ as M-elements is defined in a natural manner and we write $\mathfrak{M}_w \models \varphi[r_0, \ldots, r_n]$ if φ is satisfied in $\mathfrak{M}_w$ when interpreting x_i by r_i. The language defined by a formula φ is therefore $\{w \in \Sigma_2^+ \mid \mathfrak{M}_w \models \varphi\}$.

Let us define a language $\mathcal{L}_{SF}$ of first-order formulas ("SF" stands for star-free and will be explained by Theorem 2). The *variables* are denoted $x, y, z, \ldots$ and are ranging over M-elements. The *terms* are the variables and possibly the constants like max. The *atomic formulas* are obtained by the following rules:

1. if τ_1 and τ_2 are terms then $\tau_1 < \tau_2$ and $\tau_1 = \tau_2$ are atomic formulas
2. if τ is a term then $P_1(\tau)$ is an atomic formula.

Finally, we obtain the set $\mathcal{L}_{SF}$ of all the *formulas* by using the Boolean connectives $\neg, \wedge, \vee, \rightarrow, \leftrightarrow$ and the first-order quantifiers $(\exists x)\ldots$ and $(\forall x)\ldots$ where x is a variable. We are now able to state the Theorem of McNaughton and Papert.

Theorem 2. [10] *A language $L \subseteq \Sigma_2^+$ is star-free iff there exists a sentence φ of $\mathcal{L}_{SF}$ such that $L = \{w \in \Sigma_2^+ \mid \mathfrak{M}_w \models \varphi\}$.*

Recall that a *sentence* is a formula whose all variables are bound.

Example 2. Consider the star-free language $1^+0^* = \{1\}\overline{\overline{\emptyset}\{0\}\overline{\emptyset}}\,\overline{\overline{\emptyset}\{1\}\overline{\emptyset}}$. It is defined by the formula

$$(\exists x)[P_1(x) \wedge (\forall y)(x < y \rightarrow P_1(y)) \wedge (\forall y)(y < x \rightarrow \neg P_1(y))]. \quad (1)$$

To understand formula (1), one has to consider x as the first position where a "1" occurs.

3 Numeration Systems

Definition 1. *A* linear numeration system *U is a strictly increasing sequence $(U_n)_{n\in\mathbb{N}}$ of integers such that $U_0 = 1$, $\sup \frac{U_{n+1}}{U_n}$ is bounded and satisfying for all $n \in \mathbb{N}$ a linear recurrence relation*

$$U_{n+k} = c_{k-1}U_{n+k-1} + \cdots + c_0 U_n,\ c_i \in \mathbb{Z},\ c_0 \neq 0.$$

The *normalized representation* of $x \in \mathbb{N}$ is the word $\rho_U(x) = w_n \cdots w_0$ computed through the greedy algorithm [5] and satisfying

$$x = \sum_{\ell=0}^{n} w_\ell U_\ell.$$

A set $X \subseteq \mathbb{N}$ is *U-recognizable* if $\rho_U(X)$ is a regular language. In the same way, $X \subseteq \mathbb{N}$ is *U-star-free* if $\rho_U(X)$ is a star-free language.

Remark 2. Notice that we allow leading zeroes in $\rho_U(x)$. Indeed, considering leading zeroes does not change the U-star-free property of a set.

In the following, we shall only consider linear numeration systems $(U_n)_{n\in\mathbb{N}}$ whose characteristic polynomial is the minimal polynomial of a Pisot number [1]. (A *Pisot number* is an algebraic integer $\theta > 1$ whose all conjugates have modulus less than one.) For instance, the k-ary system and the Fibonacci system are of this kind. The function $V_U(x)$ is the greatest U_ℓ appearing in the normalized representation of x with a non-zero coefficient w_ℓ. With this definition, we can state the following result.

Theorem 3. [1] *If $U = (U_n)_{n\in\mathbb{N}}$ is a linear numeration system whose characteristic polynomial is the minimal polynomial of a Pisot number then $X \subseteq \mathbb{N}$ is U-recognizable iff there exists a first-order formula φ of $\langle \mathbb{N}, +, V_U\rangle$ having a single free variable such that*

$$X = \{x \in \mathbb{N} \mid \langle \mathbb{N}, +, V_U\rangle \models \varphi(x)\}.$$

For the special case of integer bases, the interested reader can also see the nice survey [2].

Example 3. Consider the binary system defined by the sequence $U_n = 2^n$ (we write V_k instead of V_U to refer to the base k). The set of even integers is 2-recognizable and defined in the Presburger arithmetic $\langle \mathbb{N}, +\rangle$ by the formula $\varphi(x) \equiv (\exists y)(x = y + y)$. In the same way, the powers of 2 are also 2-recognizable and defined by $\varphi(x) \equiv V_2(x) = x$.

We obtain a logical characterization of the U-star-free sets of integers using Theorems 2 and 3. One more time, for the sake of simplicity, we only consider the binary system. Therefore, we replace "U" by "2" in terms like U-recognizability and notation like ρ_U to refer to the base.

Let us introduce the binary relation $\epsilon_2(x, y)$ defined by "y is a power of 2 occurring in the normalized 2-representation of x". As an example $(74, 8)$ belongs to ϵ_2 because $\rho_2(74) = 100\underline{1}010$ but $(74, 16)$ and $(74, 31)$ do not (indeed 31 is not even a power of 2). Observe also that (x, x) belongs to ϵ_2 if and only if x is a power of 2.

Remark 3. The structures $\langle \mathbb{N}, +, V_2\rangle$ and $\langle \mathbb{N}, +, \epsilon_2\rangle$ are equivalent (i.e., for any formula $\varphi(n)$ of $\langle \mathbb{N}, +, V_2\rangle$ there exists a formula $\varphi'(n)$ of $\langle \mathbb{N}, +, \epsilon_2\rangle$ such that $\{n \in \mathbb{N} \mid \langle \mathbb{N}, +, V_2\rangle \models \varphi(n)\} = \{n \in \mathbb{N} \mid \langle \mathbb{N}, +, \epsilon_2\rangle \models \varphi'(n)\}$ and conversely). So Theorem 3 can be restated in term of $\langle \mathbb{N}, +, \epsilon_2\rangle$.

We now introduce a subset $\mathcal{L}_{2,n}$ of formulas $\varphi(n)$ in $\langle \mathbb{N}, +, \epsilon_2\rangle$ defined as follows. The aim of this construction is that a formula in $\mathcal{L}_{2,n}$ has to be used to define a 2-star-free set of integers.

3.1 Syntax of Logical Formulas in $\mathcal{L}_{2,n}$

The *variables* are ranging over $\mathbb{N}$ and denoted $n, x, y, z, \ldots$ (n is dedicated to be the only free variable). The only *terms* are the variables. The *atomic formulas* are obtained with the following rules:

1. If x and y are variables ($\neq n$) then $x < y$ and $x = y$ are atomic formulas.
2. if x is a variable ($\neq n$) then $\epsilon_2(n, x)$ is an atomic formula.

If φ is a formula whose x is a free variable ($x \neq n$) then

$$(\exists x)_2\varphi \equiv (\exists x)(\epsilon_2(x, x) \wedge \varphi) \text{ and } (\forall x)_2\varphi \equiv (\forall x)(\epsilon_2(x, x) \wedge \varphi)$$

are *formulas.* To obtain formulas, we can also use the usual Boolean connectives $\neg, \wedge, \vee, \rightarrow, \leftrightarrow$ either for formulas or atomic formulas. We are now able to define $\mathcal{L}_{2,n}$. If φ is a formula in which the only free variable is (possibly) n then φ is a formula of $\mathcal{L}_{2,n}$.

Remark 4. Notice that the free variable n appears only in the formulas of $\mathcal{L}_{2,n}$ through the relation $\epsilon_2(n, x)$.

Proposition 1. *A set $X \subseteq \mathbb{N}$ is 2-star-free iff it is definable by a formula $\varphi(n)$ of $\mathcal{L}_{2,n}$.*

We sketch the proof of Proposition 1 on an example. Let $X \subseteq \mathbb{N}$ be defined by the following formula of $\mathcal{L}_{2,n}$,

$$(\exists x)_2[\epsilon_2(n, x) \wedge (\forall y)_2(x < y \rightarrow \epsilon_2(n, y)) \wedge (\forall y)_2(y < x \rightarrow \neg\epsilon_2(n, y))]$$

Let us proceed to some syntaxical transformations, $\epsilon_2(n, x)$ (resp. $(\exists x)_2$, $(\forall x)_2$) is replaced by $P_1(x')$ (resp. $(\exists x')$, $(\forall x')$). Finally any occurrence of a variable x is replaced by x'. Thus we obtain

$$(\exists x')[P_1(x') \wedge (\forall y')(x' < y' \rightarrow P_1(y')) \wedge (\forall y')(y' < x' \rightarrow \neg P_1(y'))]$$

which is a formula of $\mathcal{L}_{SF}$ defining a star-free language L. Due to the definition of $\mathcal{L}_{2,n}$, it is clear that $L = \rho_2(X)$.

If we adapt $\mathcal{L}_{SF}$ to a larger alphabet $\{0, \ldots, c\}$ by introducing predicates $P_2, \ldots, P_c$ and if we extend $\mathcal{L}_{2,n}$ to $\mathcal{L}_{U,n}$ using $\epsilon_{j,U}(x, y)$ to say that "y is a U_ℓ occurring in the normalized U-representation of x with the coefficient $j \neq 0$" then Proposition 1 can be restated in a more general framework. (Details are left to the reader.)

Proposition 2. *Let U be a linear numeration system whose characteristic polynomial is the minimal polynomial of a Pisot number. If $\rho_U(\mathbb{N})$ is star-free and defined by the formula $\chi_\mathbb{N}$ of $\mathcal{L}_{SF}$ then a set $X \subseteq \mathbb{N}$ is U-star-free if and only if X is definable by a first-order formula of $\mathcal{L}_{U,n}$ of the form $\varphi \wedge \chi'_\mathbb{N}$ where φ is a formula of $\mathcal{L}_{U,n}$ and the formula $\chi'_\mathbb{N}$ of $\mathcal{L}_{U,n}$ depends only on the syntax of the formula $\chi_\mathbb{N}$.*

The only difference with the binary case is that a word is not necessarily a normalized U-representation and thus we have to introduce a new condition with $\chi_\mathbb{N}$. For instance, in the Fibonacci system, over the alphabet $\{0, 1\}$ two consecutive 1's are not allowed in a normalized representation and this is rendered by a formula of the form

$$\chi_\mathbb{N} \equiv (\forall x)(\forall y)[(\exists z)(x < z < y) \vee \neg(P_1(x) \wedge P_1(y))].$$

From this formula, we can easily derive a new formula of $\mathcal{L}_{U,n}$

$$\chi'_{\mathbb{N}}(n) \equiv (\forall x)_2(\forall y)_2[(\exists z)_2(x < z < y) \vee \neg(\epsilon_{1,U}(n,x) \wedge \epsilon_{1,U}(n,y))].$$

(To obtain this latter formula, we have used the same syntaxical transformations as sketched in the proof of Proposition 1.)

We can now sketch the proof of Proposition 2. If there exists a formula $\varphi(n)$ in $\mathcal{L}_{U,n}$ such that $X \subset \mathbb{N}$ is defined by $\Psi(n) \equiv \varphi(n) \wedge \chi'_{\mathbb{N}}(n)$. Then we can proceed to the same kind of syntaxical transformations on Ψ as in the sketch of the proof of Proposition 1. For instance, $\epsilon_{j,U}(n,x)$ has to be replaced by $P_j(x')$. Proceeding this way, we obtain a formula of $\mathcal{L}_{SF}$ (extended to the alphabet $\{0,\dots,c\}$) defining a language $L \subset \{0,\dots,c\}^*$. Thanks to the formula $\chi_{\mathbb{N}}$ (which is trivially derived from $\chi'_{\mathbb{N}}$), any word in L is a normalized U-representation and so it is clear that $L = \rho_U(X)$. For the converse, let $\rho_U(X)$ be defined by a formula α of $\mathcal{L}_{SF}$. It is clear that $\rho_U(X)$ is also defined by the formula $\alpha \wedge \chi_{\mathbb{N}}$ because $\rho_U(X) \subset \rho_U(\mathbb{N})$. If we proceed backwards to the now usual syntaxical transformations on $\alpha \wedge \chi_{\mathbb{N}}$, we obtain a formula of $\mathcal{L}_{U,n}$ of the required form and defining X.

4 Base Dependence

In this section, we consider only integer base numeration systems and we study the base dependence of the star-free property. We show that the sets of integers are classified into four categories.

Proposition 3. *Let $k, m \geq 2$. A set $X \subseteq \mathbb{N}$ is k-star-free if and only if it is k^m-star-free.*

Proof. Let us first show that if $X \subseteq \mathbb{N}$ is k^m-star-free then X is k-star-free. Assume that $\rho_{k^m}(X)$ is obtained by an extended regular expression over the alphabet $\Sigma_{k^m} = \{0,\dots,k^m - 1\}$ without star operation. In this expression, one can replace each occurrence of a letter $j \in \Sigma_{k^m}$ with the word $0^{m-l}\rho_k(j)$ ($l = |\rho_k(j)|$) of length m. Since we only use concatenation product, the resulting expression defining a language $L \subset \{0,\dots,k-1\}^*$ is still star-free and it is clear that $0^*\rho_k(X) = 0^*L$.

Assume now that $X \subseteq \mathbb{N}$ is k-star-free. Using Theorem 1, the minimal automaton $\mathcal{M} = (Q, q_0, \Sigma_k, \delta, F)$ of $0^*\rho_k(X)$ is permutation free. From $\mathcal{M}$, we build a new automaton $\mathcal{M}' = (Q, q_0, \Sigma_{k^m}, \delta', F)$ having the same set of states and the same initial and final states. The transition function δ' of $\mathcal{M}'$ is defined as follows. For each $j \in \Sigma_{k^m}$, $p, q \in Q$, let $w = 0^{m-l}\rho_k(j)$ where $l = |\rho_k(j)|$, then $\delta'(p,j) = q$ if and only if $\delta(p,w) = q$. One can easily check that the minimal automaton of the language L accepted by $\mathcal{M}'$ is permutation free and that $L = 0^*\rho_{k^m}(X)$. □

4.1 The Case of Ultimately Periodic Sets

A finite union of arithmetic progressions can be written as $\cup_{j=1}^{q}(r_j + s\mathbb{N}) \cup F$ where F is a finite set and s is the l.c.m. of the periods of the different progressions. Since aperiodicity is preserved up to finite modifications of a language, we can forget the finite set F and assume that $r_j < s$. Union of aperiodic sets being again aperiodic, we shall consider a single set of the form $r + s\mathbb{N}$ with $r < s$.

Proposition 4. *Let $r + s\mathbb{N}$ be such that $r < s$, $s > 1$ and the factorization of s as a product of primes is of the form $s = p_1^{\alpha_1} \cdots p_k^{\alpha_k}$, $\alpha_i > 0$, $p_i \neq p_j$. If $P = \Pi_{j=1}^k p_j$ then $r + s\,\mathbb{N}$ is (iP)-star-free for any integer $i > 0$.*

Proof. Let $\alpha = \sup_{j=1,\dots,k} \alpha_j$. It is clear that $(iP)^{\alpha+n}$ is a multiple of s for all integers $n \geq 0$ and $i > 0$. So in the (iP)-ary expansion of an integer the digits corresponding to those powers of iP provide the decomposition with multiples of s. To obtain an element of $r + s\,\mathbb{N}$, we thus have to focus on the last α digits corresponding to the powers $1, iP, \dots, (iP)^{\alpha-1}$ of weakest weight. Consider the finite set

$$Y = \{r + ns \mid n \in \mathbb{N} \text{ and } r + ns < (iP)^\alpha\}.$$

For each $y_j \in Y$, $j = 1, \dots, t$, consider the word $\rho_{iP}(y_j)$ preceded by some zeroes to obtain a word $y'_j \in \Sigma_{iP}^*$ of length α. To conclude the proof, observe that the language made up of the (iP)-ary expansions of the elements in $r + s\,\mathbb{N}$ is $\Sigma_{iP}^* \{y'_1, \dots, y'_t\}$ which is a definite language. (Recall that $L \subseteq \Sigma^*$ is a *definite language* if there are finite languages A and B such that $L = \Sigma^* A \cup B$.) □

Remark 5. The situation of Proposition 4 cannot be improved. Indeed with the previous notations, consider an integer Q which is a product of some but not all the prime factors appearing in s. For any $n \in \mathbb{N}$, Q^n is not a multiple of P and the sequence $(Q^n \bmod s)_{n \in \mathbb{N}}$ is ultimately periodic. Therefore $\rho_Q(r + s\mathbb{N})$ is regular but not star-free because, due to this periodicity, the corresponding automaton is not counter-free. As an example, one can check that $6\mathbb{N}$ is neither 2-star-free nor 3-star-free because the automata accepting $0^*\rho_2(6\mathbb{N})$ and $0^*\rho_3(6\mathbb{N})$ are not permutation free as shown in Fig. 3. For instance, consider the minimal automaton of $0^*\rho_3(6\mathbb{N})$. We denote by q_0 its initial state and by p the other non-terminal state. We see that the word 1 makes a non-trivial permutation of the set of states $\{q_0, p\}$. Indeed, $\delta(q_0, 1) = p$ and $\delta(p, 1) = q_0$.

To summarize the situation, the sets of integers can be classified into four categories:

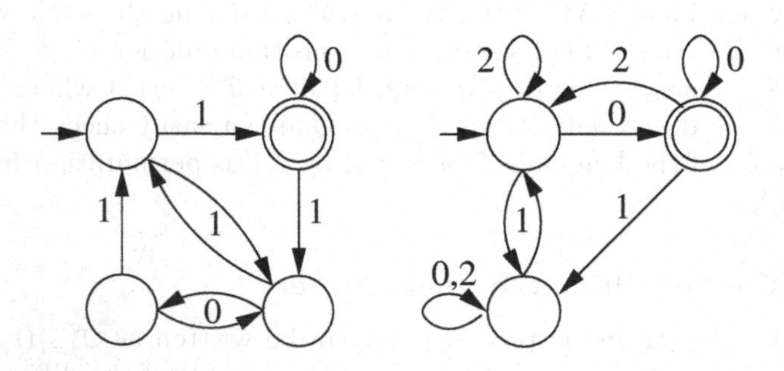

Fig. 3. Automata for $0^*\rho_2(6\mathbb{N})$ and $0^*\rho_3(6\mathbb{N})$

1. The finite and cofinite sets are k-star-free for any $k > 1$.
2. The ultimately periodic sets of period $s = p_1^{\alpha_1} \cdots p_j^{\alpha_j} > 1$ are (iP)-star-free for $P = \Pi_{k=1}^{j} p_k$ and any $i > 0$. In particular, these sets are P^m-star-free for $m \geq 1$.
3. Thanks to Cobham's theorem [3], if a k-recognizable set X is not a finite union of arithmetic progressions then X is only k^m-recognizable for $m \geq 1$ (k being simple[1]). So if a k-star-free set X is not ultimately periodic then X is only k^m-star-free for $m \geq 1$ (k being simple).
4. Finally, there are sets which are not k-star-free for any $k > 1$.

5 *k*-adic Number Systems

The k-adic numeration system is built upon the sequence $U_n = k^n$ but representations are written over the alphabet $\{1, \ldots, k\}$ instead of $\{0, \ldots, k-1\}$. It can be shown that each integer has a unique k-adic representation (see [11] for an exposition on k-adic number systems). For instance, the use of k-adic system may be relevant to remove the ambiguity due to the presence of leading zeroes in a k-ary representation. Indeed, 0 is not a valid digit in a k-adic representation (see for instance [7, p. 303] for a relation to L systems). We have the following proposition.

Proposition 5. *Let $k \geq 2$. A set $X \subseteq \mathbb{N}$ is k-star-free iff the language of the k-adic representations of the elements in X is star-free.*

Proof. Let us consider a language $\widehat{\nu_k}$ of pairs of words of the same length. The idea is that the first component is the k-adic representation of x and the second component is the k-ary representation of the same x. Since the k-ary representation is the greatest in lexicographical ordering, we allow leading zeroes in the first component to obtain words of the same length. If we consider the reversal of $\widehat{\nu_k}$, we have

$$\widehat{\nu_k}^R = \{(u, v) \in \{1, \ldots, k\}^* 0^* \times \{0, \ldots, k-1\}^* : |u| = |v|, \pi_k(u^R) = \pi_k(v^R)\}$$

where $\pi_k(w_n \cdots w_0) = \sum_{\ell=0}^{n} w_\ell\, p^\ell$. It is well-known that this language is regular [6]. The trim minimal automaton (the sink has not been represented) of $\widehat{\nu_k}^R$ is given in Fig. 4 and is clearly permutation free. So $\widehat{\nu_k}$ is a star-free language. Hence we obtain the result after applying the correct canonical homomorphisms of projection which preserve the star-free property. □

Acknowledgments

The author warmly thanks Antonio Restivo which has suggested this work during the thematic term *Semigroups, Algorithms, Automata and Languages* in Coimbra.

[1] Being multiplicatively dependent is an equivalence relation over $\mathbb{N}$, the smallest element in an equivalence class is said to be *simple*. For instance, 2, 3, 5, 6, 7, 10, 11 are simple.

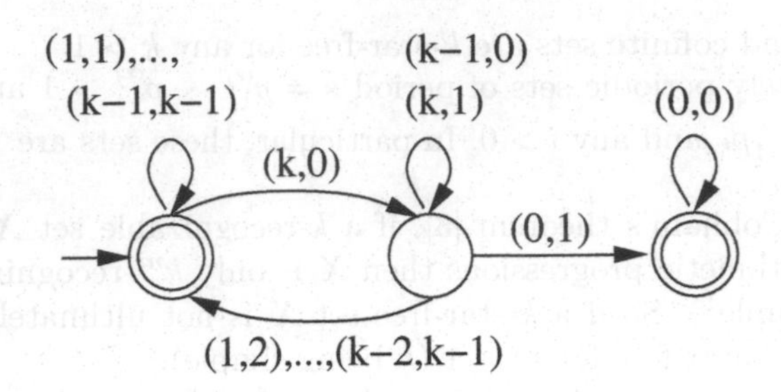

Fig. 4. From k-adic to k-ary representation.

References

1. V. Bruyère, G. Hansel, Bertrand numeration systems and recognizability. Latin American Theoretical INformatics (Valparaíso, 1995). *Theoret. Comput. Sci.* **181** (1997), 17–43.
2. V. Bruyère, G. Hansel, C. Michaux, R. Villemaire, Logic and p-recognizable sets of integers. Journées Montoises (Mons, 1992). *Bull. Belg. Math. Soc. Simon Stevin* **1** (1994), 191–238.
3. A. Cobham, On the base-dependence of sets of numbers recognizable by finite automata. *Math. Systems Theory* **3** (1969) 186–192.
4. S. Eilenberg, *Automata, languages, and machines.* Vol. A. Pure and Applied Mathematics, Vol. 58. Academic Press, New York, (1974).
5. A. S. Fraenkel, Systems of numeration. *Amer. Math. Monthly* **92** (1985), 105–114.
6. C. Frougny, Representations of numbers and finite automata. *Math. Systems Theory* **25** (1992), 37–60.
7. L. Kari, G. Rozenberg, A. Salomaa, L systems. in *Handbook of formal languages,* Vol. 1, 253–328, Springer, Berlin, (1997).
8. A. de Luca, A. Restivo, Representations of integers and laguage theory, Mathematical foundations of computer science (Prague, 1984), 407–415, *Lecture Notes in Comput. Sci.* **176**, Springer, Berlin, (1984).
9. A. de Luca, A. Restivo, Star-free sets of integers. *Theoret. Comput. Sci.* **43** (1986), 265–275.
10. R. McNaughton, S. Papert, *Counter-free automata.* M.I.T. Research Monograph, No. 65. The M.I.T. Press, Cambridge, Mass.-London, (1971).
11. A. Salomaa, *Formal languages.* Academic Press, New York, (1973).
12. M. P. Schützenberger, On finite monoids having only trivial subgroups. *Information and Control* **8** (1965), 190-194.
13. W. Thomas, Classifying regular events in symbolic logic. *J. Comput. System Sci.* **25** (1982), 360–376.

Towards a Cardinality Theorem for Finite Automata

Till Tantau

Technische Universität Berlin, Fakultät für Elektrotechnik und Informatik,
10623 Berlin, Germany,
tantau@cs.tu-berlin.de

Abstract. Kummer's cardinality theorem states that a language is recursive if a Turing machine can exclude for any n words one of the $n+1$ possibilities for the number of words in the language. This paper gathers evidence that the cardinality theorem might also hold for finite automata. Three reasons are given. First, Beigel's nonspeedup theorem also holds for finite automata. Second, the cardinality theorem for finite automata holds for $n=2$. Third, the restricted cardinality theorem for finite automata holds for all n.

1 Introduction

How difficult is it to compute the function $\#_A^n$ for a given language A? This function takes n words as input and yields the number of words in A as output. This counting problem, raised in this general form by Gasarch [6], plays an important role in a variety of proofs both in complexity theory [18,11,21,8,14] and recursion theory [16,17,4].

One way of quantifying the complexity of $\#_A^n$ is to consider its enumeration complexity. For the enumeration complexity of a function f we ask for the smallest m such that f is m-enumerable. This notion, which was first defined by Cai and Hemaspaandra [5] in the context of polynomial-time computations and which was only later transfered to recursive computations, is defined as follows. A function f, taking n-tuples of words as input, is *m-enumerable* if there exists a Turing machine that on input $x_1, \dots, x_n$ starts a possibly infinite computation during which it prints words onto an output tape. At most m different words may be printed and one of them must be $f(x_1, \dots, x_n)$.

Intuitively, the larger m, the easier it should be to m-enumerate $\#_A^n$. This intuition is wrong. Kummer's cardinality theorem, see below, states that even n-enumerating $\#_A^n$ is just as hard as deciding A. Intriguingly, the intuition *is* correct for polynomial-time computations and the work of Gasarch, Hoene, and Nickelsen [6,10,19] shows that a polynomial-time version of the cardinality theorem does not hold.

Cardinality Theorem ([16]). *If $\#_A^n$ is n-enumerable, then A is recursive.*

K. Diks et al. (Eds): MFSC 2002, LNCS 2420, pp. 625–636, 2002.

1.1 History of the Cardinality Theorem

The proof of the cardinality theorem combines ideas from different areas. Several less general results were already proved when Kummer wrote his paper 'A Proof of Beigel's Cardinality Conjecture' [16]. The title of Kummer's paper refers to the fact that Richard Beigel [3] was the first to conjecture the cardinality theorem as a generalisation of his so-called nonspeedup theorem. In the following formulation of the nonspeedup theorem, χ^n_A denotes the n-fold characteristic function of A. Note that the nonspeedup theorem is a consequence of the cardinality theorem.

Nonspeedup Theorem ([3]). *If χ^n_A is n-enumerable, then A is recursive.*

Owings [20] succeeded in proving the cardinality theorem for $n = 2$. For larger n he could only show that if $\#^n_A$ is n-enumerable, then A is recursive in the halting problem. Harizanov et al. [7] have formulated a restricted cardinality theorem, see below, whose proof is somewhat simpler than the proof of the full cardinality theorem.

Restricted Cardinality Theorem ([7]). *If $\#^n_A$ is n-enumerable via a Turing machine that never enumerates both* 0 *and n simultaneously, then A is recursive.*

1.2 Finite Automata and the Cardinality Theorem

In this paper I gather evidence that the cardinality theorem might also hold for finite automata, see Conjecture 1.1 below.

The conjecture refers to the notion of n-enumerability by finite automata, which was introduced in [22] and which is defined as follows. A function f, taking n-tuples of words as input, is *m-fa-enumerable* if there exists a finite automaton for which for every input tuple $(x_1, \ldots, x_n)$ the output attached to the last state reached is a set of at most m values that contains $f(x_1, \ldots, x_n)$. The different components of the tuple are put onto n different tapes, shorter words padded with blanks at the end, and the automaton scans the tapes synchronously. A more detailed definition of fa-enumerability is given at the beginning of the next section.

Conjecture 1.1. If $\#^n_A$ is n-fa-enumerable, then A is regular.

The following three results support the conjecture:

1. The nonspeedup theorem also holds for finite automata, see Fact 2.1.
2. The conjecture holds for $n = 2$, see Theorem 2.2.
3. The restricted form of the conjecture holds for all n, see Theorem 2.3.

Together these results bring us as near to a proof of Conjecture 1.1 as did the results in recursion theory before Kummer's breakthrough proof.

1.3 Reformulations of the Cardinality Theorem

Results on cardinality computations are 'separation results in disguise'. The above results can be reformulated purely in terms of separation and disjointness of certain relations, which will be called regular. A *regular relation* is a relation on words that can be accepted by a multi-tape automata with synchronously read tapes. Two relations A and B are *fa-separable* if there exists a regular relation C such that $A \subseteq C \subseteq \bar{B}$. Relations $A_1, \ldots, A_k$ are *disjoint* if $A_1 \cap \cdots \cap A_k$ is empty. We will only refer to this notion of disjointness in the following, not to the more restrictive and more common notion of pairwise disjointness.

We will see that the condition '$\#_A^2$ is 2-fa-enumerable' is equivalent to 'there exist disjoint regular supersets of $A \times A$, $A \times \bar{A}$, and $\bar{A} \times \bar{A}$'. The condition '$\#_A^n$ is n-fa-enumerable via a finite automaton that never enumerates both 0 and n simultaneously' is equivalent to '$A^{(n)}$ and $\bar{A}^{(n)}$ are fa-separable'. Here $A^{(n)}$ denotes the set of n-tuples of pairwise distinct elements of A. These equivalences allow us to reformulate the last two supporting results for Conjecture 1.1, Theorems 2.2 and 2.3, as follows.

2′. If there exist disjoint regular supersets of $A \times A$, $A \times \bar{A}$, and $\bar{A} \times \bar{A}$, then A is regular, see Theorem 3.2.
3′. If $A^{(n)}$ and $\bar{A}^{(n)}$ are fa-separable, then A is regular, see Theorem 3.3.

Both statements are deceptively innocent-looking and the reader is invited to try to prove them without referring to Theorems 2.2 and 2.3.

The statements are 'optimal' in the sense that if we slightly relax the imposed conditions, then the set A need no longer be regular. Indeed, A can even become *nonrecursive* as the following two results show.

- There exists a nonrecursive set A for which there exist disjoint regular supersets of $A \times A$, $A \times \bar{A}$, $\bar{A} \times A$, and $\bar{A} \times \bar{A}$, see Theorem 3.4.
- For each $n \geq 2$ there exist disjoint, recursively inseparable sets A and B such that $A^{(n)}$ and $B^{(n)}$ are fa-separable, see Theorem 3.5.

The idea of the construction for the last result can be extended to show that there exist disjoint, recursively inseparable sets that are $(3,5)$-fa-separable, see Theorem 4.1 and Section 4 for definitions. The existence of such sets is somewhat surprising, since an old result of Kinber [15, Theorem 2] states that such sets do not exist.

This paper is organised as follows. In Section 2 we prove the cardinality theorem for finite automata for $n = 2$ and the restricted cardinality theorem for finite automata for all n. In Section 3 we reformulate the finite automata versions of the cardinality theorem in terms of regular relations. In Section 4 we construct a counter-example to the above-mentioned result of Kinber [15, Theorem 2].

2 Finite Automata Versions of the Cardinality Theorem

This section presents the three results mentioned in the introduction that support Conjecture 1.1.

We first fix notations. The i-th bit of a bitstring $b \in \{0,1\}^n$ will be denoted $b[i]$. Let $b[i_1,\dots,i_k] := b[i_1]\dots b[i_k]$. For a set P of bitstrings let $P[i_1,\dots,i_k] := \{b[i_1,\dots,i_k] \mid b \in P\}$. The cardinality of a set P will be denoted $|P|$. The *n-fold characteristic function* χ_A^n of a language $A \subseteq \Sigma^*$ maps each word tuple $(x_1,\dots,x_n)$ to the bitstring $b \in \{0,1\}^n$ for which $b[i] = 1$ iff $x_i \in A$. The function $\#_A^n$ maps each word tuple $(x_1,\dots,x_n)$ to $\bigl|\{x_1,\dots,x_n\} \cap A\bigr|$. Note that if words $x_1,\dots,x_n$ are not pairwise different, then $\#_A^n(x_1,\dots,x_n) < n$.

Finite automata compute functions as follows. As done in [1,2,22] we use deterministic automata that instead of just accepting or rejecting a word may produce many different outputs, depending on the type of the last state reached. Formally, instead of a set of accepting states the automata are equipped with output functions $\gamma\colon Q \to \Gamma$ that assign an output $\gamma(q) \in \Gamma$ to each state $q \in Q$. The *output* $M(x)$ of an automaton M on input x is the value $\gamma(q)$ attached to the last state q reached by M upon input x. An automaton M *computes a function* f if $f(x) = M(x)$ for all x.

To fa-compute a function that takes n-tuples of words as input we use n input tapes. Each word is put onto a different tape, shorter words padded with blanks at the end such that all words have the same length. Then the automaton scans the words *synchronously*, meaning that in each step all heads advance exactly one symbol. Equivalently, one may also think of the input words being fed to a single-tape automaton in an interleaved fashion: first the first symbols of all input words, then the second symbols, then the third symbols, and so forth.

A function f, taking n-tuples of words as input, is *m-fa-enumerable* if there exists a deterministic n-tape automaton M such that for all words $x_1,\dots,x_n$ the set $M(x_1,\dots,x_n)$ has size at most m and contains $f(x_1,\dots,x_n)$.

Fact 2.1 ([22]). *If χ_A^n is n-fa-enumerable, then A is regular.*

Theorem 2.2. *If $\#_A^2$ is 2-fa-enumerable, then A is regular.*

Proof. If $\#_A^2$ is 2-fa-enumerable, then χ_A^2 is 3-fa-enumerable via an automaton M that always outputs one of the three sets $\{00,01,10\}$, $\{00,11\}$, and $\{01,10,11\}$.

Consider an automaton N that gets *three* input words x, y, and z and computes (in parallel) what M would output on the three inputs (x,y), (x,z), and (y,z). The automaton N outputs a set P of possibilities for $\chi_A^3(x,y,z)$ defined as follows: $P := \bigl\{b \in \{0,1\}^3 \mid b[1,2] \in M(x,y), b[1,3] \in M(x,z), b[2,3] \in M(y,z)\bigr\}$.

We now employ an argument that is similar to Kadin's [13] easy-hard arguments. Easy-hard arguments have been used in complexity theory in different proofs, see for example [9]. In such an argument one shows that either all words in Σ^* are *easy* (in a sense to be defined), in which case the language A is, well, easy; or there exist some hard words, which allow us to decide all *other* words, provided we know the characteristic values of the hard words.

Let us call a pair (x,y) of words *easy* if there exists a word z such that $Q := N(x,y,z)[1,2]$ has size at most 2. The word z will be called an *advisor* for (x,y). If a pair is not easy, that is, if there does not exist an advisor for it, we call the pair *hard*. We now distinguish two cases, depending on the existence of certain hard pairs.

Case 1. Suppose there exists a hard pair (x, y) with $\chi_A(x) \neq \chi_A(y)$, that is, $\chi_A^2(x, y) = 01$ or $\chi_A^2(x, y) = 10$. We only consider the case $\chi_A^2(x, y) = 01$, the other case is symmetric. We claim that A can be decided by an automaton that works as follows: on input of a single word z it computes $R := N(x, y, z)$. As we will argue in the following paragraph, R will contain at most one bitstring that begins with 01. But then we can output the last bit of this bitstring, which must be the correct value of $\chi_A(z)$. Thus A is regular.

To see that R contains only one bitstring starting with 01, suppose we had both $010 \in R$ and $011 \in R$. Then $000 \notin R$, since otherwise $R[2, 3] \supseteq \{10, 11, 00\}$, contradicting the assumption that one possibility has been excluded for $\#_A^2(y, z)$. Likewise, $101 \notin R$ and also $111 \notin R$, since otherwise $R[1, 3] \supseteq \{00, 01, 11\}$. Since (x, y) was a hard pair, either $R[1, 2] = \{00, 01, 10\}$ or $R[1, 2] = \{01, 10, 11\}$. In the first case, since $000 \notin R$ and $00 \in R[1, 2]$, we must have $001 \in R$. Likewise, since $101 \notin R$ and $10 \in R[1, 2]$, we must have $100 \in R$. But then $R \supseteq \{010, 011, 001, 100\}$ and thus $R[2, 3] \supseteq \{10, 11, 01, 00\}$, a contradiction. Similarly, in the second case we must have $100 \in R$ and $110 \in R$ and thus $R \supseteq \{010, 011, 100, 110\}$, which yields $R[2, 3] \supseteq \{10, 11, 00\}$, also a contradiction. This shows that R contains only one bitstring starting with 01.

Case 2. We must now show that A is regular if $\chi_A(x) = \chi_A(y)$ for every hard pair (x, y). The rough idea is as follows. Our aim is to show that χ_A^2 is 2-fa-enumerable, because then A is regular by Fact 2.1. So we wish to output a set of at most two possibilities for the characteristic string of any two input words. On input of two words x and y we first check whether the pair (x, y) is hard. How such a check can be performed is explained later. If the pair is hard, by assumption we know that $\chi_A(x) = \chi_A(y)$ and we can output the set $\{00, 11\}$. Otherwise, the pair is easy. In this case we know that there exists an advisor z such that $Q := N(x, y, z)[1, 2]$ has size at most 2. Thus if we can find the advisor, we can output Q. How an advisor can be found is explained next.

To find the advisor z, we employ a method that is also used in [22] for the proof of Fact 2.1. The idea is to first construct a *nondeterministic automaton I with ε-moves.* This automaton has two input tapes, on which it finds two words x and y. As its first step it nondeterministically branches to all states the automaton N reaches when it reads the first symbols of x and y on its first and second tape and some arbitrary symbol on the third tape (including the blank symbol, which corresponds to guessing the end of the word on the third tape). In the second step I branches to the states that N reaches upon then reading the second symbols of x and y plus some arbitrary symbol on the third tape and so on. When the ends of both x and y have been reached, I may go on simulating N using ε-moves: it nondeterministically branches to the states N would reach reading blanks on the first two tapes and some arbitrary symbol on the third tape. Note that on some nondeterministic path the automaton I will reach the state reached by N upon input (x, y, z).

We turn I into a deterministic automaton J using the standard power set construction on its states. Upon input of a pair (x, y) the automaton J will end its computation in a 'power state' p that is exactly the set of all states reached by I upon input (x, y). We define the output attached to p in J as the intersection of all sets $\gamma(q)[1,2]$ with $q \in p$, where γ is the output function of N. This ensures that $\chi^2_A(x, y)$ is always an element of $J(x, y)$. Due to the existence of the advisor z, there must exist a state $q \in p$ such that $\gamma(q)[1,2]$ has size at most 2. Thus the intersection that is output by J has size at most 2.

It remains to show how we can check whether (x, y) is a hard pair. Consider the automaton I once more and once more turn it into a deterministic automaton J'. Only this time we have a closer look at the outputs $\gamma(q)$ attached to the states $q \in p$ of the power state p. We check whether for one of these outputs $\gamma(q)[1,2]$ has size at most 2. This is the case iff (x, y) is an easy pair. □

The next proof adapts several ideas that were previously used in a new proof by Austinat et al. [2, Proposition 1] of an old result of Kinber. Kinber's original proof turns out to be wrong, see Section 4.

Theorem 2.3. *If $\#^n_A$ is n-fa-enumerable via a finite automaton that never enumerates both 0 and n simultaneously, then A is regular.*

Proof. We prove the claim by induction on n. For $n = 1$ the claim is trivial. So suppose the claim has already been shown for $n - 1$.

Let $\#^n_A$ be n-fa-enumerable via a finite automaton M that never enumerates both 0 and n simultaneously. Then on input of pairwise different words $x_1, \ldots, x_n$ the output of M either misses 0, which can be interpreted as '$\{x_1, \ldots, x_n\}$ intersects A', or it misses n, which can be interpreted as '$\{x_1, \ldots, x_n\}$ intersects $\bar{A}$'.

Similarly to the proof of Theorem 2.2 we now distinguish two cases, depending on the existence of certain hard words. Let us call a tuple $(y_1, \ldots, y_n)$ an *advisor* for a tuple $(x_1, \ldots, x_{n-1})$, if all of these words are pairwise different and if M makes the following $n + 1$ claims: it claims '$\{y_1, \ldots, y_n\}$ intersects $\bar{A}$' and for $i \in \{1, \ldots, n\}$ it claims '$\{x_1, \ldots, x_{n-1}, y_i\}$ intersects A'. Note that an advisor can only, but need not, exist if at least one x_i is in A. Let us call a tuple $(x_1, \ldots, x_{n-1})$ of pairwise different words *easy* if (a) at least one x_i is not in A or (b) there exists an advisor for it, and let us call the tuple *hard* if neither (a) nor (b) holds.

Case 1. Suppose that there exists a hard tuple $(x_1, \ldots, x_{n-1})$. Since (a) does not hold for it, all x_i are in A. Consider an automaton N that works as follows. On input y it simulates M on the input $(x_1, \ldots, x_{n-1}, y)$ and if the output is '$\{x_1, \ldots, x_{n-1}, y\}$ intersects A', it accepts, otherwise rejects. We show that $\mathrm{L}(N)$, the set of all words accepted by N, is a finite variation of A, which is hence regular. For $y \in A$ the automaton M *must* output '$\{x_1, \ldots, x_{n-1}, y\}$ intersects A' and thus $y \in \mathrm{L}(N)$, except possibly for $y \in \{x_1, \ldots, x_{n-1}\}$. For $y \notin A$, since (b) does not hold, the automata M can output '$\{x_1, \ldots, x_{n-1}, y\}$ intersects A' at most $n - 1$ times. Thus $y \notin \mathrm{L}(N)$ whenever $y \notin A$, except for these finitely many exceptions.

Case 2. Suppose all tuples of pairwise different words are easy. We show that $\#^{n-1}_A$ is $(n - 1)$-fa-enumerable via an automaton M' that never outputs 0 and

$n-1$ simultaneously. Then A is regular by the induction hypothesis. So let n input words $x_1,\dots,x_{n-1}$ be given. If they are not pairwise different, we can immediately output the set $\{0,\dots,n-2\}$. Otherwise, via the detour of a nondeterministic automaton, we search for an advisor as in the proof of Theorem 2.2. If we find one, we output '$\{x_1,\dots,x_{n-1}\}$ intersects A'. The existence of the advisor ensures that this output is correct. If we fail to find an advisor, which can only happen because (a) holds, we output '$\{x_1,\dots,x_{n-1}\}$ intersect $\bar{A}$'. Clearly this output is also correct. □

3 Regular Relations and Cardinality Computations

The introduction claims that 'results on cardinality computations are separation results in disguise'. This section explains what is meant by this claim. It starts with a lemma that shows how cardinality computations can be reformulated in terms of separation and disjointness of regular relations. Then we apply this lemma and show how the previously obtained results can be reformulated without referring to cardinality computations. At the end of the section, we show that the obtained results are optimal in certain senses.

Recall that we called a relation R *regular* if its characteristic function χ_R, defined by $\chi_R(x_1,\dots,x_n)=1$ iff $(x_1,\dots,x_n)\in R$, is fa-computable. We called two relations A and B *fa-separable* if there exists a regular relation C such that $A\subseteq C\subseteq \bar{B}$.

Lemma 3.1. *For every language A the following equivalences hold:*

1. *The function $\#^n_A$ is n-fa-enumerable via an automaton that never enumerates both 0 and n simultaneously, iff $A^{(n)}$ and $\bar{A}^{(n)}$ are fa-separable.*
2. *The function $\#^2_A$ is 2-fa-enumerable, iff there exist disjoint regular supersets of $A\times A$, $A\times\bar{A}$, and $\bar{A}\times\bar{A}$.*
3. *The function χ^2_A is 3-fa-enumerable, iff there exist disjoint regular supersets of $A\times A$, $A\times\bar{A}$, $\bar{A}\times A$, and $\bar{A}\times\bar{A}$.*

Proof. The first equivalence follows easily from the definitions.

For the second equivalence, first assume that there exist disjoint regular supersets $R_2\supseteq A\times A$, $R_1\supseteq A\times\bar{A}$, and $R_0\supseteq\bar{A}\times\bar{A}$. An automaton that witnesses the 2-fa-enumerability of $\#^2_A$ works as follows: on input (x,y) with $x\neq y$ it outputs the set Q that contains 0 if both $(x,y)\in R_0$ and $(y,x)\in R_0$, that contains 1 if either $(x,y)\in R_1$ or $(y,x)\in R_1$, and that contains 2 if both $(x,y)\in R_2$ and $(y,x)\in R_2$. Note that $|Q|\le 2$. Suppose $x\notin A$ and $y\notin A$. Then $(x,y)\in R_0$ and $(y,x)\in R_0$. Thus Q will contain 0 as required. Likewise, if $x\in A$ and $y\in A$ then $2\in Q$. Finally, if $\chi_A(x)\neq\chi_A(y)$, then either $(x,y)\in R_1$ or $(y,x)\in R_1$ and thus $1\in Q$. In all cases we have $\#^2_A(x,y)\in Q$. For the second direction, let $\#^2_A$ be 2-fa-enumerated by M. For $i\in\{0,1,2\}$ define sets $S_i:=\{(x,y)\mid i\in M(x,y)\}$ and let $\Delta:=\{(x,x)\mid x\in\Sigma^*\}$. The desired disjoint supersets are given by $S_2\cup\Delta\supseteq A\times A$, $S_1\setminus\Delta\supseteq A\times\bar{A}$, and $S_0\cup\Delta\supseteq\bar{A}\times\bar{A}$.

For the third equivalence, let R_{11}, R_{10}, R_{01}, and R_{00} denote disjoint supersets of $A\times A$, $A\times\bar{A}$, $\bar{A}\times A$, and $\bar{A}\times\bar{A}$, respectively. Then χ^2_A can be 3-fa-enumerated by an automaton M that works as follows: on input (x, y) it checks (in parallel) for which b we have $(x, y) \in R_b$. It then outputs the set of all such b. Clearly, $\chi_A^2(x, y)$ will be an element of this set and this set will have size at most 3. For the second direction, if χ_A^2 is 3-fa-enumerable via M, we can define R_b with $b \in \{00, 01, 10, 11\}$ as the set of all pairs (x, y) such that $b \in M(x, y)$. These sets have the required properties. □

Lemma 3.1 allows us to reformulate Theorems 2.2 and 2.3 as follows.

Theorem 3.2. *If there exist disjoint regular supersets of* $A \times A$, $A \times \bar{A}$, *and* $\bar{A} \times \bar{A}$, *then* A *is regular.*

Theorem 3.3. *If* $A^{(n)}$ *and* $\bar{A}^{(n)}$ *are fa-separable, then* A *is regular.*

In the following we prove that both results are 'optimal'. For the first theorem this means that if we add a superset of $\bar{A} \times A$ to the list of disjoint sets, then the theorem fails—even quite dramatically as Theorem 3.4 shows. For the second theorem this means that we cannot replace $\bar{A}$ by an arbitrary set B.

Theorem 3.4. *There exists a nonrecursive set* A *such that there exist disjoint regular supersets of* $A \times A$, $A \times \bar{A}$, $\bar{A} \times A$, *and* $\bar{A} \times \bar{A}$.

Proof. As shown in [1,22], for every infinite bitstring b the set A of all bitstrings that are lexicographically smaller than b has the property that χ_A^2 is 3-fa-enumerable. Such a set A is also called a *standard left cut* in the literature. By Lemma 3.1, since χ_A^2 is 3-fa-enumerable, there exist of disjoint regular supersets of $A \times A$, $A \times \bar{A}$, $\bar{A} \times A$, and $\bar{A} \times \bar{A}$. Since there exist uncountably many standard left cuts [12], there must exist a nonrecursive one. □

Theorem 3.5. *For each* $n \geq 2$ *there exist disjoint, recursively inseparable sets* A *and* B *such that* $A^{(n)}$ *and* $B^{(n)}$ *are fa-separable.*

Proof. For an infinite bitstring b let A_b denote the set of all nonempty prefixes of b and let B_b denote the set of all nonempty prefixes of b with the last bit toggled. Then any two words in A_b are comparable with respect to the prefix ordering $\sqsubseteq$, whereas no two different words in B_b are comparable with respect to $\sqsubseteq$. Thus for every bitstring b the relation $A_b^{(2)}$ is a subset of $\sqsubseteq$, whereas $B_b^{(2)}$ is a subset of the complement of $\sqsubseteq$. In particular, $A_b^{(n)}$ and $B_b^{(n)}$ are fa-separable for every b and all $n \geq 2$.

We construct an infinite bitstring b such that $A := A_b$ and $B := B_b$ are not recursively separable. This bitstring is constructed in stages. Let $(M_i)_{i \in \mathbb{N}}$ be an enumeration of all Turing machines (the enumeration need not be effective). In stage i we guarantee that $\mathrm{L}(M_i)$, the set of all words accepted by M_i, does not separate A_b and B_b, that is, either $A_b \not\subseteq \mathrm{L}(M_i)$ or $B_b \not\subseteq \overline{\mathrm{L}(M_i)}$.

Suppose we have already constructed $b^i := b[1, \dots, i]$ and must now decide how to define $b[i+1]$. We check whether both $b^i 0 \in \mathrm{L}(M_i)$ and $b^i 1 \notin \mathrm{L}(M_i)$ hold. If this is the case, let $b[i+1] := 1$, which will ensure both $A_b \not\subseteq \mathrm{L}(M_i)$ and $B_b \not\subseteq \overline{\mathrm{L}(M_i)}$. If this is not the case, let $b[i+1] := 0$, which will ensure either $A_b \not\subseteq \mathrm{L}(M_i)$ or $B_b \not\subseteq \overline{\mathrm{L}(M_i)}$. In either case we guarantee that $\mathrm{L}(M_i)$ does not separate A_b and B_b. □

4 Counter-example to a Theorem of Kinber

In this section we extend the ideas used in the proof of Theorem 3.5 to construct a counter-example to an old result of Kinber [15, Theorem 2]. Kinber's theorem states that if two sets are (m,n)-fa-separable for $m > n/2$, then they are fa-separable. This claim is wrong, since we will construct sets that are $(3,5)$-fa-separable, but not even recursively separable. Before we prove this, we first review Kinber's notion of (m,n)-fa-separability, which is a generalisation of fa-separability.

For two sets A and B let us call a pair (x,b), consisting of a word $x \in \Sigma^*$ and a bit b, *bad* if $b = 1$ and $x \in B$ and if $b = 0$ and $x \in A$. Two disjoint sets A and B are called *(m,n)-separable*, respectively *(m,n)-fa-separable*, if there exists an n-ary function f computable by a Turing machine, respectively a finite automaton, that on input of any n pairwise different words $x_1,\dots,x_n$ outputs a bitstring $b \in \{0,1\}^n$ such that at most $n-m$ pairs $(x_i, b[i])$ are bad. The intuition behind this definition is that an (m,n)-separating function must output 1 for words in A and 0 for words in B and it may make up to $n-m$ mistakes. Words that are neither in A nor in B play no role. Note that sets are $(1,1)$-fa-separable iff they are fa-separable.

Kinber shows that for all $m < n$ there are (m,n)-separable sets that are not recursively separable. He claims that the situation for finite automata is different, since he claims that for $m > n/2$ all (m,n)-fa-separable sets are fa-separable. Theorem 4.1 shows that this claim is wrong.

Theorem 4.1. *There exist* $(3,5)$*-fa-separable, but not recursively separable sets.*

Proof. We show that the recursively inseparable sets A and B constructed in the proof of Theorem 3.5 are $(3,5)$-fa-separable. Recall that these sets were of the form $A = A_b$ and $B = B_b$ for some infinite bitstring b. We must construct an automaton M that on input of five words $x_1,\dots,x_5$ will claim '$x_i \in A$' or '$x_i \in B$' for each i, such that among the claims for words in $A \cup B$ at most 2 are wrong.

Let $X := \{x_1,\dots,x_5\}$. Let y_i denote the word x_i without the last bit (if x_i is the empty string, then $x_i \notin A \cup B$ and we can ignore it). We will say that x_i is *associated* with y_i. Let us call two words x_i and x_j *siblings* if they are associated with the same vertex $y_i = y_j$. Let $Y := \{y_1,\dots,y_5\}$.

The automaton scans the forest structure of $(Y,\sqsubseteq)$, that is, for each pair (i,j) it finds out whether $y_i \sqsubseteq y_j$ holds. Then it considers all branches in the tree $(Y,\sqsubseteq)$ for which at least three words are associated with the vertices on this branch. Given such a branch, let y denote the last vertex on this branch. Then all vertices on this branch are prefixes of y. The automaton assigns outputs to some of the input words according to the following rule: for each $i \in \{1,\dots,5\}$, if y_i is a proper prefix of y and $x_i \sqsubseteq y$ we claim '$x_i \in A$' and if y_i is a proper prefix of y and $x_i \not\sqsubseteq y$ we claim '$x_i \in B$'. Since a word may be associated with a vertex that lies on more than one branch, the just given rule may assign conflicting outputs to a word x_i. Also, we may not have assigned any output to x_i. In either case the automaton outputs '$x_i \in A$'. Note that in both cases this ensures that if x_i has a sibling x_j, the automaton also outputs '$x_j \in A$'.

According to the construction, the output of the automaton for a word $x_i \in A \cup B$ can be *incorrect* only if y_i is not a proper prefix of the last vertex of one of the above-mentioned branches, or if two of these branches 'split' exactly at y_i. Note furthermore that if $x_i \in A \cup B$ has a sibling, at least one output will be correct for the sibling pair.

We now argue that the described procedure $(3, 5)$-fa-separates A and B. Let $X' := X \cap (A \cup B)$ be the words for which our algorithm must produce a correct output with an error margin of 2. Since for $|X'| \leq 2$ we can output anything, the interesting cases are $3 \leq |X'| \leq 5$.

For $|X'| = 5$, there can only be a mistake for one word associated with the top vertex. There cannot be any splits. Thus we can make at most one mistake.

For $|X'| = 4$, there can also be only one mistake for one word associated with the top vertex, but there can be another mistake caused by a split earlier on the branch to which the words in X' are associated. In total, we can make at most two mistakes.

For $|X'| = 3$, if a sibling pair is associated with any vertex on the branch, at least one output is correct and we are done. Otherwise, again one mistake is possible for the word associated with the top vertex. If there is no split at the root vertex, we make at most one additional mistake at the 'middle' vertex. So assume that there is a split at the root vertex. Then *two* additional input words must be associated with the branch leading away in the wrong direction from the root (since we considered only branches to which at least three words are associated). But then there cannot be another split at the middle vertex of our main branch and the output for this middle element must be correct. □

Although Theorem 2 of Kinber's paper fails, corollaries of this theorem can still be true. For example, Kinber's claim is true if instead of arbitrary disjoint sets A and B we consider A and $\bar{A}$: if a set and its complement are (m, n)-fa-separable for $m > n/2$, then it is regular. Austinat et al. [2] were the first to give a (correct) proof of this corollary. The result can also easily be derived from the more general Theorem 3.3: if A and $\bar{A}$ are (m, n)-fa-separable for $m > n/2$, then $A^{(n)}$ and $\bar{A}^{(n)}$ are fa-separable via majority voting and thus A is regular.

5 Conclusion

This paper raises two questions that I would like to recommend for further research. First, does the cardinality theorem hold for finite automata? Second, for which m and n do there exist (m, n)-fa-separable, but not fa-separable sets?

One promising approach to prove the cardinality theorem for finite automata seems to be the employment of proof techniques used in the recursive setting. This approach was successfully taken in [22] to prove the nonspeedup theorem for finite automata. Unfortunately, the transferal of the proofs appears to be highly nontrivial. For example, Kummer's proof of the cardinality theorem is based on the following *r.e. tree lemma*: if a tree is recursively enumerable (r.e.) and some finite tree cannot be embedded into it, then all its branches are recursive. It is not clear what the correct transferal of this lemma to finite automata might be.

The idea of using regular relations to reformulate the cardinality theorem for finite automata can also be applied to the original cardinality theorem. Using a similar argument as in the proof of Lemma 3.1, one can show that a set A is recursive iff there exist disjoint r.e. supersets of $A \times A$, $A \times \bar{A}$, and $\bar{A} \times \bar{A}$. More generally, let $A^{(k,n)}$ denote the set of all tuples $(x_1, \ldots, x_n)$ of pairwise different words such that exactly k of them are in A. Then the cardinality theorem can be reformulated as follows: a set A is recursive iff there exists disjoint r.e. supersets of $A^{(0,n)}, \ldots, A^{(n,n)}$. In this relational formulation we can also ask whether the cardinality theorem holds in different contexts. For example, we can ask whether a set A must be context-free if there exist disjoint context-free supersets of $A \times A$, $A \times \bar{A}$, and $\bar{A} \times \bar{A}$.

Acknowledgements

I would like to thank Holger Austinat and Ulrich Hertrampf for helpful discussions and pointers to the literature.

References

1. H. Austinat, V. Diekert, and U. Hertrampf. A structural property of regular frequency computations. *Theoretical Comput. Sci.*, to appear 2002.
2. H. Austinat, V. Diekert, U. Hertrampf, and H. Petersen. Regular frequency computations. In *Proc. RIMS Symposium on Algebraic Syst., Formal Languages and Computation*, volume 1166 of *RIMS Kokyuroku*, pages 35–42, Research Inst. for Math. Sci., Kyoto University, Japan, 2000.
3. R. Beigel. *Query-Limited Reducibilities.* PhD thesis, Stanford University, USA, 1987.
4. R. Beigel, W. Gasarch, M. Kummer, G. Martin, T. McNicholl, and F. Stephan. The complexity of ODD_n^A. *J. Symbolic Logic*, 65(1):1–18, 2000.
5. J. Cai and L. Hemachandra. Enumerative counting is hard. *Inform. Computation*, 82(1):34–44, 1989.
6. W. Gasarch. Bounded queries in recursion theory: A survey. In *Proc. 6th Structure in Complexity Theory Conf.*, pages 62–78, 1991. IEEE Computer Society Press.
7. V. Harizanov, M. Kummer, and J. Owings. Frequency computations and the cardinality theorem. *J. Symbolic Logic*, 52(2):682–687, 1992.
8. L. Hemachandra. The strong exponential hierarchy collapses. *J. Comput. Syst. Sci.*, 39(3):299–322, 1989.
9. E. Hemaspaandra, L. Hemaspaandra, and H. Hempel. A downward collapse in the polynomial hierarchy. *SIAM J. Comput.*, 28(2):383–393, 1998.
10. A. Hoene and A. Nickelsen. Counting, selecting, and sorting by query-bounded machines. In *Proc. 10th Symposium on Theoretical Aspects of Comput. Sci.*, volume 665 of *Lecture Notes in Comput. Sci.*, pages 196–205. Springer-Verlag, 1993.
11. N. Immerman. Nondeterministic space is closed under complementation. *SIAM J. Comput.*, 17(5):935–938, 1988.
12. C. Jockusch, Jr. *Reducibilities in Recursive Function Theory.* PhD thesis, Massachusetts Inst. of Technology, USA, 1966.

13. J. Kadin. The polynomial time hierarchy collapses if the boolean hierarchy collapses. *SIAM J. Comput.*, 17(6):1263–1282, 1988.
14. J. Kadin. $\mathrm{P}^{\mathrm{NP}[O(\log n)]}$ and sparse Turing-complete sets for NP. *J. Comput. Syst. Sci.*, 39(3):282–298, 1989.
15. E. Kinber. Frequency computations in finite automata. *Cybernetics*, 2:179–187, 1976.
16. M. Kummer. A proof of Beigel's cardinality conjecture. *J. Symbolic Logic*, 57(2): 677–681, 1992.
17. M. Kummer and F. Stephan. Effective search problems. *Math. Logic Quarterly*, 40(2):224–236, 1994.
18. S. Mahaney. Sparse complete sets for NP: Solution of a conjecture of Berman and Hartmanis. *J. Comput. Syst. Sci.*, 25(2):130–143, 1982.
19. A. Nickelsen. On polynomially $\mathcal{D}$-verbose sets. In *Proc. 14th Symposium on Theoretical Aspects of Comput. Sci.*, volume 1200 of *Lecture Notes in Comput. Sci.*, pages 307–318. Springer-Verlag, 1997.
20. J. Owings, Jr. A cardinality version of Beigel's nonspeedup theorem. *J. Symbolic Logic*, 54(3):761–767, 1989.
21. R. Szelepcsényi. The method of forced enumeration for nondeterministic automata. *Acta Informatica*, 26(3):279–284, 1988.
22. T. Tantau. Comparing verboseness for finite automata and Turing machines. In *Proc. 19th Symposium on Theoretical Aspects of Comput. Sci.*, volume 2285 of *Lecture Notes in Comput. Sci.*, pages 465–476. Springer-Verlag, 2002.

An Approximation Semantics for the Propositional Mu-Calculus*

Roger Villemaire

Département d'informatique, Université du Québec à Montréal,
C.P. 8888, succ. centre-ville, Montréal (Québec), Canada H3C 3P8,
villemaire.roger@uqam.ca

Abstract. We give a new semantics for the propositional μ-calculus, prove that it is equivalent to the standard semantics and use it in order to propose a constraint based axiomatization of the μ-calculus. We show that our axiomatization is sound relatively to our semantics. Completeness of the system is the topic of ongoing research but we give a simple example of how our system could be used to construct a finite counter-example of an unprovable sentence.

1 Introduction

The propositional μ-calculus is an extension of propositional logic with modalities and fixpoints (greatest and smallest) which was introduced by [Ko] to describe properties of computational systems. Its importance does not only rely on the fact that this logic is built from natural computational notion such as action (change of state) and recursion (fixpoint) but also that it subsumes most of the other propositional modal logics such as PDL [FiLa], PDLΔ [St], Process Logic [HoKoPa], and CTL* [EmHa] [Da]. Furthermore the propositional μ-calculus preserves the characterisation of bisimulation equivalence [HeMi]. Finally on the practical side most of the tools used today for specification or verification use some kind of fragment of the μ-calculus to specify properties of systems.

[EmJu] have shown that the propositional μ-calculus is decidable in exponential time and [Ko] [Wa1], [Wa2] gave axiomatizations. In later years there has been a lot of interest in model-checking, i.e. verifying if a specific μ-calculus sentence holds in a specific state (see for instance [StWa], [Cl], [Wi] [EmJuSi]) and these methods have proven to be of great practical importance.

Even if we have decision procedures and axiomatizations for the propositional μ-calculus, all known ones use automata theoretic methods. This fact limits the possibility to apply these methods inside tools. On the other hand model-checking does not use the full flexibility of logic. For instance to show that a state q satisfy a μ-calculus sentence φ can be seen as showing that $\psi \to \varphi$ for ψ a mu-calculus sentence describing q up to bisimulation. But instead of showing $\psi \to \varphi$ for ψ describing q up to bisimulation one could also use only a property ψ true for q, in fact ψ could be a property satisfied by a wide range

* This research was supported by the canadian N.S.E.R.C/C.R.S.N.G.

K. Diks et al. (Eds): MFSC 2002, LNCS 2420, pp. 637–649, 2002.

of descriptions. So instead of doing verification by describing a specification and verifying properties on it, one could describe some of the properties one garantees, leaving open a wide range of behaviors and still be able to show that some other useful property is a consequence. This approach could be helpful in verification of concurrent systems. But in order to realize such a project one needs algorithms to manipulate μ-calculus sentences either to show that they are true, or if not, give a finite counter-example. One hopes to get methods which rely on much more elementary methods than those of automata on trees in order to use these methods in practical implementations.

In order to aim at this objective we introduce in this paper an approximation semantics for the μ-calculus. We show that our semantics does not change the notion of truth (no more and no less μ-calculus sentences are true) and we give an axiomatization of the μ-calculus which is sound for our semantics. We illustrate our proof system by some examples and we conclude with an illustration of how our proof system could be used to build a counter-example to an unprovable sentence. Of course to be of practical significance our system and method should be proved to be complete, i.e. able to produce a finite counter-example for any unprovable sentence. We don't have a general proof of this, but it is the topic of ongoing research. Nevertheless we think that our new semantics is interesting in its own and should stimulate research on proof systems which do not rely on involved automata theoretic methods.

The paper is structured as follows. First we explain our approximation semantics and show that it does not change the notion of truth, then we give an axiomatization and show that it is sound in our semantics, we then give some examples of use of this proof system and we finally give an illustration of how our proof system could be used to build a counter-example to an unprovable sentence.

2 Approximation Semantics

A μ-calculus formula is build from the constant *True*, propositional constants $P, Q, \ldots$, propositional variables $X, Y, Z, \ldots$, the connectors $\wedge, \neg$, the modality $\langle a \rangle$ where a is an action from the actions set $a, b, c, \ldots$ and least fix-point $\mu X.\varphi(X)$ where X must appear positively, i.e. within an even number of negations, in $\varphi(X)$.

A μ-calculus sentence is a formula with no free variables, i.e. not bounded by fixpoints.

The standard semantics is to interpret a sentence φ as set of states $[\varphi]$ of a finite transition system (e.g. a finite set of states and a finite set of transition between states labelled by actions) in the following way. The constant *True* is interpreted as the set of all states. Other constants are interpreted as arbitrary sets of states, $\wedge$ as the intersection, $\neg$ as the complement, $\langle a \rangle \varphi$ as the set $\langle a \rangle [\varphi] = \{q : \text{state} \; ; \text{there exists } q \xrightarrow{a} q' \text{ with } q' \in [\varphi]\}$. Finally $\mu X.\varphi(X)$ is interpreted as the least fix-point of the operator $\varphi(X)$ which exists by the Tarski-Knaster Theorem.

We will call the interpretation $[-]$ the *standard valuation.* Others usual connectors can be define in the standard way, e.g. $\varphi \vee \psi \equiv \neg(\neg\varphi \wedge \neg\psi)$, $\varphi \rightarrow \psi \equiv \neg\varphi \vee \psi$, $[a]\varphi \equiv \neg\langle a\rangle\neg\varphi$ and greatest fix-point $\nu X.\varphi(X) \equiv \neg(\mu X.\neg\varphi(\neg X))$. It is often convenient to assume that there exists for any set of states S a corresponding propositional constant which we will also denote by S.

We will restrict ourself to *well-named* formulas, i.e. formulas such that for any subformula $\mu X.\varphi(X)$, X is not bounded by a fix-point in $\varphi(X)$. Since renaming fix-point variables does not change the meaning of a sentence this does not reduce the logic.

Following [Ko] we introducc

Definition 1. *For F a finite set of sentences the inductive closure (denoted $Cl(F)$) is the smallest set of sentences containing F such that*
a) if $\varphi \wedge \psi \in Cl(F)$ then $\varphi, \psi \in Cl(F)$
b) if $\neg\varphi \in Cl(F)$ then $\varphi \in Cl(F)$
c) if $\langle a\rangle\varphi \in Cl(F)$ then $\varphi \in Cl(F)$
d) if $\mu X.\varphi(X) \in Cl(F)$ then $\varphi\{X/\mu X.\varphi(X)\} \in Cl(F)$

The inductive closure of F is finite and of size linear in the size of F.

We will say that a set of sentences is *inductively closed* if it is equal to its inductive closure.

We now introduce the fundamental notion of our approximation semantics

Definition 2. *Let T be a transition system and $[-]$ be the standard valuation for T and let F be an inductively closed set of sentences. An inductive valuation v for F on T is a pair of interpretations $[-]^v_L$, $[-]^v_U$ mapping sentences of F to set of states of T such that*

PL) $[P]^v_L \subseteq [P]$ *PU)* $[P] \subseteq [P]^v_U$
$\wedge$L) $[\varphi \wedge \psi]^v_L \subseteq [\varphi]^v_L \cap [\psi]^v_L$ *$\wedge$U)* $[\varphi]^v_U \cap [\psi]^v_U \subseteq [\varphi \wedge \psi]^v_U$
$\neg$L) $[\neg\varphi]^v_L \subseteq ([\varphi]^v_U)^c$ *$\neg$U)* $([\varphi]^v_L)^c \subseteq [\neg\varphi]^v_U$
where c denotes the complement operation.
$\langle\rangle$L) $[\langle a\rangle\varphi]^v_L \subseteq \langle a\rangle[\varphi]^v_L$ *$\langle\rangle$U)* $\langle a\rangle[\varphi]^v_U \subseteq [\langle a\rangle\varphi]^v_U$
μL) $[\mu X.\varphi(X)]^v_L \subseteq [\varphi\{X/\mu X.\varphi(X)\}]^v_L$ *μU)* $[\varphi\{X/\mu X.\varphi(X)\}]^v_U \subseteq [\mu X.\varphi(X)]^v_U$
$[-]^v_L$ is called the lower valuation and $[-]^v_U$ the upper valuation.

The motivation (which we prove later but only for some inductive valuations satisfying a computational property which we will later introduce) is that if a state is in the lower valuation then it is in the standard one so we are sure that it satisfies this formula in the standard sense. On the other hand if a state is not in the upper valuation then it is not in the standard one so we are sure that it does not satisfy the formula. Hence the lower valuation approximates the standard valuation from below and the upper valuation approximates it from above.

Note that the previous definition just depends on the standard valuation of propositional constants.

Example 1. The pair formed of $[-]^v_L = [-]$ and $[-]^v_U = [-]$ (i.e. twice the standard valuation) is an inductive valuation. We will call this inductive valuation the *standard inductive valuation* and denote it by $\top$.

Example 2. The pair formed of $[-]^v_L = \emptyset$ and $[-]^v_U =$ set of all states (i.e. a lower valuation sending every sentence to the empty set and an upper valuation sending every sentence to the set of all states) form an inductive valuation. We will denote this inductive valuation by $\perp$.

We will need to compare inductive valuations in order to have a notion of one being a better (closer) approximation than another one.

Definition 3. *Let T be a transition system, F be an inductively closed set of sentences and v, w be inductive valuations for F on T. We say that v is smaller than w ($v \leq w$) if*

a) For any sentence φ of F, $[\varphi]^v_L \subseteq [\varphi]^w_L$
b) For any sentence φ of F, $[\varphi]^w_U \subseteq [\varphi]^v_U$

The idea is that w is a better (closer) approximation to the standard valuation than v.

Remark 1. $\perp$ is the smallest inductive valuation. The standard inductive valuation is not the greatest inductive valuation, since interpreting μ in both the lower and the upper valuation as the **greatest** fix-point operator will also give an inductive valuation. But we will show that for computationally interesting inductive valuations we indeed have this inequality.

Definition 4. *Let T be a transition system, F be an inductively closed set of sentences and v, w be inductive valuations for F on T. We say that v is equal to w ($v = w$) if in a) and b) of the previous definition we replace inclusion by equality.*

We will say that $v < w$ if $v \leq w$ and $v \neq w$.

We now need a notion of computation to constraint how inductive valuations are constructed.

Definition 5. *Let T be a transition system and F be an inductively closed set of sentences. An inductive run for F on T is a sequence of inductive valuations for F on T, $\perp = v_0 < v_1 < \ldots < v_n$ such that for all $\mu X.\varphi(X)$ in F and all $i \neq 0$*

$$[\mu X.\varphi(X)]^{v_i}_L \subseteq [\varphi\{X/\mu X.\varphi(X)\}]^{v_{i-1}}_L$$

The idea here is that using the Tarski-Knaster Theorem the standard valuation of $\mu X.\varphi(X)$ is computed by the recursion $\emptyset \subseteq \varphi(\emptyset) \subseteq \varphi(\varphi(\emptyset)) \subseteq \ldots$ so in order to have an inductive semantics which is an approximation of the standard one, we need a way to restrain the growth of $[\mu X.\varphi(X)]^v_L$.

We will now show that an inductive valuation which is part of an inductive run is smaller than the standard inductive valuation.

Definition 6. *We say that a variable of a formula is positive (negative) if it appear only within an even (odd) number of negations. A polarized variable of a formula is a variable which is either positive or negative. We say that a formula is polarized if all its variables are polarized.*

Definition 7. *Let $\varphi(X_1,\ldots,X_n)$ be a polarized formula and $\eta = \{X_1/\mu X_1.\varphi_1, \ldots, X_n/\mu X_n.\varphi_n\}$ be a substitution.*

a) We will denote by $\eta^v_{\varphi,L}$ the substitution $\{X_1/A_1,\ldots,X_n/A_n\}$, where $A_i = [\mu X_i.\varphi_i]^v_L$ for X_i positive in $\varphi(X_1,\ldots,X_n)$ and $A_i = [\mu X_i.\varphi_i]^v_U$ for X_i negative in $\varphi(X_1,\ldots,X_n)$.

b) We will denote by $\eta^v_{\varphi,U}$ the substitution $\{X_1/B_1,\ldots,X_n/B_n\}$, where $B_i = [\mu X_i.\varphi_i]^v_U$ for X_i positive in $\varphi(X_1,\ldots,X_n)$ and $B_i = [\mu X_i.\varphi_i]^v_L$ for X_i negative in $\varphi(X_1,\ldots,X_n)$.

Lemma 1. *Let T be a transition system and F be an inductively closed set of sentences, $\varphi(X_1,\ldots,X_n)$ be a polarized formula and $\eta = \{X_1/\mu X_1.\varphi_1,\ldots, X_n/\mu X_n.\varphi_n\}$ be a substitution. Let also $\bot = v_0 < v_1 < \ldots < v_n$ be an inductive run.*

Then

a) $[\varphi\eta]^{v_i}_L \subseteq [\varphi\eta^{v_i}_{\varphi,L}]$, for $i = 0,\ldots,n$.

b) $[\varphi\eta^{v_i}_{\varphi,U}] \subseteq [\varphi\eta]^{v_i}_U$

Proof. We show this result by induction on i and then on the structure of the sentence φ.

The case $i = 0$ is clear by definition of $\bot$. Now suppose the result is true for i, we will show that it also holds for $i+1$. This is done by induction on the structure of φ.

The case $\varphi \equiv X_i$ follows from the definition of $\eta^{v_i}_{\varphi,L}$ and of $\eta^{v_i}_{\varphi,U}$ and the definition of inductive valuation. The case $\varphi \equiv P$ for a propositional constant P follows from the definition of inductive valuation.

Let us consider negation. Let use first prove case a). By the definition of inductive valuation we have that $[\neg\psi\eta]^{v_{i+1}}_L \subseteq ([\psi\eta]^{v_{i+1}}_U)^c$ which is included in $[\psi\eta^{v_{i+1}}_{\psi,U}]^c$ by induction hypothesis (on the structure of the sentence for case b)). Finally this last set is equal to $[\neg\psi\eta^{v_{i+1}}_{\psi,U}] = [\neg\psi\eta^{v_{i+1}}_{\neg\psi,L}]$. The case b) is similar.

The proof for φ a conjunction or of the form $\langle a\rangle\psi$ being straightforward by induction on the structure of the sentence, we are left with the case of a fixpoint.

Let us first look at the case a). Since we restricted ourself to well-named sentences we have that X is not in $X_1,\ldots,X_n$. By definition of inductive run we have that $[\mu X.\psi\eta]^{v_{i+1}}_L \subseteq ([\psi\{X/\mu X.\psi\eta,\eta\}]^{v_i}_L)$ where $\{X/\mu X.\psi\eta,\eta\}$ is the substitution obtained from η by adding $X/\mu X.\psi\eta$ to it. By induction hypothesis on i this last set is included in $[\psi\{X/[\mu X.\psi\eta]^{v_i}_L, \eta^{v_i}_{\mu X.\psi,L}\}]$. Now again by induction hypothesis on i, we have that $[\mu X.\psi\eta]^{v_i}_L \subseteq [\mu X.\psi\eta^{v_i}_{\mu X.\psi\eta,L}]$ and therefore $[\psi\{X/[\mu X.\psi\eta]^{v_i}_L, \eta^{v_i}_{\mu X.\psi,L}\}] \subseteq [\psi\{X/\mu X.\psi\eta^{v_i}_{\mu X.\psi\eta,L}, \eta^{v_i}_{\mu X.\psi\eta,L}\}] = [\mu X.\psi\eta^{v_i}_{\mu X.\varphi\eta,L}]$, which is included in $[\mu X.\psi\eta^{v_{i+1}}_{\mu X.\varphi\eta,L}]$ since $v_i \leq v_{i+1}$. Finally for the case b) we have that by definition of inductive run $[\mu X.\psi\eta]^{v_{i+1}}_U \supseteq [\psi\{X/\mu X.\psi\eta,\eta\}]^{v_{i+1}}_U$ which contains by induction on the structure of the sentence, $[\psi\{X/[\mu X.\psi\eta]^{v_{i+1}}_U$, $\eta^{v_{i+1}}_{\mu X.\psi,U}\}] = [\psi\eta^{v_{i+1}}_{\mu X.\psi,U}\{X/[\mu X.\varphi\eta]^{v_{i+1}}_U]$. This show that $B = [\mu X.\psi\eta]^{v_{i+1}}_U$ is a pre-smallest fixpoint for $\Psi(X) \equiv \psi\eta^{v_{i+1}}_{\mu X.\psi,U}$(i.e. $B \supseteq [\Psi(B)]$) so B contains the least fix-point, i.e. $[\mu X.\psi\eta^{v_{i+1}}_{\mu X.\psi,U}] \subseteq [\mu X.\psi\eta]^{v_{i+1}}_U$, which ends the proof.

Taking η to be the empty substitution, the previous lemma has the following consequence.

Theorem 1. *Let T be a transition system and F be an inductively closed set of sentences. Let $\bot = v_0 < v_1 < \ldots < v_n$ be an inductive run for F on T. Then $v_i \leq \top$, for $i = 0, \ldots, n$.*

So as said before we now have that if a state is in the lower valuation *for a valuation of an inductive run* then it is in the standard one so we are sure that it satisfies this formula in the standard sense. On the other hand if a state is not in the upper valuation *for a valuation of an inductive run* then it is not in the standard one so we are sure that it does not satisfy the formula. Hence the lower valuation approximates the standard valuation from below and the upper valuation approximates it from above *for a valuation of an inductive run.*

Now we are left with the question of whether an inductive run can reach the standard valuation. We need this in order to show that we are really able to approximate the inductive standard valuation. This is what the following result shows.

Theorem 2. *Let T be a transition system and F be an inductively closed set of sentences. There exists an inductive run for F on T, $\bot = v_0 < v_1 < \ldots < v_n$ such that $v_n = \top$.*

Proof. We will first define a class of inductive valuations and then we will use some of the valuations of this class to build an inductive run reaching the standard inductive valuation.

First of all extend F to the smallest set G of formulas containing F and closed under

a) if $\varphi \wedge \psi \in G$ then $\varphi, \psi \in G$
b) if $\neg\varphi \in G$ then $\varphi \in G$
c) if $\langle a\rangle\varphi \in G$ then $\varphi \in G$
d) if $\mu X.\varphi(X) \in G$ then $\varphi\{X/\mu X.\varphi(X)\} \in G$
e) if $\mu X.\varphi(X) \in G$ and S is a set of states of T then then $\varphi\{X/S\} \in G$.
f) if $\mu X.\varphi(X) \in G$ then $\varphi \in G$.

G is again a finite set of size linear in the product of the size of F and of the set of subsets of states of T.

Let f be a function sending fixpoint sentences of G to sets of states of T. Let $\varphi(X_1, \ldots, X_n)$ be a polarized formula of G and let $\eta = \{X_1/\mu X_1.\varphi_1, \ldots, X_n/\mu X_n.\varphi_n\}$ be a substitution with $\mu X_1.\varphi_1, \ldots, \mu X_n.\varphi_n$ sentences of G. We will now define by induction on φ two formulas of G called $L^{f,\eta}(\varphi)$ and $U^{f,\eta}(\varphi)$.

PL) $L^{f,\eta}(P) = P$ PU) $U^{f,\eta}(P) = P$
XL) $L^{f,\eta}(X) = X$ XU) $U^{f,\eta}(X) = X$
$\wedge$L) $L^{f,\eta}(\varphi \wedge \varphi) = L^{f,\eta}(\varphi) \wedge L^{f,\eta}(\psi)$ $\wedge$U) $U^{f,\eta}(\varphi \wedge \varphi) = U^{f,\eta}(\varphi) \wedge U^{f,\eta}(\psi)$
<>L) $L^{f,\eta}(\langle a\rangle\varphi) = \langle a\rangle(L^{f,\eta}(\varphi))$ <>U) $U^{f,\eta}(\langle a\rangle\varphi) = \langle a\rangle(U^{f,\eta}(\varphi))$
$\neg$L) $L^{f,\eta}(\neg\varphi) = \neg U^{f,\eta}(\varphi))$ $\neg$U) $U^{f,\eta}(\neg\varphi) = \neg L^{f,\eta}(\varphi))$
μL) $L^{f,\eta}(\mu X.\varphi) = \mu X.L^{f,\{X/\mu X.\varphi\}\eta}(\varphi) \wedge f(\mu X.\varphi\eta)$
μU) $U^{f,\eta}(\mu X.\varphi) = \mu X.U^{f,\{X/\mu X.\varphi\}\eta}(\varphi)$

where $\{X/\mu X.\varphi\}\eta$ is the composition of substitutions, i.e. it means $\{X/\mu X.\varphi\eta, \eta\}$.

We now associate with f a pair of functions v_f defined by $[\varphi]_L^{v_f} = [L^{f,\epsilon}(\varphi)]$ and $[\varphi]_U^{v_f} = [U^{f,\epsilon}(\varphi)]$ where ϵ is the empty substitution.

We now introduce a few lemmas inside this proof to simplify the presentation.

Lemma 2. *The above v_f are inductive valuations.*

Proof. proof The statement PL) and PU) of the definition of inductive valuation follow from the corresponding clauses above, since $[P]_L^{v_f} = [L^{f,\epsilon}(P)] = [P]$ and $[P]_U^{v_f} = [U^{f,\epsilon}(P)] = [P]$

In the same way the statement $\neg$L) of the definition of inductive valuation follows from the above $\neg$L) since $[\neg\varphi]_L^{v_f} = [L^{f,\epsilon}(\neg\varphi)] = [\neg U^{f,\epsilon}(\varphi)] = ([\varphi]_U^{v_f})^c$.

Similar arguments show the cases $\neg$U), $\wedge$L), $\wedge$L), $\wedge$U) and $\langle\rangle$L) and $\langle\rangle$U).

For the case μL) we have that $[\mu X.\varphi]_L^{v_f} = [L^{f,\epsilon}(\mu X.\varphi)] = [\mu X.L^{f,\{X/\mu X.\varphi\}}(\varphi) \wedge f(\mu X.\varphi)] = [(L^{f,\{X/\mu X.\varphi\}}(\varphi)\{X/\mu X.L^{f,\{X/\mu X.\varphi\}}(\varphi) \wedge f(\mu X.\varphi)\}) \wedge f(\mu X.\varphi)] \subseteq [L^{f,\epsilon}(\varphi\{X/\mu X.\varphi\})] = [\varphi\{X/\mu X.\varphi\}]_L^{v_f}$

Finally for μU) we have $[\mu X.\varphi]_U^{v_f} = [U^{f,\epsilon}(\mu X.\varphi)] = [\mu X.U^{f,\{X/\mu X.\varphi\}}(\varphi)] = [U^{f,\{X/\mu X.\varphi\}}(\varphi)\{X/\mu X.U^{f,\{X/\mu X.\varphi\}}(\varphi)\}] = [U^{f,\epsilon}(\varphi\{X/\mu X.\varphi\})] = [\varphi\{X/\mu X.\varphi]_U^{v_f}$. This completes the proof of the lemma.

We will now define functions $f_0, \ldots, f_n$ such that $\bot < v_{f_0} < v_{f_1} < \ldots < v_{f_n}$ is an inductive run and v_{f_n} is the standard inductive valuation.

Define the f_i recursively as follows.

a) f_0 to be the constant function $\emptyset$

b) $f_i(\mu X.\varphi) = [\varphi\{X/\mu X.\varphi\}]_L^{v_{f_{i-1}}}$

If we show that $\bot \leq v_{f_0} \leq v_{f_1} \leq \ldots \leq v_{f_n}$, then since $[\mu X.\varphi]_L^{v_{f_i}} = [L^{f_i,\epsilon}(\mu X.\varphi)] = [\mu X.L^{f_i,\{X/\mu X.\varphi\}}(\varphi) \wedge f_i(\mu X.\varphi)] \subseteq [f_i(\mu X.\varphi)] \subseteq [\varphi\{X/\mu X.\varphi\}]_L^{v_{f_{i-1}}}$ we will have that taking all v_{f_i} up to the first n such that $v_{f_n} = v_{f_{n+1}}$ (which exists since G is finite) $\bot < v_{f_0} < v_{f_1} < \ldots < v_{f_n}$ is also an inductive run.

Lemma 3. *With the above definition, $v_{f_{i-1}} \leq v_{f_i}$ for $i = 1, \ldots, n$.*

Proof. (Sketch) Note that if $f_{i-1}(\mu X.\varphi) \subseteq f_i(\mu X.\varphi)$ for any sentence $\mu X.\varphi$, then by induction on the structure of formulas we have that $v_{f_{i-1}} \leq v_{f_i}$. On the other hand if $v_{f_{i-1}} \leq v_{f_i}$ then $f_i(\mu X.\varphi) \subseteq f_{i+1}(\mu X.\varphi)$ for any sentence $\mu X.\varphi$. Since $f_0(\mu X.\varphi) = \emptyset \subseteq f_1$, the proof of the lemma is completed.

Now take n to by the smallest n such that $v_{f_n} = v_{f_{n+1}}$. In order to prove that $v_{f_n} = \top$, we prove :

Lemma 4. *With the above definition, the following hold*

a) $[\varphi\eta]_L^{v_{f_n}} = [\varphi\eta_{\varphi,L}^{v_{f_n}}]$

and

b) $[\varphi\eta_{\varphi,U}^{v_{f_n}}] = [\varphi\eta]_U^{v_{f_n}}$

Proof. (Sketch) We show this by induction on φ.

The only difficult cases are of μ, let us show here these cases.

For μL) since by lemma 2.1 $[\mu X.\varphi\eta]_L^{v_{f_n}} \subseteq [\mu X.\varphi\eta_{\varphi,L}^{v_{f_n}}]$ to have equality it is sufficient to show that $[\mu X.\varphi\eta]_L^{v_{f_n}}$ is a pre-least-fixpoint of $[\varphi\eta_{\varphi,L}^{v_{f_n}}(X)]$.

Let us now show this. We have that

$$[\mu X.\varphi\eta]_L^{v_{f_n}} = [\mu X.\varphi\eta]_L^{v_{f_{n+1}}} =$$

$$[\mu X.L^{f_{n+1},\{X/\mu X.\varphi\}}(\varphi\eta) \wedge f_{n+1}(\mu X\varphi\eta))] =$$

$$[L^{f_{n+1},\{X/\mu X.\varphi\}}(\varphi\eta)\{X/\mu X.L^{f_{n+1},\{X/\mu X.\varphi\}}(\varphi\eta) \wedge f_{n+1}(\mu X\varphi\eta)\}] \cap [f_{n+1}(\mu X\varphi\eta))]$$
$$= [L^{f_{n+1},\epsilon}(\varphi\eta\{X/\mu X.\varphi\})] \cap [\varphi\eta\{X/\mu X.\varphi\eta\})]_L^{v_{f_n}} = [\varphi\eta\{X/\mu X.\varphi\eta\})]_L^{v_{f_n}}$$

which by induction rule is equal to $[\varphi\eta_{\varphi,L}^{v_{f_n}}\{X/[\mu X.\varphi\eta]_L^{v_{f_n}}\}]$.

For μU) we use the fact that $[\mu X.\varphi\eta]_U^{v_{f_n}} = [\mu X.U^{f_n,\epsilon}(\varphi\eta)] = \bigcap_{[S]\supseteq[U^{f_n,\epsilon}(\varphi\eta)\{X/S\}]}[S]$ which since $[U^{f_n,\epsilon}(\varphi\eta)\{X/S\}] = [\varphi\eta\{X/S\}]_U^{v_{f_n}}$ is equal to $\bigcap_{[S]\supseteq[\varphi\eta\{X/S\}]_U^{v_{f_n}}}[S]$. Since by induction hypothesis $[\varphi\eta\{X/S\}]_U^{f_n} = [\varphi\eta_{\varphi,U}^{v_{f_n}}\{X/S\}]$ it follows that $[\mu X.\varphi\eta]_U^{f_n} = [\mu X.\varphi\eta_{\varphi,U}^{v_{f_n}}]$. This completes the proof of the lemma and also the proof of the theorem.

3 Sequent Calculus

We introduce for the propositional μ-calculus a double-sided sequent calculus.

Definition 8. *A context sentence is a μ-calculus sentence with a upper-script representing the stage of the computation for this sentence, for example φ^α. The upper-script is called a context.*

Definition 9. *A context sequent on A a set of contexts, is a double sided sequent of contexts sentences whose context are taken for A. For example $\varphi_1^{\alpha_1}, \ldots, \varphi_n^{\alpha_n} \vdash \psi_1^{\beta_1}, \ldots, \psi_m^{\beta_m}$.*

The interpretation of a double sequent is the classical one. A sequent is true if any state of a transition system which satisfy all the sentences on the left must also satisfy at least one formula on the right. Sequent calculus for propositional calculus give a proof system which is analogous to the transform of a sentence into conjunctive normal form. For the μ-calculus it does not seem to be possible to give such a simple deductive system, therefore we consider a proof system with context sequents and also hypotheses which we will now introduce.

Definition 10. *A hypothesis is a finite set of context sequents with a partial order on the contextes such that for any sequent of the hypothesis this order is total on its contexts.*

Definition 11. *A context valuation for the contextes $\alpha_1, \ldots, \alpha_n$ is a mapping V sending an α_i to inductive valuation $V(\alpha_i)$.*

Definition 12. *A context valuation V satisfies a sequent $\varphi_1^{\alpha_1}, \ldots, \varphi_n^{\alpha_n} \vdash \psi_1^{\beta_1}, \ldots, \psi_m^{\beta_m}$ if $[\varphi_1]_L^{V(\alpha_1)} \cap \cdots \cap [\varphi_n]_L^{V(\alpha_n)} \subseteq [\psi_1]_U^{V(\beta_1)} \cup \cdots \cup [\psi_m]_U^{V(\beta_m)}$*

Remark 2. If V and V' are context valuations such that $V(\alpha_i) \leq V'(\alpha_i)$ for $i = 1, \dots, n$ and $V(\beta_j) \leq V'(\beta_j)$ for $j = 1, \dots, m$ then if V' satisfies $\varphi_1^{\alpha_1}, \dots, \varphi_n^{\alpha_n} \vdash \psi_1^{\beta_1}, \dots, \psi_m^{\beta_m}$ then V satisfy it too.

Definition 13. *A context valuation V satisfy a hypothesis H if the following properties are satisfied*

a) if $\alpha \leq \beta$ in the partial order of H then $V(\alpha) \leq V(\beta)$

b) if $\alpha_1 \leq \dots \leq \alpha_n$ is a chain in the partial order of H then $V(\alpha_1) \leq \dots \leq V(\alpha_n)$ can be expanded (adding more inductive valuations in the chain) to an inductive run.

c) V satisfy every sequent of the hypothesis.

Definition 14. *A context sequent is the consequence of a hypothesis if any context valuation which satisfy this hypothesis also satisfy the context sequent.*

Theorem 3. *A context sequent is the consequence of a hypothesis containing no context sequent if and only if it is true for the standard inductive valuation.*

Proof. Follows from Theorem 2.1 and Theorem 2.2.

We now give a proof system sound relative to this notion of consequence.

Definition 15. *A context graph is a finite transition system whose nodes are context sequents and a hypothesis, such that for any context sequent appearing in this transition system, the order is a total order on the contexts of this sequent.*

Definition 16. *A context sequent $\varphi_1^{\alpha_1}, \dots, \varphi_n^{\alpha_n} \vdash \psi_1^{\beta_1}, \dots, \psi_m^{\beta_m}$ is smaller than a hypothesis if there exists a sequent $\varphi_1^{\gamma_1}, \dots, \varphi_n^{\gamma_n} \vdash \psi_1^{\delta_1}, \dots, \psi_m^{\delta_m}$ in the hypothesis such that $\alpha_i \leq \gamma_i$ and $\beta_j \leq \delta_j$ for $i = 1, \dots, n$ and $j = 1, \dots, m$.*

Remark 3. If a context valuation V satisfies the Hypothesis of a context graph then it satisfies also all sequents smaller than a Hypothesis.

Definition 17. *A proof graph is a context graph with a distinguished node called the root built using the rules in Fig. 1.*

In these rules the sequents on the right are called the *antecedents* and the sequent on the left is called the *consequent.*

Definition 18. *A sequent is said to be provable from Hypothesis H if there exists a proof graph having this sequent as its root, built using the above rules (starting with H as hypothesis set) which is a tree, whose leaves are all consequents of some rule without antecedent.*

Theorem 4. *If a sequent is provable from hypothesis H then it is a consequence of H.*

Proof. This is soundness of our system. Every rule can be shown sound in the same way as for classical propositional calculus, except rule μ-Left. For this one first completes the sequence of valuations associated with the contexts to an inductive run and then applies the definition of inductive run.

¬-right	$\Gamma \vdash (\neg\varphi)^\alpha, \Gamma'$ if		$\Gamma, \varphi^\alpha \vdash \Gamma'$
¬-left	$\Gamma, (\neg\psi)^\alpha \vdash \Gamma'$ if		$\Gamma \vdash \psi^\alpha, \Gamma'$
∧-right	$\Gamma \vdash (\varphi \wedge \varphi')^\alpha, \Gamma'$if	$\Gamma \vdash \varphi^\alpha, \Gamma'$ and	$\Gamma \vdash (\varphi')^\alpha, \Gamma'$
∧-left	$\Gamma, (\psi \wedge \psi')^\alpha \vdash \Gamma'$if		$\Gamma, \psi^\alpha, (\psi')^\alpha \vdash \Gamma'$
True-right	$\Gamma \vdash \text{True}^\alpha, \Gamma'$		
True-left	$\Gamma, \text{True}^\alpha \vdash \Gamma'$ if		$\Gamma \vdash \Gamma'$

Axiom $\Gamma \vdash \Gamma'$ for Γ and Γ' such that $\Gamma \cap \Gamma' \neq \emptyset$

Hypothesis $\Gamma \vdash \Gamma'$ if $\Gamma \vdash \Gamma'$ is a smaller than a hypothesis

$\langle a \rangle$ $(\langle a \rangle\varphi)^\alpha, \Gamma \vdash \Gamma'$ if $(\varphi)^\alpha \vdash \Sigma'$, where $\Sigma' = \{(\psi)^\beta; (\langle a \rangle\psi)^\beta \in \Gamma'\}$.

μ-right $\Gamma \vdash (\mu X.\varphi)^\alpha, \Gamma'$ if $\Gamma \vdash (\varphi\{X/\mu X.\varphi\})^\alpha, \Gamma'$

μ-left $\Gamma, (\mu X.\varphi)^\alpha \vdash \Gamma'$ if $\Gamma, (\mu X.\varphi)^{\alpha'} \vdash \Gamma'$ and $\Gamma, (\varphi\{X/\mu X.\varphi\})^\beta \vdash \Gamma'$ where α' is the predecessor of α in this sequent, β is a new context symbol in this proof and $\Gamma, (\mu X.\varphi)^\beta \vdash \Gamma'$ has been added to the hypothesis of this proof.

Fig. 1.

4 Some Examples of the Proof System

Example 3. Let un first show that $(\mu X.\langle a \rangle X)^\alpha \vdash$, which means that $(\mu X.\langle a \rangle X)$ is empty, is provable in our system.

First apply rule μ-Left to

$$(\mu X.\langle a \rangle X)^\alpha \vdash$$

to get

$$(\langle a \rangle(\mu X.\langle a \rangle X))^\beta \vdash$$

where $\beta \leq \alpha$ and $(\mu X.\langle a \rangle X)^\beta \vdash$ is added to the hypothesis.

Then apply rule $\langle\rangle$ for action a to get

$$(\mu X.\langle a \rangle X)^\beta \vdash$$

Now this last sequent is smaller than a hypothesis so applying rule Hypothesis we have a proof.

Example 4. In the same way $(\mu X.X)^\alpha \vdash$ is provable, since applying rule μ-Left to it gives $((\mu X.X))^\beta \vdash$ where $\beta \leq \alpha$ and $(\mu X.X)^\beta \vdash$ is added to the hypothesis. Again rule Hypothesis give a proof.

Example 5. Moving to a somewhat more interesting example let us consider

$$(\mu X.\langle a \rangle True \vee [c]X)^\alpha \vdash (\mu X.\langle a \rangle True \vee \langle b \rangle True \vee [c]X)^\alpha$$

Applying rule μ-left give

$$(\langle a\rangle True \vee [c](\mu X.\langle a\rangle True \vee [c]X))^{\beta} \vdash (\mu X.\langle a\rangle True \vee \langle b\rangle True \vee [c]X)^{\alpha}$$

where $(\mu X.\langle a\rangle True \vee [c]X)^{\beta} \vdash (\mu X.\langle a\rangle True \vee \langle b\rangle True \vee [c]X)^{\alpha}$ is added to the hypothesis.

Now using $\vee$-left (in fact we gets such a rule translating $\vee$ into a $\wedge$, we will use this here to get the proof simpler) we have now to show that

$$(\langle a\rangle True)^{\beta} \vdash (\mu X.\langle a\rangle True \vee \langle b\rangle True \vee [c]X)^{\alpha}$$

and

$$([c](\mu X.\langle a\rangle True \vee [c]X))^{\beta} \vdash (\mu X.\langle a\rangle True \vee \langle b\rangle True \vee [c]X)^{\alpha}$$

Using μ-right, $\vee$-right and then Axiom on the first one get rid of it. For the second one applies μ-right to get

$$\begin{aligned}&([c](\mu X.\langle a\rangle True \vee [c]X))^{\beta}\\ &\quad\vdash (\langle a\rangle True \vee \langle b\rangle True \vee [c](\mu X.\langle a\rangle True \vee \langle b\rangle True \vee [c]X))^{\alpha}\end{aligned}$$

again $\vee$-right gives

$$\begin{aligned}&([c](\mu X.\langle a\rangle True \vee [c]X))^{\beta}\\ &\quad\vdash (\langle a\rangle True)^{\alpha}, (\langle b\rangle True)^{\alpha}, ([c](\mu X.\langle a\rangle True \vee \langle b\rangle True \vee [c]X))^{\alpha}\end{aligned}$$

Now translating $[c]$ into $\neg\langle c\rangle\neg$ and applying $\neg$-Left and $\neg$-Right we get

$$\begin{gathered}(\langle c\rangle\neg(\mu X.\langle a\rangle True \vee \langle b\rangle True \vee [c]X))^{\alpha} \vdash\\ (\langle a\rangle True)^{\alpha}, (\langle b\rangle True)^{\alpha}, (\langle c\rangle\neg(\mu X.\langle a\rangle True \vee [c]X))^{\beta}\end{gathered}$$

Finally applying $\langle\rangle$ for c and again $\neg$-left and $\neg$-right we just have to use Hypothesis to finish the proof.

Example 6. We would like to finish with a very simple example of how our system could be used to construct a counter-example for some non-provable sequent.

Take

$$(\nu X.\langle a\rangle X)^{\alpha} \vdash$$

and convert it replacing the greatest fixpoint by its definition give us the following sequent.

$$(\neg\mu X.\neg\langle a\rangle\neg X)^{\alpha} \vdash \tag{1}$$

Applying $\neg$-left we get

$$\vdash (\mu X.\neg\langle a\rangle\neg X)^{\alpha} \tag{2}$$

Applying μ-right we get

$$\vdash (\neg\langle a\rangle\neg(\mu X.\neg\langle a\rangle\neg X))^{\alpha} \tag{3}$$

Applying ¬-right we get

$$(4) \qquad ((\langle a\rangle\neg(\mu X.\neg\langle a\rangle\neg X))^{\alpha} \vdash$$

Applying $\langle\rangle$ on a we get back to (1).

Now in order to build a counter-example to this sequent we must give a transition system and an inductive valuation corresponding to α which does not satisfy (1). Take this proof graph to be the transition system with the application $\langle\rangle$ as an a labelled transition. Merge the sequent (1),(2),(3) and (4) since we get one from the other by applying another rule than $\langle\rangle$. So we get a unique state with a a loop on it. Finally take the lower valuation of a sentence to be the set of states which contains a sequent having it on the left and the upper valuation as being the complement of the set of states which contains a sequent having it on the right. The application of the rules give that this is indeed is an inductive valuation.

This is of course not a proof that our system will always produce a finite counter-example. It is only the motivation why I think that this proof system is worth studying.

References

[Cl] R. Cleaveland, Tableau-Based Model Checking in the Propositional Mu-Calculus, Acta Informatica 27, 725-747 (1990).

[Da] M. Dam, CTL* and ECTL* as fragments of the modal μ-calculus, Theoretical Computer Science 126 (1994) 77-96.

[EmHa] E.A. Emerson , J. Halpern. "Sometimes" and "not never" revisited: On branching versus linear time temporal logic. Journal of the ACM, 33:175-211, 1986.

[EmJu] E. A. Emerson, C. S. Jutla. The complexity of tree automata and logics of programs. 29th IEEE Symp. on Foundation of Computer Science, 1988.

[EmJuSi] E.A. Emerson, C.S. Jutla, A.P. Sistla, On model-checking for fragments of the μ-calculus, CAV'93, LNCS 697.

[FiLa] M.J. Fischer, R.E. Ladner. Propositional dynamic logic of regular programs, Journal of Computer and System Sciences, 18:194-211, 1979

[HeMi] M. Hennessy, R. Milner. Algebraic laws for nondeterminism and concurrency, Journal of the ACM 32 (1985) 137-162.

[HoKoPa] D. Harel, D. Kozen, R. Parikh. Process logic: Expressiveness, decdability and completeness. Journal of Computer and System Sciences, 25:144-201, 1982.

[Ko] D. Kozen. Results on the propositional mu-calculus, Theoretical Computer Science, 27:333-354, 1983.

[St] R. S. Streett,. Propositional dynamic logic of looping and converse is elementary decidable. Information and Computation, 54:121-141, 1982.

[StWa] C. Stirling, D. Walker, Local model checking in the modal mu-calculus, Theoretical Computer Science, 89 (1991) 161-177.

[Wal] I. Walukiewicz, A Complete Deductive System for the mu-Calculus PhD-thesis, Warsaw University, 1994.

[Wa2] I. Walukiewicz, Completeness of Kozen's Axiomatisation of the Propositional μ-calculus, LICS'95.

[Wi] G. Winskel, A note on model checking the modal ν-calculus, ICALP'89, LNCS 372.

Author Index

Lecture Notes in Computer Science

For information about Vols. 1–2341
please contact your bookseller or Springer-Verlag

Vol. 2342: I. Horrocks, J. Hendler (Eds.), The Semantic Web – ISCW 2002. Proceedings, 2002. XVI, 476 pages. 2002.

Vol. 2345: E. Gregori, M. Conti, A.T. Campbell, G. Omidyar, M. Zukerman (Eds.), NETWORKING 2002. Proceedings, 2002. XXVI, 1256 pages. 2002.

Vol. 2346: H. Unger, T. Böhme, A. Mikler (Eds.), Innovative Internet Computing Systems. Proceedings, 2002. VIII, 251 pages. 2002.

Vol. 2347: P. De Bra, P. Brusilovsky, R. Conejo (Eds.), Adaptive Hypermedia and Adaptive Web-Based Systems. Proceedings, 2002. XV, 615 pages. 2002.

Vol. 2348: A. Banks Pidduck, J. Mylopoulos, C.C. Woo, M. Tamer Ozsu (Eds.), Advanced Information Systems Engineering. Proceedings, 2002. XIV, 799 pages. 2002.

Vol. 2349: J. Kontio, R. Conradi (Eds.), Software Quality – ECSQ 2002. Proceedings, 2002. XIV, 363 pages. 2002.

Vol. 2350: A. Heyden, G. Sparr, M. Nielsen, P. Johansen (Eds.), Computer Vision – ECCV 2002. Proceedings, Part I. XXVIII, 817 pages. 2002.

Vol. 2351: A. Heyden, G. Sparr, M. Nielsen, P. Johansen (Eds.), Computer Vision – ECCV 2002. Proceedings, Part II. XXVIII, 903 pages. 2002.

Vol. 2352: A. Heyden, G. Sparr, M. Nielsen, P. Johansen (Eds.), Computer Vision – ECCV 2002. Proceedings, Part III. XXVIII, 919 pages. 2002.

Vol. 2353: A. Heyden, G. Sparr, M. Nielsen, P. Johansen (Eds.), Computer Vision – ECCV 2002. Proceedings, Part IV. XXVIII, 841 pages. 2002.

Vol. 2355: M. Matsui (Ed.), Fast Software Encryption. Proceedings, 2001. VIII, 169 pages. 2001.

Vol. 2356: R. Kohavi, B.M. Masand, M. Spiliopoulou, J. Srivastava (Eds.), WEBKDD 2002 – Mining Log Data Across All Customers Touch Points. Proceedings, 2001. XI, 167 pages. 2002. (Subseries LNAI).

Vol. 2358: T. Hendtlass, M. Ali (Eds.), Developments in Applied Artificial Intelligence. Proceedings, 2002 XIII, 833 pages. 2002. (Subseries LNAI).

Vol. 2359: M. Tistarelli, J. Bigun, A.K. Jain (Eds.), Biometric Authentication. Proceedings, 2002. X, 197 pages. 2002.

Vol. 2360: J. Esparza, C. Lakos (Eds.), Application and Theory of Petri Nets 2002. Proceedings, 2002. X, 445 pages. 2002.

Vol. 2361: J. Blieberger, A. Strohmeier (Eds.), Reliable Software Technologies – Ada-Europe 2002. Proceedings, 2002 XIII, 367 pages. 2002.

Vol. 2362: M. Tanabe, P. van den Besselaar, T. Ishida (Eds.), Digital Cities II. Proceedings, 2001. XI, 399 pages. 2002.

Vol. 2363: S.A. Cerri, G. Gouardères, F. Paraguaçu (Eds.), Intelligent Tutoring Systems. Proceedings, 2002. XXVIII, 1016 pages. 2002.

Vol. 2364: F. Roli, J. Kittler (Eds.), Multiple Classifier Systems. Proceedings, 2002. XI, 337 pages. 2002.

Vol. 2365: J. Daemen, V. Rijmen (Eds.), Fast Software Encryption. Proceedings, 2002. XI, 277 pages. 2002.

Vol. 2366: M.-S. Hacid, Z.W. Raś, D.A. Zighed, Y. Kodratoff (Eds.), Foundations of Intelligent Systems. Proceedings, 2002. XII, 614 pages. 2002. (Subseries LNAI).

Vol. 2367: J. Fagerholm, J. Haataja, J. Järvinen, M. Lyly. P. Råback, V. Savolainen (Eds.), Applied Parallel Computing. Proceedings, 2002. XIV, 612 pages. 2002.

Vol. 2368: M. Penttonen, E. Meineche Schmidt (Eds.), Algorithm Theory – SWAT 2002. Proceedings, 2002. XIV, 450 pages. 2002.

Vol. 2369: C. Fieker, D.R. Kohel (Eds.), Algorithmic Number Theory. Proceedings, 2002. IX, 517 pages. 2002.

Vol. 2370: J. Bishop (Ed.), Component Deployment. Proceedings, 2002. XII, 269 pages. 2002.

Vol. 2371: S. Koenig, R.C. Holte (Eds.), Abstraction, Reformulation, and Approximation. Proceedings, 2002. XI, 349 pages. 2002. (Subseries LNAI).

Vol. 2372: A. Pettorossi (Ed.), Logic Based Program Synthesis and Transformation. Proceedings, 2001. VIII, 267 pages. 2002.

Vol. 2373: A. Apostolico, M. Takeda (Eds.), Combinatorial Pattern Matching. Proceedings, 2002. VIII, 289 pages. 2002.

Vol. 2374: B. Magnusson (Ed.), ECOOP 2002 – Object-Oriented Programming. XI, 637 pages. 2002.

Vol. 2375: J. Kivinen, R.H. Sloan (Eds.), Computational Learning Theory. Proceedings, 2002. XI, 397 pages. 2002. (Subseries LNAI).

Vol. 2377: A. Birk, S. Coradeschi, T. Satoshi (Eds.), RoboCup 2001: Robot Soccer World Cup V. XIX, 763 pages. 2002. (Subseries LNAI).

Vol. 2378: S. Tison (Ed.), Rewriting Techniques and Applications. Proceedings, 2002. XI, 387 pages. 2002.

Vol. 2379: G.J. Chastek (Ed.), Software Product Lines. Proceedings, 2002. X, 399 pages. 2002.

Vol. 2380: P. Widmayer, F. Triguero, R. Morales, M. Hennessy, S. Eidenbenz, R. Conejo (Eds.), Automata, Languages and Programming. Proceedings, 2002. XXI, 1069 pages. 2002.

Vol. 2381: U. Egly, C.G. Fermüller (Eds.), Automated Reasoning with Analytic Tableaux and Related Methods. Proceedings, 2002. X, 341 pages. 2002 .(Subseries LNAI).

Vol. 2382: A. Halevy, A. Gal (Eds.), Next Generation Information Technologies and Systems. Proceedings, 2002. VIII, 169 pages. 2002.

Vol. 2383: M.S. Lew, N. Sebe, J.P. Eakins (Eds.), Image and Video Retrieval. Proceedings, 2002. XII, 388 pages. 2002.

Vol. 2384: L. Batten, J. Seberry (Eds.), Information Security and Privacy. Proceedings, 2002. XII, 514 pages. 2002.

Vol. 2385: J. Calmet, B. Benhamou, O. Caprotti, L. Henocque, V. Sorge (Eds.), Artificial Intelligence, Automated Reasoning, and Symbolic Computation. Proceedings, 2002. XI, 343 pages. 2002. (Subseries LNAI).

Vol. 2386: E.A. Boiten, B. Möller (Eds.), Mathematics of Program Construction. Proceedings, 2002. X, 263 pages. 2002.

Vol. 2387: O.H. Ibarra, L. Zhang (Eds.), Computing and Combinatorics. Proceedings, 2002. XIII, 606 pages. 2002.

Vol. 2388: S.-W. Lee, A. Verri (Eds.), Pattern Recognition with Support Vector Machines. Proceedings, 2002. XI, 420 pages. 2002.

Vol. 2389: E. Ranchhod, N.J. Mamede (Eds.), Advances in Natural Language Processing. Proceedings, 2002. XII, 275 pages. 2002. (Subseries LNAI).

Vol. 2391: L.-H. Eriksson, P.A. Lindsay (Eds.), FME 2002: Formal Methods – Getting IT Right. Proceedings, 2002. XI, 625 pages. 2002.

Vol. 2392: A. Voronkov (Ed.), Automated Deduction – CADE-18. Proceedings, 2002. XII, 534 pages. 2002. (Subseries LNAI).

Vol. 2393: U. Priss, D. Corbett, G. Angelova (Eds.), Conceptual Structures: Integration and Interfaces. Proceedings, 2002. XI, 397 pages. 2002. (Subseries LNAI).

Vol. 2395: G. Barthe, P. Dybjer, L. Pinto, J. Saraiva (Eds.), Applied Semantics. IX, 537 pages. 2002.

Vol. 2396: T. Caelli, A. Amin, R.P.W. Duin, M. Kamel, D. de Ridder (Eds.), Structural, Syntactic, and Statistical Pattern Recognition. Proceedings, 2002. XVI, 863 pages. 2002.

Vol. 2398: K. Miesenberger, J. Klaus, W. Zagler (Eds.), Computers Helping People with Special Needs. Proceedings, 2002. XXII, 794 pages. 2002.

Vol. 2399: H. Hermanns, R. Segala (Eds.), Process Algebra and Probabilistic Methods. Proceedings, 2002. X, 215 pages. 2002.

Vol. 2401: P.J. Stuckey (Ed.), Logic Programming. Proceedings, 2002. XI, 486 pages. 2002.

Vol. 2402: W. Chang (Ed.), Advanced Internet Services and Applications. Proceedings, 2002. XI, 307 pages. 2002.

Vol. 2403: Mark d'Inverno, M. Luck, M. Fisher, C. Preist (Eds.), Foundations and Applications of Multi-Agent Systems. Proceedings, 1996-2000. X, 261 pages. 2002. (Subseries LNAI).

Vol. 2404: E. Brinksma, K.G. Larsen (Eds.), Computer Aided Verification. Proceedings, 2002. XIII, 626 pages. 2002.

Vol. 2405: B. Eaglestone, S. North, A. Poulovassilis (Eds.), Advances in Databases. Proceedings, 2002. XII, 199 pages. 2002.

Vol. 2406: C. Peters, M. Braschler, J. Gonzalo, M. Kluck (Eds.), Evaluation of Cross-Language Information Retrieval Systems. Proceedings, 2001. X, 601 pages. 2002.

Vol. 2407: A.C. Kakas, F. Sadri (Eds.), Computational Logic: Logic Programming and Beyond. Part I. XII, 678 pages. 2002. (Subseries LNAI).

Vol. 2408: A.C. Kakas, F. Sadri (Eds.), Computational Logic: Logic Programming and Beyond. Part II. XII, 628 pages. 2002. (Subseries LNAI).

Vol. 2409: D.M. Mount, C. Stein (Eds.), Algorithm Engineering and Experiments. Proceedings, 2002. VIII, 207 pages. 2002.

Vol. 2410: V.A. Carreño, C.A. Muñoz, S. Tahar (Eds.), Theorem Proving in Higher Order Logics. Proceedings, 2002. X, 349 pages. 2002.

Vol. 2412: H. Yin, N. Allinson, R. Freeman, J. Keane, S. Hubbard (Eds.), Intelligent Data Engineering and Automated Learning – IDEAL 2002. Proceedings, 2002. XV, 597 pages. 2002.

Vol. 2413: K. Kuwabara, J. Lee (Eds.), Intelligent Agents and Multi-Agent Systems. Proceedings, 2002. X, 221 pages. 2002. (Subseries LNAI).

Vol. 2414: F. Mattern, M. Naghshineh (Eds.), Pervasive Computing. Proceedings, 2002. XI, 298 pages. 2002.

Vol. 2415: J. Dorronsoro (Ed.), Artificial Neural Networks – ICANN 2002. Proceedings, 2002. XXVIII, 1382 pages. 2002.

Vol. 2417: M. Ishizuka, A. Sattar (Eds.), PRICAI 2002: Trends in Artificial Intelligence. Proceedings, 2002. XX, 623 pages. 2002. (Subseries LNAI).

Vol. 2418: D. Wells, L. Williams (Eds.), Extreme Programming and Agile Methods – XP/Agile Universe 2002. Proceedings, 2002. XII, 292 pages. 2002.

Vol. 2419: X. Meng, J. Su, Y. Wang (Eds.), Advances in Web-Age Information Management. Proceedings, 2002. XV, 446 pages. 2002.

Vol. 2420: K. Diks, W. Rytter (Eds.), Mathematical Foundations of Computer Science 2002. Proceedings, 2002. XII, 652 pages. 2002.

Vol. 2421: L. Brim, P. Jančar, M. Křetínský, A. Kučera (Eds.), CONCUR 2002 – Concurrency Theory. Proceedings, 2002. XII, 611 pages. 2002.

Vol. 2423: D. Lopresti, J. Hu, R. Kashi (Eds.), Document Analysis Systems V. Proceedings, 2002. XIII, 570 pages. 2002.

Vol. 2430: T. Elomaa, H. Mannila, H. Toivonen (Eds.), Machine Learning: ECML 2002. Proceedings, 2002. XIII, 532 pages. 2002. (Subseries LNAI).

Vol. 2431: T. Elomaa, H. Mannila, H. Toivonen (Eds.), Principles of Data Mining and Knowledge Discovery. Proceedings, 2002. XIV, 514 pages. 2002. (Subseries LNAI).

Vol. 2436: J. Fong, R.C.T. Cheung, H.V. Leong, Q. Li (Eds.), Advances in Web-Based Learning. Proceedings, 2002. XIII, 434 pages. 2002.

Vol. 2440: J.M. Haake, J.A. Pino (Eds.), Groupware – CRIWG 2002. Proceedings, 2002. XII, 285 pages. 2002.

Vol. 2442: M. Yung (Ed.), Advances in Cryptology – CRYPTO 2002. Proceedings, 2002. XIV, 627 pages. 2002.

Vol. 2444: A. Buchmann, F. Casati, L. Fiege, M.-C. Hsu, M.-C. Shan (Eds.), Technologies for E-Services. Proceedings, 2002. X, 171 pages. 2002.